Triangles and Angles

Right Triangle

Triangle has one 90° (right) angle.

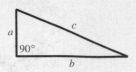

Pythagorean Formula
(*for right triangles*)

$a^2 + b^2 = c^2$

Right Angle

Measure is 90°.

Isosceles Triangle

Two sides are equal.

$AB = BC$

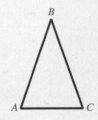

Straight Angle

Measure is 180°.

Equilateral Triangle

All sides are equal.

$AB = BC = CA$

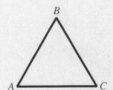

Complementary Angles

The sum of the measures of two complementary angles is 90°.

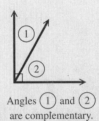

Angles ① and ② are complementary.

Sum of the Angles of Any Triangle

$A + B + C = 180°$

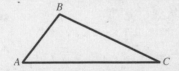

Supplementary Angles

The sum of the measures of two supplementary angles is 180°.

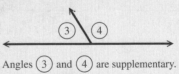

Angles ③ and ④ are supplementary.

Similar Triangles

Corresponding angles are equal; corresponding sides are proportional.

$A = D, B = E, C = F$

$$\frac{AB}{DE} = \frac{AC}{DF} = \frac{BC}{EF}$$

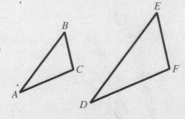

Vertical Angles

Vertical angles have equal measures.

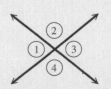

Angle ① = Angle ③

Angle ② = Angle ④

Beginning Algebra

10th EDITION

MARGARET L. LIAL
American River College

JOHN HORNSBY
University of New Orleans

TERRY McGINNIS

PEARSON
Addison Wesley

Boston   San Francisco   New York
London   Toronto   Sydney   Tokyo   Singapore   Madrid
Mexico City   Munich   Paris   Cape Town   Hong Kong   Montreal

Publisher: Greg Tobin
Editor in Chief: Maureen O'Connor
Senior Project Editor: Lauren Morse
Assistant Editor: Caroline Case
Senior Managing Editor: Karen Wernholm
Senior Production Supervisor: Kathleen A. Manley
Senior Designer: Dennis Schaefer
Photo Researcher: Beth Anderson
Digital Assets Manager: Marianne Groth
Media Producer: Sharon Tomasulo
Software Development: John O'Brien, MathXL; Ted Hartman, TestGen
Marketing Manager: Michelle Renda
Marketing Assistant: Alexandra Waibel
Senior Author Support/Technology Specialist: Joe Vetere
Senior Prepress Supervisor: Caroline Fell
Rights and Permissions Advisor: Dana Weightman
Manufacturing Manager: Evelyn Beaton
Media Buyer: Ginny Michaud
Text Design: IKO Ink
Production Coordination and Composition: WestWords/PMG
Illustrations: Network Graphics

Cover photo: Blazing Autumn © Copyright Lorraine Cota Manley

Photo Credits: see page I-8

Many of the designations used by manufacturers and sellers to distinguish their products are claimed as trademarks. Where those designations appear in this book, and Addison-Wesley was aware of a trademark claim, the designations have been printed in initial caps or all caps.

Library of Congress Cataloging-in-Publication Data
Lial, Margaret L.
    Beginning algebra.—10th ed. / Margaret L. Lial, John Hornsby, Terry McGinnis.
       p.   cm.
    Includes index.
    ISBN 0-321-43726-8
    1. Algebra.   I. Hornsby, E. John.   II. McGinnis, Terry.   III. Title.

QA152.3.L5   2007
512.9—dc22

          2006049439

6 7 8 9 10—DOW—11 10 09

Contents

List of Applications

Preface

The tenth edition of *Beginning Algebra* continues our ongoing commitment to provide the best possible text and supplements package to help instructors teach and students succeed. To that end, we have tried to address the diverse needs of today's students through a more open design, updated figures and graphs, helpful features, careful explanations of topics, and a comprehensive package of supplements and study aids. We have also taken special care to respond to the suggestions of users and reviewers and have added many new examples and exercises based on their feedback. Students who have never studied algebra—as well as those who require further review of basic algebraic concepts before taking additional courses in mathematics, business, science, nursing, or other fields—will benefit from the text's student-oriented approach.

This text is part of a series that includes the following books:

- *Intermediate Algebra,* Tenth Edition, by Lial, Hornsby, and McGinnis
- *Beginning and Intermediate Algebra,* Fourth Edition, by Lial, Hornsby, and McGinnis
- *Algebra for College Students,* Sixth Edition, by Lial, Hornsby, and McGinnis

Key Features

We believe students and instructors will welcome the following helpful features.

Enhanced Annotated Instructor's Edition   For easier reference, margin answers in the *Annotated Instructor's Edition* are now given in a single-column format whenever possible. In addition, the authors have added approximately 35 new Teaching Tips and more than 100 new and updated Classroom Examples.

NEW   *Tab Your Way to Success!*   A "Tab Your Way to Success!" guide provides students with color-coded Post-It® tabs to mark important pages of the text that they may need to return to for review work, test preparation, or instructor help.

Chapter Openers   New and updated chapter openers feature real-world applications of mathematics that are relevant to students and tied to specific material within the chapters. Examples of topics include Americans' personal savings rate, the Olympics, and student credit card debt. (See pages 1, 95, and 189—Chapters 1, 2, and 3.)

Real-Life Applications   We are always on the lookout for interesting data to use in real-life applications. As a result, we have included many new or updated examples and exercises from fields such as business, pop culture, sports, the life sciences, and technology that show the relevance of algebra to daily life. (See pages 190, 446, and 522.) A comprehensive List of Applications appears at the beginning of the text. (See pages vii–x.)

Figures, Photos, and **NEW** *Hand-Drawn Graphs*   Today's students are more visually oriented than ever. Thus, we have made a concerted effort to include mathematical figures, diagrams, tables, and graphs, including the new "hand-drawn" style of graphs, whenever possible. (See pages 138, 190, and 204.) Many of the graphs also use a style similar to that seen by students in today's print and electronic media. Photos have been incorporated to enhance applications in examples and exercises. (See pages 161, 236, and 595.)

Emphasis on Problem Solving   Introduced in Chapter 2, our six-step problem-solving method is integrated throughout the text. The six steps, *Read, Assign a Variable, Write an Equation, Solve, State the Answer,* and *Check,* are emphasized in boldface type and repeated in examples and exercises to reinforce the problem-solving process for students. (See pages 118, 300, and 447.) **PROBLEM-SOLVING HINT** boxes provide students with helpful problem-solving tips and strategies. (See pages 119, 124, and 150.)

Learning Objectives   Each section begins with clearly stated, numbered objectives, and the included material is directly keyed to these objectives so that students know exactly what is covered in each section. (See pages 96, 328, and 548.)

Cautions and Notes   One of the most popular features of previous editions, **CAUTION** and **NOTE** boxes warn students about common errors and emphasize important ideas throughout the exposition. (See pages 16, 110, and 120.) Highlighted in bright yellow, the text design makes them easy to spot.

NEW   *Pointers*   Pointers from the authors have been added to examples and provide students with important on-the-spot reminders and warnings about common pitfalls. (See pages 77, 135, and 193.)

Connections   Connections boxes provide connections to the real world or to other mathematical concepts, historical background, and thought-provoking questions for writing, class discussion, or group work. (See pages 125, 349, and 555.)

Now Try Exercises   To actively engage students in the learning process, each example concludes with a reference to one or more parallel exercises from the corresponding exercise set. In this way, students are able to immediately apply and reinforce the concepts and skills presented in the examples. These Now Try exercises are now marked with gray screens in the exercise sets so they can be easily spotted. Using the new Video Lectures on CD or DVD with Solution Clips, students can watch an instructor work through the complete solution to one Now Try problem for every example in the text. Exercises with a solution on video are marked with a CD icon ◉ in the exercise sets. (See pages 38, 126, and 199.)

Ample and Varied Exercise Sets   One of the most commonly mentioned strengths of this text is its exercise sets. The text contains a wealth of exercises to provide students with opportunities to practice, apply, connect, and extend the algebraic concepts and skills they are learning. Numerous illustrations, tables, graphs, and photos have been added to the exercise sets to help students visualize the problems they are solving. Problem types include writing ✐ , estimation, graphing calculator ▦ , and challenging "brain buster" exercises that go beyond the examples as well as applications and multiple-choice, matching, true/false, and fill-in-the-blank problems.

NEW • *Concept Check* exercises facilitate mathematical thinking and conceptual understanding. (See pages 38, 242, and 350.)

NEW • *WHAT WENT WRONG?* exercises ask students to identify typical errors in solutions and work the problems correctly. (See pages 226, 343, and 592.)

NEW • **PREVIEW EXERCISES**, brought back from earlier editions by popular request, *review* previously-studied concepts and *preview* skills needed for the upcoming section. (See pages 102, 195, and 203.)

• *Relating Concepts Exercises*   These sets of exercises help students tie together topics and develop problem-solving skills as they compare and contrast ideas, identify and describe patterns, and extend concepts to new situations. (See pages 174, 268, and 451.) These exercises make great collaborative activities for pairs or small groups of students.

• *Summary Exercises*   Based on user feedback, every chapter includes at least one set of in-chapter summary exercises. These special exercise sets provide students with the all-important *mixed* review problems they need to master topics. Summaries of solution methods or additional examples are often included. (See pages 66, 298, and 430.)

• *Technology Insights Exercises*   We assume that all students of this text have access to scientific calculators. *While graphing calculators are not required for this text,* some students may go on to courses that use them. For this reason, we have included Technology Insights exercises in selected exercise sets. These exercises provide an opportunity for students to interpret typical results seen on graphing calculator screens. Actual calculator screens from the Texas Instruments TI-83/84 Plus graphing calculator are featured. (See pages 229, 283, and 647.)

Group Activities   Appearing at the end of each chapter, these real-data activities allow students to apply the mathematical content of the chapter in a collaborative setting. (See pages 176, 316, and 533.)

Ample Opportunity for Review   Each chapter concludes with a Chapter Summary that features Key Terms, New Symbols, Test Your Word Power, and a Quick Review of each section's content with additional examples. A comprehensive set of Chapter Review Exercises, keyed to individual sections, is included, as are Mixed Review Exercises and a Chapter Test. Beginning with Chapter 2, each chapter concludes with a set of Cumulative Review Exercises that cover material going back to Chapter 1. The new Pass the Test: Chapter Test Solutions on Video with Interactive Chapter Summaries CD includes many helpful review resources based on the Chapter Summary. (See pages 177, 261, and 317.)

Test Your Word Power   To help students understand and master mathematical vocabulary, this feature can be found in each Chapter Summary. Key terms from the chapter are presented along with four possible definitions in a multiple-choice format. Answers and examples illustrating each term are provided. An interactive version of Test Your Word Power is available on the Pass the Test CD. (See pages 84, 177, and 261.)

Glossary   A comprehensive glossary of key terms from throughout the text is included at the back of the book. (See pages G-1 to G-6.)

What content changes have been made?

▶ A primary focus of this revision of the text was to polish and enhance individual presentations of topics and exercise sets, based on user and reviewer feedback, and we have worked hard to do this throughout the book. Some of the specific content changes include the following:

- There are approximately 1087 new and updated exercises, including many problems that focus on drill, skill development, and review. These include new Concept Check exercises, What Went Wrong? problems, and Preview Exercises.

- When a new type of graph is introduced (Sections 3.2, 4.5, 5.4, and 9.5), a new "hand-drawn" graph style is used to simulate what a student might actually sketch on graph paper.

- Real-world data in over 300 applications have been updated.

- Chapter 3 includes a new set of summary exercises on linear equations and graphs. All chapters now include a set of these popular mixed-review problems.

- The presentation of the following topics has been expanded:

    Review of fractions (Section 1.1)
    Applications from geometry (Section 2.5 and 5.1)
    Graphing linear equations in two variables (Section 3.2)
    Slope (Section 3.3)
    Slope-intercept form of the equation of a line (Section 3.4)
    Scientific notation and applications (Section 5.3)

What supplements are available?

▶ For a comprehensive list of the supplements and study aids that accompany *Beginning Algebra,* Tenth Edition, see pages xvi and xvii.

Acknowledgments

▶ The comments, criticisms, and suggestions of users, nonusers, instructors, and students have positively shaped this textbook over the years, and we are most grateful for the many responses we have received. Thanks to the following people for their review work, feedback, assistance at various meetings, and additional media contributions:

Barbara Aaker, *Community College of Denver*
Kim Bennekin, *Georgia Perimeter College*
Dixie Blackinton, *Weber State University*
Callie Daniels, *St. Charles Community College*
Cheryl Davids, *Central Carolina Technical College*
Chris Diorietes, *Fayetteville Technical Community College*
Sylvia Dreyfus, *Meridian Community College*
LaTonya Ellis, *Bishop State Community College*
Beverly Hall, *Fayetteville Technical Community College*
Sandee House, *Georgia Perimeter College*
Joe Howe, *St. Charles Community College*
Lynette King, *Gadsden State Community College*
Linda Kodama, *Kapi´olani Community College*
Carlea McAvoy, *South Puget Sound Community College*
James Metz, *Kapi´olani Community College*
Jean Millen, *Georgia Perimeter College*
Molly Misko, *Gadsden State Community College*

Jane Roads, *Moberly Area Community College*
Melanie Smith, *Bishop State Community College*
Erik Stubsten, *Chattanooga State Technical Community College*
Tong Wagner, *Greenville Technical College*
Sessia Wyche, *University of Texas at Brownsville*

Special thanks are due all those instructors at Broward Community College for their insightful comments.

Over the years, we have come to rely on an extensive team of experienced professionals. Our sincere thanks go to these dedicated individuals at Addison-Wesley, who worked long and hard to make this revision a success: Greg Tobin, Maureen O'Connor, Lauren Morse, Michelle Renda, Caroline Case, Alexandra Waibel, Kathy Manley, Dennis Schaefer, and Sharon Smith.

Abby Tanenbaum did an outstanding job helping us revise traditional and real data applications. Melena Fenn provided excellent production work. Thanks are due Jeff Cole, who supplied accurate, helpful solutions manuals, and Jim Ball, who provided the comprehensive Printed Test Bank. We are most grateful to Lucie Haskins for yet another accurate, useful index; Becky Troutman for preparing the comprehensive List of Applications; and Gary Williams, Cathy Zucco-Teveloff, and Perian Herring for accuracy checking page proofs.

As an author team, we are committed to the goal stated earlier in this Preface—to provide the best possible text and supplements package to help instructors teach and students succeed. We are most grateful to all those over the years who have aspired to this goal with us. As we continue to work toward it, we would welcome any comments or suggestions you might have. Please feel free to send your comments via e-mail to math@aw.com.

Margaret L. Lial
John Hornsby
Terry McGinnis

STUDENT SUPPLEMENTS

Student's Solutions Manual
- Provides detailed solutions to the odd-numbered section-level exercises and summary exercises and to all Relating Concepts, Chapter Review, Chapter Test, and Cumulative Review Exercises

ISBNs: 0-321-44471-X and 978-0-321-44471-4

NEW Video Lectures on CD with Solution Clips
NEW Video Lectures on DVD with Solution Clips
- Complete set of digitized videos on CD-ROM (or DVD) for students to use at home or on campus
- Includes a full lecture for each section of the text
- Students can also choose to watch an instructor work the solution to exercises that have been correlated to all examples from the text (one exercise for each example)
- Each exercise that has a video solution available is denoted in the exercise sets by a CD icon 🌐
- Optional captioning in English and Spanish is available for the lecture portion of this product (Video Lectures on CD only)

CD ISBNs: 0-321-47458-9 and 978-0-321-47458-2
DVD ISBNs: 0-321-44792-1 and 978-0-321-44792-0

NEW Pass the Test: Chapter Test Solutions on Video with Interactive Chapter Summaries on CD
Included with each Student Edition of the book, this CD-ROM contains:
- Interactive "Key Terms" with definitions
- Interactive "Test Your Word Power"
- Summary lectures for each key concept from the "Quick Review" for each chapter
- Video footage of an instructor working through the complete solutions for all chapter test problems

Additional Skill and Drill Manual
- Provides additional practice and test preparation for students

ISBNs: 0-321-44793-X and 978-0-321-44793-7

NEW MathXL® Tutorials on CD
- Provides algorithmically generated practice exercises that correlate at the objective level to the content of the text
- Every exercise is accompanied by an example and a guided solution, and selected exercises may also include a video clip
- The software provides helpful feedback and can generate printed summaries of students' progress

ISBNs: 0-321-44111-7 and 978-0-321-44111-9

INSTRUCTOR SUPPLEMENTS

Annotated Instructor's Edition
- Provides answers to all text exercises in color next to the corresponding problems, along with teaching tips and extra examples for the classroom
- Icons identify writing 📝 and calculator 🖩 exercises

ISBNs: 0-321-44786-7 and 978-0-321-44786-9

Instructor's Solutions Manual
- Provides complete solutions to all text exercises
- **NEW** Now includes solutions to all Classroom Examples

ISBNs: 0-321-44107-9 and 978-0-321-44107-2

Instructor and Adjunct Support Manual
- Includes resources designed to help both new and adjunct faculty with course preparation and classroom management
- Offers helpful teaching tips correlated to the sections of the text

ISBNs: 0-321-44950-9 and 978-0-321-44950-4

NEW Online Lesson Plans
- Lesson plans for each section of the book
- Worksheets covering additional topics
- Correlation to California Mathematics Content Standards for Algebra 1

Printed Test Bank and Instructor's Resource Guide
- The test bank contains two diagnostic pretests, four free-response and two multiple-choice test forms per chapter, and two final exams
- The resource guide contains additional practice exercises for most objectives of every section and a conversion guide from the ninth to the tenth edition

ISBNs: 0-321-44627-5 and 978-0-321-44627-5

Online Answer Book
- Provides answers to all the exercises in the text

NEW PowerPoint Lecture Slides
- Presents key concepts and definitions from the text
- Provides complete solutions to all Classroom Examples from the Annotated Instructor's Edition

NEW Active Learning Lecture Slides
- Multiple choice questions are available for each section of the book, allowing instructors to quickly assess mastery of material in class
- Available in PowerPoint, these slides can be used with classroom response systems

STUDENT SUPPLEMENTS

Addison-Wesley Math Tutor Center

- Staffed by qualified mathematics instructors
- Provides tutoring on examples and odd-numbered exercises from the textbook through a registration number with a new textbook or purchased separately
- Accessible via toll-free telephone, toll-free fax, e-mail, or the Internet
www.aw-bc/tutorcenter

InterAct Math Tutorial Website www.interactmath.com

- Get practice and tutorial help online!
- Retry an exercise as many times as you like with new values each time for unlimited practice and mastery
- Every exercise is accompanied by an interactive guided solution that gives you helpful feedback when an incorrect answer is entered
- View the steps of a worked-out sample problem similar to the one you're working on

INSTRUCTOR SUPPLEMENTS

TestGen®

- **NEW** Now includes a pre-made-test for each chapter that has been correlated problem-by-problem to the chapter tests in the book
- Enables instructors to build, edit, print, and administer tests using a computerized bank of questions developed to cover all text objectives
- Algorithmically based, TestGen® allows instructors to create multiple but equivalent versions of the same question or test with the click of a button
- Instructors can also modify test bank questions or add new questions
- Tests can be printed or administered online
- Available on a dual-platform Windows/Macintosh CD-ROM

ISBNs: 0-321-44945-2 and 978-0-321-44945-0

Adjunct Support Center

The Math Adjunct Support Center is staffed by qualified mathematics instructors with over 50 years combined experience at both the community college and university level. Assistance is provided for faculty in the following areas:

- Suggested syllabus consultation
- Tips on using materials packaged with your book
- Book-specific content assistance
- Teaching suggestions including advice on classroom strategies

www.aw-bc.com/tutorcenter/math-adjunct.html

Available for Students and Instructors

MathXL®   MathXL® is a powerful online homework, tutorial, and assessment system that accompanies your Addison-Wesley textbook in mathematics or statistics. With MathXL, instructors can create, edit, and assign online homework and tests using algorithmically generated exercises correlated at the objective level to the textbook. They can also create and assign their own online exercises and import TestGen tests for added flexibility. All student work is tracked in MathXL's online gradebook. Students can take chapter tests in MathXL and receive personalized study plans based on their test results. The study plan diagnoses weaknesses and links students directly to tutorial exercises for the objectives they need to study and retest. Students can also access supplemental video clips directly from selected exercises. MathXL is available to qualified adopters. For more information, visit our website at www.mathxl.com, or contact your Addison-Wesley sales representative.

MyMathLab   MyMathLab is a series of text-specific, easily customizable online courses for Addison-Wesley textbooks in mathematics and statistics. MyMathLab is powered by CourseCompass™—Pearson Education's online teaching and learning environment—and by MathXL®—our online homework, tutorial, and assessment system. MyMathLab gives instructors the tools they need to deliver all or a portion of their course online, whether students are in a lab setting or working from home. MyMathLab provides a rich and flexible set of course materials, featuring free-response exercises that are algorithmically generated for unlimited practice and mastery. Students can also use online tools, such as video lectures, animations, and a multimedia textbook, to independently improve their understanding and performance. Instructors can use MyMathLab's homework and test managers to select and assign online exercises correlated directly to the textbook, and they can also create and assign their own online exercises and import TestGen tests for added flexibility. MyMathLab's online gradebook—designed specifically for mathematics and statistics—automatically tracks students' homework and test results and gives the instructor control over how to calculate final grades. Instructors can also add offline (paper-and-pencil) grades to the MathXL gradebook. MyMathLab is available to qualified adopters. For more information, visit our Web site at www.mymathlab.com or contact your Addison-Wesley sales representative.

Feature Walkthrough

CHAPTER 2

Linear Equations and Inequalities in One Variable

In 1896, 241 competitors from 14 countries gathered in Athens, Greece, for the first modern Olympic Games. In 2004, the XXVIII Olmpic Summer Games returned to Athens as a truly international event, attracting 11,099 athletes from 202 countries.

One ceremonial aspect of the games is the flying of the Olympic flag with its five interlocking rings of different colors on a white background. First introduced at the 1920 Games in Antwerp, Belgium, the rings on the flag symbolize unity among the nations of Africa, the Americas, Asia, Australia, and Europe. (*Source:* www.athens2004.com; *Microsoft Encarta Encyclopedia.*)

Throughout this chapter we use *linear equations* to solve applications about the Olympics.

Chapter Opener

Each chapter opens with an application and section outline. The application in the opener is tied to material presented later in the chapter.

---

96 CHAPTER 2 Linear Equations and Inequalities in One Variable

2.1 The Addition Property of Equality

OBJECTIVES

1. Identify linear equations.
2. Use the addition property of equality.
3. Simplify and then use the addition property of equality.

Recall from **Section 1.3** that an *equation* is a statement asserting that two algebraic expressions are equal. The simplest type of equation is a *linear equation.*

OBJECTIVE 1 Identify linear equations.

Linear Equation in One Variable

A **linear equation in one variable** can be written in the form
$$Ax + B = C,$$
for real numbers A, B, and C, with $A \neq 0$.

For example,
$$4x + 9 = 0, \quad 2x - 3 = 5, \quad \text{and} \quad x = 7 \qquad \text{Linear equations}$$
are linear equations in one variable (x). The final two can be written in the specified form with the use of properties developed in this chapter. However,
$$x^2 + 2x = 5, \quad \frac{1}{x} = 6, \quad \text{and} \quad |2x + 6| = 0 \qquad \text{Nonlinear equations}$$
are *not* linear equations.

As we saw in **Section 1.3**, a *solution* of an equation is a number that makes the equation true when it replaces the variable. An equation is solved by finding its **solution set**, the set of all solutions. Equations that have exactly the same solution sets are **equivalent equations**. A linear equation in x is solved by using a series of steps to produce a simpler equivalent equation of the form

For example,
$$4x + 9 = 0, \quad 2x - 3 = 5, \quad \text{and} \quad x = 7 \qquad \text{Linear equations}$$
are linear equations in one variable (x). The final two can be written in the specified form with the use of properties developed in this chapter. However,
$$x^2 + 2x = 5, \quad \frac{1}{x} = 6, \quad \text{and} \quad |2x + 6| = 0 \qquad \text{Nonlinear equations}$$
are *not* linear equations.

As we saw in **Section 1.3**, a *solution* of an equation is a number that makes the equation true when it replaces the variable. An equation is solved by finding its **solution set**, the set of all solutions. Equations that have exactly the same solution sets are **equivalent equations**. A linear equation in x is solved by using a series of steps to produce a simpler equivalent equation of the form
$$x = \text{a number} \qquad \text{or} \qquad \text{a number} = x.$$

OBJECTIVE 2 Use the addition property of equality. In the equation $x - 5 = 2$, both $x - 5$ and 2 represent the same number because that is the meaning of the equals sign. To solve the equation, we change the left side from $x - 5$ to just x. We do this by adding 5 to $x - 5$. We use 5 because 5 is the opposite (additive inverse) of -5, and $-5 + 5 = 0$. To keep the two sides equal, we must also add 5 to the right side.
$$
\begin{aligned}
x - 5 &= 2 & &\text{Given equation} \\
x - 5 + 5 &= 2 + 5 & &\text{Add 5 to each side.} \\
x + 0 &= 7 & &\text{Additive inverse property} \\
x &= 7 & &\text{Additive identity property}
\end{aligned}
$$
The solution of the given equation is 7. We check by replacing x with 7 in the original equation.

Check:
$$
\begin{aligned}
x - 5 &= 2 & &\text{Original equation} \\
7 - 5 &= 2 \quad ? & &\text{Let } x = 7. \\
2 &= 2 & &\text{True}
\end{aligned}
$$

Learning Objectives

Each section opens with a highlighted list of clearly stated, numbered learning objectives. These learning objectives are restated throughout the section where appropriate for reinforcement.

Cautions

Students are warned of common errors through the use of **Caution** boxes throughout the text.

Notes

Important ideas are emphasized in **Note** boxes that appear throughout the text.

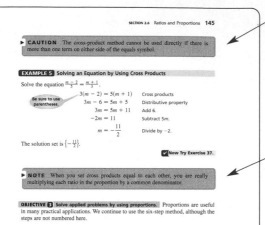

▶ **CAUTION** The cross-product method cannot be used directly if there is more than one term on either side of the equals symbol.

EXAMPLE 5 Solving an Equation by Using Cross Products

Solve the equation $\frac{m-2}{5} = \frac{m+1}{3}$.

Be sure to use parentheses.

$$3(m-2) = 5(m+1) \quad \text{Cross products}$$
$$3m - 6 = 5m + 5 \quad \text{Distributive property}$$
$$3m = 5m + 11 \quad \text{Add 6.}$$
$$-2m = 11 \quad \text{Subtract 5m.}$$
$$m = -\frac{11}{2} \quad \text{Divide by } -2.$$

The solution set is $\left\{ -\frac{11}{2} \right\}$.

✔ Now Try Exercise 37.

▶ **NOTE** When you set cross products equal to each other, you are really multiplying each ratio in the proportion by a common denominator.

OBJECTIVE 3 Solve applied problems by using proportions. Proportions are useful in many practical applications. We continue to use the six-step method, although the steps are not numbered here.

Now Try Exercises

Now Try Exercises are found after each example to encourage students to work exercises in the exercise sets that parallel the example just studied. **NEW**—Students can watch an instructor working through the complete solution to one *Now Try* problem for every example in the text on the new Video Lectures on CD or DVD with Solution Clips.

Classroom Examples and Teaching Tips

The *Annotated Instructor's Edition* provides answers to all text exercises and Group Activities in color in the margin or next to the corresponding exercise. *Classroom Examples* are also included to provide instructors with examples that are different from those that students have in their textbooks. Solutions to the Classroom Examples are found in the Instructor's Solutions Manual or in the PowerPoint Lecture Slides. *Teaching Tips* offer guidance on presenting the material at hand.

Pointers

NEW Examples have been made even more student-friendly with **pointers** from the authors that provide on-the-spot reminders and warnings about common pitfalls.

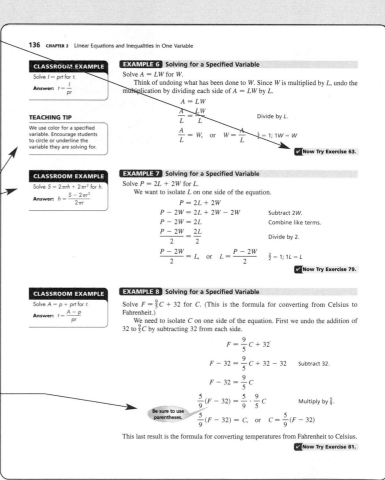

CLASSROOM EXAMPLE

Solve $I = prt$ for t.

Answer: $t = \frac{I}{pr}$

TEACHING TIP

We use color for a specified variable. Encourage students to circle or underline the variable they are solving for.

EXAMPLE 6 Solving for a Specified Variable

Solve $A = LW$ for W.

Think of undoing what has been done to W. Since W is multiplied by L, undo the multiplication by dividing each side of $A = LW$ by L.

$$A = LW$$
$$\frac{A}{L} = \frac{LW}{L} \quad \text{Divide by } L.$$
$$\frac{A}{L} = W, \quad \text{or} \quad W = \frac{A}{L} \qquad \frac{L}{L} = 1; \ 1W = W$$

✔ Now Try Exercise 63.

CLASSROOM EXAMPLE

Solve $S = 2\pi rh + 2\pi r^2$ for h.

Answer: $h = \frac{S - 2\pi r^2}{2\pi r}$

EXAMPLE 7 Solving for a Specified Variable

Solve $P = 2L + 2W$ for L.

We want to isolate L on one side of the equation.

$$P = 2L + 2W$$
$$P - 2W = 2L + 2W - 2W \quad \text{Subtract } 2W.$$
$$P - 2W = 2L \quad \text{Combine like terms.}$$
$$\frac{P - 2W}{2} = \frac{2L}{2} \quad \text{Divide by 2.}$$
$$\frac{P - 2W}{2} = L, \quad \text{or} \quad L = \frac{P - 2W}{2} \qquad \frac{2}{2} = 1; \ 1L = L$$

✔ Now Try Exercise 79.

CLASSROOM EXAMPLE

Solve $A = p + prt$ for t.

Answer: $t = \frac{A - p}{pr}$

EXAMPLE 8 Solving for a Specified Variable

Solve $F = \frac{9}{5}C + 32$ for C. (This is the formula for converting from Celsius to Fahrenheit.)

We need to isolate C on one side of the equation. First we undo the addition of 32 to $\frac{9}{5}C$ by subtracting 32 from each side.

$$F = \frac{9}{5}C + 32$$
$$F - 32 = \frac{9}{5}C + 32 - 32 \quad \text{Subtract 32.}$$
$$F - 32 = \frac{9}{5}C$$
$$\frac{5}{9}(F - 32) = \frac{5}{9} \cdot \frac{9}{5}C \quad \text{Multiply by } \frac{5}{9}.$$

Be sure to use parentheses.

$$\frac{5}{9}(F - 32) = C, \quad \text{or} \quad C = \frac{5}{9}(F - 32)$$

This last result is the formula for converting temperatures from Fahrenheit to Celsius.

✔ Now Try Exercise 81.

Connections

Connections boxes provide connections to the real world or to other mathematical concepts, historical background, and offer thought-provoking questions for writing or class discussion.

Writing Exercises

Writing exercises abound in the Lial series through the Connections boxes and also in the exercise sets (as marked with a pencil icon ✎).

NEW Preview Exercises

Preview Exercises have been added to the end of each section exercise set to help students transition from one section to the next.

PREVIEW EXERCISES

Solve each equation. See **Section 2.2.**

87. $\frac{x}{12} = \frac{12}{72}$ **88.** $\frac{x}{15} = \frac{144}{60}$ **89.** $0.06x = 300$ **90.** $0.4x = 80$

91. $\frac{3}{4}x = 21$ **92.** $-\frac{5}{6}x = 30$ **93.** $-3x = \frac{1}{4}$ **94.** $4x = \frac{1}{3}$

OBJECTIVES

1. Learn the six steps for solving applied problems.
2. Solve problems involving unknown numbers.
3. Solve problems involving sums of quantities.
4. Solve problems involving supplementary and complementary angles.
5. Solve problems involving consecutive integers.

OBJECTIVE 1 Learn the six steps for solving applied problems. We now look at how algebra is used to solve applied problems. While there is no one specific method that enables you to solve all kinds of applied problems, the following six-step method is often applicable.

Solving an Applied Problem

Step 1 **Read** the problem carefully until you understand what is given and what is to be found.
Step 2 **Assign a variable** to represent the unknown value, using diagrams or tables as needed. Write down what the variable represents. If necessary, express any other unknown values in terms of the variable.
Step 3 **Write an equation** using the variable expression(s).
Step 4 **Solve** the equation.
Step 5 **State the answer.** Does it seem reasonable?
Step 6 **Check** the answer in the words of the *original* problem.

Problem Solving

The Lial *six-step problem-solving method* is introduced in Chapter 2 and is then continually reinforced in examples, exercises, and problem-solving hint boxes throughout the text.

EXAMPLE 4 Finding the Height of a Triangular Sail

The area of a triangular sail of a sailboat is 126 ft². (Recall that "ft²" means "square feet.") The base of the sail is 12 ft. Find the height of the sail.

Step 1 **Read.** We must find the height of the triangular sail.
Step 2 **Assign a variable.** Let $h =$ the height of the sail, in feet. See Figure 8.
Step 3 **Write an equation.** The formula for the area of a triangle is $A = \frac{1}{2}bh$, where A is the area, b is the base, and h is the height. Using the information given in the problem, we substitute 126 for A and 12 for b in the formula.

$$A = \frac{1}{2}bh$$
$$126 = \frac{1}{2}(12)h \qquad A = 126,\ b = 12$$

Step 4 **Solve.**
$$126 = 6h \qquad \text{Multiply.}$$
$$21 = h \qquad \text{Divide by 6.}$$

Step 5 **State the answer.** The height of the sail is 21 ft.
Step 6 **Check** to see that the values $A = 126$, $b = 12$, and $h = 21$ satisfy the formula for the area of a triangle.

✓ Now Try Exercise 51.

FIGURE 8

Summary Exercises

Summary Exercises appear in all chapters to provide students with *mixed* practice problems needed to master topics.

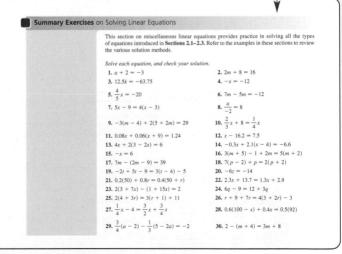

Summary Exercises on Solving Linear Equations

This section on miscellaneous linear equations provides practice in solving all the types of equations introduced in **Sections 2.1–2.3.** Refer to the examples in these sections to review the various solution methods.

Solve each equation, and check your solution.

1. $a + 2 = -3$ **2.** $2m + 8 = 16$
3. $12.5k = -63.75$ **4.** $-x = -12$
5. $\frac{4}{5}x = -20$ **6.** $7m - 5m = -12$
7. $5x - 9 = 4(x - 3)$ **8.** $\frac{a}{-2} = 8$
9. $-3(m - 4) + 2(5 + 2m) = 29$ **10.** $\frac{2}{3}x + 8 = \frac{1}{4}x$
11. $0.08x + 0.06(x + 9) = 1.24$ **12.** $x - 16.2 = 7.5$
13. $4x + 2(3 - 2x) = 6$ **14.** $-0.3x + 2.1(x - 4) = -6.6$
15. $-x = 6$ **16.** $3(m + 5) - 1 + 2m = 5(m + 2)$
17. $7m - (2m - 9) = 39$ **18.** $7(p - 2) + p = 2(p + 2)$
19. $-2t + 5t - 9 = 3(t - 4) - 5$ **20.** $-6z = -14$
21. $0.2(50) + 0.8r = 0.4(50 + r)$ **22.** $2.3x + 13.7 = 1.3x + 2.9$
23. $2(3 + 7x) - (1 + 15x) = 2$ **24.** $6q - 9 = 12 + 3q$
25. $2(4 + 3r) = 3(r + 1) + 11$ **26.** $r + 9 + 7r = 4(3 + 2r) - 3$
27. $\frac{1}{4}x - 4 = \frac{3}{2}x + \frac{3}{4}x$ **28.** $0.6(100 - x) + 0.4x = 0.5(92)$
29. $\frac{3}{4}(a - 2) - \frac{1}{3}(5 - 2a) = -2$ **30.** $2 - (m + 4) = 3m + 8$

Ample and Varied Exercise Sets

This text contains over 7,700 exercises, including over 2,100 review exercises, plus numerous conceptual and writing exercises that go beyond the examples. Multiple-choice, matching, true/false, and completion exercises help to provide variety. Exercises suitable for graphing calculator use are marked with an icon. **NEW** Students can watch an instructor work through the complete solution to all exercises marked with a CD icon on the Video Lectures on CD or DVD with Solution Clips.

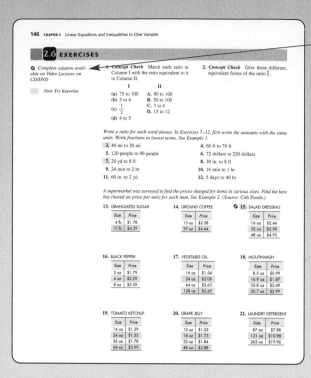

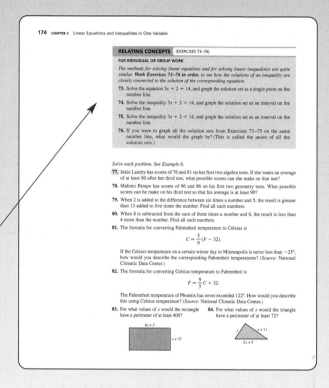

Relating Concepts

Found in selected exercise sets, these exercises tie together topics and highlight the relationships among various concepts and skills.

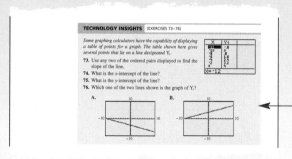

Technology Insights

Technology Insights exercises are found in selected exercise sets throughout the text. These exercises illustrate the power of graphing calculators and provide an opportunity for students to interpret typical results seen on graphing calculator screens. (A graphing calculator is *not* required to complete these exercises).

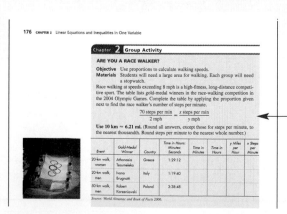

Group Activities

Appearing at the end of each chapter, these activities allow students to work collaboratively to solve a problem related to chapter material.

Ample Opportunity for Review

One of the most popular features of the Lial textbooks is the extensive and well thought-out end-of-chapter material. At the end of each chapter, students will find:

Key Terms and New Symbols that are keyed back to the appropriate section for easy reference and study. **NEW** Look for interactive Key Terms with Definitions on the Pass the Test CD.

Test Your Word Power to help students understand and master mathematical vocabulary; key terms from the chapter are presented with four possible definitions in multiple-choice format. **NEW** Look for an interactive version of *Test Your Word Power* on the Pass the Test CD.

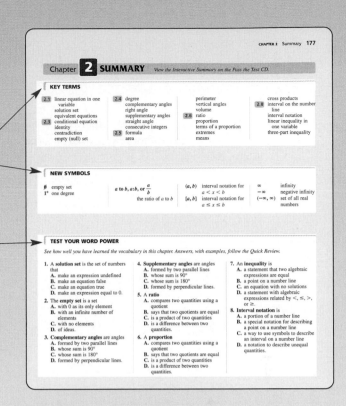

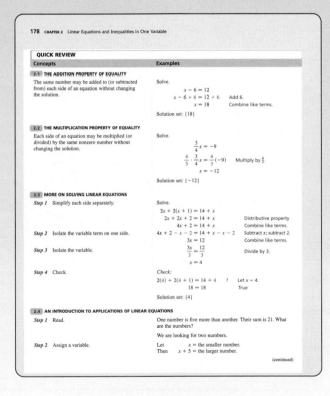

Quick Review sections give students main concepts from the chapter (referenced back to the appropriate section) and an adjacent example of each concept. **NEW** Quick Review Summary Lectures are available for each concept on the Pass The Test CD.

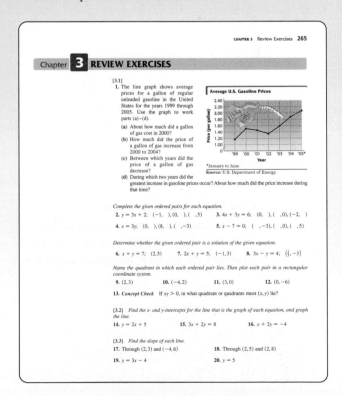

Review Exercises are keyed to the appropriate sections so that students can refer to examples of that type of problem if they need help.

Mixed Review Exercises require students to solve problems without the help of section references.

MIXED REVIEW EXERCISES

Solve.

69. $\frac{x}{7} = \frac{x-5}{2}$ **70.** $I = prt$ for r **71.** $-2x > -4$

72. $2k - 5 = 4k + 13$ **73.** $0.05x + 0.02x = 4.9$ **74.** $2 - 3(x - 5) = 4 + x$

75. $9x - (7x + 2) = 3x + (2 - x)$ **76.** $\frac{1}{3}s + \frac{1}{2}s + 7 = \frac{5}{6}s + 5 + 2$

77. *Concept Check* A student solved $3 - (8 + 4x) = 2x + 7$ and gave the solution set {6}. Verify that this answer is incorrect by checking it in the equation. Then explain the error and give the correct solution set. (*Hint:* The error involves the subtraction sign.)

78. Athletes in vigorous training programs can eat 50 calories per day for every 2.2 lb of body weight. To the nearest hundred, how many calories can a 175-lb athlete consume per day? (*Source: The Gazette,* Cedar Rapids Iowa, March 23, 2002.)

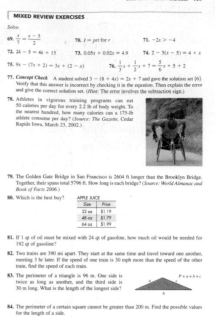

79. The Golden Gate Bridge in San Francisco is 2604 ft longer than the Brooklyn Bridge. Together, their spans total 5796 ft. How long is each bridge? (*Source: World Almanac and Book of Facts 2006.*)

80. Which is the best buy?

APPLE JUICE	
Size	Price
32 oz	$1.19
48 oz	$1.79
64 oz	$1.99

81. If 1 qt of oil must be mixed with 24 qt of gasoline, how much oil would be needed for 192 qt of gasoline?

82. Two trains are 390 mi apart. They start at the same time and travel toward one another, meeting 3 hr later. If the speed of one train is 30 mph more than the speed of the other train, find the speed of each train.

83. The perimeter of a triangle is 96 m. One side is twice as long as another, and the third side is 30 m long. What is the length of the longest side? $P = a + b + c$

84. The perimeter of a certain square cannot be greater than 200 m. Find the possible values for the length of a side.

---

186 CHAPTER 2 Linear Equations and Inequalities in One Variable

Chapter **2** TEST

View the complete solutions to all Chapter Test exercises on the Pass the Test CD.

Solve each equation.

1. $5x + 9 = 7x + 21$ **2.** $-\frac{4}{7}x = -12$

3. $7 - (x - 4) = -3x + 2(x + 1)$ **4.** $0.6(x + 20) + 0.8(x - 10) = 46$

5. $-8(2x + 4) = -4(4x + 8)$

Solve each problem.

6. In the 2005 baseball season, the St. Louis Cardinals won the most games of any major league team. The Cardinals won 24 less than twice as many games as they lost. They played 162 regular-season games. How many wins and losses did the Cardinals have? (*Source:* mlb.com)

7. Three islands in the Hawaiian island chain are Hawaii (the Big Island), Maui, and Kauai. Together, their areas total 5300 mi². The island of Hawaii is 3293 mi² larger than the island of Maui, and Maui is 177 mi² larger than Kauai. What is the area of each island?

8. Find the measure of an angle if its supplement measures 10° more than three times its complement.

9. The formula for the perimeter of a rectangle is $P = 2L + 2W$.
   (a) Solve for W.
   (b) If $P = 116$ and $L = 40$, find the value of W.

10. Find the measure of each marked angle.

$(3x + 15)°$ $(4x - 5)°$

Solve each proportion.

11. $\frac{z}{8} = \frac{12}{16}$ **12.** $\frac{x+5}{3} = \frac{x-3}{4}$

Solve each problem.

13. Which is the better buy for processed cheese slices, 8 slices for $2.19 or 12 slices for $3.30?

14. The distance between Milwaukee and Boston is 1050 mi. On a certain map, this distance is represented by 42 in. On the same map, Seattle and Cincinnati are 92 in. apart. What is the actual distance between Seattle and Cincinnati?

Chapter Tests help students practice for the real thing. **NEW** The Pass the Test CD in the back of the text offers video of an instructor working through the complete solution for every exercise from the chapter tests.

Cumulative Review Exercises gather various types of exercises from preceding chapters to help students remember and retain what they are learning throughout the course.

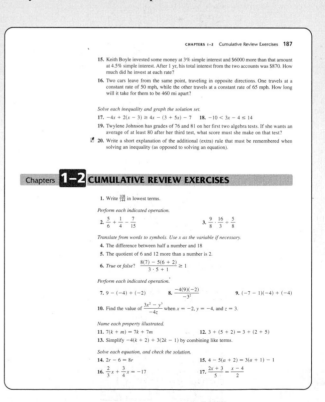

CHAPTERS 1–2 Cumulative Review Exercises **187**

15. Keith Boyle invested some money at 3% simple interest and $6000 more than that amount at 4.5% simple interest. After 1 yr, his total interest from the two accounts was $870. How much did he invest at each rate?

16. Two cars leave from the same point, traveling in opposite directions. One travels at a constant rate of 50 mph, while the other travels at a constant rate of 65 mph. How long will it take for them to be 460 mi apart?

Solve each inequality and graph the solution set.

17. $-4x + 2(x - 3) \geq 4x - (3 + 5x) - 7$ **18.** $-10 < 3x - 4 \leq 14$

19. Twylene Johnson has grades of 76 and 81 on her first two algebra tests. If she wants an average of at least 80 after her third test, what score must she make on that test?

20. Write a short explanation of the additional (extra) rule that must be remembered when solving an inequality (as opposed to solving an equation).

Chapters **1–2** CUMULATIVE REVIEW EXERCISES

1. Write $\frac{108}{124}$ in lowest terms.

Perform each indicated operation.

2. $\frac{5}{6} + \frac{1}{4} - \frac{7}{15}$ **3.** $\frac{9}{8} \cdot \frac{16}{3} \div \frac{5}{8}$

Translate from words to symbols. Use x as the variable if necessary.

4. The difference between half a number and 18

5. The quotient of 6 and 12 more than a number is 2.

6. *True or false?* $\frac{8(7) - 5(6 + 2)}{3 \cdot 5 + 1} \geq 1$

Perform each indicated operation.

7. $9 - (-4) + (-2)$ **8.** $\frac{-4(9)(-2)}{-3^2}$ **9.** $(-7 - 1)(-4) + (-4)$

10. Find the value of $\frac{3x^2 - y^3}{-4z}$ when $x = -2$, $y = -4$, and $z = 3$.

Name each property illustrated.

11. $7(k + m) = 7k + 7m$ **12.** $3 + (5 + 2) = 3 + (2 + 5)$

13. Simplify $-4(k + 2) + 3(2k - 1)$ by combining like terms.

Solve each equation, and check the solution.

14. $2r - 6 = 8r$ **15.** $4 - 5(a + 2) = 3(a + 1) - 1$

16. $\frac{2}{3}x + \frac{3}{4}x = -17$ **17.** $\frac{2x+3}{5} = \frac{x-4}{2}$

Tab Your Way to Success
Use these tabs to mark important sections and pages.

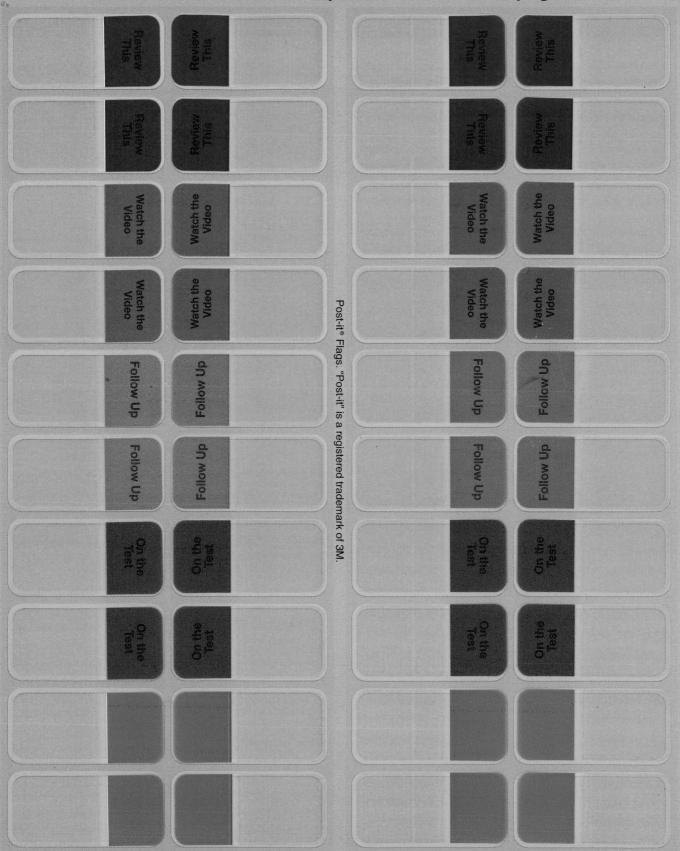

Post-it® Flags. "Post-it" is a registered trademark of 3M.

TAKE CHARGE OF YOUR LEARNING

These tabs offer several ways to be successful in your mathematics class by taking charge of your own learning.

Review This
Use this tab to identify the most important information in each chapter so that you can refer to these pages easily.

Watch the Video
Tab an exercise for which you would like to see the complete solution worked out on video.

Follow Up
Tab any material you need extra help on or wish to discuss with your instructor.

On the Test
Mark material that your instructor emphasizes or material that you expect will be on the test.

Blank Tab
Use this tab to mark any information you might like to refer to again easily.

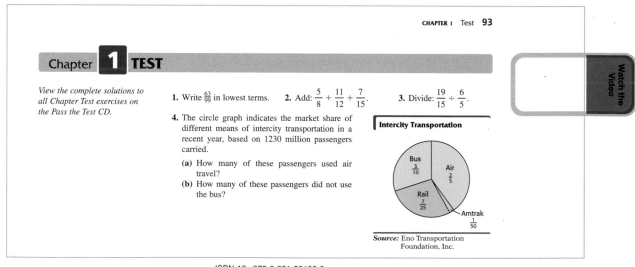

ISBN-13: 978-0-321-50135-6
ISBN-10:   0-321-50135-7

NEW *Pass the Test: Chapter Test Solutions on Video with Interactive Chapter Summaries on CD,* included with each new copy of the book, is based on the end-of-chapter material from the book and offers the following tools for each chapter:

- Interactive *Key Terms* with definitions
- Interactive *Test Your Word Power* vocabulary exercises
- Summary lectures for each key concept from the *Quick Review*
- Video footage of an instructor working through the complete solutions for all chapter test problems.

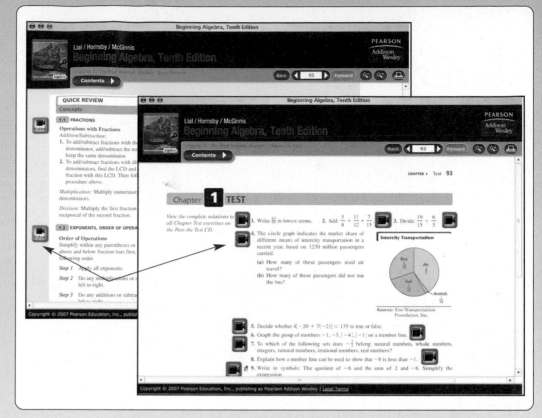

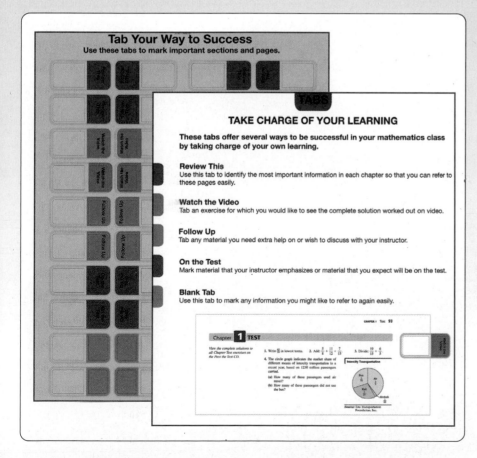

NEW *Tab Your Way to Success* appears at the front of the text. This page of color-coded Post-It® tabs makes it easy for students to flag pages they want to return to for review, test preparation, or instructor help.

The Real Number System

The personal savings rate of Americans has declined steadily since 1984, when it stood at 10.8% of after-tax income. In 2005, Americans spent everything they earned and more, causing the personal savings rate to drop to -0.5%, the first time there has been a negative personal savings rate since 1933. That year, Americans depleted their savings to cope with the job layoffs and business failures of the Great Depression. (*Source:* Commerce Department.)

In this chapter, we examine *positive* and *negative numbers* and apply them to situations such as the personal savings rate of Americans in Exercise 117 of Section 1.5.

1.1 Fractions

In everyday life, the numbers seen most often are the **natural numbers,**

$$1, 2, 3, 4, \ldots,$$

the **whole numbers,**

$$0, 1, 2, 3, 4, \ldots,$$

and **fractions,** such as

$$\frac{1}{2}, \quad \frac{2}{3}, \quad \text{and} \quad \frac{15}{7}.$$

The parts of a fraction are named as follows:

$$\text{Fraction bar} \rightarrow \frac{4}{7}. \begin{array}{l} \leftarrow \text{Numerator} \\ \leftarrow \text{Denominator} \end{array}$$

The fraction bar represents division $\left(\frac{a}{b} = a \div b\right)$ and also serves as a grouping symbol.

If the numerator of a fraction is less than the denominator, we call the fraction a **proper fraction.** A proper fraction has a value less than 1. If the numerator is greater than or equal to the denominator, the fraction is an **improper fraction** and has a value greater than or equal to 1. An improper fraction that is greater than 1 is often written as a **mixed number.** For example,

$$\text{Improper fraction} \rightarrow \frac{12}{5} \quad \text{may be written as} \quad 2\frac{2}{5}. \leftarrow \text{Mixed number}$$

OBJECTIVE 1 Learn the definition of *factor.* In the statement $2 \times 9 = 18$, the numbers 2 and 9 are called **factors** of 18. Other factors of 18 include 1, 3, 6, and 18. The result of the multiplication, 18, is called the **product.** We can represent the product of two numbers in several ways. For example, the following all represent the product of 6 and 3:

$$6 \times 3, \quad 6 \cdot 3, \quad (6)(3), \quad 6(3), \quad (6)3. \quad \textbf{Products}$$

The number 18 is **factored** by writing it as the product of two or more numbers. For example, 18 can be factored as

$$6 \cdot 3, \quad 18 \cdot 1, \quad 9 \cdot 2, \quad \text{or} \quad 3 \cdot 3 \cdot 2.$$

In algebra, a raised dot $\cdot$ is often used instead of the $\times$ symbol to indicate multiplication because $\times$ may be confused with the letter x.

A natural number greater than 1 is **prime** if it has only itself and 1 as factors. "Factors" are understood here to mean natural number factors.

$$2, 3, 5, 7, 11, 13, 17, 19, 23, 29, 31, 37 \quad \textbf{First dozen prime numbers}$$

A natural number greater than 1 that is not prime is called a **composite number.** Some examples follow.

$$4, 6, 8, 9, 10, 12 \quad \textbf{Composite numbers}$$

By agreement, the number 1 is neither prime nor composite.

Sometimes we must find all **prime factors** of a number—those factors which are prime numbers. For example, the only prime factors of 18 are 2 and 3.

EXAMPLE 1 Factoring Numbers

Write each number as the product of prime factors.

(a) 35

Write 35 as the product of the prime fractors 5 and 7, or as

$$35 = 5 \cdot 7.$$

(b) 24

One way to begin is to divide by the smallest prime, 2, to get

$$24 = 2 \cdot 12.$$

Now divide 12 by 2 to find factors of 12.

$$24 = 2 \cdot 2 \cdot 6$$

Since 6 can be written as $2 \cdot 3$,

$$24 = 2 \cdot 2 \cdot 2 \cdot 3. \quad \leftarrow \text{All factors are prime.}$$

✔ **Now Try Exercises 9 and 19.**

▶ **NOTE** We need not start with the smallest prime factor, as shown in Example 1(b). In fact, no matter which prime factor we start with, we will *always* obtain the same prime factorization.

OBJECTIVE 2 Write fractions in lowest terms. A fraction is in **lowest terms** when the numerator and denominator have no factors in common (other than 1). We use the **basic principle of fractions** to write a fraction in lowest terms.

Basic Principle of Fractions

If the numerator and denominator of a fraction are multiplied or divided by the same nonzero number, the value of the fraction is not changed.

For example, $\frac{12}{16}$ can be written in lowest terms as follows:

$$\frac{12}{16} = \frac{3 \cdot 4}{4 \cdot 4} = \frac{3}{4} \cdot \frac{4}{4} = \frac{3}{4} \cdot 1 = \frac{3}{4}. \qquad \frac{4}{4} = 1$$

This procedure uses the rule for multiplying fractions (covered in the next objective) and the multiplication property of 1 (covered in **Section 1.7**).

To write a fraction in lowest terms, use these steps.

Writing a Fraction in Lowest Terms

Step 1 Write the numerator and the denominator as the product of prime factors.

Step 2 Divide the numerator and the denominator by the **greatest common factor,** the product of all factors common to both.

EXAMPLE 2 Writing Fractions in Lowest Terms

Write each fraction in lowest terms.

(a) $\dfrac{10}{15} = \dfrac{2 \cdot 5}{3 \cdot 5} = \dfrac{2 \cdot 1}{3 \cdot 1} = \dfrac{2}{3}$

The factored form shows that 5 is the greatest common factor of 10 and 15. Dividing both numerator and denominator by 5 gives $\frac{10}{15}$ in lowest terms as $\frac{2}{3}$.

(b) $\dfrac{15}{45} = \dfrac{3 \cdot 5}{3 \cdot 3 \cdot 5} = \dfrac{1 \cdot 1}{3 \cdot 1 \cdot 1} = \dfrac{1}{3}$

The factored form shows that 3 and 5 are the prime factors of both 15 and 45. Dividing both 15 and 45 by $3 \cdot 5 = 15$ gives $\frac{15}{45}$ in lowest terms as $\frac{1}{3}$.

✔ **Now Try Exercises 25 and 27.**

We can simplify this process by finding the greatest common factor in the numerator and denominator by inspection. For instance, in Example 2(b), we can use 15 rather than $3 \cdot 5$.

$$\frac{15}{45} = \frac{15}{3 \cdot 15} = \frac{1}{3 \cdot 1} = \frac{1}{3}$$

Remember to write 1 in the numerator.

▶ **CAUTION** When writing fractions like $\frac{15}{45}$ from Example 2(b) in lowest terms, be sure to include the factor 1 in the numerator or an error may result.

OBJECTIVE 3 Multiply and divide fractions.

Multiplying Fractions

If $\dfrac{a}{b}$ and $\dfrac{c}{d}$ are fractions, then $\qquad \dfrac{a}{b} \cdot \dfrac{c}{d} = \dfrac{a \cdot c}{b \cdot d}.$

That is, to multiply two fractions, multiply their numerators and then multiply their denominators.

EXAMPLE 3 Multiplying Fractions

Find each product, and write it in lowest terms.

(a) $\dfrac{3}{8} \cdot \dfrac{4}{9} = \dfrac{3 \cdot 4}{8 \cdot 9}$ Multiply numerators.
Multiply denominators.

$\qquad = \dfrac{3 \cdot 4}{2 \cdot 4 \cdot 3 \cdot 3}$ Factor the denominator.

$\qquad = \dfrac{1}{2 \cdot 3}$ Divide numerator and denominator by $3 \cdot 4$, or 12.

Remember to write 1 in the numerator.

$\qquad = \dfrac{1}{6}$ Lowest terms

Think: $4 \cdot 5 = 20$, and $20 + 1 = 21$, so $5\frac{1}{4} = \frac{21}{4}$.

(b) $2\dfrac{1}{3} \cdot 5\dfrac{1}{4} = \dfrac{7}{3} \cdot \dfrac{21}{4}$ Write as improper fractions.

Think: $3 \cdot 2 = 6$, and $6 + 1 = 7$, so $2\frac{1}{3} = \frac{7}{3}$.

$= \dfrac{7 \cdot 21}{3 \cdot 4}$ Multiply numerators.
Multiply denominators.

$= \dfrac{7 \cdot 3 \cdot 7}{3 \cdot 4}$ Factor the numerator.

$= \dfrac{49}{4},$ or $12\dfrac{1}{4}$ Write in lowest terms
and as a mixed number.

✔ **Now Try Exercises 35 and 41.**

▶ **NOTE** Some students prefer to divide out any common factors before multiplying. For example,

$$\dfrac{3}{8} \cdot \dfrac{4}{9} = \dfrac{3}{2 \cdot 4} \cdot \dfrac{4}{3 \cdot 3} \qquad \text{Example 3(a)}$$

$$= \dfrac{1}{2 \cdot 3} = \dfrac{1}{6}. \qquad \text{The same answer results.}$$

Two fractions are **reciprocals** of each other if their product is 1. For example,

$$\dfrac{3}{4} \cdot \dfrac{4}{3} = \dfrac{12}{12} = 1,$$

so $\frac{3}{4}$ and $\frac{4}{3}$ are reciprocals, as are $\frac{1}{6}$ and 6 $\left(\text{since } 6 = \frac{6}{1}\right)$. Because division is the opposite (or inverse) of multiplication, we use reciprocals to divide fractions.

Dividing Fractions

If $\dfrac{a}{b}$ and $\dfrac{c}{d}$ are fractions, then $\dfrac{a}{b} \div \dfrac{c}{d} = \dfrac{a}{b} \cdot \dfrac{d}{c}.$

That is, to divide by a fraction, multiply by its reciprocal.

We will explain why this method works in **Chapter 7.** However, as an example, we know that $20 \div 10 = 2$ and $20 \cdot \frac{1}{10} = 2$. The answer to a division problem is called a **quotient.** For example, the quotient of 20 and 10 is 2.

EXAMPLE 4 Dividing Fractions

Find each quotient, and write it in lowest terms.

(a) $\dfrac{3}{4} \div \dfrac{8}{5} = \dfrac{3}{4} \cdot \dfrac{5}{8} = \dfrac{3 \cdot 5}{4 \cdot 8} = \dfrac{15}{32}$

Multiply by the reciprocal of the second fraction.

(b) $\dfrac{3}{4} \div \dfrac{5}{8} = \dfrac{3}{4} \cdot \dfrac{8}{5} = \dfrac{3 \cdot 8}{4 \cdot 5} = \dfrac{3 \cdot 4 \cdot 2}{4 \cdot 5} = \dfrac{6}{5},$ or $1\dfrac{1}{5}$

> Remember to write in lowest terms.

(c) $\dfrac{5}{8} \div 10 = \dfrac{5}{8} \div \dfrac{10}{1} = \dfrac{5}{8} \cdot \dfrac{1}{10} = \dfrac{5 \cdot 1}{8 \cdot 10} = \dfrac{5 \cdot 1}{8 \cdot 5 \cdot 2} = \dfrac{1}{16}$

Write 10 as $\frac{10}{1}$.

(d) $1\dfrac{2}{3} \div 4\dfrac{1}{2} = \dfrac{5}{3} \div \dfrac{9}{2}$ Write as improper fractions.

$= \dfrac{5}{3} \cdot \dfrac{2}{9}$ Multiply by the reciprocal of the second fraction.

$= \dfrac{10}{27}$ Multiply numerators.
Multiply denominators.

> ✔ **Now Try Exercises 43, 45, and 49.**

OBJECTIVE 4 **Add and subtract fractions.** The result of adding two numbers is called the **sum** of the numbers. For example, $2 + 3 = 5$, so 5 is the sum of 2 and 3.

Adding Fractions

If $\dfrac{a}{b}$ and $\dfrac{c}{b}$ are fractions, then $\dfrac{a}{b} + \dfrac{c}{b} = \dfrac{a + c}{b}.$

That is, to find the sum of two fractions having the *same* denominator, add the numerators and keep the *same* denominator.

EXAMPLE 5 **Adding Fractions with the Same Denominator**

Find each sum, and write it in lowest terms.

(a) $\dfrac{3}{7} + \dfrac{2}{7} = \dfrac{3 + 2}{7} = \dfrac{5}{7}$ Add numerators; keep the same denominator.

(b) $\dfrac{2}{10} + \dfrac{3}{10} = \dfrac{2 + 3}{10} = \dfrac{5}{10} = \dfrac{1}{2}$ Write in lowest terms.

> ✔ **Now Try Exercise 55.**

If the fractions to be added do not have the same denominators, we can still use the preceding method, but only *after* we rewrite the fractions with a common denominator. For example, to rewrite $\frac{3}{4}$ as a fraction with denominator 32, that is,

$$\dfrac{3}{4} = \dfrac{?}{32},$$

we must find the number that can be multiplied by 4 to give 32. Since $4 \cdot 8 = 32$, we use the number 8.

$$\dfrac{3}{4} = \dfrac{3 \cdot 8}{4 \cdot 8} = \dfrac{24}{32}$$ Multiply numerator and denominator by 8.

Finding the Least Common Denominator

To add or subtract fractions with different denominators, find the **least common denominator (LCD)** as follows:

Step 1 Factor each denominator.

Step 2 For the LCD, use every factor that appears in any factored form. If a factor is repeated, use the largest number of repeats in the LCD.

EXAMPLE 6 Adding Fractions with Different Denominators

Find each sum, and write it in lowest terms.

(a) $\dfrac{4}{15} + \dfrac{5}{9}$

To find the least common denominator, first factor both denominators.

$$15 = 5 \cdot 3 \qquad \text{and} \qquad 9 = 3 \cdot 3$$

Since 5 and 3 appear as factors, and 3 is a factor of 9 twice, the LCD is

$$\overset{\displaystyle 15 \quad\ \ 9}{\underset{\displaystyle 5 \cdot 3 \cdot 3,}{\wedge\ \ \wedge}} \qquad \text{or} \qquad 45.$$

Write each fraction with 45 as denominator.

$$\frac{4}{15} = \frac{4 \cdot 3}{15 \cdot 3} = \frac{12}{45} \qquad \text{and} \qquad \frac{5}{9} = \frac{5 \cdot 5}{9 \cdot 5} = \frac{25}{45}$$

Now add the two equivalent fractions.

$$\frac{4}{15} + \frac{5}{9} = \frac{12}{45} + \frac{25}{45} = \frac{37}{45}$$

(b) $3\dfrac{1}{2} + 2\dfrac{3}{4}$

We add mixed numbers by using either of two methods.

Method 1 Write the mixed numbers as improper fractions.

$$3\frac{1}{2} + 2\frac{3}{4} = \frac{7}{2} + \frac{11}{4} \qquad\qquad \text{Change to improper fractions.}$$

$$= \frac{14}{4} + \frac{11}{4} \qquad\qquad \text{Get a common denominator.}$$

$$= \frac{25}{4}, \quad \text{or} \quad 6\frac{1}{4} \qquad \text{Add; write as a mixed number.}$$

Method 2 Write $3\frac{1}{2}$ as $3\frac{2}{4}$. Then add vertically.

$$\left.\begin{array}{l} 3\dfrac{1}{2} = 3\dfrac{2}{4} \\[2mm] + \ 2\dfrac{3}{4} = 2\dfrac{3}{4} \end{array}\right\} \quad \begin{array}{l}\text{Add the whole numbers and} \\ \text{the fractions separately.}\end{array}$$

$$5\frac{5}{4} = 5 + 1\frac{1}{4} = 6\frac{1}{4}, \quad \text{or} \quad \frac{25}{4}$$

✔ **Now Try Exercises 57 and 59.**

The **difference** between two numbers is found by subtracting the numbers. For example, $9 - 5 = 4$, so the difference between 9 and 5 is 4. Subtraction of fractions is similar to addition.

Subtracting Fractions

If $\dfrac{a}{b}$ and $\dfrac{c}{b}$ are fractions, then $\qquad \dfrac{a}{b} - \dfrac{c}{b} = \dfrac{a - c}{b}.$

That is, to find the difference between two fractions having the *same* denominator, subtract the numerators and keep the *same* denominator.

EXAMPLE 7 Subtracting Fractions

Find each difference, and write it in lowest terms.

(a) $\dfrac{15}{8} - \dfrac{3}{8} = \dfrac{15 - 3}{8}$ Subtract numerators; keep the same denominator.

$= \dfrac{12}{8} = \dfrac{3}{2},$ or $1\dfrac{1}{2}$ Write in lowest terms and as a mixed number.

(b) $\dfrac{7}{18} - \dfrac{4}{15}$

Here, $18 = 2 \cdot 3 \cdot 3$ and $15 = 3 \cdot 5$, so the LCD is $2 \cdot 3 \cdot 3 \cdot 5 = 90$.

$$\dfrac{7}{18} - \dfrac{4}{15} = \dfrac{7 \cdot 5}{2 \cdot 3 \cdot 3 \cdot 5} - \dfrac{4 \cdot 2 \cdot 3}{2 \cdot 3 \cdot 3 \cdot 5} = \dfrac{35}{90} - \dfrac{24}{90} = \dfrac{11}{90}$$

(c) $\dfrac{15}{32} - \dfrac{11}{45}$

Since $32 = 2 \cdot 2 \cdot 2 \cdot 2 \cdot 2$ and $45 = 3 \cdot 3 \cdot 5$, there are no common factors, and the LCD is $32 \cdot 45 = 1440$.

$$\dfrac{15}{32} - \dfrac{11}{45} = \dfrac{15 \cdot 45}{32 \cdot 45} - \dfrac{11 \cdot 32}{45 \cdot 32}$$ Get a common denominator.

$$= \dfrac{675}{1440} - \dfrac{352}{1440}$$

$$= \dfrac{323}{1440}$$ Subtract numerators; keep the common denominator.

(d) $4\dfrac{1}{2} - 1\dfrac{3}{4} = \dfrac{9}{2} - \dfrac{7}{4}$ Change to improper fractions.

$= \dfrac{18}{4} - \dfrac{7}{4}$ Get a common denominator.

$= \dfrac{11}{4},$ or $2\dfrac{3}{4}$ Subtract; write as a mixed number.

Alternatively, we could use the vertical method.

$$4\frac{1}{2} = 4\frac{2}{4} = 3\frac{6}{4} \qquad 4\frac{2}{4} = 3 + 1 + \frac{2}{4} = 3 + \frac{4}{4} + \frac{2}{4} = 3\frac{6}{4}$$

$$-1\frac{3}{4} = 1\frac{3}{4} = 1\frac{3}{4}$$

$$2\frac{3}{4}, \quad \text{or} \quad \frac{11}{4}$$

> ✔ **Now Try Exercises 63, 65, and 69.**

OBJECTIVE 5 Solve applied problems that involve fractions.

EXAMPLE 8 Adding Fractions to Solve an Applied Problem

The diagram in Figure 1 appears in the book *Woodworker's 39 Sure-Fire Projects.* It is the front view of a corner bookcase/desk. Find the height of the bookcase/desk from the floor to the top of the writing surface.

We must add the following measures (in the figure, ″ means inches):

$$\frac{3}{4}, \quad 4\frac{1}{2}, \quad 9\frac{1}{2}, \quad \frac{3}{4}, \quad 9\frac{1}{2}, \quad \frac{3}{4}, \quad 4\frac{1}{2}.$$

We begin by changing $4\frac{1}{2}$ to $4\frac{2}{4}$ and $9\frac{1}{2}$ to $9\frac{2}{4}$, since the common denominator is 4. Then we use Method 2 from Example 6(b).

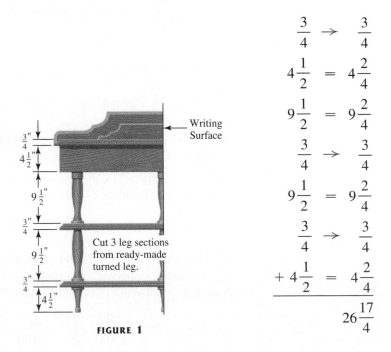

$$\frac{3}{4} \rightarrow \frac{3}{4}$$
$$4\frac{1}{2} = 4\frac{2}{4}$$
$$9\frac{1}{2} = 9\frac{2}{4}$$
$$\frac{3}{4} \rightarrow \frac{3}{4}$$
$$9\frac{1}{2} = 9\frac{2}{4}$$
$$\frac{3}{4} \rightarrow \frac{3}{4}$$
$$+ 4\frac{1}{2} = 4\frac{2}{4}$$
$$26\frac{17}{4}$$

FIGURE 1

Since $\frac{17}{4} = 4\frac{1}{4}$, $26\frac{17}{4} = 26 + 4\frac{1}{4} = 30\frac{1}{4}$. The height is $30\frac{1}{4}$ in.

> ✔ **Now Try Exercise 77.**

OBJECTIVE 6 Interpret data in a circle graph. A **circle graph,** or **pie chart,** is often used to give a pictorial representation of data. A circle is used to indicate the total of all the categories represented. The circle is divided into sectors, or wedges (like pieces of pie), whose sizes show the relative magnitudes of the categories. The sum of all the fractional parts must be 1 (for 1 whole circle).

EXAMPLE 9 Using a Circle Graph to Interpret Information

In November 2005, there were about 970 million Internet users worldwide. The circle graph in Figure 2 shows the approximate fractions of these users living in various regions of the world.

Worldwide Internet Users By Region

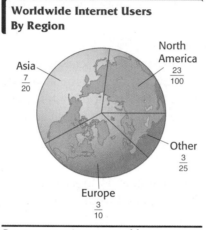

Source: www.internetworldstats.com

FIGURE 2

(a) Which region had the largest share of Internet users in November 2005? What was that share?

In the circle graph, the sector for Asia is the largest, so Asia had the largest share of Internet users, $\frac{7}{20}$.

(b) Estimate the number of Internet users in North America in November 2005.

A share of $\frac{23}{100}$ can be rounded to $\frac{25}{100}$, or $\frac{1}{4}$, and the total number of Internet users, 970 million, can be rounded to 1000 million (1 billion). We multiply $\frac{1}{4}$ by 1000. The number of Internet users in North America would be about

$$\frac{1}{4}(1000) = 250 \text{ million.}$$

(c) How many actual Internet users were there in North America in November 2005?

To find the answer, we multiply the actual fraction from the graph for North America, $\frac{23}{100}$, by the number of users, 970 million:

$$\frac{23}{100}(970) = \frac{23}{100} \cdot \frac{970}{1} = \frac{22{,}310}{100} = 223\frac{1}{10}.$$

Thus, $223\frac{1}{10}$ million, or 223,100,000, people in North America used the Internet. This number is reasonable, given our estimate in part (b).

✔ Now Try Exercises 87 and 89.

1.1 EXERCISES

Concept Check *Decide whether each statement is* true *or* false. *If it is false, say why.*

1. In the fraction $\frac{3}{7}$, 3 is the numerator and 7 is the denominator.

2. The mixed number equivalent of $\frac{41}{5}$ is $8\frac{1}{5}$.

3. The fraction $\frac{17}{51}$ is in lowest terms.

4. The reciprocal of $\frac{8}{2}$ is $\frac{4}{1}$.

5. The product of 8 and 2 is 10.

6. The difference between 12 and 2 is 6.

Identify each number as prime, composite, *or* neither. *If the number is composite, write it as the product of prime factors. See Example 1.*

7. 19   **8.** 31   ⊙ **9.** 64   **10.** 99

11. 3458   **12.** 1025   **13.** 1   **14.** 0

15. 30   **16.** 40   **17.** 500   **18.** 700

19. 124   **20.** 120   **21.** 29   **22.** 83

Write each fraction in lowest terms. See Example 2.

23. $\frac{8}{16}$ **24.** $\frac{4}{12}$ ⊙ **25.** $\frac{15}{18}$ **26.** $\frac{16}{20}$

27. $\frac{18}{90}$ **28.** $\frac{16}{64}$ **29.** $\frac{144}{120}$ **30.** $\frac{132}{77}$

31. *Concept Check* One of the following is the correct way to write $\frac{16}{24}$ in lowest terms. Which one is it?

A. $\frac{16}{24} = \frac{8+8}{8+16} = \frac{8}{16} = \frac{1}{2}$ **B.** $\frac{16}{24} = \frac{4 \cdot 4}{4 \cdot 6} = \frac{4}{6}$

C. $\frac{16}{24} = \frac{8 \cdot 2}{8 \cdot 3} = \frac{2}{3}$ **D.** $\frac{16}{24} = \frac{14+2}{21+3} = \frac{2}{3}$

32. *Concept Check* For the fractions $\frac{p}{q}$ and $\frac{r}{s}$, which one of the following can serve as a common denominator?

A. $q \cdot s$ **B.** $q + s$ **C.** $p \cdot r$ **D.** $p + r$

Find each product or quotient, and write it in lowest terms. See Examples 3 and 4.

33. $\frac{4}{5} \cdot \frac{6}{7}$ **34.** $\frac{5}{9} \cdot \frac{10}{7}$ ⊙ **35.** $\frac{1}{10} \cdot \frac{12}{5}$ **36.** $\frac{6}{11} \cdot \frac{2}{3}$

37. $\frac{15}{4} \cdot \frac{8}{25}$ **38.** $\frac{4}{7} \cdot \frac{21}{8}$ **39.** $3\frac{1}{4} \cdot 1\frac{2}{3}$ **40.** $2\frac{2}{3} \cdot 5\frac{4}{5}$

41. $2\frac{3}{8} \cdot 3\frac{1}{5}$ **42.** $3\frac{3}{5} \cdot 7\frac{1}{6}$ ⊙ **43.** $\frac{5}{4} \div \frac{3}{8}$ **44.** $\frac{7}{6} \div \frac{9}{10}$

45. $\frac{32}{5} \div \frac{8}{15}$ **46.** $\frac{24}{7} \div \frac{6}{21}$ **47.** $\frac{3}{4} \div 12$ **48.** $\frac{2}{5} \div 30$

49. $2\frac{1}{2} \div 1\frac{5}{7}$ **50.** $2\frac{2}{9} \div 1\frac{2}{5}$ **51.** $2\frac{5}{8} \div 1\frac{15}{32}$ **52.** $2\frac{3}{10} \div 7\frac{4}{5}$

▨ **53.** Write a summary explaining how to multiply and divide two fractions. Give examples.

▨ **54.** Write a summary explaining how to add and subtract two fractions. Give examples.

Find each sum or difference, and write it in lowest terms. See Examples 5–7.

55. $\dfrac{7}{12} + \dfrac{1}{12}$ **56.** $\dfrac{3}{16} + \dfrac{5}{16}$ **57.** $\dfrac{5}{9} + \dfrac{1}{3}$ **58.** $\dfrac{4}{15} + \dfrac{1}{5}$

59. $3\dfrac{1}{8} + 2\dfrac{1}{4}$ **60.** $4\dfrac{2}{3} + 2\dfrac{1}{6}$ **61.** $3\dfrac{1}{4} + 1\dfrac{4}{5}$ **62.** $5\dfrac{3}{4} + 3\dfrac{1}{3}$

63. $\dfrac{13}{15} - \dfrac{3}{15}$ **64.** $\dfrac{11}{12} - \dfrac{3}{12}$ **65.** $\dfrac{7}{12} - \dfrac{1}{9}$ **66.** $\dfrac{11}{16} - \dfrac{1}{12}$

67. $4\dfrac{3}{4} - 1\dfrac{2}{5}$ **68.** $8\dfrac{4}{5} - 7\dfrac{4}{9}$ **69.** $6\dfrac{1}{4} - 5\dfrac{1}{3}$ **70.** $5\dfrac{1}{3} - 2\dfrac{1}{2}$

Solve each problem. See Example 8.

Use the table to answer Exercises 71 and 72.

	Microwave	Stove Top		
Servings	1	1	4	6
Water	$\frac{3}{4}$ cup	1 cup	3 cups	4 cups
Grits	3 Tbsp	3 Tbsp	$\frac{3}{4}$ cup	1 cup
Salt (optional)	Dash	Dash	$\frac{1}{4}$ tsp	$\frac{1}{2}$ tsp

Source: Package of Quaker Quick Grits.

71. How many cups of water would be needed for eight microwave servings?

72. How many tsp of salt would be needed for five stove-top servings? (*Hint:* 5 is halfway between 4 and 6.)

Oversized drivers have resulted in the use of longer wooden golf tees. As a result, the Pride Golf Tee Company, the only U.S. manufacturer of wooden golf tees, has created the Professional Tee System, shown in the figure. Use the information given to work Exercises 73 and 74. (Source: The Gazette, Cedar Rapids, Iowa, August 27, 2005.)

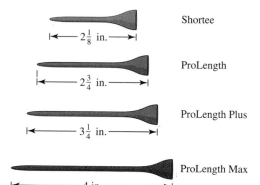

73. Find the difference in length between the ProLength Plus and the once-standard Shortee.

74. The ProLength Max tee is the longest tee allowed by the U.S. Golf Association's *Rules of Golf.* How much longer is the ProLength Max than the Shortee?

75. A hardware store sells a 40-piece socket wrench set. The measure of the largest socket is $\frac{3}{4}$ in. The measure of the smallest is $\frac{3}{16}$ in. What is the difference between these measures?

76. Two sockets in a socket wrench set have measures of $\frac{9}{16}$ in. and $\frac{3}{8}$ in. What is the difference between these two measures?

77. A business owner has decided to expand by buying a piece of property next to his store. The property has an irregular shape, with five sides, as shown in the figure. Find the total distance around the piece of property. (This distance is called the **perimeter** of the figure.)

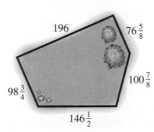

Measurements in feet

78. Find the perimeter of the triangle in the figure.

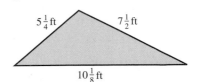

79. A piece of board is $15\frac{5}{8}$ in. long. If it must be divided into three pieces of equal length, how long must each piece be?

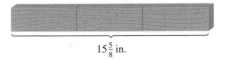

$15\frac{5}{8}$ in.

80. Sondra Braeseker's favorite recipe for barbecue sauce calls for $2\frac{1}{3}$ cups of tomato sauce. The recipe makes enough barbecue sauce to serve seven people. How much tomato sauce is needed for one serving?

81. A cake recipe calls for $1\frac{3}{4}$ cups of sugar. A caterer has $15\frac{1}{2}$ cups of sugar on hand. How many cakes can he make?

82. An upholsterer needs $2\frac{1}{4}$ yd of fabric to cover a chair. How many chairs can be covered with $23\frac{2}{3}$ yd of fabric?

83. It takes $2\frac{3}{8}$ yd of fabric to make a costume for a school play. How much fabric would be needed for seven costumes?

84. A cookie recipe calls for $2\frac{2}{3}$ cups of sugar. How much sugar would be needed to make four batches of cookies?

85. First published in 1953, the digest-sized *TV Guide* has recently been changed to a full-sized magazine. The new magazine is 3 in. wider than the old guide. What is the difference in their heights? (*Source: TV Guide.*)

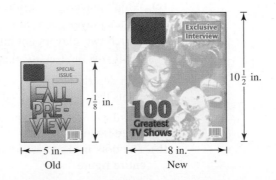

86. Under existing standards, most of the holes in Swiss cheese must have diameters between $\frac{11}{16}$ and $\frac{13}{16}$ in. To accommodate new high-speed slicing machines, the U.S. Department of Agriculture wants to reduce the minimum size to $\frac{3}{8}$ in. How much smaller is $\frac{3}{8}$ in. than $\frac{11}{16}$ in.? (*Source:* U.S. Department of Agriculture.)

Approximately 34 million people living in the United States in 2004 were born in other countries. The circle graph gives the fractional number from each region of birth for these people. Use the graph to answer each question. See Example 9.

87. What fractional part of the foreign-born population was from other regions?

88. What fractional part of the foreign-born population was from Latin America or Asia?

89. How many (in millions) were born in Europe?

U.S. Foreign-Born Population By Region of Birth

Other

Latin America $\frac{27}{50}$

Asia $\frac{1}{4}$

Europe $\frac{7}{50}$

Source: U.S. Census Bureau.

90. At the conclusion of the Addison-Wesley softball league season, batting statistics for five players were as follows:

Player	At-Bats	Hits	Home Runs
Maureen O'Connor	40	9	2
Lauren Morse	36	12	3
Ron Hampton	11	5	1
Greg Tobin	16	8	0
Joe Vetere	20	10	2

Use the table to answer each question. Estimate as necessary.

(a) Which player got a hit in exactly $\frac{1}{3}$ of his or her at-bats?

(b) Which player got a hit in just less than $\frac{1}{2}$ of his or her at-bats?

(c) Which player got a home run in just less than $\frac{1}{10}$ of his or her at-bats?

(d) Which player got a hit in just less than $\frac{1}{4}$ of his or her at-bats?

(e) Which two players got hits in exactly the same fractional parts of their at-bats? What was the fractional part, expressed in lowest terms?

91. For each description, write a fraction in lowest terms that represents the region described.

(a) The dots in the rectangle as a part of the dots in the entire figure

(b) The dots in the triangle as a part of the dots in the entire figure

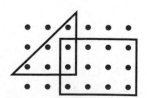

(c) The dots in the overlapping region of the triangle and the rectangle as a part of the dots in the triangle alone

(d) The dots in the overlapping region of the triangle and the rectangle as a part of the dots in the rectangle alone

92. *Concept Check* Estimate the best approximation for the following sum:

$$\frac{14}{26} + \frac{98}{99} + \frac{100}{51} + \frac{90}{31} + \frac{13}{27}.$$

A. 6   **B.** 7   **C.** 5   **D.** 8

1.2 Exponents, Order of Operations, and Inequality

OBJECTIVES

1 Use exponents.

2 Use the rules for order of operations.

3 Use more than one grouping symbol.

4 Know the meanings of $\neq$, $<$, $>$, $\leq$, and $\geq$.

5 Translate word statements to symbols.

6 Write statements that change the direction of inequality symbols.

7 Interpret data in a bar graph.

OBJECTIVE 1 Use exponents. In the prime factored form of 81, written

$$81 = 3 \cdot 3 \cdot 3 \cdot 3,$$

the factor 3 appears four times. In algebra, repeated factors are written with an *exponent*. For example, in $3 \cdot 3 \cdot 3 \cdot 3$, the number 3 appears as a factor four times, so the product is written as 3^4 and is read "3 to the fourth power."

$$\underbrace{3 \cdot 3 \cdot 3 \cdot 3}_{\text{4 factors of 3}} = 3^4 \quad \overset{\text{Exponent}}{\underset{\text{Base}}{}}$$

The number 4 is the **exponent,** or **power,** and 3 is the **base** in the **exponential expression** 3^4. A natural number exponent, then, tells how many times the base is used as a factor. *A number raised to the first power is simply that number.* For example,

$$5^1 = 5 \quad \text{and} \quad \left(\frac{1}{2}\right)^1 = \frac{1}{2}.$$

EXAMPLE 1 Evaluating Exponential Expressions

Find the value of each exponential expression.

(a) $5^2 = \underbrace{5 \cdot 5}_{} = 25$

　　　　5 is used as a factor 2 times.

　　Read 5^2 as "5 squared" or "the square of 5."

(b) $6^3 = \underbrace{6 \cdot 6 \cdot 6}_{} = 216$

　　　　6 is used as a factor 3 times.

　　Read 6^3 as "6 cubed" or "the cube of 6."

(c) $2^5 = 2 \cdot 2 \cdot 2 \cdot 2 \cdot 2 = 32$　　2 is used as a factor 5 times.

　　Read 2^5 as "2 to the fifth power."

(d) $\left(\frac{2}{3}\right)^3 = \frac{2}{3} \cdot \frac{2}{3} \cdot \frac{2}{3} = \frac{8}{27}$　　$\frac{2}{3}$ is used as a factor 3 times.

✔ **Now Try Exercises 5 and 17.**

> ▶ **CAUTION** *Squaring, or raising a number to the second power, is* **not** *the same as doubling the number.* For example,
>
> $$3^2 \quad \text{means} \quad 3 \cdot 3, \quad not \quad 2 \cdot 3.$$
>
> Thus $3^2 = 9$, *not* 6. Similarly, cubing, or raising a number to the third power, does *not* mean tripling the number.

OBJECTIVE 2 **Use the rules for order of operations.** Many problems involve more than one operation. To indicate the order in which the operations should be performed, we often use **grouping symbols.** If no grouping symbols are used, we apply the rules for order of operations.

Consider the expression $5 + 2 \cdot 3$. To show that the multiplication should be performed before the addition, we can use parentheses to write

$$5 + (2 \cdot 3), \quad \text{which equals} \quad 5 + 6, \quad \text{or} \quad 11.$$

If addition is to be performed first, the parentheses should group $5 + 2$ as follows:

$$(5 + 2) \cdot 3, \quad \text{which equals} \quad 7 \cdot 3, \quad \text{or} \quad 21.$$

Other grouping symbols used in more complicated expressions are brackets [ ], braces { }, and fraction bars. (For example, in $\frac{8 - 2}{3}$, the expression $8 - 2$ is considered to be grouped in the numerator.)

To work problems with more than one operation, we use the following **order of operations.** This order is used by most calculators and computers.

Order of Operations

If grouping symbols are present, simplify within them, innermost first (and above and below fraction bars separately), in the following order:

Step 1 Apply all **exponents.**

Step 2 Do any **multiplications** or **divisions** in the order in which they occur, working from left to right.

Step 3 Do any **additions** or **subtractions** in the order in which they occur, working from left to right.

If no grouping symbols are present, start with Step 1.

When no operation is indicated, as in 3(7) or $(-5)(-4)$, multiplication is understood.

EXAMPLE 2 **Using the Rules for Order of Operations**

Find the value of each expression.

(a) $4 + 5 \cdot 6$

Use the rules for order of operations given in the box.

$$4 + 5 \cdot 6 \qquad \text{Be careful! Multiply first.}$$

$$= 4 + 30 \qquad \text{Multiply.}$$
$$= 34 \qquad \text{Add.}$$

(b) $9(6 + 11)$

$\quad = 9(17)$ Work inside parentheses.

$\quad = 153$ Multiply.

(c) $6 \cdot 8 + 5 \cdot 2$

$\quad = 48 + 10$ Multiply, working from left to right.

$\quad = 58$ Add.

> Start here.

(d) $2(5 + 6) + 7 \cdot 3$

$\quad = 2(11) + 7 \cdot 3$ Work inside parentheses.

$\quad = 22 + 21$ Multiply.

$\quad = 43$ Add.

> $2^3 = 2 \cdot 2 \cdot 2$, not $2 \cdot 3$.

(e) $9 - 2^3 + 5$

$\quad = 9 - 2 \cdot 2 \cdot 2 + 5$ Apply the exponent.

$\quad = 9 - 8 + 5$ Multiply.

$\quad = 1 + 5$ Subtract.

$\quad = 6$ Add.

> Think: $3^3 = 3 \cdot 3 \cdot 3$

(f) $72 \div 2 \cdot 3 + 4 \cdot 2^3 - 3^3$

$\quad = 72 \div 2 \cdot 3 + 4 \cdot 8 - 27$ Apply the exponents.

$\quad = 36 \cdot 3 + 4 \cdot 8 - 27$ Divide.

$\quad = 108 + 32 - 27$ Multiply.

$\quad = 140 - 27$ Add.

$\quad = 113$ Subtract.

Multiplications and divisions are done from left to right *as they appear;* then additions and subtractions are done from left to right, *as they appear.*

> ✔ **Now Try Exercises 25, 31, 39, and 41.**

OBJECTIVE 3 Use more than one grouping symbol. An expression with double (or *nested*) parentheses, such as $2(8 + 3(6 + 5))$, can be confusing. For clarity, we often use brackets, [   ], in place of one pair of parentheses.

EXAMPLE 3 Using Brackets and Fraction Bars as Grouping Symbols

Simplify each expression.

> Start here.

(a) $2[8 + 3(6 + 5)]$

$\quad = 2[8 + 3(11)]$ Add inside parentheses.

$\quad = 2[8 + 33]$ Multiply inside brackets.

$\quad = 2[41]$ Add inside brackets.

$\quad = 82$ Multiply.

(b) $\dfrac{4(5 + 3) + 3}{2(3) - 1}$

The expression can be written as the quotient

$$[4(5 + 3) + 3] \div [2(3) - 1],$$

which shows that the fraction bar groups the numerator and denominator separately. Simplify both numerator and denominator, and then divide if possible.

$$\dfrac{4(5 + 3) + 3}{2(3) - 1}$$

$$= \dfrac{4(8) + 3}{2(3) - 1} \qquad \text{Work inside parentheses.}$$

$$= \dfrac{32 + 3}{6 - 1} \qquad \text{Multiply.}$$

$$= \dfrac{35}{5} \qquad \text{Add and subtract.}$$

$$= 7 \qquad \text{Divide.}$$

✔ **Now Try Exercises 45 and 47.**

OBJECTIVE 4 **Know the meanings of ≠, <, >, ≤, and ≥.** So far, we have used the symbols for the operations of arithmetic $(+, -, \cdot, \div)$ and the symbol for equality $(=)$. The symbols $\neq, <, >, \leq,$ and $\geq$ are used to express an **inequality,** a statement that two expressions are not equal. The equality symbol with a slash through it, $\neq$, means "is not equal to." For example,

$$7 \neq 8. \qquad \text{7 is not equal to 8.}$$

If two numbers are not equal, then one of the numbers must be less than the other. The symbol $<$ represents "is less than," so

$$7 < 8. \qquad \text{7 is less than 8.}$$

The word *is* in the phrase "is less than" is a verb, which actually makes $7 < 8$ a sentence. This is *not* the case for expressions like $7 + 8, 7 - 8,$ and so on.

The symbol $>$ means "is greater than." For example,

$$8 > 2. \qquad \text{8 is greater than 2.}$$

To keep the meanings of the symbols $<$ and $>$ clear, remember that the symbol always points to the lesser number.

$$\text{Lesser number} \rightarrow \; \mathbf{8} < 15$$

$$15 > \mathbf{8} \leftarrow \text{Lesser number}$$

Two other symbols, $\leq$ and $\geq$, also represent the idea of inequality. The symbol $\leq$ means "is less than or equal to," so

$$5 \leq 9. \qquad \text{5 is less than or equal to 9.}$$

If either the $<$ part or the $=$ part is true, then the inequality $\leq$ is true. The statement $5 \leq 9$ is true, since $5 < 9$ is true. Also, $8 \leq 8$ is true, since $8 = 8$ is true. But it is not true that $13 \leq 9$ because neither $13 < 9$ nor $13 = 9$ is true.

The symbol $\geq$ means "is greater than or equal to." Again,

$$9 \geq 5 \qquad \text{9 is greater than or equal to 5.}$$

is true because $9 > 5$ is true.

EXAMPLE 4 Using Inequality Symbols

Determine whether each statement is *true* or *false*.

(a) $6 \neq 5 + 1$

The statement is false because 6 *is equal to* $5 + 1$.

(b) $5 + 3 < 19$

Since $5 + 3$ represents a number (8) that is less than 19, this statement is true.

(c) $15 \leq 20 \cdot 2$

The statement $15 \leq 20 \cdot 2$ is true, since $15 < 40$.

(d) $25 \geq 30$

Both $25 > 30$ and $25 = 30$ are false. Because of this, $25 \geq 30$ is false.

(e) $12 \geq 12$

Since $12 = 12$, this statement is true.

(f) $9 < 9$

Since $9 = 9$, this statement is false.

(g) $\dfrac{6}{15} \geq \dfrac{2}{3}$

To compare fractions, write them with a common denominator. Here, 15 is a common denominator and $\frac{2}{3} = \frac{10}{15}$. Now decide whether $\frac{6}{15} \geq \frac{10}{15}$ is true or false. Both statements $\frac{6}{15} > \frac{10}{15}$ and $\frac{6}{15} = \frac{10}{15}$ are false; therefore, $\frac{6}{15} \geq \frac{2}{3}$ is false.

✔ **Now Try Exercises 53 and 55.**

OBJECTIVE 5 Translate word statements to symbols. An important part of algebra deals with translating words into algebraic notation.

EXAMPLE 5 Translating from Words to Symbols

Write each word statement in symbols.

(a) Twelve equals ten plus two.

$$12 = 10 + 2$$

(b) Nine is less than ten.

$$9 < 10$$

(c) Fifteen is not equal to eighteen.

$$15 \neq 18$$

(d) Seven is greater than four.

$$7 > 4$$

(e) Thirteen is less than or equal to forty.

$$13 \leq 40$$

(f) Eleven is greater than or equal to eleven.

$$11 \geq 11$$

✔ **Now Try Exercises 67 and 69.**

▶ **OBJECTIVE ⑥ Write statements that change the direction of inequality symbols.** Any statement with $<$ can be converted to one with $>$, and any statement with $>$ can be converted to one with $<$. We do this by reversing the order of the numbers and the direction of the symbol. For example,

Interchange numbers.

$$6 < 10 \qquad \text{becomes} \qquad 10 > 6.$$

Reverse symbol.

EXAMPLE 6 Converting between Inequality Symbols

Parts (a)–(d) each show a statement written in two equally correct ways. In each inequality, the inequality symbol points toward the lesser number.

(a) $9 < 16 \qquad 16 > 9$ **(b)** $5 > 2 \qquad 2 < 5$

(c) $3 \leq 8 \qquad 8 \geq 3$ **(d)** $12 \geq 5 \qquad 5 \leq 12$

✔ **Now Try Exercise 83.**

Here is a summary of the symbols discussed in this section.

Symbol	Meaning	Example
$=$	Is equal to	$0.5 = \frac{1}{2}$ means 0.5 is equal to $\frac{1}{2}$.
$\neq$	Is not equal to	$3 \neq 7$ means 3 is not equal to 7.
$<$	Is less than	$6 < 10$ means 6 is less than 10.
$>$	Is greater than	$15 > 14$ means 15 is greater than 14.
$\leq$	Is less than or equal to	$4 \leq 8$ means 4 is less than or equal to 8.
$\geq$	Is greater than or equal to	$1 \geq 0$ means 1 is greater than or equal to 0.

▶ **CAUTION** The equality and inequality symbols are used to write mathematical *sentences.* They differ from the symbols for operations $(+, -, \cdot, \text{and} \div)$ discussed earlier, which are used to write mathematical *expressions* that represent a number.

For example, compare the sentence

$$4 < 10,$$

which gives the relationship between 4 and 10, with the expression

$$4 + 10,$$

which tells how to operate on 4 and 10 to get the number 14.

OBJECTIVE ⑦ Interpret data in a bar graph. **Bar graphs** are often used to summarize data in a concise manner.

EXAMPLE 7 Interpreting Inequality Concepts by Means of a Bar Graph

The bar graph in Figure 3 shows the federal budget outlays for national defense over the 10-year period 1995–2004.

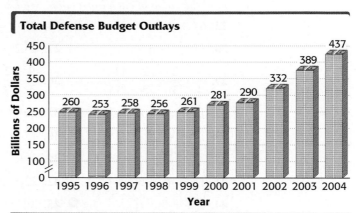

Source: U.S. Office of Management and Budget.

FIGURE 3

(a) Which years had outlays greater than 300 billion dollars?

Look for numbers greater than 300 at the tops of the columns. In 2002, 2003, and 2004, the outlays were 332 billion, 389 billion, and 437 billion dollars, respectively.

(b) Which years had outlays greater than 450 billion dollars?

Since the tops of all the bars are below the line representing 450 billion, there were no such years. This is reinforced by the fact that all of the numbers at the tops of the bars are less than 450 (billion).

✔ **Now Try Exercise 93.**

1.2 EXERCISES

◐ *Complete solution available on Video Lectures on CD/DVD*

▨ *Now Try Exercise*

Concept Check *Decide whether each statement is* true *or* false. *If it is false, explain why.*

1. When evaluated, $4 + 3(8 - 2)$ is equal to 42.

2. $3^3 = 9$

3. The statement "4 is 12 less than 16" is interpreted as $4 = 12 - 16$.

4. The statement "6 is 4 less than 10" is interpreted as $6 < 10 - 4$.

Find the value of each exponential expression. See Example 1.

◐ **5.** 7^2 **6.** 4^2 **7.** 12^2 **8.** 14^2

9. 4^3 **10.** 5^3 **11.** 10^3 **12.** 11^3

13. 3^4 **14.** 6^4 **15.** 4^5 **16.** 3^5

17. $\left(\dfrac{2}{3}\right)^4$ **18.** $\left(\dfrac{3}{4}\right)^3$ **19.** $(0.4)^3$ **20.** $(0.5)^4$

✎ **21.** Explain in your own words how to evaluate a power of a number, such as 6^3.

✎ **22.** Explain why any power of 1 must be equal to 1.

Find the value of each expression. See Examples 2 and 3.

23. $64 \div 4 \cdot 2$ **24.** $250 \div 5 \cdot 2$ 🌐 **25.** $13 + 9 \cdot 5$

26. $11 + 7 \cdot 6$ **27.** $25.2 - 12.6 \div 4.2$ **28.** $12.4 - 9.3 \div 3.1$

29. $\dfrac{1}{4} \cdot \dfrac{2}{3} + \dfrac{2}{5} \cdot \dfrac{11}{3}$ **30.** $\dfrac{9}{4} \cdot \dfrac{2}{3} + \dfrac{4}{5} \cdot \dfrac{5}{3}$ **31.** $9 \cdot 4 - 8 \cdot 3$

32. $11 \cdot 4 + 10 \cdot 3$ **33.** $20 - 4 \cdot 3 + 5$ **34.** $18 - 7 \cdot 2 + 6$

35. $10 + 40 \div 5 \cdot 2$ **36.** $12 + 64 \div 8 - 4$ **37.** $18 - 2(3 + 4)$

38. $30 - 3(4 + 2)$ **39.** $3(4 + 2) + 8 \cdot 3$ **40.** $9(1 + 7) + 2 \cdot 5$

41. $18 - 4^2 + 3$ **42.** $22 - 2^3 + 9$ **43.** $5[3 + 4(2^2)]$

44. $6[2 + 8(3^3)]$ 🌐 **45.** $3^2[(11 + 3) - 4]$ **46.** $4^2[(13 + 4) - 8]$

47. $\dfrac{6(3^2 - 1) + 8}{8 - 2^2}$ **48.** $\dfrac{2(8^2 - 4) + 8}{29 - 3^3}$

49. $\dfrac{4(6 + 2) + 8(8 - 3)}{6(4 - 2) - 2^2}$ **50.** $\dfrac{6(5 + 1) - 9(1 + 1)}{5(8 - 6) - 2^3}$

✎ **51.** Explain how you would use the order of operations to simplify $4 + 3(2^2 - 1)^3$.

52. ***Concept Check*** In evaluating $(4^2 + 3^3)^4$, what is the *last* exponent that would be applied?

First simplify both sides of each inequality. Then tell whether the given statement is true *or* false. *See Examples 2–4.*

🌐 **53.** $9 \cdot 3 - 11 \le 16$ **54.** $6 \cdot 5 - 12 \le 18$

55. $5 \cdot 11 + 2 \cdot 3 \le 60$ **56.** $9 \cdot 3 + 4 \cdot 5 \ge 48$

57. $0 \ge 12 \cdot 3 - 6 \cdot 6$ **58.** $10 \le 13 \cdot 2 - 15 \cdot 1$

59. $45 \ge 2[2 + 3(2 + 5)]$ **60.** $55 \ge 3[4 + 3(4 + 1)]$

61. $[3 \cdot 4 + 5(2)] \cdot 3 > 72$ **62.** $2 \cdot [7 \cdot 5 - 3(2)] \le 58$

63. $\dfrac{3 + 5(4 - 1)}{2 \cdot 4 + 1} \ge 3$ **64.** $\dfrac{7(3 + 1) - 2}{3 + 5 \cdot 2} \le 2$

65. $3 \ge \dfrac{2(5 + 1) - 3(1 + 1)}{5(8 - 6) - 4 \cdot 2}$ **66.** $7 \le \dfrac{3(8 - 3) + 2(4 - 1)}{9(6 - 2) - 11(5 - 2)}$

Write each word statement in symbols. See Example 5.

67. Fifteen is equal to five plus ten.   **68.** Twelve is equal to twenty minus eight.

🌐 **69.** Nine is greater than five minus four.   **70.** Ten is greater than six plus one.

71. Sixteen is not equal to nineteen.   **72.** Three is not equal to four.

73. One-half is less than or equal to two-fourths.

74. One-third is less than or equal to three-ninths.

Write each statement in words and decide whether it is true *or* false.

75. $7 < 19$ **76.** $9 < 10$ **77.** $3 \neq 6$

78. $9 \neq 13$ **79.** $8 \geq 11$ **80.** $4 \leq 2$

81. *Concept Check*   Construct a true statement that involves an addition on the left side, the symbol $\geq$, and a multiplication on the right side.

82. Construct a false statement that involves subtraction on the left side, the symbol $\leq$, and a division on the right side. Then tell why the statement is false and how it could be changed to become true.

Write each statement with the inequality symbol reversed while keeping the same meaning. See Example 6.

83. $5 < 30$ **84.** $8 > 4$ **85.** $2.5 \geq 1.3$ **86.** $4.1 \leq 5.3$

87. *Concept Check*   What English-language phrase is used to express the fact that one person's age *is less than* another person's age?

88. *Concept Check*   What English-language phrase is used to express the fact that one person's height *is greater than* another person's height?

One way to measure a person's cardiofitness is to calculate how many METs, or metabolic units, he or she can reach at peak exertion. One MET is the amount of energy used when sitting quietly. To calculate ideal METs, we can use the expression

$$14.7 - \text{age} \cdot 0.13 \qquad \text{For women}$$

or

$$14.7 - \text{age} \cdot 0.11. \qquad \text{For men}$$

(*Source: New England Journal of Medicine, August 2005.*)

89. A 40-yr-old woman wishes to calculate her ideal MET.

   (a) Write the expression, using her age.
   (b) Calculate her ideal MET. (*Hint:* Use the rules for order of operations.)
   (c) Researchers recommend that a person reach approximately 85% of his or her MET when exercising. Calculate 85% of the ideal MET from part (b). Then refer to the following table. What activity can the woman do that is approximately this value?

Activity	METs	Activity	METs
Golf (with cart)	2.5	Skiing (water or downhill)	6.8
Walking (3 mph)	3.3	Swimming	7.0
Mowing lawn (power mower)	4.5	Walking (5 mph)	8.0
Ballroom or square dancing	5.5	Jogging	10.2
Cycling	5.7	Skipping rope	12.0

Source: Harvard School of Public Health.

90. Repeat parts (a)–(c) of Exercise 89 for a 55-yr-old man.

91. Repeat parts (a)–(c) of Exercise 89, using your age.

92. The table shows the number of pupils per teacher in U.S. public schools in selected states.

State	Pupils per Teacher
Alaska	16.7
Texas	14.7
California	20.5
Wyoming	12.5
Maine	12.3
Idaho	17.8
Missouri	13.9

Source: National Center for Education Statistics.

(a) Which states had a figure greater than 13.9?

(b) Which states had a figure that was at most 14.7?

(c) Which states had a figure not less than 13.9?

The bar graph shows world coal production by year for the years 1997 through 2003. Use the graph to answer Exercises 93 and 94. See Example 7.

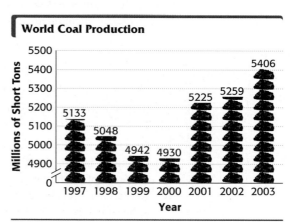

World Coal Production

Source: U.S. Energy Information Administration.

93. In three of the years represented, coal production was less than production in the previous year. What were these three years?

94. What was the first year in which production was greater than 5200 million short tons?

Concept Check *Insert one pair of parentheses so that the left side of each equation is equal to the right side.*

95. $3 \cdot 6 + 4 \cdot 2 = 60$

96. $2 \cdot 8 - 1 \cdot 3 = 42$

97. $10 - 7 - 3 = 6$

98. $15 - 10 - 2 = 7$

99. $8 + 2^2 = 100$

100. $4 + 2^2 = 36$

1.3   Variables, Expressions, and Equations

OBJECTIVES

1 Evaluate algebraic expressions, given values for the variables.

2 Translate word phrases to algebraic expressions.

3 Identify solutions of equations.

4 Identify solutions of equations from a set of numbers.

5 Distinguish between *expressions* and *equations*.

A **variable** is a symbol, usually a letter such as x, y, or z, used to represent any unknown number. An **algebraic expression** is a sequence of numbers, variables, operation symbols, and/or grouping symbols formed according to the rules of algebra.

$$x + 5, \quad 2m - 9, \quad 8p^2 + 6(p - 2) \qquad \text{Algebraic expressions}$$

In $2m - 9$, the $2m$ means $2 \cdot m$, the product of 2 and m; $8p^2$ represents the product of 8 and p^2. Also, $6(p - 2)$ means the product of 6 and $p - 2$.

OBJECTIVE 1   Evaluate algebraic expressions, given values for the variables.   An algebraic expression has different numerical values for different values of the variables.

EXAMPLE 1   Evaluating Expressions

Find the value of each algebraic expression when $m = 5$.

(a) $8m$

$$= 8 \cdot m$$
$$= 8 \cdot 5 \qquad \text{Let } m = 5.$$
$$= 40 \qquad \text{Multiply.}$$

(b) $3m^2$

$$= 3 \cdot m^2$$
$$= 3 \cdot 5^2 \qquad \text{Let } m = 5. \qquad \boxed{5^2 = 5 \cdot 5}$$
$$= 3 \cdot 25 \qquad \text{Square 5.}$$
$$= 75 \qquad \text{Multiply.}$$

✔ **Now Try Exercises 15 and 17.**

▶ **CAUTION**   In Example 1(b), $3m^2$ means $3 \cdot m^2$, **not** $3m \cdot 3m$. **Unless parentheses are used, the exponent refers only to the variable or number just before it.** To write $3m \cdot 3m$ with exponents, use parentheses: $(3m)^2$.

EXAMPLE 2   Evaluating Expressions

Find the value of each expression when $x = 5$ and $y = 3$.

(a) $\qquad\qquad 2x + 7y$

Follow the rules for order of operations.

$$= 2 \cdot 5 + 7 \cdot 3 \qquad \text{Let } x = 5 \text{ and } y = 3.$$
$$= 10 + 21 \qquad \text{Multiply.}$$
$$= 31 \qquad \text{Add.}$$

(b) $\dfrac{9x - 8y}{2x - y}$

$$= \frac{9 \cdot 5 - 8 \cdot 3}{2 \cdot 5 - 3} \qquad \text{Let } x = 5 \text{ and } y = 3.$$
$$= \frac{45 - 24}{10 - 3} \qquad \text{Multiply.}$$
$$= \frac{21}{7}, \quad \text{or} \quad 3 \qquad \text{Subtract, then divide.}$$

(c) $x^2 - 2y^2$

$3^2 = 3 \cdot 3$

$= 5^2 - 2 \cdot 3^2$ Let $x = 5$ and $y = 3$.

$5^2 = 5 \cdot 5$ $= 25 - 2 \cdot 9$ Apply the exponents.

$= 25 - 18$ Multiply.

$= 7$ Subtract.

✔ **Now Try Exercises 27, 35, and 37.**

OBJECTIVE 2 Translate word phrases to algebraic expressions.

▶ **PROBLEM-SOLVING HINT**   Sometimes variables must be used to change word phrases into algebraic expressions. This process will be important later in solving applied problems.

EXAMPLE 3 Using Variables to Write Word Phrases as Algebraic Expressions

Write each word phrase as an algebraic expression, using x as the variable.

(a) The sum of a number and 9

"Sum" is the answer to an addition problem. The given phrase translates as

$$x + 9 \quad \text{or} \quad 9 + x.$$

(b) 7 minus a number

"Minus" indicates subtraction, so the translation is

$$7 - x.$$

Note that $x - 7$ would not be correct, because we cannot subtract in either order and get the same result.

(c) A number subtracted **from 12**

Since a number is subtracted *from* 12, write this as

$$12 - x.$$

Compare this result with "12 subtracted from a number," which is $x - 12$.

(d) The product of 11 and a number

$$11 \cdot x, \quad \text{or} \quad 11x$$

(e) 5 divided by a number

$$5 \div x, \quad \text{or} \quad \frac{5}{x}$$

$\frac{x}{5}$ is not correct here.

(f) The product of 2 and the difference between a number and 8

We are multiplying 2 times another number. This number is the difference between a number and 8, written $x - 8$. Using parentheses around this difference, we find that the final expression is

$$2(x - 8).$$

✔ **Now Try Exercises 43, 49, and 53.**

▶ **CAUTION** Notice that, in translating the words "the difference between a number and 8" in Example 3(f), the order is kept the same: $x - 8$. "The difference between 8 and a number" would be written $8 - x$.

OBJECTIVE 3 Identify solutions of equations. An **equation** is a statement that two algebraic expressions are equal. *Therefore, an equation always includes the equality symbol, =.*

$$x + 4 = 11, \qquad 2y = 16, \qquad 4p + 1 = 25 - p,$$
$$z^2 = 4, \qquad \frac{3}{4}x + \frac{1}{2} = 0, \qquad 4(m - 0.5) = 2m \quad \Bigg\}\quad \text{Equations}$$

To **solve** an equation means to find the values of the variable that make the equation true. Such values of the variable are called the **solutions** of the equation.

EXAMPLE 4 Deciding whether a Number Is a Solution of an Equation

Decide whether the given number is a solution of the equation.

(a) $5p + 1 = 36;\quad 7$

$$5p + 1 = 36$$
$$5 \cdot 7 + 1 = 36 \qquad ? \qquad \text{Let } p = 7.$$
$$35 + 1 = 36 \qquad ?$$
$$36 = 36 \qquad\qquad \text{True}$$

Be careful! Multiply first.

The number 7 is a solution of the equation.

(b) $9m - 6 = 32;\quad 4$

$$9m - 6 = 32$$
$$9 \cdot 4 - 6 = 32 \qquad ? \qquad \text{Let } m = 4.$$
$$36 - 6 = 32 \qquad ?$$
$$30 = 32 \qquad\qquad \text{False}$$

The number 4 is not a solution of the equation.

✔ **Now Try Exercises 59 and 61.**

OBJECTIVE 4 Identify solutions of equations from a set of numbers. A **set** is a collection of objects. In mathematics, these objects are most often numbers. The objects that belong to the set, called **elements** of the set, are written between **braces.** For example, the set containing the numbers 1, 2, 3, 4, and 5 is written as

$$\{1, 2, 3, 4, 5\}.$$

For more information about sets, see **Appendix C** at the back of this book.

In some cases, the set of numbers from which the solutions of an equation must be chosen is specifically stated. One way of determining solutions is the direct substitution of all possible replacements. The ones that lead to a true statement are solutions.

EXAMPLE 5 Finding a Solution from a Given Set

Write each word statement as an equation. Use x as the variable. Then find all solutions of the equation from the set

$$\{0, 2, 4, 6, 8, 10\}.$$

(a) The sum of a number and four is six.

The word *is* suggests "equals." If x represents the unknown number, then translate as follows:

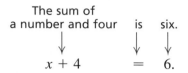

The sum of
a number and four   is   six.

$$x + 4 \qquad = \qquad 6.$$

Try each number from the given set $\{0, 2, 4, 6, 8, 10\}$, in turn, to see that 2 is the only solution of $x + 4 = 6$.

(b) Nine more than five times a number is 49.

Use x to represent the unknown number. Start with $5x$ and then add 9 to it. The word *is* translates as $=$.

$$5x + 9 = 49$$

Try each number from $\{0, 2, 4, 6, 8, 10\}$. The solution is 8, since $5 \cdot 8 + 9 = 49$.

(c) The sum of a number and 12 is equal to four times the number.

If x represents the number, "the sum of a number and 12," is represented by $x + 12$. The translation is

$$x + 12 = 4x.$$

Trying each replacement in the equation leads to a true statement only when $x = 4$, since $4 + 12 = 4(4)$.

✔ **Now Try Exercises 69 and 73.**

OBJECTIVE 5 Distinguish between *expressions* and *equations.* Students often have trouble distinguishing between equations and expressions. *An equation is a sentence— it has something on the left side, an = sign, and something on the right side. An expression is a phrase that represents a number.*

$\underbrace{4x + 5}$ $=$ $\underbrace{9}$	$4x + 5$
Left side ↑ Right side	↑
Equation	Expression
(to solve)	(to simplify or evaluate)

EXAMPLE 6 Distinguishing between Equations and Expressions

Decide whether each of the following is an *equation* or an *expression.*

(a) $2x - 5y$

There is no equals sign, so this is an expression.

(b) $2x = 5y$

Because there is an equals sign with something on either side of it, this is an equation.

✔ **Now Try Exercises 77 and 81.**

1.3 EXERCISES

Concept Check *Fill in each blank with the correct response.*

1. If $x = 3$, then the value of $x + 7$ is _____.

2. If $x = 1$ and $y = 2$, then the value of $4xy$ is _____.

3. The sum of 12 and x is represented by the expression _____. If $x = 9$, the value of that expression is _____.

4. If x can be chosen from the set $\{0, 1, 2, 3, 4, 5\}$, the only solution of $x + 5 = 9$ is _____.

5. Will the equation $x = x + 4$ ever have a solution? _____

6. $2x + 3$ is an _____, while $2x + 3 = 8$ is an _____.
   (equation/expression)   (equation/expression)

✍ *In Exercises 7–12, give a short explanation.*

7. Explain why $2x^3$ is not the same as $2x \cdot 2x \cdot 2x$.

8. Why are "5 less than a number" and "5 is less than a number" translated differently?

9. When evaluating the expression $4x^2$ for $x = 3$, explain why 3 must be squared *before* multiplying by 4.

10. What value of x would cause the expression $2x + 3$ to equal 9? Explain your reasoning.

11. There are many pairs of values of x and y for which $2x + y$ will equal 6. Name two such pairs and describe how you determined them.

12. Suppose that, for the equation $3x - y = 9$, the value of x is given as 4. What would be the corresponding value of y? How do you know this?

Find the value if (a) $x = 4$ and (b) $x = 6$. See Example 1.

13. $x + 9$ **14.** $x - 1$ **15.** $5x$ **16.** $7x$ ⊕ **17.** $4x^2$

18. $5x^2$ **19.** $\dfrac{x + 1}{3}$ **20.** $\dfrac{x - 2}{5}$ **21.** $\dfrac{3x - 5}{2x}$ **22.** $\dfrac{4x - 1}{3x}$

23. $3x^2 + x$ **24.** $2x + x^2$ **25.** $6.459x$ **26.** $0.74x^2$

Find the value if (a) $x = 2$ and $y = 1$ and (b) $x = 1$ and $y = 5$. See Example 2.

⊕ **27.** $8x + 3y + 5$ **28.** $4x + 2y + 7$ **29.** $3(x + 2y)$ **30.** $2(2x + y)$

31. $x + \dfrac{4}{y}$ **32.** $y + \dfrac{8}{x}$ **33.** $\dfrac{x}{2} + \dfrac{y}{3}$ **34.** $\dfrac{x}{5} + \dfrac{y}{4}$

35. $\dfrac{2x + 4y - 6}{5y + 2}$ **36.** $\dfrac{4x + 3y - 1}{x}$ **37.** $2y^2 + 5x$ **38.** $6x^2 + 4y$

39. $\dfrac{3x + y^2}{2x + 3y}$ **40.** $\dfrac{x^2 + 1}{4x + 5y}$ **41.** $0.841x^2 + 0.32y^2$ **42.** $0.941x^2 + 0.2y^2$

Write each word phrase as an algebraic expression, using x as the variable. See Example 3.

⊕ **43.** Twelve times a number   **44.** Nine times a number

45. Seven added to a number   **46.** Thirteen added to a number

47. Two subtracted from a number   **48.** Eight subtracted from a number

49. A number subtracted from seven

50. A number subtracted from fourteen

51. The difference between a number and 6

52. The difference between 6 and a number

53. 12 divided by a number

54. A number divided by 12

55. The product of 6 and four less than a number

56. The product of 9 and five more than a number

57. Suppose that the directions on a test read "Solve the following expressions." How would you politely correct the person who wrote these directions? What alternative directions might you suggest?

58. In the phrase "Four more than the product of a number and 6," does the word *and* signify the operation of addition? Explain.

Decide whether the given number is a solution of the equation. See Example 4.

59. $5m + 2 = 7$; 1

60. $3r + 5 = 8$; 1

61. $2y + 3(y - 2) = 14$; 3

62. $6a + 2(a + 3) = 14$; 2

63. $6p + 4p + 9 = 11$; $\dfrac{1}{5}$

64. $2x + 3x + 8 = 20$; $\dfrac{12}{5}$

65. $3r^2 - 2 = 46$; 4

66. $2x^2 + 1 = 19$; 3

67. $\dfrac{z + 4}{2 - z} = \dfrac{13}{5}$; $\dfrac{1}{3}$

68. $\dfrac{x + 6}{x - 2} = \dfrac{37}{5}$; $\dfrac{13}{4}$

Write each word statement as an equation. Use x as the variable. Find all solutions from the set {2, 4, 6, 8, 10}. See Example 5.

69. The sum of a number and 8 is 18.

70. A number minus three equals 1.

71. Sixteen minus three-fourths of a number is 13.

72. The sum of six-fifths of a number and 2 is 14.

73. One more than twice a number is 5.

74. The product of a number and 3 is 6.

75. Three times a number is equal to 8 more than twice the number.

76. Twelve divided by a number equals $\frac{1}{3}$ times that number.

Identify each as an expression *or an* equation. *See Example 6.*

77. $3x + 2(x - 4)$

78. $5y - (3y + 6)$

79. $7t + 2(t + 1) = 4$

80. $9r + 3(r - 4) = 2$

81. $x + y = 3$

82. $x + y - 3$

A **mathematical model** *is an equation that describes the relationship between two quantities. For example, the life expectancy of Americans can be approximated by the equation*

$$y = 0.212x - 347,$$

where x is a year between 1943 and 2005 and y is age in years. (Source: Centers for Disease Control and Prevention.)

Use this model to approximate life expectancy (to the nearest tenth of a year) in each of the following years.

83. 1943

84. 1960

85. 1980

86. 2005

1.4 Real Numbers and the Number Line

OBJECTIVES

1 Classify numbers and graph them on number lines.

2 Tell which of two real numbers is less than the other.

3 Find additive inverses and absolute values of real numbers.

4 Interpret the meanings of real numbers from a table of data.

OBJECTIVE ❶ Classify numbers and graph them on number lines. In **Section 1.1,** we introduced two important sets of numbers: the *natural numbers* and the *whole numbers.*

Natural Numbers*

$\{1, 2, 3, 4, \ldots\}$ is the set of **natural numbers.**

Whole Numbers

$\{0, 1, 2, 3, 4, \ldots\}$ is the set of **whole numbers.**

▶ **NOTE** The three dots show that the list of numbers continues in the same way indefinitely.

These numbers, along with many others, can be represented on a **number line** like the one in Figure 4. We draw a number line by choosing any point on the line and labeling it 0. Then we choose any point to the right of 0 and label it 1. The distance between 0 and 1 gives a unit of measure used to locate other points, as shown in Figure 4. The points labeled in Figure 4 correspond to the first few whole numbers.

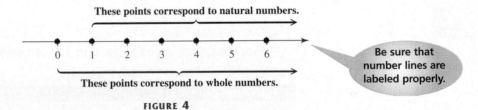

FIGURE 4

The natural numbers are located to the right of 0 on the number line. But numbers may also be placed to the left of 0. For each natural number, we can place a corresponding number to the left of 0. These numbers, written -1, -2, -3, -4, and so on, are shown in Figure 5 on the next page. Each is the **opposite,** or **negative,** of a natural number. The natural numbers, their opposites, and 0 form a new set of numbers called the *integers.*

Integers

$\{\ldots, -3, -2, -1, 0, 1, 2, 3, \ldots\}$ is the set of **integers.**

---

*The symbols { and } are braces used in conjunction with sets. See **Appendix C.**

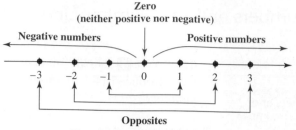

FIGURE 5

Positive numbers and *negative numbers* are called **signed numbers.** There are many practical applications of negative numbers. For example, a Fahrenheit temperature on a cold January day might be −10°, and a business that spends more than it takes in has a negative "profit" (a loss).

EXAMPLE 1 **Using Negative Numbers in Applications**

Use an integer to express the number in boldface italics in each application.

(a) The lowest Fahrenheit temperature ever recorded in meteorological records was *129*° below zero at Vostok, Antarctica, on July 21, 1983. (*Source: World Almanac and Book of Facts 2006.*)

Use −129 because "below zero" indicates a negative number.

(b) The shore surrounding the Dead Sea is *1340* ft below sea level. (*Source: Microsoft Encarta Encyclopedia.*)

Again, "below sea level" indicates a negative number, −1340.

✔ **Now Try Exercises 1 and 3.**

Fractions, introduced in **Section 1.1,** are examples of *rational numbers.*

Rational Numbers

$\{x \mid x$ is a quotient of two integers, with denominator not $0\}$ is the set of **rational numbers.**

(Read the part in the braces as "the set of all numbers x such that x is a quotient of two integers, with denominator not 0.")

▶ **NOTE** The set symbolism used in the definition of rational numbers,

$$\{x \mid x \text{ has a certain property}\},$$

is called **set-builder notation.** This notation is convenient to use when it is not possible to list all the elements of a set.

Since any integer can be written as the quotient of itself and 1, *all integers are also rational numbers.* For example, $-5 = \frac{-5}{1}$. A decimal number that comes to an end (terminates), such as 0.23, is a rational number. For example, $0.23 = \frac{23}{100}$. Decimal numbers that repeat in a fixed block of digits, such as $0.3333\ldots = 0.\overline{3}$ and $0.454545\ldots = 0.\overline{45}$, are also rational numbers. For example, $0.\overline{3} = \frac{1}{3}$.

To **graph** a number, we place a dot on the number line at the point that corresponds to the number. The number is called the **coordinate** of the point. Think of the graph of a set of numbers as a picture of the set. Figure 6 shows a number line with the graphs of several rational numbers.

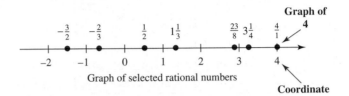

Graph of selected rational numbers

FIGURE 6

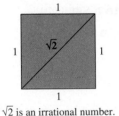

$\sqrt{2}$ is an irrational number.

FIGURE 7

Although many numbers are rational, not all are. For example, a square that measures one unit on a side has a diagonal whose length is the square root of 2, written $\sqrt{2}$. See Figure 7. It can be shown that $\sqrt{2}$ cannot be written as a quotient of integers. Because of this, $\sqrt{2}$ is not rational; it is *irrational*. Other examples of *irrational numbers* are $\sqrt{3}$, $\sqrt{7}$, $-\sqrt{10}$, and π (the ratio of the *circumference* of a circle to its *diameter*).

Irrational Numbers

$\{x \mid x \text{ is a nonrational number represented by a point on the number line}\}$ is the set of **irrational numbers.**

The decimal form of an irrational number neither terminates nor repeats. Irrational numbers are discussed in **Chapter 8.**

Both rational and irrational numbers can be represented by points on the number line and together form the set of *real numbers.*

Real Numbers

$\{x \mid x \text{ is a rational or an irrational number}\}$ is the set of **real numbers.**

An example of a number that is not a real number is the square root of a negative number, such as $\sqrt{-5}$. These numbers are discussed in **Chapter 9.**

Two ways to represent the relationships among the various types of real numbers are shown in Figure 8 on the next page. Notice that every real number is either a rational number or an irrational number.

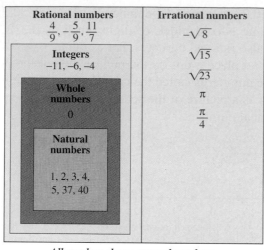

All numbers shown are real numbers.

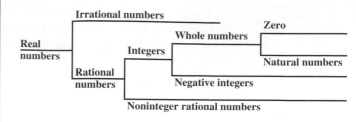

FIGURE 8

EXAMPLE 2 Determining whether a Number Belongs to a Set

List the numbers in the following set that belong to each set of numbers:

$$\left\{-5, -\frac{2}{3}, 0, \sqrt{2}, 3\frac{1}{4}, 5, 5.8\right\}.$$

(a) Natural numbers
The only natural number in the set is 5.

(b) Whole numbers
The whole numbers consist of the natural numbers and 0. So the elements of the set that are whole numbers are 0 and 5.

(c) Integers
The integers in the set are -5, 0, and 5.

(d) Rational numbers
The rational numbers are -5, $-\frac{2}{3}$, 0, $3\frac{1}{4}$, 5, and 5.8, since each of these numbers *can* be written as the quotient of two integers. For example, $3\frac{1}{4} = \frac{13}{4}$ and $5.8 = \frac{58}{10}$.

(e) Irrational numbers
The only irrational number in the set is $\sqrt{2}$.

(f) Real numbers
All the numbers in the set are real numbers.

✔ **Now Try Exercise 19.**

OBJECTIVE 2 Tell which of two real numbers is less than the other.   Given any two whole numbers, you probably can tell which number is less than the other. But what happens with two negative numbers, as in the set of integers? Positive numbers decrease as the corresponding points on the number line go to the left. For example, $8 < 12$ because 8 is to the left of 12 on the number line. This ordering is extended to all real numbers by definition.

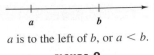

a is to the left of b, or $a < b$.

FIGURE 9

Ordering of Real Numbers

For any two real numbers a and b, ***a is less than b*** if a is to the left of b on the number line. See Figure 9.

This means that any negative number is less than 0, and any negative number is less than any positive number. Also, 0 is less than any positive number.

EXAMPLE 3   Determining the Order of Real Numbers

Is it true that $-3 < -1$?

To decide whether the statement is true, locate both numbers, -3 and -1, on a number line, as shown in Figure 10. Since -3 is to the left of -1 on the number line, -3 is less than -1. The statement $-3 < -1$ is true.

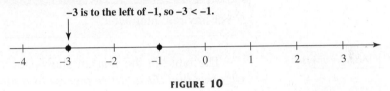

−3 is to the left of −1, so −3 < −1.

FIGURE 10

✔ **Now Try Exercise 53.**

▶ **NOTE**   In **Section 1.2**, we saw how it is possible to rewrite a statement involving $<$ as an equivalent statement involving $>$. The question in Example 3 can also be worded as follows: Is it true that $-1 > -3$? This is, of course, also a true statement.

We can also say that, for any two real numbers a and b, ***a is greater than b*** if a is to the right of b on the number line.

OBJECTIVE **3** **Find additive inverses and absolute values of real numbers.**   By a property of the real numbers, for any real number x (except 0), there is exactly one number on the number line the same distance from 0 as x, but on the opposite side of 0. For example, Figure 11 shows that the numbers 1 and -1 are each the same distance from 0, but on opposite sides of 0. The numbers 1 and -1 are called *additive inverses,* or *opposites,* of each other.

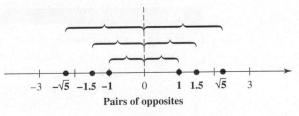

Pairs of opposites

FIGURE 11

Additive Inverse

The **additive inverse** of a number x is the number that is the same distance from 0 on the number line as x, but on the opposite side of 0.

The additive inverse of a number can be indicated by writing the symbol $-$ in front of the number. With this symbol, the additive inverse of 7 is written -7. The additive inverse of -3 is written $-(-3)$ and can be read "the opposite of -3" or "the negative of -3." Figure 11 suggests that 3 is the additive inverse of -3. A number can have only one additive inverse, so the symbols 3 and $-(-3)$ must represent the same number, which means that

$$-(-3) = 3.$$

This idea can be generalized.

Double Negative Rule*

For any real number x, $-(-x) = x.$

Number	Additive Inverse
-4	$-(-4)$ or 4
0	0
19	-19
$-\dfrac{2}{3}$	$\dfrac{2}{3}$
0.52	-0.52

The table in the margin shows several numbers and their additive inverses. *It suggests that the additive inverse of a number is found by changing the sign of the number.* An important property of additive inverses will be studied in **Section 1.7:** $a + (-a) = (-a) + a = 0$, for all real numbers a.

As previously mentioned, additive inverses are numbers that are the same distance from 0 on the number line. We can also express this idea by saying that a number and its additive inverse have the same *absolute value*. The **absolute value** of a real number can be defined as the distance between 0 and the number on the number line. The symbol for the absolute value of the number x is $|x|$, **read "the absolute value of x."** For example, the distance between 2 and 0 on the number line is 2 units, so

$$|2| = 2.$$

Because the distance between -2 and 0 on the number line is also 2 units,

$$|-2| = 2.$$

Distance is a physical measurement, which is never negative. *Therefore, the absolute value of a number is never negative.* For example,

$$|12| = 12 \quad \text{and} \quad |-12| = 12,$$

since both 12 and -12 lie at a distance of 12 units from 0 on the number line. Also, since 0 is a distance of 0 units from 0, $|0| = 0$.

In symbols, the absolute value of x is defined as follows.

Absolute Value

For any real number x,

$$|x| = \begin{cases} x & \text{if } x \geq 0 \\ -x & \text{if } x < 0. \end{cases}$$

---

*This rule is justified by interpreting $-(-x)$ as $-1 \cdot (-x) = [(-1)(-1)]x = 1 \cdot x = x$. This interpretation requires concepts covered in later sections of the chapter.

By this definition, if x is a positive number or 0, then its absolute value is x itself. For example, since 8 is a positive number, $|8| = 8$. *If x is a negative number, then its absolute value is the additive inverse of x.* This means that if $x = -8$, then $|-8| = -(-8) = 8$, since the additive inverse of -8 is 8.

> ▶ **CAUTION** The definition of absolute value can be confusing if it is not read carefully. The "$-x$" in the second part of the definition *does not* represent a negative number. Since x is negative in the second part, $-x$ represents the opposite of a negative number—that is, a positive number. *The absolute value of a number is never negative.*

EXAMPLE 4 Finding the Absolute Value

Simplify by finding the absolute value.

(a) $|5| = 5$ **(b)** $|-5| = -(-5) = 5$

(c) $-|5| = -(5) = -5$ **(d)** $-|-5| = -(5) = -5$

(e) $|8 - 2| = |6| = 6$ **(f)** $-|8 - 2| = -|6| = -6$

✔ **Now Try Exercises 35, 37, and 39.**

Parts (e) and (f) of Example 4 show that absolute value bars are also grouping symbols. We perform any operations that appear inside absolute value symbols before finding the absolute value.

OBJECTIVE 4 Interpret the meanings of real numbers from a table of data. The next example shows how signed numbers can be used to interpret data.

EXAMPLE 5 Interpreting Data

The Producer Price Index (PPI) is the oldest continuous statistical series published by the Bureau of Labor Statistics. It measures the average changes in prices received by producers of all commodities produced in the United States.

The table shows the percent change in the Producer Price Index for selected commodities from 2002 to 2003 and from 2003 to 2004. Use the table to answer each question.

Commodity	Change from 2002 to 2003	Change from 2003 to 2004
Farm products	12.6	10.6
Gasoline	23.3	24.7
Machinery and equipment	−0.8	0.2
Iron and steel	6.5	33.7
Electronic components and accessories	−1.7	−2.2

Source: U.S. Bureau of Labor Statistics.

(a) What commodity in which year represents the greatest percent decrease?

We must find the negative number with the greatest absolute value. The number that satisfies this condition is −2.2; the greatest percent decrease was shown by electronic components and accessories from 2003 to 2004.

(b) Which commodity in which year represents the least change?

In this case, we must find the number (either positive, negative, or zero) with the least absolute value. From 2003 to 2004, machinery and equipment showed the least change, an increase of 0.2%.

✔ **Now Try Exercises 65 and 67.**

1.4 EXERCISES

🌐 *Complete solution available on Video Lectures on CD/DVD*

▪ *Now Try Exercise*

In Exercises 1–4, use an integer to express each number in boldface italics representing a change. In Exercises 5–8, use a rational number. See Example 1.

🌐 **1.** Between July 1, 2004, and July 1, 2005, the population of the United States increased by approximately ***2,845,000.*** (*Source:* U.S. Census Bureau.)

2. From 2000 to 2005, the mean SAT verbal score for Massachusetts residents increased by ***9,*** while the mathematics score increased by ***14.*** (*Source:* The College Board.)

3. From 1995 to 2005, the number of cable TV systems in the United States went from 11,218 to 8409, representing a decrease of ***2809.*** (*Source: Television and Cable Factbook.*)

4. In 1935, there were 15,295 banks in the United States. By 2004, the number was 8975, representing a decrease of ***6320*** banks. (*Source:* Federal Deposit Insurance Corporation.)

5. Enrollment at Kirkwood Community College declined ***2.4***% from fall 2004 to fall 2005. Enrollment at Des Moines Area Community College rose ***5.2***% during this same period. (*Source:* Iowa Department of Education.)

6. Fortune Brands, Inc., lost $***891.5*** million in the third quarter of 1999. In the first quarter of 2005, the company earned $***796.3*** million. (*Source:* Fortune Brands, Inc.)

7. On Monday, January 9, 2006, the Dow Jones Industrial Average (DJIA) closed at 11,011.90. On the previous Friday, it had closed at 10,959.31. Thus, on Monday, it closed up ***52.59*** points. (*Source: The Washington Post.*)

8. On Thursday, January 12, 2006, the NASDAQ closed at 2316.69. On the previous day, it had closed at 2331.36. Thus, on Thursday, it closed down ***14.67*** points. (*Source: The Washington Post.*)

Concept Check *In Exercises 9–14, give a number that satisfies the given condition.*

9. An integer between 3.5 and 4.5

10. A rational number between 3.8 and 3.9

11. A whole number that is not positive and is less than 1

12. A whole number greater than 4.5

13. An irrational number that is between $\sqrt{11}$ and $\sqrt{13}$

14. A real number that is neither negative nor positive

Concept Check *In Exercises 15–18, decide whether each statement is* true *or* false.

15. Every natural number is positive. **16.** Every whole number is positive.

17. Every integer is a rational number. **18.** Every rational number is a real number.

For Exercises 19 and 20, see Example 2. List all numbers from each set that are

(a) *natural numbers;* **(b)** *whole numbers;* **(c)** *integers;*
(d) *rational numbers;* **(e)** *irrational numbers;* **(f)** *real numbers.*

19. $\left\{-9, -\sqrt{7}, -1\frac{1}{4}, -\frac{3}{5}, 0, \sqrt{5}, 3, 5.9, 7\right\}$

20. $\left\{-5.3, -5, -\sqrt{3}, -1, -\frac{1}{9}, 0, 1.2, 1.8, 3, \sqrt{11}\right\}$

21. Explain in your own words the different sets of numbers introduced in this section, and give an example of each kind.

22. What two possible situations exist for the decimal representation of a rational number?

Graph each group of numbers on a number line. See Figures 5 and 6.

23. $0, 3, -5, -6$

24. $2, 6, -2, -1$

25. $-2, -6, -4, 3, 4$

26. $-5, -3, -2, 0, 4$

27. $\frac{1}{4}, 2\frac{1}{2}, -3\frac{4}{5}, -4, -1\frac{5}{8}$

28. $5\frac{1}{4}, 4\frac{5}{9}, -2\frac{1}{3}, 0, -3\frac{2}{5}$

29. *Concept Check* Match each expression in Column I with its value in Column II. Choices in Column II may be used once, more than once, or not at all.

I	II
(a) $|-7|$	**A.** 7
(b) $-(-7)$	**B.** -7
(c) $-|-7|$	**C.** Neither A nor B
(d) $-|-(-7)|$	**D.** Both A and B

30. *Concept Check* Fill in the blanks with the correct values: The opposite of -2 is _____, while the absolute value of -2 is _____. The additive inverse of -2 is _____, while the additive inverse of the absolute value of -2 is _____.

Find **(a)** *the opposite (or additive inverse) of each number and* **(b)** *the absolute value of each number.*

31. -4 **32.** -8 **33.** 6 **34.** 11

Simplify by finding the absolute value. See Example 4.

35. $|-6|$ **36.** $|-15|$ **37.** $-\left|-\frac{2}{3}\right|$

38. $-\left|-\frac{4}{5}\right|$ **39.** $|6 - 3|$ **40.** $-|6 - 3|$

Select the lesser of the two given numbers. See Examples 3 and 4.

41. $-12, -4$

42. $-9, -14$

43. $-8, -7$

44. $-15, -16$

45. $3, |-4|$

46. $5, |-2|$

47. $|-3.5|, |-4.5|$

48. $|-8.9|, |-9.8|$

49. $-|-6|, -|-4|$

50. $-|-2|, -|-3|$

51. $|5 - 3|, |6 - 2|$

52. $|7 - 2|, |8 - 1|$

Decide whether each statement is true *or* false. *See Examples 3 and 4.*

53. $-5 < -2$

54. $-8 > -2$

55. $-4 \leq -(-5)$

56. $-6 \leq -(-3)$

57. $|-6| < |-9|$

58. $|-12| < |-20|$

59. $-|8| > |-9|$

60. $-|12| > |-15|$

61. $-|-5| \geq -|-9|$

62. $-|-12| \leq -|-15|$

63. $|6 - 5| \geq |6 - 2|$

64. $|13 - 8| \leq |7 - 4|$

The table shows the percent change in the Producer Price Index (PPI) for selected industries from 2002 to 2003 and from 2003 to 2004. Use the table to answer Exercises 65–68. See Example 5.

Industry	Change from 2002 to 2003	Change from 2003 to 2004
Book publishers	3.7	3.8
Telephone apparatus manufacturing	-3.5	-5.1
Construction machinery manufacturing	1.4	3.1
Petroleum refineries	25.9	25.0
Electronic computer manufacturing	-19.6	-12.3

Source: U.S. Bureau of Labor Statistics.

65. Which industry in which year represents the greatest percentage increase?

66. Which industry in which year represents the greatest percentage decrease?

67. Which industry in which year represents the least change?

68. Which industries represent a decrease for both years?

Concept Check *Give three numbers between* -6 *and* 6 *that satisfy each given condition.*

69. Positive real numbers but not integers

70. Real numbers but not positive numbers

71. Real numbers but not whole numbers

72. Rational numbers but not integers

73. Real numbers but not rational numbers

74. Rational numbers but not negative numbers

75. Students often say "Absolute value is always positive." Is this true? Explain.

76. ***Concept Check*** *True* or *false:* If a is negative, then $|a| = -a$.

1.5 Adding and Subtracting Real Numbers

OBJECTIVES

1 Add two numbers with the same sign.

2 Add positive and negative numbers.

3 Use the definition of subtraction.

4 Use the rules for order of operations with real numbers.

5 Interpret words and phrases involving addition and subtraction.

6 Use signed numbers to interpret data.

In this and the next section, we extend the rules for operations with positive numbers to the negative numbers.

OBJECTIVE 1 Add two numbers with the same sign.   Recall from **Section 1.1** that the answer to an addition problem is called a **sum.** A number line can be used to illustrate adding real numbers.

EXAMPLE 1 Adding Numbers on a Number Line

(a) Use a number line to find the sum $2 + 3$.

Add the positive numbers 2 and 3 on the number line by starting at 0 and drawing an arrow 2 units to the *right,* as shown in Figure 12. This arrow represents the number 2 in the sum $2 + 3$. Then, from the right end of that arrow, draw another arrow 3 units to the right. The number below the end of this second arrow is 5, so $2 + 3 = 5$.

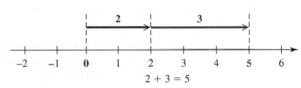

$2 + 3 = 5$

FIGURE 12

(b) Use a number line to find the sum $-2 + (-4)$. (Parentheses are placed around the -4 to avoid the confusing use of $+$ and $-$ next to each other.)

To add the negative numbers -2 and -4 on the number line, we start at 0 and draw an arrow 2 units to the *left,* as shown in Figure 13. From the left end of the first arrow, we draw a second arrow 4 units to the left. We draw the arrow to the left to represent the addition of a *negative* number. The number below the end of this second arrow is -6, so $-2 + (-4) = -6$.

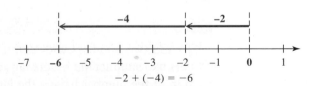

$-2 + (-4) = -6$

FIGURE 13

✔ Now Try Exercise 1.

In Example 1(b), the sum of the two negative numbers -2 and -4 is a negative number whose distance from 0 is the sum of the distance of -2 from 0 and the distance of -4 from 0. *That is, the sum of two negative numbers is the negative of the sum of their absolute values.*

$$-2 + (-4) = -\big(|-2| + |-4|\big) = -(2 + 4) = -6$$

Adding Numbers with the Same Sign

To add two numbers with the *same* sign, add the absolute values of the numbers. The sum has the same sign as the given numbers.

Example: $-4 + (-3) = -7$

EXAMPLE 2 Adding Two Negative Numbers

Find each sum.

(a) $-2 + (-9) = -(|-2| + |-9|) = -(2 + 9) = -11$

(b) $-8 + (-12) = -20$ **(c)** $-15 + (-3) = -18$

✔ **Now Try Exercise 7.**

OBJECTIVE 2 Add positive and negative numbers. We can use a number line to illustrate the sum of a positive number and a negative number.

EXAMPLE 3 Adding Numbers with Different Signs

Use a number line to find the sum $-2 + 5$.

    We find the sum $-2 + 5$ on the number line by starting at 0 and drawing an arrow 2 units to the left. From the left end of this arrow, we draw a second arrow 5 units to the right, as shown in Figure 14. The number below the end of the second arrow is 3, so $-2 + 5 = 3$.

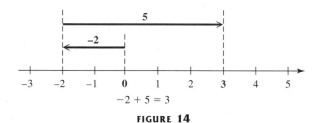

FIGURE 14

✔ **Now Try Exercise 3.**

Adding Numbers with Different Signs

To add two numbers with *different* signs, find the absolute values of the numbers and subtract the smaller absolute value from the larger. Give the answer the sign of the number having the larger absolute value.

Example: $-12 + 6 = -6$

For instance, to add -12 and 5, find their absolute values:

$$|-12| = 12 \quad \text{and} \quad |5| = 5.$$

Then find the difference between these absolute values: $12 - 5 = 7$. The sum will be negative, since $|-12| > |5|$, so the final answer is

$$-12 + 5 = -7.$$

While a number line is useful in showing the rules for addition, it is important to be able to find sums mentally.

EXAMPLE 4 Adding Mentally

Check each answer, trying to work the addition mentally. If you have trouble, use a number line.

(a) $7 + (-4) = 3$ **(b)** $-8 + 12 = 4$

(c) $-\dfrac{1}{2} + \dfrac{1}{8} = -\dfrac{4}{8} + \dfrac{1}{8} = -\dfrac{3}{8}$

> Remember to find a common denominator.

(d) $\dfrac{5}{6} + \left(-1\dfrac{1}{3}\right) = \dfrac{5}{6} + \left(-\dfrac{4}{3}\right) = \dfrac{5}{6} + \left(-\dfrac{8}{6}\right) = -\dfrac{3}{6} = -\dfrac{1}{2}$

(e) $-4.6 + 8.1 = 3.5$ **(f)** $-16 + 16 = 0$ **(g)** $42 + (-42) = 0$

✔ **Now Try Exercises 13, 15, and 27.**

Parts (f) and (g) in Example 4 suggest that the sum of a number and its additive inverse is 0. That is always true, and this property is discussed further in **Section 1.7.**
The rules for adding signed numbers are summarized as follows.

Adding Signed Numbers

Same sign Add the absolute values of the numbers. The sum has the same sign as the given numbers.

Different signs Find the absolute values of the numbers and subtract the smaller absolute value from the larger. Give the answer the sign of the number having the larger absolute value.

OBJECTIVE 3 Use the definition of subtraction. Recall from **Section 1.1** that the answer to a subtraction problem is called a **difference.** We can illustrate the subtraction of 4 from 7, written $7 - 4$, with a number line. As seen in Figure 15, we begin at 0 and draw an arrow 7 units to the right. From the right end of this arrow, we draw an arrow 4 units to the left. The number at the end of the second arrow shows that $7 - 4 = 3$.

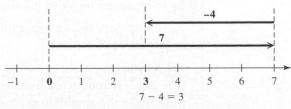

$7 - 4 = 3$

FIGURE 15

The procedure used to find the difference $7 - 4$ is exactly the same procedure that would be used to find the sum $7 + (-4)$, so

$$7 - 4 = 7 + (-4).$$

This equation suggests that *subtracting* a positive number from a larger positive number is the same as *adding* the additive inverse of the smaller number to the larger. This result leads to the definition of subtraction for all real numbers.

Definition of Subtraction

For any real numbers x and y,

$$x - y = x + (-y).$$

That is, to *subtract y from x, add the additive inverse* (or opposite) of y to x.

EXAMPLE 5 Using the Definition of Subtraction

Subtract.

Change − to +.

No change ⟶

Additive inverse of 3

(a) $12 - 3 = 12 + (-3) = 9$

(b) $5 - 7 = 5 + (-7) = -2$

(c) $-6 - 9 = -6 + (-9) = -15$

Change − to +.

No change ⟶

Additive inverse of −5

(d) $-3 - (-5) = -3 + (5) = 2$

(e) $\dfrac{4}{3} - \left(-\dfrac{1}{2}\right) = \dfrac{4}{3} + \dfrac{1}{2} = \dfrac{8}{6} + \dfrac{3}{6} = \dfrac{11}{6},$ or $1\dfrac{5}{6}$

✔ **Now Try Exercises 43, 51, and 55.**

Uses of the Symbol −

We use the symbol − for three purposes:

1. to represent subtraction, as in $9 - 5 = 4$;

2. to represent negative numbers, such as -10, -2, and -3;

3. to represent the opposite (or negative) of a number, as in "the opposite (or negative) of 8 is -8."

We may see more than one use of − in the same problem, such as $-6 - (-9)$, where -9 is subtracted from -6. The meaning of the − symbol depends on its position in the algebraic expression.

OBJECTIVE 4 Use the rules for order of operations with real numbers. In problems that have grouping symbols, first perform all operations inside the parentheses and brackets, as in **Section 1.2.** Work from the inside out.

EXAMPLE 6 Adding and Subtracting with Grouping Symbols

Perform each indicated operation.

Start here.

(a) $-6 - [2 - (8 + 3)]$

$= -6 - [2 - 11]$ Add.

$= -6 - [2 + (-11)]$ Definition of subtraction

$= -6 - [-9]$ Add.

$= -6 + (9)$ Definition of subtraction

$= 3$ Add.

(b) $5 + [(-3 - 2) - (4 - 1)]$

$= 5 + [(-3 + (-2)) - 3]$

$= 5 + [(-5) - 3]$

$= 5 + [(-5) + (-3)]$

$= 5 + [-8]$

$= -3$

(c) $\dfrac{2}{3} - \left[\dfrac{1}{12} - \left(-\dfrac{1}{4}\right)\right]$

$= \dfrac{8}{12} - \left[\dfrac{1}{12} - \left(-\dfrac{3}{12}\right)\right]$ Find a common denominator.

$= \dfrac{8}{12} - \left[\dfrac{1}{12} + \dfrac{3}{12}\right]$ Definition of subtraction

$= \dfrac{8}{12} - \dfrac{4}{12}$ Add.

$= \dfrac{4}{12}$ Subtract.

$= \dfrac{1}{3}$ Lowest terms

(d) $\qquad |4 - 7| + 2|6 - 3|$

$= |-3| + 2|3|$ Work within absolute value bars.

$= 3 + 2 \cdot 3$ Evaluate absolute values.

$= 3 + 6$ Multiply.

Be careful! Multiply first.

$= 9$ Add.

✔ **Now Try Exercises 65, 75, and 79.**

OBJECTIVE 5 Interpret words and phrases involving addition and subtraction. The word *sum* indicates addition. The table lists other words and phrases that indicate addition in problem solving.

Word or Phrase	Example	Numerical Expression and Simplification
Sum of	The *sum of* −3 and 4	−3 + 4 = 1
Added to	5 *added to* −8	−8 + 5 = −3
More than	12 *more than* −5	−5 + 12 = 7
Increased by	−6 *increased by* 13	−6 + 13 = 7
Plus	3 *plus* 14	3 + 14 = 17

EXAMPLE 7 Interpreting Words and Phrases Involving Addition

Write a numerical expression for each phrase and simplify the expression.

(a) The sum of −8 and 4 and 6

$$-8 + 4 + 6$$

To simplify, add in order from left to right, to obtain

$$-4 + 6, \quad \text{or} \quad 2.$$

(b) 3 more than −5, increased by 12

$$-5 + 3 + 12 \quad \text{simplifies to} \quad -2 + 12, \quad \text{or} \quad 10.$$

✔ **Now Try Exercises 83 and 87.**

The word *difference* indicates subtraction. Other words and phrases that indicate subtraction in problem solving are given in the table.

Word, Phrase, or Sentence	Example	Numerical Expression and Simplification
Difference between	The *difference between* −3 and −8	−3 − (−8) = −3 + 8 = 5
Subtracted from	12 *subtracted from* 18	18 − 12 = 6
From…, subtract….	From 12, subtract 8.	12 − 8 = 12 + (−8) = 4
Less	6 *less* 5	6 − 5 = 1
Less than	6 *less than* 5	5 − 6 = 5 + (−6) = −1
Decreased by	9 *decreased by* −4	9 − (−4) = 9 + 4 = 13
Minus	8 *minus* 5	8 − 5 = 3

▶ **CAUTION** In subtracting two numbers, be careful to write them in the correct order, because, in general,

$$a - b \neq b - a.$$

For example,

$$5 - 3 \neq 3 - 5.$$

Think carefully before interpreting an expression involving subtraction.

EXAMPLE 8 Interpreting Words and Phrases Involving Subtraction

Write a numerical expression for each phrase and simplify the expression.

(a) The difference between −8 and 5

When "difference between" is used, write the numbers in the order they are given.* The expression is

$$-8 - 5, \quad \text{which simplifies to} \quad -8 + (-5), \quad \text{or} \quad -13.$$

(b) 4 subtracted from the sum of 8 and −3

Here, addition is also used, as indicated by the word *sum*. First, add 8 and −3. Next, subtract 4 from this sum. The expression is

$$[8 + (-3)] - 4, \quad \text{which simplifies to} \quad 5 - 4, \quad \text{or} \quad 1.$$

(c) 4 less than −6

Here, 4 must be taken *from* −6, so write −6 first.

Be careful with order. 　　　$-6 - 4$ 　 simplifies to 　 $-6 + (-4)$, 　 or 　 -10.

Notice that "4 less than −6" differs from "4 *is less than* −6." The second of these is symbolized $4 < -6$ (which is a false statement).

(d) 8, decreased by 5 less than 12

First, write "5 less than 12" as $12 - 5$. Next, subtract $12 - 5$ from 8.

$$8 - (12 - 5) \quad \text{simplifies to} \quad 8 - 7, \quad \text{or} \quad 1.$$

✔ **Now Try Exercises 91 and 97.**

EXAMPLE 9 Solving a Problem Involving Subtraction

The record-high temperature in the United States is 134° Fahrenheit, recorded at Death Valley, California, in 1913. The record low is −80°F, at Prospect Creek, Alaska, in 1971. See Figure 16. What is the difference between these highest and lowest temperatures? (*Source: National Climatic Data Center.*)

We must subtract the lowest temperature from the highest temperature.

$$134 - (-80) = 134 + 80 \qquad \text{Definition of subtraction}$$
$$= 214 \qquad \text{Add.}$$

The difference between the two temperatures is 214°F.

✔ **Now Try Exercise 113.**

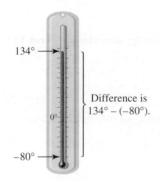

134°

Difference is
$134° - (-80°)$.

0°

−80°

FIGURE 16

OBJECTIVE 6 Use signed numbers to interpret data.

EXAMPLE 10 Using a Signed Number to Interpret Data

The bar graph in Figure 17 on the next page gives the Producer Price Index (PPI) for crude materials between 1999 and 2004.

---

*In some cases, people interpret "the difference between" (at least for two positive numbers) to represent the larger minus the smaller. However, we will not do so in this book.

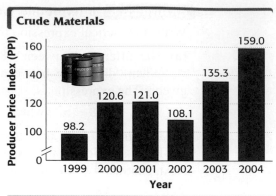

Source: U.S. Bureau of Labor Statistics.

FIGURE 17

(a) Use a signed number to represent the change in the PPI from 2003 to 2004.

To find this change, we start with the index number for 2004 and then subtract from it the index number for 2003.

$$\underbrace{159.0}_{\text{2004 index}} - \underbrace{135.3}_{\text{2003 index}} = \underbrace{+23.7}_{\substack{\text{A positive number} \\ \text{indicates an increase.}}}$$

(b) Use a signed number to represent the change in the PPI from 2001 to 2002.

$$\underbrace{108.1}_{\text{2002 index}} - \underbrace{121.0}_{\text{2001 index}} = \underbrace{108.1 + (-121.0) = -12.9}_{\substack{\text{A negative number} \\ \text{indicates a decrease.}}}$$

✔ **Now Try Exercises 99 and 101.**

1.5 EXERCISES

Complete solution available on Video Lectures on CD/DVD

Now Try Exercise

Concept Check *Fill in each blank with the correct response.*

1. The sum of two negative numbers will always be a _____ number.
(positive/negative)
Give a number-line illustration using the sum $-2 + (-3)$.

2. The sum of a number and its opposite will always be _____.

3. If I am adding a positive number and a negative number, and the negative number has the larger absolute value, the sum will be a _____ number. Give
(positive/negative)
a number-line illustration using the sum $-4 + 2$.

4. To simplify the expression $8 + [-2 + (-3 + 5)]$, I should begin by adding _____ and _____, according to the rule for order of operations.

5. Explain in words how to add signed numbers. Consider the various cases and give examples.

6. Explain in words how to subtract signed numbers.

Find each sum. See Examples 1–6.

7. $-6 + (-2)$ **8.** $-8 + (-3)$ **9.** $-3 + (-9)$

10. $-11 + (-5)$ **11.** $5 + (-3)$ **12.** $11 + (-8)$

13. $6 + (-8)$ **14.** $3 + (-7)$ **15.** $-3.5 + 12.4$

16. $-12.5 + 21.3$ **17.** $4 + [13 + (-5)]$ **18.** $6 + [2 + (-13)]$

19. $8 + [-2 + (-1)]$ **20.** $12 + [-3 + (-4)]$ **21.** $-2 + [5 + (-1)]$

22. $-8 + [9 + (-2)]$ **23.** $-6 + [6 + (-9)]$ **24.** $-3 + [11 + (-8)]$

25. $[(-9) + (-3)] + 12$ **26.** $[(-8) + (-6)] + 10$ **27.** $-\dfrac{1}{6} + \dfrac{2}{3}$

28. $-\dfrac{6}{25} + \dfrac{19}{20}$ **29.** $\dfrac{5}{8} + \left(-\dfrac{17}{12}\right)$ **30.** $\dfrac{9}{10} + \left(-\dfrac{3}{5}\right)$

31. $2\dfrac{1}{2} + \left(-3\dfrac{1}{4}\right)$ **32.** $-4\dfrac{3}{8} + 6\dfrac{1}{2}$

33. $-6.1 + [3.2 + (-4.8)]$ **34.** $-9.4 + [-5.8 + (-1.4)]$

35. $[-3 + (-4)] + [5 + (-6)]$ **36.** $[-8 + (-3)] + [-7 + (-6)]$

37. $[-4 + (-3)] + [8 + (-1)]$ **38.** $[-5 + (-9)] + [16 + (-21)]$

39. $[-4 + (-6)] + [(-3) + (-8)] + [12 + (-11)]$

40. $[-2 + (-11)] + [12 + (-2)] + [18 + (-6)]$

Find each difference. See Examples 1–6.

41. $3 - 6$ **42.** $7 - 12$ **43.** $5 - 9$

44. $8 - 13$ **45.** $-6 - 2$ **46.** $-11 - 4$

47. $-9 - 5$ **48.** $-12 - 15$ **49.** $6 - (-3)$

50. $12 - (-2)$ **51.** $-6 - (-2)$ **52.** $-7 - (-5)$

53. $2 - (3 - 5)$ **54.** $-3 - (4 - 11)$ **55.** $\dfrac{1}{2} - \left(-\dfrac{1}{4}\right)$

56. $\dfrac{1}{3} - \left(-\dfrac{4}{3}\right)$ **57.** $-\dfrac{3}{4} - \dfrac{5}{8}$ **58.** $-\dfrac{5}{6} - \dfrac{1}{2}$

59. $\dfrac{5}{8} - \left(-\dfrac{1}{2} - \dfrac{3}{4}\right)$ **60.** $\dfrac{9}{10} - \left(\dfrac{1}{8} - \dfrac{3}{10}\right)$ **61.** $3.4 - (-8.2)$

62. $5.7 - (-11.6)$ **63.** $-6.4 - 3.5$ **64.** $-4.4 - 8.6$

Perform each indicated operation. See Examples 1–6.

65. $(4 - 6) + 12$ **66.** $(3 - 7) + 4$ **67.** $(8 - 1) - 12$

68. $(9 - 3) - 15$ **69.** $6 - (-8 + 3)$ **70.** $8 - (-9 + 5)$

71. $2 + (-4 - 8)$ **72.** $6 + (-9 - 2)$

73. $|-5 - 6| + |9 + 2|$ **74.** $|-4 + 8| + |6 - 1|$

75. $|-8 - 2| - |-9 - 3|$ **76.** $|-4 - 2| - |-8 - 1|$

77. $-9 + [(3 - 2) - (-4 + 2)]$ **78.** $-8 - [(-4 - 1) + (9 - 2)]$

79. $-3 + [(-5 - 8) - (-6 + 2)]$ **80.** $-4 + [(-12 + 1) - (-1 - 9)]$

81. $-9.1237 + [(-4.8099 - 3.2516) + 11.27903]$

82. $-7.6247 - [(-3.9928 + 1.42773) - (-2.80981)]$

Write a numerical expression for each phrase and simplify. See Examples 7 and 8.

83. The sum of -5 and 12 and 6

84. The sum of -3 and 5 and -12

85. 14 added to the sum of -19 and -4

86. -2 added to the sum of -18 and 11

87. The sum of -4 and -10, increased by 12

88. The sum of -7 and -13, increased by 14

89. $\frac{2}{7}$ more than the sum of $\frac{5}{7}$ and $-\frac{9}{7}$

90. 1.85 more than the sum of -1.25 and -4.75

91. The difference between 4 and -8

92. The difference between 7 and -14

93. 8 less than -2

94. 9 less than -13

95. The sum of 9 and -4, decreased by 7

96. The sum of 12 and -7, decreased by 14

97. 12 less than the difference between 8 and -5

98. 19 less than the difference between 9 and -2

The bar graph shows federal budget outlays for the U.S. Treasury Department for the years 2002 through 2005. Use a signed number to represent the change in outlay for each period. See Example 10.

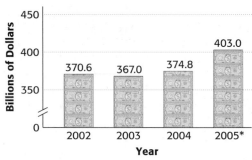

*Estimated

Source: U.S. Office of Management and Budget.

99. 2002 to 2003

100. 2003 to 2004

101. 2004 to 2005

102. 2002 to 2005

The two tables show the heights of some selected mountains and the depths of some selected trenches. Use the information given to answer Exercises 103–108.

Mountain	Height (in feet)
Foraker	17,400
Wilson	14,246
Pikes Peak	14,110

Trench	Depth (in feet, as a negative number)
Philippine	−32,995
Cayman	−24,721
Java	−23,376

Source: World Almanac and Book of Facts 2006.

103. What is the difference between the height of Mt. Foraker and the depth of the Philippine Trench?

104. What is the difference between the height of Pikes Peak and the depth of the Java Trench?

105. How much deeper is the Cayman Trench than the Java Trench?

106. How much deeper is the Philippine Trench than the Cayman Trench?

107. How much higher is Mt. Wilson than Pikes Peak?

108. If Mt. Wilson and Pikes Peak were stacked one on top of the other, how much higher would they be than Mt. Foraker?

Solve each problem. See Example 9.

109. On the basis of census population projections for 2020, New York will lose 5 seats in the U.S. House of Representatives, Pennsylvania will lose 4 seats, and Ohio will lose 3. Write a signed number that represents the total number of seats these three states are projected to lose. (*Source:* Population Reference Bureau.)

110. Michigan is projected to lose 3 seats in the U.S. House of Representatives and Illinois 2 in 2020. The states projected to gain the most seats are California with 9, Texas with 5, Florida with 3, Georgia with 2, and Arizona with 2. Write a signed number that represents the algebraic sum of these changes. (*Source:* Population Reference Bureau.)

111. The largest change in temperature ever recorded within a 24-hr period occurred in Browning, Montana, on January 23–24, 1916. The temperature fell 100°F from a starting temperature of 44°F. What was the low temperature during this period? (*Source: Guinness World Records 2006.*)

112. The lowest temperature ever recorded in Tennessee was −32°F. The highest temperature ever recorded there was 145°F more than the lowest. What was this highest temperature? (*Source:* National Climatic Data Center.)

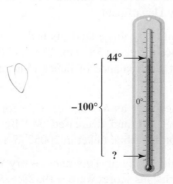

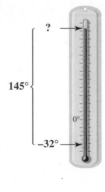

113. The lowest temperature ever recorded in Illinois was −36°F on January 5, 1999. The lowest temperature ever recorded in Utah was set on February 1, 1985, and was 33°F lower than Illinois's record low. What is the record low temperature for Utah? (*Source:* National Climatic Data Center.)

114. The top of Mt. Whitney, visible from Death Valley, has an altitude of 14,494 ft above sea level. The bottom of Death Valley is 282 ft below sea level. Using 0 as sea level, find the difference between these two elevations. (*Source: World Almanac and Book of Facts 2006.*)

115. The surface, or rim, of a canyon is at altitude 0. On a hike down into the canyon, a party of hikers stops for a rest at 130 m below the surface. The hikers then descend another 54 m. Write the new altitude as a signed number.

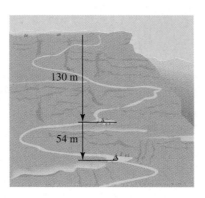

116. A pilot announces to the passengers that the current altitude of their plane is 34,000 ft. Because of turbulence, the pilot is forced to descend 2100 ft. Write the new altitude as a signed number.

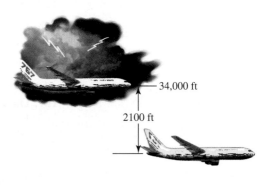

117. In 1984, Americans saved 10.8% of their after-tax incomes. In 2005, they saved −0.5%. (*Source:* Commerce Department.)

    (a) Find the difference between these two amounts.

    (b) How could Americans have a negative personal savings rate in 2005?

118. In 2000, the federal budget had a surplus of $236 billion. In 2004, the federal budget had a deficit of $413 billion. Find the difference between these amounts. (*Source:* Treasury Department.)

119. In 1998, undergraduate college students had an average credit card balance of $1879. The average balance increased $869 by 2000 and then dropped $579 by 2004. What was the average credit card balance of undergraduate college students in 2004? (*Source:* Nellie Mae.)

120. In 1999, companies paid an average of $243 for an airline ticket. This average price increased $16 by 2001 and decreased $40 by 2005. What was the average price companies paid for an airline ticket in 2005? (*Source:* American Express.)

121. Deronn Bowen enjoys playing Triominoes every Wednesday night. Last Wednesday, on four successive turns, his scores were −19, 28, −5, and 13. What was his final score for the four turns?

122. Dorothy Easley also enjoys playing Triominoes. On five successive turns, her scores were −13, 15, −12, 24, and 14. What was her total score for the five turns?

123. In August, Amy Loshak began with a checking account balance of $904.89. Her checks and deposits for August are as follows:

Checks	Deposits
$35.84	$85.00
$26.14	$120.76
$3.12	

Assuming no other transactions, what was her account balance at the end of August?

124. In September, Jeff Guild began with a checking account balance of $904.89. His checks and deposits for September are as follows:

Checks	Deposits
$41.29	$80.59
$13.66	$276.13
$84.40	

Assuming no other transactions, what was his account balance at the end of September?

125. Angie Matthews owes $870.00 on her MasterCard account. She returns two items costing $35.90 and $150.00 and receives credit for these on the account. Next, she makes a purchase of $82.50 and then two more purchases of $10.00 each. She makes a payment of $500.00. She then incurs a finance charge of $37.23. How much does she still owe?

126. Sanford Geraci owes $679.00 on his Visa account. He returns three items costing $36.89, $29.40, and $113.55 and receives credit for these on the account. Next, he makes purchases of $135.78 and $412.88 and two purchases of $20.00 each. He makes a payment of $400. He then incurs a finance charge of $24.57. How much does he still owe?

Concept Check In Exercises 127–130, suppose that x represents a positive number and y represents a negative number. Determine whether the given expression must represent a positive number or a negative number.

127. $x - y$ **128.** $y - x$ **129.** $x + |y|$ **130.** $y - |x|$

1.6 Multiplying and Dividing Real Numbers

OBJECTIVES

1 Find the product of a positive number and a negative number.

2 Find the product of two negative numbers.

3 Identify factors of integers.

(continued)

The result of multiplication is called the **product.** We already know how to multiply positive numbers, and we know that the product of two positive numbers is positive. We also know that the product of 0 and any positive number is 0, so we extend that property to all real numbers.

Multiplication by Zero

For any real number x, $x \cdot 0 = 0.$

OBJECTIVE ❶ Find the product of a positive number and a negative number. To define the product of a positive and a negative number so that the result is consistent with the multiplication of two positive numbers, look at the following pattern.

$$3 \cdot 5 = 15$$
$$3 \cdot 4 = 12$$
$$3 \cdot 3 = 9$$
$$3 \cdot 2 = 6 \qquad \text{The products}$$
$$3 \cdot 1 = 3 \qquad \text{decrease by 3.}$$
$$3 \cdot 0 = 0$$
$$3 \cdot (-1) = ?$$

What should $3(-1)$ equal? The product $3(-1)$ represents the sum

$$-1 + (-1) + (-1) = -3,$$

so the product should be -3. Also,

$$3(-2) = -2 + (-2) + (-2) = -6$$

and
$$3(-3) = -3 + (-3) + (-3) = -9.$$

These results maintain the pattern in the list, which suggests the following rule.

Multiplying Numbers with Different Signs

For any positive real numbers x and y,

$$x(-y) = -(xy) \qquad \text{and} \qquad (-x)y = -(xy).$$

That is, the product of two numbers with opposite signs is negative.

Examples: $6(-3) = -18$ and $(-6)3 = -18$

EXAMPLE 1 Multiplying a Positive Number and a Negative Number

Find each product, using the multiplication rule given in the box.

(a) $8(-5) = -(8 \cdot 5) = -40$ **(b)** $(-5)4 = -(5 \cdot 4) = -20$

(c) $-9\left(\dfrac{1}{3}\right) = -\left(9 \cdot \dfrac{1}{3}\right) = -3$ **(d)** $6.2(-4.1) = -(6.2 \cdot 4.1) = -25.42$

✔ **Now Try Exercise 11.**

OBJECTIVE ❷ Find the product of two negative numbers. The product of two positive numbers is positive, and the product of a positive and a negative number is negative. What about the product of two negative numbers? Look at another pattern.

$$-5(4) = -20$$
$$-5(3) = -15$$
$$-5(2) = -10 \qquad \text{The products}$$
$$-5(1) = -5 \qquad \text{increase by 5.}$$
$$-5(0) = 0$$
$$-5(-1) = ?$$

The numbers on the left of the equals sign (in color) decrease by 1 for each step down the list. The products on the right increase by 5 for each step down the list. To maintain this pattern, $-5(-1)$ should be 5 more than $-5(0)$, or 5 more than 0, so

$$-5(-1) = 5.$$

The pattern continues with

$$-5(-2) = 10$$
$$-5(-3) = 15$$
$$-5(-4) = 20$$
$$-5(-5) = 25,$$

and so on, which suggests the next rule.

Multiplying Two Negative Numbers

For any positive real numbers x and y,

$$-x(-y) = xy.$$

That is, the product of two negative numbers is positive.

Example: $-5(-4) = 20$

EXAMPLE 2 Multiplying Two Negative Numbers

Find each product, using the multiplication rule given in the box.

(a) $-9(-2) = 9 \cdot 2 = 18$ **(b)** $-6(-12) = 6 \cdot 12 = 72$

(c) $-8(-1) = 8 \cdot 1 = 8$ **(d)** $-\dfrac{2}{3}\left(-\dfrac{3}{2}\right) = \dfrac{2}{3} \cdot \dfrac{3}{2} = 1$

✔ **Now Try Exercise 13.**

The following box summarizes multiplying signed numbers.

Multiplying Signed Numbers

The product of two numbers having the *same* sign is *positive*.

The product of two numbers having *different* signs is *negative*.

OBJECTIVE 3 Identify factors of integers. In **Section 1.1,** the definition of a *factor* was given for whole numbers. For example, since $9 \cdot 5 = 45$, both 9 and 5 are factors of 45. The definition can be extended to integers. If the product of two integers is a third integer, then each of the two integers is a *factor* of the third. For example, $-3(-6) = 18$, so -3 and -6 are factors of 18. The table on the next page shows several integers and the factors of those integers.

Integer	18	20	15	7	1
Pairs of factors	1, 18	1, 20	1, 15	1, 7	1, 1
	2, 9	2, 10	3, 5	−1, −7	−1, −1
	3, 6	4, 5	−1, −15		
	−1, −18	−1, −20	−3, −5		
	−2, −9	−2, −10			
	−3, −6	−4, −5			

✔ **Now Try Exercise 29.**

Number	Multiplicative Inverse (Reciprocal)
4	$\dfrac{1}{4}$
$0.3 = \dfrac{3}{10}$	$\dfrac{10}{3}$
−5	$\dfrac{1}{-5}$ or $-\dfrac{1}{5}$
$-\dfrac{5}{8}$	$-\dfrac{8}{5}$
0	None
1	1
−1	−1

OBJECTIVE 4   Use the reciprocal of a number to apply the definition of division.   In **Section 1.5,** we saw that the difference between two numbers is found by adding the additive inverse of the second number to the first. Similarly, the *quotient* of two numbers is found by *multiplying* by the *reciprocal,* or *multiplicative inverse.* By definition, since

$$8 \cdot \frac{1}{8} = \frac{8}{8} = 1 \qquad \text{and} \qquad \frac{5}{4} \cdot \frac{4}{5} = \frac{20}{20} = 1,$$

the reciprocal or multiplicative inverse of 8 is $\frac{1}{8}$ and of $\frac{5}{4}$ is $\frac{4}{5}$.

Reciprocal or Multiplicative Inverse

Pairs of numbers whose product is 1 are called **reciprocals,** or **multiplicative inverses,** of each other.

The table in the margin shows several numbers and their multiplicative inverses. Why is there no multiplicative inverse for the number 0? Suppose that k is to be the multiplicative inverse of 0. Then $k \cdot 0$ should equal 1. But $k \cdot 0 = 0$ for any number k. Since there is no value of k that is a solution of the equation $k \cdot 0 = 1$, the following statement can be made:

0 has no multiplicative inverse.

Definition of Division

For any real numbers x and y, with $y \neq 0$, $\qquad \dfrac{x}{y} = x \cdot \dfrac{1}{y}.$

That is, to divide two numbers, multiply the first by the reciprocal, or multiplicative inverse, of the second.

The definition of division indicates that y, the number to divide by, cannot be 0. Since 0 has no multiplicative inverse,

$\frac{x}{0}$ **is not a number, and** *division by 0 is undefined.* **If a division problem involves division by 0, write "undefined."**

> ▶ **NOTE** While division *by* 0 $\left(\text{that is, } \frac{x}{0}\right)$ is undefined, we may divide 0 by any nonzero number. In fact, if $y \neq 0$, then
>
> $$\frac{0}{y} = 0.$$

✔ **Now Try Exercise 9.**

Since division is defined in terms of multiplication, all the rules for multiplying signed numbers also apply to dividing them.

EXAMPLE 3 Using the Definition of Division

Find each quotient, using the definition of division.

> Remember to write in lowest terms.

(a) $\dfrac{12}{3} = 12 \cdot \dfrac{1}{3} = 4$
  (b) $\dfrac{-10}{2} = -10 \cdot \dfrac{1}{2} = -5$

(c) $\dfrac{-1.47}{-7} = -1.47\left(-\dfrac{1}{7}\right) = 0.21$
  (d) $-\dfrac{2}{3} \div \left(-\dfrac{4}{5}\right) = -\dfrac{2}{3} \cdot \left(-\dfrac{5}{4}\right) = \dfrac{5}{6}$

(e) $\dfrac{0}{-10} = 0$ $\dfrac{0}{y} = 0$ $(y \neq 0)$
  (f) $\dfrac{-10}{0}$ Undefined

✔ **Now Try Exercises 33, 35, 45, and 47.**

When dividing fractions, multiplying by the reciprocal works well. However, using the definition of division directly with integers is awkward. It is easier to divide in the usual way and then determine the sign of the answer.

Dividing Signed Numbers

The quotient of two numbers having the *same* sign is *positive.*
The quotient of two numbers having *different* signs is *negative.*

Examples: $\dfrac{-15}{-5} = 3,$ $\dfrac{15}{-5} = -3,$ and $\dfrac{-15}{5} = -3$

EXAMPLE 4 Dividing Signed Numbers

Find each quotient.

(a) $\dfrac{8}{-2} = -4$
  (b) $\dfrac{-4.5}{-0.09} = 50$

> Remember to write in lowest terms.

(c) $-\dfrac{1}{8} \div \left(-\dfrac{3}{4}\right) = -\dfrac{1}{8} \cdot \left(-\dfrac{4}{3}\right) = \dfrac{1}{6}$

✔ **Now Try Exercises 39, 41, and 43.**

From the definitions of multiplication and division of real numbers,

$$\frac{-40}{8} = -40 \cdot \frac{1}{8} = -5 \quad \text{and} \quad \frac{40}{-8} = 40\left(\frac{1}{-8}\right) = -5, \quad \text{so} \quad \frac{-40}{8} = \frac{40}{-8}.$$

Based on this example, the quotient of a positive and a negative number can be expressed in any of the following three forms.

> For any positive real numbers x and y, $\quad \dfrac{-x}{y} = \dfrac{x}{-y} = -\dfrac{x}{y}.$

Similarly, the quotient of two negative numbers can be expressed as a quotient of two positive numbers.

> For any positive real numbers x and y, $\quad \dfrac{-x}{-y} = \dfrac{x}{y}.$

OBJECTIVE 5 Use the rules for order of operations when multiplying and dividing signed numbers.

EXAMPLE 5 Using the Rules for Order of Operations

Perform each indicated operation.

(a) $-9(2) - (-3)(2)$

$= -18 - (-6)$ Multiply.

$= -18 + 6$ Definition of subtraction

$= -12$ Add.

(b) $-6(-2) - 3(-4)$

$= 12 - (-12)$

$= 12 + 12$

$= 24$

(c) $-5(-2 - 3)$

$= -5(-5)$

$= 25$

(d) $\dfrac{5(-2) - 3(4)}{2(1 - 6)}$

$= \dfrac{-10 - 12}{2(-5)}$ Simplify the numerator and denominator separately.

$= \dfrac{-22}{-10} = \dfrac{11}{5}$ Remember to write in lowest terms.

✔ **Now Try Exercises 53 and 67.**

We summarize the rules for operations with signed numbers on the next page.

Operations with Signed Numbers

Addition

Same sign Add the absolute values of the numbers. The sum has the same sign as the given numbers.

$$-4 + (-6) = -10$$

Different signs Find the absolute values of the numbers and subtract the smaller absolute value from the larger. Give the answer the sign of the number having the larger absolute value.

$$4 + (-6) = -(6 - 4) = -2$$

Subtraction

Add the additive inverse (or opposite) of the second number to the first number.

$$8 - (-3) = 8 + 3 = 11$$

Multiplication and Division

Same sign The product or quotient of two numbers with the same sign is positive.

$$-5(-6) = 30 \qquad \text{and} \qquad \frac{-36}{-12} = 3$$

Different signs The product or quotient of two numbers with different signs is negative.

$$-5(6) = -30 \qquad \text{and} \qquad \frac{18}{-6} = -3$$

Division by 0 is undefined.

OBJECTIVE 6 Evaluate expressions involving variables.

EXAMPLE 6 Evaluating Expressions for Numerical Values

Evaluate each expression, given that $x = -1$, $y = -2$, and $m = -3$.

(a)

$$(3x + 4y)(-2m)$$

$$= [3(-1) + 4(-2)][-2(-3)] \qquad \text{Substitute the given values for the variables.}$$

> Use parentheses around substituted negative values to avoid errors.

$$= [-3 + (-8)][6] \qquad \text{Multiply.}$$

$$= [-11]6 \qquad \text{Add inside the brackets.}$$

$$= -66 \qquad \text{Multiply.}$$

(b) $2x^2 - 3y^2$

$$= 2(-1)^2 - 3(-2)^2 \qquad \text{Substitute.}$$

$$= 2(1) - 3(4) \qquad \text{Apply the exponents.}$$

$$= 2 - 12 \qquad \text{Multiply.}$$

$$= -10 \qquad \text{Subtract.}$$

(c) $\dfrac{4y^2 + x}{m}$

$= \dfrac{4(-2)^2 + (-1)}{-3}$ Substitute.

$= \dfrac{4(4) + (-1)}{-3}$ Apply the exponent.

$= \dfrac{16 + (-1)}{-3}$ Multiply.

$= \dfrac{15}{-3},$ or -5 Add, then divide.

✔ **Now Try Exercises 79 and 87.**

OBJECTIVE 7 **Interpret words and phrases involving multiplication and division.** The word *product* refers to multiplication. The table gives other key words and phrases that indicate multiplication in problem solving.

Word or Phrase	Example	Numerical Expression and Simplification
Product of	The *product of* -5 and -2	$-5(-2) = 10$
Times	13 *times* -4	$13(-4) = -52$
Twice (meaning "2 times")	*Twice* 6	$2(6) = 12$
Of (used with fractions)	$\dfrac{1}{2}$ *of* 10	$\dfrac{1}{2}(10) = 5$
Percent of	12% *of* -16	$0.12(-16) = -1.92$
As much as	$\dfrac{2}{3}$ *as much as* 30	$\dfrac{2}{3}(30) = 20$

EXAMPLE 7 **Interpreting Words and Phrases Involving Multiplication**

Write a numerical expression for each phrase and simplify the expression.

(a) The product of 12 and the sum of 3 and -6

Here, 12 is multiplied by "the sum of 3 and -6." The expression is

$$12[3 + (-6)], \quad \text{which simplifies to} \quad 12[-3], \quad \text{or} \quad -36.$$

(b) Twice the difference between 8 and -4

$$2[8 - (-4)] \quad \text{simplifies to} \quad 2[12], \quad \text{or} \quad 24.$$

(c) Two-thirds of the sum of -5 and -3

$$\frac{2}{3}[-5 + (-3)] \quad \text{simplifies to} \quad \frac{2}{3}[-8], \quad \text{or} \quad -\frac{16}{3}.$$

(d) 15% of the difference between 14 and -2

$$0.15[14 - (-2)] \quad \text{simplifies to} \quad 0.15[16], \quad \text{or} \quad 2.4.$$

Remember that 15% = 0.15.

(e) Double the product of 3 and 4

$$2 \cdot (3 \cdot 4) \quad \text{simplifies to} \quad 2(12), \quad \text{or} \quad 24.$$

✔ **Now Try Exercises 91, 95, and 103.**

The word *quotient* refers to division. In algebra, quotients are usually represented with a fraction bar; the symbol ÷ is seldom used. The table gives some key phrases associated with division.

Phrase	Example	Numerical Expression and Simplification
Quotient of	The *quotient of* −24 and 3	$\dfrac{-24}{3} = -8$
Divided by	−16 *divided by* −4	$\dfrac{-16}{-4} = 4$
Ratio of	The *ratio of* 2 to 3	$\dfrac{2}{3}$

EXAMPLE 8 Interpreting Words and Phrases Involving Division

Write a numerical expression for each phrase and simplify the expression.

(a) The quotient of 14 and the sum of −9 and 2

"Quotient" indicates division. The number 14 is the numerator and "the sum of −9 and 2" is the denominator. The expression is

$$\frac{14}{-9+2}, \quad \text{which simplifies to} \quad \frac{14}{-7}, \quad \text{or} \quad -2.$$

(b) The product of 5 and −6, divided by the difference between −7 and 8

The numerator of the fraction representing the division is found by multiplying 5 and −6. The denominator is found by subtracting −7 and 8. The expression is

$$\frac{5(-6)}{-7-8}, \quad \text{which simplifies to} \quad \frac{-30}{-15}, \quad \text{or} \quad 2.$$

✔ **Now Try Exercise 97.**

OBJECTIVE 8 Translate simple sentences into equations. In this section and the previous one, important words and phrases involving the four operations of arithmetic have been introduced. We can use these words and phrases to interpret sentences that translate into equations.

EXAMPLE 9 Translating Sentences into Equations

Write each sentence in symbols, using x as the variable. Then guess or use trial and error to find the solution, which comes from the list of integers between −12 and 12, inclusive.

(a) Three times a number is −18.

The word *times* indicates multiplication, and the word *is* translates as the equals sign (=).

$$3x = -18 \qquad 3 \cdot x = 3x$$

Since the integer between −12 and 12, inclusive, that makes this statement true is −6, the solution of the equation is −6.

(b) The sum of a number and 9 is 12.

$$x + 9 = 12$$

Since 3 + 9 = 12, the solution of this equation is 3.

(c) The difference between a number and 5 is 0.

$$x - 5 = 0$$

Since $5 - 5 = 0$, the solution of this equation is 5.

(d) The quotient of 24 and a number is -2.

$$\frac{24}{x} = -2$$

Here, x must be a negative number, since the numerator is positive and the quotient is negative. Since $\frac{24}{-12} = -2$, the solution is -12.

✔ **Now Try Exercises 109 and 113.**

▶ **CAUTION** Notice the distinction between the problems found in Examples 7 and 8 and those in Example 9. In Examples 7 and 8, the *phrases* translate as *expressions,* while in Example 9, the *sentences* translate as *equations.* *Remember that an equation is a sentence with something on the left side, an = sign, and something on the right side. An expression is a phrase.*

$$\frac{5(-6)}{-7 - 8}$$
Expression

$$3x = -18$$
Equation

1.6 EXERCISES

Concept Check   *Fill in each blank with one of the following:* greater than 0, less than 0, equal to 0.

1. The product or the quotient of two numbers with the same sign is _____.

2. The product or the quotient of two numbers with different signs is _____.

3. If three negative numbers are multiplied together, the product is _____.

4. If two negative numbers are multiplied together and then their product is divided by a negative number, the result is _____.

5. If a negative number is squared and the result is added to a positive number, the final answer is _____.

6. The reciprocal of a negative number is _____.

7. If three positive numbers, five negative numbers, and zero are multiplied together, the product is _____.

8. The fifth power of a negative number is _____.

🌐 **9.** *Concept Check*   Complete this statement: The quotient formed by any nonzero number divided by 0 is _____, and the quotient formed by 0 divided by any nonzero number is _____. Give an example of each quotient.

10. Concept Check   Which expression is undefined?

A. $\dfrac{5-5}{5+5}$ **B.** $\dfrac{5+5}{5+5}$ **C.** $\dfrac{5-5}{5-5}$ **D.** $\dfrac{5-5}{5}$

Find each product. See Examples 1 and 2.

11. $3(-4)$ | **12.** $-3(4)$ | **13.** $-3(-4)$ | **14.** $-2(-8)$

15. $-10(-12)$ | **16.** $9(-5)$ | **17.** $3(-11)$ | **18.** $3(-15)$

19. $-6.8(0.35)$ | **20.** $-4.6(0.24)$ | **21.** $-\dfrac{3}{8}\cdot\left(-\dfrac{10}{9}\right)$ | **22.** $-\dfrac{5}{4}\cdot\left(-\dfrac{5}{8}\right)$

23. $\dfrac{2}{15}\left(-1\dfrac{1}{4}\right)$ | **24.** $\dfrac{3}{7}\left(-1\dfrac{5}{9}\right)$ | **25.** $-8\left(-\dfrac{3}{4}\right)$ | **26.** $-6\left(-\dfrac{5}{3}\right)$

Find all integer factors of each number. See Objective 3.

27. 32 | **28.** 36 | **29.** 40 | **30.** 50 | **31.** 31 | **32.** 17

Find each quotient. See Examples 3 and 4.

33. $\dfrac{15}{5}$ | **34.** $\dfrac{25}{5}$ | **35.** $\dfrac{-30}{6}$ | **36.** $\dfrac{-28}{14}$

37. $\dfrac{-28}{-4}$ | **38.** $\dfrac{-35}{-7}$ | **39.** $\dfrac{96}{-16}$ | **40.** $\dfrac{38}{-19}$

41. $-\dfrac{4}{3}\div\left(-\dfrac{1}{8}\right)$ | **42.** $-\dfrac{5}{6}\div\left(-\dfrac{15}{7}\right)$ | **43.** $\dfrac{-8.8}{2.2}$ | **44.** $\dfrac{-4.6}{-0.23}$

45. $\dfrac{0}{-2}$ | **46.** $\dfrac{0}{-8}$ | **47.** $\dfrac{12.4}{0}$ | **48.** $\dfrac{6.5}{0}$

Perform each indicated operation. See Example 5.

49. $7-3\cdot6$ | **50.** $8-2\cdot5$ | **51.** $-10-(-4)(2)$

52. $-11-(-3)(6)$ | **53.** $-7(3-8)$ | **54.** $-5(4-7)$

55. $(12-14)(1-4)$ | **56.** $(8-9)(4-12)$ | **57.** $(7-10)(10-4)$

58. $(5-12)(19-4)$ | **59.** $(-2-8)(-6)+7$ | **60.** $(-9-4)(-2)+10$

61. $3(-5)+|3-10|$ | **62.** $4(-8)+|4-15|$ | **63.** $\dfrac{-5(-6)}{9-(-1)}$

64. $\dfrac{-12(-5)}{7-(-5)}$ | **65.** $\dfrac{-21(3)}{-3-6}$ | **66.** $\dfrac{-40(3)}{-2-3}$

67. $\dfrac{-10(2)+6(2)}{-3-(-1)}$ | **68.** $\dfrac{-12(4)+5(3)}{-14-(-3)}$ | **69.** $\dfrac{3^2-4^2}{7(-8+9)}$

70. $\dfrac{5^2-7^2}{2(3+3)}$ | **71.** $\dfrac{8(-1)-|(-4)(-3)|}{-6-(-1)}$ | **72.** $\dfrac{-27(-2)-|6\cdot4|}{-2(3)-2(2)}$

73. $\dfrac{-13(-4)-(-8)(-2)}{(-10)(2)-4(-2)}$ | **74.** $\dfrac{-5(2)+[3(-2)-4]}{-3-(-1)}$

75. Concept Check   If x and y are both replaced by *negative* numbers, is the value of $4x+8y$ positive or negative?

76. Concept Check   Repeat Exercise 75, but replace x and y with *positive* numbers.

Evaluate each expression if x = 6, y = −4, and a = 3. See Example 6.

77. $5x - 2y + 3a$

78. $6x - 5y + 4a$

79. $(2x + y)(3a)$

80. $(5x - 2y)(-2a)$

81. $\left(\dfrac{1}{3}x - \dfrac{4}{5}y\right)\left(-\dfrac{1}{5}a\right)$

82. $\left(\dfrac{5}{6}x + \dfrac{3}{2}y\right)\left(-\dfrac{1}{3}a\right)$

83. $(-5 + x)(-3 + y)(3 - a)$

84. $(6 - x)(5 + y)(3 + a)$

85. $-2y^2 + 3a$

86. $5x - 4a^2$

87. $\dfrac{2y^2 - x}{a + 10}$

88. $\dfrac{xy + 8a}{x - y}$

Write a numerical expression for each phrase and simplify. See Examples 7 and 8.

89. The product of −9 and 2, added to 9

90. The product of 4 and −7, added to −12

91. Twice the product of −1 and 6, subtracted from −4

92. Twice the product of −8 and 2, subtracted from −1

93. Nine subtracted from the product of 1.5 and −3.2

94. Three subtracted from the product of 4.2 and −8.5

95. The product of 12 and the difference between 9 and −8

96. The product of −3 and the difference between 3 and −7

97. The quotient of −12 and the sum of −5 and −1

98. The quotient of −20 and the sum of −8 and −2

99. The sum of 15 and −3, divided by the product of 4 and −3

100. The sum of −18 and −6, divided by the product of 2 and −4

101. Two-thirds of the difference between 8 and −1

102. Three-fourths of the sum of −8 and 12

103. 20% of the product of −5 and 6

104. 30% of the product of −8 and 5

105. The sum of $\frac{1}{2}$ and $\frac{5}{8}$, times the difference between $\frac{3}{5}$ and $\frac{1}{3}$

106. The sum of $\frac{3}{4}$ and $\frac{1}{2}$, times the difference between $\frac{2}{3}$ and $\frac{1}{6}$

107. The product of $-\frac{1}{2}$ and $\frac{3}{4}$, divided by $-\frac{2}{3}$

108. The product of $-\frac{2}{3}$ and $-\frac{1}{5}$, divided by $\frac{1}{7}$

Write each statement in symbols, using x as the variable. Then guess or use trial and error to find the solution, which comes from the set of integers between −12 and 12, inclusive. See Example 9.

109. The quotient of a number and 3 is −3.

110. The quotient of a number and 4 is −1.

111. 6 less than a number is 4.

112. 7 less than a number is 2.

113. When 5 is added to a number, the result is −5.

114. When 6 is added to a number, the result is −3.

*To find the **average** of a group of numbers, we add the numbers and then divide the sum by the number of terms added. For example, to find the average of 14, 8, 3, 9, and 1, we add them and then divide by 5.*

$$\dfrac{14 + 8 + 3 + 9 + 1}{5} = \dfrac{35}{5} = 7 \leftarrow \text{Average}$$

115. Find the average of 23, 18, 13, −4, and −8.

116. Find the average of 18, 12, 0, −4, and −10.

117. What is the average of all integers between −10 and 14, inclusive?

118. What is the average of all even integers between −18 and 4, inclusive?

119. If the average of a group of numbers is 0, what is the sum of all the numbers?

120. Suppose that in a group of numbers, some are positive and some are negative. Under what conditions will the average be a positive number? Under what conditions will the average be negative?

*The operation of division is used in **divisibility tests**. A divisibility test allows us to determine whether a given number is divisible (without remainder) by another number.*

121. An integer is divisible by 2 if its last digit is divisible by 2, and not otherwise. Show that

   (a) 3,473,986 is divisible by 2 and
   (b) 4,336,879 is not divisible by 2.

122. An integer is divisible by 3 if the sum of its digits is divisible by 3, and not otherwise. Show that

   (a) 4,799,232 is divisible by 3 and
   (b) 2,443,871 is not divisible by 3.

123. An integer is divisible by 4 if its last two digits form a number divisible by 4, and not otherwise. Show that

   (a) 6,221,464 is divisible by 4 and
   (b) 2,876,335 is not divisible by 4.

124. An integer is divisible by 5 if its last digit is divisible by 5, and not otherwise. Show that

   (a) 3,774,595 is divisible by 5 and
   (b) 9,332,123 is not divisible by 5.

125. An integer is divisible by 6 if it is divisible by both 2 and 3, and not otherwise. Show that

   (a) 1,524,822 is divisible by 6 and
   (b) 2,873,590 is not divisible by 6.

126. An integer is divisible by 8 if its last three digits form a number divisible by 8, and not otherwise. Show that

   (a) 2,923,296 is divisible by 8 and
   (b) 7,291,623 is not divisible by 8.

127. An integer is divisible by 9 if the sum of its digits is divisible by 9, and not otherwise. Show that

   (a) 4,114,107 is divisible by 9 and
   (b) 2,287,321 is not divisible by 9.

128. An integer is divisible by 12 if it is divisible by both 3 and 4, and not otherwise. Show that

   (a) 4,253,520 is divisible by 12 and
   (b) 4,249,474 is not divisible by 12.

Summary Exercises on Operations with Real Numbers

Operations with Signed Numbers

Addition

Same sign   Add the absolute values of the numbers. The sum has the same sign as the given numbers.

Different signs   Find the absolute values of the numbers, and subtract the smaller absolute value from the larger. Give the answer the sign of the number having the larger absolute value.

Subtraction

Add the additive inverse (or opposite) of the second number to the first number.

Multiplication and Division

Same sign   The product or quotient of two numbers with the same sign is positive.

Different signs   The product or quotient of two numbers with different signs is negative.

Division by 0 is undefined.

Perform each indicated operation.

1. $14 - 3 \cdot 10$

2. $-3(8) - 4(-7)$

3. $(3 - 8)(-2) - 10$

4. $-6(7 - 3)$

5. $7 - (-3)(2 - 10)$

6. $-4[(-2)(6) - 7]$

7. $(-4)(7) - (-5)(2)$

8. $-5[-4 - (-2)(-7)]$

9. $40 - (-2)[8 - 9]$

10. $\dfrac{5(-4)}{-7 - (-2)}$

11. $\dfrac{-3 - (-9 + 1)}{-7 - (-6)}$

12. $\dfrac{5(-8 + 3)}{13(-2) + (-7)(-3)}$

13. $\dfrac{6^2 - 8}{-2(2) + 4(-1)}$

14. $\dfrac{16(-8 + 5)}{15(-3) + (-7 - 4)(-3)}$

15. $\dfrac{9(-6) - 3(8)}{4(-7) + (-2)(-11)}$

16. $\dfrac{2^2 + 4^2}{5^2 - 3^2}$

17. $\dfrac{(2 + 4)^2}{(5 - 3)^2}$

18. $\dfrac{4^3 - 3^3}{-5(-4 + 2)}$

19. $\dfrac{-9(-6) + (-2)(27)}{3(8 - 9)}$

20. $|-4(9)| - |-11|$

21. $\dfrac{6(-10 + 3)}{15(-2) - 3(-9)}$

22. $\dfrac{3^2 - 5^2}{(-9)^2 - 9^2}$

23. $\dfrac{(-10)^2 + 10^2}{-10(5)}$

24. $-\dfrac{3}{4} \div \left(-\dfrac{5}{8}\right)$

25. $\dfrac{1}{2} \div \left(-\dfrac{1}{2}\right)$

26. $\dfrac{8^2 - 12}{(-5)^2 + 2(6)}$

27. $\left[\dfrac{5}{8} - \left(-\dfrac{1}{16}\right)\right] + \dfrac{3}{8}$

28. $\left(\dfrac{1}{2} - \dfrac{1}{3}\right) - \dfrac{5}{6}$

29. $-0.9(-3.7)$

30. $-5.1(-0.2)$

31. $-3^2 - 2^2$

32. $|-2(3) + 4| - |-2|$

33. $40 - (-2)[-5 - 3]$

Evaluate each expression if $x = -2$, $y = 3$, and $a = 4$.

34. $-x + y - 3a$

35. $(x + 6)^3 - y^3$

36. $(x - y) - (a - 2y)$

37. $\left(\dfrac{1}{2}x + \dfrac{2}{3}y\right)\left(-\dfrac{1}{4}a\right)$

38. $\dfrac{2x + 3y}{a - xy}$

39. $\dfrac{x^2 - y^2}{x^2 + y^2}$

40. $-x^2 + 3y$

41. $\left(\dfrac{x}{y}\right)^3$

42. $\left(\dfrac{a}{x}\right)^2$

1.7 Properties of Real Numbers

OBJECTIVES

1 Use the commutative properties.

2 Use the associative properties.

3 Use the identity properties.

4 Use the inverse properties.

5 Use the distributive property.

If you were asked to find the sum

$$3 + 89 + 97,$$

you might mentally add $3 + 97$ to get 100 and then add $100 + 89$ to get 189. While the rule for order of operations says to add from left to right, we may change the order of the terms and group them in any way we choose without affecting the sum.

These are examples of shortcuts that we use in everyday mathematics. Such shortcuts are justified by the basic properties of addition and multiplication, discussed in this section. In these properties, a, b, and c represent real numbers.

OBJECTIVE 1 Use the commutative properties. The word *commute* means to go back and forth. Many people commute to work or to school. If you travel from home to work and follow the same route from work to home, you travel the same distance each time. The **commutative properties** say that if two numbers are added or multiplied in any order, the result is the same.

Commutative Properties

$$a + b = b + a \qquad \text{Addition}$$
$$ab = ba \qquad \text{Multiplication}$$

EXAMPLE 1 Using the Commutative Properties

Use a commutative property to complete each statement.

(a) $-8 + 5 = 5 +$ _____

By the commutative property of addition, the missing number is -8, since $-8 + 5 = 5 + (-8)$.

(b) $(-2)7 =$ _____(-2)

By the commutative property of multiplication, the missing number is 7, since $(-2)7 = 7(-2)$.

✔ **Now Try Exercises 1 and 3.**

OBJECTIVE 2 Use the associative properties. When we *associate* one object with another, we think of those objects as being grouped together. The **associative properties** say that when we add or multiply three numbers, we can group the first two together or the last two together and get the same answer.

Associative Properties

$$(a + b) + c = a + (b + c) \qquad \text{Addition}$$
$$(ab)c = a(bc) \qquad \text{Multiplication}$$

EXAMPLE 2 Using the Associative Properties

Use an associative property to complete each statement.

(a) $8 + (-1 + 4) = (8 + \underline{\quad}) + 4$

The missing number is -1.

(b) $[2 \cdot (-7)] \cdot 6 = 2 \cdot \underline{\quad}$

The completed expression on the right should be $2 \cdot [(-7) \cdot 6]$.

✔ **Now Try Exercises 5 and 7.**

By the associative property of addition, the sum of three numbers will be the same no matter how the numbers are "associated" in groups. For this reason, parentheses can be left out in many addition problems. For example, both

$$(-1 + 2) + 3 \quad \text{and} \quad -1 + (2 + 3)$$

can be written as

$$-1 + 2 + 3.$$

In the same way, parentheses also can be left out of many multiplication problems.

EXAMPLE 3 Distinguishing between the Associative and Commutative Properties

(a) Is $(2 + 4) + 5 = 2 + (4 + 5)$ an example of the associative property or the commutative property?

The order of the three numbers is the same on both sides of the equals sign. The only change is in the *grouping,* or association, of the numbers. Therefore, this is an example of the associative property.

(b) Is $6(3 \cdot 10) = 6(10 \cdot 3)$ an example of the associative property or the commutative property?

The same numbers, 3 and 10, are grouped on each side. On the left, the 3 appears first, but on the right, the 10 appears first. Since the only change involves the *order* of the numbers, this statement is an example of the commutative property.

(c) Is $(8 + 1) + 7 = 8 + (7 + 1)$ an example of the associative property or the commutative property?

In the statement, both the order and the grouping are changed. On the left, the order of the three numbers is 8, 1, and 7. On the right, it is 8, 7, and 1. On the left, the 8 and 1 are grouped; on the right, the 7 and 1 are grouped. Therefore, both the associative and the commutative properties are used.

✔ **Now Try Exercises 15 and 23.**

EXAMPLE 4 Using the Commutative and Associative Properties

Find the sum $23 + 41 + 2 + 9 + 25$.

The commutative and associative properties make it possible to choose pairs of numbers whose sums are easy to find.

$$23 + 41 + 2 + 9 + 25 = (41 + 9) + (23 + 2) + 25$$
$$= 50 + 25 + 25$$
$$= 100$$

✔ **Now Try Exercise 37.**

OBJECTIVE 3 Use the identity properties. If a child wears a costume on Halloween, the child's appearance is changed, but his or her *identity* is unchanged. The identity of a real number is left unchanged when identity properties are applied. The **identity properties** say that the sum of 0 and any number equals that number, and the product of 1 and any number equals that number.

Identity Properties

$$a + 0 = a \quad \text{and} \quad 0 + a = a \quad \text{Addition}$$
$$a \cdot 1 = a \quad \text{and} \quad 1 \cdot a = a \quad \text{Multiplication}$$

The number 0 leaves the identity, or value, of any real number unchanged by addition. For this reason, 0 is called the **identity element for addition,** or the **additive identity.** Since multiplication by 1 leaves any real number unchanged, 1 is the **identity element for multiplication,** or the **multiplicative identity.**

EXAMPLE 5 Using the Identity Properties

These statements are examples of the identity properties.

(a) $-3 + 0 = -3$ Addition   **(b)** $1 \cdot \dfrac{1}{2} = \dfrac{1}{2}$ Multiplication

> ✔ **Now Try Exercise 21.**

We use the identity property for multiplication to write fractions in lowest terms and to find common denominators.

EXAMPLE 6 Using the Identity Property to Simplify Expressions

Simplify.

(a) $\dfrac{49}{35} = \dfrac{7 \cdot 7}{5 \cdot 7}$ Factor.

$\phantom{\dfrac{49}{35}} = \dfrac{7}{5} \cdot \dfrac{7}{7}$ Write as a product.

$\phantom{\dfrac{49}{35}} = \dfrac{7}{5} \cdot 1$ Divide.

$\phantom{\dfrac{49}{35}} = \dfrac{7}{5}$ Identity property

(b) $\dfrac{3}{4} + \dfrac{5}{24} = \dfrac{3}{4} \cdot 1 + \dfrac{5}{24}$ Identity property

$\phantom{\dfrac{3}{4} + \dfrac{5}{24}} = \dfrac{3}{4} \cdot \dfrac{6}{6} + \dfrac{5}{24}$ Use $1 = \frac{6}{6}$ to get a common denominator.

$\phantom{\dfrac{3}{4} + \dfrac{5}{24}} = \dfrac{18}{24} + \dfrac{5}{24}$ Multiply.

$\phantom{\dfrac{3}{4} + \dfrac{5}{24}} = \dfrac{23}{24}$ Add.

> ✔ **Now Try Exercise 27.**

OBJECTIVE 4 Use the inverse properties. Each day before you go to work or school, you probably put on your shoes before you leave. Before you go to sleep at night, you probably take them off, and this leads to the same situation that existed before you put them on. These operations from everyday life are examples of *inverse* operations.

The **inverse properties** of addition and multiplication lead to the additive and multiplicative identities, respectively. Recall that $-a$ is the **additive inverse,** or **opposite,** of a and $\frac{1}{a}$ is the **multiplicative inverse,** or **reciprocal,** of the nonzero number a. The sum of the numbers a and $-a$ is 0, and the product of the nonzero numbers a and $\frac{1}{a}$ is 1.

Inverse Properties

$$a + (-a) = 0 \quad \text{and} \quad -a + a = 0 \qquad \text{Addition}$$

$$a \cdot \frac{1}{a} = 1 \quad \text{and} \quad \frac{1}{a} \cdot a = 1 \quad (a \neq 0) \qquad \text{Multiplication}$$

EXAMPLE 7 Using the Inverse Properties

These statements are examples of the inverse properties.

(a) $-\dfrac{1}{2} + \dfrac{1}{2} = 0$ **(b)** $4 + (-4) = 0$ **(c)** $-0.75 + \dfrac{3}{4} = 0$

(d) $\dfrac{2}{5} \cdot \dfrac{5}{2} = 1$ **(e)** $-5\left(-\dfrac{1}{5}\right) = 1$ **(f)** $4(0.25) = 1$

✔ **Now Try Exercise 19.**

EXAMPLE 8 Using Properties to Simplify an Expression

Simplify $-2x + 10 + 2x$.

$$-2x + 10 + 2x$$
$$= (-2x + 10) + 2x \qquad \text{Order of operations}$$
$$= [10 + (-2x)] + 2x \qquad \text{Commutative property}$$
$$= 10 + [(-2x) + 2x] \qquad \text{Associative property}$$
$$= 10 + 0 \qquad \text{Inverse property}$$
$$= 10 \qquad \text{Identity property}$$

Note that for *any* value of x, $-2x$ and $2x$ are additive inverses; that is why we can use the inverse property in this simplification.

✔ **Now Try Exercise 43.**

► **NOTE** The detailed procedure shown in Example 8 is seldom used in practice. We include the example to show how the properties of this section apply, even though steps may be skipped when we actually do the simplification.

OBJECTIVE 5 Use the distributive property. The everyday meaning of the word *distribute* is "to give out from one to several." An important property of real number operations involves this idea. Look at the value of the following expressions:

$$2(5 + 8), \quad \text{which equals} \quad 2(13), \quad \text{or} \quad 26$$
$$2(5) + 2(8), \quad \text{which equals} \quad 10 + 16, \quad \text{or} \quad 26.$$

Since both expressions equal 26,

$$2(5 + 8) = 2(5) + 2(8).$$

This result is an example of the *distributive property,* the only property involving *both* addition and multiplication. With this property, a product can be changed to a sum or difference. This idea is illustrated by the divided rectangle in Figure 18.

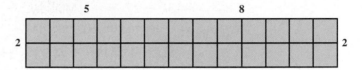

The area of the left part is 2(5) = 10.
The area of the right part is 2(8) = 16.
The total area is 2(5 + 8) = 2(13) = 26,
or the total area is 2(5) + 2(8) = 10 + 16 = 26.
Thus, 2(5 + 8) = 2(5) + 2(8).

FIGURE 18

The **distributive property** says that multiplying a number a by a sum of numbers $b + c$ gives the same result as multiplying a by b and a by c and then adding the two products.

Distributive Property

$$a(b + c) = ab + ac \qquad \text{and} \qquad (b + c)a = ba + ca$$

As the arrows show, the a outside the parentheses is "distributed" over the b and c inside. The distributive property is also valid for multiplication over subtraction.

$$a(b - c) = ab - ac \qquad \text{and} \qquad (b - c)a = ba - ca$$

The distributive property can be extended to more than two numbers.

$$a(b + c + d) = ab + ac + ad$$

The distributive property can be used "in reverse." For example, we can write

$$ac + bc = (a + b)c.$$

EXAMPLE 9 Using the Distributive Property

Use the distributive property to rewrite each expression.

(a) $\quad 5(9 + 6)$

$$= 5 \cdot 9 + 5 \cdot 6 \qquad \text{Distributive property}$$
$$= 45 + 30 \qquad \text{Multiply.}$$
$$= 75 \qquad \text{Add.}$$

Multiply first.

(b) $4(x + 5 + y)$

$= 4x + 4 \cdot 5 + 4y$ Distributive property

$= 4x + 20 + 4y$ Multiply.

(c) $-2(x + 3)$

$= -2x + (-2)(3)$ Distributive property

$= -2x - 6$ Multiply.

(d) $3(k - 9)$

$= 3[k + (-9)]$ Definition of subtraction

$= 3k + 3(-9)$ Distributive property

$= 3k - 27$ Multiply.

(e) $8(3r + 11t + 5z)$

$= 8(3r) + 8(11t) + 8(5z)$ Distributive property

$= (8 \cdot 3)r + (8 \cdot 11)t + (8 \cdot 5)z$ Associative property

$= 24r + 88t + 40z$ Multiply.

(f) $6 \cdot 8 + 6 \cdot 2$

$= 6(8 + 2)$ Distributive property

$= 6(10)$ Add.

$= 60$ Multiply.

(g) $4x - 4m$

$= 4(x - m)$

(h) $6x - 12$

$= 6 \cdot x - 6 \cdot 2$

$= 6(x - 2)$

> ✔ **Now Try Exercises 57, 63, 65, and 69.**

▶ **CAUTION** In practice, we often omit the first step in Example 9(d), where we rewrote the subtraction as addition of the additive inverse, and instead write

$3(k - 9)$

$= 3k - 3(9)$ Be careful not to make a sign error.

$= 3k - 27.$

The symbol $-a$ may be interpreted as $-1 \cdot a$. Similarly, when a negative sign precedes an expression within parentheses, it may also be interpreted as the factor -1. Thus, we can use the distributive property to remove (or clear) parentheses from expressions such as $-(2y + 3)$.

$-(2y + 3)$

$= -1 \cdot (2y + 3)$ $-a = -1 \cdot a$

$= -1 \cdot 2y + (-1) \cdot 3$ Distributive property

$= -2y - 3$ Multiply.

EXAMPLE 10 Using the Distributive Property to Remove Parentheses

Write each expression without parentheses.

(a) $-(7r - 8)$

$$= -1(7r - 8) \qquad -a = -1 \cdot a$$
$$= -1(7r) + (-1)(-8) \qquad \text{Distributive property}$$
$$= -7r + 8 \qquad \text{Multiply.}$$

(b) $-(-9w + 2)$

$$= -1(-9w + 2)$$
$$= -1(-9w) - 1(2)$$
$$= 9w - 2$$

(c) $-(-x - 3y + 6z)$

$$= -1(-1x - 3y + 6z)$$
$$= -1(-1x) - 1(-3y) - 1(6z)$$
$$= x + 3y - 6z$$

✔ **Now Try Exercises 75 and 79.**

▶ **NOTE** In expressions like those in Example 10, we can also interpret the negative sign in front of the parentheses to mean the *opposite* of each of the terms within the parentheses.

Here is a summary of the properties of real numbers discussed in this section.

Properties of Addition and Multiplication

For any real numbers a, b, and c, the following properties hold.

Commutative Properties $a + b = b + a \qquad ab = ba$

Associative Properties $(a + b) + c = a + (b + c)$
$$(ab)c = a(bc)$$

Identity Properties   There is a real number 0 such that
$$a + 0 = a \qquad \text{and} \qquad 0 + a = a.$$
There is a real number 1 such that
$$a \cdot 1 = a \qquad \text{and} \qquad 1 \cdot a = a.$$

Inverse Properties   For each real number a, there is a single real number $-a$ such that
$$a + (-a) = 0 \qquad \text{and} \qquad (-a) + a = 0.$$
For each nonzero real number a, there is a single real number $\frac{1}{a}$ such that
$$a \cdot \frac{1}{a} = 1 \qquad \text{and} \qquad \frac{1}{a} \cdot a = 1.$$

Distributive Properties $a(b + c) = ab + ac \qquad (b + c)a = ba + ca$

1.7 EXERCISES

Use the commutative or the associative property to complete each statement. State which property is used. See Examples 1 and 2.

1. $-12 + 6 = 6 +$ _____

2. $8 + (-4) = -4 +$ _____

3. $-6 \cdot 3 =$ _____ $\cdot (-6)$

4. $-12 \cdot 6 = 6 \cdot$ _____

5. $(4 + 7) + 8 = 4 + ($ _____ $+ 8)$

6. $(-2 + 3) + 6 = -2 + ($ _____ $+ 6)$

7. $8 \cdot (3 \cdot 6) = ($ _____ $\cdot 3) \cdot 6$

8. $6 \cdot (4 \cdot 2) = (6 \cdot$ _____ $) \cdot 2$

9. *Concept Check* Match each item in Column I with the correct choice(s) from Column II. Choices may be used once, more than once, or not at all.

I

(a) Identity element for addition

(b) Identity element for multiplication

(c) Additive inverse of a

(d) Multiplicative inverse, or reciprocal, of the nonzero number a

(e) The number that is its own additive inverse

(f) The two numbers that are their own multiplicative inverses

(g) The only number that has no multiplicative inverse

(h) An example of the associative property

(i) An example of the commutative property

(j) An example of the distributive property

II

A. $(5 \cdot 4) \cdot 3 = 5 \cdot (4 \cdot 3)$

B. 0

C. $-a$

D. -1

E. $5 \cdot 4 \cdot 3 = 60$

F. 1

G. $(5 \cdot 4) \cdot 3 = 3 \cdot (5 \cdot 4)$

H. $5(4 + 3) = 5 \cdot 4 + 5 \cdot 3$

I. $\dfrac{1}{a}$

10. Explain the difference between the commutative and associative properties.

Decide whether each statement is an example of the commutative, associative, identity, inverse, *or* distributive property. *See Examples 1, 2, 3, 5, 6, 7, and 9.*

11. $7 + 18 = 18 + 7$

12. $13 + 12 = 12 + 13$

13. $5 \cdot (13 \cdot 7) = (5 \cdot 13) \cdot 7$

14. $-4 \cdot (2 \cdot 6) = (-4 \cdot 2) \cdot 6$

15. $-6 + (12 + 7) = (-6 + 12) + 7$

16. $(-8 + 13) + 2 = -8 + (13 + 2)$

17. $-6 + 6 = 0$

18. $12 + (-12) = 0$

19. $\dfrac{2}{3}\left(\dfrac{3}{2}\right) = 1$

20. $\dfrac{5}{8}\left(\dfrac{8}{5}\right) = 1$

21. $2.34 + 0 = 2.34$

22. $-8.456 + 0 = -8.456$

23. $(4 + 17) + 3 = 3 + (4 + 17)$

24. $(-8 + 4) + 12 = 12 + (-8 + 4)$

25. $6(x + y) = 6x + 6y$

26. $14(t + s) = 14t + 14s$

27. $-\dfrac{5}{9} = -\dfrac{5}{9} \cdot \dfrac{3}{3} = -\dfrac{15}{27}$

28. $\dfrac{13}{12} = \dfrac{13}{12} \cdot \dfrac{7}{7} = \dfrac{91}{84}$

29. $5(2x) + 5(3y) = 5(2x + 3y)$

30. $3(5t) - 3(7r) = 3(5t - 7r)$

31. Write a paragraph explaining in your own words the identity and inverse properties of addition and multiplication.

32. Write a paragraph explaining in your own words the distributive property of multiplication with respect to addition. Give examples.

33. *Concept Check* The following conversation actually took place between one of the authors of this book and his son, Jack, when Jack was 4 years old:

DADDY: "Jack, what is $3 + 0$?"
JACK: "3."
DADDY: "Jack, what is $4 + 0$?"
JACK: "4. And Daddy, *string* plus zero equals *string*!"

What property of addition did Jack recognize? (Jack is now 21 and a mathematics major in college.)

34. *Concept Check* The distributive property holds for multiplication with respect to addition. Is there a distributive property for addition with respect to multiplication? If not, give an example to show why.

Find each sum. See Example 4.

35. $97 + 13 + 3 + 37$ **36.** $49 + 199 + 1 + 1$

37. $1999 + 2 + 1 + 8$ **38.** $2998 + 3 + 2 + 17$

39. $159 + 12 + 141 + 88$ **40.** $106 + 8 + (-6) + (-8)$

41. $843 + 627 + (-43) + (-27)$ **42.** $1846 + 1293 + (-46) + (-93)$

Simplify each expression. See Examples 7 and 8.

43. $6t + 8 - 6t + 3$ **44.** $9r + 12 - 9r + 1$ **45.** $\dfrac{2}{3}x - 11 + 11 - \dfrac{2}{3}x$

46. $\dfrac{1}{5}y + 4 - 4 - \dfrac{1}{5}y$ **47.** $\left(\dfrac{9}{7}\right)(-0.38)\left(\dfrac{7}{9}\right)$ **48.** $\left(\dfrac{4}{5}\right)(-0.73)\left(\dfrac{5}{4}\right)$

49. $t + (-t) + \dfrac{1}{2}(2)$ **50.** $w + (-w) + \dfrac{1}{4}(4)$

51. *Concept Check* Evaluate $25 - (6 - 2)$ and evaluate $(25 - 6) - 2$. Do you think subtraction is associative?

52. *Concept Check* Evaluate $180 \div (15 \div 3)$ and evaluate $(180 \div 15) \div 3$. Do you think division is associative?

53. *Concept Check* Suppose that a student simplifies the expression $-3(4 - 6)$ as follows:

$$-3(4 - 6)$$
$$= -3(4) - 3(6)$$
$$= -12 - 18$$
$$= -30.$$

The student made a very common error. *WHAT WENT WRONG?* Work the problem correctly.

54. Explain how the procedure of changing $\frac{3}{4}$ to $\frac{9}{12}$ requires the use of the multiplicative identity element, 1.

Rewrite each expression. Simplify if possible. See Example 9.

55. $5(9 + 8)$ **56.** $6(11 + 8)$ **57.** $4(t + 3)$

58. $5(w + 4)$ **59.** $-8(r + 3)$ **60.** $-11(x + 4)$

61. $-5(y - 4)$ **62.** $-9(g - 4)$ **63.** $-\dfrac{4}{3}(12y + 15z)$

64. $-\dfrac{2}{5}(10b + 20a)$ **65.** $8z + 8w$ **66.** $4s + 4r$

67. $7(2v) + 7(5r)$ **68.** $13(5w) + 13(4p)$ **69.** $8(3r + 4s - 5y)$

70. $2(5u - 3v + 7w)$ **71.** $-3(8x + 3y + 4z)$ **72.** $-5(2x - 5y + 6z)$

73. $5x + 15$ **74.** $9p + 18$

Write each expression without parentheses. See Example 10.

75. $-(4t + 3m)$ **76.** $-(9x + 12y)$ **77.** $-(-5c - 4d)$

78. $-(-13x - 15y)$ **79.** $-(-q + 5r - 8s)$ **80.** $-(-z + 5w - 9y)$

81. *Concept Check*   The operations of "getting out of bed" and "taking a shower" are not commutative. Give an example of another pair of everyday operations that are not commutative.

82. *Concept Check*   The phrase "dog biting man" has two different meanings, depending on how the words are associated:

$$(\text{dog biting}) \text{ man} \qquad \text{dog (biting man)}$$

Give another example of a three-word phrase that has different meanings depending on how the words are associated.

RELATING CONCEPTS   (EXERCISES 83–86)

FOR INDIVIDUAL OR GROUP WORK

In Section 1.6, we used a pattern to see that the product of two negative numbers is a positive number. In these exercises, we show another justification for determining the sign of the product of two negative numbers. **Work Exercises 83–86 in order.**

83. Evaluate the expression $-3[5 + (-5)]$ by using the rules for order of operations.

84. Write the expression in Exercise 83 after using the distributive property. Do not simplify the products.

85. The product $-3(5)$ should be part of the answer you wrote for Exercise 84. On the basis of the results in **Section 1.6,** what is this product?

86. In Exercise 83, you should have obtained 0 as an answer. Now consider the following, using the results of Exercises 83 and 85.

$$-3[5 + (-5)] = -3(5) + (-3)(-5)$$
$$0 = -15 + ?$$

The question mark represents the product $(-3)(-5)$. When added to -15, it must give a sum of 0. Therefore, how must we interpret $(-3)(-5)$?

Concept Check   *In Exercises 87 and 88, consider the given statement.*
(a) *Is this statement true?*
(b) *If it is not true, tell which property you think was erroneously applied.*

87. $-2(5 \cdot 7) = (-2 \cdot 5) \cdot (-2 \cdot 7)$ **88.** $4 + (3 \cdot 5) = (4 + 3) \cdot (4 + 5)$

1.8 Simplifying Expressions

OBJECTIVES

1 Simplify expressions.

2 Identify terms and numerical coefficients.

3 Identify like terms.

4 Combine like terms.

5 Simplify expressions from word phrases.

OBJECTIVE 1 Simplify expressions. We now simplify expressions, using the properties of addition and multiplication introduced in **Section 1.7**.

EXAMPLE 1 Simplifying Expressions

Simplify each expression.

(a) $4x + 8 + 9$ simplifies to $4x + 17$.

(b) $4(3m - 2n)$

$$= 4(3m) - 4(2n) \qquad \text{Distributive property}$$
$$= (4 \cdot 3)m - (4 \cdot 2)n \qquad \text{Associative property}$$
$$= 12m - 8n$$

(c) $\quad 6 + 3(4k + 5)$

Don't start by adding!

$$= 6 + 3(4k) + 3(5) \qquad \text{Distributive property}$$
$$= 6 + (3 \cdot 4)k + 3(5) \qquad \text{Associative property}$$
$$= 6 + 12k + 15 \qquad \text{Multiply.}$$
$$= 6 + 15 + 12k \qquad \text{Commutative property}$$
$$= 21 + 12k \qquad \text{Add.}$$

(d) $\quad 5 - (2y - 8)$

$$= 5 - 1(2y - 8) \qquad -a = -1 \cdot a$$
$$= 5 - 1(2y) - 1(-8) \qquad \text{Distributive property}$$

Be careful with signs.

$$= 5 - 2y + 8 \qquad \text{Multiply.}$$
$$= 5 + 8 - 2y \qquad \text{Commutative property}$$
$$= 13 - 2y \qquad \text{Add.}$$

✔ **Now Try Exercises 7 and 9.**

The steps using the commutative and associative properties will not be shown in the rest of the examples, but you should be aware that they are usually involved.

Term	Numerical Coefficient
$-7y$	-7
$34r^3$	34
$-26x^5yz^4$	-26
$-k = -1k$	-1
$r = 1r$	1
$\dfrac{3x}{8} = \dfrac{3}{8}x$	$\dfrac{3}{8}$
$\dfrac{x}{3} = \dfrac{1x}{3} = \dfrac{1}{3}x$	$\dfrac{1}{3}$

OBJECTIVE 2 Identify terms and numerical coefficients. A **term** is a number, a variable, or a product or quotient of numbers and variables raised to powers,* such as

$$9x, \quad 15y^2, \quad -3, \quad -8m^2n, \quad \frac{2}{p}, \quad \text{and} \quad k. \qquad \text{Terms}$$

In the term $9x$, the **numerical coefficient,** or simply **coefficient,** of the variable x is 9. In the term $-8m^2n$, the numerical coefficient of m^2n is -8. Additional examples are shown in the table in the margin.

✔ **Now Try Exercises 13 and 19.**

---

*Another name for certain terms, **monomial**, is introduced in **Chapter 5**.

▶ **CAUTION** It is important to be able to distinguish between *terms* and *factors*. For example, in the expression $8x^3 + 12x^2$, there are two *terms*, $8x^3$ and $12x^2$. Terms are separated by a $+$ or $-$ sign. On the other hand, in the one-term expression $(8x^3)(12x^2)$, $8x^3$ and $12x^2$ are *factors*. Factors are multiplied.

OBJECTIVE 3 Identify like terms. Terms with exactly the same variables that have the same exponents are **like terms.** For example, $9m$ and $4m$ have the same variable and are like terms. Also, $6x^3$ and $-5x^3$ are like terms. The terms $-4y^3$ and $4y^2$ have different exponents and are **unlike terms.** Here are some additional examples.

$5x$ and $-12x$ $3x^2y$ and $5x^2y$ Like terms
$4xy^2$ and $5xy$ $-7w^3z^3$ and $2xz^3$ Unlike terms

✔ **Now Try Exercises 23 and 29.**

OBJECTIVE 4 Combine like terms. Recall the distributive property:

$$x(y + z) = xy + xz.$$

As seen in **Section 1.7,** this statement can also be written "backward" as

$$xy + xz = x(y + z).$$

This form of the distributive property may be used to find the sum or difference of like terms. For example,

$$3x + 5x = (3 + 5)x = 8x.$$

Using the distributive property in this way is called **combining like terms.**

EXAMPLE 2 Combining Like Terms

Combine like terms in each expression.

(a) $-9m + 5m$
$= (-9 + 5)m$
$= -4m$

(b) $6r + 3r + 2r$
$= (6 + 3 + 2)r$
$= 11r$

(c) $4x + x$
$= 4x + 1x$ $x = 1x$
$= (4 + 1)x$
$= 5x$

(d) $16y^2 - 9y^2$
$= (16 - 9)y^2$
$= 7y^2$

(e) $32y + 10y^2$ cannot be combined because $32y$ and $10y^2$ are unlike terms. We cannot use the distributive property here to combine coefficients.

✔ **Now Try Exercises 37, 41, and 53.**

▶ **CAUTION** *Remember that only like terms may be combined.*

EXAMPLE 3 **Simplifying Expressions Involving Like Terms**

Simplify each expression.

(a) $14y + 2(6 + 3y)$

$$= 14y + 2(6) + 2(3y) \qquad \text{Distributive property}$$
$$= 14y + 12 + 6y \qquad \text{Multiply.}$$
$$= 20y + 12 \qquad \text{Combine like terms.}$$

(b) $\qquad 9k - 6 - 3(2 - 5k)$ *Be careful with signs.*

$$= 9k - 6 - 3(2) - 3(-5k) \qquad \text{Distributive property}$$
$$= 9k - 6 - 6 + 15k \qquad \text{Multiply.}$$
$$= 24k - 12 \qquad \text{Combine like terms.}$$

(c) $\qquad -(2 - r) + 10r$

$$= -1(2 - r) + 10r \qquad -a = -1 \cdot a$$
$$= -1(2) - 1(-r) + 10r \qquad \text{Distributive property}$$
Be careful with signs. $\quad = -2 + 1r + 10r \qquad \text{Multiply.}$
$$= -2 + 11r \qquad \text{Combine like terms.}$$

(d) $100[0.03(x + 4)]$

$$= [(100)(0.03)](x + 4) \qquad \text{Associative property}$$
$$= 3(x + 4) \qquad \text{Multiply.}$$
$$= 3x + 12 \qquad \text{Distributive property}$$

(e) $5(2a - 6) - 3(4a - 9)$

$$= 10a - 30 - 12a + 27 \qquad \text{Distributive property}$$
$$= -2a - 3 \qquad \text{Combine like terms.}$$

(f) $-\dfrac{2}{3}(x - 6) - \dfrac{1}{6}x$

$$= -\frac{2}{3}x - \frac{2}{3}(-6) - \frac{1}{6}x \qquad \text{Distributive property}$$

$$= -\frac{2}{3}x + 4 - \frac{1}{6}x \qquad \text{Multiply.}$$

$$= -\frac{4}{6}x + 4 - \frac{1}{6}x \qquad \text{Get a common denominator.}$$

$$= -\frac{5}{6}x + 4 \qquad \text{Combine like terms.}$$

✔ **Now Try Exercises 57, 59, 61, and 65.**

▶ **NOTE** Examples 2 and 3 suggest that like terms may be combined by adding or subtracting the coefficients of the terms and keeping the same variable factors.

OBJECTIVE 5 Simplify expressions from word phrases. Earlier we saw how to translate words, phrases, and statements into expressions and equations. Now we can simplify translated expressions by combining like terms.

EXAMPLE 4 Translating Words to a Mathematical Expression

Translate to a mathematical expression and simplify: The sum of 9, five times a number, four times the number, and six times the number.

The word "sum" indicates that the terms should be added. Use x to represent the number. Then the phrase translates as

$$9 + 5x + 4x + 6x, \quad \text{Write with symbols.}$$

which simplifies to

$$9 + 15x. \quad \text{Combine like terms.}$$

✔ **Now Try Exercise 77.**

▶ **CAUTION** In Example 4, we are dealing with an expression to be simplified, *not* an equation to be solved.

1.8 EXERCISES

● *Complete solution available on Video Lectures on CD/DVD*

Now Try Exercise

Concept Check *In Exercises 1–4, choose the letter of the correct response.*

1. Which is true for all real numbers x?

A. $6 + 2x = 8x$ **B.** $6 - 2x = 4x$

C. $6x - 2x = 4x$ **D.** $3 + 8(4x - 6) = 11(4x - 6)$

2. Which is an example of a pair of like terms?

A. $6t, 6w$ **B.** $-8x^2y, 9xy^2$ **C.** $5ry, 6yr$ **D.** $-5x^2, 2x^3$

3. Which is an example of a term with numerical coefficient 5?

A. $5x^3y^7$ **B.** x^5 **C.** $\dfrac{x}{5}$ **D.** 5^2xy^3

4. Which is a correct translation for "six times a number, subtracted from the product of eleven and the number" (if x represents the number)?

A. $6x - 11x$ **B.** $11x - 6x$ **C.** $(11 + x) - 6x$ **D.** $6x - (11 + x)$

Simplify each expression. See Example 1.

5. $4r + 19 - 8$ **6.** $7t + 18 - 4$ ● **7.** $5 + 2(x - 3y)$

8. $8 + 3(s - 6t)$ **9.** $-2 - (5 - 3p)$ **10.** $-10 - (7 - 14r)$

11. $6 + (4 - 3x) - 8$ **12.** $-12 + (7 - 8x) + 6$

In each term, give the numerical coefficient of the variable(s). See Objective 2.

● **13.** $-12k$ **14.** $-23y$ **15.** $5m^2$ **16.** $-3n^6$ **17.** xw

18. pq **19.** $-x$ **20.** $-t$ **21.** $\dfrac{x}{5}$ **22.** $\dfrac{2x}{3}$

Identify each group of terms as like *or* unlike. *See Objective 3.*

23. $8r, -13r$ **24.** $-7a, 12a$ **25.** $5z^4, 9z^3$ **26.** $8x^5, -10x^3$

27. $4, 9, -24$ **28.** $7, 17, -83$ **29.** x, y **30.** t, s

31. *Concept Check* A student simplified the expression $7x - 2(3 - 2x)$ as shown.

$$7x - 2(3 - 2x)$$
$$= 7x - 2(3) - 2(2x)$$
$$= 7x - 6 - 4x$$
$$= 3x - 6$$

WHAT WENT WRONG? Find the correct simplified answer.

32. There is an old saying, "You can't add apples and oranges." Explain how this saying can be applied to the goal of Objective 4 in this section.

Simplify each expression. See Examples 1–3.

33. $9y + 8y$ **34.** $15m + 12m$

35. $-4a - 2a$ **36.** $-3z - 9z$

37. $12b + b$ **38.** $30x + x$

39. $2k + 9 + 5k + 6$ **40.** $2 + 17z + 1 + 2z$

41. $-5y + 3 - 1 + 5 + y - 7$ **42.** $2k - 7 - 5k + 7k - 3 - k$

43. $-2x + 3 + 4x - 17 + 20$ **44.** $r - 6 - 12r - 4 + 6r$

45. $16 - 5m - 4m - 2 + 2m$ **46.** $6 - 3z - 2z - 5 + z - 3z$

47. $-10 + x + 4x - 7 - 4x$ **48.** $-p + 10p - 3p - 4 - 5p$

49. $1 + 7x + 11x - 1 + 5x$ **50.** $-r + 2 - 5r + 3 + 4r$

51. $-\dfrac{4}{3} + 2t + \dfrac{1}{3}t - 8 - \dfrac{8}{3}t$ **52.** $-\dfrac{5}{6} + 8x + \dfrac{1}{6}x - 7 - \dfrac{7}{6}$

53. $6y^2 + 11y^2 - 8y^2$ **54.** $-9m^3 + 3m^3 - 7m^3$

55. $2p^2 + 3p^2 - 8p^3 - 6p^3$ **56.** $5y^3 + 6y^3 - 3y^2 - 4y^2$

57. $2(4x + 6) + 3$ **58.** $4(6y - 9) + 7$

59. $100[0.05(x + 3)]$ **60.** $100[0.06(x + 5)]$

61. $-4(y - 7) - 6$ **62.** $-5(t - 13) - 4$

63. $-\dfrac{4}{3}(y - 12) - \dfrac{1}{6}y$ **64.** $-\dfrac{7}{5}(t - 15) - \dfrac{1}{2}t$

65. $-5(5y - 9) + 3(3y + 6)$ **66.** $-3(2t + 4) + 8(2t - 4)$

67. $-3(2r - 3) + 2(5r + 3)$ **68.** $-4(5y - 7) + 3(2y - 5)$

69. $8(2k - 1) - (4k - 3)$ **70.** $6(3p - 2) - (5p + 1)$

71. $-2(-3k + 2) - (5k - 6) - 3k - 5$ **72.** $-2(3r - 4) - (6 - r) + 2r - 5$

73. $-4(-3k + 3) - (6k - 4) - 2k + 1$ **74.** $-5(8j + 2) - (5j - 3) - 3j + 17$

75. $-7.5(2y + 4) - 2.9(3y - 6)$ **76.** $8.4(6t - 6) + 2.4(9 - 3t)$

Translate each phrase into a mathematical expression. Use x as the variable. Combine like terms when possible. See Example 4.

77. Five times a number, added to the sum of the number and three

78. Six times a number, added to the sum of the number and six

79. A number multiplied by -7, subtracted from the sum of 13 and six times the number

80. A number multiplied by 5, subtracted from the sum of 14 and eight times the number

81. Six times a number added to -4, subtracted from twice the sum of three times the number and 4 (*Hint: Twice* means two times.)

82. Nine times a number added to 6, subtracted from triple the sum of 12 and 8 times the number (*Hint: Triple* means three times.)

83. Write the expression $9x - (x + 2)$ in words, as in Exercises 77–82.

84. Write the expression $2(3x + 5) - 2(x + 4)$ in words, as in Exercises 77–82.

RELATING CONCEPTS (EXERCISES 85–88)

FOR INDIVIDUAL OR GROUP WORK

A manufacturer has fixed costs of $1000 *to produce widgets. Each widget costs* $5 *to make. The fixed cost to produce gadgets is* $750, *and each gadget costs* $3 *to make.* **Work Exercises 85–88 in order.**

85. Write an expression for the cost to make x widgets. (*Hint:* The cost will be the sum of the fixed cost and the cost per item times the number of items.)

86. Write an expression for the cost to make y gadgets.

87. Use your answers from Exercises 85 and 86 to write an expression for the total cost to make x widgets and y gadgets.

88. Simplify the expression you wrote in Exercise 87.

Chapter **1** Group Activity

COMPARING FLOOR PLANS OF HOUSES

Objective Use arithmetic skills to make comparisons.

This activity explores perimeters and areas of different-shaped homes. Floor plans for three different homes are as follows:

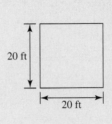
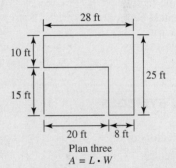

Plan one
$A = \pi r^2$
$C = 2\pi r$

Plan two
$A = s^2$

Plan three
$A = L \cdot W$

A. As a group, look at the dimensions of the given floor plans. Considering only the dimensions, which plan do you think has the greatest area?

B. Now have each student in your group pick one floor plan. For each plan, find the following. (Round all answers to the nearest whole number. In Plan one, let $\pi = 3.14$.)

   1. The area of the plan

   2. The perimeter or circumference (the distance around the outside) of the plan

C. Share your findings with the group and answer the following questions.

   1. What did you determine about the areas of the three floor plans?

   2. Which plan has the smallest perimeter? Which has the largest perimeter?

   3. Why do you think that houses with round floor plans might be more energy efficient?

   4. What advantages do you think houses with square or rectangular floor plans have? Why do you think floor plans with these shapes are most common in homes today?

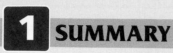

Chapter **1** SUMMARY　*View the Interactive Summary on the Pass the Test CD.*

KEY TERMS

1.1 natural numbers
whole numbers
fractions
numerator
denominator
proper fraction
improper fraction
mixed number
factor
product
factored
prime number
composite number
prime factors
greatest common
　factor
lowest terms

basic principle of
　fractions
reciprocal
quotient
sum
least common
　denominator (LCD)
difference
1.2 exponent (power)
base
exponential expression
grouping symbols
order of operations
inequality
1.3 variable
algebraic expression
equation

solution
set
element
1.4 number line
integers
signed numbers
rational numbers
set-builder notation
graph
coordinate
irrational numbers
real numbers
additive inverse
　(opposite)
absolute value
1.6 multiplicative inverse
　(reciprocal)

1.7 commutative property
associative property
identity property
identity element for
　addition (additive
　identity)
identity element for
　multiplication
　(multiplicative
　identity)
inverse property
distributive property
1.8 term
numerical coefficient
like terms
unlike terms
combining like terms

NEW SYMBOLS

a^n　n factors of a

[ ]　brackets

$=$　is equal to

$\neq$　is not equal to

$<$　is less than

$>$　is greater than

$\leq$　is less than or equal
　to

$\geq$　is greater than or
　equal to

{ }　set braces

$\{x \mid x$ **has a certain
　property**$\}$
　set-builder notation

$-x$　the additive inverse, or
　opposite, of x

$|x|$　absolute value of x

$\dfrac{1}{x}$　the multiplicative
　inverse, or reciprocal,
　of the nonzero
　number x

$a(b), (a)b, (a)(b), a \cdot b,$
　or ab　a times b

$a \div b, \dfrac{a}{b},$ **or** a/b
　a divided by b

TEST YOUR WORD POWER

See how well you have learned the vocabulary in this chapter. Answers, with examples, follow the Quick Review.

1. A **factor** is
　A. the answer in an addition problem
　B. the answer in a multiplication problem
　C. one of two or more numbers that are added to get another number
　D. one of two or more numbers that are multiplied to get another number.

2. A number is **prime** if
　A. it cannot be factored

B. it has just one factor
C. it has only itself and 1 as factors
D. it has at least two different factors.

3. An **exponent** is
　A. a symbol that tells how many numbers are being multiplied
　B. a number raised to a power
　C. a number that tells how many times a factor is repeated

D. one of two or more numbers that are multiplied.

4. A **variable** is
　A. a symbol used to represent an unknown number
　B. a value that makes an equation true
　C. a solution of an equation
　D. the answer in a division problem.

(continued)

5. An **integer** is
  A. a positive or negative number
  B. a natural number, its opposite, or zero
  C. any number that can be graphed on a number line
  D. the quotient of two numbers.

6. A **coordinate** is
  A. the number that corresponds to a point on a number line
  B. the graph of a number
  C. any point on a number line

D. the distance from 0 on a number line.

7. The **absolute value** of a number is
  A. the graph of the number
  B. the reciprocal of the number
  C. the opposite of the number
  D. the distance between 0 and the number on a number line.

8. A **term** is
  A. a numerical factor
  B. a number, a variable, or a product or quotient of numbers and variables raised to powers

C. one of several variables with the same exponents
  D. a sum of numbers and variables raised to powers.

9. A **numerical coefficient** is
  A. the numerical factor of the variable(s) in a term
  B. the number of terms in an expression
  C. a variable raised to a power
  D. the variable factor in a term.

QUICK REVIEW

Concepts	Examples

1.1 FRACTIONS

Operations with Fractions

Addition/Subtraction:
1. To add/subtract fractions with the same denominator, add/subtract the numerators and keep the same denominator.
2. To add/subtract fractions with different denominators, find the LCD and write each fraction with this LCD. Then follow the procedure above.

Multiplication: Multiply numerators and multiply denominators.

Division: Multiply the first fraction by the reciprocal of the second fraction.

Perform each operation.

$$\frac{2}{5} + \frac{7}{5} = \frac{2+7}{5} = \frac{9}{5}, \text{ or } 1\frac{4}{5}$$

$$\frac{2}{3} - \frac{1}{2} = \frac{4}{6} - \frac{3}{6} \qquad \text{6 is the LCD.}$$
$$= \frac{4-3}{6} = \frac{1}{6}$$

$$\frac{4}{3} \cdot \frac{5}{6} = \frac{20}{18} = \frac{10}{9}, \text{ or } 1\frac{1}{9}$$

$$\frac{6}{5} \div \frac{1}{4} = \frac{6}{5} \cdot \frac{4}{1} = \frac{24}{5}, \text{ or } 4\frac{4}{5}$$

1.2 EXPONENTS, ORDER OF OPERATIONS, AND INEQUALITY

Order of Operations
Simplify within any parentheses or brackets and above and below fraction bars first, in the following order.

Step 1 Apply all exponents.

Step 2 Do any multiplications or divisions from left to right.

Step 3 Do any additions or subtractions from left to right.

If no grouping symbols are present, start with Step 1.

Simplify $36 - 4(2^2 + 3)$. **Start inside the parentheses.**

$$36 - 4(2^2 + 3)$$
$$= 36 - 4(4 + 3)$$
$$= 36 - 4(7)$$
$$= 36 - 28$$
$$= 8$$

(continued)

Concepts	Examples

1.3 VARIABLES, EXPRESSIONS, AND EQUATIONS

Evaluate an expression with a variable by substituting a given number for the variable.

Evaluate $2x + y^2$ if $x = 3$ and $y = -4$.

$$2x + y^2$$
$$= 2(3) + (-4)^2$$
$$= 6 + 16 \quad \text{Use parentheses to avoid errors.}$$
$$= 22$$

Values of a variable that make an equation true are solutions of the equation.

Is 2 a solution of $5x + 3 = 18$?

$$5(2) + 3 = 18 \quad ? \quad \text{Let } x = 2.$$
$$13 = 18 \qquad \text{False}$$

2 is not a solution.

1.4 REAL NUMBERS AND THE NUMBER LINE

Ordering Real Numbers
a is less than b if a is to the left of b on the number line.

Graph -2, 0, and 3.

$$-2 < 3 \qquad 3 > 0 \qquad 0 < 3$$

The additive inverse of x is $-x$.

$$-(5) = -5 \qquad -(-7) = 7 \qquad -0 = 0$$

The absolute value of x, $|x|$, is the distance between x and 0 on the number line.

$$|13| = 13 \qquad |0| = 0 \qquad |-5| = 5$$

1.5 ADDING AND SUBTRACTING REAL NUMBERS

Adding Signed Numbers
To add two numbers with the *same* sign, add their absolute values. The sum has that same sign.

Add.
$$9 + 4 = 13$$
$$-8 + (-5) = -13$$

To add two numbers with *different* signs, subtract their absolute values. The sum has the sign of the number with larger absolute value.

$$7 + (-12) = -5$$
$$-5 + 13 = 8$$

Definition of Subtraction
$$x - y = x + (-y)$$

Subtract.
$$-3 - 4 = -3 + (-4) = -7$$
$$-2 - (-6) = -2 + 6 = 4$$
$$13 - (-8) = 13 + 8 = 21$$

1.6 MULTIPLYING AND DIVIDING REAL NUMBERS

Multiplying and Dividing Signed Numbers
The product (or quotient) of two numbers having the *same sign* is *positive;* the product (or quotient) of two numbers having *different signs* is *negative.*

Multiply or divide.

$$6 \cdot 5 = 30 \qquad -7(-8) = 56 \qquad \frac{20}{4} = 5$$

$$\frac{-24}{-6} = 4 \qquad -6(5) = -30 \qquad 6(-5) = -30$$

$$\frac{-18}{9} = -2 \qquad \frac{49}{-7} = -7$$

(continued)

Concepts	Examples

Definition of Division

$$\frac{x}{y} = x \cdot \frac{1}{y}, \quad y \neq 0$$

$$\frac{10}{2} = 10 \cdot \frac{1}{2} = 5$$

Division by 0 is undefined. 0 divided by a nonzero number equals 0.

$\frac{5}{0}$ is undefined. $\frac{0}{5} = 0$

1.7   PROPERTIES OF REAL NUMBERS

Commutative Properties

$$a + b = b + a$$
$$ab = ba$$

$$7 + (-1) = -1 + 7$$
$$5(-3) = (-3)5$$

Associative Properties

$$(a + b) + c = a + (b + c)$$
$$(ab)c = a(bc)$$

$$(3 + 4) + 8 = 3 + (4 + 8)$$
$$[-2(6)]4 = -2[(6)4]$$

Identity Properties

$$a + 0 = a \qquad 0 + a = a$$
$$a \cdot 1 = a \qquad 1 \cdot a = a$$

$$-7 + 0 = -7 \qquad 0 + (-7) = -7$$
$$9 \cdot 1 = 9 \qquad 1 \cdot 9 = 9$$

Inverse Properties

$$a + (-a) = 0 \qquad -a + a = 0$$
$$a \cdot \frac{1}{a} = 1 \qquad \frac{1}{a} \cdot a = 1 \quad (a \neq 0)$$

$$7 + (-7) = 0 \qquad -7 + 7 = 0$$
$$-2\left(-\frac{1}{2}\right) = 1 \qquad -\frac{1}{2}(-2) = 1$$

Distributive Properties

$$a(b + c) = ab + ac$$
$$(b + c)a = ba + ca$$
$$a(b - c) = ab - ac$$

$$5(4 + 2) = 5(4) + 5(2)$$
$$(4 + 2)5 = 4(5) + 2(5)$$
$$9(5 - 4) = 9(5) - 9(4)$$

1.8   SIMPLIFYING EXPRESSIONS

Only like terms may be combined. We use the distributive property to combine like terms.

$$-3y^2 + 6y^2 + 14y^2$$
$$= (-3 + 6 + 14)y^2$$
$$= 17y^2$$

$$4(3 + 2x) - 6(5 - x)$$
$$= 4(3) + 4(2x) - 6(5) - 6(-x)$$
$$= 12 + 8x - 30 + 6x$$
$$= 14x - 18$$

Answers to Test Your Word Power

1. D; *Example:* Since $2 \times 5 = 10$, the numbers 2 and 5 are factors of 10; other factors of 10 are $-10, -5, -2, -1, 1,$ and 10.

2. C; *Examples:* 2, 3, 11, 41, 53

3. C; *Example:* In 2^3, the number 3 is the exponent (or power), so 2 is a factor three times; $2^3 = 2 \cdot 2 \cdot 2 = 8$.

4. A; *Examples: a, b, c*

5. B; *Examples:* $-9, 0, 6$

6. A; *Example:* The point graphed 3 units to the right of 0 on a number line has coordinate 3.

7. D; *Examples:* $|2| = 2$ and $|-2| = 2$

8. B; *Examples:* $6, \frac{x}{2}, -4ab^2$

9. A; *Examples:* The term 3 has numerical coefficient 3, $8z$ has numerical coefficient 8, and $-10x^4y$ has numerical coefficient -10.

Chapter **1** REVIEW EXERCISES

[1.1]* *Perform each indicated operation.*

1. $\dfrac{8}{5} \div \dfrac{32}{15}$

2. $\dfrac{3}{8} + 3\dfrac{1}{2} - \dfrac{3}{16}$

3. The circle graph illustrates how 400 people responded to a survey that asked "Do you think that the CEOs of corporations guilty of accounting fraud should go to jail?" What fractional part of the group did not have an opinion?

4. Based on the graph in Exercise 3, how many people responded "yes"?

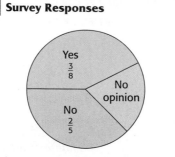

Survey Responses

[1.2] *Find the value of each exponential expression.*

5. 5^4

6. $\left(\dfrac{3}{5}\right)^3$

7. $(0.02)^2$

8. $(0.1)^3$

Find the value of each expression.

9. $8 \cdot 5 - 13$

10. $7[3 + 6(3^2)]$

11. $\dfrac{9(4^2 - 3)}{4 \cdot 5 - 17}$

12. $\dfrac{6(5 - 4) + 2(4 - 2)}{3^2 - (4 + 3)}$

Tell whether each statement is true *or* false.

13. $12 \cdot 3 - 6 \cdot 6 \le 0$

14. $3[5(2) - 3] > 20$

15. $9 \le 4^2 - 8$

Write each word statement in symbols.

16. Thirteen is less than seventeen.

17. Five plus two is not equal to ten.

18. The bar graph on the next page shows the number of people naturalized in the United States (that is, made citizens of the U.S.) for the years 1995 to 2004.

  (a) In which years were *fewer than* 600 thousand people naturalized?
  (b) In which years were *at least* 700 thousand people naturalized?
  (c) How many people were naturalized in the five years having the largest numbers of naturalizations?

---

*For help with the Review Exercises in this text, refer to the appropriate section given in brackets.

People Naturalized

Source: U.S. Department of Homeland Security.

[1.3] *Evaluate each expression if x = 6 and y = 3.*

19. $2x + 6y$ **20.** $4(3x - y)$ **21.** $\dfrac{x}{3} + 4y$ **22.** $\dfrac{x^2 + 3}{3y - x}$

Write each word phrase as an algebraic expression, using x as the variable.

23. Six added to a number

24. A number subtracted from eight

25. Nine subtracted from six times a number

26. Three-fifths of a number added to 12

Decide whether the given number is a solution of the given equation.

27. $5x + 3(x + 2) = 22;\quad 2$

28. $\dfrac{t + 5}{3t} = 1;\quad 6$

Write each word statement as an equation. Use x as the variable. Then find the solution from the set $\{0, 2, 4, 6, 8, 10\}$.

29. Six less than twice a number is 10.

30. The product of a number and 4 is 8.

[1.4] *Graph each group of numbers on a number line.*

31. $-4, -\dfrac{1}{2}, 0, 2.5, 5$

32. $-2, |-3|, -3, |-1|$

Classify each number, using the sets natural numbers, whole numbers, integers, rational numbers, irrational numbers, *and* real numbers.

33. $\dfrac{4}{3}$

34. $\sqrt{6}$

Select the lesser number in each pair.

35. $-10, 5$ **36.** $-8, -9$ **37.** $-\dfrac{2}{3}, -\dfrac{3}{4}$ **38.** $0, -|23|$

Decide whether each statement is true *or* false.

39. $12 > -13$ **40.** $0 > -5$ **41.** $-9 < -7$ **42.** $-13 \geq -13$

*For each number, **(a)** find the opposite of the number and **(b)** find the absolute value of the number.*

43. -9 **44.** 0 **45.** 6 **46.** $-\dfrac{5}{7}$

Simplify.

47. $|-12|$ **48.** $-|3|$ **49.** $-|-19|$ **50.** $-|9-2|$

[1.5] *Perform each indicated operation.*

51. $-10+4$ **52.** $14+(-18)$ **53.** $-8+(-9)$

54. $\dfrac{4}{9}+\left(-\dfrac{5}{4}\right)$ **55.** $-13.5+(-8.3)$ **56.** $(-10+7)+(-11)$

57. $[-6+(-8)+8]+[9+(-13)]$ **58.** $(-4+7)+(-11+3)+(-15+1)$

59. $-7-4$ **60.** $-12-(-11)$

61. $5-(-2)$ **62.** $-\dfrac{3}{7}-\dfrac{4}{5}$

63. $2.56-(-7.75)$ **64.** $(-10-4)-(-2)$

65. $(-3+4)-(-1)$ **66.** $-(-5+6)-2$

Write a numerical expression for each phrase and simplify the expression.

67. 19 added to the sum of -31 and 12   **68.** 13 more than the sum of -4 and -8

69. The difference between -4 and -6 **70.** Five less than the sum of 4 and -8

Find the solution of each equation from the set $\{-3, -2, -1, 0, 1, 2, 3\}$ by guessing or by trial and error.

71. $x+(-2)=-4$ **72.** $12+x=11$

Solve each problem.

73. Like many people, Otis Taylor neglects to keep up his checkbook balance. When he finally balanced his account, he found that the balance was $-\$23.75$, so he deposited $\$50.00$. What is his new balance?

74. The low temperature in Yellowknife, in the Canadian Northwest Territories, one January day was $-26°$F. It rose $16°$ that day. What was the high temperature?

75. Mike O'Hanian owed a friend $\$28$. He repaid $\$13$, but then borrowed another $\$14$. What positive or negative amount represents his present financial status?

76. If the temperature drops $7°$ below its previous level of $-3°$, what is the new temperature?

77. Peyton Manning of the Indianapolis Colts passed for a gain of 8 yd, was sacked for a loss of 12 yd, and then threw a 42 yd touchdown pass. What positive or negative number represents the total net yardage for the plays?

78. On Monday, February 13, 2006, the Dow Jones Industrial Average closed at 10,892.32, down 26.73 from the previous Friday. What was the closing value the previous Friday? (*Source: Times Picayune.*)

[1.6]   *Perform each indicated operation.*

79. $(-12)(-3)$ **80.** $15(-7)$ **81.** $-\dfrac{4}{3}\left(-\dfrac{3}{8}\right)$ **82.** $(-4.8)(-2.1)$

83. $5(8 - 12)$ **84.** $(5 - 7)(8 - 3)$ **85.** $2(-6) - (-4)(-3)$

86. $3(-10) - 5$ **87.** $\dfrac{-36}{-9}$ **88.** $\dfrac{220}{-11}$

89. $-\dfrac{1}{2} \div \dfrac{2}{3}$ **90.** $-33.9 \div (-3)$ **91.** $\dfrac{-5(3) - 1}{8 - 4(-2)}$

92. $\dfrac{5(-2) - 3(4)}{-2[3 - (-2)] - 1}$ **93.** $\dfrac{10^2 - 5^2}{8^2 + 3^2 - (-2)}$ **94.** $\dfrac{(0.6)^2 + (0.8)^2}{(-1.2)^2 - (-0.56)}$

Evaluate each expression if $x = -5$, $y = 4$, and $z = -3$.

95. $6x - 4z$ **96.** $5x + y - z$ **97.** $5x^2$ **98.** $z^2(3x - 8y)$

Write a numerical expression for each phrase and simplify the expression.

99. Nine less than the product of -4 and 5   **100.** Five-sixths of the sum of 12 and -6

101. The quotient of 12 and the sum of 8 and -4 **102.** The product of -20 and 12, divided by the difference between 15 and -15

Write each sentence in symbols, using x as the variable, and find the solution by guessing or by trial and error. All solutions come from the list of integers between -12 and 12.

103. 8 times a number is -24.   **104.** The quotient of a number and 3 is -2.

Find the average of each group of numbers.

105. 26, 38, 40, 20, 4, 14, 96, 18   **106.** $-12, 28, -36, 0, 12, -10$

[1.7]   *Decide whether each statement is an example of the commutative, associative, identity, inverse, or distributive property.*

107. $6 + 0 = 6$ **108.** $5 \cdot 1 = 5$

109. $-\dfrac{2}{3}\left(-\dfrac{3}{2}\right) = 1$ **110.** $17 + (-17) = 0$

111. $5 + (-9 + 2) = [5 + (-9)] + 2$ **112.** $w(xy) = (wx)y$

113. $3x + 3y = 3(x + y)$ **114.** $(1 + 2) + 3 = 3 + (1 + 2)$

Use the distributive property to rewrite each expression. Simplify if possible.

115. $7y + 14$ **116.** $-12(4 - t)$ **117.** $3(2s) + 3(5y)$ **118.** $-(-4r + 5s)$

119. Evaluate $25 - (5 - 2)$ and $(25 - 5) - 2$. Use this example to explain why subtraction is not associative.

120. Evaluate $180 \div (15 \div 5)$ and $(180 \div 15) \div 5$. Use this example to explain why division is not associative.

[1.8] *Combine like terms whenever possible.*

121. $2m + 9m$

122. $15p^2 - 7p^2 + 8p^2$

123. $5p^2 - 4p + 6p + 11p^2$

124. $-2(3k - 5) + 2(k + 1)$

125. $7(2m + 3) - 2(8m - 4)$

126. $-(2k + 8) - (3k - 7)$

Translate each phrase into a mathematical expression. Use x to represent the number, and combine like terms when possible.

127. Seven times a number, subtracted from the product of -2 and three times the number

128. A number multiplied by 8, added to the sum of 5 and four times the number

MIXED REVIEW EXERCISES*

Perform each indicated operation.

129. $[(-2) + 7 - (-5)] + [-4 - (-10)]$

130. $\left(-\dfrac{5}{6}\right)^2$

131. $\dfrac{6(-4) + 2(-12)}{5(-3) + (-3)}$

132. $\dfrac{3}{8} - \dfrac{5}{12}$

133. $\dfrac{8^2 + 6^2}{7^2 + 1^2}$

134. $-16(-3.5) - 7.2(-3)$

135. $2\dfrac{5}{6} - 4\dfrac{1}{3}$

136. $-8 + [(-4 + 17) - (-3 - 3)]$

137. $-\dfrac{12}{5} \div \dfrac{9}{7}$

138. $(-8 - 3) - 5(2 - 9)$

139. $5x^2 - 12y^2 + 3x^2 - 9y^2$

140. $-4(2t + 1) - 8(-3t + 4)$

141. Write a sentence or two explaining the special considerations involving 0 in division.

142. "Two negatives give a positive" is often heard from students. Is this correct? Use more precise language in explaining what it means.

143. Use x as the variable and write an expression for "the product of 5, and the sum of a number and 7." Then use the distributive property to rewrite the expression.

144. The highest temperature ever recorded in Iowa was 118°F at Keokuk on July 20, 1934. The lowest temperature ever recorded in the state was at Elkader on February 3, 1996, and was 165° lower than the highest temperature. What is the record low temperature for Iowa? (*Source:* National Climatic Data Center.)

---

*The order of exercises in this final group does not correspond to the order in which topics occur in the chapter. This random ordering should help you prepare for the chapter test in yet another way.

Chapter **1** TEST

View the complete solutions to all Chapter Test exercises on the Pass the Test CD.

1. Write $\frac{63}{99}$ in lowest terms.

2. Add: $\frac{5}{8} + \frac{11}{12} + \frac{7}{15}$.

3. Divide: $\frac{19}{15} \div \frac{6}{5}$.

4. The circle graph indicates the market share of different means of intercity transportation in a recent year, based on 1230 million passengers carried.

   (a) How many of these passengers used air travel?

   (b) How many of these passengers did not use the bus?

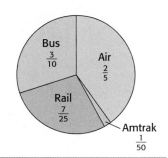

Intercity Transportation

Bus $\frac{3}{10}$ · Air $\frac{2}{5}$ · Rail $\frac{7}{25}$ · Amtrak $\frac{1}{50}$

Source: Eno Transportation Foundation, Inc.

5. Decide whether $4[-20 + 7(-2)] \leq 135$ is true or false.

6. Graph the group of numbers $-1, -3, |-4|, |-1|$ on a number line.

7. To which of the following sets does $-\frac{2}{3}$ belong: natural numbers, whole numbers, integers, rational numbers, irrational numbers, real numbers?

8. Explain how a number line can be used to show that -8 is less than -1.

9. Write in symbols: The quotient of -6 and the sum of 2 and -8. Simplify the expression.

Perform each indicated operation.

10. $-2 - (5 - 17) + (-6)$

11. $-5\frac{1}{2} + 2\frac{2}{3}$

12. $-6 - [-7 + (2 - 3)]$

13. $4^2 + (-8) - (2^3 - 6)$

14. $(-5)(-12) + 4(-4) + (-8)^2$

15. $\dfrac{-7 - (-6 + 2)}{-5 - (-4)}$

16. $\dfrac{30(-1 - 2)}{-9[3 - (-2)] - 12(-2)}$

Find the solution of each equation from the set $\{-6, -4, -2, 0, 2, 4, 6\}$ by guessing or by trial and error.

17. $-x + 3 = -3$

18. $-3x = -12$

Evaluate each expression, given $x = -2$ and $y = 4$.

19. $3x - 4y^2$

20. $\dfrac{5x + 7y}{3(x + y)}$

Solve each problem.

21. The highest elevation in Argentina is Mt. Aconcagua, which is 6960 m above sea level. The lowest point in Argentina is the Valdeś Peninsula, 40 m below sea level. Find the difference between the highest and lowest elevations.

22. For a certain system of rating relief pitchers, 3 points are awarded for a save, 3 points are awarded for a win, 2 points are subtracted for a loss, and 2 points are subtracted for a blown save. If Mariano Rivera of the New York Yankees has 4 saves, 3 wins, 2 losses, and 1 blown save, how many points does he have?

23. The bar graph shows public secondary school (grades 7–12) enrollment in millions for selected years from 1980 to 2003 in the United States. Use a signed number to represent the change in enrollment for each period.

(a) 1980 to 1985
(b) 1985 to 1990
(c) 1995 to 2000
(d) 2000 to 2003

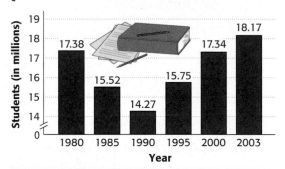

Public Secondary School Enrollment

Source: National Education Association.

Match each property in Column I with the example of it in Column II.

I	II
24. Commutative property	**A.** $3x + 0 = 3x$
25. Associative property	**B.** $(5 + 2) + 8 = 8 + (5 + 2)$
26. Inverse property	**C.** $-3(x + y) = -3x + (-3y)$
27. Identity property	**D.** $-5 + (3 + 2) = (-5 + 3) + 2$
28. Distributive property	**E.** $-\dfrac{5}{3}\left(-\dfrac{3}{5}\right) = 1$

29. What property is used to show that $3(x + 1) = 3x + 3$?

30. Consider the expression $-6[5 + (-2)]$.

(a) Evaluate it by first working within the brackets.
(b) Evaluate it by using the distributive property.
✎ (c) Why must the answers in items (a) and (b) be the same?

Simplify by combining like terms.

31. $8x + 4x - 6x + x + 14x$
32. $5(2x - 1) - (x - 12) + 2(3x - 5)$

Linear Equations and Inequalities in One Variable

In 1896, 241 competitors from 14 countries gathered in Athens, Greece, for the first modern Olympic Games. In 2004, the XXVIII Olmpic Summer Games returned to Athens as a truly international event, attracting 11,099 athletes from 202 countries.

One ceremonial aspect of the games is the flying of the Olympic flag with its five interlocking rings of different colors on a white background. First introduced at the 1920 Games in Antwerp, Belgium, the rings on the flag symbolize unity among the nations of Africa, the Americas, Asia, Australia, and Europe. (*Source:* www.athens2004.com; *Microsoft Encarta Encyclopedia.*)

Throughout this chapter we use *linear equations* to solve applications about the Olympics.

2.1 The Addition Property of Equality

OBJECTIVES

1 Identify linear equations.

2 Use the addition property of equality.

3 Simplify and then use the addition property of equality.

Recall from **Section 1.3** that an *equation* is a statement asserting that two algebraic expressions are equal. The simplest type of equation is a *linear equation.*

OBJECTIVE 1 Identify linear equations.

> **Linear Equation in One Variable**
>
> A **linear equation in one variable** can be written in the form
> $$Ax + B = C,$$
> for real numbers A, B, and C, with $A \neq 0$.

For example,
$$4x + 9 = 0, \quad 2x - 3 = 5, \quad \text{and} \quad x = 7 \qquad \text{Linear equations}$$
are linear equations in one variable (x). The final two can be written in the specified form with the use of properties developed in this chapter. However,
$$x^2 + 2x = 5, \quad \frac{1}{x} = 6, \quad \text{and} \quad |2x + 6| = 0 \qquad \text{Nonlinear equations}$$
are *not* linear equations.

As we saw in **Section 1.3,** a *solution* of an equation is a number that makes the equation true when it replaces the variable. An equation is solved by finding its **solution set,** the set of all solutions. Equations that have exactly the same solution sets are **equivalent equations.** A linear equation in x is solved by using a series of steps to produce a simpler equivalent equation of the form
$$x = \text{a number} \qquad \text{or} \qquad \text{a number} = x.$$

OBJECTIVE 2 Use the addition property of equality. In the equation $x - 5 = 2$, both $x - 5$ and 2 represent the same number because that is the meaning of the equals sign. To solve the equation, we change the left side from $x - 5$ to just x. We do this by adding 5 to $x - 5$. We use 5 because 5 is the opposite (additive inverse) of -5, and $-5 + 5 = 0$. To keep the two sides equal, we must also add 5 to the right side.

$$
\begin{aligned}
x - 5 &= 2 &&\text{Given equation} \\
x - 5 + 5 &= 2 + 5 &&\text{Add 5 to each side.} \\
x + 0 &= 7 &&\text{Additive inverse property} \\
x &= 7 &&\text{Additive identity property}
\end{aligned}
$$

The solution of the given equation is 7. We check by replacing x with 7 in the original equation.

Check:
$$
\begin{aligned}
x - 5 &= 2 &&\text{Original equation} \\
7 - 5 &= 2 \quad ? &&\text{Let } x = 7. \\
2 &= 2 &&\text{True}
\end{aligned}
$$

Since the final equation is true, 7 checks as the solution and $\{7\}$ is the solution set.

To solve the equation, we added the same number to each side. The **addition property of equality** justifies this step.

Addition Property of Equality

If A, B, and C are real numbers, then the equations

$$A = B \qquad \text{and} \qquad A + C = B + C$$

are equivalent equations.

That is, we can add the same number to each side of an equation without changing the solution.

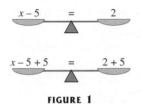

FIGURE 1

In this property, C represents a real number. Any quantity that represents a real number can be added to each side of an equation to obtain an equivalent equation.

▶ **NOTE** Equations can be thought of in terms of a balance. Thus, adding the same quantity to each side does not affect the balance. See Figure 1.

EXAMPLE 1 Using the Addition Property of Equality

Solve $x - 16 = 7$.

Our goal is to get an equivalent equation of the form $x =$ a number. To do this, we use the addition property of equality to add 16 to each side.

$$x - 16 = 7$$
$$x - 16 + 16 = 7 + 16 \qquad \text{Add 16 to each side.}$$
$$x = 23 \qquad \text{Combine like terms.}$$

Check: Substitute 23 for x in the *original* equation.

$$x - 16 = 7 \qquad \text{Original equation}$$
$$23 - 16 = 7 \quad ? \quad \text{Let } x = 23.$$
$$7 = 7 \qquad \text{True}$$

Since a true statement results, $\{23\}$ is the solution set.

✔ **Now Try Exercise 7.**

▶ **CAUTION** Do NOT write the solution set in Example 1 as $\{x = 23\}$. This

EXAMPLE 2 Using the Addition Property of Equality

Solve $x - 2.9 = -6.4$.

We use the addition property of equality to isolate x on the left side.

$$x - 2.9 = -6.4$$
$$x - 2.9 + 2.9 = -6.4 + 2.9 \qquad \text{Add 2.9 to each side.}$$
$$x = -3.5$$

Check:

$$x - 2.9 = -6.4 \qquad \text{Original equation}$$

$$-3.5 - 2.9 = -6.4 \quad ? \quad \text{Let } x = -3.5.$$

$$-6.4 = -6.4 \qquad \text{True}$$

Since a true statement results, the solution set is $\{-3.5\}$.

✔ **Now Try Exercise 21.**

The addition property of equality says that the same number may be *added* to each side of an equation. In **Section 1.5,** subtraction was defined as addition of the opposite. Thus, we can also use the following rule when solving an equation.

> The same number may be subtracted from each side of an equation without changing the solution.

EXAMPLE 3 Using the Addition Property of Equality

Solve $-7 = x + 22$.

Here, the variable x is on the right side of the equation. To isolate x on the right, we must eliminate the 22 by subtracting 22 from each side.

$$-7 = x + 22$$

$$-7 - 22 = x + 22 - 22 \qquad \text{Subtract 22 from each side.}$$

$$-29 = x, \quad \text{or} \quad x = -29$$

Check:

$$-7 = x + 22 \qquad \text{Original equation}$$

$$-7 = -29 + 22 \quad ? \quad \text{Let } x = -29.$$

$$-7 = -7 \qquad \text{True}$$

Thus, the solution set is $\{-29\}$.

✔ **Now Try Exercise 17.**

▶ **CAUTION** The final line of the check does *not* give the solution to the problem, only a confirmation that the solution found is correct.

EXAMPLE 4 Subtracting a Variable Expression

Solve $\frac{3}{5}k + 17 = \frac{8}{5}k$.

$$\frac{3}{5}k + 17 = \frac{8}{5}k \qquad \text{Original equation}$$

$$\frac{3}{5}k + 17 - \frac{3}{5}k = \frac{8}{5}k - \frac{3}{5}k \qquad \text{Subtract } \tfrac{3}{5}k \text{ from each side.}$$

$$17 = 1k \qquad \text{Combine like terms; } \tfrac{5}{5}k = 1k.$$

$$17 = k \qquad \text{Multiplicative identity property}$$

(From now on we will skip the step that changes $1k$ to k.) Check the solution by replacing k with 17 in the original equation. The solution set is $\{17\}$.

✔ **Now Try Exercise 31.**

What happens if we solve the equation in Example 4 by first subtracting $\frac{8}{5}k$ from each side?

$$\frac{3}{5}k + 17 = \frac{8}{5}k$$

$\frac{3}{5}k + 17 - \frac{8}{5}k = \frac{8}{5}k - \frac{8}{5}k$ Subtract $\frac{8}{5}k$ from each side.

$17 - k = 0$ Combine like terms; $-\frac{5}{5}k = -1k = -k$.

$17 - k - 17 = 0 - 17$ Subtract 17 from each side.

$-k = -17$ Combine like terms; additive inverse

This result gives the value of $-k$, but not of k itself. However, it does say that the additive inverse of k is -17, which means that k must be 17, the same result we obtained in Example 4.

$$k = 17$$

(This result can also be justified by the multiplication property of equality, covered in **Section 2.2.**) We make the following generalization:

If a is a number and $-x = a$, then $x = -a$.

OBJECTIVE 3 Simplify and then use the addition property of equality.

EXAMPLE 5 Simplifying an Equation before Solving

Solve $3k - 12 + k + 2 = 5 + 3k + 2$.

Begin by combining like terms on each side of the equation.

$$3k - 12 + k + 2 = 5 + 3k + 2$$
$$4k - 10 = 7 + 3k$$

Next, get all terms that contain variables on the same side of the equation and all terms without variables on the other side. One way to start is to subtract $3k$ from each side.

$4k - 10 - 3k = 7 + 3k - 3k$ Subtract $3k$ from each side.

$k - 10 = 7$ Combine like terms.

$k - 10 + 10 = 7 + 10$ Add 10 to each side.

$k = 17$ Combine like terms.

Check: Substitute 17 for k in the original equation.

$3k - 12 + k + 2 = 5 + 3k + 2$ Original equation

$3(17) - 12 + 17 + 2 = 5 + 3(17) + 2$?   Let $k = 17$.

$51 - 12 + 17 + 2 = 5 + 51 + 2$?   Multiply.

$58 = 58$ True

The check results in a true statement, so the solution set is $\{17\}$.

✔ **Now Try Exercise 49.**

EXAMPLE 6 Using the Distributive Property to Simplify an Equation

Solve $3(2 + 5x) - (1 + 14x) = 6$.

Be sure to distribute to all terms within the parentheses.

$$3(2 + 5x) - (1 + 14x) = 6$$
$$3(2 + 5x) - 1(1 + 14x) = 6 \qquad -(1 + 14x) = -1(1 + 14x)$$

$3(2) + 3(5x) - 1(1) - 1(14x) = 6$	Distributive property
$6 + 15x - 1 - 14x = 6$	Multiply.
$x + 5 = 6$	Combine like terms.
$x + 5 - 5 = 6 - 5$	Subtract 5 from each side.
$x = 1$	Combine like terms.

Check by substituting 1 for x in the original equation. The solution set is $\{1\}$.

✔ **Now Try Exercise 57.**

▶ **CAUTION** *Be careful to apply the distributive property correctly* in a problem like that in Example 6, or a sign error may result.

2.1 EXERCISES

🌐 *Complete solution available on Video Lectures on CD/DVD*

Now Try Exercise

1. *Concept Check* Which pairs of equations are equivalent equations?

A. $x + 2 = 6$ and $x = 4$ **B.** $10 - x = 5$ and $x = -5$
C. $x + 3 = 9$ and $x = 6$ **D.** $4 + x = 8$ and $x = -4$

2. *Concept Check* Decide whether each of the following is an *expression* or an *equation*. If it is an expression, simplify it. If it is an equation, solve it.

(a) $5x + 8 - 4x + 7$ **(b)** $-6y + 12 + 7y - 5$
(c) $5x + 8 - 4x = 7$ **(d)** $-6y + 12 + 7y = -5$

📝 **3.** In your own words, state the addition property of equality. Give an example.

📝 **4.** Explain how to check a solution of an equation.

Solve each equation, and check your solution. See Examples 1–4.

5. $x - 4 = 8$ **6.** $x - 8 = 9$ 🌐 **7.** $x - 12 = 19$

8. $x - 15 = 25$ **9.** $x - 5 = -8$ **10.** $x - 7 = -9$

11. $r + 9 = 13$ **12.** $x + 6 = 10$ **13.** $x + 26 = 17$

14. $x + 45 = 24$ **15.** $7 + r = -3$ **16.** $8 + k = -4$

🌐 **17.** $2 = p + 15$ **18.** $3 = z + 17$ **19.** $-2 = x - 12$

20. $-6 = x - 21$ 🌐 **21.** $x - 8.4 = -2.1$ **22.** $x - 15.5 = -5.1$

23. $t + 12.3 = -4.6$ **24.** $x + 21.5 = -13.4$ **25.** $3x + 7 = 2x + 4$

26. $9x - 1 = 8x + 4$ **27.** $8t + 6 = 7t + 6$ **28.** $13t + 9 = 12t + 9$

29. $-4x + 7 = -5x + 9$ **30.** $-6x + 3 = -7x + 10$ 🌐 **31.** $\dfrac{2}{5}w - 6 = \dfrac{7}{5}w$

32. $-\dfrac{2}{7}z + 2 = \dfrac{5}{7}z$ **33.** $5.6x + 2 = 4.6x$ **34.** $9.1x - 5 = 8.1x$

35. $3p = 2p$ **36.** $8b = 7b$ **37.** $1.2y - 4 = 0.2y - 4$

38. $7.7r + 6 = 6.7r + 6$ **39.** $\dfrac{1}{2}x + 2 = -\dfrac{1}{2}x$ **40.** $\dfrac{1}{5}x - 7 = -\dfrac{4}{5}x$

41. $3x + 7 - 2x = 0$ **42.** $5x + 4 - 4x = 0$ **43.** $0.4x + 3 = 0.3x$

44. *Concept Check* Which of the following are *not* linear equations in one variable?

 A. $x^2 - 5x + 6 = 0$ **B.** $x^3 = x$ **C.** $3x - 4 = 0$ **D.** $7x - 6x = 3 + 9x$

45. Define a linear equation in one variable in words.

46. Refer to the definition of a linear equation in one variable given in this section. Why is the restriction $A \neq 0$ necessary?

Solve each equation, and check your solution. See Examples 5 and 6.

47. $5t + 3 + 2t - 6t = 4 + 12$ **48.** $4x + 3x - 6 - 6x = 10 + 3$

49. $6x + 5 + 7x + 3 = 12x + 4$ **50.** $4x - 3 - 8x + 1 = -5x + 9$

51. $5.2q - 4.6 - 7.1q = -0.9q - 4.6$ **52.** $-4.0x + 2.7 - 1.6x = -4.6x + 2.7$

53. $\dfrac{5}{7}x + \dfrac{1}{3} = \dfrac{2}{5} - \dfrac{2}{7}x + \dfrac{2}{5}$ **54.** $\dfrac{6}{7}s - \dfrac{3}{4} = \dfrac{4}{5} - \dfrac{1}{7}s + \dfrac{1}{6}$

55. $(5y + 6) - (3 + 4y) = 10$ **56.** $(8r - 3) - (7r + 1) = -6$

57. $2(p + 5) - (9 + p) = -3$ **58.** $4(k - 6) - (3k + 2) = -5$

59. $-6(2b + 1) + (13b - 7) = 0$ **60.** $-5(3w - 3) + (1 + 16w) = 0$

61. $10(-2x + 1) = -19(x + 1)$ **62.** $2(2 - 3r) = -5(r - 3)$

63. $-2(8p + 2) - 3(2 - 7p) - 2(4 + 2p) = 0$

64. $-5(1 - 2z) + 4(3 - z) - 7(3 + z) = 0$

65. $4(7x - 1) + 3(2 - 5x) - 4(3x + 5) = -6$

66. $9(2m - 3) - 4(5 + 3m) - 5(4 + m) = -3$

67. *Concept Check* Write an equation that requires the use of the addition property of equality, in which 6 must be added to each side to solve the equation and the solution is a negative number.

68. *Concept Check* Write an equation that requires the use of the addition property of equality, in which $\frac{1}{2}$ must be subtracted from each side and the solution is a positive number.

Write an equation using the information given in the problem. Use x as the variable. Then solve the equation.

69. Three times a number is 17 more than twice the number. Find the number.

70. One added to three times a number is three less than four times the number. Find the number.

71. If six times a number is subtracted from seven times the number, the result is -9. Find the number.

72. If five times a number is added to three times the number, the result is the sum of seven times the number and 9. Find the number.

Most of the exercise sets in the rest of the book end with brief sets of "Preview Exercises." These exercises are designed to help you review ideas introduced earlier, as well as preview ideas needed for the next few sections. If you need help with these Preview Exercises, look in the section or sections indicated.

PREVIEW EXERCISES

Simplify each expression. See **Section 1.8.**

73. $\dfrac{2}{3}\left(\dfrac{3}{2}\right)$ 　　　　　 **74.** $\dfrac{5}{6}\left(\dfrac{6}{5}\right)$ 　　　　　 **75.** $-\dfrac{5}{4}\left(-\dfrac{4}{5}x\right)$

76. $-\dfrac{9}{7}\left(-\dfrac{7}{9}x\right)$ 　　 **77.** $9\left(\dfrac{r}{9}\right)$ 　　　　　 **78.** $6\left(\dfrac{t}{6}\right)$

2.2 The Multiplication Property of Equality

OBJECTIVES

1 Use the multiplication property of equality.

2 Combine terms in equations, and then use the multiplication property of equality.

OBJECTIVE 1 Use the multiplication property of equality. The addition property of equality alone is not enough to solve some equations, such as $3x + 2 = 17$.

$$3x + 2 = 17$$
$$3x + 2 - 2 = 17 - 2 \qquad \text{Subtract 2 from each side.}$$
$$3x = 15 \qquad \text{Combine like terms.}$$

Notice that the coefficient of x on the left side is 3, not 1 as desired. Another property is needed to change $3x = 15$ to an equation of the form

$$x = \text{a number.}$$

If $3x = 15$, then $3x$ and 15 represent the same number. Multiplying $3x$ and 15 by the same number will also result in an equality. The **multiplication property of equality** states that we can multiply each side of an equation by the same nonzero number without changing the solution.

Multiplication Property of Equality

If A, B, and C $(C \neq 0)$ represent real numbers, then the equations

$$A = B \qquad \text{and} \qquad AC = BC$$

are equivalent equations.

　　　That is, we can multiply each side of an equation by the same nonzero number without changing the solution.

This property can be used to solve $3x = 15$. The $3x$ on the left must be changed to $1x$, or x, instead of $3x$. To isolate x, we multiply each side of the equation by $\frac{1}{3}$. We use $\frac{1}{3}$ because $\frac{1}{3}$ is the reciprocal of 3 and $\frac{1}{3} \cdot 3 = \frac{3}{3} = 1$.

$$3x = 15$$

$$\frac{1}{3}(3x) = \frac{1}{3} \cdot 15 \qquad \text{Multiply each side by } \tfrac{1}{3}.$$

$$\left(\frac{1}{3} \cdot 3\right)x = \frac{1}{3} \cdot 15 \qquad \text{Associative property}$$

$$1x = 5 \qquad \text{Multiplicative inverse property}$$

$$x = 5 \qquad \text{Multiplicative identity property}$$

We check by substituting 5 for x in the original equation. The solution set of the equation is $\{5\}$.

Just as the addition property of equality permits *subtracting* the same number from each side of an equation, the multiplication property of equality permits *dividing* each side of an equation by the same nonzero number. For example, the equation $3x = 15$, which we just solved by multiplying each side by $\tfrac{1}{3}$, could also be solved by dividing each side by 3.

$$3x = 15$$

$$\frac{3x}{3} = \frac{15}{3} \qquad \text{Divide each side by 3.}$$

$$x = 5$$

We can divide each side of an equation by the same nonzero number without changing the solution. ***Do not, however, divide each side by a variable, as that may result in losing a valid solution.***

> ▶ **NOTE** In practice, it is usually easier to multiply on each side if the coefficient of the variable is a fraction, and divide on each side if the coefficient is an integer. For example, to solve
>
> $$-\frac{3}{4}x = 12,$$
>
> it is easier to multiply by $-\tfrac{4}{3}$, the reciprocal of $-\tfrac{3}{4}$, than to divide by $-\tfrac{3}{4}$. On the other hand, to solve
>
> $$-5x = -20,$$
>
> it is easier to divide by -5 than to multiply by $-\tfrac{1}{5}$.

EXAMPLE 1 **Dividing Each Side of an Equation by a Nonzero Number**

Solve $25x = 30$.

Transform the equation so that x (instead of $25x$) is on the left by using the multiplication property of equality. Divide each side by 25, the coefficient of x.

$$25x = 30$$

$$\frac{25x}{25} = \frac{30}{25} \qquad \text{Divide each side by 25.}$$

$$x = \frac{30}{25} = \frac{6}{5} \qquad \text{Write in lowest terms.}$$

Check: Substitute $\frac{6}{5}$ for x in the original equation.

$$25x = 30 \qquad \text{Original equation}$$
$$\frac{25}{1}\left(\frac{6}{5}\right) = 30 \qquad ? \qquad \text{Let } x = \tfrac{6}{5}.$$
$$30 = 30 \qquad \text{True}$$

The solution set is $\left\{\frac{6}{5}\right\}$.

✔ **Now Try Exercise 31.**

EXAMPLE 2 Solving an Equation with Decimals

Solve $-2.1x = 6.09$.

$$-2.1x = 6.09$$
$$\frac{-2.1x}{-2.1} = \frac{6.09}{-2.1} \qquad \text{Divide each side by } -2.1.$$
$$x = -2.9 \qquad \text{Divide; use a calculator.}$$

Check:
$$-2.1x = 6.09 \qquad \text{Original equation}$$
$$-2.1(-2.9) = 6.09 \qquad ? \qquad \text{Let } x = -2.9.$$
$$6.09 = 6.09 \qquad \text{True}$$

The solution set is $\{-2.9\}$.

✔ **Now Try Exercise 39.**

EXAMPLE 3 Using the Multiplication Property of Equality

Solve $\dfrac{x}{4} = 3$.

Replace $\frac{x}{4}$ by $\frac{1}{4}x$, since dividing by 4 is the same as multiplying by $\frac{1}{4}$. To isolate x on the left, multiply each side by 4, the reciprocal of the coefficient of x.

$$\frac{x}{4} = 3$$

In practice, these steps are often combined.

$$\frac{1}{4}x = 3 \qquad \text{Change } \tfrac{x}{4} \text{ to } \tfrac{1}{4}x.$$
$$4 \cdot \frac{1}{4}x = 4 \cdot 3 \qquad \text{Multiply by 4.}$$
$$x = 12 \qquad \text{Multiplicative inverse property; multiplicative identity property}$$

Check that 12 is the solution.

Check:
$$\frac{x}{4} = 3 \qquad \text{Original equation}$$
$$\frac{12}{4} = 3 \qquad ? \qquad \text{Let } x = 12.$$
$$3 = 3 \qquad \text{True}$$

The solution set is $\{12\}$.

✔ **Now Try Exercise 43.**

EXAMPLE 4 **Using the Multiplication Property of Equality**

Solve $\frac{3}{4}h = 6$.

To isolate h on the left, multiply each side of the equation by $\frac{4}{3}$. Use $\frac{4}{3}$ because $\frac{4}{3} \cdot \frac{3}{4}h = 1 \cdot h = h$.

$$\frac{3}{4}h = 6$$

$$\frac{4}{3}\left(\frac{3}{4}h\right) = \frac{4}{3} \cdot 6 \qquad \text{Multiply by } \tfrac{4}{3}.$$

$$1 \cdot h = \frac{4}{3} \cdot \frac{6}{1} \qquad \text{Multiplicative inverse property}$$

$$h = 8 \qquad \begin{array}{l}\text{Multiplicative identity property;}\\ \text{multiply fractions.}\end{array}$$

Check: $\qquad \frac{3}{4}h = 6 \qquad$ Original equation

$$\frac{3}{4}(8) = 6 \qquad ? \qquad \text{Let } h = 8.$$

$$6 = 6 \qquad \text{True}$$

The solution set is $\{8\}$.

✔ **Now Try Exercise 47.**

In **Section 2.1,** we obtained the equation $-k = -17$ in our alternative solution to Example 4. We reasoned that since this equation says that the additive inverse (or opposite) of k is -17, then k must equal 17. We can also use the multiplication property of equality to obtain the same result as detailed in the next example.

EXAMPLE 5 **Using the Multiplication Property of Equality when the Coefficient of the Variable Is -1**

Solve $-k = -17$.

On the left side, change $-k$ to k by first writing $-k$ as $-1 \cdot k$.

$$-k = -17$$

$$-1 \cdot k = -17 \qquad -k = -1 \cdot k$$

$$-1(-1 \cdot k) = -1(-17) \qquad \text{Multiply each side by } -1.$$

$$[-1(-1)] \cdot k = 17 \qquad \text{Associative property; multiply.}$$

$$1 \cdot k = 17 \qquad \text{Multiplicative inverse property}$$

$$k = 17 \qquad \text{Multiplicative identity property}$$

Check: $\qquad -k = -17 \qquad$ Original equation

$$-(17) = -17 \qquad ? \qquad \text{Let } k = 17.$$

$$-17 = -17 \qquad \text{True}$$

The solution, 17, checks, so $\{17\}$ is the solution set.

✔ **Now Try Exercise 53.**

OBJECTIVE 2   Combine terms in equations, and then use the multiplication property of equality.

EXAMPLE 6   Combining Terms in an Equation before Solving

Solve $5m + 6m = 33$.

$$5m + 6m = 33$$
$$11m = 33 \qquad \text{Combine like terms.}$$
$$\frac{11m}{11} = \frac{33}{11} \qquad \text{Divide by 11.}$$
$$m = 3$$

Check this proposed solution. The solution set is $\{3\}$.

✔ **Now Try Exercise 59.**

CONNECTIONS

The use of algebra to solve equations and applied problems is very old. The 3600-year-old Rhind Papyrus includes the following "word problem": "Aha, its whole, its seventh, it makes 19." This brief sentence describes the equation

$$x + \frac{x}{7} = 19.$$

The solution of this equation is $16\frac{5}{8}$. The word *algebra* is from the work *Hisab al-jabr m'al muquabalah*, written in the ninth century by Muhammed ibn Musa Al-Khowarizmi. The title means "the science of transposition and cancellation." From Latin versions of Khowarizmi's text, "al-jabr" became the broad term covering the art of equation solving.

2.2   EXERCISES

Complete solution available on Video Lectures on CD/DVD

Now Try Exercise

1. In your own words, state the multiplication property of equality. Give an example.

2. In the statement of the multiplication property of equality in this section, there is a restriction that $C \neq 0$. What would happen if you multiplied each side of an equation by 0?

3. **Concept Check**   Which equation does *not* require the use of the multiplication property of equality?

   A. $3x - 5x = 6$ **B.** $-\frac{1}{4}x = 12$ **C.** $5x - 4x = 7$ **D.** $\frac{x}{3} = -2$

4. **Concept Check**   Tell whether you would use the addition or multiplication property of equality to solve each equation.

   (a) $3x = 12$ **(b)** $3 + x = 12$ **(c)** $-x = 4$ **(d)** $-12 = 6 + x$

5. A student tried to solve the equation $4x = 8$ by dividing each side by 8. Why is this not the correct procedure for solving this equation?

6. State how you would find the solution of a linear equation if your next-to-last step reads "$-x = 5$."

Concept Check *By what number is it necessary to multiply both sides of each equation to isolate x on the left side? Do not actually solve these equations.*

7. $\frac{2}{3}x = 8$

8. $\frac{4}{5}x = 6$

9. $\frac{x}{10} = 3$

10. $\frac{x}{100} = 8$

11. $-\frac{9}{2}x = -4$

12. $-\frac{8}{3}x = -11$

13. $-x = 0.36$

14. $-x = 0.29$

Concept Check *By what number is it necessary to divide both sides of each equation to isolate x on the left side? Do not actually solve these equations.*

15. $6x = 5$

16. $7x = 10$

17. $-4x = 13$

18. $-13x = 6$

19. $0.12x = 48$

20. $0.21x = 63$

21. $-x = 23$

22. $-x = 49$

Solve each equation, and check your solution. See Examples 1–5.

23. $5x = 30$

24. $7x = 56$

25. $2m = 15$

26. $3m = 10$

27. $3a = -15$

28. $5k = -70$

29. $-3x = 12$

30. $-4x = 36$

31. $10t = -36$

32. $4s = -34$

33. $-6x = -72$

34. $-8x = -64$

35. $2r = 0$

36. $5x = 0$

37. $0.2t = 8$

38. $0.9x = 18$

39. $-2.1m = 25.62$

40. $-3.9a = 31.2$

41. $\frac{1}{4}x = -12$

42. $\frac{1}{5}p = -3$

43. $\frac{z}{6} = 12$

44. $\frac{x}{5} = 15$

45. $\frac{x}{7} = -5$

46. $\frac{k}{8} = -3$

47. $\frac{2}{7}p = 4$

48. $\frac{3}{8}x = 9$

49. $-\frac{5}{6}t = -15$

50. $-\frac{3}{4}k = -21$

51. $-\frac{7}{9}c = \frac{3}{5}$

52. $-\frac{5}{6}d = \frac{4}{9}$

53. $-x = 12$

54. $-t = 14$

55. $-x = -\frac{3}{4}$

56. $-x = -\frac{1}{2}$

57. $-0.3x = 9$

58. $-0.5x = 20$

Solve each equation, and check your solution. See Example 6.

59. $4x + 3x = 21$

60. $9x + 2x = 121$

61. $3r - 5r = 10$

62. $9p - 13p = 24$

63. $5m + 6m - 2m = 63$

64. $11r - 5r + 6r = 168$

65. $-6x + 4x - 7x = 0$

66. $-5x + 4x - 8x = 0$

67. $9w - 5w + w = -3$

68. $10x - 6x + 3x = -4$

69. $\frac{1}{3}x - \frac{1}{4}x + \frac{1}{12}x = 3$

70. $\frac{2}{5}x + \frac{1}{10}x - \frac{1}{20}x = 18$

71. *Concept Check* Write an equation that requires the use of the multiplication property of equality, where each side must be multiplied by $\frac{2}{3}$ and the solution is a negative number.

72. *Concept Check* Write an equation that requires the use of the multiplication property of equality, where each side must be divided by 100 and the solution is not an integer.

Write an equation using the information given in the problem. Use x as the variable. Then solve the equation.

73. When a number is multiplied by 4, the result is 6. Find the number.

74. When a number is multiplied by -4, the result is 10. Find the number.

75. When a number is divided by -5, the result is 2. Find the number.

76. If twice a number is divided by 5, the result is 4. Find the number.

PREVIEW EXERCISES

*Simplify each expression. See **Section 1.8.***

77. $-(3m + 5)$

78. $-4(-1 + 6x)$

79. $4(-5 + 2p) - 3(p - 4)$

80. $2(4k - 7) - 4(-k + 3)$

*Solve each equation. See **Section 2.1.***

81. $4x + 5 + 2x = 7x$

82. $2x + 5x - 3x + 4 = 3x + 2$

2.3   More on Solving Linear Equations

OBJECTIVES

1 Learn and use the four steps for solving a linear equation.

2 Solve equations with fractions or decimals as coefficients.

3 Solve equations with no solution or infinitely many solutions.

4 Write expressions for two related unknown quantities.

OBJECTIVE 1   Learn and use the four steps for solving a linear equation. We solve more complicated equations using the following four-step method.

Solving a Linear Equation

Step 1   **Simplify each side separately.** Clear parentheses, using the distributive property if needed, and combine like terms.

Step 2   **Isolate the variable term on one side.** Use the addition property if necessary so that the variable term is on one side of the equation and a number is on the other.

Step 3   **Isolate the variable.** Use the multiplication property if necessary to get the equation in the form $x = $ a number. (Other letters may be used for variables.)

Step 4   **Check.** Substitute the proposed solution into the *original* equation to see if a true statement results.

EXAMPLE 1   Using the Four Steps to Solve an Equation

Solve $3x + 4 - 2x - 7 = 4x + 3$.

Step 1 $3x + 4 - 2x - 7 = 4x + 3$

$x - 3 = 4x + 3$ Combine like terms.

Step 2 $x - 3 + 3 = 4x + 3 + 3$ Add 3.

$x = 4x + 6$ Combine like terms.

$x - 4x = 4x + 6 - 4x$ Subtract 4x.

$-3x = 6$ Combine like terms.

Step 3 $\dfrac{-3x}{-3} = \dfrac{6}{-3}$ Divide by -3.

$x = -2$ $\frac{-3}{-3} = 1; \; 1x = x$

Step 4 Substitute -2 for x in the original equation to check.

$$3x + 4 - 2x - 7 = 4x + 3 \qquad \text{Original equation}$$
$$3(-2) + 4 - 2(-2) - 7 = 4(-2) + 3 \quad ? \quad \text{Let } x = -2.$$
$$-6 + 4 + 4 - 7 = -8 + 3 \quad ? \quad \text{Multiply.}$$
$$-5 = -5 \qquad \text{True}$$

The solution set of the equation is $\{-2\}$.

✔ **Now Try Exercise 7.**

▶ **NOTE** In Step 2 of Example 1, we added and subtracted the terms in such a way that the variable term ended up on the left side of the equation. Choosing differently would have put the variable term on the right side of the equation. Either way, the same solution results.

EXAMPLE 2 **Using the Four Steps to Solve an Equation**

Solve $4(k - 3) - k = k - 6$.

Step 1 Before combining like terms, use the distributive property to simplify $4(k - 3)$.

$$4(k - 3) - k = k - 6$$
$$4(k) + 4(-3) - k = k - 6 \qquad \text{Distributive property}$$
$$4k - 12 - k = k - 6 \qquad \text{Multiply.}$$
$$3k - 12 = k - 6 \qquad \text{Combine like terms.}$$

Step 2
$$3k - 12 - k = k - 6 - k \qquad \text{Subtract } k.$$
$$2k - 12 = -6 \qquad \text{Combine like terms.}$$
$$2k - 12 + 12 = -6 + 12 \qquad \text{Add 12.}$$
$$2k = 6 \qquad \text{Combine like terms.}$$

Step 3
$$\frac{2k}{2} = \frac{6}{2} \qquad \text{Divide by 2.}$$
$$k = 3$$

Step 4 Check by substituting 3 for k in the original equation. Work inside the parentheses first.

$$4(k - 3) - k = k - 6 \qquad \text{Original equation}$$
$$4(3 - 3) - 3 = 3 - 6 \quad ? \quad \text{Let } k = 3.$$
$$4(0) - 3 = 3 - 6 \quad ?$$
$$0 - 3 = 3 - 6 \quad ?$$
$$-3 = -3 \qquad \text{True}$$

The solution set of the equation is $\{3\}$.

✔ **Now Try Exercise 9.**

EXAMPLE 3 Using the Four Steps to Solve an Equation

Solve $8a - (3 + 2a) = 3a + 1$.

Step 1 Clear parentheses using the distributive property.

$$8a - (3 + 2a) = 3a + 1$$
$$8a - 1(3 + 2a) = 3a + 1 \qquad \text{Multiplicative identity property}$$
$$8a - 3 - 2a = 3a + 1 \qquad \text{Distributive property}$$
$$6a - 3 = 3a + 1$$

Be careful with signs.

Step 2
$$6a - 3 - 3a = 3a + 1 - 3a \qquad \text{Subtract } 3a.$$
$$3a - 3 = 1$$
$$3a - 3 + 3 = 1 + 3 \qquad \text{Add 3.}$$
$$3a = 4$$

Step 3
$$\frac{3a}{3} = \frac{4}{3} \qquad \text{Divide by 3.}$$
$$a = \frac{4}{3}$$

Step 4 Check this solution in the original equation.

$$8a - (3 + 2a) = 3a + 1 \qquad \text{Original equation}$$
$$8\left(\frac{4}{3}\right) - \left[3 + 2\left(\frac{4}{3}\right)\right] = 3\left(\frac{4}{3}\right) + 1 \quad ? \qquad \text{Let } a = \tfrac{4}{3}.$$
$$\frac{32}{3} - \left[3 + \frac{8}{3}\right] = 4 + 1 \qquad ?$$
$$\frac{32}{3} - \left[\frac{9}{3} + \frac{8}{3}\right] = 5 \qquad ?$$
$$\frac{32}{3} - \frac{17}{3} = 5 \qquad ?$$
$$5 = 5 \qquad \text{True}$$

The check shows that $\left\{\frac{4}{3}\right\}$ is the solution set.

✔ **Now Try Exercise 11.**

▶ **CAUTION** Be very careful with signs when solving an equation like the one in Example 3. When clearing parentheses in the expression

$$8a - (3 + 2a),$$

remember that the $-$ sign acts like a factor of -1 and affects the sign of *every* term within the parentheses. Thus,

$$8a - (3 + 2a) = 8a + (-1)(3 + 2a)$$
$$= 8a - 3 - 2a.$$

Change to $-$ in *both* terms.

EXAMPLE 4 Using the Four Steps to Solve an Equation

Solve $4(8 - 3t) = 32 - 8(t + 2)$.

Step 1 $\qquad\qquad 4(8 - 3t) = 32 - 8(t + 2)$ Be careful with signs.

$\qquad\qquad\qquad 32 - 12t = 32 - 8t - 16$ Distributive property

$\qquad\qquad\qquad 32 - 12t = 16 - 8t$

Step 2 $\qquad 32 - 12t + 12t = 16 - 8t + 12t$ Add 12t.

$\qquad\qquad\qquad\qquad 32 = 16 + 4t$

$\qquad\qquad\quad 32 - 16 = 16 + 4t - 16$ Subtract 16.

$\qquad\qquad\qquad\qquad 16 = 4t$

Step 3 $\qquad\qquad\qquad \dfrac{16}{4} = \dfrac{4t}{4}$ Divide by 4.

$\qquad\qquad\qquad 4 = t, \quad \text{or} \quad t = 4$

Step 4 *Check:* $\quad 4(8 - 3t) = 32 - 8(t + 2)$ Original equation

$\qquad\qquad 4(8 - 3 \cdot 4) = 32 - 8(4 + 2)$?   Let $t = 4$.

$\qquad\qquad\quad 4(8 - 12) = 32 - 8(6)$?

$\qquad\qquad\qquad 4(-4) = 32 - 48$?

$\qquad\qquad\qquad\quad -16 = -16$ True

Since a true statement results, the solution set is $\{4\}$.

✔ **Now Try Exercise 13.**

OBJECTIVE 2 **Solve equations with fractions or decimals as coefficients.** We clear an equation of fractions by multiplying each side by the least common denominator (LCD) of all the fractions in the equation. It is a good idea to do this to avoid messy computations.

▶ **CAUTION** When clearing an equation of fractions, be sure to multiply every term on each side of the equation by the LCD.

EXAMPLE 5 Solving an Equation with Fractions as Coefficients

Solve $\dfrac{2}{3}x - \dfrac{1}{2}x = -\dfrac{1}{6}x - 2$.

The LCD of all the fractions in the equation is 6, so multiply each side by 6.

$$\frac{2}{3}x - \frac{1}{2}x = -\frac{1}{6}x - 2$$

$$6\left(\frac{2}{3}x - \frac{1}{2}x\right) = 6\left(-\frac{1}{6}x - 2\right) \qquad \text{Multiply by 6.}$$

$$6\left(\frac{2}{3}x\right) + 6\left(-\frac{1}{2}x\right) = 6\left(-\frac{1}{6}x\right) + 6(-2) \qquad \text{Distributive property}$$

$$4x - 3x = -x - 12$$

Now use the four steps to solve the equivalent equation

$$4x - 3x = -x - 12.$$

Step 1 $\qquad\qquad x = -x - 12$ $\qquad$ Combine like terms.

Step 2 $\qquad\qquad x + x = -x - 12 + x$ $\qquad$ Add x.

$\qquad\qquad\qquad 2x = -12$ $\qquad$ Combine like terms.

Step 3 $\qquad\qquad \dfrac{2x}{2} = \dfrac{-12}{2}$ $\qquad$ Divide by 2.

$\qquad\qquad\qquad x = -6$

Step 4 $\quad$ *Check:* $\quad \dfrac{2}{3}x - \dfrac{1}{2}x = -\dfrac{1}{6}x - 2$ $\qquad$ Original equation

$$\dfrac{2}{3}(-6) - \dfrac{1}{2}(-6) = -\dfrac{1}{6}(-6) - 2 \quad ? \quad \text{Let } x = -6.$$

$$-4 + 3 = 1 - 2 \qquad ?$$

$$-1 = -1 \qquad \text{True}$$

The solution set of the equation is $\{-6\}$.

✔ **Now Try Exercise 23.**

The multiplication property is also used to clear an equation of decimals.

EXAMPLE 6 **Solving an Equation with Decimals as Coefficients**

Solve $0.1t + 0.05(20 - t) = 0.09(20)$.

The decimals are expressed as tenths and hundredths. Choose the least exponent on 10 needed to eliminate the decimals; in this case, use $10^2 = 100$. A number can be multiplied by 100 by moving the decimal point two places to the right.

$$0.10t + 0.05(20 - t) = 0.09(20) \qquad 0.1 = 0.10$$

$$10t + 5(20 - t) = 9(20) \qquad \text{Multiply by 100.}$$

Now use the four steps.

Step 1 $\qquad 10t + 5(20) + 5(-t) = 180$ $\qquad$ Distributive property

$\qquad\qquad 10t + 100 - 5t = 180$

$\qquad\qquad\qquad 5t + 100 = 180$ $\qquad$ Combine like terms.

Step 2 $\qquad 5t + 100 - 100 = 180 - 100$ $\qquad$ Subtract 100.

$\qquad\qquad\qquad\qquad 5t = 80$ $\qquad$ Combine like terms.

Step 3 $\qquad\qquad \dfrac{5t}{5} = \dfrac{80}{5}$ $\qquad$ Divide by 5.

$\qquad\qquad\qquad t = 16$

Step 4 $\quad$ Check that $\{16\}$ is the solution set by substituting 16 for t in the original equation.

✔ **Now Try Exercise 29.**

OBJECTIVE 3 Solve equations with no solution or infinitely many solutions. Each equation that we have solved so far has had exactly one solution. An equation with exactly one solution is a **conditional equation** because it is only true under certain conditions. Sometimes equations may have no solution or infinitely many solutions. (The four steps are not identified in these examples. See if you can identify them.)

EXAMPLE 7 Solving an Equation That Has Infinitely Many Solutions

Solve $5x - 15 = 5(x - 3)$.

$$5x - 15 = 5(x - 3) \qquad \text{Original equation}$$
$$5x - 15 = 5x - 15 \qquad \text{Distributive property}$$
$$5x - 15 - 5x = 5x - 15 - 5x \qquad \text{Subtract } 5x.$$
$$-15 = -15$$
$$-15 + 15 = -15 + 15 \qquad \text{Add 15.}$$
$$0 = 0 \qquad \text{True}$$

Solution set: {all real numbers}

The variable has "disappeared." Since the last statement ($0 = 0$) is true, *any* real number is a solution. (We could have predicted this from the line in the solution that says $5x - 15 = 5x - 15$, which is certainly true for *any* value of x.) An equation with both sides exactly the same, like $0 = 0$, is called an **identity.** An identity is true for all replacements of the variables. As shown above, write the solution set as *{all real numbers}*.

✔ **Now Try Exercise 17.**

▶ **CAUTION** When solving an equation like the one in Example 7, do not write {0} as the solution set. There are infinitely many other solutions.

EXAMPLE 8 Solving an Equation That Has No Solution

Solve $2x + 3(x + 1) = 5x + 4$.

$$2x + 3(x + 1) = 5x + 4 \qquad \text{Original equation}$$
$$2x + 3x + 3 = 5x + 4 \qquad \text{Distributive property}$$
$$5x + 3 = 5x + 4$$
$$5x + 3 - 5x = 5x + 4 - 5x \qquad \text{Subtract } 5x.$$
$$3 = 4 \qquad \text{False}$$

Solution set: ∅ — This is the symbol for the empty set (or null set).

Again, the variable has disappeared, but this time a false statement ($3 = 4$) results. When this happens in solving an equation, it indicates that the equation has no solution and is called a **contradiction.** Its solution set is the **empty set,** or **null set,** symbolized **∅.**

✔ **Now Try Exercise 21.**

▶ **CAUTION** **DO NOT** write {∅} to represent the empty set.

The following table summarizes the solution sets of the three types of equations presented so far.

Type of Equation	Final Equation in Solution	Number of Solutions	Solution Set
Conditional	$x =$ a number	One	{a number}
Identity	A true statement with no variable, such as $0 = 0$	Infinite	{all real numbers}
Contradiction	A false statement with no variable, such as $3 = 4$	None	$\emptyset$

OBJECTIVE 4 Write expressions for two related unknown quantities.

▶ **PROBLEM-SOLVING HINT** Often we are given a problem in which the sum of two quantities is a particular number, and we are asked to find the values of the two quantities. Example 9 shows how to express the unknown quantities in terms of a single variable.

EXAMPLE 9 Translating a Phrase into an Algebraic Expression

Perform each translation.

(a) Two numbers have a sum of 23. If one of the numbers is represented by k, find an expression for the other number.

First, suppose that the sum of two numbers is 23 and one of the numbers is 10. How would you find the other number? You would subtract 10 from 23 to get 13.

$$23 - 10 = 13$$

So instead of using 10 as one of the numbers, use k as stated in the problem. The other number would be obtained in the same way. You must subtract k from 23. Therefore, an expression for the other number is

$$23 - k.$$

(b) Two numbers have a product of 24. If one of the numbers is represented by k, find an expression for the other number.

The word *product* refers to multiplication. To obtain 24, the number k must be multiplied by

$$\frac{24}{k}.$$

✔ **Now Try Exercises 47 and 49.**

▶ **CAUTION** Since the sum of the two numbers in Example 9(a) is 23, the expression for the other number must be $23 - k$, *not* $k - 23$. To check, find the sum of the two numbers:

$$k + (23 - k) = 23, \quad \text{as required.}$$

2.3 EXERCISES

1. In your own words, give the four steps used to solve a linear equation. Give an example to demonstrate the steps.

2. After working correctly through several steps of the solution of a linear equation, a student obtains the equation $7x = 3x$. Then the student divides each side by x to get $7 = 3$ and gives $\emptyset$ as the answer. Is this correct? If not, explain why.

3. *Concept Check*   Which one of the following equations does *not* have {all real numbers} as its solution set?

A. $5x = 4x + x$ **B.** $2(x + 6) = 2x + 12$ **C.** $\dfrac{1}{2}x = 0.5x$ **D.** $3x = 2x$

4. *Concept Check*   Based on the discussion in this section, if an equation has decimals or fractions as coefficients, what additional step will make the work easier?

Solve each equation, and check your solution. See Examples 1–4, 7, and 8.

5. $3x + 8 = 5x + 10$

6. $10p + 6 = 12p - 4$

7. $12h - 5 = 11h + 5 - h$

8. $-4x - 1 = -5x + 1 + 3x$

9. $3(4x + 2) + 5x = 30 - x$

10. $5(2m + 3) - 4m = 8m + 27$

11. $-2p + 7 = 3 - (5p + 1)$

12. $4x + 9 = 3 - (x - 2)$

13. $6(3w + 5) = 2(10w + 10)$

14. $4(2x - 1) = -6(x + 3)$

15. $6(4x - 1) = 12(2x + 3)$

16. $6(2x + 8) = 4(3x - 6)$

17. $3(2x - 4) = 6(x - 2)$

18. $3(6 - 4x) = 2(-6x + 9)$

19. $7r - 5r + 2 = 5r - r$

20. $9p - 4p + 6 = 7p - 3p$

21. $11x - 5(x + 2) = 6x + 5$

22. $6x - 4(x + 1) = 2x + 4$

Solve each equation by clearing fractions or decimals. Check your solution. See Examples 5 and 6.

23. $\dfrac{3}{5}t - \dfrac{1}{10}t = t - \dfrac{5}{2}$

24. $-\dfrac{2}{7}r + 2r = \dfrac{1}{2}r + \dfrac{17}{2}$

25. $-\dfrac{1}{4}(x - 12) + \dfrac{1}{2}(x + 2) = x + 4$

26. $\dfrac{1}{9}(p + 18) + \dfrac{1}{3}(2p + 3) = p + 3$

27. $\dfrac{2}{3}k - \left(k - \dfrac{1}{2}\right) = \dfrac{1}{6}(k - 51)$

28. $-\dfrac{5}{6}q - (q - 1) = \dfrac{1}{4}(-q + 80)$

29. $0.2(60) + 0.05x = 0.10(60 + x)$

30. $0.30(30) + 0.15x = 0.20(30 + x)$

31. $1.00x + 0.05(12 - x) = 0.10(63)$

32. $0.92x + 0.98(12 - x) = 0.96(12)$

33. $0.6(10,000) + 0.8x = 0.72(10,000 + x)$

34. $0.2(5000) + 0.3x = 0.25(5000 + x)$

Solve each equation, and check your solution. See Examples 1–8.

35. $10(2x - 1) = 8(2x + 1) + 14$

36. $9(3k - 5) = 12(3k - 1) - 51$

37. $-(4x + 2) - (-3x - 5) = 3$

38. $-(6k - 5) - (-5k + 8) = -3$

39. $\frac{1}{2}(x + 2) + \frac{3}{4}(x + 4) = x + 5$

40. $\frac{1}{3}(x + 3) + \frac{1}{6}(x - 6) = x + 3$

41. $0.1(x + 80) + 0.2x = 14$

42. $0.3(x + 15) + 0.4(x + 25) = 25$

43. $4(x + 8) = 2(2x + 6) + 20$

44. $4(x + 3) = 2(2x + 8) - 4$

45. $9(v + 1) - 3v = 2(3v + 1) - 8$

46. $8(t - 3) + 4t = 6(2t + 1) - 10$

Write the answer to each problem in terms of the variable. See Example 9.

47. Two numbers have a sum of 11. One of the numbers is q. What expression represents the other number?

48. Two numbers have a sum of 26. One of the numbers is r. What expression represents the other number?

49. The product of two numbers is 9. One of the numbers is k. What expression represents the other number?

50. The product of two numbers is -3. One of the numbers is m. What expression represents the other number?

51. A football player gained x yards rushing. On the next down, he gained 7 yd. What expression represents the number of yards he gained altogether?

52. A football player gained y yards on a punt return. On the next return, he gained 4 yd. What expression represents the number of yards he gained altogether?

53. A baseball player got 65 hits one season. He got h of the hits in one game. What expression represents the number of hits he got in the rest of the games?

54. A hockey player scored 42 goals in one season. He scored n goals in one game. What expression represents the number of goals he scored in the rest of the games?

55. Monica is a years old. What expression represents her age 12 yr from now? 2 yr ago?

56. Chandler is b years old. What expression represents his age 3 yr ago? 5 yr from now?

57. Tom has r quarters. Express the value of the quarters in cents.

58. Jean has y dimes. Express the value of the dimes in cents.

59. A bank teller has t dollars, all in $5 bills. What expression represents the number of $5 bills the teller has?

60. A clerk has v dollars, all in $20 bills. What expression represents the number of $20 bills the clerk has?

61. A plane ticket costs b dollars for an adult and d dollars for a child. Find an expression that represents the total cost for 3 adults and 2 children.

62. A concert ticket costs c dollars for an adult and f dollars for a child. Find an expression that represents the total cost for 4 adults and 6 children.

PREVIEW EXERCISES

*Write each phrase as a mathematical expression using x as the variable. See **Sections 1.3, 1.5, 1.6, and 1.8.***

63. A number added to -6

64. The sum of a number and twice the number

65. A number decreased by 9

66. The difference between -5 and a number

67. The quotient of -6 and a nonzero number

68. A number divided by 17

69. The product of 12 and the difference between a number and 9

70. The quotient of 9 more than a number and 6 less than the number

Summary Exercises on Solving Linear Equations

This section on miscellaneous linear equations provides practice in solving all the types of equations introduced in **Sections 2.1–2.3.** Refer to the examples in these sections to review the various solution methods.

Solve each equation, and check your solution.

1. $a + 2 = -3$

2. $2m + 8 = 16$

3. $12.5k = -63.75$

4. $-x = -12$

5. $\dfrac{4}{5}x = -20$

6. $7m - 5m = -12$

7. $5x - 9 = 4(x - 3)$

8. $\dfrac{a}{-2} = 8$

9. $-3(m - 4) + 2(5 + 2m) = 29$

10. $\dfrac{2}{3}x + 8 = \dfrac{1}{4}x$

11. $0.08x + 0.06(x + 9) = 1.24$

12. $x - 16.2 = 7.5$

13. $4x + 2(3 - 2x) = 6$

14. $-0.3x + 2.1(x - 4) = -6.6$

15. $-x = 6$

16. $3(m + 5) - 1 + 2m = 5(m + 2)$

17. $7m - (2m - 9) = 39$

18. $7(p - 2) + p = 2(p + 2)$

19. $-2t + 5t - 9 = 3(t - 4) - 5$

20. $-6z = -14$

21. $0.2(50) + 0.8r = 0.4(50 + r)$

22. $2.3x + 13.7 = 1.3x + 2.9$

23. $2(3 + 7x) - (1 + 15x) = 2$

24. $6q - 9 = 12 + 3q$

25. $2(4 + 3r) = 3(r + 1) + 11$

26. $r + 9 + 7r = 4(3 + 2r) - 3$

27. $\dfrac{1}{4}x - 4 = \dfrac{3}{2}x + \dfrac{3}{4}x$

28. $0.6(100 - x) + 0.4x = 0.5(92)$

29. $\dfrac{3}{4}(a - 2) - \dfrac{1}{3}(5 - 2a) = -2$

30. $2 - (m + 4) = 3m + 8$

2.4 An Introduction to Applications of Linear Equations

OBJECTIVE 1 Learn the six steps for solving applied problems. We now look at how algebra is used to solve applied problems. While there is no one specific method that enables you to solve all kinds of applied problems, the following six-step method is often applicable.

Solving an Applied Problem

Step 1 **Read** the problem carefully until you understand what is given and what is to be found.

Step 2 **Assign a variable** to represent the unknown value, using diagrams or tables as needed. Write down what the variable represents. If necessary, express any other unknown values in terms of the variable.

Step 3 **Write an equation** using the variable expression(s).

Step 4 **Solve** the equation.

Step 5 **State the answer.** Does it seem reasonable?

Step 6 **Check** the answer in the words of the *original* problem.

The third step in solving an applied problem is often the hardest. To translate the problem into an equation, write the given phrases as mathematical expressions. Replace any words that mean *equal* or *same* with an $=$ sign. Other forms of the verb "to be," such as *is, are, was,* and *were,* also translate this way. The $=$ sign leads to an equation to be solved.

OBJECTIVE 2 Solve problems involving unknown numbers. Some of the simplest applied problems involve unknown numbers.

EXAMPLE 1 Finding the Value of an Unknown Number

If 4 is multiplied by a number decreased by 7, the product is 100. Find the number.

Step 1 **Read** the problem carefully. We must find an unknown number.

Step 2 **Assign a variable** to represent the unknown quantity. In this problem, we are asked to find a number, so we write

$$\text{Let } x = \text{the number.}$$

There are no other unknown quantities to find.

Step 3 **Write an equation.**

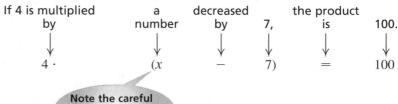

If 4 is multiplied by	a number	decreased by	7,	the product is	100.
↓	↓	↓	↓	↓	↓
$4 \cdot$	$(x$	$-$	$7)$	$=$	100

Note the careful use of parentheses around $x - 7$.

Step 4 **Solve** the equation.

$$4(x - 7) = 100$$
$$4x - 28 = 100 \qquad \text{Distributive property}$$
$$4x - 28 + 28 = 100 + 28 \qquad \text{Add 28.}$$
$$4x = 128 \qquad \text{Combine like terms.}$$
$$\frac{4x}{4} = \frac{128}{4} \qquad \text{Divide by 4.}$$
$$x = 32$$

Step 5 **State the answer.** The number is 32.

Step 6 **Check.** When 32 is decreased by 7, we get $32 - 7 = 25$. If 4 is multiplied by 25, we get 100, as required. The answer, 32, is correct.

✔ **Now Try Exercise 5.**

OBJECTIVE ③ Solve problems involving sums of quantities. A common type of problem in elementary algebra involves finding two quantities when the sum of the quantities is known. In Example 9 of **Section 2.3,** we prepared for this type of problem by writing mathematical expressions for two related unknown quantities.

▶ **PROBLEM-SOLVING HINT** In general, to solve problems involving sums of quantities, choose a variable to represent one of the unknowns. Then represent the other quantity in terms of the *same* variable, using information from the problem. Write an equation based on the words of the problem.

EXAMPLE 2 **Finding Numbers of Olympic Medals**

In the 2004 Olympic Games in Athens, Greece, the United States won 40 more medals than China. The two countries won a total of 166 medals. How many medals did each country win? (*Source:* U.S. Olympic Committee.)

Step 1 **Read** the problem. We are given information about the total number of medals and asked to find the number each country won.

Step 2 **Assign a variable.**

Let $x =$ the number of medals China won.

Then $x + 40 =$ the number of medals the United States won.

Step 3 **Write an equation.**

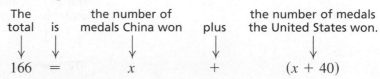

The total	is	the number of medals China won	plus	the number of medals the United States won.
↓	↓	↓	↓	↓
166	=	x	+	$(x + 40)$

Step 4 **Solve the equation.**

$$166 = x + (x + 40)$$

$$166 = 2x + 40 \qquad \text{Combine like terms.}$$

$$166 - 40 = 2x + 40 - 40 \qquad \text{Subtract 40.}$$

$$126 = 2x \qquad \text{Combine like terms.}$$

$$\frac{126}{2} = \frac{2x}{2} \qquad \text{Divide by 2.}$$

$$x = 63$$

Step 5 **State the answer.** The variable x represents the number of medals China won, so China won 63 medals. Then the number of medals the United States won is $x + 40 = 63 + 40 = 103$.

Step 6 **Check.** Since the United States won 103 medals and China won 63, the total number of medals was $103 + 63 = 166$. Because $103 - 63 = 40$, the United States won 40 more medals than China. This information agrees with what is given in the problem, so the answer checks.

✔ **Now Try Exercise 11.**

▶ **NOTE** The problem in Example 2 could also be solved by letting x represent the number of medals the United States won. Then $x - 40$ would represent the number of medals China won. The equation would be

$$166 = x + (x - 40).$$

The solution of this equation is 103, which is the number of U.S. medals. The number of Chinese medals would be $103 - 40 = 63$. The answers are the same, whichever approach is used.

EXAMPLE 3 Finding the Number of Orders for Tea

The owner of Terry's Coffeehouse found that on one day the number of orders for tea was $\frac{1}{3}$ the number of orders for coffee. If the total number of orders for the two drinks was 76, how many orders were placed for tea?

Step 1 **Read** the problem. It asks for the number of orders for tea.

Step 2 **Assign a variable.** Because of the way the problem is stated, let the variable represent the number of orders for coffee.

Let x = the number of orders for coffee.

Then $\frac{1}{3}x$ = the number of orders for tea.

Step 3 **Write an equation.** Use the fact that the total number of orders was 76.

The total	is	orders for coffee	plus	orders for tea.
↓	↓	↓	↓	↓
76	=	x	+	$\frac{1}{3}x$

Step 4 **Solve** the equation.

$$76 = \frac{4}{3}x \qquad \begin{array}{l} x = 1x = \frac{3}{3}x; \\ \text{Combine like terms.} \end{array}$$

$$\frac{3}{4}(76) = \frac{3}{4}\left(\frac{4}{3}x\right) \qquad \text{Multiply by } \tfrac{3}{4}.$$

Be careful! This is *not* the answer.

$$57 = x$$

Step 5 **State the answer.** In this problem, ***x* does not represent the quantity that we are asked to find.** The number of orders for tea was $\frac{1}{3}x$. So $\frac{1}{3}(57) = 19$ is the number of orders for tea.

Step 6 **Check.** The number of orders for coffee (x) was 57 and the number for tea was 19; 19 is one-third of 57, and $19 + 57 = 76$. Since this agrees with the information given in the problem, the answer is correct.

✔ **Now Try Exercise 23.**

> **PROBLEM-SOLVING HINT** In Example 3, it was easier to let the variable represent the quantity that was *not* specified. This required extra work in Step 5 to find the number of orders for tea. In some cases, this approach is easier than letting the variable represent the quantity that we are asked to find.

EXAMPLE 4 Analyzing a Gasoline–Oil Mixture

A lawn trimmer uses a mixture of gasoline and oil. The mixture contains 16 oz of gasoline for each ounce of oil. If the tank holds 68 oz of the mixture, how many ounces of oil and how many ounces of gasoline does it require when it is full?

Step 1 **Read** the problem. We must find how many ounces of oil and gasoline are needed to fill the tank.

Step 2 **Assign a variable.**

Let $x =$ the number of ounces of oil required.

Then $16x =$ the number of ounces of gasoline required.

Step 3 **Write an equation.**

Amount of gasoline	plus	amount of oil	is	total amount in tank.
↓	↓	↓	↓	↓
$16x$	$+$	x	$=$	68

Step 4 **Solve.**

$$17x = 68 \qquad \text{Combine like terms.}$$

$$\frac{17x}{17} = \frac{68}{17} \qquad \text{Divide by 17.}$$

$$x = 4$$

Step 5 **State the answer.** The lawn trimmer requires 4 oz of oil, and $16(4) = 64$ oz of gasoline when full.

Step 6 **Check.** Since $4 + 64 = 68$, and 64 is 16 times 4, the answer checks.

✔ **Now Try Exercise 25.**

> ▶ **PROBLEM-SOLVING HINT** Sometimes it is necessary to find three unknown quantities in an applied problem. Frequently, the three unknowns are compared in *pairs*. When this happens, it is usually easiest to let the variable represent the unknown found in both pairs.

EXAMPLE 5 Dividing a Board into Pieces

The instructions for a woodworking project call for three pieces of wood. The longest piece must be twice the length of the middle-sized piece, and the shortest piece must be 10 in. shorter than the middle-sized piece. Sue Costa has a board 70 in. long that she wishes to use. How long can each piece be?

Step 1 **Read** the problem. There will be three answers.

Step 2 **Assign a variable.** Since the middle-sized piece appears in both pairs of comparisons, let x represent the length, in inches, of the middle-sized piece. We have

$$x = \text{the length of the middle-sized piece,}$$
$$2x = \text{the length of the longest piece, and}$$
$$x - 10 = \text{the length of the shortest piece.}$$

A sketch is helpful here. See Figure 2.

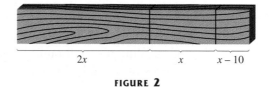

FIGURE **2**

Step 3 **Write an equation.**

Longest	plus	middle-sized	plus	shortest	is	total length.
↓	↓	↓	↓	↓	↓	↓
$2x$	$+$	x	$+$	$(x - 10)$	$=$	70

Step 4 **Solve.**

$$4x - 10 = 70 \qquad \text{Combine like terms.}$$
$$4x - 10 + 10 = 70 + 10 \qquad \text{Add 10.}$$
$$4x = 80 \qquad \text{Combine like terms.}$$
$$\frac{4x}{4} = \frac{80}{4} \qquad \text{Divide by 4.}$$
$$x = 20$$

Step 5 **State the answer.** The middle-sized piece is 20 in. long, the longest piece is $2(20) = 40$ in. long, and the shortest piece is $20 - 10 = 10$ in. long.

Step 6 **Check.** The sum of the lengths is 70 in. All conditions of the problem are satisfied.

✔ **Now Try Exercise 29.**

OBJECTIVE 4 Solve problems involving supplementary and complementary angles. An angle can be measured by a unit called the **degree** (°), which is $\frac{1}{360}$ of a complete rotation. Two angles whose sum is 90° are said to be **complementary,** or complements of each other. An angle that measures 90° is a **right angle.** Two angles whose sum is 180° are said to be **supplementary,** or supplements of each other. One angle *supplements* the other to form a **straight angle** of 180°. See Figure 3. If x represents the degree measure of an angle, then

$$90 - x \text{ represents the degree measure of its complement,}$$

and $\qquad 180 - x$ represents the degree measure of its supplement.

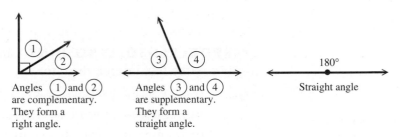

Angles ① and ② are complementary. They form a right angle.

Angles ③ and ④ are supplementary. They form a straight angle.

180°

Straight angle

FIGURE 3

EXAMPLE 6 Finding the Measure of an Angle

Find the measure of an angle whose supplement is 10° more than twice its complement.

Step 1 **Read** the problem. We are to find the measure of an angle, given information about its complement and its supplement.

Step 2 **Assign a variable.**

Let $\qquad\qquad x =$ the degree measure of the angle.

Then $\qquad 90 - x =$ the degree measure of its complement,

and $\qquad 180 - x =$ the degree measure of its supplement.

Step 3 **Write an equation.**

Supplement	is	10	more than	twice	its complement.
↓	↓	↓	↓	↓	↓
$180 - x$	$=$	10	$+$	$2 \cdot$	$(90 - x)$

Be sure to use parentheses around $90 - x$.

Step 4 **Solve.**

$$180 - x = 10 + 180 - 2x \qquad \text{Distributive property}$$
$$180 - x = 190 - 2x \qquad \text{Combine like terms.}$$
$$180 - x + 2x = 190 - 2x + 2x \qquad \text{Add } 2x.$$
$$180 + x = 190 \qquad \text{Combine like terms.}$$
$$180 + x - 180 = 190 - 180 \qquad \text{Subtract 180.}$$
$$x = 10$$

Step 5 **State the answer.** The measure of the angle is 10°.

Step 6 **Check.** The complement of 10° is 80° and the supplement of 10° is 170°. 170° is equal to 10° more than twice 80° ($170 = 10 + 2(80)$ is true); therefore, the answer is correct.

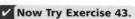 **Now Try Exercise 43.**

OBJECTIVE ⑤ **Solve problems involving consecutive integers.** Two integers that differ by 1 are called **consecutive integers.** For example, 3 and 4, 6 and 7, and −2 and −1 are pairs of consecutive integers. In general, if x represents an integer, $x + 1$ represents the next larger consecutive integer.

Consecutive *even* integers, such as 8 and 10, differ by 2. Similarly, consecutive *odd* integers, such as 9 and 11, also differ by 2. In general, if x represents an even integer, $x + 2$ represents the next larger consecutive even integer. The same holds true for odd integers; that is, if x is an odd integer, $x + 2$ is the next larger odd integer. In this book, we list consecutive integers in increasing order when solving applications.

▶ **PROBLEM-SOLVING HINT** In solving consecutive integer problems, if x = the first integer, then, for any

two consecutive integers, use	$x,\ x + 1$;
two consecutive *even* integers, use	$x,\ x + 2$;
two consecutive *odd* integers, use	$x,\ x + 2$.

EXAMPLE 7 Finding Consecutive Integers

Two pages that face each other in this book have 249 as the sum of their page numbers. What are the page numbers?

Step 1 **Read** the problem. Because the two pages face each other, they must have page numbers that are consecutive integers.

Step 2 **Assign a variable.**

Let x = the lesser page number.

Then $x + 1$ = the greater page number.

Step 3 **Write an equation.** Because the sum of the page numbers is 249, the equation is

$$x + (x + 1) = 249.$$

Step 4 **Solve.**

$$2x + 1 = 249 \quad \text{Combine like terms.}$$
$$2x = 248 \quad \text{Subtract 1.}$$
$$x = 124 \quad \text{Divide by 2.}$$

Step 5 **State the answer.** The lesser page number is 124, and the greater page number is $124 + 1 = 125$. (Your book is opened to these two pages.)

Step 6 **Check.** The sum of 124 and 125 is 249. The answer is correct.

✔ **Now Try Exercise 51.**

In the final example, we do not number the steps. See if you can identify them.

EXAMPLE 8 Finding Consecutive Odd Integers

If the lesser of two consecutive odd integers is doubled, the result is 7 more than the greater of the two integers. Find the two integers.

Let x be the lesser integer. Since the two numbers are consecutive *odd* integers, then $x + 2$ is the greater. Now we write an equation.

If the lesser is doubled,	the result is	7	more than	the greater.
↓	↓	↓	↓	↓
$2x$	$=$	7	$+$	$(x + 2)$

$$2x = 9 + x \qquad \text{Combine like terms.}$$
$$x = 9 \qquad \text{Subtract } x.$$

The lesser integer is 9 and the greater is $9 + 2 = 11$. When 9 is doubled, we get 18, which is 7 more than the greater odd integer, 11. The answers are correct.

✔ **Now Try Exercise 49.**

CONNECTIONS

George Polya (1888–1985), a native of Budapest, Hungary, wrote the modern classic *How to Solve It.* In this book, he proposed a four-step process for problem solving:

Step 1 Understand the problem.   *Step 2* Devise a plan.

Step 3 Carry out the plan.   *Step 4* Look back and check.

For Discussion or Writing

Compare Polya's four-step process with the six steps given in this section. Identify which of the steps in our list match Polya's four steps. Trial and error is also a useful problem-solving tool. Where does this method fit into Polya's steps?

2.4 EXERCISES

⊙ *Complete solution available on Video Lectures on CD/DVD*

▢ *Now Try Exercise*

1. *Concept Check* Suppose that a problem requires you to find the number of cars on a dealer's lot. Which one of the following would *not* be a reasonable answer? Justify your response.

   A. 0   **B.** 45   **C.** 1   **D.** $6\frac{1}{2}$

2. *Concept Check* Suppose that a problem requires you to find the number of hours a lightbulb is on during a day. Which one of the following would *not* be a reasonable answer? Justify your response.

   A. 0   **B.** 4.5   **C.** 13   **D.** 25

3. *Concept Check* Suppose that a problem requires you to find the distance traveled in miles. Which one of the following would *not* be a reasonable answer? Justify your response.

   A. -10 **B.** 1.8   **C.** $10\frac{1}{2}$ **D.** 50

4. *Concept Check* Suppose that a problem requires you to find the time in minutes. Which one of the following would *not* be a reasonable answer? Justify your response.

A. 0 **B.** 10.5 **C.** −5 **D.** 90

Solve each problem. See Example 1.

5. If 2 is added to five times a number, the result is equal to 5 more than four times the number. Find the number.

6. If four times a number is added to 8, the result is three times the number added to 5. Find the number.

7. If 2 is subtracted from a number and this difference is tripled, the result is 6 more than the number. Find the number.

8. If 3 is added to a number and this sum is doubled, the result is 2 more than the number. Find the number.

9. The sum of three times a number and 7 more than the number is the same as the difference between −11 and twice the number. What is the number?

10. If 4 is added to twice a number and this sum is multiplied by 2, the result is the same as if the number is multiplied by 3 and 4 is added to the product. What is the number?

Solve each problem. See Example 2.

11. The number of drive-in movie screens has declined steadily in the United States since the 1960s. California and New York were two of the states with the most screens remaining in 2001. California had 11 more than New York, and there were 107 in all in the two states. How many drive-in movie screens remained in each state? (*Source:* National Association of Theatre Owners.)

12. As of 2005, the two most highly watched episodes in the history of television were the final episode of *M*A*S*H*, broadcast on February 23, 1983, and the "Who Shot J. R.?" episode of *Dallas*, broadcast on November 21, 1980. The total number of viewers for these two episodes was about 91 million, with 9 million more people watching the *M*A*S*H* episode than the *Dallas* one. How many people watched each show? (*Source:* Nielsen Media Research.)

13. During the 109th Congress (2005–2006), the U.S. Senate had a total of 99 Democrats and Republicans. There were 11 more Republicans than Democrats. How many Democrats and Republicans were there in the Senate? (*Source: World Almanac and Book of Facts 2006.*)

14. The total number of Democrats and Republicans in the U.S. House of Representatives during the 109th Congress was 434. There were 30 more Republicans than Democrats. How many members of each party were there? (*Source: World Almanac and Book of Facts 2006.*)

15. Bruce Springsteen and the E Street Band generated top revenue on the concert circuit in 2003. Springsteen and second-place Céline Dion together took in $196.4 million from ticket sales. If Céline Dion took in $35.4 million less than Bruce Springsteen and the E Street Band, how much revenue did each generate? (*Source: Parade,* February 15, 2004.)

16. The Toyota Camry was the top-selling passenger car in the United States in 2004, followed by the Honda Accord. Honda Accord sales were 40 thousand less than Toyota Camry sales, and 814 thousand of the two types of cars were sold. How many of each make of car were sold? (*Source:* Ward's Communications.)

17. In the 2004–2005 NBA regular season, the Phoenix Suns won two more than three times as many games as they lost. The Suns played 82 games. How many wins and losses did the team have? (*Source: World Almanac and Book of Facts 2006.*)

18. In the 2005 regular baseball season, the Chicago White Sox won 27 fewer than twice as many games as they lost. They played 162 regular-season games. How many wins and losses did the team have? (*Source: World Almanac and Book of Facts 2006.*)

Solve each problem. See Examples 3 and 4.

19. The value of a "Mint State-63" (uncirculated) 1950 Jefferson nickel minted at Denver is $\frac{8}{7}$ the value of a 1945 nickel in similar condition minted at Philadelphia. Together, the total value of the two coins is $15.00. What is the value of each coin? (*Source:* Yeoman, R., *A Guide Book of United States Coins,* edited by K. Bressett, 56th edition, 2003.)

20. The world's most populous countries are China and India. As of mid-2005, the combined population of these two countries was estimated at 2.4 billion. If there were about $\frac{4}{5}$ as many people living in India as China, what was the population of each country, to the nearest tenth of a billion? (*Source:* U.S. Census Bureau.)

21. On March 8, 2003, the world's largest taco was made in the city of Mexicali, Baja California, Mexico. The taco contained approximately 1 kg of onion for every 6.6 kg of grilled steak. The total weight of these two ingredients was 617.6 kg. To the nearest tenth of a kilogram, how many kilograms of onions and how many kilograms of grilled beef were used to make the taco? (*Source: Guinness World Records.*)

22. Arnie Waldman has chosen a workout that combines weight training and aerobics. If he does 30 min of weight training, followed by a 30-min aerobics class, he will burn a total of 374 calories. If he burns $\frac{12}{5}$ as many calories doing aerobics as doing weight training, how many calories will he burn in each activity?

23. In one day, a store sold $\frac{8}{5}$ as many DVDs as CDs. The total number of DVDs and CDs sold that day was 273. How many DVDs were sold?

24. U.S. five-cent coins are made from a combination of two metals: nickel and copper. For every pound of nickel, 3 lb of copper are used. How many pounds of copper would be needed to make 560 lb of five-cent coins? (*Source:* The United States Mint.)

25. A bakery makes a special whole-grain bread using two kinds of flour: whole wheat and rye. The recipe for this bread calls for 1 oz of rye flour for every 4 oz of whole-wheat flour. How many ounces of each kind of flour should be used to make a loaf of bread weighing 32 oz?

26. Juliet Carl read the label on the bottle of an over-the-counter medication she is taking and found that the medication contains 9 mg of active ingredients for every 1 mg of inert ingredients. How much of each kind of ingredient would be contained in a single 250-mg caplet?

Solve each problem. See Example 5.

27. Al Moser is an office manager. His responsibilities include booking airline tickets for the business trips that employees of his company need to make. In one week, he booked 55 tickets, divided among three airlines. He booked 7 more tickets on American Airlines than United Airlines. On Southwest Airlines, he booked 4 more than twice as many tickets as on United. How many tickets did he book on each airline?

28. In her job as a mathematics textbook editor, Lauren Morse works 7.5 hr a day. She spent a recent day making telephone calls, writing e-mails, and attending meetings. On that day, she spent twice as much time attending meetings as making telephone calls and spent 0.5 hr longer writing e-mails than making telephone calls. How many hours did she spend on each task?

29. Nagaraj Nanjappa has a party-length submarine sandwich 59 in. long. He wants to cut it into three pieces so that the middle piece is 5 in. longer than the shortest piece and the shortest piece is 9 in. shorter than the longest piece. How long should the three pieces be?

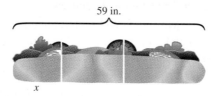

59 in.

x

30. The United States earned a total of 103 medals at the 2004 Athens Olympics. The number of gold medals earned was 6 more than the number of bronze medals. The number of silver medals earned was 10 more than the number of bronze medals. How many of each kind of medal did the United States earn? (*Source:* U.S. Olympic Committee.)

31. Venus is 31.2 million mi farther from the sun than Mercury, while Earth is 57 million mi farther from the sun than Mercury. If the total of the distances from these three planets to the sun is 196.2 million mi, how far away from the sun is Mercury? (All distances given here are *mean* (*average*) distances.) (*Source: The New York Times Almanac 2006.*)

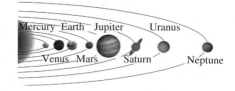

32. Together, Saturn, Jupiter, and Uranus have a total of 137 known satellites (moons). Jupiter has 16 more satellites than Saturn, and Uranus has 20 fewer satellites than Saturn. How many known satellites does Uranus have? (*Source: The New York Times Almanac 2006.*)

33. The sum of the measures of the angles of any triangle is 180°. In triangle *ABC*, angles *A* and *B* have the same measure, while the measure of angle *C* is 60° greater than each of *A* and *B*. What are the measures of the three angles?

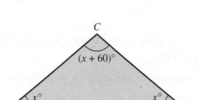

34. In triangle *ABC*, the measure of angle *A* is 141° more than the measure of angle *B*. The measure of angle *B* is the same as the measure of angle *C*. Find the measure of each angle. (*Hint:* See Exercise 33.)

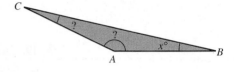

Concept Check   *Answer each question.*

35. If the sum of two numbers is *k* and one of the numbers is *m*, how can you express the other number?

36. If the product of two numbers is *r* and one of the numbers is *s* ($s \neq 0$), how can you express the other number?

37. Is there an angle whose supplement is equal to its complement? If so, what is the measure of the angle?

38. Is there an angle that is equal to its supplement? Is there an angle that is equal to its complement? If the answer is yes to either question, give the measure of the angle.

39. If *x* represents an integer, how can you express the next smaller consecutive integer in terms of *x*?

40. If *x* represents an even integer, how can you express the next smaller even integer in terms of *x*?

Solve each problem. See Example 6.

41. Find the measure of an angle whose complement is four times its measure.

42. Find the measure of an angle whose supplement is three times its measure.

43. Find the measure of an angle whose supplement measures 39° more than twice its complement.

44. Find the measure of an angle whose supplement measures 38° less than three times its complement.

45. Find the measure of an angle such that the difference between the measures of its supplement and three times its complement is 10°.

46. Find the measure of an angle such that the sum of the measures of its complement and its supplement is 160°.

Solve each problem. See Examples 7 and 8.

47. The numbers on two consecutively numbered gym lockers have a sum of 137. What are the locker numbers?

48. The sum of two consecutive checkbook check numbers is 357. Find the numbers.

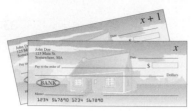

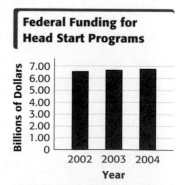

49. Find two consecutive even integers such that the lesser added to three times the greater gives a sum of 46.

50. Find two consecutive odd integers such that twice the greater is 17 more than the lesser.

51. Two pages that are back-to-back in this book have 203 as the sum of their page numbers. What are the page numbers?

52. Two houses on the same side of the street have house numbers that are consecutive even integers. The sum of the integers is 58. What are the two house numbers?

53. When the lesser of two consecutive integers is added to three times the greater, the result is 43. Find the integers.

54. If five times the lesser of two consecutive integers is added to three times the greater, the result is 59. Find the integers.

55. If the sum of three consecutive even integers is 60, what is the first of the three even integers? (*Hint:* If x and $x + 2$ represent the first two consecutive even integers, how would you represent the third consecutive even integer?)

56. If the sum of three consecutive odd integers is 69, what is the third of the three odd integers?

57. If 6 is subtracted from the third of three consecutive odd integers and the result is multiplied by 2, the answer is 23 less than the sum of the first and twice the second of the integers. Find the integers.

58. If the first and third of three consecutive even integers are added, the result is 22 less than three times the second integer. Find the integers.

Apply the ideas of this section to solve Exercises 59 and 60, which are based on the graphs.

59. In 2003, the federal funding for Head Start programs increased by $0.13 billion from the previous year. The increase from 2003 to 2004 was $0.10 billion. Over the three-year period 2002–2004, the total funding was $19.98 billion. What was the federal Head Start funding for each of these years? (*Source:* U.S. Department of Health and Human Services.)

Federal Funding for Head Start Programs

(Bar graph: Billions of Dollars vs. Year — 2002, 2003, 2004)

60. Boatbuilding, iron foundries, and amusement parks and arcades are three of the industries with the highest rates of occupational injuries. The graph shows the number of nonfatal injuries per 1000 full-time workers in 2004 in these three industries. In a typical group of 1000 workers from each of these industries, there were 30 more injuries in iron foundries (I) than in amusement parks and arcades (A), and there were 12 more injuries in amusement parks and arcades than in boatbuilding (B). Among these 3000 workers, there were 387 nonfatal occupational injuries. How many of the injuries took place in each industry? (*Source:* U.S. Bureau of Labor Statistics.)

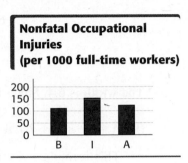

PREVIEW EXERCISES

*Use the given values to evaluate each expression. See **Section 1.3.***

61. LW; $\quad L = 6, W = 4$

62. rt; $\quad r = 25, t = 4.5$

63. $2L + 2W$; $\quad L = 8, W = 2$

64. prt; $\quad p = 4000, r = 0.04, t = 2$

65. $\dfrac{1}{2}Bh$; $\quad B = 27, h = 6$

66. $\dfrac{1}{2}h(b + B)$; $\quad h = 10, b = 4, B = 12$

2.5 Formulas and Applications from Geometry

OBJECTIVES

1 Solve a formula for one variable, given the values of the other variables.

2 Use a formula to solve an applied problem.

3 Solve problems involving vertical angles and straight angles.

4 Solve a formula for a specified variable.

Many applied problems can be solved with *formulas*. A **formula** is an equation in which variables are used to describe a relationship. For example, formulas exist for geometric figures such as squares and circles, for distance, for money earned on bank savings, and for converting English measurements to metric measurements.

$$P = 4s, \quad A = \pi r^2, \quad I = prt, \quad F = \frac{9}{5}C + 32 \qquad \text{Formulas}$$

The formulas used in this book are given on the inside covers.

OBJECTIVE 1 Solve a formula for one variable, given the values of the other variables. Given the values of all but one of the variables in a formula, we can find the value of the remaining variable. In Example 1, we use the idea of *area*. The **area** of a plane (two-dimensional) geometric figure is a measure of the surface covered by the figure.

EXAMPLE 1 Using Formulas to Evaluate Variables

Find the value of the remaining variable in each formula.

(a) $A = LW$; $A = 64$, $L = 10$

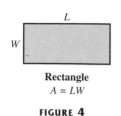

L

W

Rectangle
$A = LW$

FIGURE 4

As shown in Figure 4, this formula gives the area of a rectangle with length L and width W. Substitute the given values into the formula and then solve for W.

$$A = LW$$
$$64 = 10W \qquad \text{Let } A = 64 \text{ and } L = 10.$$
$$\frac{64}{10} = \frac{10W}{10} \qquad \text{Divide by 10.}$$
$$6.4 = W$$

The width is 6.4. Since $10(6.4) = 64$, the given area, the answer checks.

(b) $A = \dfrac{1}{2}h(b + B)$; $A = 210$, $B = 27$, $h = 10$

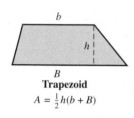

b

h

B

Trapezoid
$A = \frac{1}{2}h(b + B)$

FIGURE 5

This formula gives the area of a trapezoid with parallel sides of lengths b and B and distance h between the parallel sides. See Figure 5. Again, begin by substituting the given values into the formula.

$$A = \frac{1}{2}h(b + B)$$
$$210 = \frac{1}{2}(10)(b + 27) \qquad A = 210, h = 10, B = 27$$

Now solve for b.

$$210 = 5(b + 27) \qquad \text{Multiply.}$$
$$210 = 5b + 135 \qquad \text{Distributive property}$$
$$210 - 135 = 5b + 135 - 135 \qquad \text{Subtract 135.}$$
$$75 = 5b \qquad \text{Combine like terms.}$$
$$\frac{75}{5} = \frac{5b}{5} \qquad \text{Divide by 5.}$$
$$15 = b$$

The length of the shorter parallel side, b, is 15. Since $\frac{1}{2}(10)(15 + 27) = 210$, the given area, the answer checks.

✔ **Now Try Exercises 19 and 23.**

OBJECTIVE 2 Use a formula to solve an applied problem. Formulas are often used in applied problems. *It is a good idea to draw a sketch when a geometric figure is involved.* Examples 2 and 3 use the idea of *perimeter*. The **perimeter** of a plane (two-dimensional) geometric figure is the distance around the figure. For a polygon (e.g., a rectangle, square, or triangle), it is the sum of the lengths of its sides. We use the six steps introduced in the previous section.

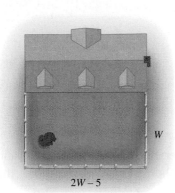

2W − 5

FIGURE 6

EXAMPLE 2 Finding the Dimensions of a Rectangular Yard

Cathleen Horne's backyard is in the shape of a rectangle. The length is 5 m less than twice the width, and the perimeter is 80 m. Find the dimensions of the yard.

Step 1   **Read** the problem. We must find the dimensions of the yard.

Step 2   **Assign a variable.** Let W = the width of the lot, in meters. Since the length is 5 meters less than twice the width, the length $L = 2W − 5$. See Figure 6.

Step 3   **Write an equation.** The formula for the perimeter of a rectangle is

$$P = 2L + 2W.$$

Perimeter   = 2 · Length   + 2 · Width

80   = 2(2W − 5)   +   2W   Substitute 2W − 5 for length L.

Step 4   **Solve** the equation.

$$80 = 4W − 10 + 2W \qquad \text{Distributive property}$$
$$80 = 6W − 10 \qquad \text{Combine like terms.}$$
$$80 + 10 = 6W − 10 + 10 \qquad \text{Add 10.}$$
$$90 = 6W \qquad \text{Combine like terms.}$$
$$\frac{90}{6} = \frac{6W}{6} \qquad \text{Divide by 6.}$$
$$15 = W$$

Step 5   **State the answer.** The width is 15 m and the length is $2(15) − 5 = 25$ m.

Step 6   **Check.** If the width of the yard is 15 m and the length is 25 m, the perimeter is $2(25) + 2(15) = 50 + 30 = 80$ m, as required.

✔ **Now Try Exercise 41.**

EXAMPLE 3 Finding the Dimensions of a Triangle

The longest side of a triangle is 3 ft longer than the shortest side. The medium side is 1 ft longer than the shortest side. If the perimeter of the triangle is 16 ft, what are the lengths of the three sides?

Step 1   **Read** the problem. We are given the perimeter of a triangle and want to find the lengths of the three sides.

Step 2   **Assign a variable.**

Let s = the length of the shortest side, in feet,

$s + 1$ = the length of the medium side, in feet, and,

$s + 3$ = the length of the longest side in feet.

See Figure 7.

Step 3   **Write an equation.** Use the formula for the perimeter of a triangle.

$$P = a + b + c$$
$$16 = s + (s + 1) + (s + 3) \qquad \text{Substitute.}$$

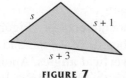

s
s + 1
s + 3

FIGURE 7

Step 4 **Solve.**

$$16 = s + (s + 1) + (s + 3)$$
$$16 = 3s + 4 \qquad \text{Combine like terms.}$$
$$12 = 3s \qquad \text{Subtract 4.}$$
$$4 = s \qquad \text{Divide by 3.}$$

Step 5 **State the answer.** Since s represents the length of the shortest side, its measure is 4 ft. The medium side measures $s + 1 = 4 + 1 = 5$ ft, and the longest side measure $s + 3 = 4 + 3 = 7$ ft.

Step 6 **Check.** The medium side, 5 ft, is 1 ft longer than the shortest side, and the longest side, 7 ft, is 3 ft longer than the shortest side. Futhermore, the perimeter is $4 + 5 + 7 = 16$ ft, as required.

✔ **Now Try Exercise 43.**

FIGURE 8

EXAMPLE 4 Finding the Height of a Triangular Sail

The area of a triangular sail of a sailboat is 126 ft². (Recall that "ft²" means "square feet.") The base of the sail is 12 ft. Find the height of the sail.

Step 1 **Read.** We must find the height of the triangular sail.

Step 2 **Assign a variable.** Let h = the height of the sail, in feet. See Figure 8.

Step 3 **Write an equation.** The formula for the area of a triangle is $A = \frac{1}{2}bh$, where A is the area, b is the base, and h is the height. Using the information given in the problem, we substitute 126 for A and 12 for b in the formula.

$$A = \frac{1}{2}bh$$

$$126 = \frac{1}{2}(12)h \qquad A = 126, b = 12$$

Step 4 **Solve.**

$$126 = 6h \qquad \text{Multiply.}$$
$$21 = h \qquad \text{Divide by 6.}$$

Step 5 **State the answer.** The height of the sail is 21 ft.

Step 6 **Check** to see that the values $A = 126, b = 12,$ and $h = 21$ satisfy the formula for the area of a triangle.

✔ **Now Try Exercise 51.**

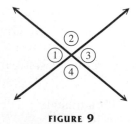

FIGURE 9

OBJECTIVE 3 Solve problems involving vertical angles and straight angles. Figure 9 shows two intersecting lines forming angles that are numbered ①, ②, ③, and ④. Angles ① and ③ lie "opposite" each other. They are called **vertical angles.** Another pair of vertical angles is ② and ④. In geometry, it is shown that vertical angles have equal measures.

Now look at angles ① and ②. When their measures are added, we get 180°, the measure of a straight angle. There are three other such pairs of angles: ② and ③, ③ and ④, and ① and ④.

The next example uses these ideas.

EXAMPLE 5 Finding Angle Measures

Refer to the appropriate figure in each part.

(a) Find the measure of each marked angle in Figure 10.

Since the marked angles are vertical angles, they have equal measures. Set $4x + 19$ equal to $6x - 5$ and solve.

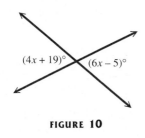

FIGURE 10

$$4x + 19 = 6x - 5$$
$$19 = 2x - 5 \qquad \text{Subtract } 4x.$$
$$24 = 2x \qquad \text{Add 5.}$$
$$12 = x \qquad \text{Divide by 2.}$$

Since $x = 12$, one angle has measure $4(12) + 19 = 67$ degrees. The other has the same measure, since $6(12) - 5 = 67$ as required. Each angle measures $67°$.

(b) Find the measure of each marked angle in Figure 11.

The measures of the marked angles must add to $180°$ because together they form a straight angle. The equation to solve is

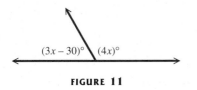

FIGURE 11

$$(3x - 30) + 4x = 180.$$
$$7x - 30 = 180 \qquad \text{Combine like terms.}$$
$$7x = 210 \qquad \text{Add 30.}$$
$$x = 30 \qquad \text{Divide by 7.}$$

Don't stop here!

To find the measures of the angles, replace x with 30 in the two expressions.

$$3x - 30 = 3(30) - 30 = 90 - 30 = 60$$
$$4x = 4(30) = 120$$

The two angle measures are $60°$ and $120°$.

✔ **Now Try Exercises 57 and 59.**

▶ **CAUTION** In Example 5, the answer is *not* the value of x. Remember to substitute the value of the variable into the expression given for each angle.

OBJECTIVE 4 **Solve a formula for a specified variable.** Sometimes it is necessary to solve a number of problems that use the same formula. For example, a surveying class might need to solve several problems that involve the formula for the area of a rectangle, $A = LW$. Suppose that in each problem the area (A) and the length (L) of a rectangle are given, and the width (W) must be found. Rather than solving for W each time the formula is used, it would be simpler to *rewrite the formula* so that it is solved for W. This process is called **solving for a specified variable** or **solving a literal equation.**

In solving a formula for a specified variable, we treat the specified variable as if it were the *only* variable in the equation and treat the other variables as if they were numbers. We use the same steps to solve the equation for the specified variable that we use to solve equations with just one variable.

EXAMPLE 6 Solving for a Specified Variable

Solve $A = LW$ for W.

Think of undoing what has been done to W. Since W is multiplied by L, undo the multiplication by dividing each side of $A = LW$ by L.

$$A = LW$$

$$\frac{A}{L} = \frac{LW}{L} \qquad\qquad \text{Divide by } L.$$

$$\frac{A}{L} = W, \quad \text{or} \quad W = \frac{A}{L} \qquad \frac{L}{L} = 1; \, 1W = W$$

✔ **Now Try Exercise 63.**

EXAMPLE 7 Solving for a Specified Variable

Solve $P = 2L + 2W$ for L.

We want to isolate L on one side of the equation.

$$P = 2L + 2W$$

$$P - 2W = 2L + 2W - 2W \qquad\qquad \text{Subtract } 2W.$$

$$P - 2W = 2L \qquad\qquad \text{Combine like terms.}$$

$$\frac{P - 2W}{2} = \frac{2L}{2} \qquad\qquad \text{Divide by 2.}$$

$$\frac{P - 2W}{2} = L, \quad \text{or} \quad L = \frac{P - 2W}{2} \qquad \frac{2}{2} = 1; \, 1L = L$$

✔ **Now Try Exercise 79.**

EXAMPLE 8 Solving for a Specified Variable

Solve $F = \frac{9}{5}C + 32$ for C. (This is the formula for converting from Celsius to Fahrenheit.)

We need to isolate C on one side of the equation. First we undo the addition of 32 to $\frac{9}{5}C$ by subtracting 32 from each side.

$$F = \frac{9}{5}C + 32$$

$$F - 32 = \frac{9}{5}C + 32 - 32 \qquad\qquad \text{Subtract 32.}$$

$$F - 32 = \frac{9}{5}C$$

$$\frac{5}{9}(F - 32) = \frac{5}{9} \cdot \frac{9}{5}C \qquad\qquad \text{Multiply by } \frac{5}{9}.$$

Be sure to use parentheses.

$$\frac{5}{9}(F - 32) = C, \quad \text{or} \quad C = \frac{5}{9}(F - 32)$$

This last result is the formula for converting temperatures from Fahrenheit to Celsius.

✔ **Now Try Exercise 81.**

EXAMPLE 9 Solving for a Specified Variable

Solve $A = \dfrac{1}{2}h(b + B)$ for B.

$$A = \frac{1}{2}h(b + B)$$

$$2A = 2 \cdot \frac{1}{2}h(b + B) \qquad \text{Multiply by 2 to clear the fraction.}$$

Multiplying 2 times $\frac{1}{2}$ here is not an application of the distributive property.

$$2A = h(b + B) \qquad 2 \cdot \frac{1}{2} = \frac{2}{2} = 1$$

$$2A = hb + hB \qquad \text{Distributive property}$$

$$2A - hb = hb + hB - hb \qquad \text{Subtract } hb.$$

$$2A - hb = hB \qquad \text{Combine like terms.}$$

$$\frac{2A - hb}{h} = \frac{hB}{h} \qquad \text{Divide by } h.$$

$$\frac{2A - hb}{h} = B, \quad \text{or} \quad B = \frac{2A - hb}{h}$$

✔ **Now Try Exercise 85.**

▶ **NOTE**   The result in Example 9 can be written in a different form as follows:

$$B = \frac{2A - hb}{h} = \frac{2A}{h} - \frac{hb}{h} = \frac{2A}{h} - b.$$

Either form is correct.

2.5 EXERCISES

Complete solution available on Video Lectures on CD/DVD

Now Try Exercise

1. In your own words, explain what is meant by each term.

    (a) Perimeter of a plane geometric figure
    (b) Area of a plane geometric figure

2. *Concept Check*   In parts (a)–(c), choose one of the following words to make the statement true: *linear, square,* or *cubic.*

    (a) If the dimensions of a plane geometric figure are given in feet, then the **area** is given in _____ feet.

    (b) If the dimensions of a rectangle are given in yards, then the **perimeter** is given in _____ yards.

    (c) If the dimensions of a pyramid are given in meters, then the **volume** is given in _____ meters.

3. *Concept Check*   If a formula has exactly five variables, how many values would you need to be given in order to find the value of any one variable?

4. *Concept Check*   The formula for changing Celsius to Fahrenheit is given in Example 8 as $F = \frac{9}{5}C + 32$. Sometimes it is seen as $F = \frac{9C}{5} + 32$. These are both correct. Why is it true that $\frac{9}{5}C$ is equal to $\frac{9C}{5}$?

Concept Check   *Decide whether the perimeter or area would be used to solve a problem concerning the measure of the quantity.*

5. Carpeting for a bedroom

6. Sod for a lawn

7. Fencing for a yard

8. Baseboards for a living room

9. Tile for a bathroom

10. Fertilizer for a garden

11. Determining the cost of replacing a linoleum floor with a wood floor

12. Determining the cost of planting rye grass in a lawn for the winter

In the following exercises, a formula is given along with the values of all but one of the variables in the formula. Find the value of the variable that is not given. If appropriate, use 3.14 as an approximation for π (pronounced "pie"). See Example 1.

13. $P = 2L + 2W$ (perimeter of a rectangle); $L = 8, W = 5$

14. $P = 2L + 2W$; $L = 6, W = 4$

15. $A = \dfrac{1}{2}bh$ (area of a triangle); $b = 8, h = 16$

16. $A = \dfrac{1}{2}bh$; $b = 10, h = 14$

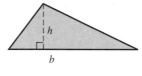

17. $P = a + b + c$ (perimeter of a triangle); $P = 12, a = 3, c = 5$

18. $P = a + b + c$; $P = 15, a = 3, b = 7$

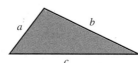

🌐 **19.** $d = rt$ (distance formula); $d = 252, r = 45$

20. $d = rt$; $d = 100, t = 2.5$

21. $I = prt$ (simple interest); $p = 7500, r = 0.035, t = 6$

22. $I = prt$; $p = 5000, r = 0.025, t = 7$

23. $A = \dfrac{1}{2}h(b + B)$ (area of a trapezoid); $A = 91, h = 7, b = 12$

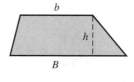

24. $A = \dfrac{1}{2}h(b + B)$; $A = 75, b = 19, B = 31$

25. $C = 2\pi r$ (circumference of a circle); $C = 16.328$

26. $C = 2\pi r$; $C = 8.164$

27. $C = 2\pi r$; $C = 20\pi$

28. $C = 2\pi r$; $C = 100\pi$

29. $A = \pi r^2$ (area of a circle); $r = 4$

30. $A = \pi r^2$; $r = 12$

31. $S = 2\pi rh$; $S = 120\pi, h = 10$

32. $S = 2\pi rh$; $S = 720\pi, h = 30$

The **volume** of a three-dimensional object is a measure of the space occupied by the object. For example, we would need to know the volume of a gasoline tank in order to know how many gallons of gasoline it would take to completely fill the tank. In the following exercises, a formula for the volume (V) of a three-dimensional object is given, along with values for the other variables. Evaluate V. (Use 3.14 as an approximation for π.) See Example 1.

33. $V = LWH$ (volume of a rectangular box); $L = 10, W = 5, H = 3$

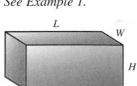

34. $V = LWH$; $L = 12, W = 8, H = 4$

35. $V = \dfrac{1}{3}Bh$ (volume of a pyramid); $B = 12$, $h = 13$

36. $V = \dfrac{1}{3}Bh$; $B = 36$, $h = 4$

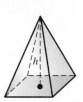

37. $V = \dfrac{4}{3}\pi r^3$ (volume of a sphere); $r = 12$

38. $V = \dfrac{4}{3}\pi r^3$; $r = 6$

Solve each perimeter problem. See Examples 2 and 3.

39. The length of a rectangle is 9 in. more than the width. The perimeter is 54 in. Find the length and the width of the rectangle.

40. The width of a rectangle is 3 ft less than the length. The perimeter is 62 ft. Find the length and the width of the rectangle.

41. The perimeter of a rectangle is 36 m. The length is 2 m more than three times the width. Find the length and the width of the rectangle.

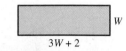

42. The perimeter of a rectangle is 36 yd. The width is 18 yd less than twice the length. Find the length and the width of the rectangle.

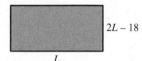

43. The longest side of a triangle is 3 in. longer than the shortest side. The medium side is 2 in. longer than the shortest side. If the perimeter of the triangle is 20 in., what are the lengths of the three sides?

44. The perimeter of a triangle is 28 ft. The medium side is 4 ft longer than the shortest side, while the longest side is twice as long as the shortest side. What are the lengths of the three sides?

45. Two sides of a triangle have the same length. The third side measures 4 m less than twice that length. The perimeter of the triangle is 24 m. Find the lengths of the three sides.

46. A triangle is such that its medium side is twice as long as its shortest side and its longest side is 7 yd less than three times its shortest side. The perimeter of the triangle is 47 yd. What are the lengths of the three sides?

*Use a formula to write an equation for each application, and then use the problem-solving method of **Section 2.4** to solve. (Use 3.14 as an approximation for π.) Formulas are found on the inside covers of this book. See Examples 2–4.*

47. In 1997, a prehistoric ceremonial site dating to about 3000 B.C. was discovered at Stanton Drew in southwestern England. The site, which is larger than Stonehenge, is a nearly perfect circle, consisting of nine concentric rings that probably held upright wooden posts. Around this timber temple is a wide, encircling ditch enclosing an area with a diameter of 443 ft. Find this enclosed area to the nearest thousand square feet. (*Hint:* Find the radius. Then use $A = \pi r^2$.) (*Source: Archaeology,* vol. 51, no. 1, Jan./Feb. 1998.)

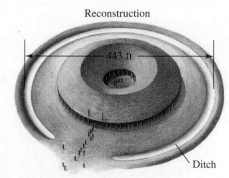

Reconstruction

443 ft

Ditch

48. The Rogers Centre (formerly the SkyDome) in Toronto, Canada, opened on June 5, 1989. It was the first stadium with a hard-shell, retractable roof. The steel dome is 630 ft in diameter. To the nearest foot, what is the circumference of this dome? (*Source:* www.ballparks.com)

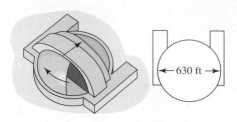

49. The largest fashion catalogue in the world was published on August 30, 2003, in Hamburg, Germany. Each of the 212 pages in the catalogue measured 1.2 m by 1.5 m. What was the perimeter of a page? What was the area? (*Source: Guinness World Records.*)

50. The world's largest sand painting was created by Buddhist monks in the Singapore Expo Hall in May 2004. The painting measured 12.24 m by 12.24 m. What was the perimeter of the sand painting? To the nearest hundredth of a square meter, what was the area? (*Source: Guinness World Records.*)

51. The area of a triangular road sign is 70 ft². If the base of the sign measures 14 ft, what is the height of the sign?

52. The largest drum ever constructed was played at the Royal Festival Hall in London in 1987. It had a diameter of 13 ft. What was the area of the circular face of the drum? (*Hint:* $A = \pi r^2$.) (*Source: Guinness World Records.*)

53. The survey plat depicted here shows two lots that form a trapezoid. The measures of the parallel sides are 115.80 ft and 171.00 ft. The height of the trapezoid is 165.97 ft. Find the combined area of the two lots. Round your answer to the nearest hundredth of a square foot.

54. Lot A in the survey plat is in the shape of a trapezoid. The parallel sides measure 26.84 ft and 82.05 ft. The height of the trapezoid is 165.97 ft. Find the area of Lot A. Round your answer to the nearest hundredth of a square foot.

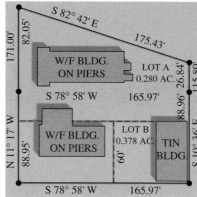

Source: Property survey in New Roads, Louisiana.

55. The U.S. Postal Service requires that any box sent by Priority Mail® have length plus girth (distance around) totaling no more than 108 in. The maximum volume that meets this condition is contained by a box with a square end 18 in. on each side. What is the length of the box? What is the maximum volume? (*Source:* United States Postal Service.)

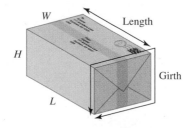

56. On March 17, 2005, a new record was set for the world's largest sandwich. The fillings of this sandwich were corned beef, cheese, lettuce, and mustard. The sandwich, made by Wild Woody's Chill and Grill in Roseville, Michigan, was 12 ft long, 12 ft wide, and $17\frac{1}{2}$ in. $\left(1\frac{11}{24}\text{ ft}\right)$ thick. What was the volume of the sandwich? (*Source: Guinness World Records.*)

Find the measure of each marked angle. See Example 5.

57.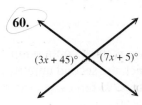
$(x + 1)°$ $(4x - 56)°$

58.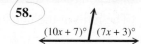
$(10x + 7)°$ $(7x + 3)°$

59.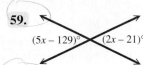
$(5x - 129)°$ $(2x - 21)°$

60.
$(3x + 45)°$ $(7x + 5)°$

61.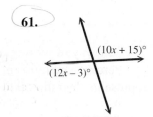
$(10x + 15)°$
$(12x - 3)°$

62.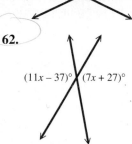
$(11x - 37)°$ $(7x + 27)°$

Solve each formula for the specified variable. See Examples 6–9.

63. $d = rt$ for t **64.** $d = rt$ for r **65.** $A = bh$ for b

66. $A = LW$ for L **67.** $C = \pi d$ for d **68.** $P = 4s$ for s

69. $V = LWH$ for H **70.** $V = LWH$ for W **71.** $I = prt$ for r

72. $I = prt$ for p **73.** $A = \frac{1}{2}bh$ for h **74.** $A = \frac{1}{2}bh$ for b

75. $V = \frac{1}{3}\pi r^2 h$ for h **76.** $V = \pi r^2 h$ for h **77.** $P = a + b + c$ for b

78. $P = a + b + c$ for a **79.** $P = 2L + 2W$ for W **80.** $A = p + prt$ for r

81. $y = mx + b$ for m **82.** $y = mx + b$ for x **83.** $Ax + By = C$ for y

84. $Ax + By = C$ for x **85.** $M = C(1 + r)$ for r **86.** $C = \frac{5}{9}(F - 32)$ for F

PREVIEW EXERCISES

*Solve each equation. See **Section 2.2.***

87. $\frac{x}{12} = \frac{12}{72}$ **88.** $\frac{x}{15} = \frac{144}{60}$ **89.** $0.06x = 300$ **90.** $0.4x = 80$

91. $\frac{3}{4}x = 21$ **92.** $-\frac{5}{6}x = 30$ **93.** $-3x = \frac{1}{4}$ **94.** $4x = \frac{1}{3}$

2.6   Ratios and Proportions

OBJECTIVE 1 Write ratios. A **ratio** is a comparison of two quantities using a quotient.

> **Ratio**
>
> The ratio of the number a to the number b $(b \neq 0)$ is written
>
> $$a \text{ to } b, \qquad a:b, \qquad \text{or} \qquad \frac{a}{b}.$$

The last way of writing a ratio is most common in algebra.

Percents are ratios in which the second number is always 100. For example, 50% represents the ratio 50 to 100, 27% represents the ratio 27 to 100, and so on.

EXAMPLE 1 Writing Word Phrases as Ratios

Write a ratio for each word phrase.

(a) The ratio of 5 hr to 3 hr is

$$\frac{5 \text{ hr}}{3 \text{ hr}} = \frac{5}{3}.$$

(b) To find the ratio of 6 hr to 3 days, first convert 3 days to hours.

$$3 \text{ days} = 3 \cdot 24 = 72 \text{ hr}$$

The ratio of 6 hr to 3 days is thus

$$\frac{6 \text{ hr}}{3 \text{ days}} = \frac{6 \text{ hr}}{72 \text{ hr}} = \frac{6}{72} = \frac{1}{12}.$$

✔ **Now Try Exercises 3 and 7.**

An application of ratios is in unit pricing, to see which size of an item offered in different sizes produces the best price per unit. To do this, set up the ratio of the price of the item to the number of units on the label. Then divide to obtain the price per unit.

EXAMPLE 2 Finding Price per Unit

The Cub Foods supermarket in Coon Rapids, Minnesota, charges the following prices for a jar of extra crunchy peanut butter.

PEANUT BUTTER

Size	Price
18 oz	$1.50
40 oz	$4.14
64 oz	$6.29

Which size is the best buy? That is, which size has the lowest unit price?

To find the best buy, write ratios comparing the price for each size of jar to the number of units (ounces) per jar. The results in the following table are rounded to the nearest thousandth.

Size	Unit Cost (dollars per ounce)
18 oz	$\dfrac{\$1.50}{18} = \$0.083$ ⟵——— The best buy
40 oz	$\dfrac{\$4.14}{40} = \$0.104$
64 oz	$\dfrac{\$6.29}{64} = \$0.098$

Because the 18-oz size produces the lowest unit cost, it is the best buy. This example shows that buying the largest size does not always provide the best buy, although it often does.

✔ **Now Try Exercise 15.**

OBJECTIVE ❷ Solve proportions. A ratio is used to compare two numbers or amounts. A **proportion** says that two ratios are equal, so it is a special type of equation. For example,

$$\frac{3}{4} = \frac{15}{20}$$

is a proportion which says that the ratios $\frac{3}{4}$ and $\frac{15}{20}$ are equal. In the proportion

$$\frac{a}{b} = \frac{c}{d} \quad (b, d \neq 0),$$

a, b, c, and d are the **terms** of the proportion. The terms a and d are called the **extremes,** and the terms b and c are called the **means.** We read the proportion $\frac{a}{b} = \frac{c}{d}$ as "a is to b as c is to d." Beginning with this proportion and multiplying each side by the common denominator, bd, gives

$$\frac{a}{b} = \frac{c}{d}$$

$$bd \cdot \frac{a}{b} = bd \cdot \frac{c}{d} \qquad \text{Multiply by } bd.$$

$$\frac{b}{b}(d \cdot a) = \frac{d}{d}(b \cdot c) \qquad \text{Associative and commutative properties}$$

$$ad = bc. \qquad \text{Commutative and identity properties}$$

We can also find the products ad and bc by multiplying diagonally.

$$\frac{a}{b} \diagdown\!\!\!\!\diagup \frac{c}{d} \begin{array}{l} bc \\[4pt] ad \end{array}$$

For this reason, ad and bc are called **cross products.**

Cross Products

If $\frac{a}{b} = \frac{c}{d}$, then the cross products ad and bc are equal.

Also, if $ad = bc$, then $\frac{a}{b} = \frac{c}{d}$ $(b, d \neq 0)$.

From this rule, if $\frac{a}{b} = \frac{c}{d}$ then $ad = bc$; that is, the product of the extremes equals the product of the means.

▶ **NOTE** If $\frac{a}{c} = \frac{b}{d}$, then $ad = cb$, or $ad = bc$. This means that the two proportions are equivalent, and

$$\text{the proportion } \frac{a}{b} = \frac{c}{d} \text{ can also be written as } \frac{a}{c} = \frac{b}{d} \quad (c, d \neq 0).$$

Sometimes one form is more convenient to work with than the other.

EXAMPLE 3 Deciding whether Proportions Are True

Decide whether each proportion is *true* or *false*.

(a) $\dfrac{3}{4} = \dfrac{15}{20}$

Check to see whether the cross products are equal.

$$4 \cdot 15 = 60$$
$$\frac{3}{4} \diagdown\!\!\!\!\!\diagup \frac{15}{20}$$
$$3 \cdot 20 = 60$$

The cross products are equal, so the proportion is true.

(b) $\dfrac{6}{7} = \dfrac{30}{32}$

The cross products are $6 \cdot 32 = 192$ and $7 \cdot 30 = 210$. The cross products are not equal, so the proportion is false.

> ✔ **Now Try Exercises 23 and 25.**

Four numbers are used in a proportion. If any three of these numbers are known, the fourth can be found.

EXAMPLE 4 Finding an Unknown in a Proportion

Solve the proportion $\frac{5}{9} = \frac{x}{63}$.

$$5 \cdot 63 = 9 \cdot x \qquad \text{Cross products must be equal.}$$
$$315 = 9x \qquad \text{Multiply.}$$
$$35 = x \qquad \text{Divide by 9.}$$

Check by substituting 35 for x in the proportion. The solution set is $\{35\}$.

> ✔ **Now Try Exercise 29.**

> **CAUTION** The cross-product method cannot be used directly if there is more than one term on either side of the equals symbol.

EXAMPLE 5 Solving an Equation by Using Cross Products

Solve the equation $\frac{m-2}{5} = \frac{m+1}{3}$.

Be sure to use parentheses.

$3(m-2) = 5(m+1)$	Cross products
$3m - 6 = 5m + 5$	Distributive property
$3m = 5m + 11$	Add 6.
$-2m = 11$	Subtract $5m$.
$m = -\dfrac{11}{2}$	Divide by -2.

The solution set is $\left\{-\frac{11}{2}\right\}$.

✔ **Now Try Exercise 37.**

> **NOTE** When you set cross products equal to each other, you are really multiplying each ratio in the proportion by a common denominator.

OBJECTIVE 3 Solve applied problems by using proportions. Proportions are useful in many practical applications. We continue to use the six-step method, although the steps are not numbered here.

EXAMPLE 6 Applying Proportions

After Lee Ann Spahr had pumped 5.0 gal of gasoline, the display showing the price read $16.60. When she finished pumping the gasoline, the price display read $48.14. How many gallons did she pump?

To solve this problem, set up a proportion, with prices in the numerators and gallons in the denominators.

Let x = the number of gallons she pumped. Then

Price ⟶ $\dfrac{\$16.60}{5.0} = \dfrac{\$48.14}{x}$ ⟵ Price
Gallons ⟶ $\quad\quad\quad\quad\quad\quad$ ⟵ Gallons

Be sure that numerators represent the same quantities and denominators represent the same quantities.

$16.60x = 5.0(48.14)$	Cross products
$16.60x = 240.70$	Multiply.
$x = 14.5$	Divide by 16.60.

She pumped 14.5 gal. Check this answer. Using a calculator to perform the arithmetic reduces the possibility of error. Notice that the way the proportion was set up uses the fact that the unit price is the same, no matter how many gallons are purchased.

✔ **Now Try Exercise 51.**

2.6 EXERCISES

Complete solution available on Video Lectures on CD/DVD

Now Try Exercise

1. *Concept Check* Match each ratio in Column I with the ratio equivalent to it in Column II.

I	II
(a) 75 to 100	**A.** 80 to 100
(b) 5 to 4	**B.** 50 to 100
(c) $\dfrac{1}{2}$	**C.** 3 to 4
(d) 4 to 5	**D.** 15 to 12

2. *Concept Check* Give three different, equivalent forms of the ratio $\frac{4}{3}$.

Write a ratio for each word phrase. In Exercises 7–12, first write the amounts with the same units. Write fractions in lowest terms. See Example 1.

3. 40 mi to 30 mi

4. 60 ft to 70 ft

5. 120 people to 90 people

6. 72 dollars to 220 dollars

7. 20 yd to 8 ft

8. 30 in. to 8 ft

9. 24 min to 2 hr

10. 16 min to 1 hr

11. 60 in. to 2 yd

12. 5 days to 40 hr

A supermarket was surveyed to find the prices charged for items in various sizes. Find the best buy (based on price per unit) for each item. See Example 2. (Source: Cub Foods.)

13. GRANULATED SUGAR

Size	Price
4 lb	$1.78
10 lb	$4.39

14. GROUND COFFEE

Size	Price
13 oz	$2.58
39 oz	$4.44

15. SALAD DRESSING

Size	Price
16 oz	$2.44
32 oz	$2.98
48 oz	$4.95

16. BLACK PEPPER

Size	Price
2 oz	$1.79
4 oz	$2.59
8 oz	$5.59

17. VEGETABLE OIL

Size	Price
16 oz	$1.54
24 oz	$2.08
64 oz	$3.63
128 oz	$5.65

18. MOUTHWASH

Size	Price
8.5 oz	$0.99
16.9 oz	$1.87
33.8 oz	$2.49
50.7 oz	$2.99

19. TOMATO KETCHUP

Size	Price
14 oz	$1.39
24 oz	$1.55
36 oz	$1.78
64 oz	$3.99

20. GRAPE JELLY

Size	Price
12 oz	$1.05
18 oz	$1.73
32 oz	$1.84
48 oz	$2.88

21. LAUNDRY DETERGENT

Size	Price
87 oz	$7.88
131 oz	$10.98
263 oz	$19.96

✎ **22.** Explain the distinction between *ratio* and *proportion*. Give examples.

Decide whether each proportion is true *or* false. *See Example 3.*

23. $\dfrac{5}{35} = \dfrac{8}{56}$ **24.** $\dfrac{4}{12} = \dfrac{7}{21}$ **25.** $\dfrac{120}{82} = \dfrac{7}{10}$

26. $\dfrac{27}{160} = \dfrac{18}{110}$ **27.** $\dfrac{\frac{1}{2}}{5} = \dfrac{1}{10}$ **28.** $\dfrac{\frac{1}{3}}{6} = \dfrac{1}{18}$

Solve each equation. See Examples 4 and 5.

29. $\dfrac{k}{4} = \dfrac{175}{20}$ **30.** $\dfrac{x}{6} = \dfrac{18}{4}$ **31.** $\dfrac{49}{56} = \dfrac{z}{8}$

32. $\dfrac{20}{100} = \dfrac{z}{80}$ **33.** $\dfrac{a}{24} = \dfrac{15}{16}$ **34.** $\dfrac{x}{4} = \dfrac{12}{30}$

35. $\dfrac{z}{2} = \dfrac{z+1}{3}$ **36.** $\dfrac{m}{5} = \dfrac{m-2}{2}$ **37.** $\dfrac{3y-2}{5} = \dfrac{6y-5}{11}$

38. $\dfrac{2r+8}{4} = \dfrac{3r-9}{3}$ **39.** $\dfrac{5k+1}{6} = \dfrac{3k-2}{3}$ **40.** $\dfrac{x+4}{6} = \dfrac{x+10}{8}$

41. $\dfrac{2p+7}{3} = \dfrac{p-1}{4}$ **42.** $\dfrac{3m-2}{5} = \dfrac{4-m}{3}$ **43.** $\dfrac{0.5x+2}{3} = \dfrac{2.25x+27}{9}$

✎ **44.** Explain why the equation $\dfrac{x+4}{3} = \dfrac{x+5}{3}$ has no solution.

Solve each problem. In Exercises 45–51, assume that all items are equally priced. See Example 6.

45. If 16 candy bars cost $20.00, how much do 24 candy bars cost?

46. If 12 ring tones cost $30.00, how much do 8 ring tones cost?

47. Eight quarts of oil cost $14.00. How much do 5 qt of oil cost?

48. Four tires cost $398.00. How much do 7 tires cost?

49. If 9 pairs of jeans cost $121.50, find the cost of 5 pairs.

50. If 7 shirts cost $87.50, find the cost of 11 shirts.

51. If 6 gal of premium unleaded gasoline costs $19.56, how much would it cost to completely fill a 15-gal tank?

52. If sales tax on a $16.00 DVD is $1.32, how much would the sales tax be on a $120.00 DVD player?

53. The distance between Kansas City, Missouri, and Denver is 600 mi. On a certain wall map, this is represented by a length of 2.4 ft. On the map, how many feet would there be between Memphis and Philadelphia, two cities that are actually 1000 mi apart?

54. The distance between Singapore and Tokyo is 3300 mi. On a certain wall map, this distance is represented by 11 in. The actual distance between Mexico City and Cairo is 7700 mi. How far apart are they on the same map?

55. A wall map of the United States has a distance of 8.5 in. between Memphis and Denver, two cities that are actually 1040 mi apart. The actual distance between St. Louis and Des Moines is 333 mi. How far apart are St. Louis and Des Moines on the map?

56. The map of the United States mentioned in the previous exercise has a distance of 8.0 in. between New Orleans and Chicago, two cities that are actually 912 mi apart. The actual distance between Milwaukee and Seattle is 1940 mi. How far apart are Milwaukee and Seattle on the map?

57. On a world globe, the distance between Capetown and Bangkok, two cities that are actually 10,080 km apart, is 12.4 in. The actual distance between Moscow and Berlin is 1610 km. How far apart are Moscow and Berlin on this globe?

58. On a world globe, the distance between Rio de Janeiro and Hong Kong, two cities that are actually 17,615 km apart, is 21.5 in. The actual distance between Paris and Stockholm is 1605 km. How far apart are Paris and Stockholm on this globe?

59. According to the directions on a bottle of Armstrong® Concentrated Floor Cleaner, for routine cleaning, $\frac{1}{4}$ cup of cleaner should be mixed with 1 gal of warm water. How much cleaner should be mixed with $10\frac{1}{2}$ gal of water?

60. The directions on the bottle mentioned in Exercise 59 also specify that, for extra-strength cleaning, $\frac{1}{2}$ cup of cleaner should be used for each gallon of water. For extra-strength cleaning, how much cleaner should be mixed with $15\frac{1}{2}$ gal of water?

61. The euro is the common currency used by most European countries, including Italy. On January 29, 2006, the exchange rate between euros and U.S. dollars was 1 euro to $1.2128. Ashley went to Rome and exchanged her U.S. currency for euros, receiving 300 euros. How much in U.S. dollars did she exchange? (*Source:* www.xe.com/ucc.)

62. If 8 U.S. dollars can be exchanged for 84.3 Mexican pesos, how many pesos can be obtained for $65? (Round to the nearest tenth.)

63. Biologists tagged 500 fish in Grand Bay on October 12. At a later date, they found 7 tagged fish in a sample of 700. Estimate the total number of fish in Grand Bay to the nearest hundred.

64. On May 25, researchers at an oxbow lake called Old River tagged 840 fish. When they returned a few weeks later, their sample of 1000 fish contained 18 that were tagged. Give an approximation of the fish population in Old River to the nearest hundred.

Two triangles are **similar** if they have the same shape (but not necessarily the same size). Similar triangles have sides that are proportional. The figure shows two similar triangles. Notice that the ratios of the corresponding sides all equal $\frac{3}{2}$:

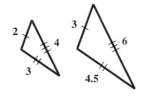

$$\frac{3}{2} = \frac{3}{2}, \qquad \frac{4.5}{3} = \frac{3}{2}, \qquad \frac{6}{4} = \frac{3}{2}.$$

If we know that two triangles are similar, we can set up a proportion to solve for the length of an unknown side.

Use a proportion to find the lengths x and y, given that each pair of triangles is similar.

65.

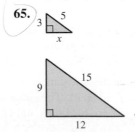

66.

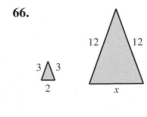

67.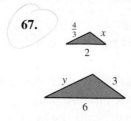

68.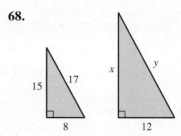

For Exercises 69 and 70, (a) draw a sketch consisting of two right triangles depicting the situation described, and (b) solve the problem. (Source: Guinness World Records.)

69. An enlarged version of the chair used by George Washington at the Constitutional Convention casts a shadow 18 ft long at the same time a vertical pole 12 ft high casts a shadow 4 ft long. How tall is the chair?

70. One of the tallest candles ever constructed was exhibited at the 1897 Stockholm Exhibition. If it cast a shadow 5 ft long at the same time a vertical pole 32 ft high cast a shadow 2 ft long, how tall was the candle?

The Consumer Price Index provides a means of determining the purchasing power of the U.S. dollar from one year to the next. Using the period from 1982 to 1984 as a measure of 100.0, the Consumer Price Index for even-numbered years from 1990 through 2004 is shown in the table. To use the Consumer Price Index to predict a price in a particular year, we set up a proportion and compare it with a known price in another year:

$$\frac{\text{price in year } A}{\text{index in year } A} = \frac{\text{price in year } B}{\text{index in year } B}.$$

Year	Consumer Price Index
1990	130.7
1992	140.3
1994	148.2
1996	156.9
1998	163.0
2000	172.2
2002	179.9
2004	188.9

Source: U.S. Bureau of Labor Statistics.

Use the Consumer Price Index figures in the table to find the amount that would be charged for the same amount of groceries that cost $120 in 1990. Give your answer to the nearest dollar.

71. in 1996   **72.** in 2000   **73.** in 2002   **74.** in 2004

PREVIEW EXERCISES

Solve, using the proper formula. See Section 2.5.

75. If an investment of $8000 earns $1280 in simple interest in 4 yr, what rate of interest is being earned?

76. If $200 earned $75 in simple interest at 5%, for how many years was the money earning interest?

77. Joel Spring earned $5700 in simple interest on a deposit of $19,000 at 3%. For how long was the money deposited?

78. What is the monetary value of 34 quarters? (*Hint:* Monetary value = number of coins × denomination.)

Solve each equation. See Section 2.3.

79. $0.15x + 0.30(3) = 0.20(3 + x)$

80. $0.20(60) + 0.05x = 0.10(60 + x)$

81. $0.92x + 0.98(12 - x) = 0.96(12)$

82. $0.10(7) + 1.00x = 0.30(7 + x)$

2.7 Further Applications of Linear Equations

OBJECTIVE 1 **Use percent in solving problems involving rates.** Recall that percent means "per hundred." Thus, percents are ratios in which the second number is always 100. For example, 50% represents the ratio 50 to 100 and 27% represents the ratio 27 to 100.

▶ **PROBLEM-SOLVING HINT** Percents are often used in problems involving mixing different concentrations of a substance or different interest rates. In each case, to get the amount of pure substance or the interest, we multiply.

Mixture Problems	Interest Problems (annual)
base × rate (%) = percentage	principal × rate (%) = interest
$b \times r = p$	$p \times r = I$

In an equation, percent is always written as a decimal. For example, 35% is written 0.35, *not* 35, and 7% is written 0.07, *not* 7.

EXAMPLE 1 Using Percents to Find Percentages

(a) If a chemist has 40 L of a 35% acid solution, then the amount of pure acid in the solution is

$$40 \qquad \times \qquad 0.35 \qquad = \qquad 14 \text{ L.}$$

Amount of solution   Rate of concentration   Amount of pure acid

(b) If $1300 is invested for one year at 7% simple interest, the amount of interest earned in the year is

$$\$1300 \qquad \times \qquad 0.07 \qquad = \qquad \$91.$$

Principal   Interest rate   Interest earned

✔ **Now Try Exercises 1 and 3.**

▶ **PROBLEM-SOLVING HINT** In the examples that follow, we use tables to organize the information in the problems. A table enables us to more easily set up an equation, which is usually the most difficult step.

OBJECTIVE 2 **Solve problems involving mixtures.** In the next example, we use percents to solve a mixture problem.

EXAMPLE 2 Solving a Mixture Problem

A chemist needs to mix 20 L of a 40% acid solution with some 70% acid solution to obtain a mixture that is 50% acid. How many liters of the 70% acid solution should be used?

Step 1 **Read** the problem. Note the percent of each solution and of the mixture.

Step 2 **Assign a variable.**

Let x = the number of liters of 70% acid solution needed.

Recall from Example 1(a) that the amount of pure acid in this solution will be given by the product of the percent of strength and the number of liters of solution, or

liters of pure acid in x liters of 70% solution = $0.70x$.

The amount of pure acid in the 20 L of 40% solution is

liters of pure acid in the 40% solution = $0.40(20) = 8$.

The new solution will contain $(x + 20)$ liters of 50% solution. The amount of pure acid in this solution is

liters of pure acid in the 50% solution = $0.50(x + 20)$.

Figure 12 illustrates this information, which is summarized in the table.

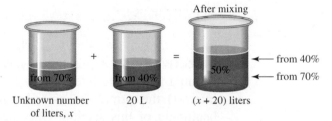

FIGURE 12

Liters of Solution	Rate (as a decimal)	Liters of Pure Acid
x	0.70	0.70x
20	0.40	0.40(20) = 8
$x + 20$	0.50	0.50($x + 20$)

Sum must equal

Step 3 **Write an equation.** The number of liters of pure acid in the 70% solution added to the number of liters of pure acid in the 40% solution will equal the number of liters of pure acid in the final mixture, so the equation is

Pure acid in 70%	plus	pure acid in 40%	is	pure acid in 50%.
↓	↓	↓	↓	↓
$0.70x$	+	$0.40(20)$	=	$0.50(x + 20)$.

Step 4 **Solve** the equation.

$$0.70x + 0.40(20) = 0.50(x + 20)$$

$$0.70x + 0.40(20) = 0.50x + 0.50(20)$$ Distributive property

$$70x + 40(20) = 50x + 50(20)$$ Multiply by 100 to clear decimals.

$$70x + 800 = 50x + 1000$$

$$20x + 800 = 1000$$ Subtract 50x.

$$20x = 200$$ Subtract 800.

$$x = 10$$ Divide by 20.

Step 5 **State the answer.** The chemist needs to use 10 L of 70% solution.

Step 6 **Check.** Since

$$0.70(10) + 0.40(20) = 7 + 8 = 15$$

and $$0.50(10 + 20) = 0.50(30) = 15,$$

the answer checks.

✔ **Now Try Exercise 15.**

▶ **NOTE** In a problem such as Example 2, the concentration of the final mixture must be *between* the concentrations of the two solutions making up the mixture.

OBJECTIVE 3 **Solve problems involving simple interest.** The next example uses the formula for simple interest, $I = prt$. Remember that when time $t = 1$ (for annual interest), the formula becomes $I = pr$, as shown in the Problem-Solving Hint at the beginning of this section. Once again, the idea of multiplying the total amount (principal) by the rate (rate of interest) gives the percentage (amount of interest).

EXAMPLE 3 **Solving a Simple Interest Problem**

Elizabeth Suco receives an inheritance. She plans to invest part of it at 6% and $2000 more than this amount at 7%. To earn $790 per year in interest, how much should she invest at each rate?

Step 1 **Read** the problem again.

Step 2 **Assign a variable.**

Let x = the amount invested at 6% (in dollars).

Then $x + 2000$ = the amount invested at 7% (in dollars).

Use a table to arrange the given information.

Amount Invested in Dollars	Rate of Interest	Interest for One Year
x	0.06	0.06x
x + 2000	0.07	0.07(x + 2000)

Step 3 **Write an equation.** Multiply amount by rate to get the interest earned. Since the total interest is to be $790, the equation is

Interest at 6%	plus	interest at 7%	is	total interest.
↓	↓	↓	↓	↓
$0.06x$	$+$	$0.07(x + 2000)$	$=$	$790.$

Step 4 **Solve** the equation. Clear parentheses; then clear decimals.

$$0.06x + 0.07x + 0.07(2000) = 790 \qquad \text{Distributive property}$$
$$6x + 7x + 7(2000) = 79{,}000 \qquad \text{Multiply by 100.}$$
$$6x + 7x + 14{,}000 = 79{,}000$$
$$13x + 14{,}000 = 79{,}000 \qquad \text{Combine like terms.}$$
$$13x = 65{,}000 \qquad \text{Subtract 14,000.}$$
$$x = 5000 \qquad \text{Divide by 13.}$$

Step 5 **State the answer.** She should invest $5000 at 6% and $5000 + $2000 = $7000 at 7%.

Step 6 **Check.** Investing $5000 at 6% and $7000 at 7% gives total interest of 0.06($5000) + 0.07($7000) = $300 + $490 = $790, as required in the original problem.

✔ **Now Try Exercise 25.**

OBJECTIVE 4 Solve problems involving denominations of money.

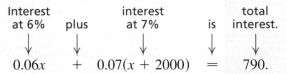

▶ **PROBLEM-SOLVING HINT** Problems that involve different denominations of money or items with different monetary values are similar to mixture and interest problems. To get the total value, we multiply.

Money Denominations Problems

number × value of one item = total value

For example, 30 dimes have a monetary value of 30($0.10) = $3. Fifteen $5 bills have a value of 15($5) = $75. A table is helpful for these problems, too.

EXAMPLE 4 Solving a Money Denominations Problem

A bank teller has 25 more $5 bills than $10 bills. The total value of the money is $200. How many of each denomination of bill does she have?

Step 1 **Read** the problem. We must find the number of each denomination of bill that the teller has.

Step 2 **Assign a variable.**

Let $x =$ the number of $10 bills.

Then $x + 25 =$ the number of $5 bills.

Organize the given information in a table.

Number of Bills	Denomination	Total Value
x	10	10x
x + 25	5	5(x + 25)

Step 3 **Write an equation.** Multiplying the number of bills by the denomination gives the monetary value. The value of the tens added to the value of the fives must be $200.

$$\underset{\substack{\text{Value of} \\ \text{tens}}}{10x} \quad \underset{\text{plus}}{+} \quad \underset{\substack{\text{value of} \\ \text{fives}}}{5(x+25)} \quad \underset{\text{is}}{=} \quad \underset{\text{\$200.}}{200}$$

Step 4 **Solve.**

$$10x + 5x + 125 = 200 \qquad \text{Distributive property}$$
$$15x + 125 = 200 \qquad \text{Combine like terms.}$$
$$15x = 75 \qquad \text{Subtract 125.}$$
$$x = 5 \qquad \text{Divide by 15.}$$

Step 5 **State the answer.** The teller has 5 tens and $5 + 25 = 30$ fives.

Step 6 **Check.** The teller has $30 - 5 = 25$ more fives than tens, and the value of the money is $5(\$10) + 30(\$5) = \$200$, as required.

✔ **Now Try Exercise 29.**

OBJECTIVE 5 **Solve problems involving distance, rate, and time.** If your car travels at an average rate of 50 mph for 2 hr, then it travels $50 \times 2 = 100$ mi. This is an example of the basic relationship between distance, rate, and time,

distance = rate × time,

given by the formula $d = rt$. By solving, in turn, for r and t in the formula, we obtain two other equivalent forms of the formula. The three forms are given here.

Distance, Rate, and Time Relationship

$$d = rt \qquad r = \frac{d}{t} \qquad t = \frac{d}{r}$$

EXAMPLE 5 **Finding Distance, Rate, or Time**

(a) The speed of sound is 1088 ft per sec at sea level at 32°F. In 5 sec under these conditions, sound travels

$$\underset{\underset{\text{Rate}}{\uparrow}}{1088} \quad \underset{\underset{\text{Time}}{\uparrow}}{\times \quad 5} \quad = \quad \underset{\underset{\text{Distance}}{\uparrow}}{5440 \text{ ft.}}$$

Here, we found distance, given rate and time, using $d = rt$.

(b) The winner of the first Indianapolis 500 race (in 1911) was Ray Harroun, driving a Marmon Wasp at an average speed of 74.59 mph. (*Source: Universal Almanac.*) To complete the 500 mi, it took him

$$\text{Distance} \rightarrow \frac{500}{74.59} = 6.70 \text{ hr} \quad \text{(rounded).} \quad \leftarrow \text{Time}$$
$$\text{Rate} \rightarrow$$

Here, we found time, given rate and distance, using $t = \frac{d}{r}$. To convert 0.70 hr to minutes, we multiply by 60 to get $0.70(60) = 42$. It took Harroun about 6 hr, 42 min, to complete the race.

(c) At the 2004 Olympic Games in Athens, Greece, Chinese swimmer Luo Xuejuan set an Olympic record of 66.64 sec in the women's 100-m breaststroke swimming event. (*Source: World Almanac and Book of Facts 2006.*) Her rate was

$$\text{Rate} = \frac{\text{Distance} \rightarrow}{\text{Time} \rightarrow} \frac{100}{66.64} = 1.50 \text{ m per sec (rounded).}$$

✔ **Now Try Exercises 37, 39, and 41.**

EXAMPLE 6 Solving a Motion Problem

Two cars leave Baton Rouge, Louisiana, at the same time and travel east on Interstate 10. One travels at a constant speed of 55 mph, and the other travels at a constant speed of 63 mph. In how many hours will the distance between them be 24 mi?

Step 1 **Read** the problem. We must find the time it will take for the distance between the cars to be 24 mi.

Step 2 **Assign a variable.** Since we are looking for time, we let $t =$ the number of hours until the distance between them is 24 mi. The sketch in Figure 13 shows what is happening in the problem.

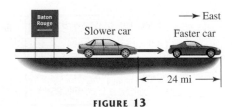

FIGURE 13

Now we construct a table, using the information given in the problem and t for the time traveled by each car. We multiply rate by time to get the expressions for distances traveled.

	Rate	× Time	= Distance	
Faster Car	63	t	$63t$	Difference is 24 mi.
Slower Car	55	t	$55t$	

The quantities $63t$ and $55t$ represent the two distances. Refer to Figure 13, and notice that the *difference* between the larger distance and the smaller distance is 24 mi.

Step 3   **Write an equation.**

$$63t - 55t = 24$$

Step 4   **Solve.** $8t = 24$ Combine like terms.

$t = 3$ Divide by 8.

Step 5   **State the answer.** It will take the cars 3 hr to be 24 mi apart.

Step 6   **Check.** After 3 hr, the faster car will have traveled $63 \times 3 = 189$ mi and the slower car will have traveled $55 \times 3 = 165$ mi. Since $189 - 165 = 24$, the conditions of the problem are satisfied.

✔ **Now Try Exercise 49.**

▶ **NOTE**   In motion problems like the one in Example 6, once you have filled in two pieces of information in each row of the table, you should automatically fill in the third piece of information, using the appropriate form of the formula relating distance, rate, and time. Set up the equation on the basis of your sketch and the information in the table.

EXAMPLE 7   **Solving a Motion Problem**

Two planes leave Memphis at the same time. One heads south to New Orleans. The other heads north to Chicago. The Chicago plane flies 50 mph faster than the New Orleans plane. In $\frac{1}{2}$ hr, the planes are 275 mi apart. What are their speeds?

Step 1   **Read** the problem carefully.

Step 2   **Assign a variable.**

Let $r =$ the speed of the slower plane.

Then $r + 50 =$ the speed of the faster plane.

Fill in a table.

	Rate	Time	Distance
Slower plane	r	$\frac{1}{2}$	$\frac{1}{2}r$
Faster plane	$r + 50$	$\frac{1}{2}$	$\frac{1}{2}(r + 50)$

Sum is 275 mi.

Step 3   **Write an equation.** As Figure 14 shows, the planes are headed in *opposite* directions. The *sum* of their distances equals 275 mi, so

$$\frac{1}{2}r + \frac{1}{2}(r + 50) = 275.$$

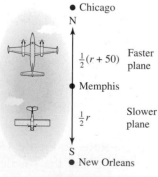

• Chicago

N

$\frac{1}{2}(r + 50)$ Faster plane

• Memphis

$\frac{1}{2}r$ Slower plane

S

• New Orleans

FIGURE 14

Step 4   **Solve.** $\dfrac{1}{2}r + \dfrac{1}{2}(r + 50) = 275$

$r + (r + 50) = 550$ Multiply by 2.

$2r + 50 = 550$ Combine like terms.

$2r = 500$ Subtract 50.

$r = 250$ Divide by 2.

Step 5 **State the answer.** The slower plane (headed south) has a speed of 250 mph. The speed of the faster plane is $250 + 50 = 300$ mph.

Step 6 **Check.** Verify that $\frac{1}{2}(250) + \frac{1}{2}(300) = 275$ mi.

✔ **Now Try Exercise 51.**

Another way to solve the problems in this section is given in **Chapter 4.**

2.7 EXERCISES

⊕ *Complete solution available on Video Lectures on CD/DVD*

☐ *Now Try Exercise*

Use the concepts of this section to answer each question. See Example 1 and the Problem-Solving Hint preceding Example 4.

⊕ **1.** How much pure alcohol is in 150 L of a 30% alcohol solution?

2. How much pure acid is in 250 mL of a 14% acid solution?

3. If $25,000 is invested for 1 yr at 3% simple interest, how much interest is earned?

4. If $10,000 is invested for 1 yr at 3.5% simple interest, how much interest is earned?

5. What is the monetary value of 35 half-dollars?

6. What is the monetary value of 283 nickels?

Concept Check   Solve each percent problem. Remember that base × rate = percentage.

7. The 2000 U.S. Census showed that the population of New Mexico was 1,819,000, with 42.1% Hispanic. What is the best estimate of the Hispanic population in New Mexico? (*Source:* U.S. Census Bureau.)

   A. 720,000   **B.** 72,000   **C.** 650,000   **D.** 36,000

8. The 2000 U.S. Census showed that the population of Alabama was 4,447,000, with 26.0% represented by African-Americans. What is the best estimate of the African-American population in Alabama? (*Source:* U.S. Census Bureau.)

   A. 500,000   **B.** 750,000   **C.** 1,100,000   **D.** 1,500,000

9. The graph shows the breakdown, by approximate percents, of the colors chosen for new 2004 model-year full-size and intermediate cars sold in the United States. If approximately 4 million of these cars were sold, about how many were each color? (*Source:* Ward's Communications.)

   (a) White   **(b)** Silver   **(c)** Blue

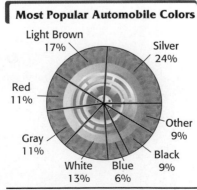

Most Popular Automobile Colors

Light Brown 17%
Silver 24%
Red 11%
Other 9%
Gray 11%
Black 9%
White 13%
Blue 6%

Source: DuPont Automotive Products.

10. An average middle-income family will spend $184,320 to raise a child born in 2004 from birth to age 17. The graph shows the breakdown, by approximate percents, for various expense categories.

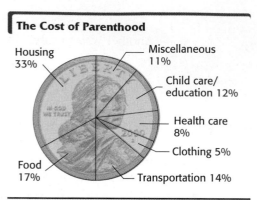

The Cost of Parenthood

Housing 33%

Miscellaneous 11%

Child care/ education 12%

Health care 8%

Clothing 5%

Transportation 14%

Food 17%

Source: U.S. Department of Agriculture.

To the nearest dollar, about how much will be spent to provide the following?

(a) Housing   **(b)** Food   **(c)** Health care

11. In 2004, the U.S. civilian labor force consisted of about 147,401,000 persons. Of this total, 8,149,000 were unemployed. To the nearest tenth, what was the percent of unemployment? (*Source:* U.S. Bureau of Labor Statistics.)

12. In 2004, the U.S. labor force (excluding agricultural employees, self-employed persons, and the unemployed) consisted of about 123,554,000 persons. Of this total, about 15,472,000 were union members. To the nearest tenth, what percent of this labor force belonged to unions? (*Source:* U.S. Bureau of Labor Statistics.)

Concept Check   *Answer each question.*

13. Suppose that a chemist is mixing two acid solutions, one of 20% concentration and the other of 30% concentration. Which one of the following concentrations could *not* be obtained?

A. 22%   **B.** 24%   **C.** 28%   **D.** 32%

14. Suppose that pure alcohol is added to a 24% alcohol mixture. Which one of the following concentrations could *not* be obtained?

A. 22%   **B.** 26%   **C.** 28%   **D.** 30%

Work each mixture problem. See Example 2.

15. How many liters of 25% acid solution must be added to 80 L of 40% solution to obtain a solution that is 30% acid?

Liters of Solution	Rate	Liters of Acid
x	0.25	$0.25x$
80	0.40	$0.40(80)$
$x + 80$	0.30	$0.30(x + 80)$

16. How many gallons of 50% antifreeze must be mixed with 80 gal of 20% antifreeze to obtain a mixture that is 40% antifreeze?

Gallons of Mixture	Rate	Gallons of Antifreeze
x	0.50	$0.50x$
80	0.20	$0.20(80)$
$x + 80$	0.40	$0.40(x + 80)$

17. A pharmacist has 20 L of a 10% drug solution. How many liters of 5% solution must be added to get a mixture that is 8%?

Liters of Solution	Rate	Liters of Pure Drug
20		20(0.10)
	0.05	
	0.08	

18. A certain metal is 20% tin. How many kilograms of this metal must be mixed with 80 kg of a metal that is 70% tin to get a metal that is 50% tin?

Kilograms of Metal	Rate	Kilograms of Pure Tin
x	0.20	
	0.70	
	0.50	

19. How many liters of a 60% acid solution must be mixed with a 75% acid solution to get 20 L of a 72% solution?

20. In a chemistry class, 12 L of a 12% alcohol solution must be mixed with a 20% solution to get a 14% solution. How many liters of the 20% solution are needed?

21. How many gallons of a 12% indicator solution must be mixed with a 20% indicator solution to get 10 gal of a 14% solution?

22. How many liters of a 10% alcohol solution must be mixed with 40 L of a 50% solution to get a 40% solution?

23. Minoxidil is a drug that has recently proven to be effective in treating male pattern baldness. Water must be added to 20 mL of a 4% minoxidil solution to dilute it to a 2% solution. How many milliliters of water should be used? (*Hint:* Water is 0% minoxidil.)

24. A pharmacist wishes to mix a solution that is 2% minoxidil. She has on hand 50 mL of a 1% solution, and she wishes to add some 4% solution to it to obtain the desired 2% solution. How much 4% solution should she add?

Work each investment problem using simple interest. See Example 3.

25. Eduardo Gomez is saving money for his college education. He deposited some money in a savings account paying 5% and $1200 less than that amount in a second account paying 4%. The two accounts produced a total of $141 interest in 1 yr. How much did he invest at each rate?

26. Roopa Patel won a prize for her work. She invested part of the money in a certificate of deposit at 4% and $3000 more than that amount in a bond paying 6%. Her annual interest income was $780. How much did Roopa invest at each rate?

27. An artist invests her earnings in two ways. Some goes into a tax-free bond paying 6%, and $6000 more than three times as much goes into mutual funds paying 5%. Her total annual interest income from the investments is $825. How much does she invest at each rate?

28. With income earned by selling the rights to his life story, an actor invests some of the money at 3% and $30,000 more than twice as much at 4%. The total annual interest earned from the investments is $5600. How much is invested at each rate?

Work each problem involving monetary values. See Example 4.

29. A coin collector has $1.70 in dimes and nickels. She has two more dimes than nickels. How many nickels does she have?

Number of Coins	Denomination	Total Value
x	0.05	0.05x
	0.10	

30. A bank teller has some $5 bills and some $20 bills. The teller has five more twenties than fives. The total value of the money is $725. Find the number of $5 bills that the teller has.

Number of Bills	Denomination	Total Value
x	5	
x + 5	20	

31. In January 2006, U.S. first-class mail rates increased to 39 cents for the first ounce, plus 24 cents for each additional ounce. If Sabrina spent $15.00 for a total of 45 stamps of these two denominations, how many stamps of each denomination did she buy? (*Source:* U.S. Postal Service.)

32. A movie theater has two ticket prices: $8 for adults and $5 for children. If the box office took in $4116 from the sale of 600 tickets, how many tickets of each kind were sold?

33. Crystal Jackson owns and operates a gourmet tea and coffee shop. Monica Hugo, who is one of Crystal's best customers, wants to buy two kinds of beans: Arabian Mocha and Colombian Decaf. If she wants twice as much Mocha as Colombian Decaf, how much of each can she buy for a total of $87.50? (Prices are listed on the sign.)

Crystal's Coffee Beans

Arabian Mocha $8.50/lb
Chocolate Mint $10.50/lb
Columbian Decaf $8.00/lb
French Roast $7.50/lb
Guatemalan Spice .. $9.50/lb
Hazelnut Decaf $10.00/lb
Italian Espresso $9.00/lb
Kona Deluxe $11.50/lb

34. In addition to offering the individual types of coffee beans, Crystal sells several special blends. The house favorite, Crystal's Special Blend, contains a combination of French Roast and Kona Deluxe beans. How many pounds of Kona should she mix with 12 lb of French Roast to get a blend to be sold for $10 a pound?

Solve each problem involving distance, rate, and time. See Example 5.

35. *Concept Check*   Which choice is the best estimate for the average speed of a bus trip of 405 mi that lasted 8.2 hr?

A. 50 mph   **B.** 30 mph   **C.** 60 mph   **D.** 40 mph

36. Suppose that an automobile averages 45 mph and travels for 30 min. Is the distance traveled $45 \times 30 = 1350$ mi? If not, explain why not, and give the correct distance.

37. A driver averaged 53 mph and took 10 hr to travel from Memphis to Chicago. What is the distance between Memphis and Chicago?

38. A small plane traveled from Warsaw to Rome, averaging 164 mph. The trip took 2 hr. What is the distance from Warsaw to Rome?

39. The winner of the 2005 Indianapolis 500 (mile) race was Dan Weldon, who drove his Honda to victory at a rate of 157.603 mph. What was his time? (*Source: World Almanac and Book of Facts 2006.*)

40. In 2005, Tony Stewart drove his Chevrolet to victory in the Brickyard 400 (mile) race. His rate was 118.782 mph. What was his time? (*Source: World Almanac and Book of Facts 2006.*)

In Exercises 41–44, find the rate on the basis of the information provided. Use a calculator and round your answers to the nearest hundredth. All events were at the 2004 Olympics. (Source: World Almanac and Book of Facts 2006.)

	Event	Participant	Distance	Time
41.	100-m hurdles, women	Joanna Hayes, USA	100 m	12.37 sec
42.	400-m hurdles, women	Fani Halkia, Greece	400 m	52.82 sec
43.	400-m hurdles, men	Felix Sánchez, Dominican Republic	400 m	47.63 sec
44.	400-m run, men	Jeremy Wariner, USA	400 m	44.00 sec

Solve each motion problem. See Examples 6 and 7.

45. Atlanta and Cincinnati are 440 mi apart. John leaves Cincinnati, driving toward Atlanta at an average speed of 60 mph. Pat leaves Atlanta at the same time, driving toward Cincinnati in her antique auto, averaging 28 mph. How long will it take them to meet?

	r	t	d
John	60	t	60t
Pat	28	t	28t

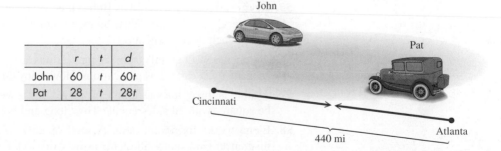

46. St. Louis and Portland are 2060 mi apart. A small plane leaves Portland, traveling toward St. Louis at an average speed of 90 mph. Another plane leaves St. Louis at the same time, traveling toward Portland and averaging 116 mph. How long will it take them to meet? See the top of the next page for a table and a figure for this problem.

	r	t	d
Plane Leaving Portland	90	t	90t
Plane Leaving St. Louis	116	t	116t

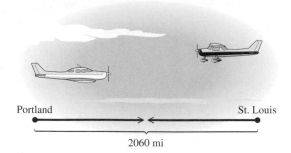

Portland   St. Louis

2060 mi

47. A train leaves Kansas City, Kansas, and travels north at 85 km per hr. Another train leaves at the same time and travels south at 95 km per hour. How long will it take before they are 315 km apart?

48. Two steamers leave a port on a river at the same time, traveling in opposite directions. Each is traveling at 22 mph. How long will it take for them to be 110 mi apart?

49. From a point on a straight road, Marco and Celeste ride bicycles in the same direction. Marco rides at 10 mph and Celeste rides at 12 mph. In how many hours will they be 15 mi apart?

50. At a given hour, two steamboats leave a city in the same direction on a straight canal. One travels at 18 mph and the other travels at 24 mph. In how many hours will the boats be 9 mi apart?

51. Two planes leave an airport at the same time, one flying east, the other flying west. The eastbound plane travels 150 mph slower. They are 2250 mi apart after 3 hr. Find the speed of each plane.

	r	t	d
Eastbound	x − 150	3	
Westbound	x	3	

52. Two trains leave a city at the same time. One travels north, and the other travels south 20 mph faster. In 2 hr, the trains are 280 mi apart. Find their speeds.

	r	t	d
Northbound	x	2	
Southbound	x + 20	2	

53. Two cars start from towns 400 mi apart and travel toward each other. They meet after 4 hr. Find the speed of each car if one travels 20 mph faster than the other.

54. Two cars leave towns 230 km apart at the same time, traveling directly toward one another. One car travels 15 km per hr slower than the other. They pass one another 2 hr later. What are their speeds?

The following problems are real "head-scratchers."

55. Kevin is three times as old as Bob. Three years ago the sum of their ages was 22 yr. How old is each now? (*Hint:* Write an expression first for the age of each now and then for the age of each three years ago.)

56. A store has 39 qt of milk, some in pint cartons and some in quart cartons. There are six times as many quart cartons as pint cartons. How many quart cartons are there? (*Hint:* 1 qt = 2 pt)

57. A table is three times as long as it is wide. If it were 3 ft shorter and 3 ft wider, it would be square (with all sides equal). How long and how wide is the table?

58. Elena works for $6 an hour. A total of 25% of her salary is deducted for taxes and insurance. How many hours must she work to take home $450?

59. Paula received a paycheck for $585 for her weekly wages less 10% deductions. How much was she paid before the deductions were made?

60. At the end of a day, the owner of a gift shop had $2394 in the cash register. This amount included sales tax of 5% on all sales. Find the amount of the sales.

2.8 Solving Linear Inequalities

OBJECTIVES

1 Graph intervals on a number line.

2 Use the addition property of inequality.

3 Use the multiplication property of inequality.

4 Solve linear inequalities by using both properties of inequality.

5 Solve applied problems by using inequalities.

6 Solve linear inequalities with three parts.

Inequalities are algebraic expressions related by

     $<$ "is less than," $\leq$ "is less than or equal to,"

     $>$ "is greater than," $\geq$ "is greater than or equal to."

We solve an inequality by finding all real number solutions of it. For example, the solution set $\{x \mid x \leq 2\}$ includes *all real numbers* that are less than or equal to 2, not just the *integers* less than or equal to 2.

OBJECTIVE 1 **Graph intervals on a number line.** A good way to show the solution set of an inequality is by graphing. We graph all the real numbers belonging to the set $\{x \mid x \leq 2\}$ by placing a square bracket at 2 on a number line and drawing an arrow extending from the bracket to the left (to represent the fact that all numbers less than 2 are also part of the graph). The graph is shown in Figure 15.

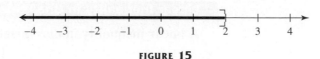

FIGURE 15

▶ **NOTE** Some texts use solid circles rather than square brackets to indicate that the endpoint is included in a number line graph. (Open circles are also used to indicate noninclusion, rather than parentheses, as described in Example 1(a).)

The set of numbers less than or equal to 2 is an example of an **interval** on the number line. To write intervals, we use **interval notation.** For example, the interval of all numbers less than or equal to 2 is written $(-\infty, 2]$. The **negative infinity** symbol $-\infty$ does not indicate a number, but shows that the interval includes all real numbers less than 2. As on the number line, the square bracket indicates that 2 is part of the solution. *A parenthesis is always used next to the infinity symbol. The set of real numbers is written in interval notation as* $(-\infty, \infty)$.

| **EXAMPLE 1** | Graphing Intervals Written in Interval Notation on a Number Line |

Write each inequality in interval notation and graph the interval.

(a) $x > -5$

The statement $x > -5$ says that x can represent any number greater than -5 but cannot equal -5. The interval is written $(-5, \infty)$. We show this interval on a graph by placing a parenthesis at -5 and drawing an arrow to the right, as in Figure 16. The parenthesis at -5 indicates that -5 is *not* part of the graph.

FIGURE 16

(b) $-1 \leq x < 3$

The statement is read "-1 is less than or equal to x *and* x is less than 3." Thus, we want the set of numbers that are *between* -1 and 3, with -1 included and 3 excluded. In interval notation, we write $[-1, 3)$, using a square bracket at -1 because -1 is part of the graph and a parenthesis at 3 because 3 is not part of the graph. The graph is shown in Figure 17.

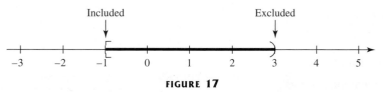

FIGURE 17

✔ **Now Try Exercises 7 and 15.**

OBJECTIVE 2 Use the addition property of inequality. Solving inequalities is similar to solving equations.

Linear Inequality in One Variable

A **linear inequality in one variable** can be written in the form

$$Ax + B < C,$$

where A, B, and C are real numbers, with $A \neq 0$.

Examples of linear inequalities in one variable are

$$x + 5 < 2, \quad y - 3 \geq 5, \quad \text{and} \quad 2k + 5 \leq 10. \qquad \text{Linear inequalities}$$

(All definitions and rules are also valid for $>$, $\leq$, and $\geq$.)

Consider the inequality $2 < 5$. If 4 is added to each side, the result is

$$2 + 4 < 5 + 4$$
$$6 < 9,$$

a true sentence. Now subtract 8 from each side:

$$2 - 8 < 5 - 8$$
$$-6 < -3.$$

The result is again a true sentence. These examples suggest the **addition property of inequality.**

Addition Property of Inequality

For any real numbers A, B, and C, the inequalities

$$A < B \qquad \text{and} \qquad A + C < B + C$$

have exactly the same solutions.

That is, the same number may be added to each side of an inequality without changing the solutions.

As with the addition property of equality, the same number may be *subtracted* from each side of an inequality.

EXAMPLE 2 Using the Addition Property of Inequality

Solve $7 + 3x > 2x - 5$.

$$7 + 3x > 2x - 5$$
$$7 + 3x - 2x > 2x - 5 - 2x \qquad \text{Subtract } 2x.$$
$$7 + x > -5 \qquad \text{Combine like terms.}$$
$$7 + x - 7 > -5 - 7 \qquad \text{Subtract } 7.$$
$$x > -12 \qquad \text{Combine like terms.}$$

The solution set is $(-12, \infty)$. Its graph is shown in Figure 18.

FIGURE 18

✔ **Now Try Exercise 23.**

▶ **NOTE** The inequality $7 + 3x > 2x - 5$ in Example 2 is equivalent to

$$2x - 5 < 7 + 3x.$$

If this inequality is solved so that the variable appears on the *right* side, the result is

$$-12 < x.$$

When this inequality is rewritten so that the variable is on the *left* side, it is

$$x > -12,$$

which is how it reads in Example 2. *Notice that the inequality symbol continues to point toward the lesser number.* Most students find it easier to graph inequalities when the variable is on the left.

OBJECTIVE 3 Use the multiplication property of inequality. The addition property of inequality cannot be used to solve an inequality such as $4x \geq 28$. This inequality requires the *multiplication property of inequality*. To see how this property works, we look at some examples.

Multiply each side of the inequality $3 < 7$ by the positive number 2.

$$3 < 7$$
$$2(3) < 2(7) \qquad \text{Multiply each side by 2.}$$
$$6 < 14 \qquad \text{True}$$

Now multiply each side of $3 < 7$ by the negative number -5.

$$3 < 7$$
$$-5(3) < -5(7) \qquad \text{Multiply each side by } -5.$$
$$-15 < -35 \qquad \text{False}$$

To get a true statement when multiplying each side by -5, we must reverse the direction of the inequality symbol.

$$3 < 7$$
$$-5(3) > -5(7) \qquad \text{Multiply by } -5; \text{ reverse the symbol.}$$
$$-15 > -35 \qquad \text{True}$$

As another example, multiply each side of $-6 < 2$ by the positive number 4.

$$-6 < 2$$
$$4(-6) < 4(2) \qquad \text{Multiply by 4.}$$
$$-24 < 8 \qquad \text{True}$$

Multiplying each side of $-6 < 2$ by -5 *and at the same time reversing the direction of the inequality symbol* gives

$$-6 < 2$$
$$-5(-6) > -5(2) \qquad \text{Multiply by } -5; \text{ reverse the symbol.}$$
$$30 > -10. \qquad \text{True}$$

In summary, the **multiplication property of inequality** has two parts.

Multiplication Property of Inequality

For any real numbers A, B, and C, with $C \neq 0$,

1. if C is *positive,* then the inequalities

$$A < B \qquad \text{and} \qquad AC < BC$$

   have exactly the same solutions;

2. if C is *negative,* then the inequalities

$$A < B \qquad \text{and} \qquad AC > BC$$

   have exactly the same solutions.

   That is, each side of an inequality may be multiplied by the same positive number without changing the solutions. *If the multiplier is negative, we must reverse the direction of the inequality symbol.*

The multiplication property of inequality also permits *division* of each side of an inequality by the same nonzero number.

   It is important to remember the differences in the multiplication property for positive and negative numbers.

1. When each side of an inequality is multiplied or divided by a positive number, the direction of the inequality symbol *does not change.* (Adding or subtracting terms on each side also does not change the symbol.)

2. When each side of an inequality is multiplied or divided by a negative number, the direction of the symbol *does change*. ***Reverse the direction of the inequality symbol only when multiplying or dividing each side by a negative number.***

EXAMPLE 3 Using the Multiplication Property of Inequality

Solve each inequality and graph the solution set.

(a) $3x < -18$

Using the multiplication property of inequality, we divide each side by 3. Since 3 is a positive number, the direction of the inequality symbol *does not* change.

$$3x < -18$$

$$\frac{3x}{3} < \frac{-18}{3} \quad \text{Divide by 3.}$$

$$x < -6$$

The solution set is $(-\infty, -6)$. The graph is shown in Figure 19.

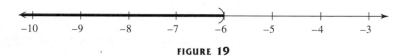

FIGURE 19

(b) $-4x \geq 8$

Here, each side of the inequality must be divided by -4, a negative number, which *does* change the direction of the inequality symbol.

$$-4x \geq 8$$

Reverse the inequality when multiplying or dividing by a negative number.

$$\frac{-4x}{-4} \leq \frac{8}{-4} \quad \text{Divide by } -4; \text{ reverse the symbol.}$$

$$x \leq -2$$

The solution set $(-\infty, -2]$ is graphed in Figure 20.

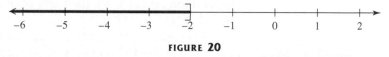

FIGURE 20

✔ **Now Try Exercises 33 and 35.**

▶ **CAUTION** Even though the number on the right side of the inequality in Example 3(a) is negative (-18), *do not reverse the direction of the inequality symbol*. Reverse the symbol only when *multiplying* or *dividing* by a negative number, as shown in Example 3(b).

OBJECTIVE 4 Solve linear inequalities by using both properties of inequality. To solve a linear inequality, follow the steps on the next page.

Solving a Linear Inequality

Step 1 **Simplify each side separately.** Use the distributive property to clear parentheses and combine like terms on each side as needed.

Step 2 **Isolate the variable terms on one side.** Use the addition property of inequality to get all terms with variables on one side of the inequality and all numbers on the other side.

Step 3 **Isolate the variable.** Use the multiplication property of inequality to change the inequality to the form $x < k$ or $x > k$, where k is a number.

Remember: Reverse the direction of the inequality symbol only when multiplying or dividing each side of an inequality by a negative number.

EXAMPLE 4 Solving a Linear Inequality

Solve $3x + 2 - 5 > -x + 7 + 2x$ and graph the solution set.

Step 1 Combine like terms and simplify.

$$3x + 2 - 5 > -x + 7 + 2x$$
$$3x - 3 > x + 7$$

Step 2 Use the addition property of inequality.

$$3x - 3 + 3 > x + 7 + 3 \qquad \text{Add 3.}$$
$$3x > x + 10$$
$$3x - x > x + 10 - x \qquad \text{Subtract } x.$$
$$2x > 10$$

Step 3 Use the multiplication property of inequality.

Because 2 is positive, keep the symbol >.

$$\frac{2x}{2} > \frac{10}{2} \qquad \text{Divide by 2.}$$
$$x > 5$$

Since 2 is positive, the direction of the inequality symbol is not changed in Step 3. The solution set is $(5, \infty)$. Its graph is shown in Figure 21.

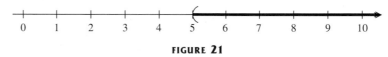

FIGURE 21

✔ **Now Try Exercise 45.**

EXAMPLE 5 Solving a Linear Inequality

Solve $5(x - 3) - 7x \geq 4(x - 3) + 9$ and graph the solution set.

Step 1 Simplify and combine like terms.

$$5(x - 3) - 7x \geq 4(x - 3) + 9$$
$$5x - 15 - 7x \geq 4x - 12 + 9 \qquad \text{Distributive property}$$
$$-2x - 15 \geq 4x - 3 \qquad \text{Combine like terms.}$$

Step 2 Use the addition property of inequality.

$$-2x - 15 - 4x \geq 4x - 3 - 4x \qquad \text{Subtract } 4x.$$
$$-6x - 15 \geq -3$$
$$-6x - 15 + 15 \geq -3 + 15 \qquad \text{Add } 15.$$
$$-6x \geq 12$$

Step 3 Divide each side by -6, a negative number.

Because -6 is negative, change $\geq$ to $\leq$.

$$\frac{-6x}{-6} \leq \frac{12}{-6} \qquad \text{Divide by } -6; \text{ reverse the symbol.}$$
$$x \leq -2$$

The solution set is $(-\infty, -2]$. Its graph is shown in Figure 22.

FIGURE 22

✔ **Now Try Exercise 49.**

OBJECTIVE 5 **Solve applied problems by using inequalities.** Until now, the applied problems that we have studied have all led to equations.

▶ **PROBLEM-SOLVING HINT** Inequalities can be used to solve applied problems involving phrases that suggest inequality. The table gives some of the more common such phrases, along with examples and translations.

Phrase	Example	Inequality
Is more than	A number *is more than* 4.	$x > 4$
Is less than	A number *is less than* -12.	$x < -12$
Is at least	A number *is at least* 6.	$x \geq 6$
Is at most	A number *is at most* 8.	$x \leq 8$

We use the same six problem-solving steps from **Section 2.4,** changing Step 3 to "Write an inequality" instead of "Write an equation."

▶ **CAUTION** Do not confuse statements such as "5 is more than a number" with the phrase "5 more than a number." The first of these is expressed as $5 > x$, while the second is expressed as $x + 5$ or $5 + x$.

The next example shows an application of algebra that is important to anyone who has ever asked, "What score can I make on my next test and have a (particular grade) in this course?" It uses the idea of finding the average of a number of grades. In general, to find the average of n numbers, add the numbers and then divide by n.

EXAMPLE 6 Finding an Average Test Score

Brent has grades of 86, 88, and 78 on his first three tests in geometry. If he wants an average of at least 80 after his fourth test, what are the possible scores he can make on that test?

Step 1 **Read** the problem again.

Step 2 **Assign a variable.** Let $x =$ Brent's score on his fourth test.

Step 3 **Write an inequality.** To find his average after four tests, add the test scores and divide by 4.

$$\underbrace{\frac{86 + 88 + 78 + x}{4}}_{\text{Average}} \underbrace{\geq}_{\text{is at least}} \underbrace{80}_{\text{80.}}$$

Step 4 **Solve.**

$$\frac{252 + x}{4} \geq 80 \qquad \text{Add the known scores.}$$

$$4\left(\frac{252 + x}{4}\right) \geq 4(80) \qquad \text{Multiply by 4.}$$

$$252 + x \geq 320$$

$$252 + x - 252 \geq 320 - 252 \qquad \text{Subtract 252.}$$

$$x \geq 68 \qquad \text{Combine like terms.}$$

Step 5 **State the answer.** He must score 68 or more on the fourth test to have an average of *at least* 80.

Step 6 **Check.**

$$\frac{86 + 88 + 78 + 68}{4} = \frac{320}{4} = 80$$

✔ **Now Try Exercise 77.**

▶ **CAUTION** Errors often occur when the phrases "at least" and "at most" appear in applied problems. Remember that

	at least	translates as	greater than or equal to
and	at most	translates as	less than or equal to.

OBJECTIVE 6 Solve linear inequalities with three parts. Inequalities that say that one number is *between* two other numbers are **three-part inequalities.** For example,

$$-3 < 5 < 7$$

says that 5 is between -3 and 7. For some applications, it is necessary to work with a three-part inequality such as

$$3 < x + 2 < 8,$$

where $x + 2$ is between 3 and 8. To solve this inequality, we subtract 2 from each of the three parts of the inequality.

$$3 - 2 < x + 2 - 2 < 8 - 2 \qquad \text{Subtract 2 from each part.}$$
$$1 < x < 6$$

The idea is to get the inequality in the form

$$\text{a number} < x < \text{another number,}$$

using "is less than." The solution set (in this case, the interval $(1, 6)$) can then easily be graphed.

▶ **CAUTION**   When inequalities have three parts, the order of the parts is important. It would be *wrong* to write an inequality as $8 < x + 2 < 3$, since this would imply that $8 < 3$, a false statement. ***In general, three-part inequalities are written so that the symbols point in the same direction and both point toward the lesser number.***

EXAMPLE 7   **Solving Three-Part Inequalities**

Solve each inequality and graph the solution set.

(a)
$$4 \le 3x - 5 < 6$$
$$4 + 5 \le 3x - 5 + 5 < 6 + 5 \qquad \text{Add 5 to each part.}$$
$$9 \le 3x < 11$$
$$\frac{9}{3} \le \frac{3x}{3} < \frac{11}{3} \qquad \text{Divide each part by 3.}$$

Remember to divide all *three* parts by 3.

$$3 \le x < \frac{11}{3}$$

The solution set is $\left[3, \frac{11}{3}\right)$. Its graph is shown in Figure 23.

FIGURE 23

(b) $-4 \le \dfrac{2}{3}m - 1 < 8$

Fractional coefficients in equations can be eliminated by multiplying each side by the least common denominator of the fractions. The same is true for inequalities.

$$3(-4) \le 3\left(\frac{2}{3}m - 1\right) < 3(8) \qquad \text{Multiply each part by 3.}$$
$$-12 \le 2m - 3 < 24 \qquad \text{Distributive property}$$
$$-12 + 3 \le 2m - 3 + 3 < 24 + 3 \qquad \text{Add 3 to each part.}$$
$$-9 \le 2m < 27$$
$$\frac{-9}{2} \le \frac{2m}{2} < \frac{27}{2} \qquad \text{Divide each part by 2.}$$
$$-\frac{9}{2} \le m < \frac{27}{2}$$

The solution set is $\left[-\frac{9}{2}, \frac{27}{2}\right)$. Its graph is shown in Figure 24.

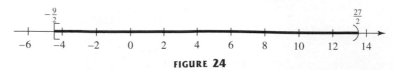

FIGURE 24

✔ **Now Try Exercises 61 and 67.**

▶ **N O T E**   The inequality in Example 7(b), $-4 \le \frac{2}{3}m - 1 < 8$, can also be solved by first adding 1 to each part and then multiplying each part by $\frac{3}{2}$. Do this and confirm that the same solution set results.

2.8 EXERCISES

🌐 *Complete solution available on Video Lectures on CD/DVD*

Now Try Exercise

📝 **1.** Explain how to determine whether to use a parenthesis or a square bracket at the endpoint when graphing an inequality on a number line.

📝 **2.** How does the graph of $x \ge -7$ differ from the graph of $x > -7$?

Write an inequality involving the variable x that describes each set of numbers graphed. See Example 1.

3.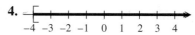

4.

5.

6.

Write each inequality in interval notation and graph the interval. See Example 1.

🌐 **7.** $k \le 4$ **8.** $r \le -11$ **9.** $x < -3$ **10.** $x < 3$

11. $t > 4$ **12.** $m > 5$ **13.** $8 \le x \le 10$ **14.** $3 \le x \le 5$

15. $0 < y \le 10$ **16.** $-3 \le x < 5$ **17.** $4 > -x > -3$ **18.** $6 \ge -x \ge 4$

19. *Concept Check*   Why is it wrong to write $3 < x < -2$ to indicate that x is between -2 and 3?

20. *Concept Check*   If $p < q$ and $r < 0$, which one of the following statements is false?

   A. $pr < qr$ **B.** $pr > qr$ **C.** $p + r < q + r$ **D.** $p - r < q - r$

Solve each inequality. Write the solution set in interval notation and graph it. See Example 2.

21. $z - 8 \ge -7$ **22.** $p - 3 \ge -11$ 🌐 **23.** $2k + 3 \ge k + 8$

24. $3x + 7 \ge 2x + 11$ **25.** $3n + 5 < 2n - 6$ **26.** $5x - 2 < 4x - 5$

27. *Concept Check*   Under what conditions must the inequality symbol be reversed when solving an inequality?

28. *Concept Check*   By what number must you *multiply* each side of $0.2x > 6$ to get just x on the left side?

29. Explain the steps you would use to solve the inequality $-5x > 20$.

30. Your friend tells you that, when solving the inequality $6x < -42$, he reversed the direction of the inequality symbol because of the -42. How would you respond?

Solve each inequality. Write the solution set in interval notation and graph it. See Example 3.

31. $3x < 18$

32. $5x < 35$

33. $2y \geq -20$

34. $6m \geq -24$

35. $-8t > 24$

36. $-7x > 49$

37. $-x \geq 0$

38. $-k < 0$

39. $-\dfrac{3}{4}r < -15$

40. $-\dfrac{7}{8}t < -14$

41. $-0.02x \leq 0.06$

42. $-0.03v \geq -0.12$

Solve each inequality. Write the solution set in interval notation and graph it. See Examples 4 and 5.

43. $5r + 1 \geq 3r - 9$

44. $6t + 3 < 3t + 12$

45. $6x + 3 + x < 2 + 4x + 4$

46. $-4w + 12 + 9w \geq w + 9 + w$

47. $-x + 4 + 7x \leq -2 + 3x + 6$

48. $14y - 6 + 7y > 4 + 10y - 10$

49. $5(x + 3) - 6x \leq 3(2x + 1) - 4x$

50. $2(x - 5) + 3x < 4(x - 6) + 1$

51. $\dfrac{2}{3}(p + 3) > \dfrac{5}{6}(p - 4)$

52. $\dfrac{7}{9}(y - 4) \leq \dfrac{4}{3}(y + 5)$

53. $4x - (6x + 1) \leq 8x + 2(x - 3)$

54. $2y - (4y + 3) > 6y + 3(y + 4)$

55. $5(2k + 3) - 2(k - 8) > 3(2k + 4) + k - 2$

56. $2(3z - 5) + 4(z + 6) \geq 2(3z + 2) + 3z - 15$

Write a three-part inequality involving the variable x that describes each set of numbers graphed. See Example 1(b).

57.

58.

59.

60.

Solve each inequality. Write the solution set in interval notation and graph it. See Example 7.

61. $-5 \leq 2x - 3 \leq 9$

62. $-7 \leq 3x - 4 \leq 8$

63. $5 < 1 - 6m < 12$

64. $-1 \leq 1 - 5q \leq 16$

65. $10 < 7p + 3 < 24$

66. $-8 \leq 3r - 1 \leq -1$

67. $-12 \leq \dfrac{1}{2}z + 1 \leq 4$

68. $-6 \leq 3 + \dfrac{1}{3}a \leq 5$

69. $1 \leq 3 + \dfrac{2}{3}p \leq 7$

70. $2 < 6 + \dfrac{3}{4}x < 12$

71. $-7 \leq \dfrac{5}{4}r - 1 \leq -1$

72. $-12 \leq \dfrac{3}{7}a + 2 \leq -4$

RELATING CONCEPTS (EXERCISES 73–76)

FOR INDIVIDUAL OR GROUP WORK

The methods for solving linear equations and for solving linear inequalities are quite similar. **Work Exercises 73–76 in order,** *to see how the solutions of an inequality are closely connected to the solution of the corresponding equation.*

73. Solve the equation $3x + 2 = 14$, and graph the solution set as a single point on the number line.

74. Solve the inequality $3x + 2 > 14$, and graph the solution set as an interval on the number line.

75. Solve the inequality $3x + 2 < 14$, and graph the solution set as an interval on the number line.

76. If you were to graph all the solution sets from Exercises 73–75 on the same number line, what would the graph be? (This is called the *union* of all the solution sets.)

Solve each problem. See Example 6.

77. Inkie Landry has scores of 76 and 81 on her first two algebra tests. If she wants an average of at least 80 after her third test, what possible scores can she make on that test?

78. Mabimi Pampo has scores of 96 and 86 on his first two geometry tests. What possible scores can he make on his third test so that his average is at least 90?

79. When 2 is added to the difference between six times a number and 5, the result is greater than 13 added to five times the number. Find all such numbers.

80. When 8 is subtracted from the sum of three times a number and 6, the result is less than 4 more than the number. Find all such numbers.

81. The formula for converting Fahrenheit temperature to Celsius is

$$C = \frac{5}{9}(F - 32).$$

If the Celsius temperature on a certain winter day in Minneapolis is never less than $-25°$, how would you describe the corresponding Fahrenheit temperatures? (*Source:* National Climatic Data Center.)

82. The formula for converting Celsius temperature to Fahrenheit is

$$F = \frac{9}{5}C + 32.$$

The Fahrenheit temperature of Phoenix has never exceeded 122°. How would you describe this using Celsius temperature? (*Source:* National Climatic Data Center.)

83. For what values of x would the rectangle have a perimeter of at least 400?

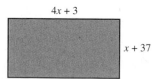

84. For what values of x would the triangle have a perimeter of at least 72?

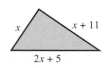

85. An international phone call costs $2.00 for the first 3 min, plus $0.30 per minute for each minute or fractional part of a minute after the first 3 min. If x represents the number of minutes of the length of the call after the first 3 min, then $2 + 0.30x$ represents the cost of the call. If Jorge has $5.60 to spend on a call, what is the maximum total time he can use the phone?

86. If the call described in Exercise 85 costs between $5.60 and $6.50, what are the possible total lengths of time for the call?

87. At the Speedy Gas'n Go, a car wash costs $4.50 and gasoline is selling for $3.20 per gallon. Lee Ann Spahr has $38.10 to spend, and her car is so dirty that she must have it washed. What is the maximum number of gallons of gasoline that she can purchase?

88. A rental company charges $15 to rent a chain saw, plus $2 per hour. Al Fabre can spend no more than $35 to clear some logs from his yard. What is the maximum amount of time he can use the rented saw?

Many applications from economics involve inequalities rather than equations. For example, a company that produces DVDs has found that revenue from the sales of the DVDs is $5 per DVD, less sales costs of $100. Production costs are $125, plus $4 per DVD. Profit (P) is given by revenue (R) less cost (C), so the company must find the production level x that makes

$$P = R - C > 0.$$

89. Write an expression for revenue R, letting x represent the production level (number of DVDs to be produced).

90. Write an expression for production costs C in terms of x.

91. Write an expression for profit P, and then solve the inequality $P = R - C > 0$.

92. Describe the solution in terms of the problem.

PREVIEW EXERCISES

*Find the value of y when (a) x = −2 and (b) x = 4. See **Sections 1.3 and 2.3**.*

93. $y = 5x + 3$ **94.** $y = 4 - 3x$ **95.** $6x - 2 = y$

96. $4x + 7y = 11$ **97.** $2x - 5y = 10$ **98.** $y + 3x = 8$

Chapter **2** **Group Activity**

ARE YOU A RACE WALKER?

Objective Use proportions to calculate walking speeds.
Materials Students will need a large area for walking. Each group will need a stopwatch.

Race walking at speeds exceeding 8 mph is a high-fitness, long-distance competitive sport. The table lists gold-medal winners in the race-walking competition in the 2004 Olympic Games. Complete the table by applying the proportion given next to find the race walker's number of steps per minute.

$$\frac{70 \text{ steps per min}}{2 \text{ mph}} = \frac{x \text{ steps per min}}{y \text{ mph}}$$

Use 10 km ≈ 6.21 mi. (Round all answers, except those for steps per minute, to the nearest thousandth. Round steps per minute to the nearest whole number.)

Event	Gold-Medal Winner	Country	Time in Hours: Minutes: Seconds	Time in Minutes	Time in Hours	y Miles per Hour	x Steps per Minute
20-km walk, women	Athanasia Tsoumeleka	Greece	1:29:12				
20-km walk, men	Ivano Brugnatti	Italy	1:19:40				
50-km walk, men	Robert Korzeniowski	Poland	3:38:48				

Source: World Almanac and Book of Facts 2006.

A. Using a stopwatch, take turns counting how many steps each member of the group takes in one minute while walking at a normal pace. Record the results in the following table. Then do it again at a fast pace. Record these results.

Name	Normal Pace		Fast Pace	
	x Steps per Minute	y Miles per Hour	x Steps per Minute	y Miles per Hour

B. Use the given proportion to convert the numbers from part A to miles per hour, and complete the table.

1. Find the average speed for the group at a normal pace and at a fast pace.

2. What is the minimum number of steps per minute you would have to take to be a race walker?

3. At a fast pace, did anyone in the group walk fast enough to be a race walker? Explain how you decided.

Chapter **2** SUMMARY *View the Interactive Summary on the Pass the Test CD.*

KEY TERMS

2.1 linear equation in one variable
solution set
equivalent equations
2.3 conditional equation
identity
contradiction
empty (null) set

2.4 degree
complementary angles
right angle
supplementary angles
straight angle
consecutive integers
2.5 formula
area

perimeter
vertical angles
volume
2.6 ratio
proportion
terms of a proportion
extremes
means

cross products
2.8 interval on the number line
interval notation
linear inequality in one variable
three-part inequality

NEW SYMBOLS

$\emptyset$ empty set
$1°$ one degree

a to b, $a{:}b$, or $\dfrac{a}{b}$
the ratio of a to b

(a, b) interval notation for $a < x < b$
$[a, b]$ interval notation for $a \le x \le b$

∞ infinity
$-\infty$ negative infinity
$(-\infty, \infty)$ set of all real numbers

TEST YOUR WORD POWER

See how well you have learned the vocabulary in this chapter. Answers, with examples, follow the Quick Review.

1. A **solution set** is the set of numbers that
   A. make an expression undefined
   B. make an equation false
   C. make an equation true
   D. make an expression equal to 0.

2. The **empty set** is a set
   A. with 0 as its only element
   B. with an infinite number of elements
   C. with no elements
   D. of ideas.

3. **Complementary angles** are angles
   A. formed by two parallel lines
   B. whose sum is 90°
   C. whose sum is 180°
   D. formed by perpendicular lines.

4. **Supplementary angles** are angles
   A. formed by two parallel lines
   B. whose sum is 90°
   C. whose sum is 180°
   D. formed by perpendicular lines.

5. A **ratio**
   A. compares two quantities using a quotient
   B. says that two quotients are equal
   C. is a product of two quantities
   D. is a difference between two quantities.

6. A **proportion**
   A. compares two quantities using a quotient
   B. says that two quotients are equal
   C. is a product of two quantities
   D. is a difference between two quantities.

7. An **inequality** is
   A. a statement that two algebraic expressions are equal
   B. a point on a number line
   C. an equation with no solutions
   D. a statement with algebraic expressions related by $<$, $\le$, $>$, or $\ge$.

8. **Interval notation** is
   A. a portion of a number line
   B. a special notation for describing a point on a number line
   C. a way to use symbols to describe an interval on a number line
   D. a notation to describe unequal quantities.

QUICK REVIEW

Concepts	Examples

2.1 THE ADDITION PROPERTY OF EQUALITY

The same number may be added to (or subtracted from) each side of an equation without changing the solution.

Solve.

$$x - 6 = 12$$
$$x - 6 + 6 = 12 + 6 \qquad \text{Add 6.}$$
$$x = 18 \qquad \text{Combine like terms.}$$

Solution set: $\{18\}$

2.2 THE MULTIPLICATION PROPERTY OF EQUALITY

Each side of an equation may be multiplied (or divided) by the same nonzero number without changing the solution.

Solve.

$$\frac{3}{4}x = -9$$
$$\frac{4}{3} \cdot \frac{3}{4}x = \frac{4}{3}(-9) \qquad \text{Multiply by } \tfrac{4}{3}.$$
$$x = -12$$

Solution set: $\{-12\}$

2.3 MORE ON SOLVING LINEAR EQUATIONS

Step 1 Simplify each side separately.

Solve.

$$2x + 2(x + 1) = 14 + x$$

$$2x + 2x + 2 = 14 + x \qquad \text{Distributive property}$$
$$4x + 2 = 14 + x \qquad \text{Combine like terms.}$$

Step 2 Isolate the variable term on one side.

$$4x + 2 - x - 2 = 14 + x - x - 2 \qquad \text{Subtract } x; \text{ subtract 2.}$$
$$3x = 12 \qquad \text{Combine like terms.}$$

Step 3 Isolate the variable.

$$\frac{3x}{3} = \frac{12}{3} \qquad \text{Divide by 3.}$$
$$x = 4$$

Step 4 Check.

Check:

$$2(4) + 2(4 + 1) = 14 + 4 \qquad ? \qquad \text{Let } x = 4.$$
$$18 = 18 \qquad \text{True}$$

Solution set: $\{4\}$

2.4 AN INTRODUCTION TO APPLICATIONS OF LINEAR EQUATIONS

Step 1 Read.

One number is five more than another. Their sum is 21. What are the numbers?

We are looking for two numbers.

Step 2 Assign a variable.

Let $\quad x$ = the smaller number.
Then $\quad x + 5$ = the larger number.

(continued)

Concepts	Examples

Step 3 Write an equation.

Step 4 Solve the equation.

$$x + (x + 5) = 21$$

$2x + 5 = 21$ Combine like terms.

$2x + 5 - 5 = 21 - 5$ Subtract 5.

$2x = 16$ Combine like terms.

$\dfrac{2x}{2} = \dfrac{16}{2}$ Divide by 2.

$x = 8$

Step 5 State the answer.

The numbers are 8 and 13.

Step 6 Check.

13 is five more than 8, and $8 + 13 = 21$. It checks.

2.5 FORMULAS AND APPLICATIONS FROM GEOMETRY

To find the value of one of the variables in a formula, given values for the others, substitute the known values into the formula.

Find L if $A = LW$, given that $A = 24$ and $W = 3$.

$24 = L \cdot 3$ $A = 24, W = 3$

$\dfrac{24}{3} = \dfrac{L \cdot 3}{3}$ Divide by 3.

$8 = L$

To solve a formula for one of the variables, isolate that variable by treating the other variables as numbers and using the steps for solving equations.

Solve $P = 2L + 2W$ for W.

$P - 2L = 2L + 2W - 2L$ Subtract $2L$.

$P - 2L = 2W$ Combine like terms.

$\dfrac{P - 2L}{2} = \dfrac{2W}{2}$ Divide by 2.

$\dfrac{P - 2L}{2} = W, \quad$ or $\quad W = \dfrac{P - 2L}{2}$

2.6 RATIOS AND PROPORTIONS

To write a ratio, express quantities in the same units.

4 ft to 8 in. $= 48$ in. to 8 in. $= \dfrac{48}{8} = \dfrac{6}{1}$

To solve a proportion, use the method of cross products.

Solve $\dfrac{x}{12} = \dfrac{35}{60}$.

$60x = 12 \cdot 35$ Cross products

$60x = 420$ Multiply.

$\dfrac{60x}{60} = \dfrac{420}{60}$ Divide by 60.

$x = 7$

Solution set: $\{7\}$

(continued)

Concepts	Examples

2.7 FURTHER APPLICATIONS OF LINEAR EQUATIONS

Step 1 Read.

A sum of money is invested at simple interest in two ways. Part is invested at 12%, and $20,000 less than that amount is invested at 10%. If the total interest for 1 yr is $9000, find the amount invested at each rate.

Step 2 Assign a variable. Make a table to help solve the problems in this section.

Let $\qquad x =$ amount invested at 12%.

Then $\quad x - 20,000 =$ amount invested at 10%.

Dollars Invested	Rate of Interest	Interest for One Year
x	0.12	0.12x
$x - 20,000$	0.10	0.10(x − 20,000)

Step 3 Write an equation.

Step 4 Solve the equation.

$$0.12x + 0.10(x - 20,000) = 9000$$

$$0.12x + 0.10x + 0.10(-20,000) = 9000 \qquad \text{Distributive property}$$

$$12x + 10x + 10(-20,000) = 900,000 \qquad \text{Multiply by 100.}$$

$$12x + 10x - 200,000 = 900,000$$

$$22x - 200,000 = 900,000$$

$$22x = 1,100,000 \qquad \text{Add 200,000.}$$

$$x = 50,000 \qquad \text{Divide by 22.}$$

Steps 5 and 6 State the answer and check the solution.

$50,000 is invested at 12% and $30,000 is invested at 10%.

The three forms of the formula relating distance, rate, and time are

$$d = rt, \quad r = \frac{d}{t}, \quad \text{and} \quad t = \frac{d}{r}.$$

Two cars leave from the same point, traveling in opposite directions. One travels at 45 mph and the other at 60 mph. How long will it take them to be 210 mi apart?

Let $t =$ time it takes for them to be 210 mi apart.

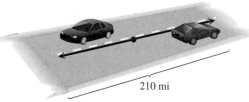

210 mi

The table gives the information from the problem.

	Rate	Time	Distance
One Car	45	t	45t
Other Car	60	t	60t

The sum of the distances, 45t and 60t, must be 210 mi.

$$45t + 60t = 210$$

$$105t = 210 \qquad \text{Combine like terms.}$$

$$t = 2 \qquad \text{Divide by 105.}$$

It will take them 2 hr to be 210 mi apart.

(continued)

Concepts	Examples

2.8 SOLVING LINEAR INEQUALITIES

Step 1 Simplify each side separately.

Step 2 Isolate the variable term on one side.

Step 3 Isolate the variable.

Be sure to reverse the direction of the inequality symbol when multiplying or dividing by a negative number.

To solve a three-part inequality such as
$$4 < 2x + 6 < 8,$$
work with all three expressions at the same time.

Solve the inequality and graph the solution set.

$$3(1 - x) + 5 - 2x > 9 - 6$$

$3 - 3x + 5 - 2x > 9 - 6$	Clear parentheses.
$8 - 5x > 3$	Combine like terms.
$8 - 5x - 8 > 3 - 8$	Subtract 8.
$-5x > -5$	Combine like terms.
$\dfrac{-5x}{-5} < \dfrac{-5}{-5}$	Divide by -5; change $>$ to $<$.
$x < 1$	

Solution set: $(-\infty, 1)$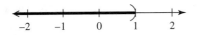

Solve.

$4 < 2x + 6 < 8$	
$4 - 6 < 2x + 6 - 6 < 8 - 6$	Subtract 6.
$-2 < 2x < 2$	Combine like terms.
$\dfrac{-2}{2} < \dfrac{2x}{2} < \dfrac{2}{2}$	Divide by 2.
$-1 < x < 1$	

Solution set: $(-1, 1)$

Answers to Test Your Word Power

1. C; *Example:* {8} is the solution set of $2x + 5 = 21$.

2. C; *Example:* The empty set $\emptyset$ is the solution set of $5x + 3 = 5x + 4$.

3. B; *Example:* Angles with measures $35°$ and $55°$ are complementary angles.

4. C; *Example:* Angles with measures $112°$ and $68°$ are supplementary angles.

5. A; *Example:* $\frac{7 \text{ in.}}{12 \text{ in.}}$, or $\frac{7}{12}$

6. B; *Example:* $\frac{2}{3} = \frac{8}{12}$

7. D; *Examples:* $x < 5, 7 + 2y \geq 11$, $-5 < 2z - 1 \leq 3$

8. C; *Examples:* $(-\infty, 5], (1, \infty), [-3, 3)$

Chapter **2** REVIEW EXERCISES

[2.1–2.3] *Solve each equation.*

1. $x - 5 = 1$

2. $x + 8 = -4$

3. $3k + 1 = 2k + 8$

4. $5k = 4k + \dfrac{2}{3}$

5. $(4r - 2) - (3r + 1) = 8$

6. $3(2x - 5) = 2 + 5x$

7. $7k = 35$

8. $12r = -48$

9. $2p - 7p + 8p = 15$

10. $\dfrac{m}{12} = -1$ **11.** $\dfrac{5}{8}k = 8$ **12.** $12m + 11 = 59$

13. $3(2x + 6) - 5(x + 8) = x - 22$ **14.** $5x + 9 - (2x - 3) = 2x - 7$

15. $\dfrac{1}{2}r - \dfrac{r}{3} = \dfrac{r}{6}$ **16.** $0.1(x + 80) + 0.2x = 14$

17. $3x - (-2x + 6) = 4(x - 4) + x$ **18.** $2(y - 3) - 4(y + 12) = -2(y + 27)$

[2.4] *Solve each problem.*

19. In 2005, Oklahoma had a total of 101 members in its House of Representatives, consisting of only Democrats and Republicans. There were 13 fewer Democrats than Republicans. How many representatives from each party were there? (*Source:* National Conference of State Legislatures, 2005.)

20. The land area of Hawaii is 5213 mi² greater than the area of Rhode Island. Together, the areas total 7637 mi². What is the area of each of the two states?

21. The height of Kegon Falls in Japan is $\frac{11}{8}$ the height of Rhaiadr Falls in Wales. The sum of the heights of these two waterfalls is 570 ft. Find the height of each. (*Source:* National Geographic Society.)

22. The height of Seven Falls in Colorado is $\frac{5}{2}$ the height of Twin Falls in Idaho. The sum of the heights is 420 ft. Find the height of each. (*Source: World Almanac and Book of Facts.*)

23. The supplement of an angle measures 10 times the measure of its complement. What is the measure of the angle?

24. Find two consecutive odd integers such that when the lesser is added to twice the greater, the result is 24 more than the greater integer.

[2.5] *A formula is given along with the values for all but one of the variables. Find the value of the variable that is not given. Use 3.14 as an approximation for* π.

25. $A = \dfrac{1}{2}bh; A = 44, b = 8$ **26.** $A = \dfrac{1}{2}h(b + B); h = 8, b = 3, B = 4$

27. $C = 2\pi r; C = 29.83$ **28.** $V = \dfrac{4}{3}\pi r^3; r = 6$

Solve each formula for the specified variable.

29. $A = bh$ for h **30.** $A = \dfrac{1}{2}h(b + B)$ for h

Find the measure of each marked angle.

31.

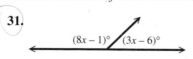

32.

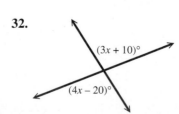

Solve each problem.

33. The perimeter of a certain rectangle is 16 times the width. The length is 12 cm more than the width. Find the width of the rectangle.

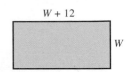

34. The Ziegfield Room in Reno, Nevada, has a circular turntable on which its showgirls dance. The circumference of the table is 62.5 ft. What is the diameter? What is the radius? What is the area? (Use $\pi = 3.14$.) (*Source: Guinness World Records.*)

35. A baseball diamond is a square with a side of 90 ft. The pitcher's mound is located 60.5 ft from home plate, as shown in the figure. Find the measures of the angles marked in the figure. (*Hint:* Recall that the sum of the measures of the angles of any triangle is 180°.)

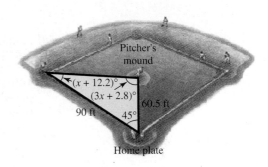

36. *Concept Check* What is wrong with the following problem? "The formula for the area of a trapezoid is $A = \frac{1}{2}h(b + B)$. If $h = 12$ and $b = 14$, find the numerical value of A."

[2.6] *Give a ratio for each word phrase, writing fractions in lowest terms.*

37. 60 cm to 40 cm

38. 5 days to 2 weeks

39. 90 in. to 10 ft

40. 3 months to 3 yr

Solve each equation.

41. $\dfrac{p}{21} = \dfrac{5}{30}$

42. $\dfrac{5 + x}{3} = \dfrac{2 - x}{6}$

Use proportions to solve each problem.

43. If 2 lb of fertilizer will cover 150 ft² of lawn, how many pounds would be needed to cover 500 ft²?

44. The tax on a $24.00 item is $2.04. How much tax would be paid on a $36.00 item?

45. The distance between two cities on a road map is 32 cm. The two cities are actually 150 km apart. The distance on the map between two other cities is 80 cm. How far apart are these cities?

46. In the 2004 Olympics held in Athens, Greece, Chinese athletes earned 63 medals. Two of every 9 medals were bronze. How many bronze medals did China earn? (*Source: World Almanac and Book of Facts 2006.*)

Find the best buy for each item.

47. CEREAL

Size	Price
15 oz	$2.69
20 oz	$3.29
25.5 oz	$3.49

48. PRUNE JUICE

Size	Price
32 oz	$1.95
48 oz	$2.89
64 oz	$3.29

[2.7] *Solve each problem.*

49. The San Francisco Giants baseball team borrowed $160 million of the $290 million needed to build a new stadium. Corporate sponsors provided the rest of the cost. What percent of the cost of the park was borrowed? (*Source:* Michael K. Ozanian, "Fields of Debt," *Forbes,* December 15, 1997.)

50. A nurse must mix 15 L of a 10% solution of a drug with some 60% solution to obtain a 20% mixture. How many liters of the 60% solution will be needed?

51. Todd Cardella invested $10,000, from which he earns an annual income of $550 per year. He invested part of the $10,000 at 5% annual interest and the remainder in bonds paying 6% interest. How much did he invest at each rate?

52. In 1846, the vessel *Yorkshire* traveled from Liverpool to New York, a distance of 3150 mi, in 384 hr. What was the *Yorkshire's* average speed? Round your answer to the nearest tenth.

53. Honey Kirk drove from Louisville to Dallas, a distance of 819 mi, averaging 63 mph. What was her driving time?

54. Two planes leave St. Louis at the same time. One flies north at 350 mph and the other flies south at 420 mph. In how many hours will they be 1925 mi apart?

[2.8] *Write each inequality in interval notation and graph it.*

55. $x \geq -4$ 56. $x < 7$ 57. $-5 \leq x < 6$

58. **Concept Check** Which one of the following inequalities requires reversal of the inequality symbol when it is solved?

     A. $4x \geq -36$ **B.** $-4x \leq 36$ **C.** $4x < 36$ **D.** $4x > 36$

Solve each inequality and graph the solution set.

59. $x + 6 \geq 3$ 60. $5x < 4x + 2$

61. $-6x \leq -18$ 62. $8(x - 5) - (2 + 7x) \geq 4$

63. $4x - 3x > 10 - 4x + 7x$

64. $3(2x + 5) + 4(8 + 3x) < 5(3x + 2) + 2x$

65. $-3 \leq 2x + 1 \leq 4$ 66. $9 < 3x + 5 \leq 20$

Solve each problem.

67. Carlotta Valdés has grades of 94 and 88 on her first two calculus tests. What possible scores on a third test will give her an average of at least 90?

68. If nine times a number is added to 6, the result is at most 3. Find all such numbers.

MIXED REVIEW EXERCISES

Solve.

69. $\dfrac{x}{7} = \dfrac{x-5}{2}$

70. $I = prt$ for r

71. $-2x > -4$

72. $2k - 5 = 4k + 13$

73. $0.05x + 0.02x = 4.9$

74. $2 - 3(x - 5) = 4 + x$

75. $9x - (7x + 2) = 3x + (2 - x)$

76. $\dfrac{1}{3}s + \dfrac{1}{2}s + 7 = \dfrac{5}{6}s + 5 + 2$

77. *Concept Check*   A student solved $3 - (8 + 4x) = 2x + 7$ and gave the solution set $\{6\}$. Verify that this answer is incorrect by checking it in the equation. Then explain the error and give the correct solution set. (*Hint:* The error involves the subtraction sign.)

78. Athletes in vigorous training programs can eat 50 calories per day for every 2.2 lb of body weight. To the nearest hundred, how many calories can a 175-lb athlete consume per day? (*Source: The Gazette,* Cedar Rapids Iowa, March 23, 2002.)

79. The Golden Gate Bridge in San Francisco is 2604 ft longer than the Brooklyn Bridge. Together, their spans total 5796 ft. How long is each bridge? (*Source: World Almanac and Book of Facts 2006.*)

80. Which is the best buy?

APPLE JUICE

Size	Price
32 oz	$1.19
48 oz	$1.79
64 oz	$1.99

81. If 1 qt of oil must be mixed with 24 qt of gasoline, how much oil would be needed for 192 qt of gasoline?

82. Two trains are 390 mi apart. They start at the same time and travel toward one another, meeting 3 hr later. If the speed of one train is 30 mph more than the speed of the other train, find the speed of each train.

83. The perimeter of a triangle is 96 m. One side is twice as long as another, and the third side is 30 m long. What is the length of the longest side?

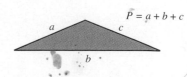

84. The perimeter of a certain square cannot be greater than 200 m. Find the possible value for the length of a side.

Chapter	**2**	TEST

*View the complete solutions
to all Chapter Test exercises
on the Pass the Test CD.*

Solve each equation.

1. $5x + 9 = 7x + 21$

2. $-\dfrac{4}{7}x = -12$

3. $7 - (x - 4) = -3x + 2(x + 1)$

4. $0.6(x + 20) + 0.8(x - 10) = 46$

5. $-8(2x + 4) = -4(4x + 8)$

Solve each problem.

6. In the 2005 baseball season, the St. Louis Cardinals won the most games of any major league team. The Cardinals won 24 less than twice as many games as they lost. They played 162 regular-season games. How many wins and losses did the Cardinals have? (*Source:* mlb.com)

7. Three islands in the Hawaiian island chain are Hawaii (the Big Island), Maui, and Kauai. Together, their areas total 5300 mi². The island of Hawaii is 3293 mi² larger than the island of Maui, and Maui is 177 mi² larger than Kauai. What is the area of each island?

8. Find the measure of an angle if its supplement measures 10° more than three times its complement.

9. The formula for the perimeter of a rectangle is $P = 2L + 2W$.

(a) Solve for W.
(b) If $P = 116$ and $L = 40$, find the value of W.

10. Find the measure of each marked angle.

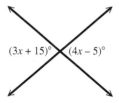

Solve each proportion.

11. $\dfrac{z}{8} = \dfrac{12}{16}$

12. $\dfrac{x + 5}{3} = \dfrac{x - 3}{4}$

Solve each problem.

13. Which is the better buy for processed cheese slices, 8 slices for $2.19 or 12 slices for $3.30?

14. The distance between Milwaukee and Boston is 1050 mi. On a certain map, this distance is represented by 42 in. On the same map, Seattle and Cincinnati are 92 in. apart. What is the actual distance between Seattle and Cincinnati?

15. Keith Boyle invested some money at 3% simple interest and $6000 more than that amount at 4.5% simple interest. After 1 yr, his total interest from the two accounts was $870. How much did he invest at each rate?

16. Two cars leave from the same point, traveling in opposite directions. One travels at a constant rate of 50 mph, while the other travels at a constant rate of 65 mph. How long will it take for them to be 460 mi apart?

Solve each inequality and graph the solution set.

17. $-4x + 2(x - 3) \geq 4x - (3 + 5x) - 7$ **18.** $-10 < 3x - 4 \leq 14$

19. Twylene Johnson has grades of 76 and 81 on her first two algebra tests. If she wants an average of at least 80 after her third test, what score must she make on that test?

✏ **20.** Write a short explanation of the additional (extra) rule that must be remembered when solving an inequality (as opposed to solving an equation).

Chapters **1-2 CUMULATIVE REVIEW EXERCISES**

1. Write $\frac{108}{144}$ in lowest terms.

Perform each indicated operation.

2. $\dfrac{5}{6} + \dfrac{1}{4} - \dfrac{7}{15}$

3. $\dfrac{9}{8} \cdot \dfrac{16}{3} \div \dfrac{5}{8}$

Translate from words to symbols. Use x as the variable if necessary.

4. The difference between half a number and 18

5. The quotient of 6 and 12 more than a number is 2.

6. *True* or *false?* $\dfrac{8(7) - 5(6 + 2)}{3 \cdot 5 + 1} \geq 1$

Perform each indicated operation.

7. $9 - (-4) + (-2)$

8. $\dfrac{-4(9)(-2)}{-3^2}$

9. $(-7 - 1)(-4) + (-4)$

10. Find the value of $\dfrac{3x^2 - y^3}{-4z}$ when $x = -2$, $y = -4$, and $z = 3$.

Name each property illustrated.

11. $7(k + m) = 7k + 7m$

12. $3 + (5 + 2) = 3 + (2 + 5)$

13. Simplify $-4(k + 2) + 3(2k - 1)$ by combining like terms.

Solve each equation, and check the solution.

14. $2r - 6 = 8r$

15. $4 - 5(a + 2) = 3(a + 1) - 1$

16. $\dfrac{2}{3}x + \dfrac{3}{4}x = -17$

17. $\dfrac{2x + 3}{5} = \dfrac{x - 4}{2}$

Solve each formula for the indicated variable.

18. $3x + 4y = 24$ for y

19. $A = P(1 + ni)$ for n

Solve each inequality. Graph the solution set.

20. $6(r - 1) + 2(3r - 5) \leq -4$

21. $-18 \leq -9z < 9$

Solve each problem.

22. For a woven hanging, Miguel Hidalgo needs three pieces of yarn, which he will cut from a 40-cm piece. The longest piece is to be three times as long as the middle-sized piece, and the shortest piece is to be 5 cm shorter than the middle-sized piece. What lengths should he cut?

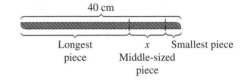

23. A fully inflated professional basketball has a circumference of 78 cm. What is the radius of a circular cross section through the center of the ball? (Use 3.14 as the approximation for π.) Round your answer to the nearest hundredth.

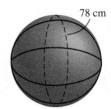

24. Piotrek Galkowski wants to increase a recipe that serves 6 to make enough for 20 people. The recipe calls for $1\frac{1}{4}$ cups of grated cheese. How much cheese will be needed to serve 20?

25. Two cars are 400 mi apart. Both start at the same time and travel toward one another. They meet 4 hr later. If the speed of one car is 20 mph faster than the other, what is the speed of each car?

Linear Equations and Inequalities in Two Variables; Functions

While U.S. consumers in general continue to pile up credit card debt, fewer undergraduate college students are carrying credit cards, and those with cards are using them less. In 2004, 76% of undergraduates carried at least one credit card, down from a peak of 83% in 2001. The average outstanding balance also dropped to $2169, from a high of $2748 in 2000. These declines are attributed to increased financial education aimed specifically at high school and college students. More work remains to further improve the financial well-being of the nation's students. (*Source:* Nellie Mae.)

In Example 7 of Section 3.2, we examine a *linear equation in two variables* that models credit card debt in the United States.

3.1 Reading Graphs; Linear Equations in Two Variables

OBJECTIVES

1 Interpret graphs.

2 Write a solution as an ordered pair.

3 Decide whether a given ordered pair is a solution of a given equation.

4 Complete ordered pairs for a given equation.

5 Complete a table of values.

6 Plot ordered pairs.

As we saw in **Chapter 1,** circle graphs (pie charts) and bar graphs provide a convenient way to organize and communicate information. Along with *line graphs*, they can be used to analyze data, make predictions, or simply to entertain us.

OBJECTIVE 1 Interpret graphs. Recall that a **bar graph** is used to show comparisons. It consists of a series of bars (or simulations of bars) arranged either vertically or horizontally. In a bar graph, values from two categories are paired with each other.

EXAMPLE 1 Interpreting a Bar Graph

The bar graph in Figure 1 shows U.S. sales of motor scooters, which have gained popularity due to their fuel efficiency. The graph compares sales in thousands.

Source: Motorcycle Industry Council.

FIGURE 1

(a) In what years were sales greater than 50 thousand?

Locate 50 on the vertical scale and follow the line across to the right. Three years—2002, 2003, and 2004—have bars that extend above the line for 50, so sales were greater than 50 thousand in those years.

(b) Estimate sales in 2000 and 2004.

Locate the top of the bar for 2000, and move horizontally across to the vertical scale to see that it is about 40. Sales in 2000 were about 40 thousand. Follow the top of the bar for 2004 across to the vertical scale to see that it lies about halfway between 80 and 90 thousand, so sales in 2004 were about 85,000.

(c) Describe the change in sales as the years progressed.

As the years progressed, sales increased steadily, from about 15 thousand in 1998 to 85 thousand in 2004.

✔ **Now Try Exercises 1 and 3.**

A line graph is used to show changes or trends in data over time. To form a **line graph,** we connect a series of points representing data with line segments.

EXAMPLE 2 Interpreting a Line Graph

The line graph in Figure 2 shows the number of U.S. households in millions that purchased real Christmas trees for the years 1999 through 2004.

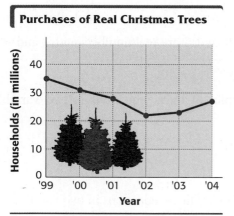

Purchases of Real Christmas Trees

Source: National Christmas Tree Association.

FIGURE 2

(a) Between which years did the number of households purchasing real Christmas trees increase?

The lines between 2002 and 2003 and between 2003 and 2004 rise, so the number of households purchasing real trees increased from 2002 to 2004.

(b) What was the general trend in the number of households purchasing real Christmas trees from 1999 through 2002? After 2002?

The line graph falls from 1999 to 2002, so the number of households purchasing real trees decreased over those years. This trend changed after 2002, when the number of households purchasing real trees increased.

(c) Estimate the number of households purchasing real Christmas trees in 1999 and 2002. About how much did this number decline between 1999 and 2002?

Move up from 1999 on the horizontal scale to the point plotted for 1999. This point is about halfway between the lines for 30 and 40 million, so about 35 million households purchased trees in 1999. Move up from 2002 on the horizontal scale to the point plotted for 2002. Then move across to the vertical scale. The point for 2002 is about 22 million.

Therefore, between 1999 and 2002, the number of households purchasing real Christmas trees declined by about

$$35 - 22 = 13 \text{ million households.}$$

✔ **Now Try Exercises 5 and 7.**

Many everyday situations, such as those illustrated in Examples 1 and 2, involve two quantities that are related. The equations and applications we discussed in **Chapter 2** had only one variable. In this chapter, we extend those ideas to *linear equations in two variables*.

Linear Equation in Two Variables

A **linear equation in two variables** is an equation that can be written in the form

$$Ax + By = C,$$

where A, B, and C are real numbers and A and B are not both 0.

Some examples of linear equations in two variables in this form, called *standard form*, are

$$3x + 4y = 9, \quad x - y = 0, \quad \text{and} \quad x + 2y = -8. \qquad \text{Linear equations in two variables}$$

▶ **NOTE** Other linear equations in two variables, such as

$$y = 4x + 5 \quad \text{and} \quad 3x = 7 - 2y,$$

are not written in standard form, but could be. We discuss the forms of linear equations in more detail in **Section 3.4.**

OBJECTIVE 2 Write a solution as an ordered pair. Recall from **Section 1.2** that a *solution* of an equation is a number that makes the equation true when it replaces the variable. For example, the linear equation in *one* variable $x - 2 = 5$ has solution 7, since replacing x with 7 gives a true statement.

A solution of a linear equation in *two* variables requires *two* numbers, one for each variable. For example, a true statement results when we replace x with 2 and y with 13 in the equation $y = 4x + 5$, since

$$13 = 4(2) + 5. \qquad \text{Let } x = 2; y = 13.$$

The pair of numbers $x = 2$ and $y = 13$ gives a solution of the equation $y = 4x + 5$. The phrase "$x = 2$ and $y = 13$" is abbreviated

$$\underset{\text{Ordered pair}}{\underbrace{(2, 13)}}$$

x-value ——↓ ↓—— *y*-value

with the x-value, 2, and the y-value, 13, given as a pair of numbers written inside parentheses. ***The x-value is always given first.*** A pair of numbers such as $(2, 13)$ is called an **ordered pair,** since the order in which the numbers are written is important. The ordered pairs $(2, 13)$ and $(13, 2)$ are *not* the same. The second pair indicates that $x = 13$ and $y = 2$. For two ordered pairs to be equal, their x-coordinates must be equal *and* their y-coordinates must be equal.

OBJECTIVE 3 Decide whether a given ordered pair is a solution of a given equation.
We substitute the x- and y-values of an ordered pair into a linear equation in two variables to see whether the ordered pair is a solution. An ordered pair that is a solution of an equation is said to *satisfy* the equation.

EXAMPLE 3 Deciding whether Ordered Pairs Are Solutions of an Equation

Decide whether each ordered pair is a solution of the equation $2x + 3y = 12$.

(a) $(3, 2)$

To see whether $(3, 2)$ is a solution of the equation $2x + 3y = 12$, substitute 3 for x and 2 for y in the equation.

$$2x + 3y = 12$$
$$2(3) + 3(2) = 12 \quad ? \quad \text{Let } x = 3; \text{ let } y = 2.$$
$$6 + 6 = 12 \quad ?$$
$$12 = 12 \quad \text{True}$$

This result is true, so $(3, 2)$ is a solution of $2x + 3y = 12$.

(b) $(-2, -7)$

$$2x + 3y = 12$$
$$2(-2) + 3(-7) = 12 \quad ? \quad \text{Let } x = -2; \text{ let } y = -7.$$

> Use parentheses to avoid errors.

$$-4 + (-21) = 12 \quad ?$$
$$-25 = 12 \quad \text{False}$$

This result is false, so $(-2, -7)$ is *not* a solution of $2x + 3y = 12$.

> ✔ **Now Try Exercises 17 and 21.**

OBJECTIVE 4 Complete ordered pairs for a given equation. Choosing a number for one variable in a linear equation makes it possible to find the value of the other variable.

EXAMPLE 4 Completing Ordered Pairs

Complete each ordered pair for the equation $y = 4x + 5$.

(a) $(7, \quad)$

> The x-value always comes first.

In this ordered pair, $x = 7$. To find the corresponding value of y, replace x with 7 in the equation.

$$y = 4x + 5$$
$$y = 4(7) + 5 \quad \text{Let } x = 7.$$
$$y = 28 + 5 \quad \text{Multiply.}$$
$$y = 33 \quad \text{Add.}$$

The ordered pair is $(7, 33)$.

(b) (, −3)

In this ordered pair, $y = -3$. Find the corresponding value of x by replacing y with -3 in the equation.

$$y = 4x + 5$$
$$-3 = 4x + 5 \qquad \text{Let } y = -3.$$
$$-8 = 4x \qquad \text{Subtract 5 from each side.}$$
$$-2 = x \qquad \text{Divide each side by 4.}$$

The ordered pair is $(-2, -3)$.

✔ **Now Try Exercises 31 and 33.**

OBJECTIVE 5 **Complete a table of values.** Ordered pairs are often displayed in a **table of values.** Although we usually write tables of values vertically, they may be written horizontally.

EXAMPLE 5 **Completing Tables of Values**

Complete the table of values for each equation. Then write the results as ordered pairs.

(a) $x - 2y = 8$

x	y
2	
10	
	0
	−2

To complete the first two ordered pairs, let $x = 2$ and then let $x = 10$.

If	$x = 2,$		If	$x = 10,$
then	$x - 2y = 8$		then	$x - 2y = 8$
becomes	$2 - 2y = 8$		becomes	$10 - 2y = 8$
	$-2y = 6$			$-2y = -2$
	$y = -3.$			$y = 1.$

Complete the last two ordered pairs by letting $y = 0$ and then $y = -2$.

If	$y = 0,$		If	$y = -2,$
then	$x - 2y = 8$		then	$x - 2y = 8$
becomes	$x - 2(0) = 8$		becomes	$x - 2(-2) = 8$
	$x - 0 = 8$			$x + 4 = 8$
	$x = 8.$			$x = 4.$

The completed table of values follows.

x	y
2	−3
10	1
8	0
4	−2

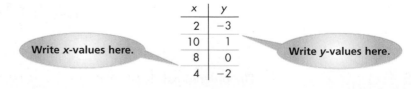

Write *x*-values here.

Write *y*-values here.

The corresponding ordered pairs are $(2, -3)$, $(10, 1)$, $(8, 0)$, and $(4, -2)$. Each ordered pair is a solution of the given equation $x - 2y = 8$.

(b) $x = 5$

x	y
	-2
	6
	3

The given equation is $x = 5$. No matter which value of y is chosen, the value of x is always 5.

x	y
5	-2
5	6
5	3

The corresponding ordered pairs are $(5, -2)$, $(5, 6)$, and $(5, 3)$.

✔ **Now Try Exercises 41 and 45.**

▶ **NOTE** We can think of $x = 5$ in Example 5(b) as an equation in two variables by rewriting $x = 5$ as $x + 0y = 5$. This form of the equation shows that, for any value of y, the value of x is 5. Similarly, $y = 4$ is the same as $0x + y = 4$.

OBJECTIVE 6 Plot ordered pairs. In **Section 2.3,** we saw that linear equations in *one* variable had either one, zero, or an infinite number of real number solutions. These solutions could be graphed on *one* number line. Every linear equation in *two* variables has an infinite number of ordered pairs as solutions. Each choice of a number for one variable leads to a particular real number for the other variable.

To graph these solutions, represented as ordered pairs (x, y), we need *two* number lines, one for each variable. The two number lines are drawn as shown in Figure 3. The horizontal number line is called the **x-axis,** and the vertical line is called the **y-axis.** Together, the x-axis and y-axis form a **rectangular coordinate system,** also called the **Cartesian coordinate system,** in honor of René Descartes (1596–1650), the French mathematician who is credited with its invention.

The coordinate system is divided into four regions, called **quadrants.** These quadrants are numbered counterclockwise, as shown in Figure 3. *Points on the axes themselves are not in any quadrant.* The point at which the x-axis and y-axis meet is called the **origin.** The origin, which is labeled 0 in Figure 3, is the point corresponding to $(0, 0)$.

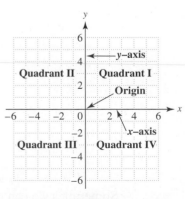

FIGURE 3

The *x*-axis and *y*-axis determine a **plane**—a flat surface illustrated by a sheet of paper. By referring to the two axes, we can associate every point in the plane with an ordered pair. The numbers in the ordered pair are called the **coordinates** of the point.

For example, locate the point associated with the ordered pair $(2, 3)$ by starting at the origin. Since the *x*-coordinate is 2, go 2 units to the right along the *x*-axis. Then, since the *y*-coordinate is 3, turn and go up 3 units on a line parallel to the *y*-axis. The point $(2, 3)$ is **plotted** in Figure 4. From now on, we will refer to the point with *x*-coordinate 2 and *y*-coordinate 3 as the point $(2, 3)$.

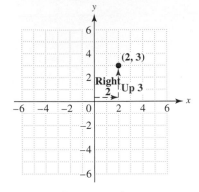

FIGURE 4

▶ **NOTE** In a plane, *both* numbers in the ordered pair are needed to locate a point. The ordered pair is a name for the point.

EXAMPLE 6 Plotting Ordered Pairs

Plot the given points in a coordinate system.

(a) $(1, 5)$ **(b)** $(-2, 3)$ **(c)** $(-1, -4)$ **(d)** $(7, -2)$

(e) $\left(\dfrac{3}{2}, 2\right)$ **(f)** $(5, 0)$ **(g)** $(0, -5)$

Figure 5 shows the graphs of the points. In part (c), locate the point $(-1, -4)$ by first going 1 unit to the left along the *x*-axis. Then turn and go 4 units down, parallel to the *y*-axis. Plot the point $\left(\frac{3}{2}, 2\right)$ in part (e) by going $\frac{3}{2}$ (or $1\frac{1}{2}$) units to the right along the *x*-axis. Then turn and go 2 units up, parallel to the *y*-axis.

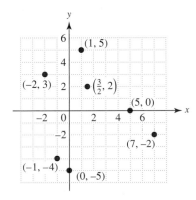

FIGURE 5

✔ Now Try Exercises 51 and 55.

Sometimes we can use a linear equation to mathematically describe, or *model*, a real-life situation, as shown in the next example.

| EXAMPLE 7 | Completing Ordered Pairs to Estimate the Number of Twin Births |

The number of twin births in the United States has increased steadily in recent years. The annual number of twin births from 1998 through 2003 can be closely approximated by the linear equation

Number of twin births ⌐   ⌐ Year

$$y = 3.563x - 7007.7,$$

which relates x, the year, and y, the number of twin births in thousands. (*Source:* Department of Health and Human Services.)

(a) Complete the table of values for the given linear equation.

x (Year)	y (Number of Twin Births, in thousands)
1998	
2000	
2003	

To find y when $x = 1998$, we substitute into the equation.

≈ means "is approximately equal to."

$$y = 3.563(1998) - 7007.7 \quad \text{Let } x = 1998.$$

$$y \approx 111 \quad \text{Use a calculator.}$$

This means that in 1998, there were about 111 thousand (or 111,000) twin births in the United States.

We substitute the years 2000 and 2003 in the same way to complete the table.

x (Year)	y (Number of Twin Births, in thousands)
1998	111
2000	118
2003	129

We can write the results from the table of values as ordered pairs (x, y). Each year x is paired with the number of twin births y (in thousands):

$$(1998, 111), \quad (2000, 118), \quad \text{and} \quad (2003, 129).$$

(b) Graph the ordered pairs found in part (a).

The ordered pairs are graphed in Figure 6 on the next page. This graph of ordered pairs of data is called a **scatter diagram.** Notice how the axes are labeled: x represents the year, and y represents the number of twin births in thousands. Different scales are used on the two axes. Here, each square represents 1 unit in the horizontal direction and 5 units in the vertical direction. Because the numbers in the first ordered pair are large, we show a break in the axes near the origin.

NUMBER OF TWIN BIRTHS

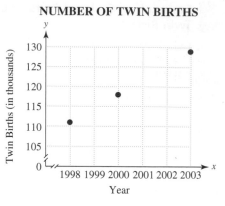

FIGURE 6

A scatter diagram enables us to tell whether two quantities are related to each other. In Figure 6, the plotted points could be connected to form a straight *line,* so the variables x (year) and y (number of twin births) have a *line*ar relationship. The increase in the number of twin births is also reflected.

✔ **Now Try Exercise 75.**

▶ **CAUTION** The equation in Example 7 is valid only for the years 1998 through 2003, because it was based on data for those years. *Do not assume that this equation would provide reliable data for other years, since the data for those years may not follow the same pattern.*

We can think of ordered pairs as representing an input value x and an output value y. If we input x into the equation, the output is y. We encounter many examples of this type of relationship every day.

- The cost to fill a tank with gasoline depends on how many gallons are needed; the number of gallons is the input, and the cost is the output.

- The distance traveled depends on the traveling time; input a time, and the output is a distance.

- The growth of a plant depends on the amount of sun it gets; the input is the amount of sun, and the output is the growth.

This idea is illustrated in Figure 7 for the data of Example 7 with an input–output "machine."

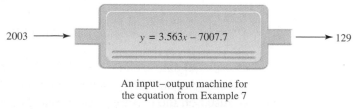

An input–output machine for
the equation from Example 7

FIGURE 7

In **Section 3.6,** we extend this idea to the concept of a *function.*

3.1 EXERCISES

Complete solution available on Video Lectures on CD/DVD

Now Try Exercise

The bar graph compares egg production in billions of eggs for six states in the year 2002. Use the graph to work Exercises 1–4. See Example 1.

1. Name the top two egg-producing states. Estimate their production.

2. Which states had egg production less than 5 billion eggs?

3. Which state had the least production? Estimate this production.

4. How did egg production in Texas (TX) compare with egg production in Iowa (IA)?

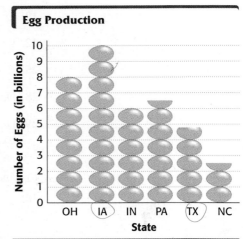

Egg Production

Source: U.S. Department of Agriculture.

The line graph shows the overall unemployment rate in the U.S. civilian labor force for the years 1998 through 2004. Use the graph to work Exercises 5–8. See Example 2.

5. Between which pairs of consecutive years did the unemployment rate decrease?

6. What was the general trend in the unemployment rate between 2000 and 2003?

7. Estimate the overall unemployment rate in 2003 and 2004. About how much did the unemployment rate decline between 2003 and 2004?

8. During which year(s) was the unemployment rate greater than 5%, but less than 6%?

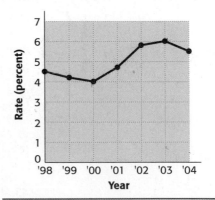

Unemployment Rate

Source: U.S. Department of Labor.

Concept Check   *Fill in each blank with the correct response.*

9. The symbol (x, y) _____ represent an ordered pair, while the symbols $[x, y]$ and
                                    (does/does not)
$\{x, y\}$ _____ represent ordered pairs.
        (do/do not)

10. The ordered pair $(3, 2)$ is a solution of the equation $2x - 5y =$ _____.

11. The point whose graph has coordinates $(-4, 2)$ is in quadrant _____.

12. The point whose graph has coordinates $(0, 5)$ lies on the _____-axis.

13. The ordered pair $(4,$ _____$)$ is a solution of the equation $y = 3$.

14. The ordered pair ($\_\_\_\_\_, -2)$ is a solution of the equation $x = 6$.

15. Define a linear equation in one variable and a linear equation in two variables, and give examples of each.

Decide whether the given ordered pair is a solution of the given equation. See Example 3.

16. $x + y = 9$; $(0, 9)$ **17.** $x + y = 8$; $(0, 8)$ **18.** $2x - y = 6$; $(4, 2)$

19. $2x + y = 5$; $(3, -1)$ **20.** $4x - 3y = 6$; $(2, 1)$ **21.** $5x - 3y = 15$; $(5, 2)$

22. $y = 3x$; $(2, 6)$ **23.** $x = -4y$; $(-8, 2)$ **24.** $x = -6$; $(-6, 5)$

25. $y = 2$; $(4, 2)$ **26.** $x + 4 = 0$; $(-6, 2)$ **27.** $x - 6 = 0$; $(4, 2)$

28. Do $(4, -1)$ and $(-1, 4)$ represent the same ordered pair? Explain.

29. Do the ordered pairs $(3, 4)$ and $(4, 3)$ correspond to the same point on the plane? Explain.

Complete each ordered pair for the equation $y = 2x + 7$. See Example 4.

30. $(2, \quad)$ **31.** $(5, \quad)$ **32.** $(\quad, 0)$ **33.** $(\quad, -3)$

Complete each ordered pair for the equation $y = -4x - 4$. See Example 4.

34. $(0, \quad)$ **35.** $(\quad, 0)$ **36.** $(\quad, 16)$ **37.** $(\quad, 24)$

38. Explain why it would be easier to find the corresponding y-value for $x = \frac{1}{3}$ than for $x = \frac{1}{7}$ in the equation $y = 6x + 2$.

39. *Concept Check* For the equation $y = mx + b$, what is the y-value corresponding to $x = 0$ for *any* value of m?

Complete each table of values. Write the results as ordered pairs. See Example 5.

40. $2x + 3y = 12$

x	y
0	
	0
	8

41. $4x + 3y = 24$

x	y
0	
	0
	4

42. $3x - 5y = -15$

x	y
0	
	0
	-6

43. $4x - 9y = -36$

x	y
	0
0	
	8

44. $x = -9$

x	y
	6
	2
	-3

45. $x = 12$

x	y
	3
	8
	0

46. $y = -6$

x	y
8	
4	
-2	

47. $y = -10$

x	y
4	
0	
-4	

48. $x - 8 = 0$

x	y
	8
	3
	0

49. $y + 2 = 0$

x	y
9	
2	
0	

50. Give the ordered pairs that correspond to the points labeled in the figure. (All coordinates are integers.) See Objective 6.

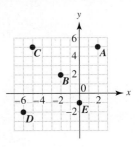

Plot each ordered pair in a rectangular coordinate system. See Example 6.

51. $(6, 2)$ **52.** $(5, 3)$ **53.** $(-4, 2)$ **54.** $(-3, 5)$ **55.** $\left(-\dfrac{4}{5}, -1\right)$

56. $\left(-\dfrac{3}{2}, -4\right)$ **57.** $(0, 4)$ **58.** $(-3, 0)$ **59.** $(4, 0)$ **60.** $(0, -3)$

Concept Check   *Fill in each blank with the word* positive *or the word* negative.

The point with coordinates (x, y) is in

61. quadrant III if x is _____ and y is _____.

62. quadrant II if x is _____ and y is _____.

63. quadrant IV if x is _____ and y is _____.

64. quadrant I if x is _____ and y is _____.

Complete each table of values and then plot the ordered pairs. See Examples 5 and 6.

65. $x - 2y = 6$

x	y
0	
	0
2	
	-1

66. $2x - y = 4$

x	y
0	
	0
1	
	-6

67. $3x - 4y = 12$

x	y
0	
	0
-4	
	-4

68. $2x - 5y = 10$

x	y
0	
	0
-5	
	-3

69. $y + 4 = 0$

x	y
0	
5	
-2	
-3	

70. $x - 5 = 0$

x	y
	1
	0
	6
	-4

71. Look at your graphs of the ordered pairs in Exercises 65–70. Describe the pattern indicated by the plotted points.

Solve each problem. See Example 7.

72. Suppose that it costs $5000 to start up a business selling snow cones. Furthermore, it costs $0.50 per cone in labor, ice, syrup, and overhead. Then the cost to make x snow cones is given by y dollars, where $y = 0.50x + 5000$. Express each of the following as an ordered pair.

   (a) When 100 snow cones are made, the cost is $5050. (*Hint:* What does x represent? What does y represent?)

   (b) When the cost is $6000, the number of snow cones made is 2000.

73. It costs a flat fee of $20 plus $5 per day to rent a pressure washer. Therefore, the cost to rent the pressure washer for x days is given by $y = 5x + 20$, where y is in dollars. Express each of the following as an ordered pair.

   (a) When the washer is rented for 5 days, the cost is $45.

   (b) I paid $50 when I returned the washer, so I must have rented it for 6 days.

74. The table shows the number of U.S. students studying abroad (in thousands) for several academic years.

U.S. STUDENTS STUDYING ABROAD

Academic Year	Number of Students (in thousands)
1998	130
1999	144
2000	154
2001	161
2002	175
2003	191

Source: Institute of International Education.

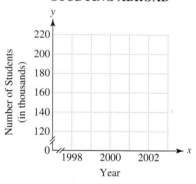

(a) Write the data from the table as ordered pairs (x, y), where x represents the year and y represents the number of U.S. students studying abroad.

(b) What does the ordered pair $(2001, 161)$ mean in the context of this problem?

(c) Make a scatter diagram of the data, using the ordered pairs from part (a) and the given grid.

(d) Describe the pattern indicated by the points on the scatter diagram. What is the trend in the number of U.S. students studying abroad?

75. The table shows the rate (in percent) at which 4-yr college students (both public and private) graduate within 5 yr.

4-YEAR COLLEGE STUDENTS GRADUATING WITHIN 5 YEARS

Year	Rate (%)
1996	53.3
1997	52.8
1998	52.1
1999	51.6
2000	51.2
2001	50.9

Source: ACT.

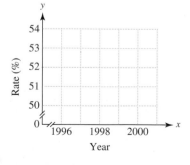

(a) Write the data from the table as ordered pairs (x, y), where x represents the year and y represents graduation rate.

(b) What does the ordered pair $(2002, 51.0)$ mean in the context of this problem?

(c) Make a scatter diagram of the data, using the ordered pairs from part (a) and the given grid.

(d) Describe the pattern indicated by the points on the scatter diagram. What is happening to rates at which 4-yr college students graduate within 5 yr?

76. The maximum benefit for the heart from exercising occurs if the heart rate is in the target heart rate zone. The lower limit of this target zone can be approximated by the linear equation

$$y = -0.65x + 143,$$

where x represents age and y represents heartbeats per minute. (*Source: The Gazette,* Cedar Rapids, Iowa, April 8, 2002.)

Age	Heartbeats (per minute)
20	
40	
60	
80	

(a) Complete the table of values for this linear equation.

(b) Write the data from the table of values as ordered pairs.

(c) Make a scatter diagram of the data. Do the points lie in an approximately linear pattern?

77. (See Exercise 76.) The upper limit of the target heart rate zone can be approximated by the linear equation

$$y = -0.85x + 187,$$

where x represents age and y represents heartbeats per minute. (*Source: The Gazette,* Cedar Rapids, Iowa, April 8, 2002.)

Age	Heartbeats (per minute)
20	
40	
60	
80	

(a) Complete the table of values for this linear equation.

(b) Write the data from the table of values as ordered pairs.

(c) Make a scatter diagram of the data. Describe the pattern indicated by the data.

78. Refer to Exercises 76 and 77. What is the target heart rate zone for age 20? age 40?

3.2 Graphing Linear Equations in Two Variables

OBJECTIVE 1 Graph linear equations by plotting ordered pairs. We know that infinitely many ordered pairs satisfy a linear equation in two variables. We find these ordered-pair solutions by choosing as many values of x (or y) as we wish and then completing each ordered pair.

For example, consider the equation $x + 2y = 7$. If we choose $x = 1$, then $y = 3$, so the ordered pair $(1, 3)$ is a solution of $x + 2y = 7$.

$$1 + 2(3) = 1 + 6 = 7 \qquad (1, 3) \text{ is a solution.}$$

This ordered pair and others that are solutions of $x + 2y = 7$ are graphed in Figure 8.

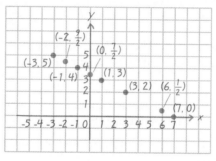

FIGURE 8

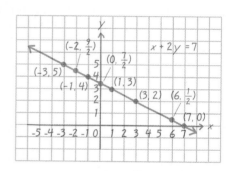

FIGURE 9

Notice that the points plotted in Figure 8 all appear to lie on a straight line, as shown in Figure 9. In fact,

every point on the line represents a solution of the equation $x + 2y = 7$, and every solution of the equation corresponds to a point on the line.

The line gives a "picture" of all the solutions of the equation $x + 2y = 7$. Only a portion of the line is shown here, but it extends indefinitely in both directions, as suggested by the arrowhead on each end. The line is called the **graph** of the equation, and the process of plotting the ordered pairs and drawing the line through the corresponding points is called **graphing.**

The preceding discussion can be generalized.

Graph of a Linear Equation

The graph of any linear equation in two variables is a straight line.

Notice that the word *line* appears in the name "*line*ar equation."

Since two distinct points determine a line, we can graph a straight line by finding any two different points on the line. However, it is a good idea to plot a third point as a check.

EXAMPLE 1 **Graphing a Linear Equation**

Graph the linear equation $4x - 5y = 20$.

At least two different points are needed to draw the graph. First let $x = 0$ and then let $y = 0$ to complete two ordered pairs.

$$4x - 5y = 20$$
$$4(0) - 5y = 20 \qquad \text{Let } x = 0.$$
$$0 - 5y = 20$$
$$-5y = 20$$
$$y = -4$$

Write each x-value first.

$$4x - 5y = 20$$
$$4x - 5(0) = 20 \qquad \text{Let } y = 0.$$
$$4x - 0 = 20$$
$$4x = 20$$
$$x = 5$$

The ordered pairs are $(0, -4)$ and $(5, 0)$. Find a third ordered pair (as a check) by choosing a number other than 0 for x or y. We choose $y = 2$.

$$4x - 5y = 20$$
$$4x - 5(2) = 20 \qquad \text{Let } y = 2.$$
$$4x - 10 = 20$$
$$4x = 30 \qquad \text{Add 10.}$$
$$x = \frac{30}{4} \qquad \text{Divide by 4.}$$
$$x = \frac{15}{2} \qquad \text{Lowest terms}$$

This gives the ordered pair $\left(\frac{15}{2}, 2\right)$, or $\left(7\frac{1}{2}, 2\right)$. Plot the three ordered pairs $(0, -4)$, $(5, 0)$, and $\left(7\frac{1}{2}, 2\right)$, and draw a line through them. This line, shown in Figure 10, is the graph of $4x - 5y = 20$.

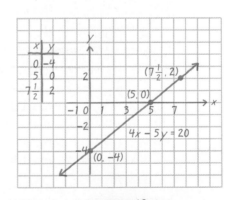

FIGURE 10

✔ **Now Try Exercise 1.**

▶ **CAUTION** When graphing a linear equation as in Example 1, all three points should lie on the same straight line. If they don't, double-check the ordered pairs you found.

EXAMPLE 2 Graphing a Linear Equation

Graph the linear equation $y = -\frac{3}{2}x + 3$.

Although this equation is not in standard form ($Ax + By = C$), it *could* be written in that form, so it is a linear equation. Two different points on the graph can be found by first letting $x = 0$ and then letting $y = 0$.

$$y = -\frac{3}{2}x + 3$$

$$y = -\frac{3}{2}(0) + 3 \qquad \text{Let } x = 0.$$

$$y = 0 + 3$$

$$y = 3$$

$$y = -\frac{3}{2}x + 3$$

$$0 = -\frac{3}{2}x + 3 \qquad \text{Let } y = 0.$$

$$\frac{3}{2}x = 3 \qquad \text{Add } \frac{3}{2}x.$$

$$x = 2 \qquad \text{Multiply by } \frac{2}{3}.$$

This gives the ordered pairs $(0, 3)$ and $(2, 0)$. To find a third point (as a check), we let $x = -2$. (Other numbers could be used, but a multiple of 2 makes multiplying by $-\frac{3}{2}$ easier.)

$$y = -\frac{3}{2}x + 3$$

$$y = -\frac{3}{2}(-2) + 3 \qquad \text{Let } x = -2.$$

$$y = 3 + 3$$

$$y = 6 \qquad \text{The ordered pair is } (-2, 6).$$

These three ordered pairs are shown in the table accompanying Figure 11. Plot the corresponding points, and then draw a line through them. This line, shown in Figure 11, is the graph of $y = -\frac{3}{2}x + 3$.

x	y
0	3
2	0
-2	6

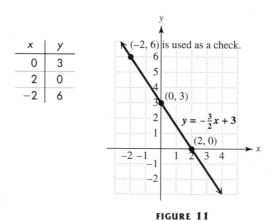

FIGURE 11

✔ **Now Try Exercise 3.**

OBJECTIVE 2 **Find intercepts.** In Figure 11, the graph intersects (crosses) the y-axis at $(0, 3)$ and the x-axis at $(2, 0)$. For this reason, $(0, 3)$ is called the **y-intercept** and $(2, 0)$ is called the **x-intercept** of the graph.

The intercepts are particularly useful for graphing linear equations. The intercepts are found by replacing, in turn, each variable with 0 in the equation and solving for the value of the other variable.

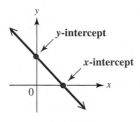

Finding Intercepts

To find the *x*-intercept, let $y = 0$ in the given equation and solve for *x*. Then $(x, 0)$ is the *x*-intercept.

To find the *y*-intercept, let $x = 0$ in the given equation and solve for *y*. Then $(0, y)$ is the *y*-intercept.

EXAMPLE 3 Finding Intercepts

Find the intercepts for the graph of $2x + y = 4$. Then draw the graph.

Find the *y*-intercept by letting $x = 0$; find the *x*-intercept by letting $y = 0$.

$2x + y = 4$	$2x + y = 4$
$2(0) + y = 4$ Let $x = 0$.	$2x + 0 = 4$ Let $y = 0$.
$0 + y = 4$	$2x = 4$
$y = 4$ *y*-intercept is $(0, 4)$.	$x = 2$ *x*-intercept is $(2, 0)$.

Write each *x*-value first.

The graph, with the two intercepts in red, is given in Figure 12. Find a third point as a check. For example, choosing $x = 4$ gives $y = -4$. Plot $(0, 4)$, $(2, 0)$, and $(4, -4)$ and draw a line through those points. This line, shown in Figure 12, is the graph of $2x + y = 4$.

x	y
0	4
2	0
4	-4

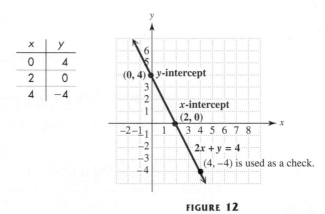

FIGURE 12

✔ **Now Try Exercise 17.**

▶ **CAUTION** When choosing *x*- or *y*-values to find ordered pairs to plot, be careful to choose so that the resulting points are not too close together. For example, using $(-1, -1)$, $(0, 0)$, and $(1, 1)$ may result in an inaccurate line. It is better to choose points whose *x*-values differ by at least 2.

OBJECTIVE 3 **Graph linear equations of the form $Ax + By = 0$.** In earlier examples, the x- and y-intercepts were used to help draw the graphs. This is not always possible. Example 4 shows what to do when the x- and y-intercepts are the same point.

EXAMPLE 4 **Graphing an Equation of the Form $Ax + By = 0$**

Graph $x - 3y = 0$.

If we let $x = 0$, then $y = 0$, giving the ordered pair $(0, 0)$. Letting $y = 0$ also gives $(0, 0)$. This is the same ordered pair, so we choose *two other values* for x or y. Choosing 2 for y gives $x - 3 \cdot 2 = 0$, or $x = 6$, so another ordered pair is $(6, 2)$. For a check point, we choose -6 for x, getting -2 for y. We use the ordered pairs $(-6, -2)$, $(0, 0)$, and $(6, 2)$ to draw the graph shown in Figure 13.

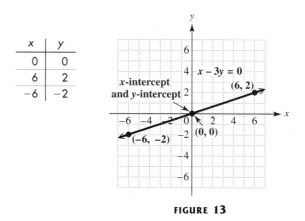

FIGURE 13

✔ **Now Try Exercise 45.**

Example 4 can be generalized as follows.

Line through the Origin
If A and B are nonzero real numbers, the graph of a linear equation of the form
$$Ax + By = 0$$
passes through the origin $(0, 0)$.

OBJECTIVE 4 **Graph linear equations of the form $y = k$ or $x = k$.** The equation $y = -4$ is a linear equation in which the coefficient of x is 0. (To see this, write $y = -4$ as $0x + y = -4$.) Also, $x = 3$ is a linear equation in which the coefficient of y is 0. These equations lead to horizontal and vertical straight lines, respectively.

EXAMPLE 5 Graphing an Equation of the Form $y = k$

Graph $y = -4$.

As the equation states, for any value of x, y is always equal to -4. To get ordered pairs that are solutions of this equation, we choose any numbers for x and always use -4 for y. Three ordered pairs that satisfy the equation are shown in the table of values. Drawing a line through these points gives the horizontal line shown in Figure 14. The y-intercept is $(0, -4)$; there is no x-intercept.

x	y
-2	-4
0	-4
3	-4

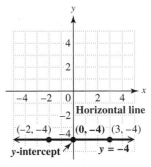

FIGURE 14

✔ **Now Try Exercise 49.**

Horizontal Line

The graph of the linear equation $y = k$, where k is a real number, is the horizontal line with y-intercept $(0, k)$ and no x-intercept.

EXAMPLE 6 Graphing an Equation of the Form $x = k$

Graph $x - 3 = 0$.

First we add 3 to each side of the equation $x - 3 = 0$ to get $x = 3$. All the ordered pairs that are solutions of this equation have an x-value of 3. Any number can be used for y. We show three ordered pairs that satisfy the equation in the table of values. Drawing a line through these points gives the vertical line shown in Figure 15. The x-intercept is $(3, 0)$; there is no y-intercept.

x	y
3	3
3	0
3	-2

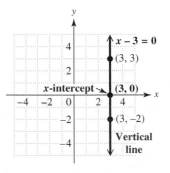

FIGURE 15

✔ **Now Try Exercise 51.**

Vertical Line

The graph of the linear equation $x = k$, where k is a real number, is the vertical line with x-intercept $(k, 0)$ and no y-intercept.

In particular, notice that the horizontal line $y = 0$ is the x-axis and the vertical line $x = 0$ is the y-axis.

▶ **CAUTION** The equations of horizontal and vertical lines are often confused with each other. Remember that the graph of $y = k$ is parallel to the x-axis and that of $x = k$ is parallel to the y-axis.

The different forms of linear equations from this section and the methods of graphing them are given in the following summary.

Graphing Straight Lines		
Equation	*To Graph*	*Example*
$y = k$	Draw a horizontal line, through $(0, k)$.	$y = -2$
$x = k$	Draw a vertical line, through $(k, 0)$.	$x = 4$
$Ax + By = 0$	The graph passes through $(0, 0)$. Get additional points that lie on the graph by choosing any value of x or y, except 0.	$x = 2y$
$Ax + By = C$, but not of the types above	Find any two points on the line. A good choice is to find the intercepts. Let $x = 0$, and find the corresponding value of y; then let $y = 0$, and find x. As a check, get a third point by choosing a value of x or y that has not yet been used.	$(4, 3)$ $(2, 0)$ $3x - 2y = 6$ $(0, -3)$

OBJECTIVE ⑤ Use a linear equation to model data.

EXAMPLE 7 Using a Linear Equation to Model Credit Card Debt

Credit card debt in the United States increased steadily from 1995 through 2003. The amount of debt y in billions of dollars can be modeled by the linear equation

$$y = 38.7x + 450,$$

where $x = 0$ represents 1995, $x = 1$ represents 1996, and so on. (*Source:* Board of Governors of the Federal Reserve System.)

(a) Use the equation to approximate credit card debt in the years 1995, 2000, and 2003.

For 1995:	$y = 38.7(0) + 450$	Replace x with 0.
	$y = 450$ billion dollars	
For 2000:	$y = 38.7(5) + 450$	$2000 - 1995 = 5$; replace x with 5.
	$y = 643.5$ billion dollars	
For 2003:	$y = 38.7(8) + 450$	$2003 - 1995 = 8$; replace x with 8.
	$y = 759.6$ billion dollars	

(b) Write the information from part (a) as three ordered pairs, and use them to graph the given linear equation.

Since x represents the year and y represents the debt, the ordered pairs are $(0, 450)$, $(5, 643.5)$, and $(8, 759.6)$. See Figure 16. (Arrowheads are not included with the graphed line, since the data are for the years 1995 to 2003 only—that is, from $x = 0$ to $x = 8$.)

U.S. CREDIT CARD DEBT

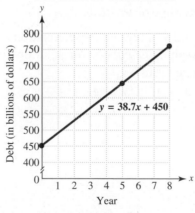

FIGURE 16

(c) Use the graph and then the equation to approximate credit card debt in 2002.

For 2002, $x = 7$. On the graph, find 7 on the horizontal axis, move up to the graphed line and then across to the vertical axis. It appears that credit card debt in 2002 was about 725 billion dollars. To use the equation, substitute 7 for x.

$$y = 38.7x + 450$$
$$y = 38.7(7) + 450 \qquad \text{Let } x = 7.$$
$$y = 720.9 \text{ billion dollars}$$

This result for 2002 is close to our estimate of 725 billion dollars from the graph.

✔ Now Try Exercise 63.

CONNECTIONS

Beginning in this chapter, we include information on the basic features of graphing calculators. The most obvious feature is their ability to graph equations. We must solve the equation for y in order to enter it into the calculator. Also, we must select an appropriate "window" for the graph. The window is determined by the minimum and maximum values of x and y. Graphing calculators have a *standard window*, often from $x = -10$ to $x = 10$ and from $y = -10$ to $y = 10$. Sometimes this is written $[-10, 10]$, $[-10, 10]$, with the x-interval shown first.

For example, to graph the equation $2x + y = 4$, discussed in Example 3, we first solve for y.

$$2x + y = 4$$
$$y = -2x + 4 \qquad \text{Subtract } 2x.$$

If we enter this equation as $y = -2x + 4$ and choose the standard window, the calculator displays the graph in Figure 17.

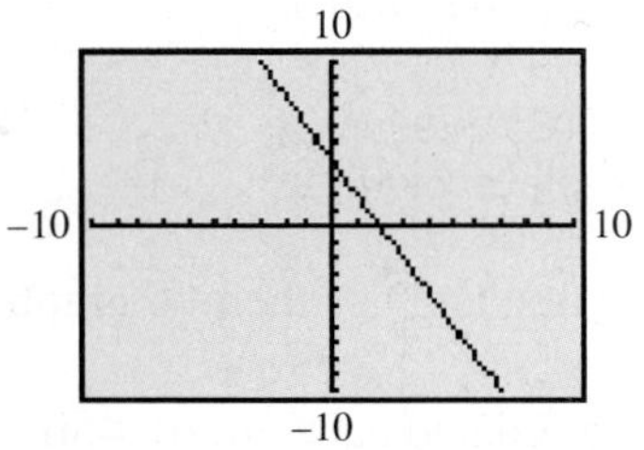

The x-value of the point on the graph where $y = 0$ (the x-intercept) gives the solution of the equation:

$$y = 0,$$

or $\qquad\qquad -2x + 4 = 0. \qquad \text{Substitute } -2x + 4 \text{ for } y.$

Since each tick mark on the x-axis represents 1, the graph shows that $(2, 0)$ is the x-intercept.

FIGURE 17

For Discussion or Writing

How would you rewrite these equations, with one side equal to 0, to enter them into a graphing calculator for solution? (It is not necessary to clear parentheses or combine terms.)

1. $3x + 4 - 2x - 7 = 4x + 3$ **2.** $5x - 15 = 3(x - 2)$

3.2 EXERCISES

Complete solution available on Video Lectures on CD/DVD

Now Try Exercise

Use the given equation to complete the given ordered pairs. Then graph each equation by plotting the points and drawing a line through them. See Examples 1 and 2.

1. $x + y = 5$
$(0, \quad), (\quad, 0), (2, \quad)$

2. $x - y = 2$
$(0, \quad), (\quad, 0), (5, \quad)$

3. $y = \dfrac{2}{3}x + 1$
$(0, \quad), (3, \quad), (-3, \quad)$

4. $y = -\dfrac{3}{4}x + 2$
$(0, \quad), (4, \quad), (-4, \quad)$

5. $3x = -y - 6$
$(0, \quad), (\quad, 0), \left(-\dfrac{1}{3}, \quad\right)$

6. $x = 2y + 3$
$(\quad, 0), (0, \quad), \left(\quad, \dfrac{1}{2}\right)$

Concept Check *In Exercises 7–12, match the information about each graph in Column I with the correct linear equation in Column II.*

I	**II**
7. The graph of the equation has *x*-intercept $(4, 0)$.	**A.** $x = 5$
8. The graph of the equation has *y*-intercept $(0, -4)$.	**B.** $y = -3$
9. The graph of the equation passes through the origin.	**C.** $2x - 5y = 8$
10. The graph of the equation is a vertical line.	**D.** $x + 4y = 0$
11. The graph of the equation is a horizontal line.	**E.** $3x + y = -4$
12. The graph of the equation passes through $(9, 2)$.	

Concept Check *In Exercises 13–16, describe what the graph of each linear equation will look like on the coordinate plane.* (Hint: *Rewrite the equation if necessary so that it is in a more recognizable form.*)

13. $3x = y - 9$ **14.** $x - 10 = 1$ **15.** $3y = -6$ **16.** $2x = 4y$

Find the x-intercept and the y-intercept for the graph of each equation. See Examples 1–6.

17. $2x - 3y = 24$ **18.** $-3x + 8y = 48$ **19.** $x + 6y = 0$ **20.** $3x - y = 0$

21. $5x - 2y = 20$ **22.** $-3x + 2y = 12$ **23.** $y = -2x + 4$ **24.** $y = 3x + 6$

25. $y = \dfrac{1}{3}x - 2$ **26.** $y = \dfrac{1}{4}x - 1$ **27.** $x - 4 = 0$ **28.** $x = 5$

29. $y = 2.5$ **30.** $y + 1.5 = 0$

31. ***Concept Check*** What is the equation of the *x*-axis? What is the equation of the *y*-axis?

32. A student attempted to graph $4x + 5y = 0$ by finding intercepts. First she let $x = 0$ and found y; then she let $y = 0$ and found x. In both cases, the resulting point was $(0, 0)$. She knew that she needed at least two different points to graph the line, but was unsure what to do next, since finding intercepts gave her only one point. Explain to her what to do next.

Graph each linear equation. See Examples 1–6.

33. $x = y + 2$ **34.** $x = -y + 6$ **35.** $x - y = 4$ **36.** $x - y = 5$

37. $2x + y = 6$ **38.** $-3x + y = -6$ **39.** $y = 2x - 5$ **40.** $y = 4x + 3$

41. $3x + 7y = 14$ **42.** $6x - 5y = 18$ **43.** $y = -\dfrac{3}{4}x + 3$ **44.** $y = -\dfrac{2}{3}x - 2$

45. $y - 2x = 0$ **46.** $y + 3x = 0$ **47.** $y = -6x$ **48.** $y = 4x$

49. $y = -1$ **50.** $y = 3$ **51.** $x + 2 = 0$ **52.** $x - 4 = 0$

53. $-3y = 15$ **54.** $-2y = 12$ **55.** $x + 2 = 8$ **56.** $x - 1 = -4$

57. Write a paragraph summarizing how to graph a linear equation in two variables.

*Recall that the maximum benefit for the heart from exercising occurs if the heart rate is in the target heart rate zone. In **Section 3.1**, Exercises 76–78, we used the linear equations*

$$y = -0.65x + 143 \quad \text{Lower limit}$$

$$y = -0.85x + 187 \quad \text{Upper limit}$$

to approximate the target heart rate zone, where x represents age and y represents heartbeats per minute. Use this information in Exercises 58–61 on the next page.

58. (a) Use the appropriate equation to find the lower limit of the target heart rate zone for ages 20, 40, 60, and 80. Write your results as ordered pairs.
(b) Graph the equation, using the data from part (a).
(c) Use the graph to estimate the lower limit for age 30.
(d) Use the linear equation to approximate the lower limit (to the nearest whole number) for age 30.
(e) How does the approximation obtained from the equation compare with the estimate from the graph?

59. (a) Use the appropriate equation to find the upper limit of the target heart rate zone for ages 20, 40, 60, and 80. Write your results as ordered pairs.
(b) Graph the equation, using the data from part (a).
(c) Use the graph to estimate the upper limit for age 30.
(d) Use the linear equation to approximate the upper limit (to the nearest whole number) for age 30.
(e) How does the approximation obtained from the equation compare with the estimate from the graph?

60. Use the results of Exercises 58(d) and 59(d) to determine the target heart rate zone for age 30.

61. Should the graphs of the target heart rate zone in Exercises 58 and 59 be used to estimate the target heart rate zone for ages below 20 or above 80? Why or why not?

Solve each problem. See Example 7.

62. The height *y* (in centimeters) of a woman is related to the length *x* of her radius (the bone from the wrist to the elbow) and is approximated by the linear equation

$$y = 3.9x + 73.5.$$

(a) Use the equation to find the approximate heights of women with radii of lengths 20 cm, 26 cm, and 22 cm.
(b) Graph the equation, using the data from part (a).
(c) Use the graph to estimate the length of the radius in a woman who is 167 cm tall. Then use the equation to find the length of the radius to the nearest centimeter. (*Hint:* Substitute for *y* in the equation.)

63. The weight *y* (in pounds) of a man taller than 60 in. can be approximated by the linear equation

$$y = 5.5x - 220,$$

where *x* is the height of the man in inches.

(a) Use the equation to approximate the weights of men whose heights are 62 in., 66 in., and 72 in.
(b) Graph the equation, using the data from part (a).
(c) Use the graph to estimate the height of a man who weighs 155 lb. Then use the equation to find the height of this man to the nearest inch. (*Hint:* Substitute for *y* in the equation.)

64. U.S. per capita consumption of cheese increased for the years 1980 through 2002 as shown in the graph on the next page. If *x* = 0 represents 1980, *x* = 5 represents 1985, and so on, per capita consumption can be modeled by the linear equation

$$y = 0.5541x + 18.60,$$

where *y* is in pounds.

(a) Use the equation to approximate cheese consumption (to the nearest tenth) in 1990, 1995, and 2002.

(b) Use the graph to estimate consumption for the same years.

(c) How do the approximations obtained from the equation compare with the estimates from the graph?

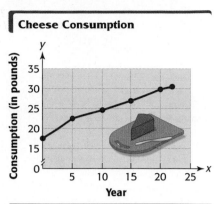

Cheese Consumption

Source: U.S. Department of Agriculture.

65. In the United States, sporting goods sales *y* (in billions of dollars) from 1997 through 2002 are modeled by the linear equation

$$y = 1.983x + 53.66,$$

where $x = 0$ corresponds to 1990, $x = 7$ corresponds to 1997, and so on.

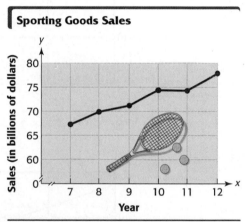

Sporting Goods Sales

Source: National Sporting Goods Association.

(a) Use the equation to approximate sporting goods sales in 1997, 2000, and 2002. Round your answers to the nearest tenth.

(b) Use the graph to estimate sales for the same years.

(c) How do the approximations obtained from the equation compare with the estimates from the graph?

66. The graph shows the value of a sport-utility vehicle (SUV) over the first 5 yr of ownership. Use the graph to do the following.

(a) Determine the initial value of the SUV.

(b) Find the *depreciation* (loss in value) from the original value after the first 3 yr.

(c) What is the annual or yearly depreciation in each of the first 5 yr?

(d) What does the ordered pair $(5, 5000)$ mean in the context of this problem?

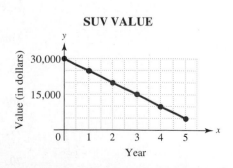

SUV VALUE

67. Demand for an item is often closely related to its price. As price increases, demand decreases, and as price decreases, demand increases. Suppose demand for a video game is 2000 units when the price is $40 and is 2500 units when the price is $30.

    (a) Let x be the price and y be the demand for the game. Graph the two given pairs of prices and demands.

    (b) Assume that the relationship is linear. Draw a line through the two points from part (a). From your graph, estimate the demand if the price drops to $20.

    (c) Use the graph to estimate the price if the demand is 3500 units.

TECHNOLOGY INSIGHTS (EXERCISES 68–72)

*In Exercises 68–71, we give the graph of a linear equation in two variables solved for y and a corresponding linear equation in one variable, with y replaced by 0. Solve the equation in one variable, using the methods of **Section 2.3**. Is the solution you get the same as the x-intercept (labeled "Zero") on the calculator screen?*

68. $8 - 2(3x - 4) - 2x = 0$

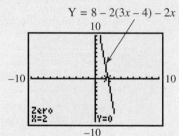

$Y = 8 - 2(3x - 4) - 2x$

69. $5(2x - 1) - 4(2x + 1) - 7 = 0$

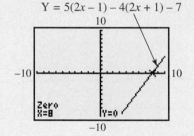

$Y = 5(2x - 1) - 4(2x + 1) - 7$

70. $0.6x - 0.1x - x + 2.5 = 0$

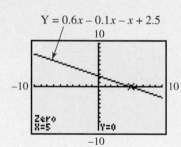

$Y = 0.6x - 0.1x - x + 2.5$

71. $-\dfrac{2}{7}x + 2x - \dfrac{1}{2}x - \dfrac{17}{2} = 0$

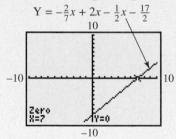

$Y = -\frac{2}{7}x + 2x - \frac{1}{2}x - \frac{17}{2}$

✎ **72.** From the results in Exercises 68–71, what can you conclude?

PREVIEW EXERCISES

*Find each quotient. See **Section 1.8**.*

73. $\dfrac{4 - 2}{8 - 5}$ **74.** $\dfrac{-3 - 5}{2 - 7}$ **75.** $\dfrac{-2 - (-4)}{3 - (-1)}$ **76.** $\dfrac{5 - (-7)}{-4 - (-1)}$

*Solve each formula for the indicated variable. See **Section 2.5**.*

77. $d = rt$ for t **78.** $A = \dfrac{1}{2}bh$ for b

79. $P = a + b + c$ for c **80.** $180 = A + B + C$ for B

3.3 The Slope of a Line

OBJECTIVES

1. Find the slope of a line, given two points.

2. Find the slope from the equation of a line.

3. Use slopes to determine whether two lines are parallel, perpendicular, or neither.

An important characteristic of the lines we graphed in **Section 3.2** is their slant, or "steepness." See Figure 18.

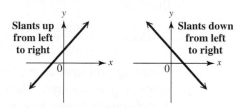

FIGURE 18

One way to measure the steepness of a line is to compare the vertical change in the line with the horizontal change while moving along the line from one fixed point to another. This measure of steepness is called the *slope* of the line.

OBJECTIVE 1 Find the slope of a line given two points. To find the steepness, or slope, of the line in Figure 19, we begin at point Q and move to point P. The vertical change, or **rise,** is the change in the y-values, which is the difference $6 - 1 = 5$ units. The horizontal change, or **run,** is the change in the x-values, which is the difference $5 - 2 = 3$ units.

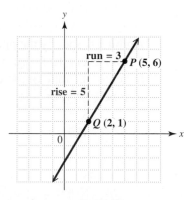

FIGURE 19

Remember from **Section 2.6** that one way to compare two numbers is by using a ratio. The **slope** is the ratio of the vertical change in y to the horizontal change in x. The line in Figure 19 has

$$\text{slope} = \frac{\text{vertical change in } y \text{ (rise)}}{\text{horizontal change in } x \text{ (run)}} = \frac{5}{3}.$$

To confirm this ratio, we can count grid squares. We start at point Q in Figure 19 and count *up* 5 grid squares to find the vertical change (rise). To find the horizontal change (run) and arrive at point P, we count to the *right* 3 grid squares. The slope is $\frac{5}{3}$, as found analytically.

EXAMPLE 1 Finding the Slope of a Line

Find the slope of the line in Figure 20.

We use the two points shown on the line. The vertical change is the difference in the y-values, or $-1 - 3 = -4$, and the horizontal change is the difference in the x-values, or $6 - 2 = 4$. Thus, the line has

$$\text{slope} = \frac{\text{change in } y \text{ (rise)}}{\text{change in } x \text{ (run)}} = \frac{-4}{4}, \quad \text{or} \quad -1.$$

Counting grid squares, we begin at point P and count *down* 4 grid squares. Then we count to the *right* 4 grid squares to reach point Q. Because we counted down, we write the vertical change as a negative number, -4 here. The slope is $\frac{-4}{4}$, or -1.

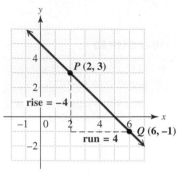

FIGURE 20

✔ **Now Try Exercise 3.**

▶ **NOTE** *The slope of a line is the same for any two points on the line.* To see this, refer to Figure 20. Find the points $(3, 2)$ and $(5, 0)$, which also lie on the line. If we start at $(3, 2)$ and count *down* 2 units and then to the *right* 2 units, we arrive at $(5, 0)$. The slope is $\frac{-2}{2}$, or -1, the same slope we found in Example 1.

The idea of slope is used in many everyday situations. See Figure 21. For example, a highway with a 10%, or $\frac{1}{10}$, grade (or slope) rises 1 m for every 10 m horizontally. Architects specify the pitch of a roof by using slope; a $\frac{5}{12}$ roof means that the roof rises 5 ft for every 12 ft that it runs in the horizontal direction. The slope of a stairwell also indicates the ratio of the vertical rise to the horizontal run. In the figure, the slope of the stairwell is $\frac{8}{12}$, or $\frac{2}{3}$.

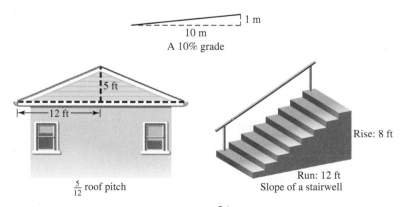

FIGURE 21

We can generalize the preceding discussion and find the slope of a line through two nonspecific points (x_1, y_1) and (x_2, y_2). (This notation is called **subscript notation.** Read x_1 as "x-sub-one" and x_2 as "x-sub-two.") See Figure 22.

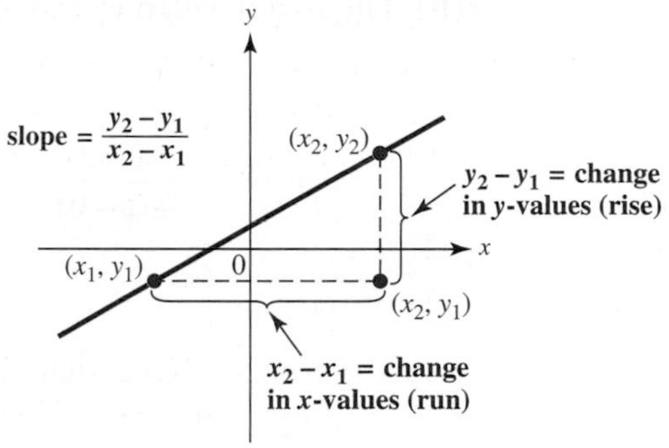

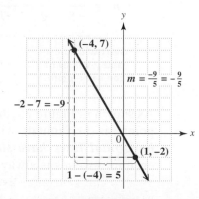

$$\text{slope} = \frac{y_2 - y_1}{x_2 - x_1}$$

$y_2 - y_1 = \text{change in } y\text{-values (rise)}$

$x_2 - x_1 = \text{change in } x\text{-values (run)}$

FIGURE 22

Moving along the line from the point (x_1, y_1) to the point (x_2, y_2), we see that y changes by $y_2 - y_1$ units. This is the vertical change (rise). Similarly, x changes by $x_2 - x_1$ units, which is the horizontal change (run). The slope of the line is the ratio of $y_2 - y_1$ to $x_2 - x_1$.

Traditionally, the letter m represents slope. The slope m of a line is defined as follows.

Slope Formula

The **slope** of the line through the points (x_1, y_1) and (x_2, y_2) is

$$m = \frac{\text{change in } y}{\text{change in } x} = \frac{y_2 - y_1}{x_2 - x_1} \qquad \text{if } x_1 \neq x_2.$$

The slope of a line tells how fast y changes for each unit of change in x; that is, *the slope gives the change in y for each unit of change in x.*

EXAMPLE 2 Finding Slopes of Lines

Find the slope of each line.

(a) The line through $(-4, 7)$ and $(1, -2)$

Use the slope formula. Let $(-4, 7) = (x_1, y_1)$ and $(1, -2) = (x_2, y_2)$. Then

$$\text{slope } m = \frac{\text{change in } y}{\text{change in } x} = \frac{y_2 - y_1}{x_2 - x_1} = \frac{-2 - 7}{1 - (-4)} = \frac{-9}{5} = -\frac{9}{5}.$$

Count grid squares in Figure 23 to confirm your calculation.

$(-4, 7)$

$m = \frac{-9}{5} = -\frac{9}{5}$

$-2 - 7 = -9$

$(1, -2)$

$1 - (-4) = 5$

FIGURE 23

(b) The line through $(-9, -2)$ and $(12, 5)$

$$m = \frac{\overset{\displaystyle y\text{-value}}{\downarrow}\,5 - (-2)}{\underset{\uparrow}{12 - (-9)}} = \frac{7}{21} = \frac{1}{3}$$

Corresponding *x*-value

See Figure 24. Note that the same slope is obtained by subtracting in reverse order.

$$m = \frac{\overset{\displaystyle y\text{-value}}{\downarrow}\,-2 - 5}{\underset{\uparrow}{-9 - 12}} = \frac{-7}{-21} = \frac{1}{3}$$

Corresponding *x*-value

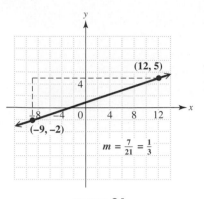

FIGURE 24

✔ **Now Try Exercises 27 and 33.**

▶ **CAUTION** *It makes no difference which point is (x_1, y_1) or (x_2, y_2); however, be consistent.* Start with the *x*- and *y*-values of one point (either one), and subtract the corresponding values of the other point.

The slopes we found for the lines in Figures 23 and 24 suggest the following.

Positive and Negative Slopes

A line with positive slope rises (slants up) from left to right.

A line with negative slope falls (slants down) from left to right.

EXAMPLE 3 **Finding the Slope of a Horizontal Line**

Find the slope of the line through $(-5, 4)$ and $(2, 4)$.

$$m = \frac{4 - 4}{-5 - 2} = \frac{0}{-7} = 0 \quad \text{Zero slope}$$

Start with the *x*- and *y*-values of the same point.

As shown in Figure 25, the line through these two points is horizontal, with equation $y = 4$. *All horizontal lines have slope 0,* since the difference in *y*-values is always 0.

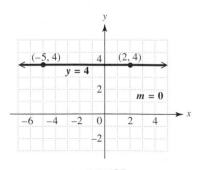

FIGURE 25

✔ **Now Try Exercise 31.**

EXAMPLE 4 Finding the Slope of a Vertical Line

Find the slope of the line through $(6, 2)$ and $(6, -4)$.

$$m = \frac{2 - (-4)}{6 - 6} = \frac{6}{0} \qquad \text{Undefined slope}$$

Since division by 0 is undefined, the slope is undefined. The graph in Figure 26 shows that the line through the given two points is vertical with equation $x = 6$. All points on a vertical line have the same x-value, so *the slope of any vertical line is undefined.*

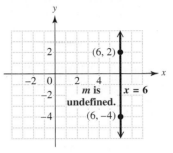

FIGURE 26

✔ **Now Try Exercise 35.**

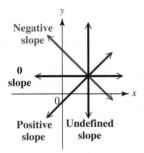

Slopes of lines

FIGURE 27

Slopes of Horizontal and Vertical Lines

Horizontal lines, with equations of the form $y = k$, have **slope 0.**

Vertical lines, with equations of the form $x = k$, have **undefined slope.**

Figure 27 summarizes the four cases for slopes of lines.

OBJECTIVE 2 Find the slope from the equation of a line. Consider the equation

$$y = -3x + 5.$$

We can find the slope of the line using any two points on the line. We get these two points by first choosing two different values of x and then finding the corresponding values of y. We choose $x = -2$ and $x = 4$.

$y = -3x + 5$	$y = -3x + 5$
$y = -3(-2) + 5$ Let $x = -2$.	$y = -3(4) + 5$ Let $x = 4$.
$y = 6 + 5$	$y = -12 + 5$
$y = 11$	$y = -7$

The ordered pairs are $(-2, 11)$ and $(4, -7)$. Now we use the slope formula.

$$m = \frac{11 - (-7)}{-2 - 4} = \frac{18}{-6} = -3$$

The slope, -3, is the same number as the coefficient of x in the given equation $y = -3x + 5$. It can be shown that this always happens, *as long as the equation is solved for y.* This fact is used to find the slope of a line from its equation.

Finding the Slope of a Line from Its Equation

Step 1   Solve the equation for y.

Step 2   The slope is given by the coefficient of x.

EXAMPLE 5 Finding Slopes from Equations

Find the slope of each line.

(a) $2x - 5y = 4$

Step 1 Solve the equation for y.

$$2x - 5y = 4$$
$$-5y = -2x + 4 \qquad \text{Subtract } 2x \text{ from each side.}$$
$$y = \frac{2}{5}x - \frac{4}{5} \qquad \text{Divide by } -5.$$

Slope ⟶

Step 2 The slope is given by the coefficient of x, so the slope is $\frac{2}{5}$.

(b) $8x + 4y = 1$

Solve the equation for y.

$$8x + 4y = 1$$
$$4y = -8x + 1 \qquad \text{Subtract } 8x.$$
$$y = -2x + \frac{1}{4} \qquad \text{Divide by } 4.$$

The slope of this line is given by the coefficient of x, which is -2.

✔**Now Try Exercise 45.**

OBJECTIVE ❸ Use slopes to determine whether two lines are parallel, perpendicular, or neither. Two lines in a plane that never intersect are **parallel.** We use slopes to tell whether two lines are parallel. For example, Figure 28 shows the graphs of the two equations $x + 2y = 4$ and $x + 2y = -6$. These lines appear to be parallel. Solving the equation $x + 2y = 4$ for y gives $y = -\frac{1}{2}x + 2$. Solving $x + 2y = -6$ for y gives $y = -\frac{1}{2}x - 3$. Both lines have slope $-\frac{1}{2}$. *Nonvertical parallel lines always have equal slopes.*

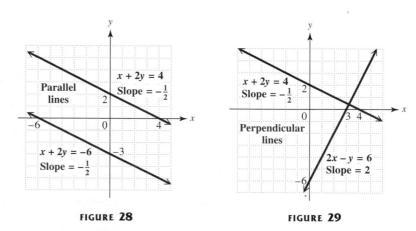

FIGURE 28 　　　　　　　　　　　 **FIGURE 29**

Figure 29 shows the graphs of $x + 2y = 4$ and $2x - y = 6$. These lines appear to be **perpendicular.** (That is, they intersect at a 90° angle.) As before, solving

$x + 2y = 4$ for y gives $y = -\frac{1}{2}x + 2$, with slope $-\frac{1}{2}$. Solving $2x - y = 6$ for y gives $y = 2x - 6$, with slope 2. The product of $-\frac{1}{2}$ and 2 is

$$-\frac{1}{2}(2) = -1.$$

Number	Negative Reciprocal
$\dfrac{3}{4}$	$-\dfrac{4}{3}$
$\dfrac{1}{2}$	$-\dfrac{2}{1}$, or -2
-6, or $-\dfrac{6}{1}$	$\dfrac{1}{6}$
-0.4, or $-\dfrac{4}{10}$	$\dfrac{10}{4}$, or 2.5

The product of each number and its negative reciprocal is −1.

This condition is true in general. ***The product of the slopes of two perpendicular lines, neither of which is vertical, is always −1.*** This means that the slopes of perpendicular lines are negative (or opposite) reciprocals—if one slope is the nonzero number a, the other is $-\frac{1}{a}$. The table in the margin shows several examples.

Slopes of Parallel and Perpendicular Lines

Two lines with the same slope are parallel.

Two lines whose slopes have a product of -1 are perpendicular.

EXAMPLE 6 Deciding whether Two Lines Are Parallel or Perpendicular

Decide whether each pair of lines is *parallel, perpendicular,* or *neither.*

(a) $x + 2y = 7$
   $-2x + y = 3$

Find the slope of each line by first solving each equation for y.

$$
\begin{array}{c|c}
\begin{aligned}
x + 2y &= 7 \\
2y &= -x + 7 \\
y &= -\frac{1}{2}x + \frac{7}{2}
\end{aligned}
&
\begin{aligned}
-2x + y &= 3 \\
y &= 2x + 3
\end{aligned}
\\
\text{Slope is } -\frac{1}{2}. & \text{Slope is } 2.
\end{array}
$$

Since the slopes are not equal, the lines are not parallel. Check the product of the slopes: $-\frac{1}{2}(2) = -1$. The two lines are perpendicular because the product of their slopes is -1, indicating that the slopes are negative reciprocals.

(b) $3x - y = 4$ $\xrightarrow{\text{Solve for } y.}$ $y = 3x - 4$
   $6x - 2y = -12$ $\longrightarrow$ $y = 3x + 6$

Both lines have slope 3, so the lines are parallel.

(c) $4x + 3y = 6$ $\xrightarrow{\text{Solve for } y.}$ $y = -\frac{4}{3}x + 2$
   $2x - y = 5$ $\longrightarrow$ $y = 2x - 5$

Here the slopes are $-\frac{4}{3}$ and 2. These lines are neither parallel nor perpendicular, because $-\frac{4}{3} \neq 2$ and $-\frac{4}{3} \cdot 2 \neq -1$.

(d) $5x - y = 1$ $\xrightarrow{\text{Solve for } y.}$ $y = 5x - 1$

$x - 5y = -10$ $\xrightarrow{\hspace{2cm}}$ $y = \dfrac{1}{5}x + 2$

The slopes are 5 and $\frac{1}{5}$. The lines are not parallel, nor are they perpendicular. $\left(\textbf{\textit{Be careful!}} \; \mathbf{5\left(\frac{1}{5}\right) = 1}, \textbf{\textit{not}} \; \mathbf{-1}.\right)$

✔ **Now Try Exercises 57, 61, and 63.**

CONNECTIONS

Because the viewing window of a graphing calculator is a rectangle, the graphs of perpendicular lines will not appear perpendicular unless appropriate intervals are used for *x* and *y*. Graphing calculators usually have a key to select a "square" window automatically. In a square window, the *x*-interval is about 1.5 times the *y*-interval because the screen is about 1.5 times as wide as it is high. The equations from Figure 29 are graphed with the standard (nonsquare) window and then with a square window.

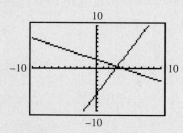
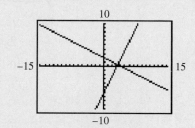

A standard (nonsquare) window
Lines do not appear perpendicular.

A square window
Lines appear perpendicular.

3.3 EXERCISES

🔘 *Complete solution available on Video Lectures on CD/DVD*

Now Try Exercise

📝 **1.** In the context of the graph of a straight line, what is meant by "rise"? What is meant by "run"?

Use the indicated points to find the slope of each line. See Example 1.

2.

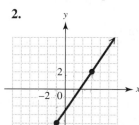

🔘 **3.**

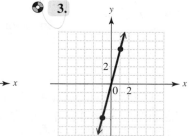

4.

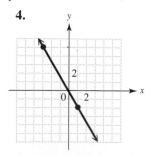

5.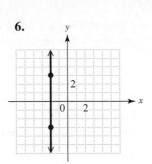

6.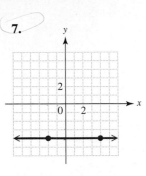

7.

8. Concept Check Look at the graph in Exercise 2 and answer the following.

(a) Start at the point $(-1, -4)$, and count vertically up to the horizontal line that goes through the other plotted point. What is this vertical change? (Remember: "Up" means positive, "down" negative.)

(b) From this new position, count horizontally to the other plotted point. What is this horizontal change? (Remember: "Right" means positive, "left" negative.)

(c) What is the quotient of the numbers found in parts (a) and (b)? What do we call this number?

(d) If we were to *start* at the point $(3, 2)$ and *end* at the point $(-1, -4)$, do you think that the answer to part (c) would be the same? Explain why or why not.

9. Concept Check Match the graph of each line in (a)–(d) with its slope in A–D.

(a)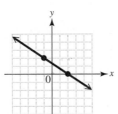

(b)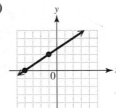

A. $\dfrac{2}{3}$

B. $\dfrac{3}{2}$

(c)

(d)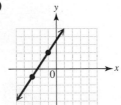

C. $-\dfrac{2}{3}$

D. $-\dfrac{3}{2}$

10. Concept Check Decide whether the line with the given slope rises from left to right, falls from left to right, is horizontal, or is vertical.

(a) $m = -4$ (b) $m = 0$ (c) m is undefined.   (d) $m = \dfrac{3}{7}$

Concept Check *On a pair of axes similar to the one shown, sketch the graph of a straight line having the indicated slope.*

11. Negative

12. Positive

13. Undefined

14. Zero

15. Explain in your own words what is meant by *slope* of a line.

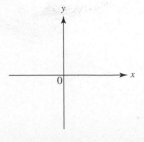

Concept Check *The figure at the right shows a line that has a positive slope (because it rises from left to right) and a positive y-value for the y-intercept (because it intersects the y-axis above the origin).*

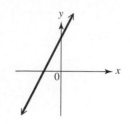

For each line in Exercises 16–21, decide whether **(a)** *the slope is positive, negative, or zero and* **(b)** *the y-value of the y-intercept is positive, negative, or zero.*

16.

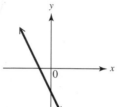

17.

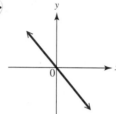

18.

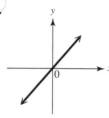

19.

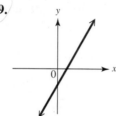

20.

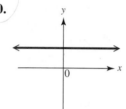

21.

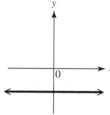

22. ***Concept Check*** What is the slope (or pitch) of this roof?

23. ***Concept Check*** What is the slope (or grade) of this hill?

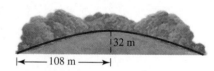

24. ***Concept Check*** A ski slope drops 25 ft vertically for every 100 horizontal feet. Which of the following express its slope? (*Hint*: There is more than one answer.)

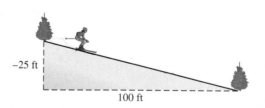

A. $-\dfrac{1}{4}$ **B.** -4 **C.** $-\dfrac{25}{100}$ **D.** -0.25 **E.** -25 **F.** -25%

25. ***Concept Check*** A student was asked to find the slope of the line through the points $(2, 5)$ and $(-1, 3)$. His answer, $-\frac{2}{3}$, was incorrect. He showed his work as

$$\frac{3-5}{2-(-1)} = \frac{-2}{3} = -\frac{2}{3}.$$

WHAT WENT WRONG? Give the correct slope.

Find the slope of the line through each pair of points. See Examples 2–4.

26. $(4, -1)$ and $(-2, -8)$

27. $(1, -2)$ and $(-3, -7)$

28. $(-8, 0)$ and $(0, -5)$

29. $(0, 3)$ and $(-2, 0)$

30. $(6, -5)$ and $(-12, -5)$

31. $(4, 3)$ and $(-6, 3)$

32. $(-4, -5)$ and $(-5, -8)$

33. $(-2, 4)$ and $(-3, 7)$

34. $(-8, 6)$ and $(-8, -1)$

35. $(-12, 3)$ and $(-12, -7)$

36. $(3.1, 2.6)$ and $(1.6, 2.1)$

37. $(4.8, 2.5)$ and $(3.6, 2.2)$

38. $\left(\dfrac{2}{3}, \dfrac{1}{2}\right)$ and $\left(\dfrac{1}{3}, -\dfrac{5}{6}\right)$

39. $\left(-\dfrac{7}{5}, \dfrac{3}{10}\right)$ and $\left(\dfrac{1}{5}, -\dfrac{1}{2}\right)$

Find the slope of each line. See Example 5.

40. $y = 2x - 3$

41. $y = 5x + 12$

42. $2y = -x + 4$

43. $4y = x + 1$

44. $-6x + 4y = 4$

45. $3x - 2y = 3$

46. $2x + 4y = 5$

47. $-3x + 2y = 5$

48. $x = -2$

49. $y = -5$

50. $y = 4$

51. $x = 6$

52. *Concept Check* What is the slope of a line whose graph is parallel to the graph of $-5x + y = -3$? Perpendicular to the graph of $-5x + y = -3$?

53. *Concept Check* What is the slope of a line whose graph is parallel to the graph of $3x + y = 7$? Perpendicular to the graph of $3x + y = 7$?

54. If two nonvertical lines are parallel, what do we know about their slopes? If two lines are perpendicular and neither is parallel to an axis, what do we know about their slopes? Why must the lines be nonvertical?

55. *Concept Check* If two lines are both vertical or both horizontal, which of the following are they?

A. Parallel   **B.** Perpendicular   **C.** Neither parallel nor perpendicular

56. *Concept Check* If a line is vertical, what is true of any line that is perpendicular to it?

For each pair of equations, give the slopes of the lines and then determine whether the two lines are parallel, perpendicular, *or* neither parallel nor perpendicular. *See Example 6.*

57. $2x + 5y = 4$
$4x + 10y = 1$

58. $-4x + 3y = 4$
$-8x + 6y = 0$

59. $8x - 9y = 6$
$8x + 6y = -5$

60. $5x - 3y = -2$
$3x - 5y = -8$

61. $3x - 2y = 6$
$2x + 3y = 3$

62. $3x - 5y = -1$
$5x + 3y = 2$

63. $5x - y = 1$
$x - 5y = -10$

64. $3x - 4y = 12$
$4x + 3y = 12$

RELATING CONCEPTS (EXERCISES 65–70)

FOR INDIVIDUAL OR GROUP WORK

Figure A gives public school enrollment (in thousands) in grades 9–12 in the United States. Figure B gives the (average) number of public school students per computer.

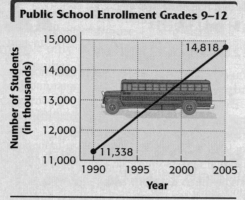

Public School Enrollment Grades 9–12

Source: U.S. Department of Education.

FIGURE A

Students Per Computer

Source: Quality Education Data, Inc.

FIGURE B

Work Exercises 65–70 in order.

65. Use the ordered pairs (1990, 11,338) and (2005, 14,818) to find the slope of the line in Figure A.

66. The slope of the line in Figure A is _____. This means that
$\qquad\qquad\qquad\qquad\qquad\qquad$ (positive/negative)
during the period represented, enrollment _____.
$\qquad\qquad\qquad\qquad\qquad\qquad\qquad\qquad$ (increased/decreased)

67. The slope of a line represents the *rate of change of the line.* On the basis of Figure A, what was the increase in students *per year* during the period shown?

68. Use the given information to find the slope, to the nearest hundredth, of the line in Figure B.

69. The slope of the line in Figure B is _____. This means that
$\qquad\qquad\qquad\qquad\qquad\qquad$ (positive/negative)
the number of students per computer _____ during the period
represented. $\qquad\qquad\qquad$ (increased/decreased)

70. On the basis of Figure B, what was the decrease in students per computer *per year* during the period shown?

71. The growth in billions of retail square footage of U.S. shopping centers is shown in the graph on the next page. This graph looks like a straight line. If the change in square footage each year is the same, then it *is* a straight line. Find the change in square footage for the years shown in the graph. (*Hint:* To find the change in square footage from 2000 to 2001, subtract the *y*-value for 2000 from the *y*-value for 2001.) Is the graph a straight line?

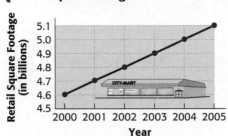

Retail Square Footage

Source: National Resource Bureau.

72. Find the slope of the line in Exercise 71 by using any two of the points shown on the line. How does the slope compare with the yearly change in square footage?

TECHNOLOGY INSIGHTS (EXERCISES 73–76)

Some graphing calculators have the capability of displaying a table of points for a graph. The table shown here gives several points that lie on a line designated Y_1.

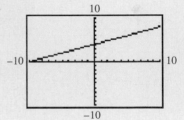

73. Use any two of the ordered pairs displayed to find the slope of the line.

74. What is the x-intercept of the line?

75. What is the y-intercept of the line?

76. Which one of the two lines shown is the graph of Y_1?

A.

B.

PREVIEW EXERCISES

Solve each equation for y. See Section 2.5.

77. $2x + 5y = 15$ **78.** $-4x + 3y = 8$ **79.** $10x = 30 + 3y$

80. $8x = 8 - 2y$ **81.** $y - (-8) = 2(x - 4)$ **82.** $y - 3 = 4[x - (-6)]$

83. $y - \left(-\dfrac{3}{5}\right) = -\dfrac{1}{2}[x - (-3)]$ **84.** $y - \left(-\dfrac{5}{8}\right) = \dfrac{3}{8}(x - 5)$

3.4 Equations of a Line

OBJECTIVES

1 Write an equation of a line by using its slope and *y*-intercept.

2 Graph a line by using its slope and a point on the line.

3 Write an equation of a line by using its slope and any point on the line.

4 Write an equation of a line by using two points on the line.

5 Find an equation of a line that fits a data set.

In **Section 3.3,** we found the slope (steepness) of a line from the equation of the line by solving the equation for *y*. In that form, the slope is the coefficient of *x*. For example, the slope of the line with equation $y = 2x + 3$ is 2, the coefficient of *x*. What does the number 3 represent? If $x = 0$, the equation becomes

$$y = 2(0) + 3 = 0 + 3 = 3.$$

Since $y = 3$ corresponds to $x = 0$, the point $(0, 3)$ is the *y*-intercept of the graph of $y = 2x + 3$. An equation such as $y = 2x + 3$ that is solved for *y* is said to be in *slope–intercept form,* because both the slope and the *y*-intercept of the line can be read directly from the equation.

Slope–Intercept Form

The **slope–intercept form** of the equation of a line with slope *m* and *y*-intercept $(0, b)$ is

$$y = mx + b.$$

Slope ⬈   ⬉ $(0, b)$ is the *y*-intercept.

Remember that the intercept in the slope–intercept form is the *y-intercept.*

▶ **NOTE**   The slope–intercept form is the most useful form for a linear equation because of the information we can determine from it. It is also the form used by graphing calculators and the one that describes a *linear function,* an important concept in mathematics.

OBJECTIVE 1 **Write an equation of a line by using its slope and *y*-intercept.**   Given the slope and *y*-intercept of a line, we can use the slope–intercept form to find an equation of the line.

EXAMPLE 1   **Finding an Equation of a Line**

Find an equation of the line with slope $\frac{2}{3}$ and *y*-intercept $(0, -1)$.

Here, $m = \frac{2}{3}$ and $b = -1$, so the equation is

Slope ⬇   ⬇ *y*-intercept is $(0, b)$.
$$y = mx + b$$

$$y = \frac{2}{3}x + (-1), \quad \text{or} \quad y = \frac{2}{3}x - 1.$$

✔ **Now Try Exercise 11.**

OBJECTIVE 2 **Graph a line by using its slope and a point on the line.**   We can use the slope and *y*-intercept to graph a line.

Graphing a Line by Using the Slope and y-Intercept

Step 1 Write the equation in slope–intercept form, if necessary, by solving for *y*.

Step 2 Identify the *y*-intercept. Graph the point $(0, b)$.

Step 3 Identify slope *m* of the line. Use the geometric interpretation of slope ("rise over run") to find another point on the graph by counting from the *y*-intercept.

Step 4 Join the two points with a line to obtain the graph.

EXAMPLE 2 Graphing a Line by Using the Slope and *y*-intercept

Graph $2x - 3y = 3$ by using the slope and *y*-intercept.

Step 1 Solve for *y* to write the equation in slope–intercept form.

$$2x - 3y = 3 \qquad \text{Given equation}$$
$$-3y = -2x + 3 \qquad \text{Subtract 2x.}$$

Slope–intercept form $\longrightarrow$ $y = \dfrac{2}{3}x - 1 \qquad \text{Divide by } -3.$

Step 2 The *y*-intercept is $(0, -1)$. Graph this point. See Figure 30.

Step 3 The slope is $\frac{2}{3}$. By the definition of slope,

$$m = \frac{\text{change in } y \text{ (rise)}}{\text{change in } x \text{ (run)}} = \frac{2}{3}.$$

Count from the *y*-intercept up 2 units and to the right 3 units to obtain the point $(3, 1)$.

Step 4 Draw the line through the points $(0, -1)$ and $(3, 1)$ to obtain the graph of $2x - 3y = 3$. See Figure 30.

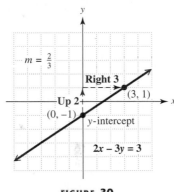

FIGURE 30

✔ **Now Try Exercise 21.**

EXAMPLE 3 Graphing a Line by Using the Slope and a Point

Graph the line through $(-2, 3)$ with slope -4.

First, locate the point $(-2, 3)$. Write the slope as

$$m = \frac{\text{change in } y \text{ (rise)}}{\text{change in } x \text{ (run)}} = -4 = \frac{-4}{1}.$$

Locate another point on the line by counting *down* 4 units (because of the negative sign) and then to the right 1 unit. Finally, draw the line through this new point *P* and the given point $(-2, 3)$. See Figure 31.

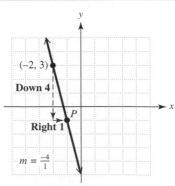

FIGURE 31

✔ **Now Try Exercise 27.**

▶ **NOTE** In Example 3, we could have written the slope as $\frac{4}{-1}$ instead. In this case, we would move up 4 units from $(-2, 3)$ and then to the left 1 unit (because of the negative sign). Verify that this produces the same line.

OBJECTIVE ③ Write an equation of a line by using its slope and any point on the line. We can use the slope–intercept form to write the equation of a line if we know the slope and any point on the line.

EXAMPLE 4 Using the Slope–Intercept Form to Write an Equation

Write an equation, in slope–intercept form, of the line having slope 4 passing through the point $(2, 5)$.

Since the line passes through the point $(2, 5)$, we can substitute $x = 2$, $y = 5$, and the given slope $m = 4$ into $y = mx + b$ and solve for b.

$$y = mx + b \qquad \text{Slope–intercept form}$$

Remember: $(0, b)$ is the y-intercept. Don't stop here.

$$5 = 4(2) + b \qquad \text{Let } x = 2, y = 5, \text{ and } m = 4.$$
$$5 = 8 + b \qquad \text{Multiply.}$$
$$-3 = b \qquad \text{Subtract 8.}$$

The y-intercept is $(0, -3)$. Using the given slope, 4, an equation of the line is

$$y = 4x - 3. \qquad \text{Slope–intercept form}$$

✔ **Now Try Exercise 37.**

There is another form that can be used to write the equation of a line. To develop this form, let m represent the slope of a line and let (x_1, y_1) represent a given point on the line. Let (x, y) represent any other point on the line. See Figure 32. Then,

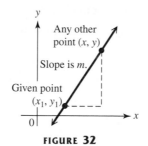

FIGURE 32

$$m = \frac{y - y_1}{x - x_1} \qquad \text{Definition of slope}$$
$$m(x - x_1) = y - y_1 \qquad \text{Multiply each side by } x - x_1.$$
$$y - y_1 = m(x - x_1). \qquad \text{Rewrite.}$$

This result is the *point–slope form* of the equation of a line.

Point–Slope Form

The **point–slope form** of the equation of a line with slope m passing through the point (x_1, y_1) is

$$\underset{\substack{\uparrow \\ \text{Given} \\ \text{point}}}{y - y_1} = \overset{\text{Slope}}{m}(x - x_1).$$

EXAMPLE 5 Using the Point–Slope Form to Write Equations

Find an equation of each line. Give the final answer in slope–intercept form.

(a) Through $(-2,4)$, with slope -3

The given point is $(-2,4)$ so $x_1 = -2$ and $y_1 = 4$. Also, $m = -3$. We substitute these values into the point–slope form to get an equation of the line, and then solve for y to write the equation in slope–intercept form.

$$y - y_1 = m(x - x_1) \qquad \text{Point–slope form}$$
$$y - 4 = -3[x - (-2)] \qquad \text{Let } y_1 = 4, m = -3, x_1 = -2.$$
$$y - 4 = -3(x + 2)$$
$$y - 4 = -3x - 6 \qquad \text{Distributive property}$$
$$y = -3x - 2 \qquad \text{Add 4.}$$

Be careful with signs.

The last equation is in slope–intercept form.

(b) Through $(4,2)$, with slope $\frac{3}{5}$

$$y - y_1 = m(x - x_1) \qquad \text{Point–slope form}$$
$$y - 2 = \frac{3}{5}(x - 4) \qquad \text{Let } y_1 = 2, m = \frac{3}{5}, x_1 = 4.$$
$$y - 2 = \frac{3}{5}x - \frac{12}{5} \qquad \text{Distributive property}$$
$$y = \frac{3}{5}x - \frac{12}{5} + \frac{10}{5} \qquad \text{Add } 2 = \frac{10}{5} \text{ to both sides.}$$
$$y = \frac{3}{5}x - \frac{2}{5} \qquad \text{Combine terms.}$$

We did not clear fractions after the substitution step, because we want the answer in slope–intercept form—that is, solved for y.

✔ **Now Try Exercise 39.**

OBJECTIVE 4 Write an equation of a line by using two points on the line. We can also use either slope–intercept form or point–slope form to find an equation of a line when two points on the line are known.

EXAMPLE 6 Finding the Equation of a Line by Using Two Points

Find an equation of the line through the points $(-2, 5)$ and $(3, 4)$. Give the final answer in slope–intercept form.

First, find the slope of the line, using the slope formula.

$$\text{slope } m = \frac{y_2 - y_1}{x_2 - x_1} = \frac{5 - 4}{-2 - 3} = \frac{1}{-5} = -\frac{1}{5}$$

Start with the x- and y-values of the same point.

Now use either $(-2, 5)$ or $(3, 4)$ and either slope–intercept or point–slope form. We choose $(3, 4)$ and point–slope form.

$$y - y_1 = m(x - x_1) \qquad \text{Point–slope form}$$

$$y - 4 = -\frac{1}{5}(x - 3) \qquad \text{Let } y_1 = 4,\ m = -\tfrac{1}{5},\ x_1 = 3.$$

$$y - 4 = -\frac{1}{5}x + \frac{3}{5} \qquad \text{Distributive property}$$

$$y = -\frac{1}{5}x + \frac{3}{5} + \frac{20}{5} \qquad \text{Add } 4 = \tfrac{20}{5} \text{ to both sides.}$$

$$y = -\frac{1}{5}x + \frac{23}{5} \qquad \text{Combine terms.}$$

The same result would be found by using $(-2, 5)$ for (x_1, y_1).

✔ **Now Try Exercise 47.**

▶ **NOTE** In Example 6, the same result would also be found by substituting the slope and either given point in slope–intercept form $y = mx + b$ and then solving for b, as in Example 4.

Many of the linear equations in **Sections 3.1–3.3** were given in the form $Ax + By = C$, called *standard form*, which we define as follows.

Standard Form

A linear equation is in **standard form** if it is written as

$$Ax + By = C,$$

where A, B, and C are integers, $A > 0$, and $B \neq 0$.

▶ **NOTE** The given definition of standard form is not the same in all texts. A linear equation can be written in this form in many different, equally correct, ways. For example,

$$3x + 4y = 12, \quad 6x + 8y = 24, \quad \text{and} \quad 9x + 12y = 36$$

all represent the same set of ordered pairs. Let us agree that $3x + 4y = 12$ is preferable to the other forms because the greatest common factor of 3, 4, and 12 is 1.

Notice also that we have defined standard form such that $A > 0$, Thus, to write an equation that has $A < 0$ in standard form, multiply each side of the equation by -1. For example,

$$-3x + 2y = -12 \qquad \text{in standard form is} \qquad 3x - 2y = 12.$$

A summary of the forms of linear equations follows.

Forms of Linear Equations

Equation	Description	Example
$x = k$	**Vertical line** Slope is undefined; x-intercept is $(k,0)$.	$x = 3$
$y = k$	**Horizontal line** Slope is 0; y-intercept is $(0,k)$.	$y = 3$
$y = mx + b$	**Slope–intercept form** Slope is m; y-intercept is $(0,b)$.	$y = \frac{3}{2}x - 6$
$y - y_1 = m(x - x_1)$	**Point–slope form** Slope is m; line passes through (x_1, y_1).	$y + 3 = \frac{3}{2}(x - 2)$
$Ax + By = C$	**Standard form** Slope is $-\frac{A}{B}$; x-intercept is $\left(\frac{C}{A},0\right)$; y-intercept is $\left(0,\frac{C}{B}\right)$.	$3x - 2y = 12$

OBJECTIVE 5 **Find an equation of a line that fits a data set.** Earlier in this chapter, we gave linear equations that modeled real data, such as number of twin births and amounts of credit card debt, and then used these equations to estimate or predict values. We can now find such an equation if the given set of data fits a linear pattern—that is, if its graph consists of points lying close to a straight line.

EXAMPLE 7 **Finding the Equation of a Line That Describes Data**

The table lists the average annual cost (in dollars) in tuition and fees for in-state students at public four-year colleges for selected years. Year 1 represents 1999, year 3 represents 2001, and so on.

Year	Cost (in dollars)
1	3362
3	3766
5	4645
7	5491

Source: The College Board.

Plot the data and find an equation that approximates it.

Letting y represent the cost in year x, we plot the data as shown in Figure 33 on the next page.

AVERAGE ANNUAL COSTS AT PUBLIC 4-YEAR COLLEGES

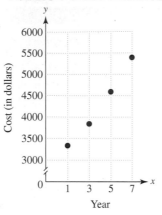

FIGURE 33

Year	Cost (in dollars)
1	3362
3	3766
5	4645
7	5491

Source: The College Board.

The points appear to lie approximately in a straight line. We can use two of the data pairs and the slope–intercept form of the equation of a line to get an equation that describes the relationship between the year and the cost. We choose the ordered pairs $(5, 4645)$ and $(7, 5491)$ from the table (repeated in the margin) and find the slope of the line through these points.

$$m = \frac{y_2 - y_1}{x_2 - x_1} \qquad \text{Slope formula}$$

$$m = \frac{5491 - 4645}{7 - 5} \qquad \text{Let } (7, 5491) = (x_2, y_2) \text{ and } (5, 4645) = (x_1, y_1).$$

$$m = 423$$

As we might expect, the slope, 423, is positive, indicating that tuition and fees *increased* \$423 each year. Now use this slope and the point $(5, 4645)$ in slope–intercept form to find an equation of the line.

$$y = mx + b \qquad \text{Slope–intercept form}$$
$$4645 = 423(5) + b \qquad \text{Substitute for } x, y, \text{ and } m.$$
$$4645 = 2115 + b \qquad \text{Multiply.}$$
$$2530 = b \qquad \text{Subtract 2115.}$$

Thus, an equation of the line is

$$y = 423x + 2530.$$

To see how well this equation approximates the ordered pairs in the data table, let $x = 3$ (for 2001) and find y.

$$y = 423x + 2530 \qquad \text{Equation of the line}$$
$$y = 423(3) + 2530 \qquad \text{Substitute 3 for } x.$$
$$y = 3799 \qquad \text{Multiply; add.}$$

The corresponding value in the table for $x = 3$ is 3766, so the equation approximates the data reasonably well. With caution, the equation could be used to predict values for years that are not included in the table.

✔ **Now Try Exercise 69.**

▶ **NOTE** In Example 7, if we had chosen two different data points, we would have gotten a slightly different equation.

3.4 EXERCISES

⊕ *Complete solution available on Video Lectures on CD/DVD*

Now Try Exercise

Concept Check *Match the description in Column I with the correct equation in Column II.*

I

1. Slope $= -2$, passes through $(4, 1)$

2. Slope $= -2$, y-intercept $(0, 1)$

3. Passes through $(0, 0)$ and $(4, 1)$

4. Passes through $(0, 0)$ and $(1, 4)$

II

A. $y = 4x$

B. $y = \frac{1}{4}x$

C. $y = -4x$

D. $y = -2x + 1$

E. $y - 1 = -2(x - 4)$

5. ***Concept Check*** Match each equation with the graph that would most closely resemble its graph.

(a) $y = x + 3$ **(b)** $y = -x + 3$ **(c)** $y = x - 3$ **(d)** $y = -x - 3$

A.

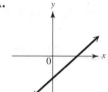

B.

C.

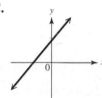

D.

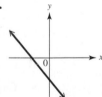

 6. Explain why the equation of a vertical line cannot be written in the form $y = mx + b$.

Concept Check *Use the geometric interpretation of slope (rise divided by run, from Section 3.3) to find the slope of each line. Then, by identifying the y-intercept from the graph, write the slope–intercept form of the equation of the line.*

7.

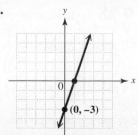

$(0, -3)$

8.

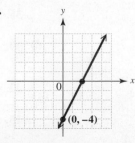

$(0, -4)$

9.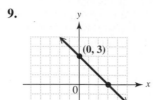

10.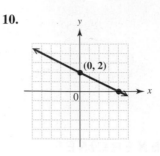

Write the equation of each line with the given slope and y-intercept. See Example 1.

11. $m = 4, (0, -3)$

12. $m = -5, (0, 6)$

13. $m = 0, (0, 3)$

14. $m = 0, (0, -4)$

15. Undefined slope, $(0, -2)$

16. Undefined slope, $(0, 5)$

Graph each equation by using the slope and y-intercept. See Example 2.

17. $y = 3x + 2$

18. $y = 4x - 4$

19. $y = -\dfrac{1}{3}x + 4$

20. $y = -\dfrac{1}{2}x + 2$

21. $2x + y = -5$

22. $3x + y = -2$

23. $4x - 5y = 20$

24. $6x - 5y = 30$

Graph each line passing through the given point and having the given slope. (In Exercises 30–33, recall the types of lines having slope 0 and undefined slope.) See Example 3.

25. $(0, 1), m = 4$

26. $(0, -5), m = -2$

27. $(1, -5), m = -\dfrac{2}{5}$

28. $(2, -1), m = -\dfrac{1}{3}$

29. $(-1, 4), m = \dfrac{2}{5}$

30. $(3, 2), m = 0$

31. $(-2, 3), m = 0$

32. $(3, -2)$, undefined slope

33. $(2, 4)$, undefined slope

34. *Concept Check* What is the common name given to a vertical line whose *x*-intercept is the origin?

35. *Concept Check* What is the common name given to a line with slope 0 whose *y*-intercept is the origin?

Write an equation for each line passing through the given point and having the given slope. Give the final answer in slope–intercept form. See Examples 4 and 5.

36. $(4, 1), m = 2$

37. $(-1, 3), m = -4$

38. $(2, 7), m = 3$

39. $(-2, 5), m = \dfrac{2}{3}$

40. $(4, 2), m = -\dfrac{1}{3}$

41. $(-4, 1), m = \dfrac{3}{4}$

42. $(6, -3), m = -\dfrac{4}{5}$

43. $(2, 1), m = \dfrac{5}{2}$

44. $(7, -2), m = -\dfrac{7}{2}$

45. *Concept Check* If a line passes through the origin and a second point whose *x*- and *y*-coordinates are equal, what is an equation of the line?

Write an equation for each line passing through the given pair of points. Give the final answer in slope–intercept form. See Example 6.

46. $(8, 5)$ and $(9, 6)$ $\quad\bullet$ **47.** $(4, 10)$ and $(6, 12)$ $\quad$ **48.** $(-1, -7)$ and $(-8, -2)$

49. $(-2, -1)$ and $(3, -4)$ $\quad$ **50.** $(0, -2)$ and $(-3, 0)$ $\quad$ **51.** $(-4, 0)$ and $(0, 2)$

52. $\left(\dfrac{1}{2}, \dfrac{3}{2}\right)$ and $\left(-\dfrac{1}{4}, \dfrac{5}{4}\right)$ $\qquad\qquad$ **53.** $\left(-\dfrac{2}{3}, \dfrac{8}{3}\right)$ and $\left(\dfrac{1}{3}, \dfrac{7}{3}\right)$

54. Describe in your own words the slope–intercept and point–slope forms of the equation of a line. Tell what information must be given to use each form to write an equation. Include examples.

RELATING CONCEPTS (EXERCISES 55–62)

FOR INDIVIDUAL OR GROUP WORK

If we think of ordered pairs of the form (C, F), then the two most common methods of measuring temperature, Celsius and Fahrenheit, can be related as follows: When $C = 0$, $F = 32$, and when $C = 100$, $F = 212$. **Work Exercises 55–62 in order.**

55. Write two ordered pairs relating these two temperature scales.

56. Find the slope of the line through the two points.

57. Use the point–slope form to find an equation of the line. (Your variables should be C and F rather than x and y.)

58. Write an equation for F in terms of C.

59. Use the equation from Exercise 58 to write an equation for C in terms of F.

60. Use the equation from Exercise 58 to find the Fahrenheit temperature when $C = 30$.

61. Use the equation from Exercise 59 to find the Celsius temperature when $F = 50$.

62. For what temperature is $F = C$?

Write an equation of the line satisfying the given conditions. Give the final answer in slope–intercept form.

63. Through $(2, -3)$; parallel to $3x = 4y + 5$

64. Through $(-1, 4)$; perpendicular to $2x + 3y = 8$

65. Perpendicular to $x - 2y = 7$; y-intercept $(0, -3)$

66. Parallel to $5x = 2y + 10$; y-intercept $(0, 4)$

*The cost y of producing x items is, in some cases, expressed as $y = mx + b$. The number b gives the **fixed cost** (the cost that is the same no matter how many items are produced), and the number m is the **variable cost** (the cost of producing an additional item). Use this information to work Exercises 67 and 68.*

67. It costs \$400 to start up a business selling snow cones. Each snow cone costs \$0.25 to produce.

(a) What is the fixed cost?

(b) What is the variable cost?

(c) Write the cost equation.

(d) What will be the cost of producing 100 snow cones, based on the cost equation?

(e) How many snow cones will be produced if the total cost is \$775?

68. It costs $2000 to purchase a copier, and each copy costs $0.02 to make.

    (a) What is the fixed cost?

    (b) What is the variable cost?

    (c) Write the cost equation.

    (d) What will be the cost of producing 10,000 copies, based on the cost equation?

    (e) How many copies will be produced if the total cost is $2600?

Solve each problem. See Example 7.

69. The table lists the average annual cost (in dollars) of tuition and fees at private four-year colleges for selected years, where year 1 represents 1997, year 3 represents 1999, and so on.

Year	Cost (in dollars)
1	13,785
3	15,518
5	17,377
7	18,950
9	21,235

Source: The College Board.

    (a) Write five ordered pairs from the data.

    (b) Plot the ordered pairs. Do the points lie approximately in a straight line?

    (c) Use the ordered pairs (3, 15,518) and (9, 21,235) to find the equation of a line that approximates the data. Write the final equation in slope–intercept form. (Round the slope to the nearest tenth.)

    (d) Use the equation from part (c) to estimate the average annual cost at private four-year colleges in 2008. (*Hint:* What is the value of x for 2008?)

70. The table gives heavy-metal nuclear waste (in thousands of metric tons) from spent reactor fuel now stored temporarily at reactor sites, awaiting permanent storage. (*Source:* "Burial of Radioactive Nuclear Waste Under the Seabed," *Scientific American,* January 1998.)

Year x	Waste y
1995	32
2000	42
2010*	61
2020*	76

*Estimated by the U.S. Department of Energy.

Let $x = 0$ represent 1995, $x = 5$ represent 2000 (since $2000 - 1995 = 5$), and so on.

    (a) For 1995, the ordered pair is $(0, 32)$. Write ordered pairs for the data for the other years given in the table.

    (b) Plot the ordered pairs (x, y). Do the points lie approximately in a straight line?

(c) Use the ordered pairs $(0, 32)$ and $(25, 76)$ to find the equation of a line that approximates the other ordered pairs. Write the equation in slope–intercept form.

(d) Use the equation from part (c) to estimate the amount of nuclear waste in 2005. (*Hint:* What is the value of x for 2005?)

TECHNOLOGY INSIGHTS (EXERCISES 71–74)

In Exercises 71 and 72, two graphing calculator views of the same line are shown. Use the displays at the bottom of the screen to find an equation of the form $y = mx + b$ for each line.

71.

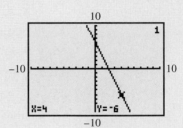

72.

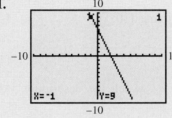

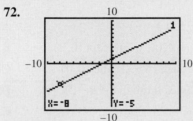

In Exercises 73 and 74, a table of values generated by a graphing calculator is shown for a line Y_1. (With graphing calculators, we use capital Y_1 and X, as the calculator does.) Use any two points to find the equation of each line. Write the final equation in slope–intercept form.

73.

X	Y₁	
-2	-.5	
-1	.25	
0	1	
1	1.75	
2	2.5	
3	3.25	
4	4	
X=-2		

74.

X	Y₁	
-4	14	
-3	10	
-2	6	
-1	2	
0	-2	
1	-6	
2	-10	
X=-4		

PREVIEW EXERCISES

*Solve each inequality and graph the solution set on a number line. See **Section 2.8.***

75. $3x + 8 > -1$

76. $\frac{1}{2}x - 3 < 2$

77. $5 - 3x \le -10$

78. $-x < 0$

79. $x - 4 \le 0$

80. $2x \ge 3x$

Summary Exercises on Linear Equations and Graphs

Graph each line, using the given information or equation.

1. $x - 2y = -4$

2. $2x + 3y = 12$

3. $m = 1$, y-intercept $(0, -2)$

4. $y = -2x + 6$

5. $m = -\frac{2}{3}$, passes through $(3, -4)$

6. Undefined slope, passes through $(-3.5, 0)$

7. $x - 4y = 0$

8. $y - 4 = -9$

9. $8x = 6y + 24$

10. $m = 1$, y-intercept $(0, -4)$

11. $5x + 2y = 10$

12. $m = -\frac{3}{4}$, passes through $(4, -4)$

13. $m = 0$, passes through $\left(0, \frac{3}{2}\right)$

14. $x + 5y = 0$

15. $y = -x + 6$

16. $4x = 3y - 24$

17. $x + 4 = 0$

18. $x - 3y = 6$

19. *Concept Check* Match the description in Column I with the correct equation in Column II.

I	**II**
(a) Slope -0.5, $b = -2$	**A.** $y = -\frac{1}{2}x$
(b) x-intercept $(4, 0)$, y-intercept $(0, 2)$	**B.** $y = -\frac{1}{2}x - 2$
(c) Passes through $(4, -2)$ and $(0, 0)$	**C.** $x - 2y = 2$
(d) $m = \frac{1}{2}$, passes through $(-2, -2)$	**D.** $x + 2y = 4$
	E. $x = 2y$

20. *Concept Check* Which of the following equations are equivalent to the equation $2x + 5y = 20$?

A. $y = -\dfrac{2}{5}x + 4$

B. $y - 2 = -\dfrac{2}{5}(x - 5)$

C. $y = \dfrac{5}{2}x - 4$

D. $2x = 5y - 20$

Write an equation for each line. Give the final answer in slope–intercept form if possible.

21. $m = -3$, $b = -6$

22. $m = \frac{3}{2}$, through $(-4, 6)$

23. Through $(1, -7)$ and $(-2, 5)$

24. Through $(0, 0)$ and $(5, 3)$

25. Through $(0, 0)$, undefined slope

26. Through $(3, 0)$ and $(0, -3)$

27. Through $(0, 0)$ and $(3, 2)$

28. $m = -2$, $b = -4$

29. Through $(5, 0)$ and $(0, -5)$

30. Through $(0, 0)$, $m = 0$

31. $m = \frac{5}{3}$, through $(-3, 0)$

32. Through $(1, -13)$ and $(-2, 2)$

3.5   Graphing Linear Inequalities in Two Variables

In **Section 3.2,** we graphed linear equations such as $2x + 3y = 6$. Now we extend this work to *linear inequalities in two variables,* such as $2x + 3y \leq 6$. (Recall that $\leq$ is read "is less than or equal to.")

Linear Inequality in Two Variables

An inequality that can be written as
$$Ax + By < C \quad \text{or} \quad Ax + By > C,$$
where A, B, and C are real numbers and A and B are not both 0, is a **linear inequality in two variables.**

The symbols $\leq$ and $\geq$ may replace $<$ and $>$ in the definition.

OBJECTIVE 1   Graph $\leq$ or $\geq$ linear inequalities.   The inequality $2x + 3y \leq 6$ means that

$$2x + 3y < 6 \qquad \text{or} \qquad 2x + 3y = 6.$$

As we found earlier, the graph of $2x + 3y = 6$ is a line. This **boundary line** divides the plane into two regions. The graph of the solutions of the inequality $2x + 3y < 6$ will include only *one* of these regions. We can find the required region by solving the given inequality for y.

$$2x + 3y \leq 6$$
$$3y \leq -2x + 6 \qquad \text{Subtract } 2x.$$
$$y \leq -\frac{2}{3}x + 2 \qquad \text{Divide by 3.}$$

From this last statement, ordered pairs in which *y is less than or equal to* $-\frac{2}{3}x + 2$ will be solutions of the inequality. Ordered pairs in which y is equal to $-\frac{2}{3}x + 2$ are on the boundary line, so pairs in which *y is less than* $-\frac{2}{3}x + 2$ will be *below* that line. (This is because, as we move *down* vertically, the y-values *decrease.*) To indicate the solutions, shade the region below the line, as in Figure 34. The shaded region, along with the boundary line, is the desired graph.

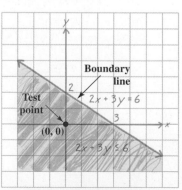

FIGURE 34

Alternatively, a test point gives a quick way to find the correct region to shade. We choose any point *not* on the boundary line. Because $(0, 0)$ is easy to substitute into an inequality, it is often a good choice. We substitute 0 for x and 0 for y in the given inequality to see whether the resulting statement is true or false.

$$2x + 3y \le 6 \qquad \text{Original inequality}$$
$$2(0) + 3(0) \le 6 \quad ? \qquad \text{Let } x = 0 \text{ and } y = 0.$$
$$0 + 0 \le 6 \quad ?$$
$$0 \le 6 \qquad \text{True}$$

Since the last statement is true, we shade the region that includes the test point $(0, 0)$. Our result agrees with that shown in Figure 34.

✔ **Now Try Exercise 19.**

OBJECTIVE 2 **Graph < or > linear inequalities.**   Inequalities that do not include the equals sign are graphed in a similar way.

EXAMPLE 1 **Graphing a Linear Inequality**

Graph $x - y > 5$.

This inequality does *not* include the equals sign. Therefore, the points on the line $x - y = 5$ do *not* belong to the graph. However, the line still serves as a boundary for two regions, one of which satisfies the inequality. To graph the inequality, first graph the equation $x - y = 5$. Use a *dashed line* to show that the points on the line are *not* solutions of the inequality $x - y > 5$. See Figure 35.

Now choose a test point to see which side of the line satisfies the inequality. Again, $(0, 0)$ is a convenient choice.

$$x - y > 5 \qquad \text{Original inequality}$$
$$0 - 0 > 5 \quad ? \qquad \text{Let } x = 0 \text{ and } y = 0.$$
$$0 > 5 \qquad \text{False}$$

Since $0 > 5$ is false, the graph of the inequality is the region that *does not* contain $(0, 0)$. Shade the *other* region, as shown in Figure 35, to obtain the required graph.

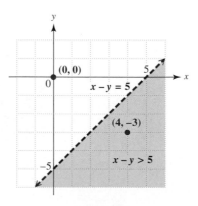

FIGURE 35

To check that the correct region is shaded, test a point in the shaded region. For example, use $(4, -3)$ from the shaded region as follows.

$$x - y > 5 \qquad \text{Original inequality}$$
$$4 - (-3) > 5 \qquad ? \qquad \text{Let } x = 4 \text{ and } y = -3.$$
$$7 > 5 \qquad \text{True}$$

Use parentheses to avoid errors.

This true statement verifies that the correct region was shaded in Figure 35.

✔ **Now Try Exercise 21.**

A summary of the steps used to graph a linear inequality in two variables follows.

Graphing a Linear Inequality

Step 1 **Graph the boundary.** Graph the line that is the boundary of the region. Use the methods of **Section 3.2.** Draw a solid line if the inequality has $\leq$ or $\geq$ because of the equality portion of the symbol; draw a dashed line if the inequality has $<$ or $>$.

Step 2 **Shade the appropriate side.** Use any point not on the line as a test point. Substitute for x and y in the *inequality.* If a true statement results, shade the side containing the test point. If a false statement results, shade the other side.

EXAMPLE 2 Graphing a Linear Inequality with a Vertical Boundary Line

Graph $x < 3$.

First, we graph $x = 3$, a vertical line going through the point $(3, 0)$. We use a dashed line (why?) and choose $(0, 0)$ as a test point.

$$x < 3 \qquad \text{Original inequality}$$
$$0 < 3 \qquad ? \qquad \text{Let } x = 0.$$
$$0 < 3 \qquad \text{True}$$

Since $0 < 3$ is true, we shade the region containing $(0, 0)$, as in Figure 36.

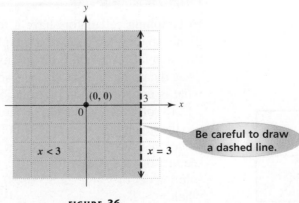

Be careful to draw a dashed line.

FIGURE 36

✔ **Now Try Exercise 25.**

OBJECTIVE 3 **Graph inequalities with a boundary through the origin.** *If the graph of an inequality has a boundary line that goes through the origin, $(0,0)$ cannot be used as a test point.*

| **EXAMPLE 3** | **Graphing a Linear Inequality with a Boundary Line through the Origin** |

Graph $x \leq 2y$.

Graph $x = 2y$, using a solid line. Some ordered pairs that can be used to graph this line are $(0,0)$, $(6,3)$, and $(4,2)$. Since $(0,0)$ is *on* the line $x = 2y$, it cannot be used as a test point. Instead, we choose $(1,3)$, a test point *off* the line.

$$x \leq 2y \qquad \text{Original inequality}$$
$$1 \leq 2(3) \quad ? \quad \text{Let } x = 1 \text{ and } y = 3.$$
$$1 \leq 6 \qquad \text{True}$$

Since $1 \leq 6$ is true, shade the side of the line containing the test point $(1,3)$. See Figure 37.

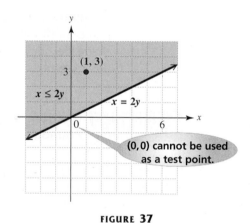

FIGURE 37

✔ **Now Try Exercise 29.**

CONNECTIONS

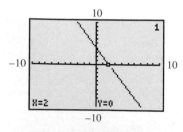

FIGURE 38

Graphing calculators have a feature that allows us to shade regions in the plane, so they can be used to graph a linear inequality in two variables. Consult the manual for specific directions for your calculator. The calculator will not draw the graph as a dashed line, so it is still necessary to understand what is and what is not included in the solution set.

To solve the inequalities in one variable, $-2x + 4 > 0$ and $-2x + 4 < 0$, we use the graph of $y = -2x + 4$ in Figure 38. For $y = -2x + 4 > 0$, we want the values of x such that $y > 0$, so that the line is *above* the x-axis. From Figure 38, we see that this is the case for $x < 2$. Thus, the solution set is $(-\infty, 2)$. Similarly, the solution set of $y = -2x + 4 < 0$ is $(2, \infty)$, because the line is *below* the x-axis when $x > 2$.

(continued)

For Discussion or Writing

1. Discuss the pros and cons of using a calculator to solve a linear inequality in one variable.
2. Use a graphing calculator to solve the following inequalities in one variable from **Section 2.8.**
   (a) $3x + 2 - 5 > -x + 7 + 2x$ (Example 4, page 168)
   (b) $3x + 2 - 5 < -x + 7 + 2x$ (Use the result from part (a).)
   (c) $4 \leq 3x - 5 < 6$ (Example 7, page 171)

3.5 EXERCISES

Complete solution available on Video Lectures on CD/DVD

Now Try Exercise

Concept Check Answer true *or* false *to each of the following.*

1. The point $(4, 0)$ lies on the graph of $3x - 4y < 12$.

2. The point $(4, 0)$ lies on the graph of $3x - 4y \leq 12$.

3. The points $(4, 1)$ and $(0, 0)$ lie on the graph of $3x - 2y \geq 0$.

4. The graph of $y > x$ does not contain points in quadrant IV.

Concept Check *The following statements were taken from articles in newspapers or magazines. Each includes a phrase that can be symbolized with one of the inequality symbols* $<, \leq, >,$ *or* $\geq$. *In Exercises 5–10, give the inequality symbol for the boldface italic words.*

5. According to the Kaiser Family Foundation, in January 2001 about one-quarter of Medicare beneficiaries spent ***more than*** $2000 a year on drugs.

6. Treating HIV-infected children and teenagers with new drugs cut the risk of death by two-thirds, to ***less than*** 1 percent annually. (*Source: Sacramento Bee,* November 22, 2001.)

7. In one study, 9 percent of pregnant women had active hepatitis, meaning that ***at most*** 9 percent of children could get it at birth. (*Source: 2000 Dow Jones & Company, Inc.*)

8. Forty percent of Americans keep ***at least*** one gun at home. (*Source:* Gallup Organization, quoted in *Reader's Digest,* March 2002.)

9. By 1937, a population of as many as a million Attwater's prairie chickens had been cut to ***less than*** 9000. (*Source: National Geographic,* March 2002.)

10. [Easter Island's] nearly 1000 statues, some *almost* 30 feet tall and weighing *as much as* 80 tons, are still an enigma. (*Source: Smithsonian,* March 2002.)

In Exercises 11–16, the straight-line boundary has been drawn. Complete the graph by shading the correct region. See Examples 1–3.

11. $x + 2y \geq 7$

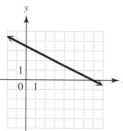

12. $2x + y \geq 5$

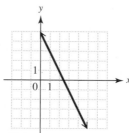

13. $-3x + 4y > 12$

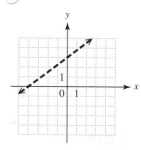

14. $x > 4$

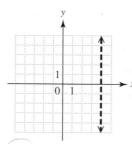

15. $y < -1$

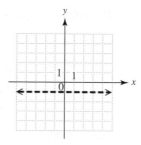

16. $x \leq 3y$

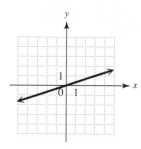

17. Explain how to determine whether to use a dashed line or a solid line when graphing a linear inequality in two variables.

18. Explain why the point $(0, 0)$ is not an appropriate choice for a test point when graphing an inequality whose boundary goes through the origin.

Graph each linear inequality. See Examples 1–3.

19. $x + y \leq 5$

20. $x + y \geq 3$

21. $2x + 3y > -6$

22. $3x + 4y < 12$

23. $y \geq 2x + 1$

24. $y < -3x + 1$

25. $x < -2$

26. $x > 1$

27. $y \leq 5$

28. $y \leq -3$

29. $y \geq 4x$

30. $y \leq 2x$

31. Explain why the graph of $y > x$ cannot lie in quadrant IV.

32. Explain why the graph of $y < x$ cannot lie in quadrant II.

TECHNOLOGY INSIGHTS (EXERCISES 33–36)

A calculator was used to generate the shaded graphs in choices A–D. Each graph has boundary line $y = 3x - 7$ or $y = -3x + 7$. Match each inequality with the appropriate choice.

33. $y \geq 3x - 7$ **34.** $y \leq 3x - 7$ **35.** $y \geq -3x + 7$ **36.** $y \leq -3x + 7$

A.
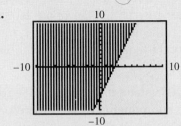

B.
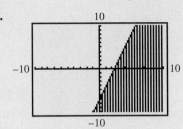

C.
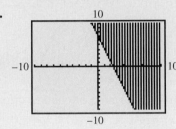

D.
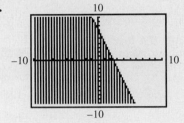

For the given information, (a) graph the inequality (here, $x \geq 0$ and $y \geq 0$, so graph only the part of the inequality in quadrant I) and (b) give some ordered pairs that satisfy the inequality.

37. A company will ship x units of merchandise to outlet I and y units of merchandise to outlet II. The company must ship a total of at least 500 units to these two outlets. The preceding information can be expressed by writing

$$x + y \geq 500.$$

38. A toy manufacturer makes stuffed bears and geese. It takes 20 min to sew a bear and 30 min to sew a goose. There is a total of 480 min of sewing time available to make x bears and y geese. These restrictions lead to the inequality

$$20x + 30y \leq 480.$$

PREVIEW EXERCISES

Find the value of $3x^2 + 8x + 5$ for each given value of x. See Section 1.3.

39. 0   **40.** -1 **41.** 4

42. $-\dfrac{5}{3}$ **43.** 1   **44.** -4

3.6 Introduction to Functions

If gasoline costs \$3.00 per gal and you buy 1 gal, then you must pay $\$3.00(1) = \$3.00$. If you buy 2 gal, your cost is $\$3.00(2) = \$6.00$; for 3 gal, your cost is $\$3.00(3) = \$9.00$, and so on. Generalizing, if x represents the number of gallons, then the cost is $\$3.00x$. If we let y represent the cost, then the equation $y = 3.00x$ *relates* the number of gallons, x, to the cost in dollars, y. The ordered pairs (x, y) that satisfy this equation form a *relation*.

OBJECTIVE 1 **Understand the definition of a relation.** In an ordered pair (x, y), x and y are called the **components** of the ordered pair. Any set of ordered pairs is called a **relation.** The set of all first components of the ordered pairs of a relation is the **domain** of the relation, and the set of all second components of the ordered pairs is the **range** of the relation.

EXAMPLE 1 Identifying Domains and Ranges of Relations Defined by Ordered Pairs

Identify the domain and range of each relation.

(a) $\{(0, 1), (2, 5), (3, 8), (4, 2)\}$

This relation has domain $\{0, 2, 3, 4\}$ and range $\{1, 2, 5, 8\}$. The correspondence between the elements of the domain and the elements of the range is shown in Figure 39.

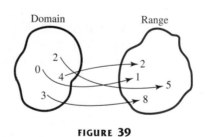

FIGURE 39

(b) $\{(3, 5), (3, 6), (3, 7), (3, 8)\}$

This relation has domain $\{3\}$ and range $\{5, 6, 7, 8\}$.

✔ **Now Try Exercises 7(a) and 9(a).**

OBJECTIVE 2 **Understand the definition of a function.** We now investigate an important type of relation called a *function*.

> **Function**
>
> A **function** is a set of ordered pairs in which each first component corresponds to exactly one second component.

By this definition, the relation in Example 1(a) is a function. But the relation in Example 1(b) is *not* a function, because the same first component, 3, corresponds to more than one second component. Notice that if the ordered pairs in Example 1(b) were interchanged, giving the relation

$$\{(5,3),(6,3),(7,3),(8,3)\},$$

the result *would* be a function. In that case, each domain element (first component) corresponds to *exactly one* range element (second component).

EXAMPLE 2 **Determining whether Relations Are Functions**

Determine whether each relation is a function.

(a) $\{(-2,4),(-1,1),(0,0),(1,1),(2,4)\}$

Notice that each first component appears once and only once. Because of this, the relation is a function.

(b) $\{(9,3),(9,-3),(4,2)\}$

The first component 9 appears in two ordered pairs and corresponds to two different second components. Therefore, this relation is not a function.

✔ **Now Try Exercises 7(b) and 9(b).**

The simple relations given in Examples 1 and 2 were defined by listing the ordered pairs or by showing the correspondence with the use of a figure. Most useful functions have an infinite number of ordered pairs and are usually defined with equations that tell how to get the second components, given the first. We have been using equations with x and y as the variables, where x represents the first component (input) and y the second component (output) in the ordered pairs.

Here are some everyday examples of functions:

1. The cost y in dollars charged by an express mail company is a function of the weight x in pounds determined by the equation $y = 1.5(x - 1) + 9$.

2. In Cedar Rapids, Iowa, the sales tax is 5% of the price of an item. The tax y on a particular item is a function of the price x, because $y = 0.05x$.

3. The distance d traveled by a car moving at a constant speed of 45 mph is a function of the time t. Thus, $d = 45t$.

As mentioned in **Section 3.1,** the function concept can be illustrated by an input–output "machine," as seen in Figure 40. It shows how the express mail company equation $y = 1.5(x - 1) + 9$ provides an output (the cost in dollars, represented by y) for a given input (the weight in pounds, given by x).

$x = 3$ ⟶ | $y = 1.5(3 - 1) + 9$ | ⟶ $y = 12$
(pounds)　　　　　　　　　　　　　　　　　　　　(dollars)
Domain　　　　　　　　　　　　　　　　　　　　　Range

An input-output (function) machine
for $y = 1.5(x - 1) + 9$

FIGURE 40

OBJECTIVE 3 Decide whether an equation defines a function. Given the graph of an equation, the definition of a function can be used to decide whether or not the graph represents a function. By the definition of a function, each *x*-value must lead to exactly one *y*-value. In Figure 41(a), the indicated *x*-value leads to two *y*-values, so this graph is not the graph of a function. A vertical line can be drawn that intersects the graph in more than one point.

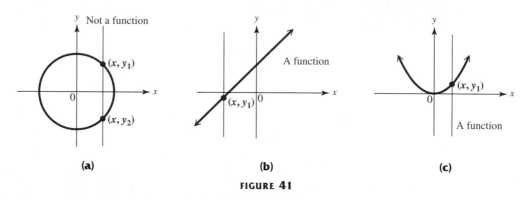

FIGURE 41

By contrast, in Figures 41(b) and 41(c) any vertical line will intersect each graph in no more than one point, so the graphs in Figures 41(b) and 41(c) are graphs of functions. This idea leads to the **vertical line test** for a function.

Vertical Line Test

If a vertical line intersects a graph in more than one point, then the graph is not the graph of a function.

As Figure 41(b) suggests, any nonvertical line is the graph of a function. For this reason, *any linear equation of the form y = mx + b defines a function.* (Recall that a vertical line has undefined slope.)

EXAMPLE 3 Deciding whether Relations Define Functions

Decide whether each relation graphed or defined here is a function.

(a)

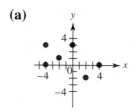

Because there are two ordered pairs with first component −4, as shown in red, this is not the graph of a function.

(b)

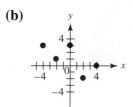

Every first component is matched with one and only one second component, and as a result, no vertical line intersects the graph in more than one point. Therefore, this is the graph of a function.

(c) $y = 2x - 9$

This linear equation is in the form $y = mx + b$. Since the graph of this equation is a line that is not vertical, the equation defines a function.

(d)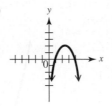

Use the vertical line test. Any vertical line will intersect the graph just once, so this is the graph of a function.

(e)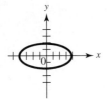

The vertical line test shows that this graph is not the graph of a function; a vertical line could intersect the graph twice.

(f) $x = 4$

The graph of $x = 4$ is a vertical line, so the equation does *not* define a function.

✔ **Now Try Exercises 11(b), 15, and 17.**

An equation in which y is squared does not usually define a function, because most x-values will lead to two y-values. This is true for *any even* power of y, such as y^2, y^4, y^6, and so on. Similarly, an equation involving $|y|$ does not usually define a function, because some x-values lead to more than one y-value.

OBJECTIVE ④ Find domains and ranges. By the definitions of domain and range given for relations, the set of all numbers that can be used as replacements for x in a function is the domain of the function, and the set of all possible values of y is the range of the function.

EXAMPLE 4 Finding the Domain and Range of Functions

Find the domain and range for each function.

(a) $y = 2x - 4$

Any number may be used for x, so the domain is the set of all real numbers. Also, any number may be used for y, so the range is also the set of all real numbers. As indicated in Figure 42, the graph of the equation is a straight line that extends infinitely in both directions, confirming that both the domain and range are $(-\infty, \infty)$.

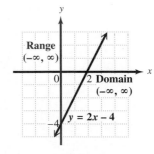

FIGURE 42

(b) $y = x^2$

Any number can be squared, so the domain is the set of all real numbers. However, since the square of a real number cannot be negative and since $y = x^2$, the values of y cannot be negative, making the range the set of all nonnegative numbers, or $[0, \infty)$ in interval notation. The ordered pairs shown in the table are used to get the graph of the function in Figure 43. While x can take any real number value, notice that y is always greater than or equal to 0.

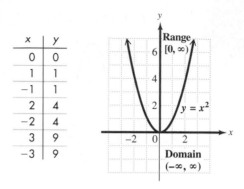

x	y
0	0
1	1
−1	1
2	4
−2	4
3	9
−3	9

FIGURE 43

✔ **Now Try Exercises 23 and 25.**

OBJECTIVE 5 Use function notation.   The letters f, g, and h are commonly used to name functions. For example, the function $y = 3x + 5$ may be written

$$f(x) = 3x + 5,$$

where $f(x)$ is read **"f of x."** The notation $f(x)$ is another way of writing y in a function. For the function defined by $f(x) = 3x + 5$, if $x = 7$, then

$$f(7) = 3 \cdot 7 + 5 \qquad \text{Let } x = 7.$$

Multiply; then add.

$$= 21 + 5 = 26.$$

Read this result, $f(7) = 26$, as "f of 7 equals 26." The notation $f(7)$ means the value of y when x is 7. The statement $f(7) = 26$ says that the value of y is 26 when x is 7. It also indicates that the point $(7, 26)$ lies on the graph of f.

Similarly, to find $f(-3)$, substitute -3 for x.

$$f(-3) = 3(-3) + 5 \qquad \text{Let } x = -3.$$

Use parentheses to avoid errors.

$$= -9 + 5 = -4$$

▶ **CAUTION**   The notation $f(x)$ does *not* mean f times x; $f(x)$ *means the value of x for the function f. It represents the y-value that corresponds to x.*

Function Notation

In the notation $f(x)$,

f is the name of the function,

x is the domain value,

and $f(x)$ is the range value y for the domain value x.

EXAMPLE 5 Using Function Notation

For the function defined by $f(x) = x^2 - 3$, find the following.

(a) $f(4)$

Substitute 4 for x.

$$f(x) = x^2 - 3$$
$$f(4) = 4^2 - 3 \qquad \text{Let } x = 4.$$

Think: $4^2 = 4 \cdot 4$

$$= 16 - 3 = 13$$

Think: $(-3)^2 = -3 \cdot (-3)$

(b) $f(0) = 0^2 - 3 = 0 - 3 = -3$ **(c)** $f(-3) = (-3)^2 - 3 = 9 - 3 = 6$

✔ **Now Try Exercise 35.**

OBJECTIVE 6 Apply the function concept in an application. Because a function assigns to each element in its domain exactly one element in its range, the function concept is used in real-data applications where two quantities are related.

EXAMPLE 6 Applying the Function Concept to Population

The resident population (in millions) of Hispanic origin in the United States for selected years is shown in the table.

HISPANIC POPULATION

Year	Population (in millions)
1998	30.8
2000	32.8
2002	37.4
2004	40.4

Source: U.S. Census Bureau.

(a) Use the table to write a set of ordered pairs that defines a function f.

If we choose the years as the domain elements and the populations as the range elements, the information in the table can be written as a set of four ordered pairs. In set notation, the function f is

$$f = \{(1998, 30.8), (2000, 32.8), (2002, 37.4), (2004, 40.4)\}.$$

(b) What is the domain of f? What is the range?

The domain is the set of years, or x-values:

$$\{1998, 2000, 2002, 2004\}.$$

The range is the set of populations, in millions, or y-values:

$$\{30.8, 32.8, 37.4, 40.4\}.$$

(c) Find $f(1998)$ and $f(2002)$.

From the table or the set of ordered pairs in part (a),

$$f(1998) = 30.8 \text{ million} \qquad \text{and} \qquad f(2002) = 37.4 \text{ million}.$$

(d) For what *x*-value does $f(x)$ equal 40.4 million? 32.8 million?
We use the table or the ordered pairs found in part (a) to determine

$$f(2004) = 40.4 \text{ million} \quad \text{and} \quad f(2000) = 32.8 \text{ million.}$$

✔ **Now Try Exercises 41 and 43.**

3.6 EXERCISES

Complete the table for the linear function defined by $f(x) = x + 2.$

x	x + 2	f(x)	(x, y)
0	2	2	(0, 2)

1. 1

2. 2

3. 3

4. *t*

5. Describe the graph of function *f* in Exercises 1–4 if the domain is {0, 1, 2, 3}.

6. Describe the graph of function *f* in Exercises 1–4 if the domain is the set of real numbers $(-\infty, \infty)$.

In Exercises 7–12, (a) give the domain and range of each relation and (b) determine whether the relation is or is not a function. See Examples 1–3.

⊙ **7.** $\{(-4, 3), (-2, 1), (0, 5), (-2, -8)\}$

8. $\{(3, 7), (1, 4), (0, -2), (-1, -1), (-2, 5)\}$

9.

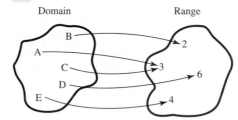

10.

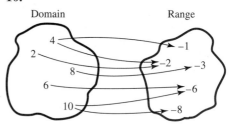

⊙ **11.**

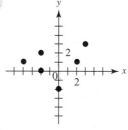

12.

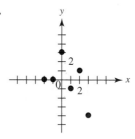

Decide whether each graph is that of a function. See Example 3.

13.

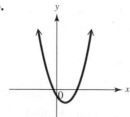

14.

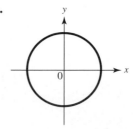

15.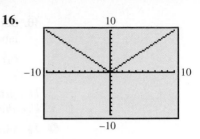

16.

Decide whether each equation defines y as a function of x. (Remember that, to be a function, every value of x must give one and only one value of y.)

17. $y = 5x + 3$

18. $y = -7x + 12$

19. $y = x^2$

20. $y = -3x^2$

21. $x = y^2$

22. $x = |y|$

Find the domain and the range for each function. See Example 4.

23. $y = 3x - 2$

24. $y = x^2 - 3$

25. $y = x^2 + 2$

26. $y = -x + 3$

27. $f(x) = \sqrt{x}$

28. $f(x) = |x|$

RELATING CONCEPTS (EXERCISES 29–32)

FOR INDIVIDUAL OR GROUP WORK

*A function defined by the equation of a line, such as $f(x) = 3x - 4$, called a **linear function,** can be graphed by replacing $f(x)$ with y and then using the methods described earlier. Let us assume that some function is written in the form $f(x) = mx + b$, for particular values of m and b. **Work Exercises 29–32 in order.***

29. If $f(2) = 4$, name the coordinates of one point on the line.

30. If $f(-1) = -4$, name the coordinates of another point on the line.

31. Use the results of Exercises 29 and 30 to find the slope of the line.

32. Use the slope–intercept form of the equation of a line to write the function in the form $f(x) = mx + b$.

---

*For each function f, find (**a**) $f(2)$, (**b**) $f(0)$, and (**c**) $f(-3)$. See Example 5.*

33. $f(x) = 4x + 3$

34. $f(x) = -3x + 5$

35. $f(x) = x^2 - x + 2$

36. $f(x) = x^3 + x$

37. $f(x) = |x|$

38. $f(x) = |x + 7|$

A calculator can be thought of as a function machine. We input a number value (from the domain), and then, by pressing the appropriate key, we obtain an output value (from the range). Use your calculator, follow the directions, and then answer each question.

39. Suppose we enter the value 4 and then take the square root (that is, activate the square root *function*, using the key labeled $\sqrt{x}$).

(**a**) What is the domain value here? (**b**) What is the range value obtained?

40. Suppose we enter the value -8 and then activate the squaring function, using the key labeled x^2.

(a) What is the domain value here? (b) What range value is obtained?

The graph shows the number of U.S. foreign-born residents (in millions) for selected years. Use the information in the graph to work Exercises 41–45. See Example 6.

41. Write the information in the graph as a set of ordered pairs. Does this set define a function?

42. Name the function g in Exercise 41. Give the domain and range of g.

43. Find $g(1980)$ and $g(1990)$.

44. For what value of x does $g(x) = 28.4$?

45. Suppose $g(2002) = 30.3$. What does this tell you in the context of the application?

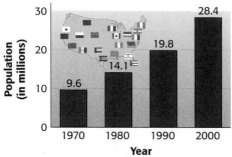

Growth of U.S. Foreign-Born Population

Source: U.S. Census Bureau.

RELATING CONCEPTS (EXERCISES 46–50)

FOR INDIVIDUAL OR GROUP WORK

The data give the percent of U.S. active-duty female military personnel during selected years from 1987 to 2003. Use the information in the table to **work Exercises 46–50 in order.**

46. (a) Plot the ordered pairs (years, percent) from the table. Give the domain and range of this relation. Do the points suggest that a linear function f would give a reasonable approximation of the data?

(b) What is $f(2000)$? For what x-value does $f(x) = 15.1$?

47. Use the first and last data pairs from Exercise 46(a) to write an equation relating the data. Give the equation in $y = mx + b$ form.

48. Use your equation from Exercise 47 to approximate the percent of active-duty female military personnel in 1993 and 2000. Round to the nearest tenth of a percent.

49. Use the second and third data pairs to write an equation relating the data. Give the equation in $y = mx + b$ form.

50. Use your equation from Exercise 49 to approximate the number of active-duty female military personnel in 1987 and 2003. Do the results in Exercise 48 or this exercise give better approximations?

ACTIVE-DUTY FEMALE
MILITARY PERSONNEL

Year	Percent
1987	10.2
1993	11.6
2000	14.4
2003	15.1

Source: U.S. Department of Defense.

TECHNOLOGY INSIGHTS (EXERCISES 51–57)

The table was generated by a graphing calculator.
The expression Y_1 *represents* $f(X)$.

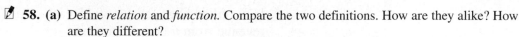

X	Y₁	
-2	-1	
-1	0	
0	1	
1	2	
2	3	
3	4	
4	5	

X= -2

51. When X = 3, Y_1 = _____.

52. What is $f(3)$?

53. When Y_1 = 2, X = _____.

54. If $f(X) = 2$, what is the value of X?

55. The points represented in the table all lie in a straight line. What is the slope of the line?

56. What is the *y*-intercept of the line?

57. Write the function in the form $y = mx + b$, for the appropriate values of *m* and *b*.

58. (a) Define *relation* and *function*. Compare the two definitions. How are they alike? How are they different?
   (b) In your own words, explain the meaning of the domain and range of a function.

PREVIEW EXERCISES

Graph each pair of equations on the same coordinate axes. See Section 3.2.

59. $2x + 3y = 12$
$4x - 2y = 8$

60. $2x - y = 4$
$x + 3y = 6$

61. $x + y = 4$
$2x = 8 - 2y$

62. $-5x + 2y = 10$
$2y = -4 + 5x$

63. $2y = -3x$
$-2x + 3y = 0$

64. $-3y = 6$
$2x = 10$

Chapter 3 Group Activity

DETERMINING BUSINESS PROFIT

Objective Use equations, tables, and graphs to make business decisions.

Graphs are used by businesses to analyze and make decisions. This activity will explore such a process.

The Parent Teacher Association of a school decides to sell bags of popcorn as a fund-raiser. The parents want to estimate their profit. Costs include $14 for popcorn (enough for 680 bags) and a total of $7 for bags to hold the popcorn.

A. Develop a profit formula.

1. From the given information, determine the total costs. (Assume that the association is buying enough supplies for 680 bags of popcorn.)

2. If the filled bags sell for $0.25 each, write an expression for the total revenue from n bags of popcorn.

3. Since Profit = Revenue − Cost, write an equation that represents the profit P for n bags of popcorn sold.

4. Work in pairs. One person should use the profit equation from part 3 to complete the table that follows. The other person should decide on an appropriate scale and graph the profit equation.

n	P
0	
	0
100	

B. Choose a different price for a bag of popcorn (between $0.20 and $0.75).

1. Write the profit equation for this cost.

2. Switch roles; that is, if you drew the graph in part A, now make a table of values for this equation. Have your partner graph the profit equation in the same coordinate system you used in part A.

C. Compare your findings and answer the following.

1. What is the break-even point (that is, when profits are 0, or revenue equals costs) for each equation?

2. What does it mean if you end up with a negative value for P?

3. If you sell all 680 bags, what will your profits be for the two different prices? Explain how you would estimate this from your graph.

4. What are the advantages and disadvantages of charging a higher price?

5. Together, decide on the price you would charge for a bag of popcorn. Explain why you chose this price.

Chapter **3** SUMMARY

View the Interactive Summary on the Pass the Test CD.

KEY TERMS

3.1 line graph
linear equation in two
  variables
ordered pair
table of values
x-axis
y-axis
rectangular (Cartesian)
  coordinate system

quadrant
origin
plane
coordinates
plot
scatter diagram
3.2 graph, graphing
y-intercept
x-intercept

3.3 rise
run
slope
subscript notation
parallel lines
perpendicular lines
3.5 linear inequality in
  two variables
boundary line

3.6 components
relation
domain
range
function
function notation

NEW SYMBOLS

(a, b) an ordered pair

m slope

(x_1, y_1) x-sub-one,
  y-sub-one

$f(x)$ function f of x

TEST YOUR WORD POWER

See how well you have learned the vocabulary in this chapter. Answers, with examples, follow the Quick Review.

1. An **ordered pair** is a pair of numbers written
   A. in numerical order between brackets
   B. between parentheses or brackets
   C. between parentheses in which order is important
   D. between parentheses in which order does not matter.

2. An **intercept** is
   A. the point where the x-axis and y-axis intersect
   B. a pair of numbers written in parentheses in which order matters
   C. one of the four regions determined by a rectangular coordinate system
   D. the point where a graph intersects the x-axis or the y-axis.

3. The **slope** of a line is
   A. the measure of the run over the rise of the line
   B. the distance between two points on the line
   C. the ratio of the change in y to the change in x along the line

D. the horizontal change compared with the vertical change of two points on the line.

4. Two lines in a plane are **parallel** if
   A. they represent the same line
   B. they never intersect
   C. they intersect at a 90° angle
   D. one has a positive slope and one has a negative slope.

5. Two lines in a plane are **perpendicular** if
   A. they represent the same line
   B. they never intersect
   C. they intersect at a 90° angle
   D. one has a positive slope and one has a negative slope.

6. A **relation** is
   A. any set of ordered pairs
   B. a set of ordered pairs in which each first component corresponds to exactly one second component
   C. two sets of ordered pairs that are related
   D. a graph of ordered pairs.

7. The **domain** of a relation is
   A. the set of all x- and y-values in the ordered pairs of the relation

B. the difference between the components in an ordered pair of the relation
   C. the set of all first components in the ordered pairs of the relation
   D. the set of all second components in the ordered pairs of the relation.

8. The **range** of a relation is
   A. the set of all x- and y-values in the ordered pairs of the relation
   B. the difference between the components in an ordered pair of the relation
   C. the set of all first components in the ordered pairs of the relation
   D. the set of all second components in the ordered pairs of the relation.

9. A **function** is
   A. any set of ordered pairs
   B. a set of ordered pairs in which each first component corresponds to exactly one second component
   C. two sets of ordered pairs that are related
   D. a graph of ordered pairs.

QUICK REVIEW

Concepts	Examples

3.1 READING GRAPHS; LINEAR EQUATIONS IN TWO VARIABLES

An ordered pair is a solution of an equation if it satisfies the equation.

Is $(2, -5)$ or $(0, -6)$ a solution of $4x - 3y = 18$?

$$4(2) - 3(-5) = 23 \neq 18 \qquad\qquad 4(0) - 3(-6) = 18$$

$(2, -5)$ is not a solution. $\qquad\qquad$ $(0, -6)$ is a solution.

If a value of either variable in an equation is given, then the other variable can be found by substitution.

Complete the ordered pair $(0, \ \ )$ for $3x = y + 4$.

$$3(0) = y + 4 \qquad \text{Let } x = 0.$$
$$0 = y + 4$$
$$-4 = y$$

The ordered pair is $(0, -4)$.

Plot the ordered pair $(-3, 4)$ by starting at the origin, going 3 units to the left, and then going 4 units up.

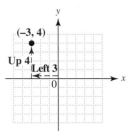

3.2 GRAPHING LINEAR EQUATIONS IN TWO VARIABLES

To graph a linear equation,

Step 1 Find at least two ordered pairs that satisfy the equation.

Step 2 Plot the corresponding points.

Step 3 Draw a straight line through the points.

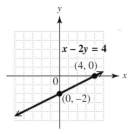

The graph of $Ax + By = 0$ passes through the origin. Find and plot another point that satisfies the equation. Then draw the line through the two points.

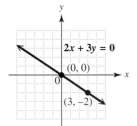

The graph of $y = k$ is a horizontal line through $(0, k)$.

The graph of $x = k$ is a vertical line through $(k, 0)$.

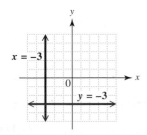

(continued)

Concepts	Examples

3.3 THE SLOPE OF A LINE

The slope of the line through (x_1, y_1) and (x_2, y_2) is

$$m = \frac{y_2 - y_1}{x_2 - x_1} \quad (x_1 \neq x_2).$$

Horizontal lines have slope 0.

Vertical lines have undefined slope.

The line through $(-2, 3)$ and $(4, -5)$ has slope

$$m = \frac{-5 - 3}{4 - (-2)} = \frac{-8}{6} = -\frac{4}{3}.$$

The line $y = -2$ has slope 0.

The line $x = 4$ has undefined slope.

To find the slope of a line from its equation, solve for y. The slope is the coefficient of x.

Find the slope of $3x - 4y = 12$.

$$-4y = -3x + 12 \qquad \text{Add } -3x.$$

$$y = \frac{3}{4}x - 3 \qquad \text{Divide by } -4.$$

Slope

Parallel lines have the same slope.

The lines $y = 3x - 1$ and $y = 3x + 4$ are parallel because both have slope 3.

The slopes of perpendicular lines are negative reciprocals. (That is, their product is -1.)

The lines $y = -3x - 1$ and $y = \frac{1}{3}x + 4$ are perpendicular because their slopes are -3 and $\frac{1}{3}$, and $-3\left(\frac{1}{3}\right) = -1$.

3.4 EQUATIONS OF A LINE

Slope–Intercept Form
$$y = mx + b$$
m is the slope.
$(0, b)$ is the y-intercept.

Find an equation of the line with slope 2 and y-intercept $(0, -5)$.

The equation is $y = 2x - 5$.

Point–Slope Form
$$y - y_1 = m(x - x_1)$$
m is the slope.
(x_1, y_1) is a point on the line.

Find an equation of the line with slope $-\frac{1}{2}$ through $(-4, 5)$.

$$y - 5 = -\frac{1}{2}[x - (-4)] \qquad \text{Substitute.}$$

$$y - 5 = -\frac{1}{2}(x + 4)$$

$$y - 5 = -\frac{1}{2}x - 2 \qquad \text{Distributive property}$$

$$y = -\frac{1}{2}x + 3 \qquad \text{Add 5.}$$

Standard Form
$$Ax + By = C$$
A, B, and C are integers, $A > 0$, and $B \neq 0$.

The equation $y = -\frac{1}{2}x + 3$ is written in standard form as

$$x + 2y = 6,$$

with $A = 1$, $B = 2$, and $C = 6$.

(continued)

Concepts	Examples

3.5 GRAPHING LINEAR INEQUALITIES IN TWO VARIABLES

Step 1 Graph the line that is the boundary of the region. Make it solid if the inequality is $\leq$ or $\geq$ because of the equality portion of the symbol; make it dashed if the inequality is $<$ or $>$.

Graph $2x + y \leq 5$.
Graph the line $2x + y = 5$.
Make it solid because of the equality portion of the symbol $\leq$.

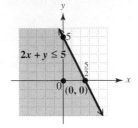

Step 2 Use any point not on the line as a test point. Substitute for x and y in the inequality. If the result is true, shade the side of the line containing the test point; if the result is false, shade the other side.

Use $(0, 0)$ as a test point.

$$2(0) + 0 \leq 5 \quad ?$$
$$0 \leq 5 \quad \text{True}$$

Shade the side of the line containing $(0, 0)$.

3.6 INTRODUCTION TO FUNCTIONS

Vertical Line Test

If a vertical line intersects a graph in more than one point, the graph is not the graph of a function.

The graph shown is not the graph of a function.

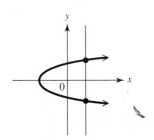

The **domain** of a function is the set of numbers that can replace x in the expression for the function. The **range** is the set of y-values that result as x is replaced by each number in the domain. To find $f(x)$ for a specific value of x, replace x by that value in the expression for the function.

Let $f(x) = x - 1$. Find
(a) the domain of f; (b) the range of f; (c) $f(-1)$.

(a) The domain is $(-\infty, \infty)$. (b) The range is $(-\infty, \infty)$.
(c) $f(-1) = -1 - 1 = -2$

Answers to Test Your Word Power

1. C; *Examples:* $(0, 3), (-3, 8), (4, 0)$

2. D; *Example:* The graph of the equation $4x - 3y = 12$ has x-intercept $(3, 0)$ and y-intercept $(0, -4)$.

3. C; *Example:* The line through $(3, 6)$ and $(5, 4)$ has slope $\dfrac{4 - 6}{5 - 3} = \dfrac{-2}{2} = -1$.

4. B; *Example:* See Figure 28 in **Section 3.3.**

5. C; *Example:* See Figure 29 in **Section 3.3.**

6. A; *Example:* $\{(0, 2), (2, 4), (3, 6), (-1, 3)\}$

7. C; *Example:* The domain in the relation in Answer 6 is the set of x-values—that is, $\{0, 2, 3, -1\}$.

8. D; *Example:* The range of the relation in Answer 6 is the set of y-values—that is, $\{2, 4, 6, 3\}$.

9. B; *Example:* The relation in Answer 6 is a function, since each x-value corresponds to exactly one y-value.

Chapter **3** REVIEW EXERCISES

[3.1]

1. The line graph shows average prices for a gallon of regular unleaded gasoline in the United States for the years 1999 through 2005. Use the graph to work parts (a)–(d).

(a) About how much did a gallon of gas cost in 2000?

(b) How much did the price of a gallon of gas increase from 2000 to 2004?

(c) Between which years did the price of a gallon of gas decrease?

(d) During which two years did the greatest increase in gasoline prices occur? About how much did the price increase during that time?

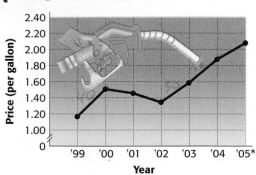

Average U.S. Gasoline Prices

*January to June

Source: U.S. Department of Energy.

Complete the given ordered pairs for each equation.

2. $y = 3x + 2$; $(-1,\ )$, $(0,\ )$, $(\ ,5)$

3. $4x + 3y = 6$; $(0,\ )$, $(\ ,0)$, $(-2,\ )$

4. $x = 3y$; $(0,\ )$, $(8,\ )$, $(\ ,-3)$

5. $x - 7 = 0$; $(\ ,-3)$, $(\ ,0)$, $(\ ,5)$

Determine whether the given ordered pair is a solution of the given equation.

6. $x + y = 7$; $(2,5)$

7. $2x + y = 5$; $(-1,3)$

8. $3x - y = 4$; $\left(\frac{1}{3},-3\right)$

Name the quadrant in which each ordered pair lies. Then plot each pair in a rectangular coordinate system.

9. $(2,3)$ **10.** $(-4,2)$ **11.** $(3,0)$ **12.** $(0,-6)$

13. *Concept Check*   If $xy > 0$, in what quadrant or quadrants must (x,y) lie?

[3.2]   *Find the x- and y-intercepts for the line that is the graph of each equation, and graph the line.*

14. $y = 2x + 5$ **15.** $3x + 2y = 8$ **16.** $x + 2y = -4$

[3.3]   *Find the slope of each line.*

17. Through $(2,3)$ and $(-4,6)$

18. Through $(2,5)$ and $(2,8)$

19. $y = 3x - 4$

20. $y = 5$

21.

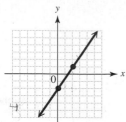

22.

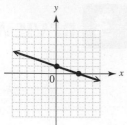

23. The line passing through these points

x	y
0	1
2	4
6	10

24. **(a)** A line parallel to the graph of
$y = 2x + 3$

(b) A line perpendicular to the graph of
$y = -3x + 3$

Decide whether each pair of lines is parallel, perpendicular, *or* neither.

25. $3x + 2y = 6$
$6x + 4y = 8$

26. $x - 3y = 1$
$3x + y = 4$

27. $x - 2y = 8$
$x + 2y = 8$

[3.4] *Write an equation for each line. Give the final answer in slope–intercept form if possible.*

28. $m = -1, b = \frac{2}{3}$

29. Through $(2, 3)$ and $(-4, 6)$

30. Through $(4, -3), m = 1$

31. Through $(-1, 4), m = \frac{2}{3}$

32. Through $(1, -1), m = -\frac{3}{4}$

33. $m = -\frac{1}{4}, b = \frac{3}{2}$

34. Slope 0, through $(-4, 1)$

35. Through $\left(\frac{1}{3}, -\frac{5}{4}\right)$, undefined slope

[3.5] *Graph each linear inequality.*

36. $3x + 5y > 9$

37. $2x - 3y > -6$

38. $x - 2y \geq 0$

[3.6] *Decide whether each relation is or is not a function. In Exercises 39 and 40, give the domain and the range.*

39. $\{(-2, 4), (0, 8), (2, 5), (2, 3)\}$

40. $\{(8, 3), (7, 4), (6, 5), (5, 6), (4, 7)\}$

41.

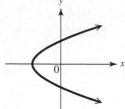

42.

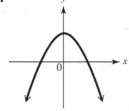

43. $2x + 3y = 12$

44. $y = x^2$

Find (a) f(2) and (b) f(−1).

45. $f(x) = 3x + 2$ **46.** $f(x) = 2x^2 - 1$ **47.** $f(x) = |x + 3|$

MIXED REVIEW EXERCISES

Concept Check   *In Exercises 48–53, match each statement to the appropriate graph or graphs in A–D. Graphs may be used more than once.*

A.
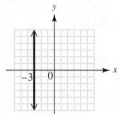

B.
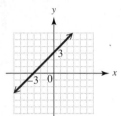

C.
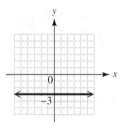

D.
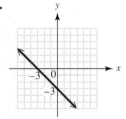

48. The line shown in the graph has undefined slope.

49. The graph of the equation has y-intercept $(0, -3)$.

50. The graph of the equation has x-intercept $(-3, 0)$.

51. The line shown in the graph has negative slope.

52. The graph is that of the equation $y = -3$.

53. The line shown in the graph has slope 1.

Find the intercepts and the slope of each line. Then graph the line.

54. $y = -2x - 5$ **55.** $x + 3y = 0$ **56.** $y - 5 = 0$

Write an equation for each line. Give the final answer in slope–intercept form.

57. $m = -\frac{1}{4}, b = -\frac{5}{4}$ **58.** Through $(8, 6), m = -3$

59. Through $(3, -5)$ and $(-4, -1)$

Graph each inequality.

60. $y < -4x$ **61.** $x - 2y \le 6$

RELATING CONCEPTS (EXERCISES 62–68)

FOR INDIVIDUAL OR GROUP WORK

*The percentages of four-year college students in public schools who earned a degree within five years of entry between 1997 and 2002 are shown in the graph. Use the graph to **work Exercises 62–68 in order.***

62. What was the percent decrease from 1997 to 2002?

63. Since the points of the graph lie approximately in a linear pattern, a straight line can be used to model the data. Will this line have positive or negative slope? Explain.

64. Write two ordered pairs for the data for 1997 and 2002.

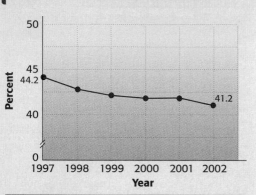

Percentages of Students Graduating Within 5 Years (Public Institutions)

Source: ACT.

65. Use the ordered pairs from Exercise 64 to find the equation of a line that models the data. Write the equation in slope–intercept form.

66. Based on the equation you found in Exercise 65, what is the slope of the line? Does it agree with your answer in Exercise 63?

67. Use the equation from Exercise 65 to approximate the percentages for 1998 through 2001, and complete the table. Round your answers to the nearest tenth.

68. Use the equation from Exercise 65 to predict the percentage for 2003. Can we be sure that this prediction is accurate?

Year	Percent
1998	
1999	
2000	
2001	

Chapter **3** TEST

View the complete solutions to all Chapter Test exercises on the Pass the Test CD.

1. Complete these ordered pairs for the equation $3x + 5y = -30$; $(0, \ )$, $(\ , 0)$, $(\ , -3)$.

2. Is $(4, -1)$ a solution of $4x - 7y = 9$?

3. How do you find the x-intercept of the graph of a linear equation in two variables? How do you find the y-intercept?

Graph each linear equation. Give the x- and y-intercepts.

4. $3x + y = 6$

5. $y - 2x = 0$

6. $x + 3 = 0$

7. $y = 1$

8. $x - y = 4$

Find the slope of each line.

9. Through $(-4, 6)$ and $(-1, -2)$

10. $2x + y = 10$

11. $x + 12 = 0$

12.

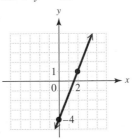

13. A line parallel to the graph of $y - 4 = 6$

Write an equation for each line. Give the final answer in slope–intercept form.

14. Through $(-1, 4)$, $m = 2$

15. The line in Exercise 12

16. Through $(2, -6)$ and $(1, 3)$

Graph each linear inequality.

17. $x + y \leq 3$

18. $3x - y > 0$

The graph shows total food and drink sales at U.S. restaurants from 1970 through 2000, where 1970 corresponds to $x = 0$. Use the graph to work Exercises 19–22.

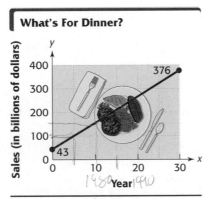

What's For Dinner?

Sales (in billions of dollars)

376

43

Year

Source: National Restaurant Association.

19. Is the slope of the line in the graph positive or negative? Explain.

20. Write two ordered pairs for the data points shown in the graph. Use them to find the slope of the line.

21. The linear equation

$$y = 11.1x + 43$$

approximates food and drink sales y in billions of dollars, where $x = 0$ represents 1970. Use the equation to approximate food and drink sales for 1990 and 1995.

22. What does the ordered pair $(30, 376)$ mean in the context of this situation?

23. Decide whether each relation represents a function. If it does, give the domain and the range.

   (a) $\{(2, 3), (2, 4), (2, 5)\}$ **(b)** $\{(0, 2), (1, 2), (2, 2)\}$

24. Use the vertical line test to determine whether the graph is that of a function.

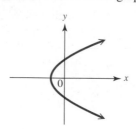

25. If $f(x) = 3x + 7$, find $f(-2)$.

Chapters **1–3** **CUMULATIVE REVIEW EXERCISES**

Perform each indicated operation.

1. $10\dfrac{5}{8} - 3\dfrac{1}{10}$

2. $\dfrac{3}{4} \div \dfrac{1}{8}$

3. $5 - (-4) + (-2)$

4. $\dfrac{(-3)^2 - (-4)(2^4)}{5(2) - (-2)^3}$

5. *True* or *false?* $\dfrac{4(3 - 9)}{2 - 6} \geq 6$

6. Find the value of $xz^3 - 5y^2$ when $x = -2$, $y = -3$, and $z = -1$.

7. What property does $3(-2 + x) = -6 + 3x$ illustrate?

8. Simplify $-4p - 6 + 3p + 8$ by combining terms.

Solve.

9. $V = \dfrac{1}{3}\pi r^2 h$ for h

10. $6 - 3(1 + a) = 2(a + 5) - 2$

11. $-(m - 3) = 5 - 2m$

12. $\dfrac{x - 2}{3} = \dfrac{2x + 1}{5}$

Solve each inequality, and graph the solutions.

13. $-2.5x < 6.5$

14. $4(x + 3) - 5x < 12$

15. $\dfrac{2}{3}x - \dfrac{1}{6}x \leq -2$

Solve each problem.

16. The gap in average annual earnings by level of education continues to increase. On the basis of the most recent statistics available, a person with a bachelor's degree can expect to earn $18,436 more each year than someone with a high school diploma. Together, the individuals would earn $61,590. How much can a person at each level of education expect to earn? (*Source:* U.S. Census Bureau.)

17. Mount Mayon in the Philippines is the most perfectly shaped conical volcano in the world. Its base is nearly a perfect circle with circumference 80 mi, and it has a height of about 8200 ft. (One mile is 5280 ft.) Find the radius of the circular base to the nearest mile. (*Hint:* This problem has some unneeded information.) (*Source: Microsoft Encarta Encyclopedia.*)

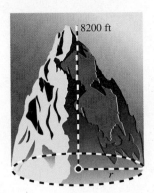

Circumference = 80 mi

18. The winning times in seconds for the women's 1000-m speed skating event in the Winter Olympics for the years 1960 through 2002 can be closely approximated by the linear equation

$$y = -0.5075x + 95.4179,$$

where x is the number of years since 1960. That is, $x = 4$ represents 1964, $x = 8$ represents 1968, and so on. (*Source: The World Almanac and Book of Facts 2006.*)

(a) Use this equation to complete the table of values. Round times to the nearest hundredth of a second.

(b) What does the ordered pair $(20, 85.27)$ mean in the context of the problem?

x	y
12	
28	
36	

19. Baby boomers are expected to inherit $10.4 trillion from their parents over the next 45 yr, an average of $50,000 each. The circle graph shows how they plan to spend their inheritances.

(a) How much of the $50,000 is expected to go toward a home purchase?

(b) How much is expected to go toward retirement?

(c) Use the answer from part (b) to estimate the amount expected to go toward paying off debts or funding children's education.

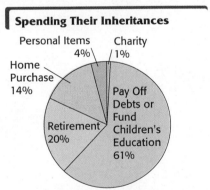

Spending Their Inheritances

Personal Items 4%
Charity 1%
Home Purchase 14%
Retirement 20%
Pay Off Debts or Fund Children's Education 61%

Source: First Interstate Bank Trust and Private Banking Group.

Consider the linear equation $-3x + 4y = 12$. Find the following.

20. The x- and y-intercepts   **21.** The slope   **22.** The graph

23. Are the lines with equations $x + 5y = -6$ and $y = 5x - 8$ *parallel, perpendicular,* or *neither*?

Write an equation for each line. Give the final answer in slope–intercept form if possible.

24. Through $(2, -5)$, slope 3   **25.** Through $(0, 4)$ and $(2, 4)$

Systems of Linear Equations and Inequalities

Although Americans continued their fascination with Hollywood and the movies in 2003, movie attendence and revenues dipped for the first time since 1991. Nonetheless, some 1.52 billion tickets were sold, and revenues exceeded $9 billion for the second year in a row. The top box office draws of the year— *The Lord of the Rings: The Return of the King* and *Finding Nemo*—attracted scores of adults and children wishing to get away from it all for a few hours. (*Source:* Exhibitor Relations Co., Nielsen EDI.)

In Exercise 13 of Section 4.4, we use a *system of linear equations* to find out how much money these top films earned.

4.1 Solving Systems of Linear Equations by Graphing

OBJECTIVES

1 Decide whether a given ordered pair is a solution of a system.

2 Solve linear systems by graphing.

3 Solve special systems by graphing.

4 Identify special systems without graphing.

5 Use a graphing calculator to solve a linear system.

A **system of linear equations,** often called a **linear system,** consists of two or more linear equations with the same variables. Examples of systems of two linear equations include

$$2x + 3y = 4 \qquad x + 3y = 1 \qquad x - y = 1$$
$$3x - y = -5 \qquad -y = 4 - 2x \qquad y = 3.$$

Linear systems

In the system on the right, think of $y = 3$ as an equation in two variables by writing it as $0x + y = 3$.

OBJECTIVE 1 Decide whether a given ordered pair is a solution of a system. A **solution of a system** of linear equations is an ordered pair that makes both equations true at the same time. A solution of an equation is said to *satisfy* the equation.

EXAMPLE 1 Determining whether an Ordered Pair Is a Solution

Decide whether the ordered pair $(4, -3)$ is a solution of each system.

(a) $x + 4y = -8$
$3x + 2y = 6$

To decide whether $(4, -3)$ is a solution of the system, substitute 4 for x and -3 for y in each equation.

$x + 4y = -8$			$3x + 2y = 6$	
$4 + 4(-3) = -8$	?		$3(4) + 2(-3) = 6$	?
$4 + (-12) = -8$	?	Multiply.	$12 + (-6) = 6$	? Multiply.
$-8 = -8$	True		$6 = 6$ True	

Because $(4, -3)$ satisfies both equations, it is a solution of the system.

(b) $2x + 5y = -7$
$3x + 4y = 2$

Again, substitute 4 for x and -3 for y in both equations.

$2x + 5y = -7$			$3x + 4y = 2$	
$2(4) + 5(-3) = -7$	?		$3(4) + 4(-3) = 2$	?
$8 + (-15) = -7$	?	Multiply.	$12 + (-12) = 2$	? Multiply.
$-7 = -7$	True		$0 = 2$	False

The ordered pair $(4, -3)$ is not a solution of this system, because it does not satisfy the second equation.

✔ **Now Try Exercises 1 and 3.**

OBJECTIVE 2 Solve linear systems by graphing. The set of all ordered pairs that are solutions of a system is its **solution set.** One way to find the solution set of a system of two linear equations is to graph both equations on the same axes. The graph of each line shows points whose coordinates satisfy the equation of that line. Any intersection point would be on both lines and would therefore be a solution of *both* equations. ***Thus, the coordinates of any point at which the lines intersect give a solution of the system.*** Because two *different* straight lines can intersect at no more than one point, there can never be more than one solution for such a system. The graph in Figure 1 shows that the solution of the system in Example 1(a) is the intersection point $(4, -3)$.

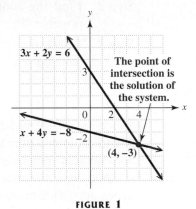

FIGURE 1

EXAMPLE 2 Solving a System by Graphing

Solve the system of equations by graphing both equations on the same axes.

$$2x + 3y = 4$$
$$3x - y = -5$$

We graph these two lines by plotting several points for each line. Recall from **Section 3.2** that the intercepts are often convenient choices. It is a good idea to use a third ordered pair as a check.

$2x + 3y = 4$

x	y
0	$\frac{4}{3}$
2	0
-2	$\frac{8}{3}$

$3x - y = -5$

x	y
0	5
$-\frac{5}{3}$	0
-2	-1

The lines in Figure 2 suggest that the graphs intersect at the point $(-1, 2)$. We check this by substituting -1 for x and 2 for y in both equations.

$$2x + 3y = 4$$
$$2(-1) + 3(2) = 4 \qquad ?$$
$$4 = 4 \qquad \text{True}$$

$$3x - y = -5$$
$$3(-1) - 2 = -5 \qquad ?$$
$$-5 = -5 \qquad \text{True}$$

Because $(-1, 2)$ satisfies both equations, the solution set of this system is $\{(-1, 2)\}$.

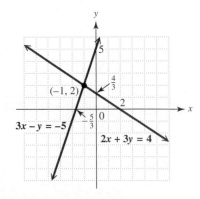

FIGURE 2

✔ **Now Try Exercise 15.**

To solve a system by graphing, follow these steps.

Solving a Linear System by Graphing

Step 1 **Graph each equation** of the system on the same coordinate axes.

Step 2 **Find the coordinates of the point of intersection** of the graphs if possible. This is the solution of the system.

Step 3 **Check** the solution in both of the original equations. Then write the solution set.

▶ **CAUTION** A difficulty with the graphing method is that it may not be possible to determine from the graph the exact coordinates of the point that represents the solution, particularly if those coordinates are not integers. The graphing method does, however, show geometrically how solutions are found and is useful when approximate answers will do.

OBJECTIVE 3 **Solve special systems by graphing.** Sometimes the graphs of the two equations in a system either do not intersect at all or are the same line.

EXAMPLE 3 **Solving Special Systems by Graphing**

Solve each system by graphing.

(a) $2x + y = 2$
$2x + y = 8$

The graphs of these lines are shown in Figure 3. The two lines are parallel and have no points in common. For such a system, there is no solution. We write the solution set as $\emptyset$.

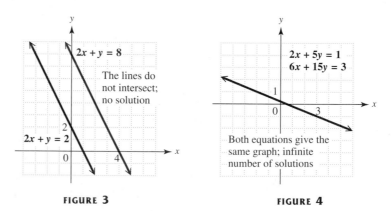

FIGURE 3   **FIGURE 4**

(b) $2x + 5y = 1$
$6x + 15y = 3$

The graphs of these two equations are the same line. See Figure 4. We can obtain the second equation by multiplying each side of the first equation by 3. In this case, every point on the line is a solution of the system, and the solution set contains an infinite number of ordered pairs, each of which satisfies both equations of the system. We write the solution set as

$$\{(x, y) \mid 2x + 5y = 1\},$$

> This is the first equation in the system. See the Note below.

read "the set of ordered pairs (x, y) such that $2x + 5y = 1$." Recall from **Section 1.4** that this notation is called **set-builder notation.** ●

> ▶ **NOTE** When a system has an infinite number of solutions, as in Example 3(b), either equation of the system could be used to write the solution set. *We prefer to use the equation (in standard form) with coefficients that are integers having no common factor (except 1).* In Example 3(b), this was the first equation of the system. This may not always be the case.

The system in Example 2 has exactly one solution. A system with at least one solution is called a **consistent system.** A system with no solution, such as the one in Example 3(a), is called an **inconsistent system.** The equations in Example 2 are **independent equations** with different graphs. The equations of the system in Example 3(b) have the same graph and are equivalent. Because they are different forms of the same equation, these equations are called **dependent equations.** Examples 2 and 3 show the three cases that may occur when solving a system of equations with two variables.

Three Cases for Solutions of Systems

1. The graphs intersect at exactly one point, which gives the (single) ordered-pair solution of the system. The **system is consistent** and the **equations are independent.** See Figure 5(a).

2. The graphs are parallel lines, so there is no solution and the solution set is ∅. The **system is inconsistent** and the **equations are independent.** See Figure 5(b).

3. The graphs are the same line. There is an infinite number of solutions, and the solution set is written in set-builder notation as $\{(x, y) \mid \underline{\hspace{2cm}}\}$, where one of the equations is written after the $\mid$ symbol. The **system is consistent** and the **equations are dependent.** See Figure 5(c).

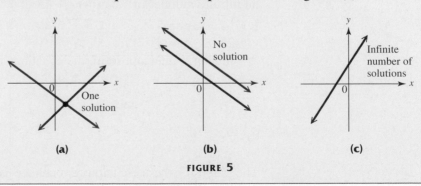

FIGURE 5

✔ **Now Try Exercises 25 and 27.**

OBJECTIVE 4 Identify special systems without graphing. Example 3 showed that the graphs of an inconsistent system are parallel lines and the graphs of a system of dependent equations are the same line. We can recognize these special kinds of systems without graphing by using slopes.

EXAMPLE 4 **Identifying the Three Cases by Using Slopes**

Describe each system without graphing. State the number of solutions.

(a) $3x + 2y = 6$
$-2y = 3x - 5$

Write each equation in slope–intercept form, $y = mx + b$, by solving for y.

$$3x + 2y = 6 \qquad\qquad -2y = 3x - 5$$

$$2y = -3x + 6 \qquad\qquad y = -\frac{3}{2}x + \frac{5}{2}$$

$$y = -\frac{3}{2}x + 3$$

Both equations have slope $-\frac{3}{2}$ but they have different y-intercepts, 3 and $\frac{5}{2}$. In **Chapter 3,** we found that lines with the same slope are parallel, so these equations have graphs that are parallel lines. The system has no solution.

(b) $2x - y = 4$

$$x = \frac{y}{2} + 2$$

Again, write the equations in slope–intercept form.

$$2x - y = 4 \qquad\qquad x = \frac{y}{2} + 2$$

$$-y = -2x + 4 \qquad\qquad \frac{y}{2} + 2 = x$$

$$y = 2x - 4 \qquad\qquad \frac{y}{2} = x - 2$$

$$y = 2x - 4$$

The equations are exactly the same; their graphs are the same line. The system has an infinite number of solutions.

(c) $x - 3y = 5$
$2x + y = 8$

In slope–intercept form, the equations are as follows.

$$x - 3y = 5 \qquad\qquad 2x + y = 8$$

$$-3y = -x + 5 \qquad\qquad y = -2x + 8$$

$$y = \frac{1}{3}x - \frac{5}{3}$$

The graphs of these equations are neither parallel nor the same line, since the slopes are different. This system has exactly one solution.

✔ **Now Try Exercises 33, 35, and 37.**

OBJECTIVE 5 Use a graphing calculator to solve a linear system.

EXAMPLE 5 Finding the Solution Set of a System with a Graphing Calculator

Solve the system from Example 1(a) by graphing with a calculator.

$$x + 4y = -8$$
$$3x + 2y = 6$$

To enter the equations in a graphing calculator, first solve each equation for y.

$x + 4y = -8$	$3x + 2y = 6$
$4y = -x - 8$	$2y = -3x + 6$
$y = -\dfrac{1}{4}x - 2$	$y = -\dfrac{3}{2}x + 3$

The calculator allows us to enter several equations to be graphed at the same time. We designate the first one Y_1 and the second one Y_2. See Figure 6(a). We graph the two equations using a standard window and then use the capability of the calculator to find the coordinates of the point of intersection of the graphs. The display at the bottom of Figure 6(b) indicates that the solution set is $\{(4, -3)\}$.

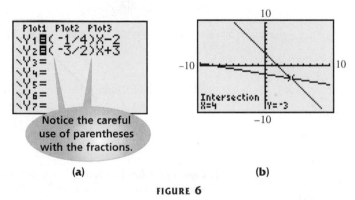

Notice the careful use of parentheses with the fractions.

(a) (b)

FIGURE 6

✔ **Now Try Exercise 53.**

4.1 EXERCISES

○ *Complete solution available on Video Lectures on CD/DVD*

Now Try Exercise

Decide whether the given ordered pair is a solution of the given system. See Example 1.

1. $(2, -3)$
$$x + y = -1$$
$$2x + 5y = 19$$

2. $(4, 3)$
$$x + 2y = 10$$
$$3x + 5y = 3$$

3. $(-1, -3)$
$$3x + 5y = -18$$
$$4x + 2y = -10$$

4. $(-9, -2)$
$$2x - 5y = -8$$
$$3x + 6y = -39$$

5. $(7, -2)$
$$4x = 26 - y$$
$$3x = 29 + 4y$$

6. $(9, 1)$
$$2x = 23 - 5y$$
$$3x = 24 + 3y$$

7. $(6, -8)$
$$-2y = x + 10$$
$$3y = 2x + 30$$

8. $(-5, 2)$
$$5y = 3x + 20$$
$$3y = -2x - 4$$

9. $(0, 0)$
$$4x + 2y = 0$$
$$x + y = 0$$

10. *Concept Check* When a student was asked to determine whether the ordered pair $(1, -2)$ is a solution of the system

$$x + y = -1$$
$$2x + y = 4,$$

he answered "yes." His reasoning was that the ordered pair satisfies the equation $x + y = -1$, since $1 + (-2) = -1$. *WHAT WENT WRONG?*

11. *Concept Check* Each ordered pair in (a)–(d) is a solution of one of the systems graphed in A–D. Because of the location of the point of intersection, you should be able to determine the correct system for each solution. Match each system from A–D with its solution from (a)–(d).

(a) $(3, 4)$ **A.**

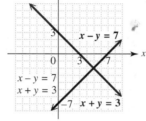

B.

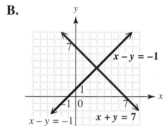

(b) $(-2, 3)$

(c) $(-3, 2)$ **C.**

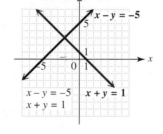

D.

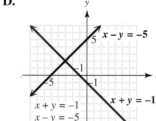

(d) $(5, -2)$

12. *Concept Check* Which ordered pair could not be a solution of the system graphed? Why is it the only valid choice?

A. $(-4, -4)$ **B.** $(-2, 2)$

C. $(-4, 4)$ **D.** $(-3, 3)$

13. *Concept Check* Which ordered pair could be a solution of the system graphed? Why is it the only valid choice?

A. $(2, 0)$ **B.** $(0, 2)$

C. $(-2, 0)$ **D.** $(0, -2)$

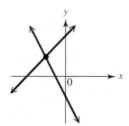

14. *Concept Check* The system $\begin{array}{l} 6x - 4y = 8 \\ 3x - 2y = 4 \end{array}$ has infinitely many solutions. Write its solution set, using set-builder notation as described in Example 3(b).

Solve each system of equations by graphing. If the system is inconsistent or the equations are dependent, say so. See Examples 2 and 3.

15. $x - y = 2$
$x + y = 6$

16. $x - y = 3$
$x + y = -1$

17. $x + y = 4$
$y - x = 4$

18. $x + y = -5$
$x - y = 5$

19. $x - 2y = 6$
$x + 2y = 2$

20. $2x - y = 4$
$4x + y = 2$

21. $3x - 2y = -3$
$-3x - y = -6$

22. $2x - y = 4$
$2x + 3y = 12$

23. $2x - 3y = -6$
$y = -3x + 2$

24. $x + 2y = 4$
$2x + 4y = 12$

25. $2x - y = 6$
$4x - 2y = 8$

26. $2x - y = 4$
$4x = 2y + 8$

27. $3x + y = 5$
$6x + 2y = 10$

28. $-3x + y = -3$
$y = x - 3$

29. $3x - 4y = 24$
$y = -\dfrac{3}{2}x + 3$

30. Solve the system $\begin{array}{l} 2x + 3y = 6 \\ x - 3y = 5 \end{array}$ by graphing. Can you check your solution? Why or why not?

31. Explain one of the drawbacks of solving a system of equations graphically.

32. Explain the three situations that may occur (regarding the number of solutions) when solving a system of two linear equations in two variables by graphing.

Without graphing, answer the following questions for each linear system. See Example 4.

(a) *Is the system inconsistent, are the equations dependent, or neither?*
(b) *Is the graph a pair of intersecting lines, a pair of parallel lines, or one line?*
(c) *Does the system have one solution, no solution, or an infinite number of solutions?*

33. $y - x = -5$
$x + y = 1$

34. $2x + y = 6$
$x - 3y = -4$

35. $x + 2y = 0$
$4y = -2x$

36. $y = 3x$
$y + 3 = 3x$

37. $5x + 4y = 7$
$10x + 8y = 4$

38. $2x + 3y = 12$
$2x - y = 4$

39. $x - 3y = 5$
$2x + y = 8$

40. $2x - y = 4$
$x = \dfrac{1}{2}y + 2$

41. $\dfrac{1}{5}x + \dfrac{1}{5}y = -1$
$-\dfrac{1}{3}x + \dfrac{1}{3}y = 1$

The numbers of daily morning and evening newspapers in the United States in selected years are shown in the graph.

42. For which years were there more evening dailies than morning dailies?

43. Estimate the year in which the number of evening and morning dailies was closest to the same. About how many newspapers of each type were there in that year?

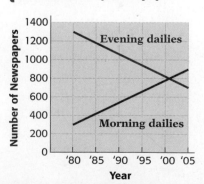

Number of Daily Newspapers

Source: Editor & Publisher International Year Book.

*An application of mathematics in economics deals with **supply and demand.** Typically, as the price of an item increases, the demand for the item decreases while the supply increases. (There are exceptions to this relationship, however.) If supply and demand can be described by straight-line equations, the point at which the lines intersect determines the **equilibrium supply** and **equilibrium demand.***

Suppose that an economist has studied the supply and demand for aluminum siding and has concluded that the price per unit, p, and the demand, x, are related by the demand equation $p = 60 - \frac{3}{4}x$, while the supply is given by the equation $p = \frac{3}{4}x$. The graphs of these two equations are shown in the figure.

SUPPLY AND DEMAND

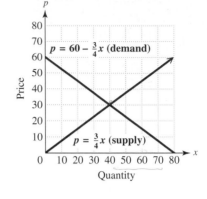

Use the graph to answer the questions in Exercises 44–46.

44. At what value of x does supply equal demand?

45. At what value of p does supply equal demand?

46. When $x > 40$, does demand exceed supply or does supply exceed demand?

47. The graph shows how college students managed their money during the years 1997 through 2004.

 (a) During what period did ATM use dominate both credit card *and* debit card use?

 (b) In what year did debit card use overtake credit card use?

 (c) In what year did debit card use overtake ATM use?

 (d) Write an ordered pair for the debit card use data in the year 1998.

 (e) Describe the trend in debit card use over this period.

How College Students Manage Their Money

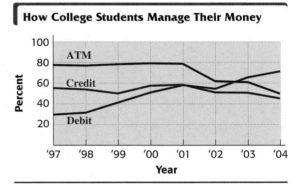

Source: Georgetown University Credit Research Center.

48. The graph shows how the average viewing hours for broadcast TV and cable/satellite TV in the United States has changed over the years 1998 through 2004.

 (a) In what year did Americans spend almost the same number of hours watching broadcast and cable/satellite TV? How many hours per year was this?

 (b) Express the point of intersection of the two graphs as an ordered pair of the form (year, hours).

 (c) During what period was the time spent watching broadcast TV almost constant?

Watching the Tube

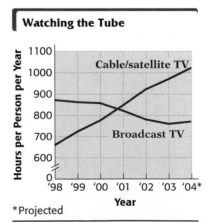

*Projected

Source: Veronis Suhler Stevenson.

✐ **(d)** Describe the trend in cable/satellite viewing time from 1998 through 2004. If a straight line were used to approximate its graph, would the line have positive, negative, or zero slope?

✐ **(e)** If a straight line were used to approximate the graph of broadcast TV viewing time from 1998 through 2004, would the line have positive, negative, or zero slope? Explain.

TECHNOLOGY INSIGHTS (EXERCISES 49–52)

Match the graphing calculator screens in choices A–D with the appropriate system in Exercises 49–52. See Example 5.

A.

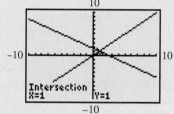

B.

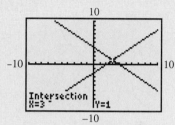

C.

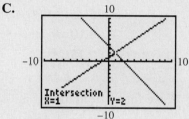

D.

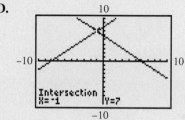

49. $x + y = 4$
$x - y = 2$

50. $x + y = 6$
$x - y = -8$

51. $2x + 3y = 5$
$x - y = 0$

52. $3x + 2y = 7$
$-x + y = 1$

 Use a graphing calculator to solve each system. See Example 5.

🌐 **53.** $3x + y = 2$
$2x - y = -7$

54. $x + 2y = -2$
$2x - y = 11$

55. $8x + 4y = 0$
$4x - 2y = 2$

56. $3x + 3y = 0$
$4x + 2y = 3$

PREVIEW EXERCISES

Solve each equation for y. See Sections 2.5 and 3.3.

57. $3x + y = 4$

58. $-2x + y = 9$

59. $9x - 2y = 4$

60. $5x - 3y = 12$

Solve each equation. Check the solution. See Section 2.3.

61. $-2(x - 2) + 5x = 10$

62. $2m - 3(4 - m) = 8$

63. $p + 4(6 - 2p) = 3$

64. $4(3 - 2k) + 3k = 12$

65. $4x - 2(1 - 3x) = 6$

66. $t + 3(2t - 4) = -13$

4.2 Solving Systems of Linear Equations by Substitution

OBJECTIVE 1 Solve linear systems by substitution. Graphing to solve a system of equations has a serious drawback. It is difficult to find an accurate solution, such as $\left(\frac{1}{3}, -\frac{5}{6}\right)$, from a graph. One algebraic method for solving a system of equations is the **substitution method.** This method is particularly useful for solving systems in which one equation is already solved, or can be solved quickly, for one of the variables.

EXAMPLE 1 Using the Substitution Method

Solve the system by the substitution method.

$$3x + 5y = 26$$
$$y = 2x$$

The second equation, $y = 2x$, is already solved for y, so we substitute $2x$ for y in the first equation.

$$3x + 5y = 26$$
$$3x + 5(2x) = 26 \qquad \text{Let } y = 2x.$$
$$3x + 10x = 26 \qquad \text{Multiply.}$$
$$13x = 26 \qquad \text{Combine like terms.}$$
$$x = 2 \qquad \text{Divide by 13.}$$

Because $x = 2$, we find y from the equation $y = 2x$ by substituting 2 for x.

$$y = 2(2) = 4 \qquad \text{Let } x = 2.$$

We check that the solution of the given system is $(2, 4)$ by substituting 2 for x and 4 for y in *both* equations.

Check:

$3x + 5y = 26$	$y = 2x$
$3(2) + 5(4) = 26 \quad ?$	$4 = 2(2) \quad ?$
$6 + 20 = 26 \quad ?$	$4 = 4 \qquad \text{True}$
$26 = 26 \qquad \text{True}$	

Since $(2, 4)$ satisfies both equations, the solution set is $\{(2, 4)\}$.

✔ **Now Try Exercise 3.**

EXAMPLE 2 Using the Substitution Method

Solve the system by the substitution method.

$$2x + 5y = 7$$
$$x = -1 - y$$

The second equation gives x in terms of y. Substitute $-1 - y$ for x in the first equation.

$$2x + 5y = 7$$
$$2(-1 - y) + 5y = 7 \qquad \text{Let } x = -1 - y.$$
$$-2 - 2y + 5y = 7 \qquad \text{Distributive property}$$

Distribute 2 to *both* −1 and −y.

$$-2 + 3y = 7 \qquad \text{Combine like terms.}$$
$$3y = 9 \qquad \text{Add 2.}$$
$$y = 3 \qquad \text{Divide by 3.}$$

To find x, substitute 3 for y in the equation $x = -1 - y$ to get

$$x = -1 - 3 = -4.$$

> Write the
> x-coordinate
> first.

Check that the solution set of the given system is $\{(-4, 3)\}$.

✔ **Now Try Exercise 5.**

▶ **CAUTION**   Be careful when you write the ordered-pair solution of a system. Even though we found y first in Example 2, *the x-coordinate is always written first in the ordered pair.*

To solve a system by substitution, follow these steps.

Solving a Linear System by Substitution

Step 1   **Solve one equation for either variable.** If one of the variables has coefficient 1 or −1, choose it, since it usually makes the substitution method easier.

Step 2   **Substitute** for that variable in the other equation. The result should be an equation with just one variable.

Step 3   **Solve** the equation from Step 2.

Step 4   **Substitute** the result from Step 3 into the equation from Step 1 to find the value of the other variable.

Step 5   **Check** the solution in both of the original equations. Then write the solution set.

EXAMPLE 3   Using the Substitution Method

Use substitution to solve the system.

$$2x = 4 - y \qquad (1)$$
$$5x + 3y = 10 \qquad (2)$$

Step 1   We must solve one of the equations for either x or y. Because the coefficient of y in equation (1) is −1, we avoid fractions by solving this equation for y.

$$2x = 4 - y \qquad (1)$$
$$y + 2x = 4 \qquad \text{Add } y.$$
$$y = -2x + 4 \qquad \text{Subtract } 2x.$$

Step 2   Now substitute $-2x + 4$ for y in equation (2).

$$5x + 3y = 10 \qquad (2)$$
$$5x + 3(-2x + 4) = 10 \qquad \text{Let } y = -2x + 4.$$

Step 3 Solve the equation from Step 2.

$$5x + 3(-2x + 4) = 10$$

$$5x - 6x + 12 = 10 \qquad \text{Distributive property}$$

$$-x + 12 = 10 \qquad \text{Combine like terms.}$$

$$-x = -2 \qquad \text{Subtract 12.}$$

$$x = 2 \qquad \text{Multiply by } -1.$$

> Distribute 3 to both $-2x$ and 4.

Step 4 Since $y = -2x + 4$ and $x = 2$,

$$y = -2(2) + 4 = 0.$$

Step 5 *Check* that $(2, 0)$ is the solution.

$2x = 4 - y$	(1)	$5x + 3y = 10$	(2)
$2(2) = 4 - 0$?		$5(2) + 3(0) = 10$?	
$4 = 4$ True		$10 = 10$ True	

Since both results are true, the solution set of the system is $\{(2, 0)\}$.

✔ **Now Try Exercise 7.**

OBJECTIVE 2 **Solve special systems by substitution.** We can solve inconsistent systems with graphs that are parallel lines and systems of dependent equations with graphs that are the same line using the substitution method.

EXAMPLE 4 **Solving an Inconsistent System by Substitution**

Use substitution to solve the system.

$$x = 5 - 2y \qquad \text{(1)}$$

$$2x + 4y = 6 \qquad \text{(2)}$$

Substitute $5 - 2y$ for x in equation (2).

$$2x + 4y = 6 \qquad \text{(2)}$$

$$2(5 - 2y) + 4y = 6 \qquad \text{Let } x = 5 - 2y.$$

$$10 - 4y + 4y = 6 \qquad \text{Distributive property}$$

$$10 = 6 \qquad \text{False}$$

This false result means that the equations in the system have graphs that are parallel lines. The system is inconsistent and has no solution, so the solution set is $\emptyset$. See Figure 7.

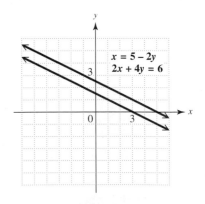

$x = 5 - 2y$
$2x + 4y = 6$

FIGURE 7

✔ **Now Try Exercise 17.**

▶ **CAUTION** It is a common error to give "false" as the solution of an inconsistent system. The correct response is ∅.

EXAMPLE 5 Solving a System with Dependent Equations by Substitution

Solve the system by the substitution method.

$$3x - y = 4 \qquad (1)$$
$$-9x + 3y = -12 \qquad (2)$$

Begin by solving equation (1) for y to get $y = 3x - 4$. Substitute $3x - 4$ for y in equation (2) and solve the resulting equation.

$$-9x + 3y = -12 \qquad (2)$$
$$-9x + 3(3x - 4) = -12 \qquad \text{Let } y = 3x - 4.$$
$$-9x + 9x - 12 = -12 \qquad \text{Distributive property}$$
$$0 = 0 \qquad \text{Add 12; combine like terms.}$$

This true result means that every solution of one equation is also a solution of the other, so the system has an infinite number of solutions. The solution set is $\{(x, y) \mid 3x - y = 4\}$. A graph of the equations of this system is shown in Figure 8.

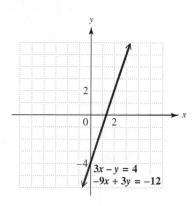

FIGURE 8

✔ **Now Try Exercise 19.**

▶ **CAUTION** It is a common error to give "true" as the solution of a system of dependent equations. Remember to give the solution set in set-builder notation using the equation in the system that is in standard form with integer coefficients that have no common factor (except 1).

OBJECTIVE 3 Solve linear systems with fractions. When a system includes an equation with fractions as coefficients, the work can be simplified. Eliminate the fractions by multiplying each side of the equation by a common denominator.

EXAMPLE 6 Using the Substitution Method with Fractions as Coefficients

Solve the system by the substitution method.

$$3x + \frac{1}{4}y = 2 \qquad (1)$$

$$\frac{1}{2}x + \frac{3}{4}y = -\frac{5}{2} \qquad (2)$$

Clear equation (1) of fractions by multiplying each side by 4.

$$4\left(3x + \frac{1}{4}y\right) = 4(2) \qquad \text{Multiply by 4.}$$

$$4(3x) + 4\left(\frac{1}{4}y\right) = 4(2) \qquad \text{Distributive property}$$

$$12x + y = 8 \qquad (3)$$

Now clear equation (2) of fractions by multiplying each side by the common denominator, 4.

$$4\left(\frac{1}{2}x + \frac{3}{4}y\right) = 4\left(-\frac{5}{2}\right) \qquad \text{Multiply by 4.}$$

$$4\left(\frac{1}{2}x\right) + 4\left(\frac{3}{4}y\right) = 4\left(-\frac{5}{2}\right) \qquad \text{Distributive property}$$

$$2x + 3y = -10 \qquad (4)$$

The given system of equations has been simplified to the equivalent system

$$12x + \ y = 8 \qquad (3)$$

$$2x + 3y = -10. \qquad (4)$$

Solve this system by the substitution method. Equation (3) can be solved for y by subtracting $12x$ from each side.

$$12x + y = 8 \qquad (3)$$

$$y = -12x + 8 \qquad \text{Subtract 12x.}$$

Now substitute this result for y in equation (4).

$$2x + 3y = -10 \qquad (4)$$

$$2x + 3(-12x + 8) = -10 \qquad \text{Let } y = -12x + 8.$$

$$2x - 36x + 24 = -10 \qquad \text{Distributive property}$$

> Distribute 3 to both −12x and 8.

$$-34x = -34 \qquad \text{Combine like terms; subtract 24.}$$

$$x = 1 \qquad \text{Divide by } -34.$$

Substitute 1 for x in $y = -12x + 8$ to get

$$y = -12(1) + 8 = -4.$$

Check by substituting 1 for x and -4 for y in both of the original equations. The solution set is $\{(1, -4)\}$.

✔ **Now Try Exercise 25.**

4.2 EXERCISES

Complete solution available on Video Lectures on CD/DVD

Now Try Exercise

1. Concept Check A student solves the system $\begin{array}{l} 5x - y = 15 \\ 7x + y = 21 \end{array}$ and finds that $x = 3$, which is correct. The student gives the solution set as $\{3\}$. Is this correct? If not, what is the solution set?

2. Concept Check A student solves the system $\begin{array}{l} x + y = 4 \\ 2x + 2y = 8 \end{array}$ and obtains the equation $0 = 0$. The student gives the solution set as $\{(0, 0)\}$. Is this correct? If not, what is the solution set?

Solve each system by the substitution method. Check each solution. See Examples 1–5.

3. $x + y = 12$
$y = 3x$

4. $x + 3y = -28$
$y = -5x$

5. $3x + 2y = 27$
$x = y + 4$

6. $4x + 3y = -5$
$x = y - 3$

7. $3x + 5y = 25$
$x - 2y = -10$

8. $5x + 2y = -15$
$2x - y = -6$

9. $3x + 4 = -y$
$2x + y = 0$

10. $2x - 5 = -y$
$x + 3y = 0$

11. $7x + 4y = 13$
$x + y = 1$

12. $3x - 2y = 19$
$x + y = 8$

13. $3x - y = 5$
$y = 3x - 5$

14. $4x - y = -3$
$y = 4x + 3$

15. $2x + y = 0$
$4x - 2y = 2$

16. $x + y = 0$
$4x + 2y = 3$

17. $2x + 8y = 3$
$x = 8 - 4y$

18. $2x + 10y = 3$
$x = 1 - 5y$

19. $2y = 4x + 24$
$2x - y = -12$

20. $2y = 14 - 6x$
$3x + y = 7$

21. When you use the substitution method, how can you tell that a system has

    (a) no solution?   **(b)** an infinite number of solutions?

22. Solve each system.

    (a) $5x - 4y = 7$
        $x = 3$

    (b) $5x - 4y = 7$
        $y = -3$

    Why are these systems easier to solve than the examples in this section?

Solve each system by the substitution method. Check each solution. See Example 6.

23. $\dfrac{1}{2}x + \dfrac{1}{3}y = 3$
$y = 3x$

24. $\dfrac{1}{4}x - \dfrac{1}{5}y = 9$
$y = 5x$

25. $\dfrac{1}{2}x + \dfrac{1}{3}y = -\dfrac{1}{3}$
$\dfrac{1}{2}x + 2y = -7$

26. $\dfrac{1}{6}x + \dfrac{1}{6}y = 1$
$-\dfrac{1}{2}x - \dfrac{1}{3}y = -5$

27. $\dfrac{x}{5} + 2y = \dfrac{8}{5}$
$\dfrac{3x}{5} + \dfrac{y}{2} = -\dfrac{7}{10}$

28. $\dfrac{x}{2} + \dfrac{y}{3} = \dfrac{7}{6}$
$\dfrac{x}{4} - \dfrac{3y}{2} = \dfrac{9}{4}$

29. $\dfrac{x}{5} + y = \dfrac{6}{5}$
$\dfrac{x}{10} + \dfrac{y}{3} = \dfrac{5}{6}$

30. $\dfrac{1}{2}x - \dfrac{1}{8}y = -\dfrac{1}{4}$
$-4x + y = 2$

31. $\dfrac{1}{6}x + \dfrac{1}{3}y = 8$
$\dfrac{1}{4}x + \dfrac{1}{2}y = 12$

32. One student solved the system $\begin{array}{l} \frac{1}{3}x - \frac{1}{2}y = 7 \\ \frac{1}{6}x + \frac{1}{3}y = 0 \end{array}$ and wrote as his answer "$x = 12$," while another solved it and wrote as her answer "$y = -6$." Who, if either, was correct? Why?

RELATING CONCEPTS (EXERCISES 33–36)

FOR INDIVIDUAL OR GROUP WORK

A system of linear equations can be used to model the cost and the revenue of a business. **Work Exercises 33–36 in order.**

33. Suppose that you start a business manufacturing and selling bicycles, and it costs you $5000 to get started. You determine that each bicycle will cost $400 to manufacture. Explain why the linear equation $y_1 = 400x + 5000$ (y_1 in dollars) gives your *total* cost of manufacturing x bicycles.

34. You decide to sell each bike for $600. What expression in x represents the revenue you will take in if you sell x bikes? Write an equation using y_2 (in dollars) to express your revenue when you sell x bikes.

35. Form a system from the two equations in Exercises 33 and 34, and then solve the system.

36. The value of x from Exercise 35 is the number of bikes it takes to *break even.* Fill in the blanks: When _____ bikes are sold, the break-even point is reached. At that point, you have spent _____ dollars and taken in _____ dollars.

 Solve each system by substitution. Then graph both lines in the standard viewing window of a graphing calculator, and use the intersection feature to support your answer. See Example 5 in **Section 4.1.** *(In Exercises 41 and 42, you will need to solve each equation for y first before graphing.)*

37. $y = 6 - x$
$y = 2x$

38. $y = 4x - 4$
$y = -3x - 11$

39. $y = -\frac{4}{3}x + \frac{19}{3}$
$y = \frac{15}{2}x - \frac{5}{2}$

40. $y = -\frac{15}{2}x + 10$
$y = \frac{25}{3}x - \frac{65}{3}$

41. $4x + 5y = 5$
$2x + 3y = 1$

42. $6x + 5y = 13$
$3x + 3y = 4$

43. *Concept Check* Assuming that there is a point of intersection and it does not appear on your screen when you solve a linear system with a graphing calculator, how can you find that point?

44. *Concept Check* Suppose that you were asked to solve the system

$$y = 1.73x + 5.28$$
$$y = -2.94x - 3.85.$$

Why would it probably be easier to solve this system with a graphing calculator than by the substitution method?

PREVIEW EXERCISES

Simplify. See Section 1.8.

45. $(14x - 3y) + (2x + 3y)$

46. $(-6x + 8y) + (6x + 2y)$

47. $(-x + 7y) + (3y + x)$

48. $(3x - 4y) + (4y - 3x)$

49. What must be added to $-4x$ to get a sum of 0?

50. What must be added to $6y$ to get a sum of 0?

51. What must $4y$ be multiplied by so that when the product is added to $8y$, the sum is 0?

52. What must $-3x$ be multiplied by so that when the product is added to $-12x$, the result is 0?

4.3   Solving Systems of Linear Equations by Elimination

OBJECTIVES

1 Solve linear systems by elimination.

2 Multiply when using the elimination method.

3 Use an alternative method to find the second value in a solution.

4 Use the elimination method to solve special systems.

OBJECTIVE 1   Solve linear systems by elimination.   An algebraic method that depends on the addition property of equality can also be used to solve systems. As mentioned earlier, adding the same quantity to each side of an equation results in equal sums:

$$\text{If } A = B, \text{ then } A + C = B + C.$$

We can take this addition a step further. Adding *equal* quantities, rather than the *same* quantity, to each side of an equation also results in equal sums:

$$\text{If } A = B \text{ and } C = D, \text{ then } A + C = B + D.$$

Using the addition property to solve systems is called the **elimination method.** With this method, the idea is to *eliminate* one of the variables. To do this, one pair of variable terms in the two equations must have coefficients that are opposites.

EXAMPLE 1   Using the Elimination Method

Use the elimination method to solve the system.

$$x + y = 5$$
$$x - y = 3$$

Each equation in this system is a statement of equality, so the sum of the right sides equals the sum of the left sides. Adding in this way gives

$$(x + y) + (x - y) = 5 + 3$$
$$2x = 8 \qquad \text{Combine like terms.}$$
$$x = 4. \qquad \text{Divide by 2.}$$

Notice that y has been eliminated. The result, $x = 4$, gives the x-value of the solution of the given system. To find the y-value of the solution, substitute 4 for x in either of the two equations of the system. Choosing the first equation, $x + y = 5$, gives

$$x + y = 5$$
$$4 + y = 5 \qquad \text{Let } x = 4.$$
$$y = 1. \qquad \text{Subtract 4.}$$

Check the solution, $(4, 1)$, by substituting 4 for x and 1 for y in both equations of the given system.

$x + y = 5$	$x - y = 3$
$4 + 1 = 5$?	$4 - 1 = 3$?
$5 = 5$ True	$3 = 3$ True

Since both results are true, the solution set of the system is $\{(4, 1)\}$.

✔ **Now Try Exercise 5.**

▶ **CAUTION** *A system is not completely solved until values for both x and y are found.* Do not stop after finding the value of only one variable. Remember to write the solution set as a set containing an ordered pair.

In general, use the following steps to solve a linear system of equations by the elimination method.

Solving a Linear System by Elimination

Step 1 **Write both equations in standard form,** $Ax + By = C$.

Step 2 **Transform the equations as needed so that the coefficients of one pair of variable terms are opposites.** Multiply one or both equations by appropriate numbers so that the sum of the coefficients of either the x- or y-terms is 0.

Step 3 **Add** the new equations to eliminate a variable. The sum should be an equation with just one variable.

Step 4 **Solve** the equation from Step 3 for the remaining variable.

Step 5 **Substitute** the result from Step 4 into either of the original equations, and solve for the other variable.

Step 6 **Check** the solution in both of the original equations. Then write the solution set.

It does not matter which variable is eliminated first. Usually, we choose the one that is more convenient to work with.

EXAMPLE 2 Using the Elimination Method

Solve the system.

$$y + 11 = 2x \quad \text{(1)}$$
$$5x = y + 26 \quad \text{(2)}$$

Step 1 Write both equations in standard form, $Ax + By = C$, to get this system:

$$-2x + y = -11 \qquad \text{Subtract } 2x \text{ and } 11.$$
$$5x - y = 26. \qquad \text{Subtract } y.$$

Step 2 Because the coefficients of y are 1 and -1, adding will eliminate y. It is not necessary to multiply either equation by a number.

Step 3 Add the two equations. This time we use vertical addition.

$$-2x + y = -11$$
$$\underline{5x - y = 26}$$
$$3x = 15 \qquad \text{Add in columns.}$$

Step 4 Solve. $\qquad\qquad\qquad x = 5 \qquad$ Divide by 3.

Step 5 Find the value of y by substituting 5 for x in either of the original equations.

$$y + 11 = 2x \qquad \text{(1)}$$
$$y + 11 = 2(5) \qquad \text{Let } x = 5.$$
$$y + 11 = 10 \qquad \text{Multiply.}$$
$$y = -1 \qquad \text{Subtract 11.}$$

Step 6 Check by substituting $x = 5$ and $y = -1$ into both of the original equations.

Check:

$y + 11 = 2x$	(1)		$5x = y + 26$	(2)
$(-1) + 11 = 2(5)$	?		$5(5) = -1 + 26$	?
$10 = 10$	True		$25 = 25$	True

Since $(5, -1)$ is a solution of *both* equations, the solution set is $\{(5, -1)\}$.

✔ **Now Try Exercise 9.**

OBJECTIVE 2 **Multiply when using the elimination method.** Sometimes we need to multiply each side of one or both equations in a system by some number before adding will eliminate a variable.

EXAMPLE 3 Multiplying Both Equations When Using the Elimination Method

Solve the system.

$$2x + 3y = -15 \quad \text{(1)}$$
$$5x + 2y = 1 \quad \text{(2)}$$

Adding the two equations gives $7x + 5y = -14$, which does not eliminate either variable. However, we can multiply each equation by a suitable number so that the coefficients of one of the two variables are opposites. For example, to eliminate x, we multiply each side of $2x + 3y = -15$ (equation (1)) by 5 and each side of $5x + 2y = 1$ (equation (2)) by -2.

$$\begin{array}{ll} 10x + 15y = -75 & \text{Multiply equation (1) by 5.} \\ \underline{-10x - 4y = -2} & \text{Multiply equation (2) by } -2. \\ 11y = -77 & \text{Add.} \\ y = -7 & \text{Divide by 11.} \end{array}$$

Find the value of x by substituting -7 for y in either equation (1) or (2).

$$\begin{array}{ll} 5x + 2(-7) = 1 & \text{Let } y = -7 \text{ in equation (2).} \\ 5x - 14 = 1 & \text{Multiply.} \\ 5x = 15 & \text{Add 14.} \\ x = 3 & \text{Divide by 5.} \end{array}$$

Check that the solution set of the system is $\{(3, -7)\}$.

✔ **Now Try Exercise 23.**

▶ **CAUTION** When using the elimination method, remember to *multiply both sides* of an equation by the same nonzero number.

OBJECTIVE 3 **Use an alternative method to find the second value in a solution.** Sometimes it is easier to find the value of the second variable in a solution by using the elimination method twice.

EXAMPLE 4 **Finding the Second Value by Using an Alternative Method**

Solve the system.

$$\begin{array}{ll} 4x = 9 - 3y & (1) \\ 5x - 2y = 8 & (2) \end{array}$$

Write equation (1) in standard form by adding $3y$ to each side to get the system

$$\begin{array}{ll} 4x + 3y = 9 & (3) \\ 5x - 2y = 8. & (2) \end{array}$$

One way to proceed is to eliminate y by multiplying each side of equation (3) by 2 and each side of equation (2) by 3 and then adding.

$$\begin{array}{ll} 8x + 6y = 18 & \text{Multiply equation (3) by 2.} \\ \underline{15x - 6y = 24} & \text{Multiply equation (2) by 3.} \\ 23x = 42 & \text{Add.} \\ x = \dfrac{42}{23} & \text{Divide by 23.} \end{array}$$

Substituting $\frac{42}{23}$ for x in one of the given equations would give y, but the arithmetic involved would be challenging. Instead, solve for y by starting again with the original equations and eliminating x. Multiply each side of equation (3) by 5 and each side of equation (2) by -4, and then add.

$$
\begin{array}{ll}
20x + 15y = 45 & \text{Multiply equation (3) by 5.} \\
\underline{-20x + 8y = -32} & \text{Multiply equation (2) by } -4. \\
23y = 13 & \text{Add.} \\
y = \dfrac{13}{23} & \text{Divide by 23.}
\end{array}
$$

Check that the solution set is $\left\{\left(\frac{42}{23}, \frac{13}{23}\right)\right\}$.

✔ **Now Try Exercise 37.**

▶ **NOTE** When the value of the first variable is a fraction, the method used in Example 4 helps avoid arithmetic errors. Of course, this method could be used to solve any system of equations.

OBJECTIVE ④ Use the elimination method to solve special systems.

EXAMPLE 5 Using the Elimination Method for an Inconsistent System or Dependent Equations

Solve each system by the elimination method.

(a) $2x + 4y = 5$
  $4x + 8y = -9$

  Multiply each side of $2x + 4y = 5$ by -2, and then add to $4x + 8y = -9$.

$$
\begin{array}{ll}
-4x - 8y = -10 & \\
\underline{4x + 8y = -9} & \\
0 = -19 & \text{False}
\end{array}
$$

The false statement $0 = -19$ indicates that the given system has solution set $\emptyset$.

(b) $3x - y = 4$
   $-9x + 3y = -12$

  Multiply each side of the first equation by 3. Then add the two equations.

$$
\begin{array}{ll}
9x - 3y = 12 & \\
\underline{-9x + 3y = -12} & \\
0 = 0 & \text{True}
\end{array}
$$

A true statement occurs when the equations are equivalent. As before, this result indicates that every solution of one equation is also a solution of the other. The solution set is $\{(x, y) \mid 3x - y = 4\}$. (See **Section 4.2,** Example 5, where the same system was solved by using substitution.)

✔ **Now Try Exercises 31 and 33.**

4.3 EXERCISES

⊙ *Complete solution available on Video Lectures on CD/DVD*

Now Try Exercise

Concept Check *Answer* true *or* false *for each statement. If false, tell why.*

1. To eliminate the *y*-terms in the system

$$2x + 12y = 7$$
$$3x + 4y = 1,$$

we should multiply the bottom equation by 3 and then add.

2. The ordered pair $(0, 0)$ *must* be a solution of a system of the form

$$Ax + By = 0$$
$$Cx + Dy = 0.$$

3. The system $\begin{array}{c} x + y = 1 \\ x + y = 2 \end{array}$ has $\emptyset$ as its solution set.

4. The ordered pair $(4, -5)$ cannot be a solution of a system that contains the equation $5x - 4y = 0$.

Solve each system by the elimination method. Check each solution. See Examples 1 and 2.

5. $x - y = -2$
$x + y = 10$

6. $x + y = 10$
$x - y = -6$

7. $2x + y = -5$
$x - y = 2$

8. $2x + y = -15$
$-x - y = 10$

9. $2y = -3x$
$-3x - y = 3$

10. $5x = y + 5$
$-5x + 2y = 0$

11. $4x - 3y = 1$
$8x = 3 + 6y$

12. $5x + 8y = 10$
$24y = -15x - 10$

13. $x + 3y = 6$
$-2x + 12 = 6y$

14. $7x + 2y = 0$
$4y = -14x$

15. $6x - y = -1$
$5y = 17 + 6x$

16. $y = 9 - 6x$
$-6x + 3y = 15$

Solve each system by the elimination method. (Hint: In Exercises 29 and 30, first clear all fractions.) Check each solution. See Examples 3–5.*

17. $2x - y = 12$
$3x + 2y = -3$

18. $x + y = 3$
$-3x + 2y = -19$

19. $x + 4y = 16$
$3x + 5y = 20$

20. $2x + y = 8$
$5x - 2y = -16$

21. $2x - 8y = 0$
$4x + 5y = 0$

22. $3x - 15y = 0$
$6x + 10y = 0$

23. $3x + 3y = 33$
$5x - 2y = 27$

24. $4x - 3y = -19$
$3x + 2y = 24$

25. $5x + 4y = 12$
$3x + 5y = 15$

26. $2x + 3y = 21$
$5x - 2y = -14$

27. $5x - 4y = 15$
$-3x + 6y = -9$

28. $4x + 5y = -16$
$5x - 6y = -20$

29. $3x = 3 + 2y$
$-\dfrac{4}{3}x + y = \dfrac{1}{3}$

30. $3x = 27 + 2y$
$x - \dfrac{7}{2}y = -25$

31. $-x + 3y = 4$
$-2x + 6y = 8$

32. $6x - 2y = 24$
$-3x + y = -12$

33. $5x - 2y = 3$
$10x - 4y = 5$

34. $3x - 5y = 1$
$6x - 10y = 4$

35. $6x - 2y = -22$
$-3x + 4y = 17$

36. $5x - 4y = -1$
$x + 8y = -9$

37. $4x = 3y - 2$
$5x + 3 = 2y$

---

*The authors thank Mitchel Levy of Broward Community College for his suggestions for this group of exercises.

38. $2x + 3y = 0$
$4x + 12 = 9y$

39. $24x + 12y = -7$
$16x - 18y = 17$

40. $9x + 4y = -3$
$6x + 6y = -7$

41. $7x - 4y = 0$
$3x = 2y$

42. *Concept Check* Write a system of the form

$$ax + by = c$$
$$dx + ey = f,$$

where $a, b, c, d, e,$ and f are consecutive integers. Solve the system, and make a conjecture.

RELATING CONCEPTS (EXERCISES 43–48)

FOR INDIVIDUAL OR GROUP WORK

*The graph shows U.S. movie attendance from 1996 through 2004. In 1996, attendance was 1339 million, as represented by the point P(1996, 1339). In 2004, attendance was 1536 million, as represented by the point Q(2004, 1536). We can find an equation of line segment PQ by using a system of equations. Then we use the equation we found to approximate the attendance in any of the years between 1996 and 2004. **Work Exercises 43–48 in order.***

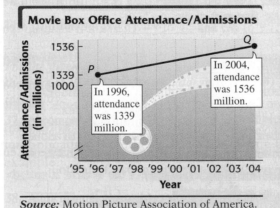

Movie Box Office Attendance/Admissions

In 1996, attendance was 1339 million.

In 2004, attendance was 1536 million.

Source: Motion Picture Association of America.

43. The line segment has an equation that can be written in the form $y = ax + b$. Using the coordinates of point P with $x = 1996$ and $y = 1339$, write an equation in the variables a and b.

44. Using the coordinates of point Q with $x = 2004$ and $y = 1536$, write a second equation in the variables a and b.

45. Write the system of equations formed from the two equations in Exercises 43 and 44, and solve the system by using the elimination method.

46. What is the equation of the segment PQ?

47. Let $x = 2002$ in the equation of Exercise 46, and solve for y. How does the result compare with the actual figure of 1639 million?

48. The actual data points for the years 1996 through 2004 do not lie in a perfectly straight line. Explain the pitfalls of relying too heavily on using the equation in Exercise 46 to approximate attendance.

PREVIEW EXERCISES

Solve each applied problem. See Sections 2.4–2.7.

49. In the 2002–2003 National Hockey League regular season, Peter Forsberg of Colorado had a total of 106 points (goals plus assists). He had 48 more assists than goals. How many goals and how many assists did he have?

50. In the 1992 Olympic Games in Barcelona, the Unified Team earned 4 more medals than the United States. Together, the two teams earned 220 medals. How many medals did each team earn?

51. In the 1992 Games, the United States had the same number of gold and bronze medals, and three fewer silver medals. How many of each kind of medal did it earn? (See Exercise 50.)

52. In the 1992 Games, the Unified Team had seven more gold than silver medals and nine more silver than bronze medals. How many of each medal did it earn? (See Exercise 50.)

53. The perimeter of a rectangle is 46 ft. The width is 7 ft less than the length. Find the width.

54. The area of a rectangle is numerically 20 more than the width, and the length is 6 cm. What is the width?

55. Carla Arriola, a cashier, has $10-bills and $20-bills. There are 6 more tens than twenties. If there are 32 bills altogether, how many of them are twenties?

56. Mitzi Fulwood traveled for 2 hr at a constant speed. Because of roadwork, she reduced her speed by 7 mph for the next 2 hr. If she traveled 206 mi, what was her speed on the first part of the trip?

Summary Exercises on Solving Systems of Linear Equations

When solving systems of linear equations, use the following guidelines to help you decide whether to use substitution or elimination.

Guidelines for Choosing a Method to Solve a System of Linear Equations

1. If one of the equations of the system is already solved for one of the variables, as in the systems

$$3x + 4y = 9 \quad \text{and} \quad -5x + 3y = 9$$
$$y = 2x - 6 \qquad\qquad x = 3y - 7,$$

the substitution method is the better choice.

2. If both equations are in standard $Ax + By = C$ form, as in

$$4x - 11y = 3$$
$$-2x + 3y = 4,$$

and none of the variables has coefficient -1 or 1, the elimination method is the better choice.

3. If one or both of the equations are in standard form and the coefficient of one of the variables is -1 or 1, as in the systems

$$3x + y = -2 \quad \text{and} \quad -x + 3y = -4$$
$$-5x + 2y = 4 \qquad\qquad 3x - 2y = 8,$$

use the elimination method, or solve for the variable with coefficient -1 or 1 and then use the substitution method.

Use the preceding guidelines to solve each problem.

1. ***Concept Check***   Assuming that you want to minimize the amount of work required, tell whether you would use the substitution or elimination method to solve each system, and why. *Do not actually solve.*

   (a) $3x + 5y = 69$
   $y = 4x$

   (b) $3x + y = -7$
   $x - y = -5$

   (c) $3x - 2y = 0$
   $9x + 8y = 7$

2. ***Concept Check***   Which one of the following systems would be easier to solve with the substitution method? Why?

$$5x - 3y = 7 \qquad 7x + 2y = 4$$
$$2x + 8y = 3 \qquad y = -3x + 1$$

In Exercises 3 and 4, (a) solve the system by the elimination method, (b) solve the system by the substitution method, and (c) tell which method you prefer for that particular system and why.

3. $4x - 3y = -8$
   $x + 3y = 13$

4. $2x + 5y = 0$
   $x = -3y + 1$

Solve each system by the method of your choice. (For Exercises 5–7, see your answers to Exercise 1.)

5. $3x + 5y = 69$
   $y = 4x$

6. $3x + y = -7$
   $x - y = -5$

7. $3x - 2y = 0$
   $9x + 8y = 7$

8. $x + y = 7$
   $x = -3 - y$

9. $6x + 7y = 4$
   $5x + 8y = -1$

10. $6x - y = 5$
    $y = 11x$

11. $4x - 6y = 10$
    $-10x + 15y = -25$

12. $3x - 5y = 7$
    $2x + 3y = 30$

13. $5x = 7 + 2y$
    $5y = 5 - 3x$

14. $4x + 3y = 1$
    $3x + 2y = 2$

15. $2x - 3y = 7$
    $-4x + 6y = 14$

16. $2x + 3y = 10$
    $-3x + y = 18$

17. $2x + 5y = 4$
    $x + y = -1$

18. $x - 3y = 7$
    $4x + y = 5$

19. $\dfrac{1}{3}x + \dfrac{1}{2}y = 4$

    $-\dfrac{1}{2}x + \dfrac{1}{4}y = -2$

Solve each system by any method. First clear all fractions.

20. $\dfrac{1}{3}x - \dfrac{1}{2}y = \dfrac{1}{6}$

    $3x - 2y = 9$

21. $\dfrac{1}{5}x + \dfrac{2}{3}y = -\dfrac{8}{5}$

    $3x - y = 9$

22. $\dfrac{1}{6}x + \dfrac{1}{6}y = 2$

    $-\dfrac{1}{2}x - \dfrac{1}{3}y = -8$

23. $\dfrac{x}{2} - \dfrac{y}{3} = 9$

    $\dfrac{x}{5} - \dfrac{y}{4} = 5$

24. $\dfrac{x}{3} - \dfrac{3y}{4} = -\dfrac{1}{2}$

    $\dfrac{x}{6} + \dfrac{y}{8} = \dfrac{3}{4}$

25. $\dfrac{x}{5} + 2y = \dfrac{16}{5}$

    $\dfrac{3x}{5} + \dfrac{y}{2} = -\dfrac{7}{5}$

4.4 Applications of Linear Systems

OBJECTIVES

1 Solve problems about unknown numbers.

2 Solve problems about quantities and their costs.

3 Solve problems about mixtures.

4 Solve problems about distance, rate (or speed), and time.

Recall from **Section 2.4** the six-step method for solving applied problems. We modify those steps slightly to allow for two variables and two equations.

Solving an Applied Problem with Two Variables

Step 1 **Read** the problem carefully until you understand what is given and what is to be found.

Step 2 **Assign variables** to represent the unknown values, using diagrams or tables as needed. Write down what each variable represents.

Step 3 **Write two equations** using both variables.

Step 4 **Solve** the system of two equations.

Step 5 **State the answer** to the problem. Is the answer reasonable?

Step 6 **Check** the answer in the words of the original problem.

OBJECTIVE 1 Solve problems about unknown numbers. Use the modified six-step method to solve problems about unknown numbers.

EXAMPLE 1 Solving a Problem about Two Unknown Numbers

In 2004, sales of sports footwear were $3551 million more than sales of sports clothing. Together, the total sales for these items amounted to $25,953 million. (*Source:* National Sporting Goods Association.) What were the sales for each?

Step 1 **Read** the problem carefully. We must find the 2004 sales (in millions of dollars) for sports footwear and sports clothing. We know how much more sports footwear sales were than sports clothing sales. Also, we know the total sales.

Step 2 **Assign variables.**

Let x = sales of sports footwear in millions of dollars,

and y = sales of sports clothing in millions of dollars.

Step 3 **Write two equations.**

$x = 3551 + y$ Sales of sports footwear were $3551 million more than sales of sports clothing.

$x + y = 25,953$ Total sales were $25,953 million.

Step 4 **Solve** the system from Step 3. We use the substitution method, since the first equation is already solved for x.

$$x + y = 25,953 \quad \text{Second equation}$$
$$(3551 + y) + y = 25,953 \quad \text{Let } x = 3551 + y.$$
$$3551 + 2y = 25,953 \quad \text{Combine like terms.}$$
$$2y = 22,402 \quad \text{Subtract 3551.}$$
Don't stop here! $\quad y = 11,201 \quad \text{Divide by 2.}$

We substitute 11,201 for y in either equation of the system to find that $x = 14,752$.

Step 5 **State the answer.** Footwear sales were $14,752 million and clothing sales were $11,201 million.

Step 6 **Check** the answer in the original problem. Since
$$14,752 - 11,201 = 3551 \quad \text{and} \quad 14,752 + 11,201 = 25,953,$$
the answer satisfies the information in the problem.

✔ **Now Try Exercise 11.**

▶ **CAUTION** If an applied problem asks for *two* values, as in Example 1, be sure to give both of them in your answer.

OBJECTIVE 2 Solve problems about quantities and their costs. We can also use a linear system to solve an applied problem involving two quantities and their costs.

EXAMPLE 2 **Solving a Problem about Quantities and Costs**

The all-time top-grossing movie, *Titanic,** earned more in Europe than in the United States. This may be because average movie prices in Europe exceed those in the United States. (*Source: Parade* magazine, September 13, 1998.)

For example, while the average movie ticket (to the nearest dollar) in 1997–1998 cost $5 in the United States, it cost an equivalent of $11 in London. Suppose that a group of 41 Americans and Londoners who paid these average prices spent a total of $307 for tickets. How many from each country were in the group?

Step 1 **Read** the problem again.

Step 2 **Assign variables.**

Let $x =$ the number of Americans in the group,

and $y =$ the number of Londoners in the group.

Summarize the information given in the problem in a table. The entries in the first two rows of the Total Value column were found by multiplying the number of tickets sold by the price per ticket.

	Number of Tickets	Price per Ticket (in dollars)	Total Value (in dollars)
Americans	x	5	$5x$
Londoners	y	11	$11y$
Total	41		307

Step 3 **Write two equations.** The total number of tickets was 41, so
$$x + y = 41. \quad \text{Total number of tickets}$$

---

*Through September 2005, *Titanic* was still number one. (*Source: Variety.*)

Since the total value was $307, the final column leads to

$$5x + 11y = 307. \quad \text{Total value of tickets}$$

These two equations form this system.

> We must write *two* equations when we use two variables.

$$x + y = 41 \qquad (1)$$
$$5x + 11y = 307 \qquad (2)$$

Step 4 **Solve** the system of equations, using the elimination method. To eliminate the *x*-terms, multiply each side of equation (1) by -5. Then add this result to equation (2).

$$
\begin{aligned}
-5x - 5y &= -205 \qquad &&\text{Multiply equation (1) by } -5. \\
\underline{5x + 11y} &= \underline{307} \qquad &&\text{(2)} \\
6y &= 102 \qquad &&\text{Add.} \\
y &= 17 \qquad &&\text{Divide by 6.}
\end{aligned}
$$

Substitute 17 for *y* in equation (1).

$$
\begin{aligned}
x + y &= 41 \qquad &&\text{(1)} \\
x + 17 &= 41 \qquad &&\text{Let } y = 17. \\
x &= 24 \qquad &&\text{Subtract 17.}
\end{aligned}
$$

Step 5 **State the answer.** There were 24 Americans and 17 Londoners in the group.

Step 6 **Check.** The sum of 24 and 17 is 41, so the number of moviegoers is correct. Since 24 Americans paid $5 each and 17 Londoners paid $11 each, the total of the admission prices is $5(24) + $11(17) = $307, which agrees with the total amount stated in the problem.

✔ **Now Try Exercise 23.**

OBJECTIVE 3 Solve problems about mixtures.   In **Section 2.7,** we solved mixture problems by using one variable. Many mixture problems can also be solved by using a system of two equations in two variables.

EXAMPLE 3 **Solving a Mixture Problem Involving Percent**

Joe Castillo, a pharmacist, needs 100 L of a 50% alcohol solution. He has on hand a 30% alcohol solution and an 80% alcohol solution, which he can mix. How many liters of each will be required to make the 100 L of a 50% alcohol solution?

Step 1 **Read** the problem. Note the percentage of each solution and of the mixture.

Step 2 **Assign variables.**

> Let x = the number of liters of 30% alcohol needed,
> and y = the number of liters of 80% alcohol needed.

Summarize the information in a table. Percents are written as decimals.

Liters of Solution	Percent (as a decimal)	Liters of Pure Alcohol
x	0.30	0.30x
y	0.80	0.80y
100	0.50	0.50(100)

Figure 9 gives an idea of what is actually happening in this problem.

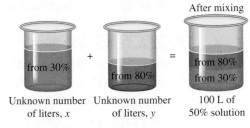

FIGURE 9

Step 3 **Write two equations.** Since the total number of liters in the final mixture will be 100, the first equation is

$$x + y = 100.$$

To find the amount of pure alcohol in each mixture, multiply the number of liters by the concentration. The amount of pure alcohol in the 30% solution added to the amount of pure alcohol in the 80% solution will equal the amount of pure alcohol in the final 50% solution. This gives the second equation,

$$0.30x + 0.80y = 0.50(100).$$

These two equations form this system:

Be sure to write *two* equations.

$$x + y = 100$$
$$0.30x + 0.80y = 50. \qquad 0.50(100) = 50$$

Step 4 **Solve** this system by the substitution method. Solving the first equation of the system for x gives $x = 100 - y$. Substitute $100 - y$ for x in the second equation.

$0.30(100 - y) + 0.80y = 50$	Let $x = 100 - y$.
$30 - 0.30y + 0.80y = 50$	Distributive property
$30 + 0.50y = 50$	Combine like terms.
$0.50y = 20$	Subtract 30.
$y = 40$	Divide by 0.50.

Distribute 0.30 to *both* 100 and −y.

Then $x = 100 - y = 100 - 40 = 60.$

Step 5 **State the answer.** The pharmacist should use 60 L of the 30% solution and 40 L of the 80% solution.

Step 6 **Check.** Since $60 + 40 = 100$ and $0.30(60) + 0.80(40) = 50$, this mixture will give 100 L of the 50% solution, as required.

✔ **Now Try Exercise 25.**

> ▶ **NOTE** The system in Example 3 could have been solved by the elimination method. Also, we could have cleared decimals by multiplying each side of the second equation by 10.

OBJECTIVE 4 Solve problems about distance, rate (or speed), and time. Problems that use the distance formula $d = rt$ were first introduced in **Section 2.7.** In many cases, these problems can be solved with systems of two linear equations. Keep in mind that setting up a table and drawing a sketch will help you solve such problems.

EXAMPLE 4 Solving a Problem about Distance, Rate, and Time

Two executives in cities 400 mi apart drive to a business meeting at a location on the line between their cities. They meet after 4 hr. Find the speed of each car if one travels 20 mph faster than the other.

Step 1 **Read** the problem carefully.

Step 2 **Assign variables.**

$$\text{Let } x = \text{the speed of the faster car,}$$
$$\text{and } y = \text{the speed of the slower car.}$$

Use the formula $d = rt$. Since each car travels for 4 hr, the time t for each car is 4. This information is shown in the table. The distance is found by using the formula $d = rt$ and the expressions already entered in the table.

	r	t	d
Faster Car	x	4	$4x$
Slower Car	y	4	$4y$

Find d, using $d = rt$.

Draw a sketch showing what is happening in the problem. See Figure 10.

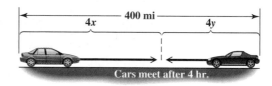

FIGURE 10

Step 3 **Write two equations.** As shown in the figure, since the total distance traveled by both cars is 400 mi, one equation is

$$4x + 4y = 400.$$

Because the faster car goes 20 mph faster than the slower car, the second equation is

$$x = 20 + y.$$

Step 4 **Solve** the system of equations.

$$4x + 4y = 400 \qquad (1)$$
$$x = 20 + y \qquad (2)$$

Use substitution. Replace x with $20 + y$ in equation (1) and solve for y.

$$4(20 + y) + 4y = 400 \qquad \text{Let } x = 20 + y.$$
$$80 + 4y + 4y = 400 \qquad \text{Distributive property}$$
$$80 + 8y = 400 \qquad \text{Combine like terms.}$$
$$8y = 320 \qquad \text{Subtract 80.}$$
$$y = 40 \qquad \text{Divide by 8.}$$

Distribute 4 to *both* 20 and *y*.

Then $x = 20 + y = 20 + 40 = 60$.

Step 5 **State the answer.** The speeds of the two cars are 40 mph and 60 mph.

Step 6 **Check.** Since each car travels for 4 hr, the total distance traveled is

$$4(60) + 4(40) = 240 + 160 = 400 \text{ mi}, \qquad \text{as required.}$$

✔ **Now Try Exercise 31.**

The problems in Examples 1–4 also could be solved by using only one variable.

4.4 EXERCISES

Concept Check *Choose the correct response in Exercises 1–7.*

1. Which expression represents the monetary value of x 20-dollar bills?

A. $\dfrac{x}{20}$ dollars   **B.** $\dfrac{20}{x}$ dollars   **C.** $(20 + x)$ dollars   **D.** $20x$ dollars

2. Which expression represents the cost of t pounds of candy that sells for $1.95 per lb?

A. $1.95t$ **B.** $\dfrac{\$1.95}{t}$ **C.** $\dfrac{t}{\$1.95}$ **D.** $1.95 + t$

3. Which expression represents the amount of interest earned on d dollars at an interest rate of 2%?

A. $2d$ dollars   **B.** $0.02d$ dollars   **C.** $0.2d$ dollars   **D.** $200d$ dollars

4. Suppose that Ira Spector wants to mix x liters of a 40% acid solution with y liters of a 35% solution to obtain 100 L of a 38% solution. One equation in a system for solving this problem is $x + y = 100$. Which one of the following is the other equation?

A. $0.35x + 0.40y = 0.38(100)$ **B.** $0.40x + 0.35y = 0.38(100)$
C. $35x + 40y = 38$ **D.** $40x + 35y = 0.38(100)$

5. According to *Natural History* magazine, the speed of a cheetah is 70 mph. If a cheetah runs for x hours, how many miles does the cheetah cover?

A. $(70 + x)$ miles   **B.** $(70 - x)$ miles   **C.** $\dfrac{70}{x}$ miles   **D.** $70x$ miles

6. What is the speed of a plane that travels at a rate of 560 mph *against* a wind of r mph?

A. $(560 + r)$ mph   **B.** $\dfrac{560}{r}$ mph   **C.** $(560 - r)$ mph   **D.** $(r - 560)$ mph

7. What is the speed of a plane that travels at a rate of 560 mph *with* a wind of r mph?

A. $\dfrac{r}{560}$ mph   **B.** $(560 - r)$ mph   **C.** $(560 + r)$ mph   **D.** $(r - 560)$ mph

8. Referring to the list of steps for solving an applied problem with two variables, write a short paragraph describing the general procedure you will use to solve the problems that follow in this exercise set.

Exercises 9 and 10 are good warm-up problems. In each case, refer to the six-step problem-solving method, fill in the blanks for Steps 2 and 3, and then complete the solution by applying Steps 4–6.

9. The sum of two numbers is 98 and the difference between them is 48. Find the two numbers.

Step 1 **Read** the problem carefully.

Step 2 **Assign variables.**
Let x = the first number and let y = _____.

Step 3 **Write two equations.**
First equation: $x + y = 98$
Second equation: _____

10. The sum of two numbers is 201 and the difference between them is 11. Find the two numbers.

Step 1 **Read** the problem carefully.

Step 2 **Assign variables.**
Let x = the first number and let y = _____.

Step 3 **Write two equations.**
First equation: $x + y = 201$
Second equation: _____

Write a system of equations for each problem, and then solve the system. See Example 1.

11. During 2004, the two top-grossing North American concert tours featured Prince and Céline Dion. Together, the two singers performed 250 shows. Céline Dion gave 58 more shows than Prince. How many shows were included in each tour? (*Source:* Pollstar.)

12. In 2005, the top two formats of U.S. commercial radio stations were country and news/talk. There were 695 fewer news/talk stations than country stations, and together they totaled 3343 stations. How many stations of each format were there? (*Source:* M Street Corporation.)

13. The two top-grossing movies of 2003 were *The Lord of the Rings: The Return of the King* and *Finding Nemo. Finding Nemo* grossed $21.4 million less than *The Lord of the Rings: The Return of the King,* and together the two films took in $700.8 million. How much did each of these movies earn? (*Source:* Nielsen EDI.)

14. The Terminal Tower in Cleveland, Ohio, is 242 ft shorter than the Key Tower, also in Cleveland. The total of the heights of the two buildings is 1658 ft. Find the heights of the buildings. (*Source: World Almanac and Book of Facts 2006.*)

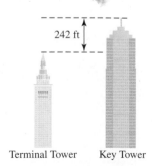

242 ft

Terminal Tower   Key Tower

*If x units of a product cost C dollars to manufacture and earn revenue of R dollars, the value of x at which the expressions for C and R are equal is called the **break-even quantity**—the number of units that produce 0 profit. In Exercises 15 and 16, **(a)** find the break-even quantity, and **(b)** decide whether the product should be produced on the basis of whether it will earn a profit. (Profit = Revenue − Cost.)*

15. $C = 85x + 900$; $R = 105x$; no more than 38 units can be sold.

16. $C = 105x + 6000$; $R = 255x$; no more than 400 units can be sold.

Write a system of equations for each problem, and then solve the system. See Example 2.

17. Jonathan, a second grader, counted the money in his piggy bank. He found that he had only quarters and dimes. When he added up his money, he found that he had 39 coins worth a total of $7.50. How many coins of each kind did he have?

Number of Coins	Value per Coin	Total Value
x	$0.25	
y	$0.10	
39		$7.50

18. Bill Kunz went to the post office to stock up on stamps. He spent $19.44 on 56 stamps, made up of a combination of 39-cent and 24-cent stamps. How many stamps of each denomination did he buy?

Number of Stamps	Denomination	Total Value
x	$0.39	
y	$0.24	
		$19.44

19. A newspaper advertised DVDs and CDs. Joyce Nemeth went shopping and bought each of her seven nephews a gift, either a DVD of the movie *Million Dollar Baby* or the latest Clay Aiken CD. The DVD cost $14.95 and the CD cost $16.88, and she spent a total of $114.30. How many DVDs and how many CDs did she buy?

20. Alan Lebovitz saw the ad mentioned in Exercise 19 and he, too, went shopping. He bought each of his five nieces a gift, either a DVD of *Eight Below* or the CD soundtrack to *Cars*. The DVD cost $14.99 and the soundtrack cost $13.88, and he spent a total of $70.51. How many DVDs and CDs did he buy?

21. Karen Walsh has twice as much money invested at 5% simple annual interest as she does at 4%. If her yearly income from these two investments is $350, how much does she have invested at each rate?

22. Charles Miller invested his textbook royalty income in two accounts, one paying 3% annual simple interest and the other paying 2% interest. He earned a total of $11 interest. If he invested three times as much in the 3% account as he did in the 2% account, how much did he invest at each rate?

23. Average movie ticket prices in the United States are, in general, lower than in other countries. It would cost $77.87 to buy three tickets in Japan plus two tickets in Switzerland. Three tickets in Switzerland plus two tickets in Japan would cost $73.83. How much does an average movie ticket cost in each of these countries? (*Source:* Business Traveler International.)

24. (See Exercise 23.) Four movie tickets in Germany plus three movie tickets in France would cost $62.27. Three tickets in Germany plus four tickets in France would cost $62.19. How much does an average movie ticket cost in each of these countries? (*Source:* Business Traveler International.)

Write a system of equations for each problem, and then solve the system. See Example 3.

25. A 40% dye solution is to be mixed with a 70% dye solution to get 120 L of a 50% solution. How many liters of the 40% and 70% solutions will be needed?

Liters of Solution	Percent (as a decimal)	Liters of Pure Dye
x	0.40	
y	0.70	
120	0.50	

26. A 90% antifreeze solution is to be mixed with a 75% solution to make 120 L of a 78% solution. How many liters of the 90% and 75% solutions will be used?

Liters of Solution	Percent (as a decimal)	Liters of Pure Antifreeze
x	0.90	
y	0.75	
120	0.78	

27. Steve Watnik wishes to mix coffee worth $6 per lb with coffee worth $3 per lb to get 90 lb of a mixture worth $4 per lb. How many pounds of the $6 and the $3 coffees will be needed?

Pounds	Dollars per Pound	Cost
x	6	
y		
90		

28. Mariana Coanda wishes to blend candy selling for $1.20 per lb with candy selling for $1.80 per lb to get a mixture that will be sold for $1.40 per lb. How many pounds of the $1.20 and the $1.80 candies should be used to get 45 lb of the mixture?

Pounds	Dollars per Pound	Cost
x		
y	1.80	
45		

29. How many pounds of nuts selling for $6 per lb and raisins selling for $3 per lb should Kelli Hammer combine to obtain 60 lb of a trail mix selling for $5 per lb?

30. Avis Proctor works at a gourmet delicatessen. She is preparing a cheese tray for a large reception. She is using some cheeses that sell for $8 per lb and others that sell for $12 per lb. How many pounds of cheese at each price should she use in order for the mixed cheeses on the tray to weigh a total of 56 lb and sell for $10.50 per lb?

Write a system of equations for each problem, and then solve the system. See Example 4.

31. Two trains start from towns 495 mi apart and travel toward each other on parallel tracks. They pass each other 4.5 hr later. If one train travels 10 mph faster than the other, find the speed of each train.

32. Two trains that are 495 mi apart travel toward each other. They pass each other 5 hr later. If one train travels half as fast as the other, what are their speeds?

33. Kansas City and Denver are 600 mi apart. Two cars start from these cities, traveling toward each other. They pass each other after 6 hr. Find the rate of each car if one travels 30 mph slower than the other.

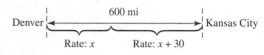

34. Toledo and Cincinnati are 200 mi apart. A car leaves Toledo traveling toward Cincinnati, and another car leaves Cincinnati at the same time, traveling toward Toledo. The car leaving Toledo averages 15 mph faster than the other car, and they pass each other after 1 hr and 36 min. What are the rates of the cars?

35. RAGBRAI®, the *Des Moines Register's* **A**nnual **G**reat **B**icycle **R**ide **A**cross **I**owa, is the longest and oldest touring bicycle ride in the world. Suppose a cyclist began the 490 mi ride on July 23, 2006 in western Iowa at the same time that a car traveling toward it left eastern Iowa. If the bicycle and the car passed each other after 7 hr and the car traveled 40 mph faster than the bicycle, find the average speed of each. (*Source:* www.ragbrai.org)

36. In 2004, Atlanta's Hartsfield Airport was the nation's busiest. Suppose two planes leave Hartsfield at the same time, one traveling east and the other traveling west. If the planes are 2100 mi apart after 2 hr and one plane travels 50 mph faster than the other, find the speed of each plane. (*Source:* Airports Council International—North America.)

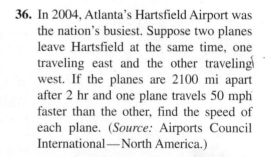

37. A boat takes 3 hr to go 24 mi upstream. It can go 36 mi downstream in the same time. Find the speed of the current and the speed of the boat in still water if x = the speed of the boat in still water and y = the speed of the current. (*Hint:* Because the current pushes the boat when it is going downstream, the speed of the boat downstream is the *sum* of the speed of the boat and the speed of the current. The current slows down the boat when it is going upstream, so the speed of the boat upstream is the *difference* of the speed of the boat and the speed of the current.)

	r	t	d
Downstream	$x + y$	3	$3(x + y)$
Upstream	$x - y$	3	

38. It takes a boat $1\frac{1}{2}$ hr to go 12 mi downstream, and 6 hr to return. Find the speed of the boat in still water and the speed of the current. Let x = the speed of the boat in still water and y = the speed of the current.

	r	t	d
Downstream	$x + y$	$\frac{3}{2}$	$\frac{3}{2}(x + y)$
Upstream	$x - y$	6	$6(x - y)$

39. If a plane can travel 440 mph into the wind and 500 mph with the wind, find the speed of the wind and the speed of the plane in still air.

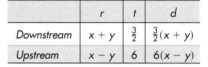

40. A small plane travels 200 mph with the wind and 120 mph against it. Find the speed of the wind and the speed of the plane in still air.

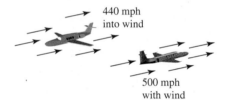

The next problems are "brain busters."

41. At the beginning of a bicycle ride for charity, Barry MacFarlane and Dane McGuckian are 30 mi apart. If they leave at the same time and ride in the same direction, Barry overtakes Dane in 6 hr. If they ride toward each other, they pass each other in 1 hr. What are their speeds?

42. Martin Peres left Farmersville in a plane at noon to travel to Exeter. Walter Wooden left Exeter in his automobile at 2 P.M. to travel to Farmersville. It is 400 mi from Exeter to Farmersville. If the sum of their speeds was 120 mph, and if they crossed paths at 4 P.M., find the speed of each.

PREVIEW EXERCISES

*Graph each linear inequality. See **Section 3.5**.*

43. $x + y \leq 4$ **44.** $2x - y > 4$ **45.** $y \geq -3x + 2$

46. $x - 3y < 6$ **47.** $2x + 4y > 8$ **48.** $3x + 2y \leq 0$

4.5 Solving Systems of Linear Inequalities

OBJECTIVES

1 Solve systems of linear inequalities by graphing.

2 Use a graphing calculator to solve a system of linear inequalities.

We graphed the solutions of a linear inequality in **Section 3.5.** Recall that, to graph the solutions of $x + 3y > 12$, for example, we first graph $x + 3y = 12$ by finding and plotting a few ordered pairs that satisfy the equation. Because the points on the line do *not* satisfy the inequality, we use a dashed line. To decide which region includes the points that are solutions, we choose a test point not on the line, such as $(0, 0)$. Substituting these values for x and y in the original inequality gives

$$x + 3y > 12$$
$$0 + 3(0) > 12 \quad ?$$
$$0 > 12. \quad \text{False}$$

This false result indicates that the solutions are those points on the side of the line that does *not* include $(0, 0)$, as shown in Figure 11.

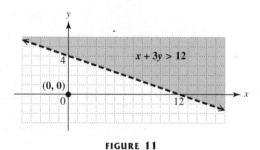

FIGURE 11

Now we use the same techniques to solve *systems* of linear inequalities.

OBJECTIVE 1 Solve systems of linear inequalities by graphing. A **system of linear inequalities** consists of two or more linear inequalities. The **solution set of a system of linear inequalities** includes all ordered pairs that make all inequalities of the system true at the same time. To solve a system of linear inequalities, use the following steps.

Solving a System of Linear Inequalities

Step 1 **Graph the inequalities.** Graph each linear inequality, using the method described in **Section 3.5.**

Step 2 **Choose the intersection.** Indicate the solution set of the system by shading the intersection of the graphs (the region where the graphs overlap).

EXAMPLE 1 Solving a System of Linear Inequalities

Graph the solution set of the system.

$$3x + 2y \le 6$$
$$2x - 5y \ge 10$$

Step 1 To graph $3x + 2y \leq 6$, graph the solid boundary line $3x + 2y = 6$ and shade the region containing $(0, 0)$, as shown in Figure 12(a). Then graph $2x - 5y \geq 10$ with the solid boundary line $2x - 5y = 10$. The test point $(0, 0)$ makes this inequality false, so shade the region on the other side of the boundary line. See Figure 12(b).

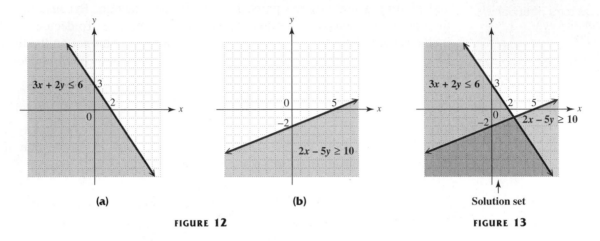

(a)	**(b)**
FIGURE 12	

Solution set
FIGURE 13

Step 2 The solution set of this system includes all points in the intersection (overlap) of the graphs of the two inequalities. As shown in Figure 13, this intersection is the gray shaded region and portions of the two boundary lines that surround it.

✔ **Now Try Exercise 5.**

▶ **NOTE** We usually do all the work on one set of axes. In the examples that follow, only one graph is shown. Be sure that the region of the final solution is clearly indicated.

EXAMPLE 2 **Solving a System of Linear Inequalities**

Graph the solution set of the system.

$$x - y > 5$$
$$2x + y < 2$$

Figure 14 shows the graphs of both $x - y > 5$ and $2x + y < 2$. Dashed lines show that the graphs of the inequalities do not include their boundary lines. The solution set of the system is the region with the gray shading. The solution set does not include either boundary line.

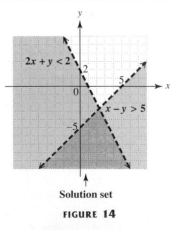

Solution set
FIGURE 14

✔ **Now Try Exercise 9.**

EXAMPLE 3 Solving a System of Three Linear Inequalities

Graph the solution set of the system.

$$4x - 3y \leq 8$$
$$x \geq 2$$
$$y \leq 4$$

Recall that $x = 2$ is a vertical line through the point $(2, 0)$, and $y = 4$ is a horizontal line through the point $(0, 4)$. The graph of the solution set is the shaded region in Figure 15, including all boundary lines.

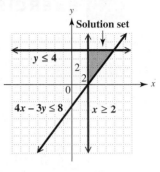

FIGURE 15

✔ **Now Try Exercise 21.**

OBJECTIVE 2 Use a graphing calculator to solve a system of linear inequalities.

 **EXAMPLE 4** Finding the Solution Set of a System of Linear Inequalities with a Graphing Calculator

Graph the solution set of the system with a calculator.

$$y < 3x + 2$$
$$y > -2x - 5$$

To graph the first inequality, we direct the calculator to shade *below* the line

$$Y_1 = 3X + 2$$

(because of the $<$ symbol). To graph the second inequality, we direct the calculator to shade *above* the line

$$Y_2 = -2X - 5$$

(because of the $>$ symbol). Figure 16(a) shows the appropriate screen for these directions on a TI-83/84 Plus calculator. Graphing the inequalities in the standard viewing window gives the screen in Figure 16(b). The crosshatched region is the intersection of the solution sets of the two individual inequalities and represents the solution set of the system.

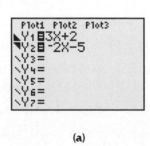

(a)   (b)

FIGURE 16

If the inequalities of a system are not solved for y as in this example, we must do so in order to enter them into the calculator. Notice that we cannot determine from the screen in Figure 16(b) whether the boundary lines are included or excluded in the solution set. For this system, neither boundary line is included.

✔ **Now Try Exercise 25.**

4.5 EXERCISES

● *Complete solution available on Video Lectures on CD/DVD*

▨ *Now Try Exercise*

Concept Check *Match each system of inequalities with the correct graph from choices A–D.*

1. $x \geq 5$
$ y \leq -3$

A.

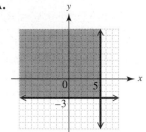

B.

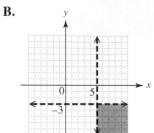

2. $x \leq 5$
$ y \geq -3$

3. $x > 5$
$ y < -3$

C.

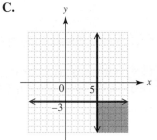

D.

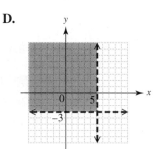

4. $x < 5$
$ y > -3$

Graph the solution set of each system of linear inequalities. See Examples 1 and 2.

● **5.** $x + y \leq 6$
$ x - y \geq 1$

6. $x + y \leq 2$
$ x - y \geq 3$

7. $4x + 5y \geq 20$
$ x - 2y \leq 5$

8. $ x + 4y \leq 8$
$2x - y \geq 4$

● **9.** $2x + 3y < 6$
$ x - y < 5$

10. $x + 2y < 4$
$ x - y < -1$

11. $y \leq 2x - 5$
$ x < 3y + 2$

12. $x \geq 2y + 6$
$ y > -2x + 4$

13. $4x + 3y < 6$
$ x - 2y > 4$

14. $3x + y > 4$
$ x + 2y < 2$

15. $x \leq 2y + 3$
$ x + y < 0$

16. $x \leq 4y + 3$
$ x + y > 0$

17. $-3x + y \geq 1$
$ 6x - 2y \geq -10$

18. $2x + 3y < 6$
$ 4x + 6y > 18$

19. $x - 3y \leq 6$
$ x \geq -4$

✏ **20.** Suppose that your friend was absent from class on the day that systems of inequalities were covered. He asked you to write a short explanation of the procedure. Do this for him.

Graph the solution set of each system. See Example 3.

● **21.** $4x + 5y < 8$
$ y > -2$
$ x > -4$

22. $x + y \geq -3$
$ x - y \leq 3$
$ y \leq 3$

23. $3x - 2y \geq 6$
$ x + y \leq 4$
$ x \geq 0$
$ y \geq -4$

24. ***Concept Check*** How can we determine that the point $(0, 0)$ is a solution of the system in Exercise 21 without actually graphing the system?

TECHNOLOGY INSIGHTS (EXERCISES 25–28)

Match each system of inequalities with its solution set. See Example 4.

A.

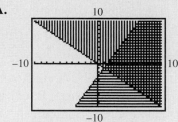

B.

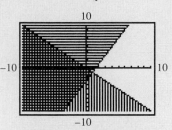

C.

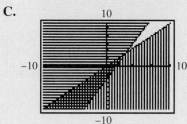

D.

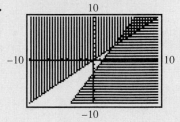

☉ 25. $y \geq x$
$y \leq 2x - 3$

26. $y \leq x$
$y \geq 2x - 3$

27. $y \geq -x$
$y \leq 2x - 3$

28. $y \leq -x$
$y \geq 2x - 3$

PREVIEW EXERCISES

*Evaluate each expression. See **Section 1.2**.*

29. $2 \cdot 2 \cdot 2 \cdot 2 \cdot 2 \cdot 2$

30. $3 \cdot 3 \cdot 3$

31. $5 \cdot 5 \cdot 5 \cdot 5$

32. $4 \cdot 4 \cdot 4 \cdot 4 \cdot 4$

33. $\dfrac{2}{3} \cdot \dfrac{2}{3} \cdot \dfrac{2}{3}$

34. $\dfrac{5}{8} \cdot \dfrac{5}{8}$

35. $(2 \cdot 2 \cdot 2)(2 \cdot 2 \cdot 2 \cdot 2)$

36. $(3 \cdot 3)(3 \cdot 3 \cdot 3)$

Chapter **4** Group Activity

TOP CONCERT TOURS

Objective Use systems of equations to analyze data.

This activity will use systems of equations to analyze the top-grossing concert tours of 2003 and 2005.

- In 2003, the top-grossing North American concert tours were those of Bruce Springsteen (and the E Street Band) and Céline Dion. Together, they grossed $196.4 million. Bruce Springsteen grossed $35.4 million more than Céline Dion. The two acts performed 192 shows, and Céline Dion performed 98 fewer shows than Bruce Springsteen. (*Source:* Pollstar.)

- In 2005, the top-grossing North American concert tours were those of The Rolling Stones and U2. Together, they grossed $300.9 million. The Rolling Stones grossed $23.1 million more than U2. The two groups performed a total of 120 shows, and U2 performed 36 more shows than The Rolling Stones. (*Source:* Pollstar.)

Using the given information, complete the table. One student should complete the information for 2003 and the other student for 2005.

A. Write a system of equations to find the total gross for each performer or group.

B. Write a system of equations to find the number of shows per year for each performer or group.

C. Use the information from Exercises A and B to find the gross per show.

Year	Group	Total Gross	Shows per Year	Gross per Show
2003	Bruce Springsteen			
2003	Céline Dion			
2005	Rolling Stones			
2005	U2			

D. Once you have completed the table, compare the results. Answer these questions.

1. What differences do you notice between 2003 and 2005?
2. Which performer or group had the highest total gross?
3. Which performer or group had the highest gross per show?

Chapter **4 SUMMARY** *View the Interactive Summary on the Pass the Test CD.*

KEY TERMS

4.1 system of linear equations
    (linear system)
solution of a system
solution set of a system

set-builder notation
consistent system
inconsistent system
independent equations

dependent equations
4.5 system of linear inequalities
solution set of a system of linear
    inequalities

TEST YOUR WORD POWER

See how well you have learned the vocabulary in this chapter. Answers, with examples, follow the Quick Review.

1. A **system of linear equations** consists of
   A. at least two linear equations with different variables
   B. two or more linear equations that have an infinite number of solutions
   C. two or more linear equations with the same variables
   D. two or more linear inequalities.

2. A **solution of a system** of linear equations is
   A. an ordered pair that makes one equation of the system true

   B. an ordered pair that makes all the equations of the system true at the same time
   C. any ordered pair that makes one or the other or both equations of the system true
   D. the set of values that make all the equations of the system false.

3. A **consistent system** is a system of equations
   A. with one solution
   B. with no solution
   C. with an infinite number of solutions

   D. that have the same graph.

4. An **inconsistent system** is a system of equations
   A. with one solution
   B. with no solution
   C. with an infinite number of solutions
   D. that have the same graph.

5. **Dependent equations**
   A. have different graphs
   B. have no solution
   C. have one solution
   D. are different forms of the same equation.

QUICK REVIEW

Concepts	Examples

4.1 SOLVING SYSTEMS OF LINEAR EQUATIONS BY GRAPHING

An ordered pair is a solution of a system if it makes all equations of the system true at the same time.

Is $(4, -1)$ a solution of the system $\begin{array}{l} x + y = 3 \\ 2x - y = 9? \end{array}$

Yes, because $4 + (-1) = 3$ and $2(4) - (-1) = 9$ are both true.

To solve a linear system by graphing,

Solve the system by graphing: $\begin{array}{l} x + y = 5 \\ 2x - y = 4. \end{array}$

Step 1 Graph each equation of the system on the same axes.

Step 2 Find the coordinates of the point of intersection.

Step 3 Check. Write the solution set.

A graphing calculator can find the solution of a system by locating the point of intersection of the graphs.

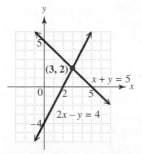

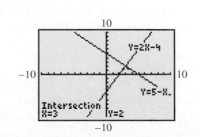

The solution $(3, 2)$ checks, so $\{(3, 2)\}$ is the solution set.

(continued)

Concepts	Examples

4.2 SOLVING SYSTEMS OF LINEAR EQUATIONS BY SUBSTITUTION

Step 1 Solve one equation for either variable.

Solve by substitution.

$$x + 2y = -5 \quad (1)$$
$$y = -2x - 1 \quad (2)$$

Equation (2) is already solved for y.

Step 2 Substitute for that variable in the other equation to get an equation in one variable.

Substitute $-2x - 1$ for y in equation (1).

$$x + 2(-2x - 1) = -5$$

Step 3 Solve the equation from Step 2.

Solve to get $x = 1$.

Step 4 Substitute the result into the equation from Step 1 to get the value of the other variable.

To find y, let $x = 1$ in equation (2).

$$y = -2(1) - 1 = -3$$

Step 5 Check. Write the solution set.

The solution, $(1, -3)$, checks, so $\{(1, -3)\}$ is the solution set.

4.3 SOLVING SYSTEMS OF LINEAR EQUATIONS BY ELIMINATION

Step 1 Write both equations in standard form, $Ax + By = C$.

Solve by elimination.

$$x + 3y = 7 \quad (1)$$
$$3x - y = 1 \quad (2)$$

Step 2 Multiply to transform the equations so that the coefficients of one pair of variable terms are opposites.

Multiply equation (1) by -3 to eliminate the x-terms.

$$\begin{array}{rl} -3x - 9y = -21 & \\ \underline{3x - y = 1} & \\ -10y = -20 & \text{Add.} \\ y = 2 & \text{Divide by } -10. \end{array}$$

Step 3 Add the equations to get an equation with only one variable.

Step 4 Solve the equation from Step 3.

Step 5 Substitute the solution from Step 4 into either of the original equations to find the value of the remaining variable.

Substitute to get the value of x.

$$\begin{array}{rl} x + 3y = 7 & (1) \\ x + 3(2) = 7 & \text{Let } y = 2. \\ x + 6 = 7 & \text{Multiply.} \\ x = 1 & \text{Subtract 6.} \end{array}$$

Step 6 Check. Write the solution set.

Since $1 + 3(2) = 7$ and $3(1) - 2 = 1$, the solution set is $\{(1, 2)\}$.

If the result of the addition step (Step 3) is a false statement, such as $0 = 4$, the graphs are parallel lines and *there is no solution. The solution set is $\emptyset$.*

$$\begin{array}{rl} x - 2y = 6 & \\ \underline{-x + 2y = -2} & \\ 0 = 4 & \text{Solution set: } \emptyset \end{array}$$

If the result is a true statement, such as $0 = 0$, the graphs are the same line, and an *infinite number of ordered pairs are solutions. The solution set is written in set-builder notation as $\{(x, y) \mid \underline{}\}$, where a form of the equation is written in the blank.*

$$\begin{array}{rl} x - 2y = 6 & \\ \underline{-x + 2y = -6} & \\ 0 = 0 & \text{Solution set: } \{(x, y) \mid x - 2y = 6\} \end{array}$$

(continued)

Concepts	Examples

4.4 APPLICATIONS OF LINEAR SYSTEMS

Use the modified six-step method.

Step 1   Read.

Step 2   Assign variables.

Step 3   Write two equations using both variables.

Step 4   Solve the system.

Step 5   State the answer.

Step 6   Check.

The sum of two numbers is 30. Their difference is 6. Find the numbers.

Let $x =$ one number; let $y =$ the other number.

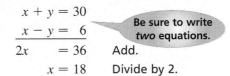

$$x + y = 30$$
$$\underline{x - y = 6}$$
$$2x = 36 \quad \text{Add.}$$
$$x = 18 \quad \text{Divide by 2.}$$

Be sure to write *two* equations.

Let $x = 18$ in the top equation: $18 + y = 30$. Solve to get $y = 12$.

The two numbers are 18 and 12.

$18 + 12 = 30$ and $18 - 12 = 6$, so the solution checks.

4.5 SOLVING SYSTEMS OF LINEAR INEQUALITIES

To solve a system of linear inequalities, graph each inequality on the same axes. (This was explained in **Section 3.5.**) The solution set of the system is formed by the overlap of the regions of the two graphs.

The shaded region shows the solution set of this system:

$$2x + 4y \geq 5$$
$$x \geq 1.$$

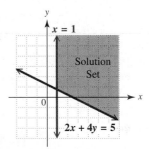

Graphing calculators also can represent solution sets of systems of inequalities.

The crosshatched region shows the solution set of this system:

$$y > 2x + 1$$
$$y < 3.$$

The boundary lines are not included.

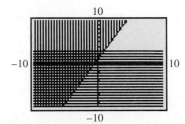

Answers to Test Your Word Power

1. C; *Example:* $2x + y = 7, \quad 3x - y = 3$

2. B; *Example:* The ordered pair $(2, 3)$ satisfies both equations of the system in Answer 1, so it is a solution of the system.

3. A; *Example:* The system in Answer 1 is consistent. The graphs of the equations intersect at exactly one point—in this case, the solution $(2,3)$.

4. B; *Example:* The equations of two parallel lines make up an inconsistent system; their graphs never intersect, so there is no solution to the system.

5. D; *Example:* The equations $4x - y = 8$ and $8x - 2y = 16$ are dependent because their graphs are the same line.

Chapter 4 REVIEW EXERCISES

[4.1] *Decide whether the given ordered pair is a solution of the given system.*

1. $(3, 4)$
$$4x - 2y = 4$$
$$5x + y = 19$$

2. $(-5, 2)$
$$x - 4y = -13$$
$$2x + 3y = 4$$

Solve each system by graphing.

3. $x + y = 4$
$$2x - y = 5$$

4. $x - 2y = 4$
$$2x + y = -2$$

5. $2x + 4 = 2y$
$$y - x = -3$$

[4.2]

6. Concept Check A student solves the system $\begin{array}{l} 2x + y = 6 \\ -2x - y = 4 \end{array}$ and gets the equation $0 = 10$. The student gives the solution set as $\{(0, 10)\}$. Is this correct? If not, what is the solution set?

7. Can a system of two linear equations in two variables have exactly three solutions? Explain.

Solve each system by the substitution method.

8. $3x + y = 7$
$$x = 2y$$

9. $2x - 5y = -19$
$$y = x + 2$$

10. $5x + 15y = 30$
$$x + 3y = 6$$

11. Concept Check After solving a system of linear equations by the substitution method, a student obtained the equation "$0 = 0$." He gave the solution set of the system as $\{(0, 0)\}$. Was his answer correct?

12. Suppose that you were asked to solve the system $\begin{array}{l} 5x - 3y = 7 \\ -x + 2y = 4 \end{array}$ by substitution. Which variable in which equation would be easiest to solve for in your first step? Why?

[4.3]

13. Concept Check Only one of the following systems does not require that we multiply one or both equations by a constant to solve the system by the elimination method. Which one is it?

A. $-4x + 3y = 7$
$$3x - 4y = 4$$

B. $5x + 8y = 13$
$$12x + 24y = 36$$

C. $2x + 3y = 5$
$$x - 3y = 12$$

D. $x + 2y = 9$
$$3x - y = 6$$

14. Concept Check For the system

$$2x + 12y = 7$$
$$3x + 4y = 1,$$

if we were to multiply the top equation by -3, by what number would we have to multiply the bottom equation to
(a) eliminate the x-terms when solving by the elimination method?
(b) eliminate the y-terms when solving by the elimination method?

Solve each system by the elimination method.

15. $2x - y = 13$
$$x + y = 8$$

16. $-4x + 3y = 25$
$$6x - 5y = -39$$

17. $3x - 4y = 9$
$$6x - 8y = 18$$

18. $2x + y = 3$
$$-4x - 2y = 6$$

[4.1–4.3] *Solve each system by any method.*

19. $2x + 3y = -5$
$3x + 4y = -8$

20. $6x - 9y = 0$
$2x - 3y = 0$

21. $x - 2y = 5$
$y = x - 7$

22. $\dfrac{x}{2} + \dfrac{y}{3} = 7$

$\dfrac{x}{4} + \dfrac{2y}{3} = 8$

23. What are the three methods of solving systems discussed in this chapter? Choose one and discuss its advantages and disadvantages.

24. *Concept Check* Why would it be easier to solve system B by the substitution method than system A?

$$\textbf{A:} \quad -5x + 6y = -7 \qquad \textbf{B:} \quad 2x + 9y = 13$$
$$2x + 5y = -5 \qquad\qquad y = 3x - 2$$

[4.4] *Solve each problem by using a system of equations.*

25. At the end of 2001, Subway topped McDonald's as the largest restaurant chain in the United States. Subway operated 148 more restaurants than McDonald's, and together the two chains had 26,346 restaurants. How many restaurants did each company operate? (*Source: USA Today,* February 4, 2002.)

26. Two of the most popular magazines in the United States are *Modern Maturity* and *Reader's Digest.* Together, the average total circulation for these two magazines during a recent six-month period was 35.6 million. *Reader's Digest*'s circulation was 5.4 million less than that of *Modern Maturity.* What were the circulation figures for each magazine? (*Source:* Audit Bureau of Circulations and Magazine Publishers of America.)

27. The two tallest buildings in Los Angeles are the U.S. Bank Tower and the Aon Center. The first of these is 11 stories taller than the second. Together, the buildings have 135 stories. How many stories tall is each building? (*Source: Los Angeles Almanac.*)

28. In the 2004 presidential election, George W. Bush received 35 more electoral votes than John Kerry. Together, the two candidates received a total of 537 electoral votes. How many electoral votes did each receive? (*Source: World Almanac and Book of Facts 2006.*)

29. The perimeter of a rectangle is 90 m. Its length is $1\frac{1}{2}$ times its width. Find the length and width of the rectangle.

30. Teresa Hodge has 20 bills, all of which are $10 or $20 bills. The total value of the money is $330. How many of each denomination does she have?

31. Christie Heinrich has candy that sells for $1.30 per lb, to be mixed with candy selling for $0.90 per lb to get 100 lb of a mix that will sell for $1 per lb. How much of each type should she use?

32. A 40% antifreeze solution is to be mixed with a 70% solution to get 90 L of a 50% solution. How many liters of the 40% and 70% solutions will be needed?

Number of Liters	Percent (as a decimal)	Amount of Pure Antifreeze
x	0.40	
y	0.70	
90	0.50	

33. Nancy Johnson invested $18,000. Part of it was invested at 3% annual simple interest, and the rest was invested at 4%. Her interest income for the first year was $650. How much did she invest at each rate?

Amount of Principal	Rate	Interest
x	0.03	
y	0.04	
$18,000	✕✕✕	

34. A certain plane flying with the wind travels 540 mi in 2 hr. Later, flying against the same wind, the plane travels 690 mi in 3 hr. Find the speed of the plane in still air and the speed of the wind.

[4.5] *Graph the solution set of each system of linear inequalities.*

35. $x + y \geq 2$
  $x - y \leq 4$

36. $y \geq 2x$
  $2x + 3y \leq 6$

37. $x + y < 3$
  $2x > y$

38. $3x - y \leq 3$
  $x \geq -2$
  $y \leq 2$

MIXED REVIEW EXERCISES

Solve each problem.

39. Letarsha Lincoln compared the monthly payments she would incur for two types of mortgages: fixed rate and variable rate. Her observations led to the graph shown.

  (a) For which years would the monthly payment be more for the fixed rate mortgage than for the variable rate mortgage?

  (b) In what year would the payments be the same, and what would those payments be?

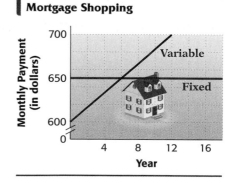

Mortgage Shopping

40. $\dfrac{2x}{3} + \dfrac{y}{4} = \dfrac{14}{3}$
  $\dfrac{x}{2} + \dfrac{y}{12} = \dfrac{8}{3}$

41. $x = y + 6$
  $2y - 2x = -12$

42. $3x + 4y = 6$
  $4x - 5y = 8$

43. $\dfrac{3x}{2} + \dfrac{y}{5} = -3$
  $4x + \dfrac{y}{3} = -11$

44. $x + y < 5$
  $x - y \geq 2$

45. $y \leq 2x$
  $x + 2y > 4$

46. The perimeter of an isosceles triangle measures 29 in. One side of the triangle is 5 in. longer than each of the two equal sides. Find the lengths of the sides of the triangle.

$y = x + 5$

47. In 2004, a total of 9.0 million people visited the Statue of Liberty and the National World War II Memorial, two popular tourist attractions. The Statue of Liberty had 1.8 million fewer visitors than the National World War II Memorial. How many visitors did each of these attractions have? (*Source:* National Park Service, Department of the Interior.)

48. Two cars leave from the same place and travel in opposite directions. One car travels 30 mph faster than the other. After $2\frac{1}{2}$ hr, they are 265 mi apart. What are the rates of the cars?

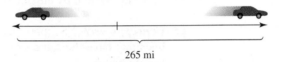

265 mi

49. A hospital bought a total of 146 bottles of glucose solution. Small bottles cost $2 each, and large ones cost $3 each. The total cost was $336. How many of each size of bottle were bought?

50. Can the problem in Exercise 46 be solved with the use of a single variable? If so, explain how it can be done, and then solve it.

Chapter **4** TEST

View the complete solutions to all Chapter Test exercises on the Pass the Test CD.

1. The graph shows a company's costs of producing computer parts and the revenue from the sale of those parts.

   (a) At what production level does the cost equal the revenue?

   (b) What is the revenue at that point?

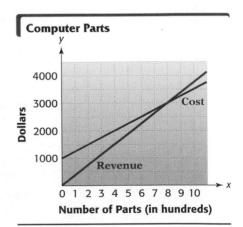

2. Decide whether each ordered pair is a solution of the system.
$$\begin{aligned} 2x + y &= -3 \\ x - y &= -9 \end{aligned}$$

   (a) $(1, -5)$ **(b)** $(1, 10)$ **(c)** $(-4, 5)$

3. Solve the system by graphing. $\quad \begin{aligned} x + 2y &= 6 \\ -2x + y &= -7 \end{aligned}$

Solve each system by substitution.

4. $2x + y = -4$
   $x = y + 7$

5. $4x + 3y = -35$
   $x + y = 0$

Solve each system by elimination.

6. $2x - y = 4$
   $3x + y = 21$

7. $4x + 2y = 2$
   $5x + 4y = 7$

8. $3x + 4y = 9$
   $2x + 5y = 13$

9. $4x + 5y = 2$
   $-8x - 10y = 6$

10. $6x - 5y = 0$
   $-2x + 3y = 0$

11. $\dfrac{6}{5}x - \dfrac{1}{3}y = -20$
   $-\dfrac{2}{3}x + \dfrac{1}{6}y = 11$

12. Solve the system by any method. $\quad \begin{aligned} 4y &= -3x + 5 \\ 6x &= -8y + 10 \end{aligned}$

13. *Concept Check*   Suppose that the graph of a system of two linear equations consists of lines that have the same slope, but different *y*-intercepts. How many solutions does the system have?

Solve each problem by using a system of equations.

14. The distance between Memphis and Atlanta is 782 mi less than the distance between Minneapolis and Houston. Together, the two distances total 1570 mi. How far is it between Memphis and Atlanta? How far is it between Minneapolis and Houston?

15. In 2004, the two most popular amusement parks in the United States were Disneyland and the Magic Kingdom at Walt Disney World. Disneyland had 1.8 million fewer visitors than the Magic Kingdom, and together they had 28.6 million visitors. How many visitors did each park have? (*Source: Amusement Business.*)

16. Susan Grody has a 25% solution of alcohol to mix with a 40% solution to get 50 L of a final mixture that is 30% alcohol. How much of each of the original solutions should she use?

17. Two cars leave from Perham, Minnesota, at the same time and travel in the same direction. One car travels one and one-third times as fast as the other. After 3 hr, they are 45 mi apart. What are the speeds of the cars?

Graph the solution set of each system of inequalities.

18. $2x + 7y \le 14$
   $x - y \ge 1$

19. $2x - y > 6$
   $4y + 12 \ge -3x$

20. *Concept Check*   Without actually graphing, determine which one of the following systems of inequalities has no solution.

A. $x \ge 4$
   $y \le 3$

B. $x + y > 4$
   $x + y < 3$

C. $x > 2$
   $y < 1$

D. $x + y > 4$
   $x - y < 3$

Chapters 1–4 CUMULATIVE REVIEW EXERCISES

1. List all integer factors of 40.

2. Find the value of the expression $\dfrac{3x^2 + 2y^2}{10y + 3}$ if $x = 1$ and $y = 5$.

Name the property that justifies each statement.

3. $5 + (-4) = (-4) + 5$ **4.** $r(s - k) = rs - rk$ **5.** $-\dfrac{2}{3} + \dfrac{2}{3} = 0$

6. Evaluate $-2 + 6[3 - (4 - 9)]$.

Solve each linear equation.

7. $2 - 3(6x + 2) = 4(x + 1) + 18$ **8.** $\dfrac{3}{2}\left(\dfrac{1}{3}x + 4\right) = 6\left(\dfrac{1}{4} + x\right)$

9. Solve the formula $P = \dfrac{kT}{V}$ for T.

Solve each linear inequality.

10. $-\dfrac{5}{6}x < 15$ **11.** $-8 < 2x + 3$

12. A survey measured public recognition of some classic advertising slogans. Complete the results shown in the table if 2500 people were surveyed.

Slogan (product or company)	Percent Recognition (nearest tenth of a percent)	Actual Number That Recognized Slogan (nearest whole number)
Please Don't Squeeze the . . . (Charmin®)	80.4%	
The Breakfast of Champions (Wheaties)	72.5%	
The King of Beers (Budweiser®)		1570
Like a Good Neighbor (State Farm)		1430

(Other slogans included "You're in Good Hands" (Allstate), "Snap, Crackle, Pop" (Rice Krispies®), and "The Un-Cola" (7-Up).)
Source: Department of Integrated Marketing Communications, Northwestern University.

Solve each problem.

13. On September 29, 2005, the U.S. Senate confirmed John G. Roberts, Jr., as the 17th Chief Justice of the United States Supreme Court. With all 100 senators voting, 56 more voted in favor of his confirmation than voted against it. How many senators voted each way? (*Source:* www.pbs.org)

14. Two angles of a triangle have the same measure. The measure of the third angle is 4° less than twice the measure of each of the equal angles. Find the measures of the three angles.

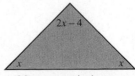

Measures are in degrees.

15. No baseball fan should be without a copy of *The Sports Encyclopedia: Baseball 2004*, by David S. Neft and Richard M. Cohen. The book provides the history of every player, team, and season from 1902 to 2003 and includes exhaustive statistics. The book has a perimeter of 37.8 in., and its width measures 2.58 in. less than its length. What are its dimensions? (*Source:* www.amazon.com)

Graph each linear equation.

16. $x - y = 4$

17. $3x + y = 6$

Find the slope of each line.

18. Through $(-5, 6)$ and $(1, -2)$

19. Perpendicular to the line $y = 4x - 3$

Find an equation for each line. Write it in slope–intercept form.

20. Through $(-4, 1)$ with slope $\frac{1}{2}$

21. Through the points $(1, 3)$ and $(-2, -3)$

22. **(a)** Write an equation of the vertical line through $(9, -2)$.
    (b) Write an equation of the horizontal line through $(4, -1)$.

Solve each system by any method.

23. $\begin{aligned} 2x - y &= -8 \\ x + 2y &= 11 \end{aligned}$

24. $\begin{aligned} 4x + 5y &= -8 \\ 3x + 4y &= -7 \end{aligned}$

25. $\begin{aligned} 3x + 5y &= 1 \\ x &= y + 3 \end{aligned}$

26. $\begin{aligned} 3x + 4y &= 2 \\ 6x + 8y &= 1 \end{aligned}$

Use a system of equations to solve each problem.

27. Admission prices at a football game were $6 for adults and $2 for children. The total value of the tickets sold was $2528, and 454 tickets were sold. How many adult and how many child tickets were sold?

Kind of Ticket	Number Sold	Cost of Each (in dollars)	Total Value (in dollars)
Adult	x	6	$6x$
Child	y		
Total	454		

28. The perimeter of a triangle measures 53 in. If two sides are of equal length and the third side measures 4 in. less than each of the equal sides, what are the lengths of the three sides?

29. A chemist needs 12 L of a 40% alcohol solution. She must mix a 20% solution and a 50% solution. How many liters of each will be required to obtain what she needs?

30. Graph the solution set of the system. $\begin{aligned} x + 2y &\le 12 \\ 2x - y &\le 8 \end{aligned}$

Exponents and Polynomials

Just how much is a *trillion*? A trillion, written 1,000,000,000,000, is a million million, or a thousand billion. A trillion seconds would last more than 31,000 years—that is, 310 centuries. The U.S. budget first exceeded $1 trillion in 1987 and topped $2 trillion for the 2003 fiscal year. (*Source: The Gazette,* Cedar Rapids, Iowa, February 5, 2002.)

In Section 5.3, we use *exponents* and *scientific notation* to write and calculate with large numbers, such as the national debt and health care expenditures.

5.1 The Product Rule and Power Rules for Exponents

OBJECTIVE 1 **Use exponents.** Recall from **Section 1.2** that in the expression 5^2, the number 5 is the *base* and 2 is the *exponent* or *power.* The expression 5^2 is called an *exponential expression.* Although we do not usually write the exponent when it is 1, in general, for any quantity a, $a^1 = a.$

EXAMPLE 1 **Using Exponents**

Write $3 \cdot 3 \cdot 3 \cdot 3$ in exponential form and evaluate.

Since 3 occurs as a factor four times, the base is 3 and the exponent is 4. The exponential expression is 3^4, read "3 to the fourth power" or simply "3 to the fourth."

$$\underbrace{3 \cdot 3 \cdot 3 \cdot 3}_{\text{4 factors of 3}} = 3^4 = 81$$

✔ Now Try Exercise 7.

EXAMPLE 2 **Evaluating Exponential Expressions**

Evaluate. Name the base and the exponent.

(a) $5^4 = 5 \cdot 5 \cdot 5 \cdot 5 = 625$

> Be careful!
> The base is 5.

(b) $-5^4 = -1 \cdot 5^4 = -1 \cdot (5 \cdot 5 \cdot 5 \cdot 5) = -625$

(c) $(-5)^4 = (-5)(-5)(-5)(-5) = 625$

Expression	Base	Exponent
5^4	5	4
-5^4	5	4
$(-5)^4$	-5	4

✔ Now Try Exercises 15 and 17.

▶ **CAUTION** Note the differences between parts (b) and (c) of Example 2. In -5^4, the absence of parentheses shows that the exponent 4 applies only to the base 5, not -5. In $(-5)^4$, the parentheses show that the exponent 4 applies to the base -5. In summary, $-a^n$ and $(-a)^n$ are not necessarily the same.

Expression	Base	Exponent	Example
$-a^n$	a	n	$-3^2 = -(3 \cdot 3) = -9$
$(-a)^n$	$-a$	n	$(-3)^2 = (-3)(-3) = 9$

OBJECTIVE 2 **Use the product rule for exponents.** By the definition of exponents,

$$2^4 \cdot 2^3 = \overbrace{(2 \cdot 2 \cdot 2 \cdot 2)}^{\text{4 factors}}\overbrace{(2 \cdot 2 \cdot 2)}^{\text{3 factors}}$$

$$= \underbrace{2 \cdot 2 \cdot 2 \cdot 2 \cdot 2 \cdot 2 \cdot 2}_{4 + 3 = 7 \text{ factors}}$$

$$= 2^7.$$

Also,
$$6^2 \cdot 6^3 = (6 \cdot 6)(6 \cdot 6 \cdot 6)$$
$$= 6 \cdot 6 \cdot 6 \cdot 6 \cdot 6$$
$$= 6^5.$$

Generalizing from these examples,

$$2^4 \cdot 2^3 = 2^{4+3} = 2^7 \quad \text{and} \quad 6^2 \cdot 6^3 = 6^{2+3} = 6^5,$$

suggests the **product rule for exponents.**

Product Rule for Exponents

For any positive integers m and n, $\quad a^m \cdot a^n = a^{m+n}.$
(Keep the same base; add the exponents.)
Example: $6^2 \cdot 6^5 = 6^{2+5} = 6^7$

▶ **CAUTION** Do not multiply the bases when using the product rule. *Keep the same base and add the exponents.* For example,
$$6^2 \cdot 6^5 = 6^7, \quad \text{not} \quad 36^7.$$

EXAMPLE 3 Using the Product Rule

Use the product rule for exponents to find each product if possible.

(a) $6^3 \cdot 6^5 = 6^{3+5} = 6^8$ **(b)** $(-4)^7(-4)^2 = (-4)^{7+2} = (-4)^9$

Keep the same base.

(c) $x^2 \cdot x = x^2 \cdot x^1 = x^{2+1} = x^3$ **(d)** $m^4 m^3 m^5 = m^{4+3+5} = m^{12}$

(e) $2^3 \cdot 3^2$

The product rule does not apply, since the bases are different.

$$2^3 \cdot 3^2 = 8 \cdot 9 = 72 \qquad \text{Evaluate } 2^3 \text{ and } 3^2; \text{ then multiply.}$$

Think: $3^2 = 3 \cdot 3$

Think: $2^3 = 2 \cdot 2 \cdot 2$

(f) $2^3 + 2^4$

The product rule does not apply to $2^3 + 2^4$, since this is a *sum*, not a *product*.

$$2^3 + 2^4 = 8 + 16 = 24 \qquad \text{Evaluate } 2^3 \text{ and } 2^4; \text{ then add.}$$

(g) $(2x^3)(3x^7)$

Since $2x^3$ means $2 \cdot x^3$ and $3x^7$ means $3 \cdot x^7$,

$$(2x^3)(3x^7) = (2 \cdot 3) \cdot (x^3 \cdot x^7) \qquad \text{Commutative and associative properties}$$
$$= 6x^{3+7} \qquad \text{Multiply; product rule}$$
$$= 6x^{10}$$

✔ **Now Try Exercises 25, 29, 31, 35, and 39.**

▶ **CAUTION** Be sure you understand the difference between *adding* and *multiplying* exponential expressions. For example,

$$8x^3 + 5x^3 = (8 + 5)x^3 = 13x^3,$$

but

$$(8x^3)(5x^3) = (8 \cdot 5)x^{3+3} = 40x^6.$$

OBJECTIVE 3 Use the rule $(a^m)^n = a^{mn}$. We can simplify an expression such as $(8^3)^2$ with the product rule for exponents.

$$(8^3)^2 = (8^3)(8^3) = 8^{3+3} = 8^6$$

The exponents in $(8^3)^2$ are multiplied to give the exponent in 8^6. As another example,

$$
\begin{aligned}
(5^2)^4 &= 5^2 \cdot 5^2 \cdot 5^2 \cdot 5^2 && \text{Definition of exponent} \\
&= 5^{2+2+2+2} && \text{Product rule} \\
&= 5^8
\end{aligned}
$$

and $2 \cdot 4 = 8$. These examples suggest **power rule (a) for exponents.**

Power Rule (a) for Exponents

For any positive integers m and n, $(a^m)^n = a^{mn}$.
(Raise a power to a power by multiplying exponents.)
Example: $(3^2)^4 = 3^{2 \cdot 4} = 3^8$

EXAMPLE 4 Using Power Rule (a)

Use power rule (a) for exponents to simplify.

(a) $(2^5)^3 = 2^{5 \cdot 3} = 2^{15}$ **(b)** $(5^7)^2 = 5^{7(2)} = 5^{14}$ **(c)** $(x^2)^5 = x^{2(5)} = x^{10}$

✔ **Now Try Exercises 43 and 45.**

OBJECTIVE 4 Use the rule $(ab)^m = a^m b^m$. We can rewrite the expression $(4x)^3$ as follows.

$$
\begin{aligned}
(4x)^3 &= (4x)(4x)(4x) && \text{Definition of exponent} \\
&= (4 \cdot 4 \cdot 4)(x \cdot x \cdot x) && \text{Commutative and associative properties} \\
&= 4^3 \cdot x^3 && \text{Definition of exponent}
\end{aligned}
$$

This example suggests **power rule (b) for exponents.**

Power Rule (b) for Exponents

For any positive integer m, $(ab)^m = a^m b^m$.
(Raise a product to a power by raising each factor to the power.)
Example: $(2p)^5 = 2^5 p^5$

EXAMPLE 5 Using Power Rule (b)

Use power rule (b) for exponents to simplify.

(a) $(3xy)^2 = 3^2x^2y^2$ Power rule (b)

$\quad\quad\quad\quad = 9x^2y^2$ $3^2 = 3 \cdot 3 = 9$

(b) $5(pq)^2 = 5(p^2q^2)$ Power rule (b)

$\quad\quad\quad\quad = 5p^2q^2$ Multiply.

(c) $3(2m^2p^3)^4 = 3[2^4(m^2)^4(p^3)^4]$ Power rule (b)

$\quad\quad\quad\quad\quad\quad = 3 \cdot 2^4 m^8 p^{12}$ Power rule (a)

$2^4 = 2 \cdot 2 \cdot 2 \cdot 2$

$\quad\quad\quad\quad\quad\quad = 48m^8p^{12}$ $3 \cdot 2^4 = 3 \cdot 16 = 48$

(d) $(-5^6)^3 = (-1 \cdot 5^6)^3$ $-a = -1 \cdot a$

$\quad\quad\quad\quad = (-1)^3 \cdot (5^6)^3$ Power rule (b)

Raise -1 $= -1 \cdot 5^{18}$ Power rule (a)

to the $= -5^{18}$

designated
power.

✔ **Now Try Exercises 49 and 53.**

▶ **CAUTION** *Power rule (b) does not apply to a sum.* For example,

$$(4x)^2 = 4^2x^2, \quad \text{but} \quad (4 + x)^2 \neq 4^2 + x^2.$$

OBJECTIVE 5 Use the rule $\left(\frac{a}{b}\right)^m = \frac{a^m}{b^m}$. Since the quotient $\frac{a}{b}$ can be written as $a\left(\frac{1}{b}\right)$, we use this fact and power rule (b) to get **power rule (c) for exponents.**

Power Rule (c) for Exponents

For any positive integer m, $\left(\dfrac{a}{b}\right)^m = \dfrac{a^m}{b^m}$ $(b \neq 0)$.

(Raise a quotient to a power by raising both numerator and denominator to the power.)

Example: $\left(\dfrac{5}{3}\right)^2 = \dfrac{5^2}{3^2}$

EXAMPLE 6 Using Power Rule (c)

Use power rule (c) for exponents to simplify.

(a) $\left(\dfrac{2}{3}\right)^5 = \dfrac{2^5}{3^5} = \dfrac{32}{243}$ **(b)** $\left(\dfrac{m}{n}\right)^3 = \dfrac{m^3}{n^3}$ $(n \neq 0)$

(c) $\left(\dfrac{1}{5}\right)^4 = \dfrac{1^4}{5^4} = \dfrac{1}{5^4} = \dfrac{1}{625}$ $1^4 = 1 \cdot 1 \cdot 1 \cdot 1 = 1$

✔ **Now Try Exercises 59 and 61.**

▶ **NOTE** In Example 6(c), we used the fact that $1^4 = 1$. *In general,* $1^n = 1$, *for any integer n.*

The rules for exponents discussed in this section are summarized in the box.

Rules for Exponents

For positive integers m and n,

		Examples
Product rule	$a^m \cdot a^n = a^{m+n}$	$6^2 \cdot 6^5 = 6^{2+5} = 6^7$
Power rules (a)	$(a^m)^n = a^{mn}$	$(3^2)^4 = 3^{2 \cdot 4} = 3^8$
(b)	$(ab)^m = a^m b^m$	$(2p)^5 = 2^5 p^5$
(c)	$\left(\dfrac{a}{b}\right)^m = \dfrac{a^m}{b^m} \quad (b \neq 0).$	$\left(\dfrac{5}{3}\right)^2 = \dfrac{5^2}{3^2}$

OBJECTIVE 6 Use combinations of rules.

EXAMPLE 7 Using Combinations of Rules

Simplify.

(a)
$$\left(\frac{2}{3}\right)^2 \cdot 2^3 = \frac{2^2}{3^2} \cdot \frac{2^3}{1} \qquad \text{Power rule (c)}$$

$$= \frac{2^2 \cdot 2^3}{3^2 \cdot 1} \qquad \text{Multiply fractions.}$$

$$= \frac{2^{2+3}}{3^2} \qquad \text{Product rule}$$

$$= \frac{2^5}{3^2} = \frac{32}{9}$$

(b)
$$(5x)^3(5x)^4 = (5x)^7 \qquad \text{Product rule}$$
$$= 5^7 x^7 \qquad \text{Power rule (b)}$$

(c)
$$(2x^2y^3)^4(3xy^2)^3 = 2^4(x^2)^4(y^3)^4 \cdot 3^3 x^3 (y^2)^3 \qquad \text{Power rule (b)}$$
$$= 2^4 x^8 y^{12} \cdot 3^3 x^3 y^6 \qquad \text{Power rule (a)}$$
$$= 2^4 \cdot 3^3 x^8 x^3 y^{12} y^6 \qquad \text{Commutative and associative properties}$$
$$= 16 \cdot 27 x^{11} y^{18} \qquad \text{Product rule}$$
$$= 432 x^{11} y^{18} \qquad \text{Multiply.}$$

> **Don't forget the factor −1.**

(d)
$$(-x^3y)^2(-x^5y^4)^3 = (-1 \cdot x^3y)^2(-1 \cdot x^5y^4)^3 \qquad -a = -1 \cdot a$$
$$= (-1)^2(x^3)^2y^2 \cdot (-1)^3(x^5)^3(y^4)^3 \qquad \text{Power rule (b)}$$
$$= (-1)^2(x^6)(y^2)(-1)^3(x^{15})(y^{12}) \qquad \text{Power rule (a)}$$
$$= (-1)^5(x^{21})(y^{14}) \qquad \text{Product rule}$$
$$= -x^{21}y^{14}$$

✔ **Now Try Exercises 65, 67, 75, and 77.**

▶ **CAUTION** In Example 7(c), do not multiply the coefficient 2 and the exponent 4.

$$(2x^2y^3)^4 = 2^4 x^{2 \cdot 4} y^{3 \cdot 4}, \quad \text{not} \quad (2 \cdot 4) x^{2 \cdot 4} y^{3 \cdot 4}.$$

OBJECTIVE 7 Use the rules for exponents in a geometry application.

EXAMPLE 8 Using Area Formulas

Find an expression that represents the area of each geometric figure.

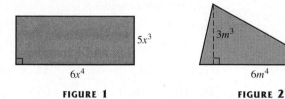

FIGURE 1   **FIGURE 2**

For Figure 1, use the formula for the area of a rectangle, $A = LW$.

$$A = (6x^4)(5x^3) \qquad \text{Area formula}$$
$$A = 6 \cdot 5 \cdot x^{4+3} \qquad \text{Product rule}$$
$$A = 30x^7$$

Figure 2 is a triangle with base $6m^4$ and height $3m^3$. Substitute these expressions into the formula for the area of a triangle, and simplify.

$$A = \frac{1}{2}bh = \frac{1}{2}(6m^4)(3m^3) = \frac{1}{2}(18m^7) = 9m^7$$

✔ **Now Try Exercise 83.**

5.1 EXERCISES

Complete solution available on Video Lectures on CD/DVD

Now Try Exercise

Concept Check *Decide whether each statement is* true *or* false. *If false, tell why.*

1. $3^3 = 9$ **2.** $(-2)^4 = 2^4$ **3.** $(a^2)^3 = a^5$ **4.** $\left(\frac{1}{4}\right)^2 = \frac{1}{4^2}$

Write each expression by using exponents. See Example 1.

5. $w \cdot w \cdot w \cdot w \cdot w \cdot w$

6. $t \cdot t \cdot t \cdot t \cdot t \cdot t \cdot t$

7. $\left(\frac{1}{2}\right)\left(\frac{1}{2}\right)\left(\frac{1}{2}\right)\left(\frac{1}{2}\right)\left(\frac{1}{2}\right)\left(\frac{1}{2}\right)$

8. $\left(-\frac{1}{4}\right)\left(-\frac{1}{4}\right)\left(-\frac{1}{4}\right)\left(-\frac{1}{4}\right)\left(-\frac{1}{4}\right)$

9. $(-4)(-4)(-4)$

10. $(-3)(-3)(-3)(-3)$

11. $(-7x)(-7x)(-7x)(-7x)$

12. $(-8p)(-8p)$

13. Explain how the expressions $(-3)^4$ and -3^4 are different.

14. Explain how the expressions $(5x)^3$ and $5x^3$ are different.

Identify the base and the exponent for each exponential expression. In Exercises 15–18, also evaluate each expression. See Example 2.

15. 3^5 **16.** 2^7 **17.** $(-3)^5$ **18.** $(-2)^7$

19. $(-6x)^4$ **20.** $(-8x)^4$ **21.** $-6x^4$ **22.** $-8x^4$

23. Explain why the product rule does not apply to the expression $5^2 + 5^3$. Then evaluate the expression by finding the individual powers and adding the results.

24. Repeat Exercise 23 for the expression $(-4)^3 + (-4)^4$.

Use the product rule, if possible, to simplify each expression. Write each answer in exponential form. See Example 3.

25. $5^2 \cdot 5^6$ **26.** $3^6 \cdot 3^7$ **27.** $4^2 \cdot 4^7 \cdot 4^3$

28. $5^3 \cdot 5^8 \cdot 5^2$ **29.** $(-7)^3(-7)^6$ **30.** $(-9)^8(-9)^5$

31. $t^3 \cdot t^8 \cdot t^{13}$ **32.** $n^5 \cdot n^6 \cdot n^9$ **33.** $(-8r^4)(7r^3)$

34. $(10a^7)(-4a^3)$ **35.** $(-6p^5)(-7p^5)$ **36.** $(-5w^8)(-9w^8)$

37. $(5x^2)(-2x^3)(3x^4)$ **38.** $(12y^3)(4y)(-3y^5)$ **39.** $3^8 + 3^9$

40. $4^{12} + 4^5$ **41.** $5^8 \cdot 3^9$ **42.** $6^3 \cdot 8^9$

Use the power rules for exponents to simplify each expression. Write each answer in exponential form. See Examples 4–6.

43. $(4^3)^2$ **44.** $(8^3)^6$ **45.** $(t^4)^5$

46. $(y^6)^5$ **47.** $(7r)^3$ **48.** $(11x)^4$

49. $(5xy)^5$ **50.** $(9pq)^6$ **51.** $(-5^2)^6$

52. $(-9^4)^8$ **53.** $(-8^3)^5$ **54.** $(-7^5)^7$

55. $8(qr)^3$ **56.** $4(vw)^5$ **57.** $\left(\dfrac{9}{5}\right)^8$

58. $\left(\dfrac{12}{7}\right)^3$ **59.** $\left(\dfrac{1}{2}\right)^3$ **60.** $\left(\dfrac{1}{3}\right)^5$

61. $\left(\dfrac{a}{b}\right)^3$ $(b \neq 0)$ **62.** $\left(\dfrac{r}{t}\right)^4$ $(t \neq 0)$

Simplify each expression. See Example 7.

63. $\left(\dfrac{5}{2}\right)^3 \cdot \left(\dfrac{5}{2}\right)^2$ **64.** $\left(\dfrac{3}{4}\right)^5 \cdot \left(\dfrac{3}{4}\right)^6$ **65.** $\left(\dfrac{9}{8}\right)^3 \cdot 9^2$

66. $\left(\dfrac{8}{5}\right)^4 \cdot 8^3$ **67.** $(2x)^9(2x)^3$ **68.** $(6y)^5(6y)^8$

69. $(-6p)^4(-6p)$ **70.** $(-13q)^3(-13q)$ **71.** $(6x^2y^3)^5$

72. $(5r^5t^6)^7$ **73.** $(x^2)^3(x^3)^5$ **74.** $(y^4)^5(y^3)^5$

75. $(2w^2x^3y)^2(x^4y)^5$ **76.** $(3x^4y^2z)^3(yz^4)^5$ **77.** $(-r^4s)^2(-r^2s^3)^5$

78. $(-ts^6)^4(-t^3s^5)^3$ **79.** $\left(\dfrac{5a^2b^5}{c^6}\right)^3$ $(c \neq 0)$ **80.** $\left(\dfrac{6x^3y^9}{z^5}\right)^4$ $(z \neq 0)$

81. *Concept Check* A student simplified $(10^2)^3$ as 1000^6. ***WHAT WENT WRONG?***

82. Explain why $(3x^2y^3)^4$ is *not* equivalent to $(3 \cdot 4)x^8y^{12}$.

Find an expression that represents the area of each figure. See Example 8. (If necessary, refer to the formulas on the inside covers. The ⌐ in the figures indicate 90° right angles.)

83.

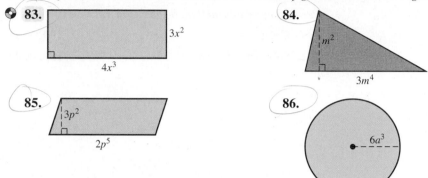

$3x^2$

$4x^3$

84.

m^2

$3m^4$

85.

$3p^2$

$2p^5$

86.

$6a^3$

Find an expression that represents the volume of each figure. (If necessary, refer to the formulas on the inside covers.)

87.

$5x^2$

$5x^2$

$5x^2$

88.

$9xy^3$

$5x^3y$

$4x^2y^4$

89. Assume that a is a number greater than 1. Arrange the following terms in order from least to greatest: $-(-a)^3$, $-a^3$, $(-a)^4$, $-a^4$. Explain how you decided on the order.

90. Devise a rule that tells whether an exponential expression with a negative base is positive or negative.

In **Chapter 2,** we used the formula for simple interest, $I = prt$, which deals with interest paid only on the principal. With **compound interest,** interest is paid on the principal and the interest earned earlier. The formula for compound interest, which involves an exponential expression, is

$$A = P(1 + r)^n,$$

where A is the amount accumulated from a principal of P dollars left untouched for n years with an annual interest rate r (expressed as a decimal).

In Exercises 91–94, use the preceding formula and a calculator to find A to the nearest cent.

91. $P = \$250$, $r = 0.04$, $n = 5$

92. $P = \$400$, $r = 0.04$, $n = 3$

93. $P = \$1500$, $r = 0.035$, $n = 6$

94. $P = \$2000$, $r = 0.025$, $n = 4$

PREVIEW EXERCISES

*Give the reciprocal of each number. See **Section 1.1**.*

95. 5   **96.** -2 **97.** $-\dfrac{1}{4}$ **98.** 0.5

*Perform each subtraction. See **Section 1.5**.*

99. $8 - (-2)$ **100.** $-2 - 8$

101. Subtract -5 from -2.   **102.** Subtract -2 from -5.

5.2   Integer Exponents and the Quotient Rule

OBJECTIVES

1 Use 0 as an exponent.

2 Use negative numbers as exponents.

3 Use the quotient rule for exponents.

4 Use combinations of rules.

In all our earlier work, exponents were positive integers. Now we want to develop a meaning for exponents that are not positive integers. Consider the following list:

$$2^4 = 16$$
$$2^3 = 8$$
$$2^2 = 4.$$

Do you see the pattern in the values? Each time we reduce the exponent by 1, the value is divided by 2 (the base). Using this pattern, we can continue the list to smaller and smaller integer exponents.

$$2^1 = 2$$
$$2^0 = 1$$
$$2^{-1} = \frac{1}{2}$$
$$2^{-2} = \frac{1}{4}$$
$$2^{-3} = \frac{1}{8}$$

From the preceding list, it appears that we should define 2^0 as 1 and negative exponents as reciprocals.

OBJECTIVE 1   Use 0 as an exponent.   The definitions of 0 and negative exponents must satisfy the rules for exponents from **Section 5.1.** For example, if $6^0 = 1$, then

$$6^0 \cdot 6^2 = 1 \cdot 6^2 = 6^2 \qquad \text{and} \qquad 6^0 \cdot 6^2 = 6^{0+2} = 6^2,$$

so that the product rule is satisfied. Check that the power rules are also valid for a 0 exponent. Thus, we define a 0 exponent as follows.

Zero Exponent

For any nonzero real number a, $\quad a^0 = 1.$

Example: $17^0 = 1$

EXAMPLE 1 Using Zero Exponents

Evaluate.

(a) $60^0 = 1$ **(b)** $(-60)^0 = 1$

(c) $-60^0 = -(1) = -1$ **(d)** $y^0 = 1 \quad (y \neq 0)$

(e) $6y^0 = 6(1) = 6 \quad (y \neq 0)$ **(f)** $(6y)^0 = 1 \quad (y \neq 0)$

✔ **Now Try Exercises 1 and 5.**

▶ **CAUTION** Look again at Example 1. In part (b), the base is -60, and since any nonzero base raised to the 0 exponent is 1, $(-60)^0 = 1$. In part (c), which could be written $-(60)^0$, the base is 60, so $-60^0 = -1$.

OBJECTIVE 2 Use negative numbers as exponents. From the lists at the beginning of this section, since $2^{-2} = \frac{1}{4}$ and $2^{-3} = \frac{1}{8}$, we can deduce that 2^{-n} should equal $\frac{1}{2^n}$. Is the product rule valid in such cases? For example, if we multiply 6^{-2} by 6^2, we get

$$6^{-2} \cdot 6^2 = 6^{-2+2} = 6^0 = 1.$$

The expression 6^{-2} behaves as if it were the reciprocal of 6^2: Their product is 1. The reciprocal of 6^2 is also $\frac{1}{6^2}$, leading us to define 6^{-2} as $\frac{1}{6^2}$. This is a particular case of the definition of negative exponents.

Negative Exponents

For any nonzero real number a and any integer n, $\quad a^{-n} = \dfrac{1}{a^n}.$

Example: $3^{-2} = \dfrac{1}{3^2}$

By definition, a^{-n} and a^n are reciprocals, since

$$a^n \cdot a^{-n} = a^n \cdot \frac{1}{a^n} = 1.$$

Because $1^n = 1$, the definition of a^{-n} can also be written

$$a^{-n} = \frac{1}{a^n} = \frac{1^n}{a^n} = \left(\frac{1}{a}\right)^n.$$

For example, $\qquad 6^{-3} = \left(\dfrac{1}{6}\right)^3 \quad$ and $\quad \left(\dfrac{1}{3}\right)^{-2} = 3^2.$

EXAMPLE 2 Using Negative Exponents

Simplify by writing with positive exponents.

(a) $3^{-2} = \dfrac{1}{3^2} = \dfrac{1}{9}$

(b) $5^{-3} = \dfrac{1}{5^3} = \dfrac{1}{125}$

(c) $\left(\dfrac{1}{2}\right)^{-3} = 2^3 = 8$ $\frac{1}{2}$ and 2 are reciprocals.

Notice that we can change the base to its reciprocal if we also change the sign of the exponent.

(d) $\left(\dfrac{2}{5}\right)^{-4} = \left(\dfrac{5}{2}\right)^4 = \dfrac{625}{16}$ $\frac{2}{5}$ and $\frac{5}{2}$ are reciprocals.

(e) $\left(\dfrac{4}{3}\right)^{-5} = \left(\dfrac{3}{4}\right)^5 = \dfrac{243}{1024}$

(f) $4^{-1} - 2^{-1} = \dfrac{1}{4} - \dfrac{1}{2} = \dfrac{1}{4} - \dfrac{2}{4} = -\dfrac{1}{4}$ Apply the exponents first, and then subtract.

> Remember to find a common denominator.

(g) $p^{-2} = \dfrac{1}{p^2}$ $(p \neq 0)$

(h) $\dfrac{1}{x^{-4}} = \dfrac{1^{-4}}{x^{-4}}$ $(x \neq 0)$ $1^{-4} = 1$

$\quad\quad = \left(\dfrac{1}{x}\right)^{-4}$ Power rule (c)

$\quad\quad = x^4$ $\frac{1}{x}$ and x are reciprocals.

✔ **Now Try Exercises 17, 19, 21, and 25.**

Consider the following:

$$\dfrac{2^{-3}}{3^{-4}} = \dfrac{\dfrac{1}{2^3}}{\dfrac{1}{3^4}} = \dfrac{1}{2^3} \div \dfrac{1}{3^4} = \dfrac{1}{2^3} \cdot \dfrac{3^4}{1} = \dfrac{3^4}{2^3}.$$ To divide by a fraction, multiply by its reciprocal.

Therefore, $\dfrac{2^{-3}}{3^{-4}} = \dfrac{3^4}{2^3}.$

Changing from Negative to Positive Exponents

For any nonzero numbers a and b and any integers m and n,

$$\dfrac{a^{-m}}{b^{-n}} = \dfrac{b^n}{a^m} \quad \text{and} \quad \left(\dfrac{a}{b}\right)^{-m} = \left(\dfrac{b}{a}\right)^m.$$

Examples: $\dfrac{3^{-5}}{2^{-4}} = \dfrac{2^4}{3^5}$ and $\left(\dfrac{4}{5}\right)^{-3} = \left(\dfrac{5}{4}\right)^3$

EXAMPLE 3 Changing from Negative to Positive Exponents

Simplify by writing with positive exponents. Assume that all variables represent nonzero real numbers.

(a) $\dfrac{4^{-2}}{5^{-3}} = \dfrac{5^3}{4^2} = \dfrac{125}{16}$

(b) $\dfrac{m^{-5}}{p^{-1}} = \dfrac{p^1}{m^5} = \dfrac{p}{m^5}$

(c) $\dfrac{a^{-2}b}{3d^{-3}} = \dfrac{bd^3}{3a^2}$

(d) $x^3y^{-4} = \dfrac{x^3y^{-4}}{1} = \dfrac{x^3}{y^4}$

(e) $\left(\dfrac{x}{2y}\right)^{-4} = \left(\dfrac{2y}{x}\right)^4 = \dfrac{2^4y^4}{x^4} = \dfrac{16y^4}{x^4}$

✔ **Now Try Exercises 31, 43, and 47.**

▶ **CAUTION** Be careful. We cannot use this rule to change negative exponents to positive exponents if the exponents occur in a *sum or difference* of terms. For example,

$$\dfrac{5^{-2} + 3^{-1}}{7 - 2^{-3}} \quad \text{would be written with positive exponents as} \quad \dfrac{\dfrac{1}{5^2} + \dfrac{1}{3}}{7 - \dfrac{1}{2^3}}.$$

OBJECTIVE 3 Use the quotient rule for exponents. How should we handle the quotient of two exponential expressions with the same base? We know that

$$\dfrac{6^5}{6^3} = \dfrac{6 \cdot 6 \cdot 6 \cdot 6 \cdot 6}{6 \cdot 6 \cdot 6} = 6^2.$$

Notice that the difference between the exponents, $5 - 3 = 2$, is the exponent in the quotient. Also,

$$\dfrac{6^2}{6^4} = \dfrac{6 \cdot 6}{6 \cdot 6 \cdot 6 \cdot 6} = \dfrac{1}{6^2} = 6^{-2}.$$

Here, $2 - 4 = -2$. These examples suggest the **quotient rule for exponents.**

Quotient Rule for Exponents

For any nonzero real number a and any integers m and n,

$$\dfrac{a^m}{a^n} = a^{m-n}.$$

(Keep the same base; subtract the exponents.)

Example: $\dfrac{5^8}{5^4} = 5^{8-4} = 5^4$

> ▶ **CAUTION** A common **error** is to write $\frac{5^8}{5^4} = 1^{8-4} = 1^4$. By the quotient rule, the quotient must have the *same base*, 5, so
>
> $$\frac{5^8}{5^4} = 5^{8-4} = 5^4.$$
>
> If you are not sure, use the definition of an exponent to write out the factors:
>
> $$\frac{5^8}{5^4} = \frac{5 \cdot 5 \cdot 5 \cdot 5 \cdot 5 \cdot 5 \cdot 5 \cdot 5}{5 \cdot 5 \cdot 5 \cdot 5} = 5^4.$$

EXAMPLE 4 Using the Quotient Rule

Simplify by writing with positive exponents. Assume that all variables represent nonzero real numbers.

(a) $\dfrac{5^8}{5^6} = 5^{8-6} = 5^2 = 25$

Keep the same base.

(b) $\dfrac{4^2}{4^9} = 4^{2-9} = 4^{-7} = \dfrac{1}{4^7}$

(c) $\dfrac{5^{-3}}{5^{-7}} = 5^{-3-(-7)} = 5^4 = 625$

Be careful with signs.

(d) $\dfrac{q^5}{q^{-3}} = q^{5-(-3)} = q^8$

(e) $\dfrac{3^2 x^5}{3^4 x^3} = \dfrac{3^2}{3^4} \cdot \dfrac{x^5}{x^3} = 3^{2-4} \cdot x^{5-3} = 3^{-2} x^2 = \dfrac{x^2}{3^2} = \dfrac{x^2}{9}$

(f) $\dfrac{(m+n)^{-2}}{(m+n)^{-4}} = (m+n)^{-2-(-4)} = (m+n)^{-2+4} = (m+n)^2 \quad (m \neq -n)$

(g) $\dfrac{7x^{-3}y^2}{2^{-1}x^2y^{-5}} = \dfrac{7 \cdot 2^1 y^2 y^5}{x^2 x^3} = \dfrac{14y^7}{x^5}$

✔ **Now Try Exercises 29, 35, and 49.**

The definitions and rules for exponents are summarized here.

Definitions and Rules for Exponents

For any integers m and n,		**Examples**
Product rule	$a^m \cdot a^n = a^{m+n}$	$7^4 \cdot 7^5 = 7^{4+5} = 7^9$
Zero exponent	$a^0 = 1 \quad (a \neq 0)$	$(-3)^0 = 1$
Negative exponent	$a^{-n} = \dfrac{1}{a^n} \quad (a \neq 0)$	$5^{-3} = \dfrac{1}{5^3}$
Quotient rule	$\dfrac{a^m}{a^n} = a^{m-n} \quad (a \neq 0)$	$\dfrac{2^2}{2^5} = 2^{2-5} = 2^{-3} = \dfrac{1}{2^3}$
Power rule (a)	$(a^m)^n = a^{mn}$	$(4^2)^3 = 4^{2 \cdot 3} = 4^6$
Power rule (b)	$(ab)^m = a^m b^m$	$(3k)^4 = 3^4 k^4$

(continued)

Definitions and Rules for Exponents (continued)

Power rule (c) $\left(\dfrac{a}{b}\right)^m = \dfrac{a^m}{b^m}$ $(b \neq 0)$ $\left(\dfrac{2}{3}\right)^2 = \dfrac{2^2}{3^2}$

Negative-to-positive rules $\dfrac{a^{-m}}{b^{-n}} = \dfrac{b^n}{a^m}$ $(a \neq 0, b \neq 0)$ $\dfrac{2^{-4}}{5^{-3}} = \dfrac{5^3}{2^4}$

$\left(\dfrac{a}{b}\right)^{-m} = \left(\dfrac{b}{a}\right)^m.$ $\left(\dfrac{4}{7}\right)^{-2} = \left(\dfrac{7}{4}\right)^2$

OBJECTIVE 4 Use combinations of rules.

EXAMPLE 5 Using Combinations of Rules

Simplify. Assume that all variables represent nonzero real numbers.

(a) $\dfrac{(4^2)^3}{4^5} = \dfrac{4^6}{4^5}$ Power rule (a)

$= 4^{6-5}$ Quotient rule

$= 4^1$

$= 4$

(b) $(2x)^3(2x)^2 = (2x)^5$ Product rule

$= 2^5 x^5$ Power rule (b)

$= 32x^5$

(c) $\left(\dfrac{2x^3}{5}\right)^{-4} = \left(\dfrac{5}{2x^3}\right)^4$ Negative-to-positive rule

$= \dfrac{5^4}{2^4 x^{12}}$ Power rules (a)–(c)

$= \dfrac{625}{16x^{12}}$

(d) $\left(\dfrac{3x^{-2}}{4^{-1}y^3}\right)^{-3} = \dfrac{3^{-3}x^6}{4^3 y^{-9}}$ Power rules (a)–(c)

$= \dfrac{x^6 y^9}{4^3 \cdot 3^3}$ Negative-to-positive rule

$= \dfrac{x^6 y^9}{1728}$

(e) $\dfrac{(4m)^{-3}}{(3m)^{-4}} = \dfrac{4^{-3}m^{-3}}{3^{-4}m^{-4}}$ Power rule (b)

$= \dfrac{3^4 m^4}{4^3 m^3}$ Negative-to-positive rule

$= \dfrac{3^4 m^{4-3}}{4^3}$ Quotient rule

$= \dfrac{3^4 m}{4^3}$

$= \dfrac{81m}{64}$

✔ **Now Try Exercises 57, 63, 65, and 67.**

5.2 EXERCISES

⊙ *Complete solution available on Video Lectures on CD/DVD*

Now Try Exercise

Decide whether each expression is equal to 0, 1, or −1. See Example 1.

1. 9^0 **2.** 5^0 **3.** $(-4)^0$ **4.** $(-10)^0$

5. -9^0 **6.** -5^0 **7.** $-(-4)^0$ **8.** $-(-10)^0$

9. $(-2)^0 - 2^0$ **10.** $(-8)^0 - 8^0$ **11.** $\dfrac{0^{10}}{10^0}$ **12.** $\dfrac{0^5}{5^0}$

Concept Check   *In Exercises 13 and 14, match each expression in Column I with the equivalent expression in Column II. Choices in Column II may be used once, more than once, or not at all. (In Exercise 13, $x \neq 0$.)*

I	II		I	II
13. (a) x^0	**A.** 0		**14. (a)** -2^{-4}	**A.** 8
(b) $-x^0$	**B.** 1		**(b)** $(-2)^{-4}$	**B.** 16
(c) $3x^0$	**C.** -1		**(c)** 2^{-4}	**C.** $-\dfrac{1}{16}$
(d) $(3x)^0$	**D.** 3		**(d)** $\dfrac{1}{2^{-4}}$	**D.** -8
(e) $-3x^0$	**E.** -3		**(e)** $\dfrac{1}{-2^{-4}}$	**E.** -16
(f) $(-3x)^0$	**F.** $\dfrac{1}{3}$		**(f)** $\dfrac{1}{(-2)^{-4}}$	**F.** $\dfrac{1}{16}$

Evaluate each expression. See Examples 1 and 2.

15. $7^0 + 9^0$ **16.** $8^0 + 6^0$ ⊙ **17.** 4^{-3} **18.** 5^{-4}

19. $\left(\dfrac{1}{2}\right)^{-4}$ **20.** $\left(\dfrac{1}{3}\right)^{-3}$ **21.** $\left(\dfrac{6}{7}\right)^{-2}$ **22.** $\left(\dfrac{2}{3}\right)^{-3}$

23. $(-3)^{-4}$ **24.** $(-4)^{-3}$ **25.** $5^{-1} + 3^{-1}$ **26.** $6^{-1} + 2^{-1}$

27. $3^{-2} - 2^{-1}$ **28.** $6^{-2} - 3^{-1}$

Simplify by writing each expression with positive exponents. Assume that all variables represent nonzero real numbers. See Examples 2–4.

⊙ **29.** $\dfrac{5^8}{5^5}$ **30.** $\dfrac{11^6}{11^3}$ ⊙ **31.** $\dfrac{3^{-2}}{5^{-3}}$ **32.** $\dfrac{4^{-3}}{3^{-2}}$ **33.** $\dfrac{5}{5^{-1}}$

34. $\dfrac{6}{6^{-2}}$ **35.** $\dfrac{x^{12}}{x^{-3}}$ **36.** $\dfrac{y^4}{y^{-6}}$ **37.** $\dfrac{1}{6^{-3}}$ **38.** $\dfrac{1}{5^{-2}}$

39. $\dfrac{2}{r^{-4}}$ **40.** $\dfrac{3}{s^{-8}}$ **41.** $\dfrac{4^{-3}}{5^{-2}}$ **42.** $\dfrac{6^{-2}}{5^{-4}}$ **43.** $p^5 q^{-8}$

44. $x^{-8} y^4$ **45.** $\dfrac{r^5}{r^{-4}}$ **46.** $\dfrac{a^6}{a^{-4}}$ **47.** $\dfrac{x^{-3}y}{4z^{-2}}$ **48.** $\dfrac{p^{-5}q^{-4}}{9r^{-3}}$

49. $\dfrac{(a+b)^{-3}}{(a+b)^{-4}}$ **50.** $\dfrac{(x+y)^{-8}}{(x+y)^{-9}}$ **51.** $\dfrac{(x+2y)^{-3}}{(x+2y)^{-5}}$ **52.** $\dfrac{(p-3q)^{-2}}{(p-3q)^{-4}}$

RELATING CONCEPTS (EXERCISES 53–56)

FOR INDIVIDUAL OR GROUP WORK

In Objective 1, we showed how 6^0 acts as 1 when it is applied to the product rule, thus motivating the definition of 0 as an exponent. We can also use the quotient rule to motivate this definition. **Work Exercises 53–56 in order.**

53. Consider the expression $\frac{25}{25}$. What is its simplest form?

54. Because $25 = 5^2$, the expression $\frac{25}{25}$ can be written as the quotient of powers of 5. Write the expression in this way.

55. Apply the quotient rule for exponents to the expression you wrote in Exercise 54. Give the answer as a power of 5.

56. Your answers in Exercises 53 and 55 must be equal because they both represent $\frac{25}{25}$. Write this equality. What definition does this result support?

Simplify by writing each expression with positive exponents. Assume that all variables represent nonzero real numbers. See Example 5.

57. $\dfrac{(7^4)^3}{7^9}$

58. $\dfrac{(5^3)^2}{5^2}$

59. $x^{-3} \cdot x^5 \cdot x^{-4}$

60. $y^{-8} \cdot y^5 \cdot y^{-2}$

61. $\dfrac{(3x)^{-2}}{(4x)^{-3}}$

62. $\dfrac{(2y)^{-3}}{(5y)^{-4}}$

❂ 63. $\left(\dfrac{x^{-1}y}{z^2}\right)^{-2}$

64. $\left(\dfrac{p^{-4}q}{r^{-3}}\right)^{-3}$

65. $(6x)^4(6x)^{-3}$

66. $(10y)^9(10y)^{-8}$

67. $\dfrac{(m^7n)^{-2}}{m^{-4}n^3}$

68. $\dfrac{(m^8n^{-4})^2}{m^{-2}n^5}$

69. $\dfrac{(x^{-1}y^2z)^{-2}}{(x^{-3}y^3z)^{-1}}$

70. $\dfrac{(a^{-2}b^{-3}c^{-4})^{-5}}{(a^2b^3c^4)^5}$

71. $\left(\dfrac{xy^{-2}}{x^2y}\right)^{-3}$

72. $\left(\dfrac{wz^{-5}}{w^{-3}z}\right)^{-2}$

The next problems are "brain busters." Simplify by writing each expression with positive exponents. Assume that all variables represent nonzero real numbers.

73. $\dfrac{(4a^2b^3)^{-2}(2ab^{-1})^3}{(a^3b)^{-4}}$

74. $\dfrac{(m^6n)^{-2}(m^2n^{-2})^3}{m^{-1}n^{-2}}$

75. $\dfrac{(2y^{-1}z^2)^2(3y^{-2}z^{-3})^3}{(y^3z^2)^{-1}}$

76. $\dfrac{(3p^{-2}q^3)^2(5p^{-1}q^{-4})^{-1}}{(p^2q^{-2})^{-3}}$

77. $\dfrac{(9^{-1}z^{-2}x)^{-1}(4z^2x^4)^{-2}}{(5z^{-2}x^{-3})^2}$

78. $\dfrac{(4^{-1}a^{-1}b^{-2})^{-2}(5a^{-3}b^4)^{-2}}{(3a^{-3}b^{-5})^2}$

79. *Concept Check*   A student simplified $\frac{16^3}{2^2}$ as shown:

$$\frac{16^3}{2^2} = \left(\frac{16}{2}\right)^{3-2} = 8^1 = 8.$$

WHAT WENT WRONG? Give the correct answer.

80. *Concept Check*   A student simplified -5^4 as shown:

$$-5^4 = (-5)^4 = 625.$$

WHAT WENT WRONG? Give the correct answer.

PREVIEW EXERCISES

Evaluate.

81. $10(6427)$ **82.** $100(72.69)$ **83.** $1000(1.23)$ **84.** $10{,}000(26.94)$

85. $34 \div 10$ **86.** $6501 \div 100$ **87.** $237 \div 1000$ **88.** $42 \div 10{,}000$

Summary Exercises on the Rules for Exponents

Simplify each expression. Use only positive exponents in your answers. Assume that all variables represent nonzero real numbers.

1. $\left(\dfrac{6x^2}{5}\right)^{12}$ **2.** $\left(\dfrac{rs^2t^3}{3t^4}\right)^6$ **3.** $(10x^2y^4)^2(10xy^2)^3$

4. $(-2ab^3c)^4(-2a^2b)^3$ **5.** $\left(\dfrac{9wx^3}{y^4}\right)^3$ **6.** $(4x^{-2}y^{-3})^{-2}$

7. $\dfrac{c^{11}(c^2)^4}{(c^3)^3(c^2)^{-6}}$ **8.** $\left(\dfrac{k^4t^2}{k^2t^{-4}}\right)^{-2}$ **9.** $5^{-1} + 6^{-1}$

10. $\dfrac{(3y^{-1}z^3)^{-1}(3y^2)}{(y^3z^2)^{-3}}$ **11.** $\dfrac{(2xy^{-1})^3}{2^3x^{-3}y^2}$ **12.** $-8^0 + (-8)^0$

13. $(z^4)^{-3}(z^{-2})^{-5}$ **14.** $\left(\dfrac{r^2st^5}{3r}\right)^{-2}$ **15.** $\dfrac{(3^{-1}x^{-3}y)^{-1}(2x^2y^{-3})^2}{(5x^{-2}y^2)^{-2}}$

16. $\left(\dfrac{5x^2}{3x^{-4}}\right)^{-1}$ **17.** $\left(\dfrac{-2x^{-2}}{2x^2}\right)^{-2}$ **18.** $\dfrac{(x^{-4}y^2)^3(x^2y)^{-1}}{(xy^2)^{-3}}$

19. $\dfrac{(a^{-2}b^3)^{-4}}{(a^{-3}b^2)^{-2}(ab)^{-4}}$ **20.** $(2a^{-30}b^{-29})(3a^{31}b^{30})$ **21.** $5^{-2} + 6^{-2}$

22. $\left[\dfrac{(x^{47}y^{23})^2}{x^{-26}y^{-42}}\right]^0$ **23.** $\left(\dfrac{7a^2b^3}{2}\right)^3$ **24.** $-(-12^0)$

25. $-(-12)^0$ **26.** $\dfrac{0^{12}}{12^0}$ **27.** $\dfrac{(2xy^{-3})^{-2}}{(3x^{-2}y^4)^{-3}}$

28. $\left(\dfrac{a^2b^3c^4}{a^{-2}b^{-3}c^{-4}}\right)^{-2}$ **29.** $(6x^{-5}z^3)^{-3}$ **30.** $(2p^{-2}qr^{-3})(2p)^{-4}$

31. $\dfrac{(xy)^{-3}(xy)^5}{(xy)^{-4}}$ **32.** $42^0 - (-12)^0$ **33.** $\dfrac{(7^{-1}x^{-3})^{-2}(x^4)^{-6}}{7^{-1}x^{-3}}$

34. $\left(\dfrac{3^{-4}x^{-3}}{3^{-3}x^{-6}}\right)^{-2}$ **35.** $(5p^{-2}q)^{-3}(5pq^3)^4$ **36.** $8^{-1} + 6^{-1}$

37. $\left[\dfrac{4r^{-6}s^{-2}t}{2r^8s^{-4}t^2}\right]^{-1}$ **38.** $(13x^{-6}y)(13x^{-6}y)^{-1}$ **39.** $\dfrac{(8pq^{-2})^4}{(8p^{-2}q^{-3})^3}$

40. $\left(\dfrac{mn^{-2}p}{m^2np^4}\right)^{-2}\left(\dfrac{mn^{-2}p}{m^2np^4}\right)^3$ **41.** $-(-3^0)^0$ **42.** $5^{-1} - 8^{-1}$

43. Concept Check *Match each expression (a)–(j) in Column I with the equivalent expression A–J in Column II. Choices in Column II may be used once, more than once, or not at all.*

I		II	
(a) $2^0 + 2^0$	**(b)** $2^1 \cdot 2^0$	**A.** 0	**B.** 1
(c) $2^0 - 2^{-1}$	**(d)** $2^1 - 2^0$	**C.** -1	**D.** 2
(e) $2^0 \cdot 2^{-2}$	**(f)** $2^1 \cdot 2^1$	**E.** $\dfrac{1}{2}$	**F.** 4
(g) $2^{-2} - 2^{-1}$	**(h)** $2^0 \cdot 2^0$	**G.** -2	**H.** -4
(i) $2^{-2} \div 2^{-1}$	**(j)** $2^0 \div 2^{-2}$	**I.** $-\dfrac{1}{4}$	**J.** $\dfrac{1}{4}$

5.3 An Application of Exponents: Scientific Notation

OBJECTIVES

1 Express numbers in scientific notation.

2 Convert numbers in scientific notation to numbers without exponents.

3 Use scientific notation in calculations.

OBJECTIVE 1 Express numbers in scientific notation. Numbers occurring in science are often extremely large (such as the distance from Earth to the sun, 93,000,000 mi) or extremely small (the wavelength of yellow-green light, approximately 0.0000006 m). Because of the difficulty of working with many zeros, scientists often express such numbers with exponents, using a form called *scientific notation*.

Scientific Notation

A number is written in **scientific notation** when it is expressed in the form

$$a \times 10^n,$$

where $1 \le |a| < 10$ and n is an integer.

In scientific notation, there is always one nonzero digit before the decimal point. This is shown in the following examples.

$3.19 \times 10^1 = 3.19 \times 10 = 31.9$	Decimal point moves 1 place to the right.
$3.19 \times 10^2 = 3.19 \times 100 = 319.$	Decimal point moves 2 places to the right.
$3.19 \times 10^3 = 3.19 \times 1000 = 3190.$	Decimal point moves 3 places to the right.
$3.19 \times 10^{-1} = 3.19 \times 0.1 = 0.319$	Decimal point moves 1 place to the left.
$3.19 \times 10^{-2} = 3.19 \times 0.01 = 0.0319$	Decimal point moves 2 places to the left.
$3.19 \times 10^{-3} = 3.19 \times 0.001 = 0.00319$	Decimal point moves 3 places to the left.

▶ **NOTE** In work with scientific notation, the times symbol, $\times$, is commonly used.

A number in scientific notation is always written with the decimal point after the first nonzero digit and then multiplied by the appropriate power of 10. For example, 56,200 is written 5.62×10^4, since

$$56,200 = 5.62 \times 10,000 = 5.62 \times 10^4.$$

Other examples include

42,000,000	written	4.2×10^7,
0.000586	written	5.86×10^{-4},
and 2,000,000,000	written	2×10^9.

It is not necessary to write 2.0.

To write a number in scientific notation, follow these steps.

Writing a Number in Scientific Notation

Step 1 Move the decimal point to the right of the first nonzero digit.

Step 2 Count the number of places you moved the decimal point.

Step 3 The number of places in Step 2 is the absolute value of the exponent on 10.

Step 4 The exponent on 10 is positive if the original number is greater than the number in Step 1; the exponent is negative if the original number is less than the number in Step 1. If the decimal point is not moved, the exponent is 0.

EXAMPLE 1 Using Scientific Notation

Write each number in scientific notation.

(a) 93,000,000

Move the decimal point to follow the first nonzero digit (the 9). Count the number of places the decimal point was moved.

$$93,000,000. \longleftarrow \text{Decimal point}$$
7 places

The number will be written in scientific notation as 9.3×10^n. To find the value of n, first compare the original number, 93,000,000, with 9.3. Since 93,000,000 is *greater* than 9.3, we must multiply by a *positive* power of 10 so that the product 9.3×10^n will equal the larger number.

Since the decimal point was moved seven places, and since n is positive,

$$93,000,000 = 9.3 \times 10^7.$$

(b) $63,200,000,000 = 6.3200000000 = 6.32 \times 10^{10}$
10 places

(c) 0.00462

Move the decimal point to the right of the first nonzero digit, and count the number of places the decimal point was moved.

$$0.00462 \quad \text{3 places}$$

Since 0.00462 is *less* than 4.62, the exponent must be *negative*.

$$0.00462 = 4.62 \times 10^{-3}$$

(d) $-0.0000762 = -7.62 \times 10^{-5}$ For negative numbers, make the result negative.
5 places

✔ **Now Try Exercises 13 and 17.**

> ▶ **NOTE** To choose the exponent when you write a number in scientific notation, think: If the original number is "large," like 93,000,000, use a *positive* exponent on 10, since positive is greater than negative. However, if the original number is "small," like 0.00462, use a *negative* exponent on 10, since negative is less than positive.

OBJECTIVE 2 Convert numbers in scientific notation to numbers without exponents.
To convert a number written in scientific notation to a number without exponents, work in reverse. *Multiplying a number by a positive power of 10 will make the number greater; multiplying by a negative power of 10 will make the number less.*

EXAMPLE 2 Writing Numbers without Exponents

Write each number without exponents.

(a) 6.2×10^3

Since the exponent is positive, we make 6.2 greater by moving the decimal point three places to the right. We attach two zeros.

$$6.2 \times 10^3 = 6.200 = 6200$$

(b) $4.283 \times 10^6 = 4.283000 = 4{,}283{,}000$ Move 6 places to the right; attach zeros as necessary.

(c) $7.04 \times 10^{-3} = 0.00704$ Move 3 places to the left.

The exponent tells the number of places and the direction that the decimal point is moved.

✔ **Now Try Exercises 21 and 29.**

OBJECTIVE 3 Use scientific notation in calculations. The next example uses scientific notation with products and quotients.

EXAMPLE 3 Multiplying and Dividing with Scientific Notation

Perform each calculation. Write answers in scientific notation and also without exponents.

(a) $(7 \times 10^3)(5 \times 10^4) = (7 \times 5)(10^3 \times 10^4)$ Commutative and associative properties

        $= 35 \times 10^7$ Multiply; product rule

        $= (3.5 \times 10^1) \times 10^7$ Write 35 in scientific notation.

        $= 3.5 \times (10^1 \times 10^7)$ Associative property

        $= 3.5 \times 10^8$ Product rule

        $= 350{,}000{,}000$ Write without exponents.

Don't stop! This number is not in scientific notation, since 35 is not between 1 and 10.

(b) $\dfrac{4 \times 10^{-5}}{2 \times 10^3} = \dfrac{4}{2} \times \dfrac{10^{-5}}{10^3} = 2 \times 10^{-8} = 0.00000002$

✔ **Now Try Exercises 35 and 47.**

> ▶ **NOTE** Multiplying or dividing numbers written in scientific notation may produce an answer in the form $a \times 10^0$. Since $10^0 = 1$, $a \times 10^0 = a$. For example,
>
> $$(8 \times 10^{-4})(5 \times 10^4) = 40 \times 10^0 = 40. \qquad 10^0 = 1$$
>
> Also, if $a = 1$, then $a \times 10^n = 10^n$. For example, we could write 1,000,000 as 10^6 instead of 1×10^6.

EXAMPLE 4 Using Scientific Notation to Solve an Application

A *nanometer* is a very small unit of measure that is equivalent to about 0.0000000000254 in. About how much would 700,000 nanometers measure in inches? (*Source: World Almanac and Book of Facts 2006.*)

Write each number in scientific notation, and then multiply.

$$700,000(0.0000000000254) = (7 \times 10^5)(2.54 \times 10^{-11}) \qquad \text{Write in scientific notation.}$$

$$= (7 \times 2.54)(10^5 \times 10^{-11})$$

$$= 17.78 \times 10^{-6} \qquad \text{Multiply; product rule}$$

Don't stop here. $\qquad = (1.778 \times 10^1) \times 10^{-6} \qquad \text{Write 17.78 in scientific notation.}$

$$= 1.778 \times 10^{-5} \qquad \text{Product rule}$$

$$= 0.00001778 \qquad \text{Write without exponents.}$$

Thus, 700,000 nanometers would measure 1.778×10^{-5} in., or 0.00001778 in.

✔ **Now Try Exercise 79.**

EXAMPLE 5 Using Scientific Notation to Solve an Application

In 2003, the national debt was $\$3.9136 \times 10^{12}$ (which is more than \$3 trillion). The population of the United States was approximately 290 million that year. About how much would each person have had to contribute in order to pay off the national debt? (*Source*: U.S. Office of Management and Budget; U.S. Census Bureau.)

Write the population in scientific notation. Then divide to obtain the per person contribution.

$$\frac{3.9136 \times 10^{12}}{290,000,000} = \frac{3.9136 \times 10^{12}}{2.9 \times 10^8} \qquad \text{Write 290 million in scientific notation.}$$

$$= \frac{3.9136}{2.9} \times 10^4 \qquad \text{Quotient rule}$$

$$= 1.3495 \times 10^4 \qquad \text{Divide; round to 4 decimal places.}$$

$$= 13,495 \qquad \text{Write without exponents.}$$

Each person would have to pay about \$13,495.

✔ **Now Try Exercise 77.**

CONNECTIONS

In 1935, Charles F. Richter devised a scale to compare the intensities, or relative power, of earthquakes. The *intensity* of an earthquake is measured relative to the intensity of a standard *zero-level* earthquake of intensity I_0. The relationship is equivalent to $I = I_0 \times 10^R$, where R is the **Richter scale** measure. For example, if an earthquake has magnitude 5.0 on the Richter scale, then its intensity is calculated as $I = I_0 \times 10^{5.0} = I_0 \times 100,000$, which is 100,000 times as intense as a zero-level earthquake. The following diagram illustrates the intensities of earthquakes and their Richter scale magnitudes.

Intensity	$I_0 \times 10^0$	$I_0 \times 10^1$	$I_0 \times 10^2$	$I_0 \times 10^3$	$I_0 \times 10^4$	$I_0 \times 10^5$	$I_0 \times 10^6$	$I_0 \times 10^7$	$I_0 \times 10^8$
Richter Scale	0	1	2	3	4	5	6	7	8

To compare two earthquakes, a ratio of the intensities is calculated. For example, to compare an earthquake that measures 8.0 on the Richter scale with one that measures 5.0, find the ratio of the intensities:

$$\frac{\text{intensity } 8.0}{\text{intensity } 5.0} = \frac{I_0 \times 10^{8.0}}{I_0 \times 10^{5.0}} = \frac{10^8}{10^5} = 10^{8-5} = 10^3 = 1000.$$

Therefore, an earthquake that measures 8.0 on the Richter scale is 1000 times as intense as one that measures 5.0.

For Discussion or Writing

The table gives Richter scale measurements for several earthquakes.

Year	Earthquake Location	Richter Scale Measurement
1964	Prince William Sound, Alaska	9.2
2004	Sumatra, Indonesia	9.0
1976	Tangshan, China	8.0
1906	San Francisco, California	7.7
2004	West Coast, Colombia	7.2

Source: U.S. Geological Survey.

1. Compare the intensity of the 2004 Sumatra earthquake with the 1976 Tangshan earthquake.

2. Compare the intensity of the 2004 Colombia earthquake with the 1964 Alaska earthquake.

3. Compare the intensity of the 2004 Sumatra earthquake with the 1906 California earthquake. (*Hint*: Use a calculator.)

4. Suppose an earthquake measures 7.2 on the Richter scale. How would the intensity of a second earthquake compare if its Richter scale measure differed by $+3.0$? By -1.0?

5.3 EXERCISES

Concept Check *Match each number written in scientific notation in Column I with the correct choice from Column II. Not all choices in Column II will be used.*

	I		II		I		II
1.	**(a)** 4.6×10^{-4}	**A.**	46,000	**2.**	**(a)** 1×10^9	**A.**	1 billion
	(b) 4.6×10^4	**B.**	460,000		**(b)** 1×10^6	**B.**	100 million
	(c) 4.6×10^5	**C.**	0.00046		**(c)** 1×10^8	**C.**	1 million
	(d) 4.6×10^{-5}	**D.**	0.000046		**(d)** 1×10^{10}	**D.**	10 billion
		E.	4600			**E.**	100 billion

Concept Check *Determine whether or not each number is written in scientific notation as defined in Objective 1. If it is not, write it as such.*

3. 4.56×10^3 **4.** 7.34×10^5 **5.** 5,600,000   **6.** 34,000

7. 0.8×10^2 **8.** 0.9×10^3 **9.** 0.004   **10.** 0.0007

▨ **11.** Explain what it means for a number to be written in scientific notation. Give examples.

▨ **12.** Explain how to multiply a number by a positive power of 10. Then explain how to multiply a number by a negative power of 10.

Write each number in scientific notation. See Example 1.

13. 5,876,000,000   **14.** 9,994,000,000   **15.** 82,350   **16.** 78,330

17. 0.000007   **18.** 0.0000004   **19.** 0.00203   **20.** 0.0000578

Write each number without exponents. See Example 2.

21. 7.5×10^5 **22.** 8.8×10^6 **23.** 5.677×10^{12} **24.** 8.766×10^9

25. 1×10^{12} **26.** 1×10^7 **27.** 6.21×10^0 **28.** 8.56×10^0

29. 7.8×10^{-4} **30.** 8.9×10^{-5} **31.** 5.134×10^{-9} **32.** 7.123×10^{-10}

*Perform the indicated operations. Write each answer **(a)** in scientific notation and **(b)** without exponents. See Example 3.*

33. $(2 \times 10^8)(3 \times 10^3)$ **34.** $(4 \times 10^7)(3 \times 10^3)$

35. $(5 \times 10^4)(3 \times 10^2)$ **36.** $(8 \times 10^5)(2 \times 10^3)$

37. $(3 \times 10^{-4})(2 \times 10^8)$ **38.** $(4 \times 10^{-3})(2 \times 10^7)$

39. $(6 \times 10^3)(4 \times 10^{-2})$ **40.** $(7 \times 10^5)(3 \times 10^{-4})$

41. $(9 \times 10^4)(7 \times 10^{-7})$ **42.** $(6 \times 10^4)(8 \times 10^{-8})$

43. $\dfrac{9 \times 10^{-5}}{3 \times 10^{-1}}$ **44.** $\dfrac{12 \times 10^{-4}}{4 \times 10^{-3}}$ **45.** $\dfrac{8 \times 10^3}{2 \times 10^2}$

46. $\dfrac{15 \times 10^4}{3 \times 10^3}$ **47.** $\dfrac{2.6 \times 10^{-3}}{2 \times 10^2}$ **48.** $\dfrac{9.5 \times 10^{-1}}{5 \times 10^3}$

49. $\dfrac{4 \times 10^5}{8 \times 10^2}$ **50.** $\dfrac{3 \times 10^9}{6 \times 10^5}$

TECHNOLOGY INSIGHTS (EXERCISES 51–56)

Graphing calculators such as the TI-83/84 Plus can display numbers in scientific notation (when in scientific mode), using the format shown in the screen on the left. For 5400, the calculator displays 5.4E3 to represent 5.4×10^3. The display 5.4E-4 means 5.4×10^{-4}. The calculator will also perform operations with numbers entered in scientific notation, as shown in the screen on the right. Notice how the rules for exponents are applied.

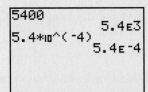

```
5400
              5.4E3
5.4*10^(-4)
              5.4E-4
```

```
(2E5)*(4E-2)
              8E3
(2E5)/(4E-2)
              5E6
(2E5)³
              8E15
```

Predict the display the calculator would give for the expression shown in each screen.

51.
```
.00000047
```

52.
```
.000021
```

53.
```
(8E5)/(4E-2)
```

54.
```
(9E-4)/(3E3)
```

55.
```
(2E6)*(2E-3)/(4E
2)
```

56.
```
(5E-3)*(1E9)/(5E
3)
```

The next problems are "brain busters." Use scientific notation to calculate the answer to each problem. Write answers in scientific notation.

57. $\dfrac{650,000,000(0.0000032)}{0.00002}$

58. $\dfrac{3,400,000,000(0.000075)}{0.00025}$

59. $\dfrac{0.00000072(0.00023)}{0.000000018}$

60. $\dfrac{0.000000081(0.000036)}{0.00000048}$

61. $\dfrac{0.0000016(240,000,000)}{0.00002(0.0032)}$

62. $\dfrac{0.000015(42,000,000)}{0.000009(0.000005)}$

Each statement comes from Astronomy! A Brief Edition *by James B. Kaler (Addison-Wesley, 1997). If the number in boldface italics is in scientific notation, write it without exponents. If the number is written without exponents, write it in scientific notation.*

63. Multiplying this view over the whole sky yields a galaxy count of more than ***10 billion.*** (page 496)

64. The circumference of the solar orbit is . . . about ***4.7 million*** km . . . . (in reference to the orbit of Jupiter, page 395)

65. The solar luminosity requires that ***2×10^9*** kg of mass be converted into energy every second. (page 327)

66. At maximum, a cosmic ray particle—a mere atomic nucleus of only ***10^{-13}*** cm across—can carry the energy of a professionally pitched baseball. (page 445)

Each statement contains a number in boldface italics. Write the number in scientific notation.

67. At the end of 2004, the total number of cellular telephone subscriptions in the world reached about ***1,341,000,000.*** (*Source:* International Telecommunications Union.)

68. In 2004, the leading U.S. advertiser was the General Motors Corporation, which spent approximately ***$4,000,000,000.*** (*Source:* Competitive Media Reporting and Publishers Information Bureau.)

69. During 2004, worldwide movie box office receipts (in U.S. dollars) totaled ***$25,240,000,000.*** (*Source:* Motion Picture Association of America.)

70. In 2004, assets of the insured commercial banks in the United States totaled about ***$8,413,000,000,000.*** (*Source:* U.S. Federal Deposit Insurance Corporation.)

Use scientific notation to calculate the answer to each problem. See Examples 3–5.

71. The body of a 150-lb person contains about 2.3×10^{-4} lb of copper. How much copper is contained in the bodies of 1200 such people?

72. In 2004, the state of Montana had about 2.8×10^4 farms with an average of 2.2×10^6 acres per farm. What was the total number of acres devoted to farmland in Montana that year? (*Source:* National Agricultural Statistics Service, U.S. Department of Agriculture.)

73. Venus is 6.68×10^7 mi from the sun. If light travels at a speed of 1.86×10^5 mi per sec, how long does it take light to travel from the sun to Venus? (*Source: World Almanac and Book of Facts 2006.*)

74. (a) The distance to Earth from Pluto is 4.58×10^9 km. In April 1983, *Pioneer 10* transmitted radio signals from Pluto to Earth at the speed of light, 3.00×10^5 km per sec. How long (in seconds) did it take for the signals to reach Earth?
   (b) How many hours did it take for the signals to reach Earth?

75. During the 2004–2005 season, Broadway shows grossed a total of 7.69×10^8 dollars. Total attendance for the season was 1.15×10^7. What was the average ticket price for a Broadway show? (*Source:* League of American Theatres and Producers.)

76. There were 6.3×10^{10} dollars spent to attend motion pictures in a recent year. Approximately 1.3×10^8 adults attended a motion picture theater at least once. What was the average amount spent per person that year? (*Source:* U.S. National Endowment for the Arts.)

77. On March 16, 2006, Congress raised the government's debt limit to 9×10^{12}. When this national debt limit is reached, about how much will it be for every man, women, and child in the country? Use 300 million as the population of the United States. (*Source: The Gazette,* Cedar Rapids, Iowa, March 17, 2006.)

78. There are 1×10^9 Social Security numbers. The population of the United States is about 3×10^8. How many Social Security numbers are available for each person? (*Source*: U.S. Census Bureau.)

79. In March 2006, astronomers using the Spitzer Space Telescope discovered a twisted double-helix nebula, a conglomeration of dust and gas stretching across the center of the Milky Way galaxy. This nebula is 25,000 light-years from Earth. If one light-year is about 6,000,000,000,000 (that is, 6 trillion) miles, about how many miles is the twisted double-helix nebula from Earth? (*Source*: http://articles.news.aol.com)

80. A computer can perform 466,000,000 calculations per second. How many calculations can it perform per minute? Per hour?

81. In 2003, the U.S. government collected about $6730 per person in taxes. If the population was 290,000,000, how much did the government collect in taxes for 2003? (*Source*: U.S. Internal Revenue Service.)

82. Pollux, one of the brightest stars in the night sky, is 33.7 light-years from Earth. If one light-year is about 6,000,000,000,000 mi, about how many miles is Pollux from Earth? (*Source: World Almanac and Book of Facts 2006.*)

83. In 2000, the population of the United States was about 281.4 million. To the nearest dollar, calculate how much each person in the United States would have had to contribute in order to make one lucky person a trillionaire (that is, to give that person $1,000,000,000,000). (*Source*: U.S. Census Bureau.)

84. In 2003, national expenditures for health care reached $1,674,000,000,000. Using 290 million as the population of the United States, about how much, to the nearest dollar, was spent on health care per person in 2003? (*Source*: U.S. Centers for Medicare and Medicaid Services.)

PREVIEW EXERCISES

*Simplify. See **Section 1.8.***

85. $-3(2x + 4) + 4(2x - 6)$

86. $-8(-3x + 7) - 4(2x + 3)$

*Evaluate each expression for $x = 3$. See **Sections 1.3 and 1.6.***

87. $2x^2 - 3x + 9$

88. $3x^2 - 3x + 2$

89. $4x^3 - 5x^2 + 2x - 6$

90. $-4x^3 + 2x^2 - 9x - 3$

5.4   Adding and Subtracting Polynomials; Graphing Simple Polynomials

OBJECTIVE 1   Identify terms and coefficients.   In **Section 1.8,** we saw that in an expression such as

$$4x^3 + 6x^2 + 5x + 8,$$

the quantities $4x^3$, $6x^2$, $5x$, and 8 are called *terms.* In the term $4x^3$, the number 4 is called the *numerical coefficient,* or simply the *coefficient,* of x^3. In the same way, 6 is the coefficient of x^2 in the term $6x^2$, and 5 is the coefficient of x in the term $5x$. The constant term 8 can be thought of as $8 \cdot 1 = 8x^0$, since $x^0 = 1$, so 8 is the coefficient in the term 8.

EXAMPLE 1   Identifying Coefficients

Name the coefficient of each term in these expressions.

(a) $x - 6x^4$

The coefficient of x is 1, because $x = 1 \cdot x$, or $1x$. The coefficient of x^4 is -6, since we can write $x - 6x^4$ as the sum $x + (-6x^4)$.

(b) $5 - v^3$

The coefficient of the term 5 is 5, because $5 = 5v^0$. By writing $5 - v^3$ as a sum, $5 + (-v^3)$, or $5 + (-1v^3)$, we can identify the coefficient of v^3 as -1.

✔ **Now Try Exercises 9 and 13.**

OBJECTIVE 2   Add like terms.   Recall from **Section 1.8** that *like terms* have exactly the same combination of variables, with the same exponents on the variables. *Only the coefficients may differ.*

$$\left.\begin{array}{l} 19m^5 \quad \text{and} \quad 14m^5 \\ 6y^9, \quad -37y^9, \quad \text{and} \quad y^9 \\ 3pq \quad \text{and} \quad -2pq \\ 2xy^2 \quad \text{and} \quad -xy^2 \end{array}\right\} \text{Examples of like terms}$$

Using the distributive property, we combine, or add, like terms by adding their coefficients.

EXAMPLE 2   Adding Like Terms

Simplify by adding like terms.

(a) $-4x^3 + 6x^3 = (-4 + 6)x^3$

$\qquad\qquad\qquad = 2x^3$ Distributive property

(b) $9x^6 - 14x^6 + x^6 = (9 - 14 + 1)x^6$ $x^6 = 1x^6$

$\qquad\qquad\qquad\quad = -4x^6$

(c) $12m^2 + 5m + 4m^2 = (12 + 4)m^2 + 5m$
$$= 16m^2 + 5m$$

(d) $3x^2y + 4x^2y - x^2y = (3 + 4 - 1)x^2y$
$$= 6x^2y$$

> ✔ **Now Try Exercises 17, 23, and 27.**

In Example 2(c), we cannot combine $16m^2$ and $5m$. These two terms are *unlike*, because the exponents on the variables are different. ***Unlike terms have different variables or different exponents on the same variables.***

> **OBJECTIVE ③   Know the vocabulary for polynomials.**   A **polynomial in** x is a term or the sum of a finite number of terms of the form ax^n, for any real number a and any whole number n. For example,

$$16x^8 - 7x^6 + 5x^4 - 3x^2 + 4 \qquad \text{Polynomial}$$

is a polynomial in x. (The 4 can be written as $4x^0$.) This polynomial is written in **descending powers** of the variable, since the exponents on x decrease from left to right. By contrast,

$$2x^3 - x^2 + \frac{4}{x} \qquad \text{Not a Polynomial}$$

is not a polynomial in x, since a variable appears in a denominator. Of course, we could define *polynomial* using any variable and not just x, as in Example 2(c). In fact, polynomials may have terms with more than one variable, as in Example 2(d).

The **degree of a term** is the sum of the exponents on the variables. For example, $3x^4$ has degree 4, while $6x^{17}$ has degree 17. The term $5x$ (or $5x^1$) has degree 1, -7 has degree 0 (since -7 can be written as $-7x^0$), and $2x^2y$ has degree $2 + 1 = 3$. (y has an exponent of 1.)

The **degree of a polynomial** is the greatest degree of any nonzero term of the polynomial. For example, $3x^4 - 5x^2 + 6$ is of degree 4, the polynomial $5x + 7$ is of degree 1, the term 3 (or $3x^0$) is of degree 0, and $x^2y + xy - 5xy^2$ is of degree 3.

Three types of polynomials are common and are given special names. A polynomial with only one term is called a **monomial.** (*Mono* means "one," as in *mono*rail.) Examples are

$$9m, \quad -6y^5, \quad a^2, \quad \text{and} \quad 6. \qquad \text{Monomials}$$

A polynomial with exactly two terms is called a **binomial.** (*Bi-* means "two," as in *bi*cycle.) Examples are

$$-9x^4 + 9x^3, \quad 8m^2 + 6m, \quad \text{and} \quad 3m^5 - 9m^2. \qquad \text{Binomials}$$

A polynomial with exactly three terms is called a **trinomial.** (*Tri-* means "three," as in *tri*angle.) Examples are

$$9m^3 - 4m^2 + 6, \quad \frac{19}{3}y^2 + \frac{8}{3}y + 5, \quad \text{and} \quad -3m^5 - 9m^2 + 2. \qquad \text{Trinomials}$$

EXAMPLE 3 Classifying Polynomials

For each polynomial, first simplify, if possible. Then give the degree and tell whether the polynomial is a *monomial*, a *binomial*, a *trinomial*, or *none of these*.

(a) $2x^3 + 5$

The polynomial cannot be simplified. It is a binomial of degree 3.

(b) $4xy - 5xy + 2xy$

Add like terms: $4xy - 5xy + 2xy = xy$, which is a monomial of degree 2.

✔ **Now Try Exercises 29 and 31.**

OBJECTIVE 4 Evaluate polynomials. A polynomial usually represents different numbers for different values of the variable.

EXAMPLE 4 Evaluating a Polynomial

Find the value of $3x^4 + 5x^3 - 4x - 4$ when $x = -2$ and when $x = 3$.

First, substitute -2 for x.

> Use parentheses to avoid errors.

$$
\begin{aligned}
3x^4 + 5x^3 - 4x - 4 &= 3(-2)^4 + 5(-2)^3 - 4(-2) - 4 && \text{Let } x = -2. \\
&= 3(16) + 5(-8) - 4(-2) - 4 && \text{Apply exponents.} \\
&= 48 - 40 + 8 - 4 && \text{Multiply.} \\
&= 12 && \text{Add and subtract.}
\end{aligned}
$$

Next, replace x with 3.

$$
\begin{aligned}
3x^4 + 5x^3 - 4x - 4 &= 3(3)^4 + 5(3)^3 - 4(3) - 4 \\
&= 3(81) + 5(27) - 12 - 4 \\
&= 243 + 135 - 12 - 4 \\
&= 362
\end{aligned}
$$

✔ **Now Try Exercise 37.**

▶ **CAUTION** Use parentheses around the numbers that are substituted for the variable in Example 4, particularly when substituting a negative number for a variable that is raised to a power, or a sign error may result.

CONNECTIONS

In **Section 3.6,** we introduced the idea of a function: For every input x, there is one output y. For example, the age y of a dog in human years can be approximated by the polynomial equation

$$y = -0.0545x^2 + 5.047x + 11.78,$$

where x is age in dog years.

(continued)

For Discussion or Writing

Use each age in dog years from the table and the given polynomial equation to approximate the age of a dog in human years. Round answers to the nearest tenth. Then compare your approximations from the equation with the actual ages given in the table.

For example, for a 1-year-old dog,

$$y = -0.0545x^2 + 5.047x + 11.78$$
$$y = -0.0545(1)^2 + 5.047(1) + 11.78 \qquad \text{Let } x = 1.$$
$$y = 16.8. \qquad\qquad\qquad\qquad\qquad \text{Nearest tenth}$$

The approximation from the equation, 16.8 human years, is greater than the actual value, 16, from the table.

Age in Dog Years, x	Age in Human Years, y
1	16
3	28
5	36
7	44
9	52
11	60
15	76

Source: American Animal Hospital Association.

DOTTY

OBJECTIVE 5   Add and subtract polynomials.   Polynomials may be added, subtracted, multiplied, and divided.

Adding Polynomials

To add two polynomials, add like terms.

EXAMPLE 5   Adding Polynomials Vertically

(a) Add $6x^3 - 4x^2 + 3$ and $-2x^3 + 7x^2 - 5$.

Write like terms in columns.

$$\begin{array}{r} 6x^3 - 4x^2 + 3 \\ -2x^3 + 7x^2 - 5 \end{array}$$

Now add, column by column.

$$\begin{array}{ccc} 6x^3 & -4x^2 & 3 \\ -2x^3 & 7x^2 & -5 \\ \hline 4x^3 & 3x^2 & -2 \end{array}$$

Add the three sums together.

$$4x^3 + 3x^2 + (-2) = 4x^3 + 3x^2 - 2 \quad \leftarrow \text{Final sum}$$

(b) Add $2x^2 - 4x + 3$ and $x^3 + 5x$.

Write like terms in columns and add column by column.

$$
\begin{array}{r}
2x^2 - 4x + 3 \\
x^3 \qquad\ \ + 5x \\
\hline
x^3 + 2x^2 + \ \ x + 3
\end{array}
$$

> Leave spaces for missing terms.

✔ **Now Try Exercises 45 and 49.**

The polynomials in Example 5 also can be added horizontally.

EXAMPLE 6 Adding Polynomials Horizontally

(a) Add $6x^3 - 4x^2 + 3$ and $-2x^3 + 7x^2 - 5$.

Combine like terms.

$(6x^3 - 4x^2 + 3) + (-2x^3 + 7x^2 - 5) = 4x^3 + 3x^2 - 2$ Same answer found in Example 5(a)

(b) Add $2x^2 - 4x + 3$ and $x^3 + 5x$.

$(2x^2 - 4x + 3) + (x^3 + 5x) = x^3 + 2x^2 - 4x + 5x + 3$ Commutative property

$\qquad\qquad\qquad\qquad\qquad = x^3 + 2x^2 + x + 3$ Combine like terms.

✔ **Now Try Exercise 61.**

In **Section 1.5,** we defined the difference $x - y$ as $x + (-y)$. (We find the difference $x - y$ by adding x and the opposite of y.) For example,

$$7 - 2 = 7 + (-2) = 5 \qquad \text{and} \qquad -8 - (-2) = -8 + 2 = -6.$$

A similar method is used to subtract polynomials.

Subtracting Polynomials

To subtract two polynomials, change all the signs in the second polynomial and add the result to the first polynomial.

EXAMPLE 7 Subtracting Polynomials

(a) Perform the subtraction $(5x - 2) - (3x - 8)$.

$(5x - 2) - (3x - 8) = (5x - 2) + [-(3x - 8)]$ Definition of subtraction

$\qquad\qquad\qquad\qquad = (5x - 2) + [-1(3x - 8)]$

$\qquad\qquad\qquad\qquad = (5x - 2) + (-3x + 8)$ Distributive property

$\qquad\qquad\qquad\qquad = 2x + 6$ Combine like terms.

(b) Subtract $6x^3 - 4x^2 + 2$ from $11x^3 + 2x^2 - 8$.
$$(11x^3 + 2x^2 - 8) - (6x^3 - 4x^2 + 2)$$

Be careful to write the problem in the correct order.

Change all signs in the second polynomial and add.
$$(11x^3 + 2x^2 - 8) + (-6x^3 + 4x^2 - 2) = 5x^3 + 6x^2 - 10$$

To check a subtraction problem, use the fact that

if $a - b = c$,   then $a = b + c$.

For example, $6 - 2 = 4$, so check by writing $6 = 2 + 4$, which is correct. Check the preceding polynomial subtraction by adding $6x^3 - 4x^2 + 2$ and $5x^3 + 6x^2 - 10$.
$$(6x^3 - 4x^2 + 2) + (5x^3 + 6x^2 - 10) = 11x^3 + 2x^2 - 8$$

Since the sum is $11x^3 + 2x^2 - 8$, the subtraction was performed correctly.

✔ **Now Try Exercises 59 and 63.**

We use vertical subtraction in **Section 5.7** when we divide polynomials.

EXAMPLE 8 **Subtracting Polynomials Vertically**

Subtract by columns to find $(14y^3 - 6y^2 + 2y - 5) - (2y^3 - 7y^2 - 4y + 6)$.

$$14y^3 - 6y^2 + 2y - 5$$
$$\underline{2y^3 - 7y^2 - 4y + 6}$$ Arrange like terms in columns.

Change all signs in the second row, and then add.

$$14y^3 - 6y^2 + 2y - \ 5$$
$$\underline{-2y^3 + 7y^2 + 4y - \ 6}$$ Change all signs.
$$12y^3 + \ y^2 + 6y - 11$$ Add.

✔ **Now Try Exercise 55.**

We add and subtract polynomials in more than one variable by combining like terms, just as with single-variable polynomials.

EXAMPLE 9 **Adding and Subtracting Polynomials with More than One Variable**

Add or subtract as indicated.

(a) $(4a + 2ab - b) + (3a - ab + b) = 4a + 2ab - b + 3a - ab + b$
$$= 7a + ab$$ Combine like terms.

(b) $(2x^2y + 3xy + y^2) - (3x^2y - xy - 2y^2)$
$$= 2x^2y + 3xy + y^2 - 3x^2y + xy + 2y^2$$
$$= -x^2y + 4xy + 3y^2$$

Be careful with signs.

✔ **Now Try Exercises 73 and 75.**

OBJECTIVE 6 **Graph equations defined by polynomials of degree 2.** In **Chapter 3,** we introduced graphs of straight lines. These graphs were defined by linear equations (which are actually polynomial equations of degree 1). By plotting points selectively, we can graph polynomial equations of degree 2.

EXAMPLE 10 Graphing Equations Defined by Polynomials of Degree 2

Graph each equation.

(a) $y = x^2$

Select values for x; then find the corresponding y-values. Selecting $x = 2$ gives

$$y = x^2 = 2^2 = 4,$$

so the point $(2, 4)$ is on the graph of $y = x^2$. (Recall that in an ordered pair such as $(2, 4)$, *the x-value comes first and the y-value second.*) We show some ordered pairs that satisfy $y = x^2$ in the table with Figure 3. If we plot the ordered pairs from the table on a coordinate system and draw a smooth curve through them, we obtain the graph shown in Figure 3.

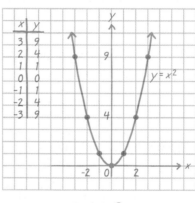

FIGURE 3

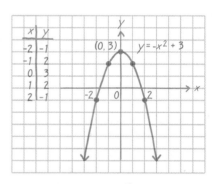

FIGURE 4

The graph of $y = x^2$ is the graph of a function, since each input x is related to just one output y. The curve in Figure 3 is called a **parabola.** The point $(0, 0)$, the *lowest* point on this graph, is called the **vertex** of the parabola. The vertical line through the vertex (the y-axis here) is called the **axis** of the parabola. The axis of a parabola is a **line of symmetry** for the graph. If the graph is folded on this line, the two halves will match.

(b) $y = -x^2 + 3$

Once again, plot points to obtain the graph. For example, if $x = -2$, then

$$y = -(-2)^2 + 3 = -4 + 3 = -1.$$

The point $(-2, -1)$ and several others are shown in the table that accompanies the graph in Figure 4. The vertex of this parabola is $(0, 3)$. Now the vertex is the *highest* point on the graph. The graph opens downward because x^2 has a negative coefficient.

✔ **Now Try Exercises 89 and 93.**

▶ **NOTE** *All polynomials of degree 2 have parabolas as their graphs.* When graphing, find points until the vertex and points on either side of it are located. (In this section, all parabolas have their vertices on the x-axis or the y-axis.)

5.4 EXERCISES

Concept Check *Fill in each blank with the correct response.*

1. In the term $7x^5$, the coefficient is _____ and the exponent is _____.

2. The expression $5x^3 - 4x^2$ has _____ term(s).
   (how many?)

3. The degree of the term $-4x^8$ is _____.

4. The polynomial $4x^2 - y^2$ _____ an example of a trinomial.
   (is/is not)

5. When $x^2 + 10$ is evaluated for $x = 4$, the result is _____.

6. $5x^{\underline{}} + 3x^3 - 7x$ is a trinomial of degree 4.

7. $3xy + 2xy - 5xy =$ _____

8. _____ is an example of a monomial with coefficient 5, in the variable x, having degree 9.

For each polynomial, determine the number of terms and name the coefficients of the terms. See Example 1.

9. $6x^4$ 　　　　　10. $-9y^5$ 　　　　　11. t^4 　　　　　12. s^7

13. $-19r^2 - r$ 　　　14. $2y^3 - y$ 　　　15. $x + 8x^2 + 5x^3$ 　　16. $v - 2v^3 - v^7$

In each polynomial, add like terms whenever possible. Write the result in descending powers of the variable. See Example 2.

17. $-3m^5 + 5m^5$ 　　　　　18. $-4y^3 + 3y^3$ 　　　　　19. $2r^5 + (-3r^5)$

20. $-19y^2 + 9y^2$ 　　　　　21. $0.2m^5 - 0.5m^2$ 　　　　　22. $-0.9y + 0.9y^2$

23. $-3x^5 + 2x^5 - 4x^5$ 　　　　　　　　24. $6x^3 - 8x^3 + 9x^3$

25. $-4p^7 + 8p^7 + 5p^9$ 　　　　　　　　26. $-3a^8 + 4a^8 - 3a^2$

27. $-4xy^2 + 3xy^2 - 2xy^2 + xy^2$ 　　　28. $3pr^5 - 8pr^5 + pr^5 + 2pr^5$

For each polynomial, first simplify, if possible, and write it in descending powers of the variable. Then give the degree of the resulting polynomial and tell whether it is a monomial, a binomial, a trinomial, or none of these. See Example 3.

29. $6x^4 - 9x$ 　　　　　　　　　　30. $7t^3 - 3t$

31. $5m^4 - 3m^2 + 6m^4 - 7m^3$ 　　　　32. $6p^5 + 4p^3 - 8p^5 + 10p^2$

33. $\dfrac{5}{3}x^4 - \dfrac{2}{3}x^4$ 　　　　　　　34. $\dfrac{4}{5}r^6 + \dfrac{1}{5}r^6$

35. $0.8x^4 - 0.3x^4 - 0.5x^4 + 7$ 　　　36. $1.2t^3 - 0.9t^3 - 0.3t^3 + 9$

Find the value of each polynomial when ***(a)*** $x = 2$ *and when* ***(b)*** $x = -1$. *See Example 4.*

37. $2x^2 - 3x - 5$ 　　　　　　　　38. $x^2 + 5x - 10$

39. $-3x^2 + 14x - 2$ 　　　　　　　40. $-2x^2 + 5x - 1$

41. $2x^5 - 4x^4 + 5x^3 - x^2$ 　　　　　42. $x^4 - 6x^3 + x^2 - 1$

Add. See Example 5.

43. $2x^2 - 4x$ 　　　　44. $-5y^3 + 3y$ 　　　⊙ 45. $3m^2 + 5m + 6$
　　$\underline{3x^2 + 2x}$ 　　　　　　$\underline{8y^3 - 4y}$ 　　　　　　$\underline{2m^2 - 2m - 4}$

46. $4a^3 - 4a^2 - 4$
$\underline{6a^3 + 5a^2 - 8}$

47. $\dfrac{2}{3}x^2 + \dfrac{1}{5}x + \dfrac{1}{6}$
$\underline{\dfrac{1}{2}x^2 - \dfrac{1}{3}x + \dfrac{2}{3}}$

48. $\dfrac{4}{7}y^2 - \dfrac{1}{5}y + \dfrac{7}{9}$
$\underline{\dfrac{1}{3}y^2 - \dfrac{1}{3}y + \dfrac{2}{5}}$

49. $9m^3 - 5m^2 + 4m - 8$ and $-3m^3 + 6m^2 - 6$

50. $12r^5 + 11r^4 - 7r^3 - 2r^2$ and $-8r^5 + 3r^3 + 2r^2$

Subtract. See Example 8.

51. $5y^3 - 3y^2$
$\underline{2y^3 + 8y^2}$

52. $-6t^3 + 4t^2$
$\underline{8t^3 - 6t^2}$

53. $12x^4 - x^2 + x$
$\underline{8x^4 + 3x^2 - 3x}$

54. $13y^5 - y^3 - 8y^2$
$\underline{7y^5 + 5y^3 + y^2}$

55. $12m^3 - 8m^2 + 6m + 7$
$\underline{-3m^3 + 5m^2 - 2m - 4}$

56. $5a^4 - 3a^3 + 2a^2 - a + 6$
$\underline{-6a^4 + a^3 - a^2 + a - 1}$

57. After reading Examples 5–8, do you have a preference regarding horizontal or vertical addition and subtraction of polynomials? Explain your answer.

58. Write a paragraph explaining how to add and subtract polynomials. Give an example using addition.

Perform each indicated operation. See Examples 6 and 7.

59. $(8m^2 - 7m) - (3m^2 + 7m - 6)$

60. $(x^2 + x) - (3x^2 + 2x - 1)$

61. $(16x^3 - x^2 + 3x) + (-12x^3 + 3x^2 + 2x)$

62. $(-2b^6 + 3b^4 - b^2) + (b^6 + 2b^4 + 2b^2)$

63. Subtract $18y^4 - 5y^2 + y$ from $7y^4 + 3y^2 + 2y$.

64. Subtract $19t^5 - 6t^3 + t$ from $8t^5 + 3t^3 + 5t$.

65. $(9a^4 - 3a^2 + 2) + (4a^4 - 4a^2 + 2) + (-12a^4 + 6a^2 - 3)$

66. $(4m^2 - 3m + 2) + (5m^2 + 13m - 4) - (16m^2 + 4m - 3)$

67. $[(8m^2 + 4m - 7) - (2m^2 - 5m + 2)] - (m^2 + m + 1)$

68. $[(9b^3 - 4b^2 + 3b + 2) - (-2b^3 - 3b^2 + b)] - (8b^3 + 6b + 4)$

69. $[(3x^2 - 2x + 7) - (4x^2 + 2x - 3)] - [(9x^2 + 4x - 6) + (-4x^2 + 4x + 4)]$

70. $[(6t^2 - 3t + 1) - (12t^2 + 2t - 6)] - [(4t^2 - 3t - 8) + (-6t^2 + 10t - 12)]$

71. *Concept Check* Without actually performing the operations, determine mentally the coefficient of the x^2-term in the simplified form of $(-4x^2 + 2x - 3) - (-2x^2 + x - 1) + (-8x^2 + 3x - 4)$.

72. *Concept Check* Without actually performing the operations, determine mentally the coefficient of the x-term in the simplified form of $(-8x^2 - 3x + 2) - (4x^2 - 3x + 8) - (-2x^2 - x + 7)$.

Add or subtract as indicated. See Example 9.

73. $(6b + 3c) + (-2b - 8c)$

74. $(-5t + 13s) + (8t - 3s)$

75. $(4x + 2xy - 3) - (-2x + 3xy + 4)$

76. $(8ab + 2a - 3b) - (6ab - 2a + 3b)$

77. $(5x^2y - 2xy + 9xy^2) - (8x^2y + 13xy + 12xy^2)$

78. $(16t^3s^2 + 8t^2s^3 + 9ts^4) - (-24t^3s^2 + 3t^2s^3 - 18ts^4)$

Find a polynomial that represents the perimeter of each rectangle or square.

79.

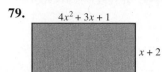

$4x^2 + 3x + 1$

$x + 2$

80.

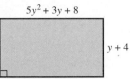

$5y^2 + 3y + 8$

$y + 4$

81.

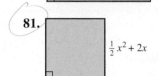

$\frac{1}{2}x^2 + 2x$

82.

$\frac{3}{4}x^2 + x$

*Find **(a)** a polynomial that represents the perimeter of each triangle and **(b)** the measures of the angles of the triangle.*

83.

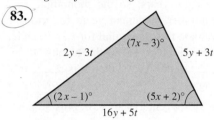

$2y - 3t$

$(7x - 3)°$

$5y + 3t$

$(2x - 1)°$

$(5x + 2)°$

$16y + 5t$

84.

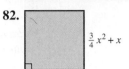

$-t^2s + 6ts$

$(8x + 3)°$

$4t^2s - 3ts^2 + 2ts$

$(6x + 7)°$

$(3x)°$

$-8t^2s + 6ts^2 + ts$

Perform each indicated operation.

85. Subtract $9x^2 - 6x + 5$ from $3x^2 - 2$.

86. Find the difference when $9x^4 + 3x^2 + 5$ is subtracted from $8x^4 - 2x^3 + x - 1$.

87. Find the difference between the sum of $5x^2 + 2x - 3$ and $x^2 - 8x + 2$ and the sum of $7x^2 - 3x + 6$ and $-x^2 + 4x - 6$.

88. Subtract the sum of $9t^3 - 3t + 8$ and $t^2 - 8t + 4$ from the sum of $12t + 8$ and $t^2 - 10t + 3$.

Graph each equation by completing the table of values. See Example 10.

89. $y = x^2 - 4$

x	y
-2	
-1	
0	
1	
2	

90. $y = x^2 - 6$

x	y
-2	
-1	
0	
1	
2	

91. $y = 2x^2 - 1$

x	y
-2	
-1	
0	
1	
2	

92. $y = 2x^2 + 2$

x	y
-2	
-1	
0	
1	
2	

93. $y = -x^2 + 4$

x	y
-2	
-1	
0	
1	
2	

94. $y = -x^2 + 2$

x	y
-2	
-1	
0	
1	
2	

95. $y = (x + 3)^2$

x	-5	-4	-3	-2	-1
y					

96. $y = (x - 4)^2$

x	2	3	4	5	6
y					

RELATING CONCEPTS	(EXERCISES 97–100)

FOR INDIVIDUAL OR GROUP WORK

As explained in the Connections box in this section, the polynomial equation

$$y = -0.0545x^2 + 5.047x + 11.78$$

gives a good approximation of the age of a dog in human years y, where x represents age in dog years. Each time we evaluate this polynomial for a value of x, we get one and only one output value y. This idea is basic to the concept of a function, one of the most important topics in mathematics.

Exercises 97–100 further illustrate the function concept with polynomials. **Work these exercises in order.**

97. It used to be thought that each dog year was about 7 human years, so the monomial $7x$ gave the number of human years for x dog years. Evaluate this monomial for $x = 9$, and then use the result to fill in the blanks: If a dog is _____ in dog years, then it is _____ in human years.

PUDDLES

98. If it costs $15 to rent a chain saw, plus $2 per day, the binomial $2x + 15$ gives the cost to rent the chain saw for x days. Evaluate this polynomial for 6, and then use the result to fill in the blanks: If the saw is rented for _____ days, the cost is _____.

99. If an object is projected upward under certain conditions, its height in feet is given by the trinomial

$$-16x^2 + 60x + 80,$$

where x is in seconds. Evaluate this polynomial for 2.5, and then use the result to fill in the blanks: If _____ seconds have elapsed, the height of the object is _____ feet.

100. The polynomial

$$1331x^2 - 9390x + 68,849$$

gives a good approximation of the number of Iowa families receiving food stamps during the period from 1996 to 2005, where $x = 0$ represents 1996. Use this polynomial to approximate the number of Iowa families receiving food stamps in 1997. (*Hint:* Any power of 1 is equal to 1, so simply add the coefficients and the constant.) (*Source:* Iowa Department of Human Services.)

PREVIEW EXERCISES

Multiply. See **Section 5.1.**

101. $(2a)(-5ab)$ **102.** $(3xz)(4x)$ **103.** $(-m^2)(m^5)$ **104.** $(2c)(3c^2)$

Multiply. See **Section 1.8.**

105. $5(x + 4)$ **106.** $-3(x^2 + 7)$ **107.** $4(2a + 6b)$ **108.** $\dfrac{1}{2}(4m - 8n)$

5.5 Multiplying Polynomials

OBJECTIVE 1 Multiply a monomial and a polynomial. As shown in **Section 5.1,** we find the product of two monomials by using the rules for exponents and the commutative and associative properties. For example,

$$-8m^6(-9n^6) = -8(-9)(m^6)(n^6) = 72m^6n^6.$$

▶ **CAUTION** *Do not confuse addition of terms with multiplication of terms.* For instance,

$$7q^5 + 2q^5 = 9q^5, \quad \text{but} \quad (7q^5)(2q^5) = 7 \cdot 2q^{5+5} = 14q^{10}.$$

To find the product of a monomial and a polynomial with more than one term, we use the distributive property and multiplication of monomials.

EXAMPLE 1 Multiplying Monomials and Polynomials

Find each product.

(a) $4x^2(3x + 5) = 4x^2(3x) + 4x^2(5)$ Distributive property

$= 12x^3 + 20x^2$ Multiply monomials.

(b) $-8m^3(4m^3 + 3m^2 + 2m - 1)$

$= -8m^3(4m^3) + (-8m^3)(3m^2)$

$+ (-8m^3)(2m) + (-8m^3)(-1)$ Distributive property

$= -32m^6 - 24m^5 - 16m^4 + 8m^3$ Multiply monomials.

✔ **Now Try Exercises 11 and 19.**

OBJECTIVE 2 Multiply two polynomials. We can use the distributive property repeatedly to find the product of any two polynomials. For example, to find the product of the polynomials $x^2 + 3x + 5$ and $x - 4$, think of $x - 4$ as a single quantity and use the distributive property as follows.

$$(x^2 + 3x + 5)(x - 4) = x^2(x - 4) + 3x(x - 4) + 5(x - 4)$$

Now use the distributive property three more times to find $x^2(x - 4)$, $3x(x - 4)$, and $5(x - 4)$.

$$x^2(x - 4) + 3x(x - 4) + 5(x - 4)$$
$$= x^2(x) + x^2(-4) + 3x(x) + 3x(-4) + 5(x) + 5(-4)$$
$$= x^3 - 4x^2 + 3x^2 - 12x + 5x - 20 \quad \text{Multiply monomials.}$$
$$= x^3 - x^2 - 7x - 20 \quad \text{Combine like terms.}$$

This example suggests the following rule.

> **Multiplying Polynomials**
>
> To multiply two polynomials, multiply each term of the second polynomial by each term of the first polynomial and add the products.

EXAMPLE 2 Multiplying Two Polynomials

Multiply $(m^2 + 5)(4m^3 - 2m^2 + 4m)$.

Multiply each term of the second polynomial by each term of the first.

$$(m^2 + 5)(4m^3 - 2m^2 + 4m)$$
$$= m^2(4m^3) + m^2(-2m^2) + m^2(4m) + 5(4m^3) + 5(-2m^2) + 5(4m)$$
$$= 4m^5 - 2m^4 + 4m^3 + 20m^3 - 10m^2 + 20m$$
$$= 4m^5 - 2m^4 + 24m^3 - 10m^2 + 20m \qquad \text{Combine like terms.}$$

✔ **Now Try Exercise 25.**

When at least one of the factors in a product of polynomials has two or more terms, the multiplication can be performed vertically.

EXAMPLE 3 Multiplying Polynomials Vertically

Multiply $(x^3 + 2x^2 + 4x + 1)(3x + 5)$ vertically.

Write the polynomials as follows.

$$\begin{array}{r} x^3 + 2x^2 + 4x + 1 \\ 3x + 5 \\ \hline \end{array}$$

It is not necessary to line up terms in columns, because any terms may be multiplied (not just like terms). Begin by multiplying each of the terms in the top row by 5.

$$\begin{array}{r} x^3 + 2x^2 + 4x + 1 \\ 3x + 5 \\ \hline 5x^3 + 10x^2 + 20x + 5 \end{array} \qquad 5(x^3 + 2x^2 + 4x + 1)$$

Notice how this process is similar to multiplication of whole numbers. Now multiply each term in the top row by $3x$.

Be careful here to place like terms in columns so they can be added.

$$\begin{array}{r} x^3 + 2x^2 + 4x + 1 \\ 3x + 5 \\ \hline 5x^3 + 10x^2 + 20x + 5 \\ 3x^4 + 6x^3 + 12x^2 + 3x \end{array} \qquad 3x(x^3 + 2x^2 + 4x + 1)$$

Add like terms.

$$\begin{array}{r} x^3 + 2x^2 + 4x + 1 \\ 3x + 5 \\ \hline 5x^3 + 10x^2 + 20x + 5 \\ 3x^4 + 6x^3 + 12x^2 + 3x \\ \hline 3x^4 + 11x^3 + 22x^2 + 23x + 5 \end{array} \longleftarrow \text{Product}$$

✔ **Now Try Exercise 29.**

| EXAMPLE 4 | **Multiplying Polynomials with Fractional Coefficients Vertically** |

Find the product of $4m^3 - 2m^2 + 4m$ and $\frac{1}{2}m^2 + \frac{5}{2}$.

$$4m^3 - 2m^2 + 4m$$
$$\frac{1}{2}m^2 + \frac{5}{2}$$

$\overline{10m^3 - 5m^2 + 10m}$ Terms of top row multiplied by $\frac{5}{2}$

$2m^5 - m^4 + 2m^3$ Terms of top row multiplied by $\frac{1}{2}m^2$

$\overline{2m^5 - m^4 + 12m^3 - 5m^2 + 10m}$ Add.

✔ **Now Try Exercise 35.**

We can use a rectangle to model polynomial multiplication. For example, to find the product

$$(2x + 1)(3x + 2),$$

label a rectangle with each term as shown next on the left. Then put the product of each pair of monomials in the appropriate box, as shown on the right.

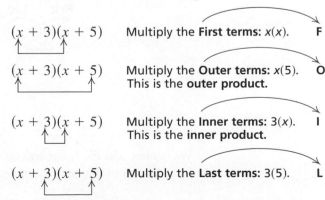

The product of the binomials is the sum of the four monomial products.

$$(2x + 1)(3x + 2) = 6x^2 + 4x + 3x + 2$$
$$= 6x^2 + 7x + 2$$

This approach can be extended to polynomials with any number of terms.

OBJECTIVE 3 **Multiply binomials by the FOIL method.** In algebra, many times the polynomials to be multiplied are binomials. For these products, the **FOIL method** reduces the rectangle method to a systematic approach without the rectangle. To develop the FOIL method, we use the distributive property to find $(x + 3)(x + 5)$.

$$(x + 3)(x + 5) = (x + 3)x + (x + 3)5$$ Distributive property
$$= x(x) + 3(x) + x(5) + 3(5)$$ Distributive property
$$= x^2 + 3x + 5x + 15$$ Multiply.
$$= x^2 + 8x + 15$$ Combine like terms.

Here is where the letters of the word FOIL originate.

$(x + 3)(x + 5)$ Multiply the **First terms:** $x(x)$.   **F**

$(x + 3)(x + 5)$ Multiply the **Outer terms:** $x(5)$.   **O**
                          This is the **outer product.**

$(x + 3)(x + 5)$ Multiply the **Inner terms:** $3(x)$.   **I**
                          This is the **inner product.**

$(x + 3)(x + 5)$ Multiply the **Last terms:** $3(5)$.   **L**

The outer product, $5x$, and the inner product, $3x$, should be added mentally to get $8x$ so that the three terms of the answer can be written without extra steps.

$$(x + 3)(x + 5) = x^2 + 8x + 15$$

A summary of the steps in the FOIL method follows.

Multiplying Binomials by the FOIL Method

Step 1 Multiply the two **First** terms of the binomials to get the first term of the answer.

Step 2 Find the **Outer** product and the **Inner** product and add them (when possible) to get the middle term of the answer.

Step 3 Multiply the two **Last** terms of the binomials to get the last term of the answer.

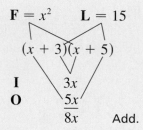

EXAMPLE 5 **Using the FOIL Method**

Use the FOIL method to find the product $(x + 8)(x - 6)$.

Step 1 **F** Multiply the first terms: $x(x) = x^2$.

Step 2 **O** Find the outer product: $x(-6) = -6x$.

  I Find the inner product: $8(x) = 8x$.

  Add the outer and inner products mentally: $-6x + 8x = 2x$.

Step 3 **L** Multiply the last terms: $8(-6) = -48$.

The product of $x + 8$ and $x - 6$ is the sum of the terms found in Steps 1–3.

$$(x + 8)(x - 6) = x^2 + 2x - 48$$

As a shortcut, this product can be found in the following manner:

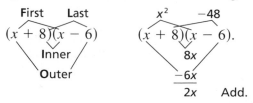

✔ **Now Try Exercise 37.**

We cannot add the inner and outer products of the FOIL method if unlike terms result, as shown in the next example.

EXAMPLE 6 Using the FOIL Method

Multiply $(9x - 2)(3y + 1)$.

First	$(9x - 2)(3y + 1)$	$27xy$
Outer	$(9x - 2)(3y + 1)$	$9x$
Inner	$(9x - 2)(3y + 1)$	$-6y$
Last	$(9x - 2)(3y + 1)$	-2

These unlike terms cannot be added.

$$\begin{array}{cccc} \text{F} & \text{O} & \text{I} & \text{L} \end{array}$$
$$(9x - 2)(3y + 1) = 27xy + 9x - 6y - 2$$

✔ **Now Try Exercise 51.**

EXAMPLE 7 Using the FOIL Method

Find each product.

$$\begin{array}{cccc} \text{F} & \text{O} & \text{I} & \text{L} \end{array}$$

(a) $(2k + 5y)(k + 3y) = 2k(k) + 2k(3y) + 5y(k) + 5y(3y)$
$$= 2k^2 + 6ky + 5ky + 15y^2 \quad \text{Multiply.}$$
$$= 2k^2 + 11ky + 15y^2 \quad \text{Combine like terms.}$$

(b) $(7p + 2q)(3p - q) = 21p^2 - pq - 2q^2 \quad \text{FOIL}$

(c) $2x^2(x - 3)(3x + 4) = 2x^2(3x^2 - 5x - 12) \quad \text{FOIL}$
$$= 6x^4 - 10x^3 - 24x^2 \quad \text{Distributive property}$$

✔ **Now Try Exercises 53 and 55.**

▶ **NOTE** Example 7(c) showed one way to multiply three polynomials. We could have multiplied $2x^2$ and $x - 3$ first and then multiplied that product and $3x + 4$.
$$2x^2(x - 3)(3x + 4) = (2x^3 - 6x^2)(3x + 4)$$
$$= 6x^4 - 10x^3 - 24x^2 \quad \text{FOIL}$$

5.5 EXERCISES

🌐 *Complete solution available on Video Lectures on CD/DVD*

Now Try Exercise

Concept Check *In Exercises 1 and 2, match each product in Column I with the correct polynomial in Column II.*

I	II
1. (a) $5x^3(6x^5)$	**A.** $125x^{15}$
(b) $-5x^5(6x^3)$	**B.** $30x^8$
(c) $(5x^5)^3$	**C.** $-216x^9$
(d) $(-6x^3)^3$	**D.** $-30x^8$

I	II
2. (a) $(x - 5)(x + 3)$	**A.** $x^2 + 8x + 15$
(b) $(x + 5)(x + 3)$	**B.** $x^2 - 8x + 15$
(c) $(x - 5)(x - 3)$	**C.** $x^2 - 2x - 15$
(d) $(x + 5)(x - 3)$	**D.** $x^2 + 2x - 15$

Find each product. See Objective 1.

3. $5y^4(3y^7)$

4. $10p^2(5p^3)$

5. $-15a^4(-2a^5)$

6. $-3m^6(-5m^4)$

7. $5p(3q^2)$

8. $4a^3(3b^2)$

9. $-6m^3(3n^2)$

10. $9r^3(-2s^2)$

Find each product. See Example 1.

11. $2m(3m + 2)$

12. $4x(5x + 3)$

13. $3p(-2p^3 + 4p^2)$

14. $4x(3 + 2x + 5x^3)$

15. $-8z(2z + 3z^2 + 3z^3)$

16. $-7y(3 + 5y^2 - 2y^3)$

17. $2y^3(3 + 2y + 5y^4)$

18. $2m^4(3m^2 + 5m + 6)$

19. $-4r^3(-7r^2 + 8r - 9)$

20. $-9a^5(-3a^6 - 2a^4 + 8a^2)$

21. $3a^2(2a^2 - 4ab + 5b^2)$

22. $4z^3(8z^2 + 5zy - 3y^2)$

23. $7m^3n^2(3m^2 + 2mn - n^3)$

24. $2p^2q(3p^2q^2 - 5p + 2q^2)$

Find each product. See Examples 2–4.

25. $(6x + 1)(2x^2 + 4x + 1)$

26. $(9y - 2)(8y^2 - 6y + 1)$

27. $(9a + 2)(9a^2 + a + 1)$

28. $(2r - 1)(3r^2 + 4r - 4)$

29. $(4m + 3)(5m^3 - 4m^2 + m - 5)$

30. $(y + 4)(3y^4 - 2y^2 + 1)$

31. $(2x - 1)(3x^5 - 2x^3 + x^2 - 2x + 3)$

32. $(2a + 3)(a^4 - a^3 + a^2 - a + 1)$

33. $(5x^2 + 2x + 1)(x^2 - 3x + 5)$

34. $(2m^2 + m - 3)(m^2 - 4m + 5)$

35. $(6x^4 - 4x^2 + 8x)\left(\dfrac{1}{2}x + 3\right)$

36. $(8y^6 + 4y^4 - 12y^2)\left(\dfrac{3}{4}y^2 + 2\right)$

Find each product. Use the FOIL method. See Examples 5–7.

37. $(m + 7)(m + 5)$

38. $(n - 1)(n + 4)$

39. $(x + 5)(x - 5)$

40. $(y + 8)(y - 8)$

41. $(2x + 3)(6x - 4)$

42. $(4m + 3)(4m + 3)$

43. $(3x - 2)(3x - 2)$

44. $(b + 8)(6b - 2)$

45. $(5a + 1)(2a + 7)$

46. $(8 - 3a)(2 + a)$

47. $(6 - 5m)(2 + 3m)$

48. $(-4 + k)(2 - k)$

49. $(5 - 3x)(4 + x)$

50. $(2m - 3n)(m + 5n)$

51. $(4x + 3)(2y - 1)$

52. $(5x + 7)(3y - 8)$

53. $(3x + 2y)(5x - 3y)$

54. $x(2x - 5)(x + 3)$

55. $3y^3(2y + 3)(y - 5)$

56. $5t^4(t + 3)(3t - 1)$

57. $-8r^3(5r^2 + 2)(5r^2 - 2)$

*Find polynomials that represent (**a**) the area and (**b**) the perimeter of each square or rectangle. (If necessary, refer to the formulas on the inside covers.)*

58.

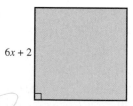

$6x + 2$

59.

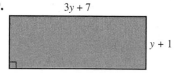

$3y + 7$

$y + 1$

60. Perform the indicated multiplications, and then describe the pattern in the products.

    (a) $(x + 4)(x - 4)$; $(y + 2)(y - 2)$; $(r + 7)(r - 7)$

    (b) $(x + 4)(x + 4)$; $(y - 2)(y - 2)$; $(r + 7)(r + 7)$

Find each product. In Exercises 71–74 and 78–80, apply the meaning of exponents.

61. $\left(3p + \dfrac{5}{4}q\right)\left(2p - \dfrac{5}{3}q\right)$

62. $\left(-x + \dfrac{2}{3}y\right)\left(3x - \dfrac{3}{4}y\right)$

63. $(x + 7)^2$ **64.** $(m + 6)^2$ **65.** $(a - 4)(a + 4)$

66. $(b - 10)(b + 10)$ **67.** $(2p - 5)^2$ **68.** $(3m + 1)^2$

69. $(5k + 3q)^2$ **70.** $(8m - 3n)^2$ **71.** $(m - 5)^3$

72. $(p + 3)^3$ **73.** $(2a + 1)^3$ **74.** $(3m - 1)^3$

75. $7(4m - 3)(2m + 1)$ **76.** $-4r(3r + 2)(2r - 5)$ **77.** $-3a(3a + 1)(a - 4)$

78. $(k + 1)^4$ **79.** $(3r - 2s)^4$ **80.** $(2z + 5y)^4$

81. $3p^3(2p^2 + 5p)(p^3 + 2p + 1)$ **82.** $5k^2(k^2 - k + 4)(k^3 - 3)$

83. $-2x^5(3x^2 + 2x - 5)(4x + 2)$ **84.** $-4x^3(3x^4 + 2x^2 - x)(-2x + 1)$

Find a polynomial that represents the area of each shaded region. In Exercises 87 and 88, leave π in your answers. (If necessary, refer to the formulas on the inside covers.)

85.   **86.**

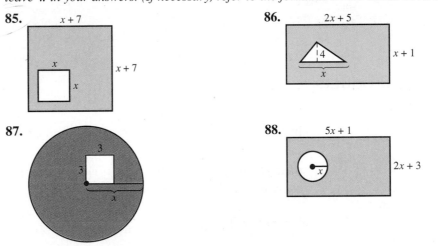

87.   **88.**

RELATING CONCEPTS (EXERCISES 89–95)

FOR INDIVIDUAL OR GROUP WORK

Work Exercises 89–95 in order.

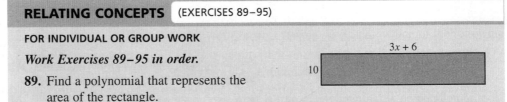

89. Find a polynomial that represents the area of the rectangle.

90. Suppose you know that the area of the rectangle is 600 yd². Use this information and the polynomial from Exercise 89 to write an equation in x, and solve it.

91. What are the dimensions of the rectangle? (Assume that all units are in yards.)

92. Suppose the rectangle represents a lawn and it costs \$3.50 per yd² to lay sod on the lawn. How much will it cost to sod the entire lawn?

93. Use the result of Exercise 91 to find the perimeter of the lawn.

94. Again, suppose the rectangle represents a lawn and it costs \$9.00 per yd to fence the lawn. How much will it cost to fence the lawn?

95. (a) Suppose that it costs k dollars per yd² to sod the lawn. Determine a polynomial in the variables x and k that represents the cost of sodding the entire lawn.

    (b) Suppose that it costs r dollars per yd to fence the lawn. Determine a polynomial in the variables x and r that represents the cost of fencing the lawn.

96. Explain the FOIL method for multiplying two binomials. Give an example.

PREVIEW EXERCISES

*Apply a power rule for exponents. See **Section 5.1.***

97. $(3m)^2$ **98.** $(5p)^2$ **99.** $(-2r)^2$

100. $(-5a)^2$ **101.** $(4x^2)^2$ **102.** $(8y^3)^2$

5.6 Special Products

OBJECTIVES

1. Square binomials.
2. Find the product of the sum and difference of two terms.
3. Find greater powers of binomials.

In this section, we develop shortcuts to find certain binomial products.

OBJECTIVE 1   Square binomials.   The square of a binomial can be found quickly by using the method suggested by Example 1.

EXAMPLE 1   Squaring a Binomial

Find $(m + 3)^2$.

$$(m + 3)(m + 3) = m^2 + 3m + 3m + 9 \quad \text{FOIL}$$
$$= m^2 + 6m + 9 \quad \text{Combine like terms.}$$

This result has the squares of the first and the last terms of the binomial:

$$m^2 = m^2 \quad \text{and} \quad 3^2 = 9.$$

The middle term is twice the product of the two terms of the binomial, since the outer and inner products are $m(3)$ and $3(m)$, and

$$m(3) + 3(m) = 2(m)(3) = 6m.$$

✔ **Now Try Exercise 1.**

This example suggests the following rules.

Square of a Binomial

The square of a binomial is a trinomial consisting of the square of the first term of the binomial, plus twice the product of the two terms, plus the square of the last term of the binomial. For x and y,

$$(x + y)^2 = x^2 + 2xy + y^2.$$

Also, $$(x - y)^2 = x^2 - 2xy + y^2.$$

EXAMPLE 2 Squaring Binomials

Square each binomial and simplify.

$$(x - y)^2 = x^2 - 2 \cdot x \cdot y + y^2$$

(a) $(5z - 1)^2 = (5z)^2 - 2(5z)(1) + (1)^2$

$$= 25z^2 - 10z + 1 \qquad (5z)^2 = 5^2 z^2 = 25z^2$$

(b) $(3b + 5r)^2 = (3b)^2 + 2(3b)(5r) + (5r)^2$

$$= 9b^2 + 30br + 25r^2$$

(c) $(2a - 9x)^2 = (2a)^2 - 2(2a)(9x) + (9x)^2$

$$= 4a^2 - 36ax + 81x^2$$

(d) $\left(4m + \dfrac{1}{2}\right)^2 = (4m)^2 + 2(4m)\left(\dfrac{1}{2}\right) + \left(\dfrac{1}{2}\right)^2$

$$= 16m^2 + 4m + \dfrac{1}{4}$$

Remember the middle term.

(e) $x(4x - 3)^2 = x(16x^2 - 24x + 9)$ Square the binomial.

$$= 16x^3 - 24x^2 + 9x \qquad \text{Distributive property}$$

✔ **Now Try Exercises 7, 9, 15, and 17.**

Notice that in the square of a sum, all of the terms are positive, as in Examples 2(b) and (d). *In the square of a difference, the middle term is negative,* as in Examples 2(a), (c), and (e).

▶ **CAUTION** A common error when squaring a binomial is to forget the middle term of the product. In general,

$$(x + y)^2 = x^2 + 2xy + y^2, \quad \text{not} \quad x^2 + y^2.$$

OBJECTIVE 2 **Find the product of the sum and difference of two terms.** In binomial products of the form $(x + y)(x - y)$, one binomial is the sum of two terms and the other is the difference of the *same* two terms. Consider $(x + 2)(x - 2)$.

$$(x + 2)(x - 2) = x^2 - 2x + 2x - 4$$
$$= x^2 - 4$$

Thus, the product of $x + y$ and $x - y$ is the difference of two squares.

Product of the Sum and Difference of Two Terms

$$(x + y)(x - y) = x^2 - y^2$$

EXAMPLE 3 Finding the Product of the Sum and Difference of Two Terms

Find each product.

(a) $(x + 4)(x - 4)$

Use the rule for the product of the sum and difference of two terms.

$$(x + 4)(x - 4) = x^2 - 4^2$$
$$= x^2 - 16$$

(b) $\left(\dfrac{2}{3} - w\right)\left(\dfrac{2}{3} + w\right) = \left(\dfrac{2}{3} + w\right)\left(\dfrac{2}{3} - w\right)$ Commutative property

$$= \left(\dfrac{2}{3}\right)^2 - w^2$$

$$= \dfrac{4}{9} - w^2$$

✔ **Now Try Exercises 25 and 37.**

EXAMPLE 4 Finding the Product of the Sum and Difference of Two Terms

Find each product.

$$(x + y)\ (x - y)\ =\ x^2\ -\ y^2$$
$$\downarrow\quad\downarrow\quad\downarrow\quad\downarrow\qquad\downarrow\qquad\downarrow$$

(a) $(5m + 3)(5m - 3) = (5m)^2 - 3^2$ Use the rule.
$$= 25m^2 - 9$$ Apply the exponents.

(b) $(4x + y)(4x - y) = (4x)^2 - y^2$ **(c)** $\left(z - \dfrac{1}{4}\right)\left(z + \dfrac{1}{4}\right) = z^2 - \dfrac{1}{16}$
$$= 16x^2 - y^2$$

(d) $p(2p + 1)(2p - 1) = p(4p^2 - 1)$
$$= 4p^3 - p$$

✔ **Now Try Exercises 33 and 41.**

OBJECTIVE 3 Find greater powers of binomials. The methods used in the previous section and this section can be combined to find greater powers of binomials.

EXAMPLE 5 Finding Greater Powers of Binomials

Find each product.

(a) $(x + 5)^3 = (x + 5)^2(x + 5)$ $a^3 = a^2 \cdot a$
$$= (x^2 + 10x + 25)(x + 5)$$ Square the binomial.
$$= x^3 + 10x^2 + 25x + 5x^2 + 50x + 125$$ Multiply polynomials.
$$= x^3 + 15x^2 + 75x + 125$$ Combine like terms.

(b) $(2y - 3)^4 = (2y - 3)^2(2y - 3)^2$ $a^4 = a^2 \cdot a^2$

$= (4y^2 - 12y + 9)(4y^2 - 12y + 9)$ Square each binomial.

$= 16y^4 - 48y^3 + 36y^2 - 48y^3 + 144y^2$ Multiply polynomials.

$ - 108y + 36y^2 - 108y + 81$

$= 16y^4 - 96y^3 + 216y^2 - 216y + 81$ Combine like terms.

✔ **Now Try Exercises 43 and 51.**

5.6 EXERCISES

🌐 **1. Concept Check**   Consider the square $(2x + 3)^2$.

     (a) What is the square of the first term, $(2x)^2$?

     (b) What is twice the product of the two terms, $2(2x)(3)$?

     (c) What is the square of the last term, 3^2?

     (d) Write the final product, which is a trinomial, using your results in parts (a)–(c).

🖊 **2.** Explain in your own words how to square a binomial. Give an example.

Find each product. See Examples 1 and 2.

3. $(m + 2)^2$ **4.** $(x + 8)^2$ **5.** $(r - 3)^2$

6. $(z - 5)^2$ 🌐 **7.** $(x + 2y)^2$ **8.** $(3m - p)^2$

9. $(5p + 2q)^2$ **10.** $(8a - 3b)^2$ **11.** $(4a + 5b)^2$

12. $(9y + z)^2$ **13.** $(7t + s)^2$ **14.** $\left(5x + \dfrac{2}{5}y\right)^2$

🌐 **15.** $\left(6m - \dfrac{4}{5}n\right)^2$ **16.** $x(2x + 5)^2$ **17.** $t(3t - 1)^2$

18. $2x(7x - 2)^2$ **19.** $-(4r - 2)^2$ **20.** $-(3y - 8)^2$

21. Concept Check   Consider the product $(7x + 3y)(7x - 3y)$.

     (a) What is the product of the first terms, $7x(7x)$?

     (b) Multiply the outer terms, $7x(-3y)$. Then multiply the inner terms, $3y(7x)$. Add the results. What is this sum?

     (c) What is the product of the last terms, $3y(-3y)$?

     (d) Write the final product, using your results in parts (a) and (c). Why is the sum found in part (b) omitted here?

🖊 **22.** Explain in your own words how to find the product of the sum and the difference of two terms. Give an example.

23. Concept Check   The square of a binomial leads to a polynomial with how many terms? The product of the sum and difference of two terms leads to a polynomial with how many terms?

Find each product. See Examples 3 and 4.

24. $(a + 8)(a - 8)$ 🌐 **25.** $(k + 5)(k - 5)$ **26.** $(2 + p)(2 - p)$

27. $(4 - 3t)(4 + 3t)$ **28.** $(2m + 5)(2m - 5)$ **29.** $(5x + 2)(5x - 2)$

30. $(3x + 4y)(3x - 4y)$ **31.** $(6a - p)(6a + p)$ **32.** $(5y + 3x)(5y - 3x)$

33. $(10x + 3y)(10x - 3y)$ **34.** $(13r + 2z)(13r - 2z)$ **35.** $(2x^2 - 5)(2x^2 + 5)$

36. $(9y^2 - 2)(9y^2 + 2)$ **37.** $\left(\dfrac{3}{4} - x\right)\left(\dfrac{3}{4} + x\right)$ **38.** $\left(7x + \dfrac{3}{7}\right)\left(7x - \dfrac{3}{7}\right)$

39. $\left(9y + \dfrac{2}{3}\right)\left(9y - \dfrac{2}{3}\right)$ **40.** $p(3p + 7)(3p - 7)$ **41.** $q(5q - 1)(5q + 1)$

42. Does $(a + b)^3$ equal $a^3 + b^3$ in general? Explain.

Find each product. See Example 5.

43. $(x + 1)^3$ **44.** $(y + 2)^3$ **45.** $(t - 3)^3$

46. $(m - 5)^3$ **47.** $(r + 5)^3$ **48.** $(p + 3)^3$

49. $(2a + 1)^3$ **50.** $(3m - 1)^3$ **51.** $(4x - 1)^4$

52. $(2x + 1)^4$ **53.** $(3r - 2t)^4$ **54.** $(2z + 5y)^4$

RELATING CONCEPTS   (EXERCISES 55–64)

FOR INDIVIDUAL OR GROUP WORK

Special products can be illustrated by using areas of rectangles. Use the figure, and **work Exercises 55–60 in order** *to justify the special product*

$$(a + b)^2 = a^2 + 2ab + b^2.$$

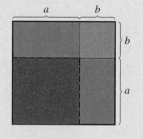

55. Express the area of the large square as the square of a binomial.

56. Give the monomial that represents the area of the red square.

57. Give the monomial that represents the sum of the areas of the blue rectangles.

58. Give the monomial that represents the area of the yellow square.

59. What is the sum of the monomials you obtained in Exercises 56–58?

60. Explain why the binomial square you found in Exercise 55 must equal the polynomial you found in Exercise 59.

To understand how the special product $(a + b)^2 = a^2 + 2ab + b^2$ *can be applied to a purely numerical problem,* **work Exercises 61–64 in order.**

61. Evaluate 35^2, using either traditional paper-and-pencil methods or a calculator.

62. The number 35 can be written as $30 + 5$. Therefore, $35^2 = (30 + 5)^2$. Use the special product for squaring a binomial with $a = 30$ and $b = 5$ to write an expression for $(30 + 5)^2$. Do not simplify at this time.

63. Use the order of operations to simplify the expression you found in Exercise 62.

64. How do the answers in Exercises 61 and 63 compare?

The special product

$$(x + y)(x - y) = x^2 - y^2$$

can be used to perform some multiplication problems. For example,

$51 \times 49 = (50 + 1)(50 - 1)$	$102 \times 98 = (100 + 2)(100 - 2)$
$= 50^2 - 1^2$	$= 100^2 - 2^2$
$= 2500 - 1$	$= 10{,}000 - 4$
$= 2499.$	$= 9996.$

Once these patterns are recognized, multiplications of this type can be done mentally. Use this method to calculate each product mentally.

65. 101×99 **66.** 103×97 **67.** 201×199

68. 301×299 **69.** $20\dfrac{1}{2} \times 19\dfrac{1}{2}$ **70.** $30\dfrac{1}{3} \times 29\dfrac{2}{3}$

Determine a polynomial that represents the area of each figure. (If necessary, refer to the formulas on the inside covers.)

71.

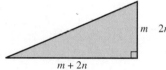

$m + 2n$, $m + 2n$

72.

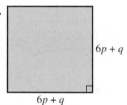

$6p + q$, $6p + q$

73.

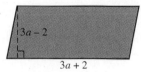

$3a - 2$, $3a + 2$

74.

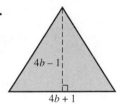

$4b - 1$, $4b + 1$

75.

$x + 2$

76.

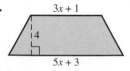

$3x + 1$, 4, $5x + 3$

In Exercises 77 and 78, refer to the figure shown here.

77. Find a polynomial that represents the volume of the cube.

78. If the value of x is 6, what is the volume of the cube?

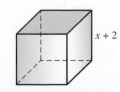

$x + 2$

PREVIEW EXERCISES

Write each product as a sum of terms. Write answers with positive exponents only. Simplify each term. See Section 1.8.

79. $\dfrac{1}{2p}(4p^2 + 2p + 8)$

80. $\dfrac{1}{5x}(5x^2 - 10x + 45)$

81. $\dfrac{1}{3m}(m^3 + 9m^2 - 6m)$

82. $\dfrac{1}{4y}(y^4 + 6y^2 + 8)$

Find each product. See Section 5.5.

83. $-3k(8k^2 - 12k + 2)$ **84.** $(3r + 5)(2r + 1)$ **85.** $(-2k + 1)(8k^2 + 9k + 3)$

Subtract. See Section 5.4.

86. $5t^2 + 2t - 6$
   $\underline{5t^2 - 3t - 9}$

87. $-4x^3 + 2x^2 - 3x + 7$
   $\underline{-4x^3 - 8x^2 + x - 4}$

88. $x^4 + 0x^3 - 4x^2 + 3x - 8$
   $\underline{x^4 - 2x^3 + 4x^2 + 6x - 7}$

5.7 Dividing Polynomials

OBJECTIVES

1 Divide a polynomial by a monomial.

2 Divide a polynomial by a polynomial.

OBJECTIVE 1 Divide a polynomial by a monomial. We add two fractions with a common denominator as follows.

$$\frac{a}{c} + \frac{b}{c} = \frac{a+b}{c}$$

In reverse, this statement gives a rule for dividing a polynomial by a monomial.

Dividing a Polynomial by a Monomial

To divide a polynomial by a monomial, divide each term of the polynomial by the monomial:

$$\frac{a+b}{c} = \frac{a}{c} + \frac{b}{c} \quad (c \neq 0).$$

Examples: $\dfrac{2+5}{3} = \dfrac{2}{3} + \dfrac{5}{3}$ and $\dfrac{x+3z}{2y} = \dfrac{x}{2y} + \dfrac{3z}{2y}$

The parts of a division problem are named here.

$$\begin{array}{c}\text{Dividend} \rightarrow \\ \text{Divisor} \rightarrow\end{array} \dfrac{12x^2 + 6x}{6x} = 2x + 1 \leftarrow \text{Quotient}$$

EXAMPLE 1 Dividing a Polynomial by a Monomial

Divide $5m^5 - 10m^3$ by $5m^2$.

Use the preceding rule, with $+$ replaced by $-$. Then use the quotient rule.

$$\frac{5m^5 - 10m^3}{5m^2} = \frac{5m^5}{5m^2} - \frac{10m^3}{5m^2} = m^3 - 2m$$

Check by multiplying: $5m^2 \cdot (m^3 - 2m) = 5m^5 - 10m^3.$ ←—Original polynomial
                                                                                               (Dividend)

Divisor Quotient

Because division by 0 is undefined, the quotient $\frac{5m^5 - 10m^3}{5m^2}$ is undefined if $m = 0$. From now on, we assume that no denominators are 0.

✔ **Now Try Exercise 15.**

EXAMPLE 2 Dividing a Polynomial by a Monomial

Divide $\dfrac{16a^5 - 12a^4 + 8a^2}{4a^3}$.

$$\frac{16a^5 - 12a^4 + 8a^2}{4a^3} = \frac{16a^5}{4a^3} - \frac{12a^4}{4a^3} + \frac{8a^2}{4a^3} \qquad \text{Divide each term by } 4a^3.$$

$$= 4a^2 - 3a + \frac{2}{a} \qquad \text{Quotient rule}$$

The quotient is not a polynomial because of the expression $\frac{2}{a}$, which has a variable in the denominator. While the sum, difference, and product of two polynomials are always polynomials, the quotient of two polynomials may not be.

Check by multiplying.

$$4a^3\left(4a^2 - 3a + \frac{2}{a}\right) = 4a^3(4a^2) + 4a^3(-3a) + 4a^3\left(\frac{2}{a}\right)$$

$$= 16a^5 - 12a^4 + 8a^2$$

✔ **Now Try Exercise 13.**

EXAMPLE 3 Dividing a Polynomial by a Monomial with a Negative Coefficient

Divide $-7x^3 + 12x^4 - 4x$ by $-4x$.

Write the polynomial in descending powers as $12x^4 - 7x^3 - 4x$ before dividing.

Write in descending powers.

$$\frac{12x^4 - 7x^3 - 4x}{-4x} = \frac{12x^4}{-4x} - \frac{7x^3}{-4x} - \frac{4x}{-4x} \qquad \text{Divide each term by } -4x.$$

$$= -3x^3 - \frac{7x^2}{-4} - (-1)$$

$$= -3x^3 + \frac{7}{4}x^2 + 1$$

Check by multiplying.

✔ **Now Try Exercise 25.**

EXAMPLE 4 Dividing a Polynomial by a Monomial

Divide $180x^4y^{10} - 150x^3y^8 + 120x^2y^6 - 90xy^4 + 100y$ by $30xy^2$.

$$\frac{180x^4y^{10} - 150x^3y^8 + 120x^2y^6 - 90xy^4 + 100y}{30xy^2}$$

$$= \frac{180x^4y^{10}}{30xy^2} - \frac{150x^3y^8}{30xy^2} + \frac{120x^2y^6}{30xy^2} - \frac{90xy^4}{30xy^2} + \frac{100y}{30xy^2}$$

$$= 6x^3y^8 - 5x^2y^6 + 4xy^4 - 3y^2 + \frac{10}{3xy}$$

✔ **Now Try Exercise 31.**

OBJECTIVE 2 Divide a polynomial by a polynomial. As shown in the box, we use a method of "long division" to divide a polynomial by a polynomial (other than a monomial). *Both polynomials must first be written in descending powers.*

Dividing Whole Numbers	Dividing Polynomials
Step 1	
Divide 6696 by 27.	Divide $8x^3 - 4x^2 - 14x + 15$ by $2x + 3$.
$27\overline{)6696}$	$2x + 3\overline{)8x^3 - 4x^2 - 14x + 15}$
Step 2	
66 divided by 27 = 2; $2 \cdot 27 = 54$.	$8x^3$ divided by $2x = 4x^2$; $4x^2(2x + 3) = 8x^3 + 12x^2$.
$\begin{array}{r} 2 \\ 27\overline{)6696} \\ 54 \end{array}$	$\begin{array}{r} 4x^2 \\ 2x + 3\overline{)8x^3 - 4x^2 - 14x + 15} \\ 8x^3 + 12x^2 \end{array}$
Step 3	
Subtract; then bring down the next digit.	Subtract; then bring down the next term.
$\begin{array}{r} 2 \\ 27\overline{)6696} \\ 54\downarrow \\ \hline 129 \end{array}$	$\begin{array}{r} 4x^2 \\ 2x + 3\overline{)8x^3 - 4x^2 - 14x + 15} \\ 8x^3 + 12x^2 \quad\downarrow \\ \hline -16x^2 - 14x \end{array}$
	(To subtract two polynomials, change the signs of the second and then add.)
Step 4	
129 divided by 27 = 4; $4 \cdot 27 = 108$.	$-16x^2$ divided by $2x = -8x$; $-8x(2x + 3) = -16x^2 - 24x$.
$\begin{array}{r} 24 \\ 27\overline{)6696} \\ 54 \\ \hline 129 \\ 108 \end{array}$	$\begin{array}{r} 4x^2 - 8x \\ 2x + 3\overline{)8x^3 - 4x^2 - 14x + 15} \\ 8x^3 + 12x^2 \\ \hline -16x^2 - 14x \\ -16x^2 - 24x \end{array}$

(continued)

Step 5

Subtract; then bring down the next digit.

$$
\begin{array}{r}
24 \\
27\overline{)6696} \\
\underline{54} \\
129 \\
\underline{108} \\
216
\end{array}
$$

Subtract; then bring down the next term.

$$
\begin{array}{r}
4x^2 - 8x \\
2x + 3\overline{)8x^3 - 4x^2 - 14x + 15} \\
\underline{8x^3 + 12x^2} \\
-16x^2 - 14x \\
\underline{-16x^2 - 24x} \\
10x + 15
\end{array}
$$

Step 6

216 divided by 27 = 8;
8 · 27 = 216.

$$
\begin{array}{r}
248 \\
27\overline{)6696} \\
\underline{54} \\
129 \\
\underline{108} \\
216 \\
\underline{216} \\
0
\end{array}
$$

6696 divided by 27 is 248.
The remainder is 0.

$10x$ divided by $2x = 5$;
$5(2x + 3) = 10x + 15.$

$$
\begin{array}{r}
4x^2 - 8x + 5 \\
2x + 3\overline{)8x^3 - 4x^2 - 14x + 15} \\
\underline{8x^3 + 12x^2} \\
-16x^2 - 14x \\
\underline{-16x^2 - 24x} \\
10x + 15 \\
\underline{10x + 15} \\
0
\end{array}
$$

$8x^3 - 4x^2 - 14x + 15$ divided by $2x + 3$ is $4x^2 - 8x + 5$. The remainder is 0.

Step 7

Check by multiplying.

$$27 \cdot 248 = 6696$$

Check by multiplying.

$$(2x + 3)(4x^2 - 8x + 5)$$
$$= 8x^3 - 4x^2 - 14x + 15$$

✔ **Now Try Exercise 51.**

EXAMPLE 5 Dividing a Polynomial by a Polynomial

Divide $\dfrac{5x + 4x^3 - 8 - 4x^2}{2x - 1}$.

The first polynomial must be written in descending powers as $4x^3 - 4x^2 + 5x - 8$. Then divide by $2x - 1$.

$$
\begin{array}{r}
2x^2 - x + 2 \\
2x - 1\overline{)4x^3 - 4x^2 + 5x - 8} \\
\underline{4x^3 - 2x^2} \\
-2x^2 + 5x \\
\underline{-2x^2 + x} \\
4x - 8 \\
\underline{4x - 2} \\
-6
\end{array}
$$

Write in descending powers.

To subtract, add the opposite.

← Remainder

$$\begin{array}{r} 2x^2 - x + 2 \\ 2x - 1 \overline{)4x^3 - 4x^2 + 5x - 8} \\ \underline{4x^3 - 2x^2} \\ -2x^2 + 5x \\ \underline{-2x^2 + x} \\ 4x - 8 \\ \underline{4x - 2} \\ -6 \end{array}$$

Step 1 $4x^3$ divided by $2x = 2x^2$; $2x^2(2x - 1) = 4x^3 - 2x^2$.

Step 2 Subtract; bring down the next term.

Step 3 $-2x^2$ divided by $2x = -x$; $-x(2x - 1) = -2x^2 + x$.

Step 4 Subtract; bring down the next term.

Step 5 $4x$ divided by $2x = 2$; $2(2x - 1) = 4x - 2$.

Step 6 Subtract. The remainder is -6. Write the remainder as the numerator of a fraction that has $2x - 1$ as its denominator. Because of the nonzero remainder, the answer is not a polynomial.

Remember to add $\frac{\text{remainder}}{\text{divisor}}$.
Don't forget the + sign.

$$\text{Dividend} \rightarrow \frac{4x^3 - 4x^2 + 5x - 8}{2x - 1} = \underbrace{2x^2 - x + 2}_{\substack{\text{Quotient} \\ \text{polynomial}}} + \underbrace{\frac{-6}{2x - 1}}_{\substack{\text{Fractional part} \\ \text{of quotient}}}$$

Divisor $\rightarrow$; $\leftarrow$ Remainder ; $\leftarrow$ Divisor

Step 7 *Check* by multiplying.

$$(2x - 1)\left(2x^2 - x + 2 + \frac{-6}{2x - 1}\right)$$

$$= (2x - 1)(2x^2) + (2x - 1)(-x) + (2x - 1)(2) + (2x - 1)\left(\frac{-6}{2x - 1}\right)$$

$$= 4x^3 - 2x^2 - 2x^2 + x + 4x - 2 - 6$$

$$= 4x^3 - 4x^2 + 5x - 8$$

✔ **Now Try Exercise 57.**

▶ **CAUTION**   Remember to include "$+ \frac{\text{remainder}}{\text{divisor}}$" as part of the answer.

EXAMPLE 6 **Dividing into a Polynomial with Missing Terms**

Divide $x^3 - 1$ by $x - 1$.

Here, the dividend, $x^3 - 1$, is missing the x^2-term and the x-term. When terms are missing, we use 0 as the coefficient for each missing term. (Zero acts as a placeholder here, just as it does in our number system.) Thus, $x^3 - 1 = x^3 + 0x^2 + 0x - 1$.

$$\begin{array}{r} x^2 + x + 1 \\ x - 1 \overline{)x^3 + 0x^2 + 0x - 1} \\ \underline{x^3 - x^2} \\ x^2 + 0x \\ \underline{x^2 - x} \\ x - 1 \\ \underline{x - 1} \\ 0 \end{array}$$

Insert placeholders for the missing terms.

The remainder is 0. The quotient is $x^2 + x + 1$. *Check* by multiplying.

$$(x^2 + x + 1)(x - 1) = x^3 - 1 \qquad \text{Quotient} \cdot \text{Divisor} = \text{Dividend}$$

✔ **Now Try Exercise 61.**

EXAMPLE 7   **Dividing by a Polynomial with Missing Terms**

Divide $x^4 + 2x^3 + 2x^2 - x - 1$ by $x^2 + 1$.

Since the divisor, $x^2 + 1$, has a missing x-term, write it as $x^2 + 0x + 1$.

Insert a placeholder for the missing term.

$$
\begin{array}{r}
x^2 + 2x + 1 \\
x^2 + 0x + 1 \overline{) x^4 + 2x^3 + 2x^2 - x - 1} \\
\underline{x^4 + 0x^3 + x^2} \\
2x^3 + x^2 - x \\
\underline{2x^3 + 0x^2 + 2x} \\
x^2 - 3x - 1 \\
\underline{x^2 + 0x + 1} \\
-3x - 2 \leftarrow \text{Remainder}
\end{array}
$$

When the result of subtracting ($-3x - 2$ here) is a polynomial of lesser degree than the divisor ($x^2 + 0x + 1$), that polynomial is the remainder. The answer is

$$x^2 + 2x + 1 + \frac{-3x - 2}{x^2 + 1}.$$

Remember to write "$+ \frac{\text{remainder}}{\text{divisor}}$."

Multiply to check that this is correct.

✔ **Now Try Exercise 71.**

EXAMPLE 8   **Dividing a Polynomial when the Quotient Has Fractional Coefficients**

Divide $4x^3 + 2x^2 + 3x + 2$ by $4x - 4$.

$$
\begin{array}{r}
\frac{6x^2}{4x} = \frac{3}{2}x \\
x^2 + \frac{3}{2}x + \frac{9}{4} \leftarrow \frac{9x}{4x} = \frac{9}{4} \\
4x - 4 \overline{) 4x^3 + 2x^2 + 3x + 2} \\
\underline{4x^3 - 4x^2} \\
6x^2 + 3x \\
\underline{6x^2 - 6x} \\
9x + 2 \\
\underline{9x - 9} \\
11
\end{array}
$$

The answer is $x^2 + \dfrac{3}{2}x + \dfrac{9}{4} + \dfrac{11}{4x - 4}$.

✔ **Now Try Exercise 75.**

5.7 EXERCISES

Now Try Exercise

Concept Check *Fill in each blank with the correct response.*

1. In the statement $\frac{6x^2 + 8}{2} = 3x^2 + 4$, _____ is the dividend, _____ is the divisor, and _____ is the quotient.

2. The expression $\frac{3x + 12}{x}$ is undefined if $x =$ _____.

3. To check the division shown in Exercise 1, multiply _____ by _____ and show that the product is _____.

4. The expression $5x^2 - 3x + 6 + \frac{2}{x}$ _____ a polynomial.
(is/is not)

5. Explain why the division problem $\frac{16m^3 - 12m^2}{4m}$ can be performed by using the methods of this section, while the division problem $\frac{4m}{16m^3 - 12m^2}$ cannot.

6. ***Concept Check*** Suppose that a polynomial in the variable x has degree 5 and it is divided by a monomial in the variable x having degree 3. What is the degree of the quotient?

Perform each division. See Examples 1–3.

7. $\frac{60x^4 - 20x^2 + 10x}{2x}$

8. $\frac{120x^6 - 60x^3 + 80x^2}{2x}$

9. $\frac{20m^5 - 10m^4 + 5m^2}{5m^2}$

10. $\frac{12t^5 - 6t^3 + 6t^2}{6t^2}$

11. $\frac{8t^5 - 4t^3 + 4t^2}{2t}$

12. $\frac{8r^4 - 4r^3 + 6r^2}{2r}$

13. $\frac{4a^5 - 4a^2 + 8}{4a}$

14. $\frac{5t^8 + 5t^7 + 15}{5t}$

Divide each polynomial by $3x^2$. See Examples 1–3.

15. $12x^5 - 9x^4 + 6x^3$

16. $24x^6 - 12x^5 + 30x^4$

17. $3x^2 + 15x^3 - 27x^4$

18. $3x^2 - 18x^4 + 30x^5$

19. $36x + 24x^2 + 6x^3$

20. $9x - 12x^2 + 9x^3$

21. $4x^4 + 3x^3 + 2x$

22. $5x^4 - 6x^3 + 8x$

Perform each division. See Examples 1–4.

23. $\frac{-27r^4 + 36r^3 - 6r^2 - 26r + 2}{-3r}$

24. $\frac{-8k^4 + 12k^3 + 2k^2 - 7k + 3}{-2k}$

25. $\frac{2m^5 - 6m^4 + 8m^2}{-2m^3}$

26. $\frac{6r^5 - 8r^4 + 10r^2}{-2r^4}$

27. $(20a^4 - 15a^5 + 25a^3) \div (5a^4)$

28. $(16y^5 - 8y^2 + 12y) \div (4y^2)$

29. $(120x^{11} - 60x^{10} + 140x^9 - 100x^8) \div (10x^{12})$

30. $(120x^{12} - 84x^9 + 60x^8 - 36x^7) \div (12x^9)$

31. $(120x^5y^4 - 80x^2y^3 + 40x^2y^4 - 20x^5y^3) \div (20xy^2)$

32. $(200a^5b^6 - 160a^4b^7 - 120a^3b^9 + 40a^2b^2) \div (40a^2b)$

33. The area of the rectangle is given by the polynomial $15x^3 + 12x^2 - 9x + 3$. What polynomial expresses the length?

34. The area of the triangle is given by the polynomial $24m^3 + 48m^2 + 12m$. What polynomial expresses the length of the base?

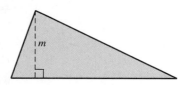

RELATING CONCEPTS (EXERCISES 35–38)

FOR INDIVIDUAL OR GROUP WORK

Our system of numeration is called a decimal system. It is based on powers of 10. *In a whole number such as* 2846, *each digit is understood to represent the number of powers of* 10 *for its place value. The* 2 *represents two thousands* (2×10^3), *the* 8 *represents eight hundreds* (8×10^2), *the* 4 *represents four tens* (4×10^1), *and the* 6 *represents six ones (or units)* (6×10^0). *In expanded form, we write*

$$2846 = (2 \times 10^3) + (8 \times 10^2) + (4 \times 10^1) + (6 \times 10^0).$$

Keeping this information in mind, **work Exercises 35–38 in order.**

35. Divide 2846 by 2, using paper-and-pencil methods: $2\overline{)2846}$.

36. Write your answer from Exercise 35 in expanded form.

37. Divide the polynomial $2x^3 + 8x^2 + 4x + 6$ by 2.

38. Compare your answers in Exercises 36 and 37. How are they similar? How are they different? For what value of x does the answer in Exercise 37 equal the answer in Exercise 36?

Perform each division. See Example 5.

39. $\dfrac{x^2 - x - 6}{x - 3}$

40. $\dfrac{m^2 - 2m - 24}{m - 6}$

41. $\dfrac{2y^2 + 9y - 35}{y + 7}$

42. $\dfrac{2y^2 + 9y + 7}{y + 1}$

43. $\dfrac{p^2 + 2p + 20}{p + 6}$

44. $\dfrac{x^2 + 11x + 16}{x + 8}$

45. $(r^2 - 8r + 15) \div (r - 3)$

46. $(t^2 + 2t - 35) \div (t - 5)$

47. $\dfrac{12m^2 - 20m + 3}{2m - 3}$

48. $\dfrac{12y^2 + 20y + 7}{2y + 1}$

49. $\dfrac{4a^2 - 22a + 32}{2a + 3}$

50. $\dfrac{9w^2 + 6w + 10}{3w - 2}$

51. $\dfrac{8x^3 - 10x^2 - x + 3}{2x + 1}$

52. $\dfrac{12t^3 - 11t^2 + 9t + 18}{4t + 3}$

53. $\dfrac{8k^4 - 12k^3 - 2k^2 + 7k - 6}{2k - 3}$

54. $\dfrac{27r^4 - 36r^3 - 6r^2 + 26r - 24}{3r - 4}$

55. $\dfrac{5y^4 + 5y^3 + 2y^2 - y - 3}{y + 1}$

56. $\dfrac{2r^3 - 5r^2 - 6r + 15}{r - 3}$

57. $\dfrac{3k^3 - 4k^2 - 6k + 10}{k - 2}$

58. $\dfrac{5z^3 - z^2 + 10z + 2}{z + 2}$

59. $\dfrac{6p^4 - 16p^3 + 15p^2 - 5p + 10}{3p + 1}$

60. $\dfrac{6r^4 - 11r^3 - r^2 + 16r - 8}{2r - 3}$

Perform each division. See Examples 5–8.

61. $(x^3 - 2x^2 - 9) \div (x - 3)$

62. $(x^3 + 2x^2 - 3) \div (x - 1)$

63. $(2x^3 + x + 2) \div (x + 1)$

64. $(3x^3 + x + 5) \div (x + 1)$

65. $\dfrac{5 - 2r^2 + r^4}{r^2 - 1}$

66. $\dfrac{4t^2 + t^4 + 7}{t^2 + 1}$

67. $\dfrac{y^3 + 1}{y + 1}$

68. $\dfrac{y^3 - 1}{y - 1}$

69. $\dfrac{a^4 - 1}{a^2 - 1}$

70. $\dfrac{a^4 - 1}{a^2 + 1}$

71. $\dfrac{x^4 - 4x^3 + 5x^2 - 3x + 2}{x^2 + 3}$

72. $\dfrac{3t^4 + 5t^3 - 8t^2 - 13t + 2}{t^2 - 5}$

73. $\dfrac{2x^5 + 9x^4 + 8x^3 + 10x^2 + 14x + 5}{2x^2 + 3x + 1}$

74. $\dfrac{4t^5 - 11t^4 - 6t^3 + 5t^2 - t + 3}{4t^2 + t - 3}$

75. $(3a^2 - 11a + 17) \div (2a + 6)$

76. $(4x^2 + 11x - 8) \div (3x + 6)$

Find a polynomial that describes each required quantity. (If necessary, refer to the formulas on the inside covers.)

77. Give the length of the rectangle.

$5x + 2$

The area is $5x^3 + 7x^2 - 13x - 6$ sq. units.

78. Find the measure of the base of the parallelogram.

$x - 1$

The area is $2x^3 + 2x^2 - 3x - 1$ sq. units.

79. If the distance traveled is $(5x^3 - 6x^2 + 3x + 14)$ mi and the rate is $(x + 1)$ mph, what is an expression for the time traveled?

80. If it costs $(4x^5 + 3x^4 + 2x^3 + 9x^2 - 29x + 2)$ dollars to fertilize a garden, and fertilizer costs $(x + 2)$ dollars per yd^2, what is an expression for the area of the garden?

PREVIEW EXERCISES

*List all positive integer factors of each number. See **Section 1.1.***

81. 18

82. 36

83. 48

84. 23

Chapter 5 | Group Activity

MEASURING THE FLIGHT OF A ROCKET

Objective Graph quadratic equations to solve an application problem.

A physics class observes the launch of two rockets, one that has an initial velocity of 48 ft per sec and another that has an initial velocity of 64 ft per sec. Use the steps that follow to determine the maximum height each rocket will achieve and how long it will take it to reach that height.

To find the height of a projectile shot vertically into the air, use the formula

$$H = -16t^2 + Vt,$$

where H = height above the launch pad in feet, t = time in seconds, and V = initial velocity.

A. One student should complete the table for the first rocket. The other student should do the same for the second rocket. Use the formula for the height of a projectile and the values provided.

Initial Velocity = 48 Feet per Second		Initial Velocity = 64 Feet per Second	
Time (seconds)	Height (feet)	Time (seconds)	Height (feet)
0		0	
0.5		0.5	
1		1	
1.5		1.5	
2		2	
2.5		2.5	
3		3	
		3.5	
		4	

B. Graph the results. You must determine the scales for the two axes on the coordinate system. Label each graph and its scale clearly.

C. Answer the following questions about each rocket.

  1. Find the time in seconds when the rocket is at height 0 ft. Why does this happen twice?

  2. What is the maximum height the rocket reached?

  3. At what time (in seconds) did the rocket reach its maximum height?

D. Compare the data from both rockets and answer the following questions.

  1. Which rocket had the greater maximum height? Explain.

  2. Which rocket reached its maximum height first? Explain.

  3. Do both rockets reach a height of 48 ft? If yes, at what times? What observations about projectiles might this situation lead to?

Chapter **5** **SUMMARY** *View the Interactive Summary on the Pass the Test CD.*

KEY TERMS

5.3 scientific notation
5.4 polynomial
descending powers
degree of a term

degree of a
   polynomial
monomial
binomial

trinomial
parabola
vertex
axis

line of symmetry
5.5 outer product
inner product
FOIL

NEW SYMBOLS

x^{-n} x to the negative n
      power

TEST YOUR WORD POWER

See how well you have learned the vocabulary in this chapter. Answers, with examples, follow the Quick Review.

1. A **polynomial** is an algebraic expression made up of
   A. a term or a finite product of terms with positive coefficients and exponents
   B. a term or a finite sum of terms with real coefficients and whole number exponents
   C. the product of two or more terms with positive exponents
   D. the sum of two or more terms with whole number coefficients and exponents.

2. The **degree of a term** is
   A. the number of variables in the term

B. the product of the exponents on the variables
C. the least exponent on the variables
D. the sum of the exponents on the variables.

3. **FOIL** is a method for
   A. adding two binomials
   B. adding two trinomials
   C. multiplying two binomials
   D. multiplying two trinomials.

4. A **binomial** is a polynomial with
   A. only one term
   B. exactly two terms
   C. exactly three terms

D. more than three terms.

5. A **monomial** is a polynomial with
   A. only one term
   B. exactly two terms
   C. exactly three terms
   D. more than three terms.

6. A **trinomial** is a polynomial with
   A. only one term
   B. exactly two terms
   C. exactly three terms
   D. more than three terms.

QUICK REVIEW

Concepts	Examples

5.1 **THE PRODUCT RULE AND POWER RULES FOR EXPONENTS**

For any integers m and n,

Product Rule $a^m \cdot a^n = a^{m+n}$

Power Rules   **(a)** $(a^m)^n = a^{mn}$

          (b) $(ab)^m = a^m b^m$

          (c) $\left(\dfrac{a}{b}\right)^m = \dfrac{a^m}{b^m}$ $(b \neq 0)$.

Perform the operations by using rules for exponents.

$$2^4 \cdot 2^5 = 2^{4+5} = 2^9$$

$$(3^4)^2 = 3^{4 \cdot 2} = 3^8$$

$$(6a)^5 = 6^5 a^5$$

$$\left(\frac{2}{3}\right)^4 = \frac{2^4}{3^4}$$

(continued)

Concepts	Examples

5.2 INTEGER EXPONENTS AND THE QUOTIENT RULE

If $a \neq 0$, then for integers m and n,

Zero Exponent $a^0 = 1$

Negative Exponent $a^{-n} = \dfrac{1}{a^n}$

Quotient Rule $\dfrac{a^m}{a^n} = a^{m-n}$

Negative-to-Positive Rules

$$\frac{a^{-m}}{b^{-n}} = \frac{b^n}{a^m} \quad (b \neq 0)$$

$$\left(\frac{a}{b}\right)^{-m} = \left(\frac{b}{a}\right)^m \quad (b \neq 0).$$

Simplify by using the rules for exponents.

$$15^0 = 1$$

$$5^{-2} = \frac{1}{5^2} = \frac{1}{25}$$

$$\frac{4^8}{4^3} = 4^{8-3} = 4^5$$

$$\frac{4^{-2}}{3^{-5}} = \frac{3^5}{4^2}$$

$$\left(\frac{6}{5}\right)^{-3} = \left(\frac{5}{6}\right)^3$$

5.3 AN APPLICATION OF EXPONENTS: SCIENTIFIC NOTATION

To write a number in scientific notation (as $a \times 10^n$, where $1 \leq |a| < 10$), move the decimal point to follow the first nonzero digit. If moving the decimal point makes the number smaller, n is positive. If it makes the number larger, n is negative. If the decimal point is not moved, n is 0.

Write in scientific notation.

$$247 = 2.47 \times 10^2$$
$$0.0051 = 5.1 \times 10^{-3}$$
$$4.8 = 4.8 \times 10^0$$

Write without exponents.

$$3.25 \times 10^5 = 325,000$$
$$8.44 \times 10^{-6} = 0.00000844$$

5.4 ADDING AND SUBTRACTING POLYNOMIALS; GRAPHING SIMPLE POLYNOMIALS

Adding Polynomials
Add like terms.

Add.

$$\begin{array}{r} 2x^2 + 5x - 3 \\ 5x^2 - 2x + 7 \\ \hline 7x^2 + 3x + 4 \end{array}$$

Subtracting Polynomials
Change the signs of the terms in the second polynomial and add the second polynomial to the first.

Subtract.

$$(2x^2 + 5x - 3) - (5x^2 - 2x + 7)$$
$$= (2x^2 + 5x - 3) + (-5x^2 + 2x - 7)$$
$$= -3x^2 + 7x - 10$$

Graphing Simple Polynomials
To graph a simple polynomial equation such as $y = x^2 - 2$, plot points near the vertex. (In this chapter, all parabolas have a vertex on the x-axis or the y-axis.)

Graph $y = x^2 - 2$.

x	y
−2	2
−1	−1
0	−2
1	−1
2	2

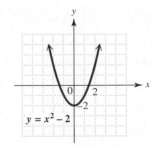

$y = x^2 - 2$

(continued)

Concepts	Examples

5.5 MULTIPLYING POLYNOMIALS

General Method for Multiplying Polynomials
Multiply each term of the first polynomial by each term of the second polynomial. Then add like terms.

Multiply.

$$
\begin{array}{r}
3x^3 - 4x^2 + 2x - 7 \\
4x + 3 \\
\hline
9x^3 - 12x^2 + 6x - 21 \\
12x^4 - 16x^3 + 8x^2 - 28x \\
\hline
12x^4 - 7x^3 - 4x^2 - 22x - 21
\end{array}
$$

FOIL Method for Multiplying Binomials

Step 1 Multiply the two **F**irst terms to get the first term of the answer.

Step 2 Find the **O**uter product and the **I**nner product, and mentally add them, when possible, to get the middle term of the answer.

Step 3 Multiply the two **L**ast terms to get the last term of the answer.

Add the terms found in Steps 1–3.

Multiply. $(2x + 3)(5x - 4)$

$$2x(5x) = 10x^2$$

$$2x(-4) + 3(5x) = 7x$$

$$3(-4) = -12$$

The product of $(2x + 3)$ and $(5x - 4)$ is $10x^2 + 7x - 12$.

5.6 SPECIAL PRODUCTS

Square of a Binomial
$$(x + y)^2 = x^2 + 2xy + y^2$$
$$(x - y)^2 = x^2 - 2xy + y^2$$

Product of the Sum and Difference of Two Terms
$$(x + y)(x - y) = x^2 - y^2$$

Multiply.

$$(3x + 1)^2 = 9x^2 + 6x + 1$$
$$(2m - 5n)^2 = 4m^2 - 20mn + 25n^2$$

$$(4a + 3)(4a - 3) = 16a^2 - 9$$

5.7 DIVIDING POLYNOMIALS

Dividing a Polynomial by a Monomial
Divide each term of the polynomial by the monomial.
$$\frac{a + b}{c} = \frac{a}{c} + \frac{b}{c}$$

Divide.

$$\frac{4x^3 - 2x^2 + 6x - 9}{2x} = 2x^2 - x + 3 - \frac{9}{2x}$$

Dividing a Polynomial by a Polynomial
Use "long division."

$$
\begin{array}{r}
2x - 5 \\
3x + 4 \overline{)6x^2 - 7x - 21} \\
\underline{6x^2 + 8x} \\
-15x - 21 \\
\underline{-15x - 20} \\
-1 \leftarrow \text{Remainder}
\end{array}
$$

The final answer is $2x - 5 + \dfrac{-1}{3x + 4}$.

Answers to Test Your Word Power

1. B; *Example:* $5x^3 + 2x^2 - 7$

2. D; *Examples:* The term 6 has degree 0, $3x$ has degree 1, $-2x^8$ has degree 8, and $5x^2y^4$ has degree 6.

3. C; *Example:* $(m + 4)(m - 3)$
    F   O   I   L
$= m(m) - 3m + 4m + 4(-3)$
$= m^2 + m - 12$

4. B; *Example:* $3t^3 + 5t$

5. A; *Examples:* -5 and $4xy^5$

6. C; *Example:* $2a^2 - 3ab + b^2$

Chapter **5** REVIEW EXERCISES

[5.1] *Use the product rule, power rules, or both to simplify each expression. Write the answers in exponential form.*

1. $4^3 \cdot 4^8$

2. $(-5)^6(-5)^5$

3. $(-8x^4)(9x^3)$

4. $(2x^2)(5x^3)(x^9)$

5. $(19x)^5$

6. $(-4y)^7$

7. $5(pt)^4$

8. $\left(\dfrac{7}{5}\right)^6$

9. $(3x^2y^3)^3$

10. $(t^4)^8(t^2)^5$

11. $(6x^2z^4)^2(x^3yz^2)^4$

12. $\left(\dfrac{2m^3n}{p^2}\right)^3$

13. Why does the product rule for exponents not apply to the expression $7^2 + 7^4$?

[5.2] *Evaluate each expression.*

14. $6^0 + (-6)^0$

15. $-(-23)^0$

16. -10^0

Simplify. Write each answer with only positive exponents. Assume that all variables represent nonzero real numbers.

17. -7^{-2}

18. $\left(\dfrac{5}{8}\right)^{-2}$

19. $(5^{-2})^{-4}$

20. $9^3 \cdot 9^{-5}$

21. $2^{-1} + 4^{-1}$

22. $\dfrac{6^{-5}}{6^{-3}}$

23. $\dfrac{x^{-7}}{x^{-9}}$

24. $\dfrac{y^4 \cdot y^{-2}}{y^{-5}}$

25. $(3r^{-2})^{-4}$

26. $(3p)^4(3p^{-7})$

27. $\dfrac{ab^{-3}}{a^4b^2}$

28. $\dfrac{(6r^{-1})^2(2r^{-4})}{r^{-5}(r^2)^{-3}}$

[5.3] *Write each number in scientific notation.*

29. 48,000,000

30. 28,988,000,000

31. 0.0000000824

Write each number without exponents.

32. 2.4×10^4

33. 7.83×10^7

34. 8.97×10^{-7}

Perform each indicated operation and write the answer without exponents.

35. $(2 \times 10^{-3}) \times (4 \times 10^5)$

36. $\dfrac{8 \times 10^4}{2 \times 10^{-2}}$

37. $\dfrac{12 \times 10^{-5} \times 5 \times 10^4}{4 \times 10^3 \times 6 \times 10^{-2}}$

Write each boldface italic number in the quote without exponents.

38. The muon, a close relative of the electron produced by the bombardment of cosmic rays against the upper atmosphere, has a half-life of 2 millionths of a second (*2×10^{-6}* s). (Excerpt from *Conceptual Physics,* 6th edition, by Paul G. Hewitt. Copyright © by Paul G. Hewitt. Published by HarperCollins College Publishers.)

39. There are 13 red balls and 39 black balls in a box. Mix them up and draw 13 out one at a time without returning any ball . . . the probability that the 13 drawings each will produce a red ball is . . . ***1.6 × 10^{-12}***. (Weaver, Warren, *Lady Luck,* Doubleday, 1963, pp. 298–299.)

Write each boldface italic number in scientific notation.

40. An electron and a positron attract each other in two ways: the electromagnetic attraction of their opposite electric charges, and the gravitational attraction of their two masses. The electromagnetic attraction is

$$\textbf{\textit{4,200,000,000,000,000,000,000,000,000,000,000,000,000,000}}$$

times as strong as the gravitational. (Asimov, Isaac, *Isaac Asimov's Book of Facts,* Bell Publishing Company, 1981, p. 106.)

41. The aircraft carrier USS John Stennis is a ***97,000***-ton nuclear powered floating city with a crew of ***5000***. (*Source:* Seelye, Katharine Q., "Staunch Allies Hard to Beat: Defense Dept., Hollywood," *New York Times,* in *Plain Dealer,* June 10, 2002.)

42. A googol is

$$\textbf{\textit{10,000,000,000,000,000,000,000,000,000,000,000,000,000,000,000,}}$$
$$\textbf{\textit{000,000,000,000,000,000,000,000,000,000,000,000,000,000,000,000.}}$$

The Web search engine Google is named after a googol. Sergey Brin, president and cofounder of Google, Inc., was a mathematics major. He chose the name Google to describe the vast reach of this search engine. (*Source: The Gazette,* Cedar Rapids, Iowa, March 2, 2001.)

43. According to Campbell, Mitchell, and Reece in *Biology Concepts and Connections* (Benjamin Cummings, 1994, p. 230), "The amount of DNA in a human cell is about ***1000*** times greater than the DNA in *E. coli.* Does this mean humans have 1000 times as many genes as the ***2000*** in *E. coli*? The answer is probably no; the human genome is thought to carry between ***50,000*** and ***100,000*** genes, which code for various proteins (as well as for tRNA and rRNA)."

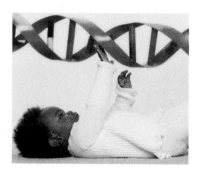

[5.4] *Combine like terms where possible in each polynomial. Write the answer in descending powers of the variable. Give the degree of the answer. Identify the polynomial as a* monomial, *a* binomial, *a* trinomial, *or* none of these.

44. $9m^2 + 11m^2 + 2m^2$

45. $-4p + p^3 - p^2 + 8p + 2$

46. $12a^5 - 9a^4 + 8a^3 + 2a^2 - a + 3$

47. $-7y^5 - 8y^4 - y^5 + y^4 + 9y$

48. $(12r^4 - 7r^3 + 2r^2) - (5r^4 - 3r^3 + 2r^2 - 1)$

49. Simplify $(5x^3y^2 - 3xy^5 + 12x^2) - (-9x^2 - 8x^3y^2 + 2xy^5)$.

Add or subtract as indicated.

50. Add.

$$-2a^3 + 5a^2$$
$$\underline{3a^3 - a^2}$$

51. Subtract.

$$6y^2 - 8y + 2$$
$$\underline{5y^2 + 2y - 7}$$

52. Subtract.

$$-12k^4 - 8k^2 + 7k$$
$$\underline{k^4 + 7k^2 - 11k}$$

Graph each equation by completing the table of values.

53. $y = -x^2 + 5$

x	-2	-1	0	1	2
y					

54. $y = 3x^2 - 2$

x	-2	-1	0	1	2
y					

[5.5] *Find each product.*

55. $(a + 2)(a^2 - 4a + 1)$

56. $(3r - 2)(2r^2 + 4r - 3)$

57. $(5p^2 + 3p)(p^3 - p^2 + 5)$

58. $(m - 9)(m + 2)$

59. $(3k - 6)(2k + 1)$

60. $(a + 3b)(2a - b)$

61. $(6k + 5q)(2k - 7q)$

62. $(s - 1)^3$

[5.6] *Find each product.*

63. $(a + 4)^2$

64. $(2r + 5t)^2$

65. $(6m - 5)(6m + 5)$

66. $(5a + 6b)(5a - 6b)$

67. $(r + 2)^3$

68. $t(5t - 3)^2$

69. Choose values for x and y to show that, in general, the following hold true.

(a) $(x + y)^2 \neq x^2 + y^2$ **(b)** $(x + y)^3 \neq x^3 + y^3$

70. Write an explanation on how to raise a binomial to the third power. Give an example.

71. Refer to Exercise 69. Suppose that you happened to let $x = 0$ and $y = 1$. Would your results be sufficient to illustrate the truth, in general, of the inequalities shown? If not, what would you need to do as your next step in working the exercise?

72. Find a polynomial that represents the volume of a cube with one side having length $(x^2 + 2)$ cm.

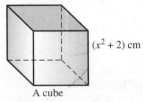

$(x^2 + 2)$ cm

A cube

73. Find a polynomial that represents the volume of a sphere with radius $(x + 1)$ in.

$(x + 1)$ in.

A sphere

[5.7] *Perform each division.*

74. $\dfrac{-15y^4}{9y^2}$

75. $\dfrac{6y^4 - 12y^2 + 18y}{6y}$

76. $(-10m^4n^2 + 5m^3n^2 + 6m^2n^4) \div (5m^2n)$

77. What polynomial, when multiplied by $6m^2n$, gives the product

$$12m^3n^2 + 18m^6n^3 - 24m^2n^2?$$

78. *Concept Check*   One of your friends in class simplified $\frac{6x^2 - 12x}{6}$ as $x^2 - 12x$. *WHAT WENT WRONG?* Give the correct answer.

Perform each division.

79. $\dfrac{2r^2 + 3r - 14}{r - 2}$

80. $\dfrac{10a^3 + 9a^2 - 14a + 9}{5a - 3}$

81. $\dfrac{x^4 - 5x^2 + 3x^3 - 3x + 4}{x^2 - 1}$

82. $\dfrac{m^4 + 4m^3 - 12m - 5m^2 + 6}{m^2 - 3}$

83. $\dfrac{16x^2 - 25}{4x + 5}$

84. $\dfrac{25y^2 - 100}{5y + 10}$

85. $\dfrac{y^3 - 8}{y - 2}$

86. $\dfrac{1000x^6 + 1}{10x^2 + 1}$

87. $\dfrac{6y^4 - 15y^3 + 14y^2 - 5y - 1}{3y^2 + 1}$

88. $\dfrac{4x^5 - 8x^4 - 3x^3 + 22x^2 - 15}{4x^2 - 3}$

MIXED REVIEW EXERCISES

Perform each indicated operation. Write answers with only positive exponents. Assume that all variables represent nonzero real numbers.

89. $5^0 + 7^0$

90. $\left(\dfrac{6r^2p}{5}\right)^3$

91. $(12a + 1)(12a - 1)$

92. 2^{-4}

93. $(8^{-3})^4$

94. $\dfrac{2p^3 - 6p^2 + 5p}{2p^2}$

95. $\dfrac{(2m^{-5})(3m^2)^{-1}}{m^{-2}(m^{-1})^2}$

96. $(3k - 6)(2k^2 + 4k + 1)$

97. $\dfrac{r^9 \cdot r^{-5}}{r^{-2} \cdot r^{-7}}$

98. $(2r + 5s)^2$

99. $(-5y^2 + 3y - 11) + (4y^2 - 7y + 15)$

100. $(2r + 5)(5r - 2)$

101. $\dfrac{2y^3 + 17y^2 + 37y + 7}{2y + 7}$

102. $(25x^2y^3 - 8xy^2 + 15x^3y) \div (10x^2y^3)$

103. $(6p^2 - p - 8) - (-4p^2 + 2p - 3)$

104. $\dfrac{3x^3 - 2x + 5}{x - 3}$

105. $(-7 + 2k)^2$

106. $\left(\dfrac{x}{y^{-3}}\right)^{-4}$

107. Find polynomials that represent the **(a)** perimeter and **(b)** area of the rectangle shown.

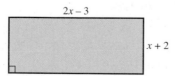

$2x - 3$

$x + 2$

108. If the side of a square has a measure represented by $5x^4 + 2x^2$, what polynomials represent its **(a)** perimeter and **(b)** area?

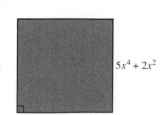

$5x^4 + 2x^2$

Chapter **5** TEST

View the complete solutions to all Chapter Test exercises on the Pass the Test CD.

Evaluate each expression.

1. 5^{-4}

2. $(-3)^0 + 4^0$

3. $4^{-1} + 3^{-1}$

4. Simplify $\dfrac{(3x^2y)^2(xy^3)^2}{(xy)^3}$. Assume that x and y are nonzero.

Simplify, and write the answer using only positive exponents. Assume that all variables represent nonzero numbers.

5. $\dfrac{8^{-1} \cdot 8^4}{8^{-2}}$

6. $\dfrac{(x^{-3})^{-2}(x^{-1}y)^2}{(xy^{-2})^2}$

7. Determine whether each expression represents a number that is *positive, negative,* or *zero.*

   (a) 3^{-4} **(b)** $(-3)^4$ **(c)** -3^4 **(d)** 3^0 **(e)** $(-3)^0 - 3^0$ **(f)** $(-3)^{-3}$

8. **(a)** Write $45{,}000{,}000{,}000$ using scientific notation.

   (b) Write 3.6×10^{-6} without using exponents.

   (c) Write the quotient without using exponents: $\dfrac{9.5 \times 10^{-1}}{5 \times 10^3}$.

9. A satellite galaxy of our own Milky Way, known as the Large Magellanic Cloud, is **1000** light-years across. A *light-year* is equal to **5,890,000,000,000** mi. (*Source:* "Images of Brightest Nebula Unveiled," *USA Today,* June 12, 2002.)

   (a) Write the two boldface italic numbers in scientific notation.

   (b) How many miles across is the Large Magellanic Cloud?

For each polynomial, combine like terms when possible and write the polynomial in descending powers of the variable. Give the degree of the simplified polynomial. Decide whether the simplified polynomial is a monomial, *a* binomial, *a* trinomial, *or* none of these.

10. $5x^2 + 8x - 12x^2$

11. $13n^3 - n^2 + n^4 + 3n^4 - 9n^2$

12. Use the table to complete a set of ordered pairs that lie on the graph of $y = 2x^2 - 4$. Then graph the equation.

x	-2	-1	0	1	2
y					

Perform each indicated operation.

13. $(2y^2 - 8y + 8) + (-3y^2 + 2y + 3) - (y^2 + 3y - 6)$

14. $(-9a^3b^2 + 13ab^5 + 5a^2b^2) - (6ab^5 + 12a^3b^2 + 10a^2b^2)$

15. Subtract.

$9t^3 - 4t^2 + 2t + 2$
$\underline{9t^3 + 8t^2 - 3t - 6}$

16. $3x^2(-9x^3 + 6x^2 - 2x + 1)$

17. $(t - 8)(t + 3)$

18. $(4x + 3y)(2x - y)$

19. $(5x - 2y)^2$

20. $(10v + 3w)(10v - 3w)$

21. $(2r - 3)(r^2 + 2r - 5)$

22. What polynomial expression represents the perimeter of this square? The area?

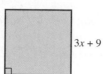

$3x + 9$

Perform each division.

23. $\dfrac{8y^3 - 6y^2 + 4y + 10}{2y}$

24. $(-9x^2y^3 + 6x^4y^3 + 12xy^3) \div (3xy)$

25. $(3x^3 - x + 4) \div (x - 2)$

Chapters 1–5 CUMULATIVE REVIEW EXERCISES

Write each fraction in lowest terms.

1. $\dfrac{28}{16}$

2. $\dfrac{55}{11}$

Perform each operation.

3. $\dfrac{2}{3} + \dfrac{1}{8}$

4. $\dfrac{7}{4} - \dfrac{9}{5}$

5. A contractor installs sheds. Each requires $1\frac{1}{4}$ yd^3 of concrete. How much concrete would be needed for 25 sheds?

6. A retailer has \$34,000 invested in her business. She finds that last year she earned 5.4% on this investment. How much did she earn?

7. List all positive integer factors of 45.

8. If $x = -2$ and $y = 4$, find the value of $\dfrac{4x - 2y}{x + y}$.

Perform each indicated operation.

9. $\dfrac{(-13 + 15) - (3 + 2)}{6 - 12}$

10. $-7 - 3[2 + (5 - 8)]$

Decide which property justifies each statement.

11. $(9 + 2) + 3 = 9 + (2 + 3)$

12. $6(4 + 2) = 6(4) + 6(2)$

13. Simplify the expression $-3(2x^2 - 8x + 9) - (4x^2 + 3x + 2)$.

Solve each equation.

14. $2 - 3(t - 5) = 4 + t$

15. $2(5h + 1) = 10h + 4$

16. $d = rt$ for r

17. $\dfrac{x}{5} = \dfrac{x - 2}{7}$

18. $\dfrac{1}{3}p - \dfrac{1}{6}p = -2$

19. $0.05x + 0.15(50 - x) = 5.50$

20. $4 - (3x + 12) = (2x - 9) - (5x - 1)$

Solve each problem.

21. A husky running the Iditarod burns $5\frac{3}{8}$ calories in exertion for every 1 calorie burned in thermo-regulation in extreme cold. According to one scientific study, a husky in top condition burns an amazing total of 11,200 calories per day. How many calories are burned for exertion, and how many are burned for regulation of body temperature? Round answers to the nearest whole number.

2006 XXXIV IDITAROD

22. If a number is subtracted from 8 and the difference is tripled, the result is three times the number. Find the number, and you will learn how many times a dolphin rests during a 24-hr period.

23. One side of a triangle is twice as long as a second side. The third side of the triangle is 17 ft long. The perimeter of the triangle cannot be more than 50 ft. Find the longest possible values for the other two sides of the triangle.

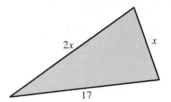

Solve each inequality.

24. $-8x \le -80$

25. $-2(x + 4) > 3x + 6$

26. $-3 \le 2x + 5 < 9$

27. Graph $y = -3x + 6$.

28. Consider the two points $(-1, 5)$ and $(2, 8)$.

    (a) Find the slope of the line joining them.

    (b) Find the equation of the line joining them.

29. Does the point $(-1, 3)$ lie within the shaded region of the graph of $y \ge x + 5$?

30. If $f(x) = x + 7$, find $f(-8)$.

Solve the system by using the method indicated.

31. $y = 2x + 5$
    $x + y = -4$ (Substitution)

32. $3x + 2y = 2$
    $2x + 3y = -7$ (Elimination)

Evaluate each expression.

33. $4^{-1} + 3^0$

34. $2^{-4} \cdot 2^5$

35. $\dfrac{8^{-5} \cdot 8^7}{8^2}$

36. Write with positive exponents only: $\dfrac{(a^{-3}b^2)^2}{(2a^{-4}b^{-3})^{-1}}$.

37. Write in scientific notation: 34,500.

38. It takes about 3.6×10^1 sec at a speed of 3.0×10^5 km per sec for light from the sun to reach Venus. How far is Venus from the sun? (*Source: World Almanac and Book of Facts 2006.*)

39. Graph $y = (x + 4)^2$, using the x-values $-6, -5, -4, -3,$ and -2 to obtain a set of points.

Perform each indicated operation.

40. $(7x^3 - 12x^2 - 3x + 8) + (6x^2 + 4) - (-4x^3 + 8x^2 - 2x - 2)$

41. $6x^5(3x^2 - 9x + 10)$

42. $(7x + 4)(9x + 3)$

43. $(5x + 8)^2$

44. $\dfrac{14x^3 - 21x^2 + 7x}{7x}$

45. $\dfrac{y^3 - 3y^2 + 8y - 6}{y - 1}$

Factoring and Applications

Wireless communication uses radio waves to carry signals and messages across distances. Cellular phones, one of the most popular forms of wireless communication, have become an invaluable tool for people to stay connected to family, friends, and work while on the go. In 2004 alone, U.S. sales of cellular phones totaled some $10,538 million as 70% of all U.S. households owned cell phones. (*Source: Microsoft Encarta Encyclopedia;* Consumer Electronics Association.)

In Exercise 33 of Section 6.6, we use a *quadratic equation* to model the number of cell phone subscribers in the United States.

6.1 The Greatest Common Factor; Factoring by Grouping

Recall from **Chapter 1** that to **factor** means "to write a quantity as a product." That is, factoring is the opposite of multiplying. For example,

$$Multiplying \qquad\qquad Factoring$$
$$6 \cdot 2 = 12, \qquad\qquad 12 = 6 \cdot 2.$$
Factors   Product   Product   Factors

Other **factored forms** of 12 are

$$-6(-2), \quad 3 \cdot 4, \quad -3(-4), \quad 12 \cdot 1, \quad \text{and} \quad -12(-1).$$

More than two factors may be used, so another factored form of 12 is $2 \cdot 2 \cdot 3$. The positive integer factors of 12 are

$$1, 2, 3, 4, 6, 12.$$

OBJECTIVE 1 **Find the greatest common factor of a list of terms.** An integer that is a factor of two or more integers is called a **common factor** of those integers. For example, 6 is a common factor of 18 and 24, since 6 is a factor of both 18 and 24. Other common factors of 18 and 24 are 1, 2, and 3. The **greatest common factor (GCF)** of a list of integers is the largest common factor of those integers. Thus, 6 is the greatest common factor of 18 and 24, since it is the largest of their common factors.

▶ **NOTE** *Factors* of a number are also *divisors* of the number. The *greatest common factor* is actually the same as the *greatest common divisor*. There are many rules for deciding what numbers divide into a given number. Here are some especially useful divisibility rules for small numbers.

A Whole Number Divisible by	Must Have the Following Property:
2	Ends in 0, 2, 4, 6, or 8
3	Sum of its digits is divisible by 3.
4	Last two digits form a number divisible by 4
5	Ends in 0 or 5
6	Divisible by both 2 and 3
8	Last three digits form a number divisible by 8
9	Sum of its digits is divisible by 9.
10	Ends in 0

Recall from **Chapter 1** that a prime number has only itself and 1 as factors. In **Section 1.1,** we factored numbers into prime factors. This is the first step in finding the greatest common factor of a list of numbers. We find the greatest common factor (GCF) of a list of numbers as follows.

Finding the Greatest Common Factor (GCF)

Step 1 **Factor.** Write each number in prime factored form.

Step 2 **List common factors.** List each prime number that is a factor of every number in the list. (If a prime does not appear in one of the prime factored forms, it cannot appear in the greatest common factor.)

Step 3 **Choose least exponents.** Use as exponents on the common prime factors the *least* exponent from the prime factored forms.

Step 4 **Multiply.** Multiply the primes from Step 3. If there are no primes left after Step 3, the greatest common factor is 1.

EXAMPLE 1 **Finding the Greatest Common Factor for Numbers**

Find the greatest common factor for each list of numbers.

(a) 30, 45

First write each number in prime factored form.

$$30 = 2 \cdot 3 \cdot 5$$
$$45 = 3 \cdot 3 \cdot 5$$

Use each prime the *least* number of times it appears in *all* the factored forms. There is no 2 in the prime factored form of 45, so there will be no 2 in the greatest common factor. The least number of times 3 appears in all the factored forms is 1, and the least number of times 5 appears is also 1. From this, the

$$GCF = 3^1 \cdot 5^1 = 15.$$

(b) 72, 120, 432

Find the prime factored form of each number.

$$72 = 2 \cdot 2 \cdot 2 \cdot 3 \cdot 3$$
$$120 = 2 \cdot 2 \cdot 2 \cdot 3 \cdot 5$$
$$432 = 2 \cdot 2 \cdot 2 \cdot 2 \cdot 3 \cdot 3 \cdot 3$$

The least number of times 2 appears in all the factored forms is 3, and the least number of times 3 appears is 1. There is no 5 in the prime factored form of either 72 or 432, so the

$$GCF = 2^3 \cdot 3^1 = 24.$$

(c) 10, 11, 14

Write the prime factored form of each number.

$$10 = 2 \cdot 5$$
$$11 = 11$$
$$14 = 2 \cdot 7$$

There are no primes common to all three numbers, so the GCF is 1.

✔ **Now Try Exercises 1 and 5.**

The greatest common factor can also be found for a list of variable terms. For example, the terms x^4, x^5, x^6, and x^7 have x^4 as the greatest common factor because each of these terms can be written with x^4 as a factor.

$$x^4 = 1 \cdot x^4, \quad x^5 = x \cdot x^4, \quad x^6 = x^2 \cdot x^4, \quad x^7 = x^3 \cdot x^4$$

> ▶ **NOTE** The exponent on a variable in the GCF is the *least* exponent that appears in *all* the common factors.

EXAMPLE 2 Finding the Greatest Common Factor for Variable Terms

Find the greatest common factor for each list of terms.

(a) $21m^7$, $-18m^6$, $45m^8$, $-24m^5$

$$21m^7 = 3 \cdot 7 \cdot m^7$$
$$-18m^6 = -1 \cdot 2 \cdot 3^2 \cdot m^6$$
$$45m^8 = 3^2 \cdot 5 \cdot m^8$$
$$-24m^5 = -1 \cdot 2^3 \cdot 3 \cdot m^5$$

First, 3 is the greatest common factor of the coefficients 21, -18, 45, and -24. The least exponent on m is 5, so the GCF of the terms is $3m^5$.

(b) x^4y^2, x^7y^5, x^3y^7, y^{15}

$$x^4y^2 = x^4 \cdot y^2$$
$$x^7y^5 = x^7 \cdot y^5$$
$$x^3y^7 = x^3 \cdot y^7$$
$$y^{15} = y^{15}$$

There is no x in the last term, y^{15}, so x will not appear in the greatest common factor. There is a y in each term, however, and 2 is the least exponent on y. The GCF is y^2.

(c) $-a^2b$, $-ab^2$

$$-a^2b = -1a^2b = -1 \cdot 1 \cdot a^2b$$
$$-ab^2 = -1ab^2 = -1 \cdot 1 \cdot ab^2$$

The factors of -1 are -1 and 1. Since $1 > -1$, the GCF is $1ab$, or simply ab.

In a list of negative terms like we have here, a negative common factor is sometimes preferable (even though it is not technically the *greatest* common factor). Thus, we might prefer $-ab$ as the common factor of these terms. (In exercises like this, either answer will be acceptable.)

✔ **Now Try Exercises 11 and 15.**

OBJECTIVE 2 **Factor out the greatest common factor.** Writing a polynomial (a sum) in factored form as a product is called **factoring.** For example, the polynomial

$$3m + 12$$

has two terms: $3m$ and 12. The greatest common factor of these two terms is 3. We can write $3m + 12$ so that each term is a product with 3 as one factor.

$$3m + 12 = 3 \cdot m + 3 \cdot 4$$

$$= 3(m + 4) \qquad \text{Distributive property}$$

The factored form of $3m + 12$ is $3(m + 4)$. This process is called **factoring out the greatest common factor.**

> ▶ **CAUTION** The polynomial $3m + 12$ is *not* in factored form when written as
>
> $$3 \cdot m + 3 \cdot 4. \qquad \text{Not in factored form}$$
>
> The *terms* are factored, but the polynomial is not. The factored form of $3m + 12$ is the *product*
>
> $$3(m + 4). \qquad \text{In factored form}$$

EXAMPLE 3 Factoring Out the Greatest Common Factor

Factor out the greatest common factor. (In part (e), use fractions in the factored form.)

(a) $5y^2 + 10y = 5y(y) + 5y(2) \qquad \text{GCF} = 5y$

$\qquad\qquad\quad = 5y(y + 2) \qquad \text{Distributive property}$

To check, multiply out the factored form: $5y(y + 2) = 5y^2 + 10y$, which is the original polynomial.

(b) $20m^5 + 10m^4 + 15m^3$

$\qquad 20m^5 + 10m^4 + 15m^3 = 5m^3(4m^2) + 5m^3(2m) + 5m^3(3) \qquad \text{GCF} = 5m^3$

$\qquad\qquad\qquad\qquad\qquad\quad = 5m^3(4m^2 + 2m + 3)$

Check: $\quad 5m^3(4m^2 + 2m + 3) = 20m^5 + 10m^4 + 15m^3 \qquad \text{Original polynomial}$

> Don't forget
> the 1.

(c) $x^5 + x^3 = x^3(x^2) + x^3(1) = x^3(x^2 + 1)$

(d) $20m^7p^2 - 36m^3p^4 = 4m^3p^2(5m^4) - 4m^3p^2(9p^2) \qquad \text{GCF} = 4m^3p^2$

$\qquad\qquad\qquad\qquad = 4m^3p^2(5m^4 - 9p^2)$

(e) $\dfrac{1}{6}n^2 + \dfrac{5}{6}n = \dfrac{1}{6}n(n) + \dfrac{1}{6}n(5) \qquad \text{GCF} = \frac{1}{6}n$

$\qquad\qquad\quad = \dfrac{1}{6}n(n + 5)$

✔ **Now Try Exercises 37, 39, 41, and 47.**

> ▶ **CAUTION** Be sure to include the 1 in a problem like Example 3(c). *Always check that the factored form can be multiplied out to give the original polynomial.*

EXAMPLE 4 Factoring Out the Greatest Common Factor

Factor out the greatest common factor.

(a) $a(a + 3) + 4(a + 3)$

The binomial $a + 3$ is the greatest common factor here.

Same

$$a(a + 3) + 4(a + 3) = (a + 3)(a + 4)$$

(b) $x^2(x + 1) - 5(x + 1) = (x + 1)(x^2 - 5)$ Factor out $x + 1$.

✔ **Now Try Exercise 55.**

OBJECTIVE 3 Factor by grouping. When a polynomial has four terms, common factors can sometimes be used to **factor by grouping.**

EXAMPLE 5 Factoring by Grouping

Factor by grouping.

(a) $2x + 6 + ax + 3a$

Group the first two terms and the last two terms, since the first two terms have a common factor of 2 and the last two terms have a common factor of a.

$$2x + 6 + ax + 3a = (2x + 6) + (ax + 3a)$$
$$= 2(x + 3) + a(x + 3)$$

The expression is still not in factored form because it is the *sum* of two terms. Now, however, $x + 3$ is a common factor and can be factored out.

$$
\begin{aligned}
2x + 6 + ax + 3a &= (2x + 6) + (ax + 3a) && \text{Group terms.} \\
&= 2(x + 3) + a(x + 3) && \text{Factor each group.} \\
&= (x + 3)(2 + a) && \text{Factor out } x + 3.
\end{aligned}
$$

The final result is in factored form because it is a *product*. Note that the goal in factoring by grouping is to get a common factor, $x + 3$ here, so that the last step is possible. Check by using the FOIL method from **Section 5.5** to multiply the binomials.

Check: $(x + 3)(2 + a) = 2x + ax + 6 + 3a$ FOIL
$$= 2x + 6 + ax + 3a,$$ Rearrange terms.

which is the original polynomial.

(b) $6ax + 24x + a + 4 = (6ax + 24x) + (a + 4)$ Group terms.
$$= 6x(a + 4) + 1(a + 4)$$ Factor each group.

Remember the 1.

$$= (a + 4)(6x + 1)$$ Factor out $a + 4$.

Check: $(a + 4)(6x + 1) = 6ax + a + 24x + 4$ FOIL
$$= 6ax + 24x + a + 4,$$ Rearrange terms.

which is the original polynomial.

(c) $2x^2 - 10x + 3xy - 15y = (2x^2 - 10x) + (3xy - 15y)$ Group terms.

$$= 2x(x - 5) + 3y(x - 5)$$ Factor each group.

$$= (x - 5)(2x + 3y)$$ Factor out $x - 5$.

Check: $(x - 5)(2x + 3y) = 2x^2 + 3xy - 10x - 15y$ FOIL

$$= 2x^2 - 10x + 3xy - 15y$$ Original polynomial

(d) $t^3 + 2t^2 - 3t - 6 = (t^3 + 2t^2) + (-3t - 6)$ Group terms.

$$= t^2(t + 2) - 3(t + 2)$$ Factor out -3 so there is a common factor $t + 2$; $-3(t + 2) = -3t - 6$.

> Be careful with signs.

$$= (t + 2)(t^2 - 3)$$ Factor out $t + 2$.

Check: $(t + 2)(t^2 - 3) = t^3 - 3t + 2t^2 - 6$ FOIL

$$= t^3 + 2t^2 - 3t - 6$$ Original polynomial

> ✔ **Now Try Exercises 69, 73, and 77.**

▶ **CAUTION**   *Be careful with signs when grouping* in a problem like Example 5(d). It is wise to check the factoring in the second step, as shown in the side comment in that example, before continuing.

Factoring by Grouping

Step 1   **Group terms.**   Collect the terms into two groups so that each group has a common factor.

Step 2   **Factor within groups.**   Factor out the greatest common factor from each group.

Step 3   **Factor the entire polynomial.**   Factor out a common binomial factor from the results of Step 2.

Step 4   **If necessary, rearrange terms.**   If Step 2 does not result in a common binomial factor, try a different grouping.

EXAMPLE 6   **Rearranging Terms before Factoring by Grouping**

Factor by grouping.

(a) $10x^2 - 12y + 15x - 8xy$

Factoring out the common factor of 2 from the first two terms and the common factor of x from the last two terms gives

$$10x^2 - 12y + 15x - 8xy = 2(5x^2 - 6y) + x(15 - 8y).$$

This does not lead to a common factor, so we try rearranging the terms. There is usually more than one way to do this.

$10x^2 - 8xy - 12y + 15x = (10x^2 - 8xy) + (-12y + 15x)$ Group terms.

$$= 2x(5x - 4y) + 3(-4y + 5x)$$ Factor each group.

$$= 2x(5x - 4y) + 3(5x - 4y)$$ Rewrite $-4y + 5x$.

$$= (5x - 4y)(2x + 3)$$ Factor out $5x - 4y$.

Check: $(5x - 4y)(2x + 3) = 10x^2 + 15x - 8xy - 12y$ FOIL

$= 10x^2 - 12y + 15x - 8xy$ Original polynomial

(b) $2xy + 12 - 3y - 8x$

We need to rearrange these terms to get two groups that each have a common factor. Trial and error suggests the following grouping:

$2xy + 12 - 3y - 8x = (2xy - 3y) + (-8x + 12)$ Group terms.

$= y(2x - 3) - 4(2x - 3)$ Factor each group.

Be careful with signs.

$= (2x - 3)(y - 4).$ Factor out $2x - 3$.

Since the quantities in parentheses in the second step must be the same, we factored out -4 rather than 4. *Check* by multiplying.

✔ **Now Try Exercises 81 and 83.**

6.1 EXERCISES

Find the greatest common factor for each list of numbers. See Example 1.

1. 40, 20, 4

2. 50, 30, 5

3. 18, 24, 36, 48

4. 15, 30, 45, 75

5. 6, 8, 9

6. 20, 22, 23

▨ **7.** How can you check your answer when you factor a polynomial?

▨ **8.** Explain how to find the greatest common factor of a list of terms. Use examples.

Find the greatest common factor for each list of terms. See Examples 1 and 2.

9. $16y, 24$

10. $18w, 27$

11. $30x^3, 40x^6, 50x^7$

12. $60z^4, 70z^8, 90z^9$

13. $12m^3n^2, 18m^5n^4, 36m^8n^3$

14. $25p^5r^7, 30p^7r^8, 50p^5r^3$

15. $-x^4y^3, -xy^2$

16. $-a^4b^5, -a^3b$

Concept Check *An expression is factored when it is written as a product, not a sum. Which of the following are not factored?*

17. $2k^2(5k)$

18. $2k^2(5k + 1)$

19. $2k^2 + (5k + 1)$

20. $(2k^2 + 1)(5k + 1)$

21. **Concept Check** Is $-xy$ a common factor of $-x^4y^3$ and $-xy^2$? If so, what is the other factor that gives $-x^4y^3$ when multiplied by $-xy$?

22. **Concept Check** Is $-a^5b^2$ a common factor of $-a^4b^5$ and $-a^3b$?

Complete each factoring.

23. $9m^4 = 3m^2(\quad)$

24. $12p^5 = 6p^3(\quad)$

25. $-8z^9 = -4z^5(\quad)$

26. $-15k^{11} = -5k^8(\quad)$

27. $6m^4n^5 = 3m^3n(\quad)$

28. $27a^3b^2 = 9a^2b(\quad)$

29. $12y + 24 = 12(\quad)$

30. $18p + 36 = 18(\quad)$

31. $10a^2 - 20a = 10a(\quad)$

32. $15x^2 - 30x = 15x(\quad)$

33. $8x^2y + 12x^3y^2 = 4x^2y(\quad)$

34. $18s^3t^2 + 10st = 2st(\quad)$

Factor out the greatest common factor. See Examples 3 and 4.

35. $27m^3 - 9m$ **36.** $36p^3 + 24p$ ◐ **37.** $16z^4 + 24z^2$ **38.** $25k^4 - 15k^2$

39. $\dfrac{1}{4}d^2 - \dfrac{3}{4}d$ **40.** $\dfrac{1}{5}z^2 + \dfrac{3}{5}z$ **41.** $12x^3 + 6x^2$ **42.** $21b^3 - 7b^2$

43. $65y^{10} + 35y^6$ **44.** $100a^5 + 16a^3$ **45.** $11w^3 - 100$ **46.** $13z^5 - 80$

47. $8m^2n^3 + 24m^2n^2$ **48.** $19p^2y - 38p^2y^3$

49. $13y^8 + 26y^4 - 39y^2$ **50.** $5x^5 + 25x^4 - 20x^3$

51. $36p^6q + 45p^5q^4 + 81p^3q^2$ **52.** $125a^3z^5 + 60a^4z^4 - 85a^5z^2$

53. $a^5 + 2a^3b^2 - 3a^5b^2 + 4a^4b^3$ **54.** $x^6 + 5x^4y^3 - 6xy^4 + 10xy$

◐ **55.** $c(x + 2) - d(x + 2)$ **56.** $r(5 - x) + t(5 - x)$

57. $m(m + 2n) + n(m + 2n)$ **58.** $3p(1 - 4p) - 2q(1 - 4p)$

Students often have difficulty when factoring by grouping because they are not able to tell when the polynomial is completely factored. For example,

$$5y(2x - 3) + 8t(2x - 3)$$

is not in factored form, because it is the *sum* of two terms: $5y(2x - 3)$ and $8t(2x - 3)$. However, because $2x - 3$ is a common factor of these two terms, the expression can now be factored as

$$(2x - 3)(5y + 8t).$$

The factored form is a *product* of two factors: $2x - 3$ and $5y + 8t$.

Concept Check   *Determine whether each expression is in factored form or is not in factored form. If it is not in factored form, factor it if possible.*

59. $8(7t + 4) + x(7t + 4)$ **60.** $3r(5x - 1) + 7(5x - 1)$

61. $(8 + x)(7t + 4)$ **62.** $(3r + 7)(5x - 1)$

63. $18x^2(y + 4) + 7(y + 4)$ **64.** $12k^3(s - 3) + 7(s + 3)$

✎ **65.** Tell why it is not possible to factor the expression in Exercise 64.

✎ **66.** Summarize the method of factoring a polynomial with four terms by grouping. Give an example.

Factor by grouping. See Examples 5 and 6.

67. $p^2 + 4p + pq + 4q$ **68.** $m^2 + 2m + mn + 2n$

◐ **69.** $a^2 - 2a + ab - 2b$ **70.** $y^2 - 6y + yw - 6w$

71. $7z^2 + 14z - az - 2a$ **72.** $5m^2 + 15mp - 2mr - 6pr$

73. $18r^2 + 12ry - 3xr - 2xy$ **74.** $8s^2 - 4st + 6sy - 3yt$

75. $3a^3 + 3ab^2 + 2a^2b + 2b^3$ **76.** $4x^3 + 3x^2y + 4xy^2 + 3y^3$

77. $1 - a + ab - b$ **78.** $6 - 3x - 2y + xy$

79. $16m^3 - 4m^2p^2 - 4mp + p^3$ **80.** $10t^3 - 2t^2s^2 - 5ts + s^3$

◐ **81.** $5m - 6p - 2mp + 15$ **82.** $y^2 + 3x - 3y - xy$

83. $18r^2 - 2ty + 12ry - 3rt$ **84.** $2b^3 + 3a^3 + 3ab^2 + 2a^2b$

85. $a^5 - 3 + 2a^5b - 6b$ **86.** $4b^3 + a^2b - 4a - ab^4$

91. Refer to Exercise 77. The answer given in the back of the book is $(1 - a)(1 - b)$. A student factored this same polynomial and got the result $(a - 1)(b - 1)$.

    (a) Is the student's answer correct?

    (b) If your answer to part (a) is *yes,* explain why these two seemingly different answers are both acceptable.

92. *Concept Check*   A student factored $18x^3y^2 + 9xy$ as $9xy(2x^2y)$. **WHAT WENT WRONG?** Factor correctly.

PREVIEW EXERCISES

Find each product. See **Section 5.5.**

93. $(x + 6)(x - 9)$ **94.** $(x - 3)(x - 6)$ **95.** $(x + 2)(x + 7)$

96. $2x(x + 5)(x - 1)$ **97.** $2x^2(x^2 + 3x + 5)$ **98.** $-5x^2(2x^2 - 4x - 9)$

6.2 Factoring Trinomials

OBJECTIVES

1 Factor trinomials with a coefficient of 1 for the squared term.

2 Factor such trinomials after factoring out the greatest common factor.

Using the FOIL method, we see that the product of the binomials $k - 3$ and $k + 1$ is

$$(k - 3)(k + 1) = k^2 - 2k - 3. \quad \text{Multiplying}$$

Suppose instead that we are given the polynomial $k^2 - 2k - 3$ and want to rewrite it as the product $(k - 3)(k + 1)$. That is,

$$k^2 - 2k - 3 = (k - 3)(k + 1). \quad \text{Factoring}$$

Recall from **Section 6.1** that this process is called factoring the polynomial. Factoring reverses or "undoes" multiplying.

OBJECTIVE 1   **Factor trinomials with a coefficient of 1 for the squared term.**   When factoring polynomials with integer coefficients, we use only integers in the factors. For example, we can factor $x^2 + 5x + 6$ by finding integers m and n such that

$$x^2 + 5x + 6 = (x + m)(x + n).$$

To find these integers m and n, we first use FOIL to multiply the two binomials on the right side of the equation:

$$(x + m)(x + n) = x^2 + nx + mx + mn.$$
$$= x^2 + (n + m)x + mn. \quad \text{Distributive property}$$

Comparing this result with $x^2 + 5x + 6$ shows that we must find integers m and n having a sum of 5 and a product of 6.

Product of m and n is 6.

$$x^2 + 5x + 6 = x^2 + (n + m)x + mn$$

Sum of m and n is 5.

Since many pairs of integers have a sum of 5, it is best to begin by listing those pairs of integers whose product is 6. Both 5 and 6 are positive, so we consider only pairs in which both integers are positive.

Factors of 6	Sums of Factors	
1, 6	$1 + 6 = 7$	
2, 3	$2 + 3 = 5$	Sum is 5.

Both pairs have a product of 6, but only the pair 2 and 3 has a sum of 5. So 2 and 3 are the required integers, and

$$x^2 + 5x + 6 = (x + 2)(x + 3).$$

Check by using the FOIL method to multiply the binomials. ***Make sure that the sum of the outer and inner products produces the correct middle term.***

Check: $(x + 2)(x + 3) = x^2 + 5x + 6$

$$2x$$
$$3x$$
$$5x \quad \text{Add.}$$

EXAMPLE 1 **Factoring a Trinomial with All Positive Terms**

Factor $m^2 + 9m + 14$.

Look for two integers whose product is 14 and whose sum is 9. List the pairs of integers whose product is 14. Then examine the sums. Again, only positive integers are needed because all signs in $m^2 + 9m + 14$ are positive.

Factors of 14	Sums of Factors	
14, 1	$14 + 1 = 15$	
7, 2	$7 + 2 = 9$	Sum is 9.

From the list, 7 and 2 are the required integers, since $7 \cdot 2 = 14$ and $7 + 2 = 9$. Thus,

$$m^2 + 9m + 14 = (m + 7)(m + 2).$$

Check: $(m + 2)(m + 7) = m^2 + 7m + 2m + 14 \quad \text{FOIL}$
$$= m^2 + 9m + 14 \quad \text{Original polynomial}$$

✔ **Now Try Exercise 25.**

▶ **NOTE** In Example 1, the answer also could have been written $(m + 2)(m + 7)$. Because of the commutative property of multiplication, the order of the factors does not matter. *Always check by multiplying.*

EXAMPLE 2 Factoring a Trinomial with a Negative Middle Term

Factor $x^2 - 9x + 20$.

We must find two integers whose product is 20 and whose sum is -9. Since the numbers we are looking for have a *positive product* and a *negative sum,* we consider only pairs of negative integers.

Factors of 20	Sums of Factors	
$-20, -1$	$-20 + (-1) = -21$	
$-10, -2$	$-10 + (-2) = -12$	
$-5, -4$	$-5 + (-4) = -9$	Sum is -9.

The required integers are -5 and -4, so
$$x^2 - 9x + 20 = (x - 5)(x - 4).$$

Check: $\quad (x - 5)(x - 4) = x^2 - 4x - 5x + 20 \qquad$ FOIL
$$= x^2 - 9x + 20 \qquad \text{Original polynomial}$$

✔ **Now Try Exercise 29.**

EXAMPLE 3 Factoring a Trinomial with Two Negative Terms

Factor $p^2 - 2p - 15$.

Find two integers whose product is -15 and whose sum is -2. If these numbers do not come to mind right away, find them (if they exist) by listing all the pairs of integers whose product is -15. Because the last term, -15, is negative, list pairs of integers with different signs.

Factors of -15	Sums of Factors	
$15, -1$	$15 + (-1) = 14$	
$-15, 1$	$-15 + 1 = -14$	
$5, -3$	$5 + (-3) = 2$	
$-5, 3$	$-5 + 3 = -2$	Sum is -2.

The required integers are -5 and 3, so
$$p^2 - 2p - 15 = (p - 5)(p + 3).$$

Check by multiplying out the factored form.

✔ **Now Try Exercise 35.**

▶ **NOTE** In Examples 1–3, notice that we listed factors in descending order (disregarding their signs) when we were looking for the required pair of integers. This helps avoid skipping the correct combination.

As shown in the next example, some trinomials cannot be factored by using only integers. We call such trinomials **prime polynomials.**

EXAMPLE 4 Deciding whether Polynomials Are Prime

Factor each trinomial if possible.

(a) $x^2 - 5x + 12$

As in Example 2, both factors must be negative to give a positive product and a negative sum. First, list all pairs of negative integers whose product is 12. Then examine the sums.

Factors of 12	Sums of Factors
$-12, -1$	$-12 + (-1) = -13$
$-6, -2$	$-6 + (-2) = -8$
$-4, -3$	$-4 + (-3) = -7$

None of the pairs of integers has a sum of -5. Therefore, the trinomial $x^2 - 5x + 12$ *cannot be factored by using only integers;* it is a *prime polynomial.*

(b) $k^2 - 8k + 11$

There is no pair of integers whose product is 11 and whose sum is -8, so $k^2 - 8k + 11$ is a prime polynomial.

✔ **Now Try Exercise 31.**

Guidelines for factoring a trinomial of the form $x^2 + bx + c$ are summarized here.

Factoring $x^2 + bx + c$

Find two integers whose product is c and whose sum is b.

1. Both integers must be positive if b and c are positive.
2. Both integers must be negative if c is positive and b is negative.
3. One integer must be positive and one must be negative if c is negative.

EXAMPLE 5 Factoring a Trinomial with Two Variables

Factor $z^2 - 2bz - 3b^2$.

Here, the coefficient of z in the middle term is $-2b$, so we need to find two expressions whose product is $-3b^2$ and whose sum is $-2b$. The expressions are $-3b$ and b, so

$$z^2 - 2bz - 3b^2 = (z - 3b)(z + b).$$

Check:
$$(z - 3b)(z + b) = z^2 + zb - 3bz - 3b^2 \qquad \text{FOIL}$$
$$= z^2 + 1bz - 3bz - 3b^2 \qquad \text{Identity property}$$
$$= z^2 - 2bz - 3b^2 \qquad \text{Combine like terms.}$$

✔ **Now Try Exercise 45.**

OBJECTIVE 2 Factor such trinomials after factoring out the greatest common factor.
The trinomial in the next example does not have a coefficient of 1 for the squared term. (In fact, there is no squared term.) However, there may be a common factor.

EXAMPLE 6 Factoring a Trinomial with a Common Factor

Factor $4x^5 - 28x^4 + 40x^3$.

First, factor out the greatest common factor, $4x^3$.

$$4x^5 - 28x^4 + 40x^3 = 4x^3(x^2 - 7x + 10)$$

Now factor $x^2 - 7x + 10$. The integers -5 and -2 have a product of 10 and a sum of -7. The completely factored form is

Include $4x^3$.

$$4x^5 - 28x^4 + 40x^3 = 4x^3(x - 5)(x - 2).$$

Check: $4x^3(x - 5)(x - 2) = 4x^3(x^2 - 7x + 10)$
$$= 4x^5 - 28x^4 + 40x^3$$

✔ **Now Try Exercise 53.**

▶ **CAUTION** *When factoring, always look for a common factor first.* Remember to include the common factor as part of the answer. As a check, multiplying out the factored form should always give the original polynomial.

6.2 EXERCISES

⊙ *Complete solution available on Video Lectures on CD/DVD*

Now Try Exercise

Concept Check In Exercises 1–4, list all pairs of integers with the given product. Then find the pair whose sum is given.

1. Product: 48;   Sum: -19

2. Product: 48;   Sum: 14

3. Product: -24;   Sum: -5

4. Product: -36;   Sum: -16

5. *Concept Check*   In factoring a trinomial in x, such as $(x + a)(x + b)$, what must be true of a and b if the coefficient of the last term of the trinomial is negative?

6. *Concept Check*   In Exercise 5, what must be true of a and b if the coefficient of the last term is positive?

7. What is meant by a *prime polynomial*?

8. How can you check your work when factoring a trinomial? Does the check ensure that the trinomial is completely factored?

9. *Concept Check*   Which one of the following is the correct factored form of $x^2 - 12x + 32$?

A. $(x - 8)(x + 4)$ **B.** $(x + 8)(x - 4)$
C. $(x - 8)(x - 4)$ **D.** $(x + 8)(x + 4)$

10. *Concept Check*   What would be the first step in factoring $2x^3 + 8x^2 - 10x$?

Complete each factoring. See Examples 1–4.

11. $p^2 + 11p + 30 = (p + 5)(\qquad)$

12. $x^2 + 10x + 21 = (x + 7)(\qquad)$

13. $x^2 + 15x + 44 = (x + 4)(\qquad)$

14. $r^2 + 15r + 56 = (r + 7)(\qquad)$

15. $x^2 - 9x + 8 = (x - 1)(\qquad)$

16. $t^2 - 14t + 24 = (t - 2)(\qquad)$

17. $y^2 - 2y - 15 = (y + 3)(\qquad)$

18. $t^2 - t - 42 = (t + 6)(\qquad)$

19. $x^2 + 9x - 22 = (x - 2)(\qquad)$

20. $x^2 + 6x - 27 = (x - 3)(\qquad)$

21. $y^2 - 7y - 18 = (y + 2)(\qquad)$

22. $y^2 - 2y - 24 = (y + 4)(\qquad)$

Factor completely. If the polynomial cannot be factored, write prime. *See Examples 1–4.*

23. $y^2 + 9y + 8$

24. $a^2 + 9a + 20$

25. $b^2 + 8b + 15$

26. $x^2 + 6x + 8$

27. $m^2 + m - 20$

28. $p^2 + 4p - 5$

29. $y^2 - 8y + 15$

30. $y^2 - 6y + 8$

31. $x^2 + 4x + 5$

32. $t^2 + 11t + 12$

33. $z^2 - 15z + 56$

34. $x^2 - 13x + 36$

35. $r^2 - r - 30$

36. $q^2 - q - 42$

37. $a^2 - 8a - 48$

38. $m^2 - 10m - 25$

39. $x^2 + 3x - 39$

40. $d^2 + 4d - 45$

41. Explain how you would factor $8 + 6x + x^2$.

42. Use your answer to Exercise 41 to factor $5 - 4x - x^2$.

Factor completely. See Examples 5 and 6.

43. $r^2 + 3ra + 2a^2$

44. $x^2 + 5xa + 4a^2$

45. $t^2 - tz - 6z^2$

46. $a^2 - ab - 12b^2$

47. $x^2 + 4xy + 3y^2$

48. $p^2 + 9pq + 8q^2$

49. $v^2 - 11vw + 30w^2$

50. $v^2 - 11vx + 24x^2$

51. $4x^2 + 12x - 40$

52. $5y^2 - 5y - 30$

53. $2t^3 + 8t^2 + 6t$

54. $3t^3 + 27t^2 + 24t$

55. $2x^6 + 8x^5 - 42x^4$

56. $4y^5 + 12y^4 - 40y^3$

57. $5m^5 + 25m^4 - 40m^2$

58. $12k^5 - 6k^3 + 10k^2$

59. $m^3n - 10m^2n^2 + 24mn^3$

60. $y^3z + 3y^2z^2 - 54yz^3$

61. Use the FOIL method from **Section 5.5** to show that $(2x + 4)(x - 3) = 2x^2 - 2x - 12$. If you are asked to factor $2x^2 - 2x - 12$ completely, why would it be incorrect to give $(2x + 4)(x - 3)$ as your answer?

62. If you are asked to factor the polynomial $3x^2 + 9x - 12$ completely, why would it be incorrect to give $(x - 1)(3x + 12)$ as your answer?

The next problems are "brain busters." Factor each polynomial.

63. $a^5 + 3a^4b - 4a^3b^2$

64. $m^3n - 2m^2n^2 - 3mn^3$

65. $y^3z + y^2z^2 - 6yz^3$

66. $k^7 - 2k^6m - 15k^5m^2$

67. $z^{10} - 4z^9y - 21z^8y^2$

68. $x^9 + 5x^8w - 24x^7w^2$

69. $(a + b)x^2 + (a + b)x - 12(a + b)$

70. $(x + y)n^2 + (x + y)n + 16(x + y)$

71. $(2p + q)r^2 - 12(2p + q)r + 27(2p + q)$

72. $(3m - n)k^2 - 13(3m - n)k + 40(3m - n)$

73. *Concept Check* What polynomial can be factored as $(a + 9)(a + 4)$?

74. *Concept Check* What polynomial can be factored as $(y - 7)(y + 3)$?

PREVIEW EXERCISES

Find each product. See Section 5.5.

75. $(2y - 7)(y + 4)$ **76.** $(3a + 2)(2a + 1)$ **77.** $(5z + 2)(3z - 2)$

78. $(4m - 3)(2m + 5)$ **79.** $(4p + 1)(2p - 3)$ **80.** $(6r - 5)(3r + 2)$

6.3   More on Factoring Trinomials

OBJECTIVES

1 Factor trinomials by grouping when the coefficient of the squared term is not 1.

2 Factor trinomials by using the FOIL method.

Trinomials such as $2x^2 + 7x + 6$, in which the coefficient of the squared term is *not* 1, are factored with extensions of the methods from the previous sections. One such method uses factoring by grouping from **Section 6.1.**

OBJECTIVE ❶ Factor trinomials by grouping when the coefficient of the squared term is not 1. Recall that a trinomial such as $m^2 + 3m + 2$ is factored by finding two numbers whose product is 2 and whose sum is 3. To factor $2x^2 + 7x + 6$, we look for two integers whose product is $2 \cdot 6 = 12$ and whose sum is 7.

$$2x^2 + 7x + 6$$

Sum is 7.

Product is $2 \cdot 6 = 12$.

By considering pairs of positive integers whose product is 12, we find the necessary integers to be 3 and 4. We use these integers to write the middle term, $7x$, as $7x = 3x + 4x$. The trinomial $2x^2 + 7x + 6$ becomes

$$2x^2 + 7x + 6 = 2x^2 + \underbrace{3x + 4x}_{7x} + 6.$$

$$= (2x^2 + 3x) + (4x + 6) \qquad \text{Group terms.}$$

$$= x(2x + 3) + 2(2x + 3) \qquad \text{Factor each group.}$$

Must be the same factor

$$2x^2 + 7x + 6 = (2x + 3)(x + 2) \qquad \text{Factor out } 2x + 3.$$

Check: $(2x + 3)(x + 2) = 2x^2 + 7x + 6$

In this example, we could have written $7x$ as $4x + 3x$. Factoring by grouping this way would give the same answer.

EXAMPLE 1 Factoring Trinomials by Grouping

Factor each trinomial.

(a) $6r^2 + r - 1$

We must find two integers with a product of $6(-1) = -6$ and a sum of 1.

Sum is 1.

$$6r^2 + r - 1 = 6r^2 + 1r - 1$$

Product is $6(-1) = -6$.

The integers are -2 and 3. We write the middle term, r, as $-2r + 3r$.

$$
\begin{aligned}
6r^2 + r - 1 &= 6r^2 - 2r + 3r - 1 && r = -2r + 3r \\
&= (6r^2 - 2r) + (3r - 1) && \text{Group terms.} \\
&= 2r(3r - 1) + 1(3r - 1) && \text{The binomials must be the same.}
\end{aligned}
$$

Remember the 1.

$$= (3r - 1)(2r + 1) \qquad\qquad \text{Factor out } 3r - 1.$$

Check: $(3r - 1)(2r + 1) = 6r^2 + r - 1$

(b) $12z^2 - 5z - 2$

Look for two integers whose product is $12(-2) = -24$ and whose sum is -5. The required integers are 3 and -8, so

$$
\begin{aligned}
12z^2 - 5z - 2 &= 12z^2 + 3z - 8z - 2 && -5z = 3z - 8z \\
&= (12z^2 + 3z) + (-8z - 2) && \text{Group terms.} \\
&= 3z(4z + 1) - 2(4z + 1) && \text{Factor each group.}
\end{aligned}
$$

Be careful with signs.

$$= (4z + 1)(3z - 2). \qquad\qquad \text{Factor out } 4z + 1.$$

Check: $(4z + 1)(3z - 2) = 12z^2 - 5z - 2$

(c) $10m^2 + mn - 3n^2$

Two integers whose product is $10(-3) = -30$ and whose sum is 1 are -5 and 6. Rewrite the trinomial with four terms.

$$
\begin{aligned}
10m^2 + mn - 3n^2 &= 10m^2 - 5mn + 6mn - 3n^2 && mn = -5mn + 6mn \\
&= 5m(2m - n) + 3n(2m - n) && \text{Group terms;} \\
& && \text{factor each group.}
\end{aligned}
$$

Check by multiplying.

$$= (2m - n)(5m + 3n) \qquad\qquad \text{Factor out } 2m - n.$$

✔ **Now Try Exercises 21, 27, and 47.**

EXAMPLE 2 Factoring a Trinomial with a Common Factor by Grouping

Factor $28x^5 - 58x^4 - 30x^3$.

First factor out the greatest common factor, $2x^3$.

$$28x^5 - 58x^4 - 30x^3 = 2x^3(14x^2 - 29x - 15)$$

To factor $14x^2 - 29x - 15$, find two integers whose product is $14(-15) = -210$ and whose sum is -29. Factoring 210 into prime factors gives

$$210 = 2 \cdot 3 \cdot 5 \cdot 7.$$

Combine the prime factors of $210 = 2 \cdot 3 \cdot 5 \cdot 7$ into pairs in different ways, using one positive and one negative (to get -210). The factors 6 and -35 have the correct sum.

$$28x^5 - 58x^4 - 30x^3 = 2x^3(14x^2 - 29x - 15)$$

Remember the common factor.

$$= 2x^3(14x^2 + 6x - 35x - 15)$$
$$= 2x^3[(14x^2 + 6x) + (-35x - 15)]$$
$$= 2x^3[2x(7x + 3) - 5(7x + 3)]$$
$$= 2x^3[(7x + 3)(2x - 5)]$$
$$= 2x^3(7x + 3)(2x - 5)$$

Check by multiplying.

✔ **Now Try Exercise 43.**

OBJECTIVE 2 **Factor trinomials by using the FOIL method.** We now show an alternative method of factoring trinomials in which the coefficient of the squared term is not 1. This method generalizes the factoring procedure explained in **Section 6.2.**

To factor $2x^2 + 7x + 6$ (the trinomial factored at the beginning of this section) by the method of **Section 6.2**, we use the FOIL method in reverse. We want to write $2x^2 + 7x + 6$ as the product of two binomials.

$$2x^2 + 7x + 6 = (\qquad)(\qquad)$$

The product of the two first terms of the binomials is $2x^2$. The possible factors of $2x^2$ are $2x$ and x or $-2x$ and $-x$. Since all terms of the trinomial are positive, we consider only positive factors. Thus, we have

$$2x^2 + 7x + 6 = (2x \qquad)(x \qquad).$$

The product of the two last terms, 6, can be factored as $1 \cdot 6, 6 \cdot 1, 2 \cdot 3$, or $3 \cdot 2$. Try each pair to find the pair that gives the correct middle term, $7x$.

$$(2x + 1)(x + 6) \qquad \text{Incorrect} \qquad \qquad (2x + 6)(x + 1) \qquad \text{Incorrect}$$
$$\underset{\overset{x}{}}{}$$
$$\underset{12x}{}$$
$$\overline{13x} \qquad \text{Add.} \qquad \qquad \qquad \overline{8x} \qquad \text{Add.}$$

with $6x$ and $2x$ shown for the right example.

Since $2x + 6 = 2(x + 3)$, the binomial $2x + 6$ has a common factor of 2, while $2x^2 + 7x + 6$ has no common factor other than 1. The product $(2x + 6)(x + 1)$ cannot be correct.

▶ **NOTE** If the terms of the original polynomial have greatest common factor 1, then all of that polynomial's binomial factors will also have GCF 1.

Now try the numbers 2 and 3 as factors of 6. Because of the presence of the common factor 2 in $2x + 2$, $(2x + 2)(x + 3)$ will not work, so we try $(2x + 3)(x + 2)$.

$$(2x + 3)(x + 2) = 2x^2 + 7x + 6 \qquad \text{Correct}$$
$$\underset{3x}{}$$
$$\underset{4x}{}$$
$$\overline{7x} \qquad \text{Add.}$$

Thus, $2x^2 + 7x + 6$ factors as

$$2x^2 + 7x + 6 = (2x + 3)(x + 2).$$

Check by multiplying out the factored form.

EXAMPLE 3 Factoring a Trinomial with All Positive Terms by Using FOIL

Factor $8p^2 + 14p + 5$.

The number 8 has several possible pairs of factors, but 5 has only 1 and 5 or -1 and -5, so begin by considering the factors of 5. Ignore the negative factors, since all coefficients in the trinomial are positive. The factors will have the form

$$(\quad + 5)(\quad + 1).$$

The possible pairs of factors of $8p^2$ are $8p$ and p or $4p$ and $2p$. Try various combinations, checking in each case to see if the middle term is $14p$.

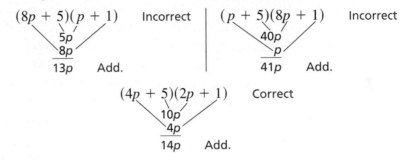

Since $14p$ is the correct middle term,

$$8p^2 + 14p + 5 = (4p + 5)(2p + 1).$$

Check: $(4p + 5)(2p + 1) = 8p^2 + 14p + 5$

✔ **Now Try Exercise 23.**

EXAMPLE 4 Factoring a Trinomial with a Negative Middle Term by Using FOIL

Factor $6x^2 - 11x + 3$.

Since 3 has only 1 and 3 or -1 and -3 as factors, it is better here to begin by factoring 3. The last term of the trinomial $6x^2 - 11x + 3$ is positive and the middle term has a negative coefficient, so we consider only negative factors. We need two negative factors, because the *product* of two negative factors is positive and their *sum* is negative, as required. Try -3 and -1 as factors of 3:

$$(\quad - 3)(\quad - 1).$$

The factors of $6x^2$ may be either $6x$ and x or $2x$ and $3x$.

$(6x - 3)(x - 1)$ Incorrect
$-3x$
$-6x$
$-9x$ Add.

$(2x - 3)(3x - 1)$ Correct
$-9x$
$-2x$
$-11x$ Add.

The factors $2x$ and $3x$ produce $-11x$, the correct middle term, so

$$6x^2 - 11x + 3 = (2x - 3)(3x - 1).$$

Check by multiplying.

✔ **Now Try Exercise 29.**

▶ **NOTE** In Example 4, we might also realize that our initial attempt to factor $6x^2 - 11x + 3$ as $(6x - 3)(x - 1)$ *cannot* be correct, since $6x - 3$ has a common factor of 3 and the original polynomial does not.

EXAMPLE 5 Factoring a Trinomial with a Negative Last Term by Using FOIL

Factor $8x^2 + 6x - 9$.

The integer 8 has several possible pairs of factors, as does -9. Since the last term is negative, one positive factor and one negative factor of -9 are needed. Since the coefficient of the middle term is relatively small, it is wise to avoid large factors such as 8 or 9. We try $4x$ and $2x$ as factors of $8x^2$, and 3 and -3 as factors of -9.

$$(4x + 3)(2x - 3) \quad \text{Incorrect}$$
$$6x$$
$$-12x$$
$$-6x \quad \text{Add.}$$

Now we try interchanging 3 and -3, since only the sign of the middle term is incorrect.

$$(4x - 3)(2x + 3) \quad \text{Correct}$$
$$-6x$$
$$12x$$
$$6x \quad \text{Add.}$$

This combination produces the correct middle term, so
$$8x^2 + 6x - 9 = (4x - 3)(2x + 3).$$

✔ **Now Try Exercise 33.**

EXAMPLE 6 Factoring a Trinomial with Two Variables

Factor $12a^2 - ab - 20b^2$.

There are several pairs of factors of $12a^2$, including
$$12a \text{ and } a, \quad 6a \text{ and } 2a, \quad \text{and} \quad 3a \text{ and } 4a,$$
just as there are many pairs of factors of $-20b^2$, including
$$20b \text{ and } -b, \quad -20b \text{ and } b, \quad 10b \text{ and } -2b, \quad -10b \text{ and } 2b,$$
$$4b \text{ and } -5b, \quad \text{and} \quad -4b \text{ and } 5b.$$

Once again, since the desired middle term is relatively small, avoid the larger factors. Try the factors $6a$ and $2a$, and $4b$ and $-5b$.
$$(6a + 4b)(2a - 5b)$$

This cannot be correct, as mentioned before, since $6a + 4b$ has a common factor while the given trinomial has GCF = 1. Try $3a$ and $4a$ with $4b$ and $-5b$.
$$(3a + 4b)(4a - 5b) = 12a^2 + ab - 20b^2 \quad \text{Incorrect}$$

Here the middle term has the wrong sign, so we interchange the signs in the factors.
$$(3a - 4b)(4a + 5b) = 12a^2 - ab - 20b^2 \quad \text{Correct}$$

✔ **Now Try Exercise 49.**

EXAMPLE 7 **Factoring Trinomials with Common Factors**

Factor each trinomial.

(a) $15y^3 + 55y^2 + 30y$

First factor out the greatest common factor, $5y$.

$$15y^3 + 55y^2 + 30y = 5y(3y^2 + 11y + 6)$$

To factor $3y^2 + 11y + 6$, try $3y$ and y as factors of $3y^2$, and 2 and 3 as factors of 6.

$$(3y + 2)(y + 3) = 3y^2 + 11y + 6 \qquad \text{Correct}$$

The completely factored form of $15y^3 + 55y^2 + 30y$ is

$$15y^3 + 55y^2 + 30y = 5y(3y + 2)(y + 3).$$

Check by multiplying.

> Remember the common factor.

(b) $-24a^3 - 42a^2 + 45a$

The common factor could be $3a$ or $-3a$. If we factor out $-3a$, the first term of the trinomial will be positive, which makes it easier to factor.

$$-24a^3 - 42a^2 + 45a = -3a(8a^2 + 14a - 15) \qquad \text{Factor out } -3a.$$
$$= -3a(4a - 3)(2a + 5) \qquad \text{Use FOIL.}$$

Check by multiplying.

✔ **Now Try Exercise 45.**

▶ **CAUTION** *Include the common factor in the final factored form.*

6.3 EXERCISES

🌐 *Complete solution available on Video Lectures on CD/DVD*

▢ *Now Try Exercise*

Factor each polynomial by grouping. (The middle term of an equivalent trinomial has already been rewritten.) See Example 1.

1. $10t^2 + 5t + 4t + 2$

2. $6x^2 + 9x + 4x + 6$

3. $15z^2 - 10z - 9z + 6$

4. $12p^2 - 9p - 8p + 6$

5. $8s^2 - 4st + 6st - 3t^2$

6. $3x^2 - 7xy + 6xy - 14y^2$

Concept Check *Complete the steps to factor each trinomial by grouping.*

7. $2m^2 + 11m + 12$

(a) Find two integers whose product is

_____ · _____ = _____

and whose sum is _____.

(b) The required integers are _____ and _____.

(c) Write the middle term, $11m$, as _____ + _____.

(d) Rewrite the given trinomial as

_____.

(e) Factor the polynomial in part (d) by grouping.

(f) Check by multiplying.

8. $6y^2 - 19y + 10$

(a) Find two integers whose product is

_____ · _____ = _____

and whose sum is _____.

(b) The required integers are _____ and _____.

(c) Write the middle term, $-19y$, as _____ + _____.

(d) Rewrite the given trinomial as

_____.

(e) Factor the polynomial in part (d) by grouping.

(f) Check by multiplying.

9. *Concept Check* Which pair of integers would be used to rewrite the middle term when one is factoring $12y^2 + 5y - 2$ by grouping?

A. $-8, 3$ **B.** $8, -3$
C. $-6, 4$ **D.** $6, -4$

10. *Concept Check* Which pair of integers would be used to rewrite the middle term when one is factoring $20b^2 - 13b + 2$ by grouping?

A. $10, 3$ **B.** $-10, -3$
C. $8, 5$ **D.** $-8, -5$

Concept Check *Decide which is the correct factored form of the given polynomial.*

11. $2x^2 - x - 1$

  A. $(2x - 1)(x + 1)$
  B. $(2x + 1)(x - 1)$

12. $3a^2 - 5a - 2$

  A. $(3a + 1)(a - 2)$
  B. $(3a - 1)(a + 2)$

13. $4y^2 + 17y - 15$

  A. $(y + 5)(4y - 3)$
  B. $(2y - 5)(2y + 3)$

14. $12c^2 - 7c - 12$

  A. $(6c - 2)(2c + 6)$
  B. $(4c + 3)(3c - 4)$

Complete each factoring. See Examples 1–7.

15. $6a^2 + 7ab - 20b^2 = (3a - 4b)(\quad\quad)$

16. $9m^2 - 3mn - 2n^2 = (3m + n)(\quad\quad)$

17. $2x^2 + 6x - 8 = 2(\quad\quad\quad\quad)$
$ = 2(\quad\quad)(\quad\quad)$

18. $3x^2 - 9x - 30 = 3(\quad\quad\quad\quad)$
$ = 3(\quad\quad)(\quad\quad)$

19. For the polynomial $12x^2 + 7x - 12$, 2 is not a common factor. Explain why the binomial $2x - 6$, then, cannot be a factor of the polynomial.

20. Factor $4k^2 + 7k - 15$ twice, using the two methods discussed in the text. Do your answers agree? Which method do you prefer?

Factor each trinomial completely. See Examples 1–7.

21. $3a^2 + 10a + 7$

22. $7r^2 + 8r + 1$

23. $2y^2 + 7y + 6$

24. $5z^2 + 12z + 4$

25. $15m^2 + m - 2$

26. $6x^2 + x - 1$

27. $12s^2 + 11s - 5$

28. $20x^2 + 11x - 3$

29. $10m^2 - 23m + 12$

30. $6x^2 - 17x + 12$

31. $8w^2 - 14w + 3$

32. $9p^2 - 18p + 8$

33. $20y^2 - 39y - 11$

34. $10x^2 - 11x - 6$

35. $3x^2 - 15x + 16$

36. $2t^2 + 13t - 18$

37. $20x^2 + 22x + 6$

38. $36y^2 + 81y + 45$

39. $24x^2 - 42x + 9$

40. $48b^2 - 74b - 10$

41. $40m^2q + mq - 6q$

42. $15a^2b + 22ab + 8b$

43. $15n^4 - 39n^3 + 18n^2$

44. $24a^4 + 10a^3 - 4a^2$

45. $15x^2y^2 - 7xy^2 - 4y^2$

46. $14a^2b^3 + 15ab^3 - 9b^3$

47. $5a^2 - 7ab - 6b^2$

48. $6x^2 - 5xy - y^2$

49. $12s^2 + 11st - 5t^2$

50. $25a^2 + 25ab + 6b^2$

51. $6m^6n + 7m^5n^2 + 2m^4n^3$

52. $12k^3q^4 - 4k^2q^5 - kq^6$

53. $5 - 6x + x^2$

54. $7 + 8x + x^2$

55. $16 + 16x + 3x^2$

56. $18 + 65x + 7x^2$

57. $-10x^3 + 5x^2 + 140x$

58. $-18k^3 - 48k^2 + 66k$

59. $12x^2 - 47x - 4$

60. $12x^2 - 19x - 10$

61. $24y^2 - 41xy - 14x^2$

62. $24x^2 + 19xy - 5y^2$

63. $36x^4 - 64x^2y + 15y^2$

64. $36x^4 + 59x^2y + 24y^2$

65. $48a^2 - 94ab - 4b^2$

66. $48t^2 - 147ts + 9s^2$

67. $10x^4y^5 + 39x^3y^5 - 4x^2y^5$

68. $14x^7y^4 - 31x^6y^4 + 6x^5y^4$

69. $36a^3b^2 - 104a^2b^2 - 12ab^2$

70. $36p^4q + 129p^3q - 60p^2q$

71. $24x^2 - 46x + 15$

72. $24x^2 - 94x + 35$

73. $24x^4 + 55x^2 - 24$

74. $24x^4 + 17x^2 - 20$

75. $24x^2 + 38xy + 15y^2$

76. $24x^2 + 62xy + 33y^2$

77. $24x^2z^4 - 113xz^2 - 35$

78. *Concept Check*   On a quiz, a student factored $3k^3 - 12k^2 - 15k$ by first factoring out the common factor $3k$ to get $3k(k^2 - 4k - 5)$. Then the student wrote

$$k^2 - 4k - 5 = k^2 - 5k + k - 5$$
$$= k(k - 5) + 1(k - 5)$$
$$= (k - 5)(k + 1). \qquad \text{Her answer}$$

WHAT WENT WRONG? What is the correct factored form?

If a trinomial has a negative coefficient for the squared term, as in $-2x^2 + 11x - 12$, it is usually easier to factor by first factoring out the common factor -1:

$$-2x^2 + 11x - 12 = -1(2x^2 - 11x + 12)$$
$$= -1(2x - 3)(x - 4).$$

Use this method to factor each trinomial. See Example 7(b).

79. $-x^2 - 4x + 21$

80. $-x^2 + x + 72$

81. $-3x^2 - x + 4$

82. $-5x^2 + 2x + 16$

83. $-2a^2 - 5ab - 2b^2$

84. $-3p^2 + 13pq - 4q^2$

85. The answer given in the back of the book for Exercise 79 is $-1(x + 7)(x - 3)$. Is $(x + 7)(3 - x)$ also a correct answer? Explain.

86. One answer for Exercise 80 is $-1(x + 8)(x - 9)$. Is $(-x - 8)(-x + 9)$ also a correct answer? Explain.

Factor each polynomial. Remember to factor out the greatest common factor as the first step.

87. $25q^2(m + 1)^3 - 5q(m + 1)^3 - 2(m + 1)^3$

88. $18x^2(y - 3)^2 - 21x(y - 3)^2 - 4(y - 3)^2$

89. $15x^2(r + 3)^3 - 34xy(r + 3)^3 - 16y^2(r + 3)^3$

90. $4t^2(k + 9)^7 + 20ts(k + 9)^7 + 25s^2(k + 9)^7$

The next problems are "brain busters." Find all integers k so that the trinomial can be factored by the methods of this section.

91. $5x^2 + kx - 1$

92. $2x^2 + kx - 3$

93. $2m^2 + km + 5$

94. $3y^2 + ky + 4$

PREVIEW EXERCISES

Find each product. See Section 5.6.

95. $(7p + 3)(7p - 3)$

96. $(3h + 5k)(3h - 5k)$

97. $\left(r^2 + \frac{1}{2}\right)\left(r^2 - \frac{1}{2}\right)$

98. $(x + 6)^2$

99. $(3t + 4)^2$

100. $\left(c - \frac{2}{3}\right)^2$

6.4 Special Factoring Techniques

OBJECTIVES

1 Factor a difference of squares.

2 Factor a perfect square trinomial.

3 Factor a difference of cubes.

4 Factor a sum of cubes.

By reversing the rules for multiplication of binomials from **Section 5.6,** we get rules for factoring polynomials in certain forms.

OBJECTIVE 1 **Factor a difference of squares.** The formula for the product of the sum and difference of the same two terms is

$$(x + y)(x - y) = x^2 - y^2.$$

Reversing this rule leads to the following special factoring rule.

Factoring a Difference of Squares

$$x^2 - y^2 = (x + y)(x - y)$$

For example,

$$m^2 - 16 = m^2 - 4^2 = (m + 4)(m - 4).$$

The following conditions must be true for a binomial to be a difference of squares:

1. Both terms of the binomial must be squares, such as

$$x^2, \qquad 9y^2, \qquad 25, \qquad 1, \qquad m^4.$$

2. The second terms of the binomials must have different signs (one positive and one negative).

EXAMPLE 1 **Factoring Differences of Squares**

Factor each binomial if possible. (In part (c), use fractions in the factored form.)

$$x^2 - y^2 = (x + y)(x - y)$$

(a) $a^2 - 49 = a^2 - 7^2 = (a + 7)(a - 7)$

(b) $y^2 - m^2 = (y + m)(y - m)$

Check by multiplying out the factored form.

(c) $z^2 - \dfrac{9}{16} = z^2 - \left(\dfrac{3}{4}\right)^2 = \left(z + \dfrac{3}{4}\right)\left(z - \dfrac{3}{4}\right)$

(d) $x^2 - 8$

Because 8 is not the square of an integer, this binomial is not a difference of squares. It is a prime polynomial.

(e) $p^2 + 16$

Since $p^2 + 16$ is a *sum* of squares, it is not equal to $(p + 4)(p - 4)$. Also, using FOIL gives

$$(p - 4)(p - 4) = p^2 - 8p + 16 \neq p^2 + 16$$

and

$$(p + 4)(p + 4) = p^2 + 8p + 16 \neq p^2 + 16,$$

so $p^2 + 16$ is a prime polynomial.

✔ **Now Try Exercises 7, 9, and 11.**

▶ **CAUTION** As Example 1(e) suggests, *after any common factor is removed, a sum of squares cannot be factored.*

EXAMPLE 2 Factoring Differences of Squares

Factor each difference of squares.

$$x^2 \; - \; y^2 \; = \; (x \; + \; y) \; (x \; - \; y)$$

(a) $25m^2 - 16 = (5m)^2 - 4^2 = (5m + 4)(5m - 4)$

(b) $49z^2 - 64t^2 = (7z)^2 - (8t)^2 = (7z + 8t)(7z - 8t)$

✔ **Now Try Exercises 13 and 21.**

▶ **NOTE** As in previous sections, you should *always check a factored form by multiplying.*

EXAMPLE 3 Factoring More Complex Differences of Squares

Factor completely.

(a) $81y^2 - 36$

First factor out the common factor, 9.

$$81y^2 - 36 = 9(9y^2 - 4) \qquad \text{Factor out 9.}$$
$$= 9[(3y)^2 - 2^2]$$
$$= 9(3y + 2)(3y - 2) \qquad \text{Difference of squares}$$

(b) $p^4 - 36 = (p^2)^2 - 6^2 = (p^2 + 6)(p^2 - 6)$

Neither $p^2 + 6$ nor $p^2 - 6$ can be factored further.

(c) $m^4 - 16 = (m^2)^2 - 4^2$

$$= (m^2 + 4)(m^2 - 4) \qquad \text{Difference of squares}$$
$$= (m^2 + 4)(m + 2)(m - 2) \qquad \text{Difference of squares again}$$

Don't stop here.

✔ **Now Try Exercises 17, 25, and 27.**

▶ **CAUTION** *Factor again when any of the factors is a difference of squares,* as in Example 3(c). Check by multiplying.

OBJECTIVE 2 Factor a perfect square trinomial. The expressions 144, $4x^2$, and $81m^6$ are called *perfect squares* because

$$144 = 12^2, \qquad 4x^2 = (2x)^2, \qquad \text{and} \qquad 81m^6 = (9m^3)^2.$$

A **perfect square trinomial** is a trinomial that is the square of a binomial. For example, $x^2 + 8x + 16$ is a perfect square trinomial because it is the square of the binomial $x + 4$:

$$x^2 + 8x + 16 = (x + 4)(x + 4) = (x + 4)^2.$$

On the one hand, a necessary condition for a trinomial to be a perfect square is that *two of its terms be perfect squares.* For this reason, $16x^2 + 4x + 15$ is not a perfect square trinomial, because only the term $16x^2$ is a perfect square.

On the other hand, even if two of the terms are perfect squares, the trinomial may not be a perfect square trinomial. For example, $x^2 + 6x + 36$ has two perfect square terms, x^2 and 36, but it is not a perfect square trinomial.

We can multiply to see that the square of a binomial gives one of the following perfect square trinomials.

Factoring Perfect Square Trinomials

$$x^2 + 2xy + y^2 = (x + y)^2$$
$$x^2 - 2xy + y^2 = (x - y)^2$$

The middle term of a perfect square trinomial is always twice the product of the two terms in the squared binomial (as shown in **Section 5.6**). Use this rule to check any attempt to factor a trinomial that appears to be a perfect square.

EXAMPLE 4   **Factoring a Perfect Square Trinomial**

Factor $x^2 + 10x + 25$.

The x^2-term is a perfect square, and so is 25. Try to factor the trinomial as

$$x^2 + 10x + 25 = (x + 5)^2.$$

To check, take twice the product of the two terms in the squared binomial.

$$2 \cdot x \cdot 5 = 10x$$

Twice   First term   Last term
of binomial   of binomial

Since $10x$ is the middle term of the trinomial, the trinomial is a perfect square and can be factored as $(x + 5)^2$. Thus,

$$x^2 + 10x + 25 = (x + 5)^2.$$

✔ **Now Try Exercise 33.**

EXAMPLE 5   **Factoring Perfect Square Trinomials**

Factor each trinomial.

(a) $x^2 - 22x + 121$

The first and last terms are perfect squares ($121 = 11^2$ or $(-11)^2$). Check to see whether the middle term of $x^2 - 22x + 121$ is twice the product of the first and last terms of the binomial $x - 11$.

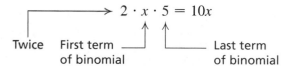

$$2 \cdot x \cdot (-11) = -22x$$

Twice   First   Last
term   term

Since twice the product of the first and last terms of the binomial is the middle term, $x^2 - 22x + 121$ is a perfect square trinomial and

$$x^2 - 22x + 121 = (x - 11)^2.$$

Same sign

Notice that the sign of the second term in the squared binomial is the same as the sign of the middle term in the trinomial.

(b) $9m^2 - 24m + 16 = (3m)^2 + 2(3m)(-4) + (-4)^2 = (3m - 4)^2$

Twice —— First —— Last
term   term

(c) $25y^2 + 20y + 16$

The first and last terms are perfect squares.

$$25y^2 = (5y)^2 \qquad \text{and} \qquad 16 = 4^2$$

Twice the product of the first and last terms of the binomial $5y + 4$ is

$$2 \cdot 5y \cdot 4 = 40y,$$

which is not the middle term of $25y^2 + 20y + 16$. This trinomial is not a perfect square. In fact, the trinomial cannot be factored even with the methods of the previous sections. It is a prime polynomial.

(d) $12z^3 + 60z^2 + 75z = 3z(4z^2 + 20z + 25)$ Factor out $3z$.

$$= 3z[(2z)^2 + 2(2z)(5) + 5^2]$$
$$= 3z(2z + 5)^2$$

✔ **Now Try Exercises 35, 43, and 49.**

▶ **NOTE**

1. The sign of the second term in the squared binomial is always the same as the sign of the middle term in the trinomial.

2. The first and last terms of a perfect square trinomial must be *positive*, because they are squares. For example, the polynomial $x^2 - 2x - 1$ cannot be a perfect square, because the last term is negative.

3. Perfect square trinomials can also be factored by using grouping or the FOIL method, although using the method of this section is often easier.

OBJECTIVE 3 **Factor a difference of cubes.** Just as we factored the difference of squares in Objective 1, we can also factor the **difference of cubes** by using the following pattern.

Factoring a Difference of Cubes

$$x^3 - y^3 = (x - y)(x^2 + xy + y^2)$$

This pattern for factoring a difference of cubes should be memorized. To see that the pattern is correct, multiply $(x - y)(x^2 + xy + y^2)$.

$$\begin{array}{r} x^2 + xy + y^2 \\ x - y \\ \hline -x^2y - xy^2 - y^3 \\ x^3 + x^2y + xy^2 \\ \hline x^3 \qquad\qquad - y^3 \end{array}$$

Notice the pattern of the terms in the factored form of $x^3 - y^3$.

- $x^3 - y^3 =$ (a binomial factor)(a trinomial factor)
- The binomial factor has the difference of the cube roots of the given terms.
- The terms in the trinomial factor are all positive.
- What you write in the binomial factor determines the trinomial factor:

$$x^3 - y^3 = (x - y)(\underset{\text{First term squared}}{x^2} + \underset{\substack{\text{positive} \\ \text{product of} \\ \text{the terms}}}{xy} + \underset{\substack{\text{second term} \\ \text{squared}}}{y^2}).$$

> **CAUTION** The polynomial $x^3 - y^3$ is not equivalent to $(x - y)^3$, because $(x - y)^3$ can also be written as
>
> $$(x - y)^3 = (x - y)(x - y)(x - y)$$
> $$= (x - y)(x^2 - 2xy + y^2),$$
>
> but
>
> $$x^3 - y^3 = (x - y)(x^2 + xy + y^2).$$

EXAMPLE 6 Factoring Differences of Cubes

Factor each polynomial.

(a) $m^3 - 125$

Let $x = m$ and $y = 5$ in the pattern for the difference of cubes.

$$x^3 - y^3 = (x - y)(x^2 + xy + y^2)$$
$$m^3 - 125 = m^3 - 5^3 = (m - 5)(m^2 + 5m + 5^2) \qquad \text{Let } x = m, y = 5.$$
$$= (m - 5)(m^2 + 5m + 25)$$

(b) $8p^3 - 27$

Since $8p^3 = (2p)^3$ and $27 = 3^3$,

$$8p^3 - 27 = (2p)^3 - 3^3$$
$$= (2p - 3)[(2p)^2 + (2p)3 + 3^2]$$
$$= (2p - 3)(4p^2 + 6p + 9).$$

(c) $4m^3 - 32 = 4(m^3 - 8)$ Factor out 4.

$$= 4(m^3 - 2^3)$$

$$= 4(m - 2)(m^2 + 2m + 4)$$

(d) $125t^3 - 216s^6 = (5t)^3 - (6s^2)^3$

$$= (5t - 6s^2)[(5t)^2 + 5t(6s^2) + (6s^2)^2]$$

$$= (5t - 6s^2)(25t^2 + 30ts^2 + 36s^4)$$

✔ **Now Try Exercises 59, 63, 69, and 77.**

▶ **CAUTION** A common error in factoring a difference of cubes, such as $x^3 - y^3 = (x - y)(x^2 + xy + y^2)$, is to try to factor $x^2 + xy + y^2$. It is easy to confuse this factor with the perfect square trinomial $x^2 + 2xy + y^2$. Because there is no 2 in $x^2 + xy + y^2$, it is unusual to be able to further factor an expression of the form $x^2 + xy + y^2$.

OBJECTIVE 4 Factor a sum of cubes. A sum of squares, such as $m^2 + 25$, cannot be factored by using real numbers, but a **sum of cubes** can be factored by the following pattern. *It, too, should be memorized.*

Factoring a Sum of Cubes

$$x^3 + y^3 = (x + y)(x^2 - xy + y^2)$$

Compare the pattern for the *sum* of cubes with the pattern for the *difference* of cubes. The only difference between them is the positive and negative signs.

$$x^3 - y^3 = (x - y)(x^2 + xy + y^2) \qquad \text{Difference of cubes}$$

Same sign   Positive   Opposite sign

$$x^3 + y^3 = (x + y)(x^2 - xy + y^2) \qquad \text{Sum of cubes}$$

Same sign   Positive   Opposite sign

Observing these relationships should help you to remember these patterns.

EXAMPLE 7 Factoring Sums of Cubes

Factor each polynomial.

(a) $k^3 + 27 = k^3 + 3^3$

$$= (k + 3)(k^2 - 3k + 3^2)$$
$$= (k + 3)(k^2 - 3k + 9)$$

(b) $8m^3 + 125n^3 = (2m)^3 + (5n)^3$

$$= (2m + 5n)[(2m)^2 - 2m(5n) + (5n)^2]$$
$$= (2m + 5n)(4m^2 - 10mn + 25n^2)$$

(c) $1000a^6 + 27b^3 = (10a^2)^3 + (3b)^3$

$$= (10a^2 + 3b)[(10a^2)^2 - (10a^2)(3b) + (3b)^2]$$
$$= (10a^2 + 3b)(100a^4 - 30a^2b + 9b^2)$$

✔ **Now Try Exercises 61, 71, and 79.**

The methods of factoring discussed in this section are summarized here.

Special Factorizations

Difference of squares	$x^2 - y^2 = (x + y)(x - y)$
Perfect square trinomials	$x^2 + 2xy + y^2 = (x + y)^2$
	$x^2 - 2xy + y^2 = (x - y)^2$
Difference of cubes	$x^3 - y^3 = (x - y)(x^2 + xy + y^2)$
Sum of cubes	$x^3 + y^3 = (x + y)(x^2 - xy + y^2)$

The *sum* of *squares* can be factored only if the terms have a common factor.

6.4 EXERCISES

1. *Concept Check* To help you factor the difference of squares, complete the following list of squares:

$1^2 = $ _____ $2^2 = $ _____ $3^2 = $ _____ $4^2 = $ _____ $5^2 = $ _____

$6^2 = $ _____ $7^2 = $ _____ $8^2 = $ _____ $9^2 = $ _____ $10^2 = $ _____

$11^2 = $ _____ $12^2 = $ _____ $13^2 = $ _____ $14^2 = $ _____ $15^2 = $ _____

$16^2 = $ _____ $17^2 = $ _____ $18^2 = $ _____ $19^2 = $ _____ $20^2 = $ _____ .

2. *Concept Check* The following powers of x are all perfect squares: $x^2, x^4, x^6, x^8, x^{10}$. On the basis of this observation, we may make a conjecture (an educated guess) that if the power of a variable is divisible by _____ (with 0 remainder), then we have a perfect square.

3. *Concept Check* To help you factor the sum or difference of cubes, complete the following list of cubes:

$1^3 = $ _____ $2^3 = $ _____ $3^3 = $ _____ $4^3 = $ _____ $5^3 = $ _____

$6^3 = $ _____ $7^3 = $ _____ $8^3 = $ _____ $9^3 = $ _____ $10^3 = $ _____ .

4. *Concept Check* The following powers of x are all perfect cubes: x^3, x^6, x^9, x^{12}, x^{15}. On the basis of this observation, we may make a conjecture that if the power of a variable is divisible by _____ (with 0 remainder), then we have a perfect cube.

5. *Concept Check* Identify each monomial as a *perfect square*, a *perfect cube*, *both of these*, or *neither of these*.

(a) $64x^6y^{12}$ (b) $125t^6$ (c) $49x^{12}$ (d) $81r^{10}$

6. *Concept Check* What must be true for x^n to be both a perfect square and a perfect cube?

Factor each binomial completely. If the binomial is prime, say so. Use your answers from Exercises 1 and 2 as necessary. (Use fractions in Exercises 9, 10, 15, and 16.) See Examples 1–3.

7. $y^2 - 25$

8. $t^2 - 16$

9. $p^2 - \dfrac{1}{9}$

10. $q^2 - \dfrac{1}{4}$

11. $m^2 + 64$

12. $k^2 + 49$

13. $9r^2 - 4$

14. $4x^2 - 9$

15. $36m^2 - \dfrac{16}{25}$

16. $100b^2 - \dfrac{4}{49}$

17. $36x^2 - 16$

18. $32a^2 - 8$

19. $196p^2 - 225$

20. $361q^2 - 400$

21. $16r^2 - 25a^2$

22. $49m^2 - 100p^2$

23. $100x^2 + 49$

24. $81w^2 + 16$

25. $p^4 - 49$

26. $r^4 - 25$

27. $x^4 - 1$

28. $y^8 - 256$

29. $p^4 - 256$

30. $16k^4 - 1$

31. *Concept Check* When a student was directed to factor $x^4 - 81$ completely, his teacher did not give him full credit for the answer $(x^2 + 9)(x^2 - 9)$. The student argued that since his answer does indeed give $x^4 - 81$ when multiplied out, he should be given full credit. **WHAT WENT WRONG?** Give the correct factored form.

32. The binomial $4x^2 + 16$ is a sum of squares that *can* be factored. How is this binomial factored? When can the sum of squares be factored?

Factor each trinomial completely. (Use fractions in Exercises 37 and 38 and decimals in Exercises 39 and 40.) See Examples 4 and 5.

33. $w^2 + 2w + 1$

34. $p^2 + 4p + 4$

35. $x^2 - 8x + 16$

36. $x^2 - 10x + 25$

37. $t^2 + t + \dfrac{1}{4}$

38. $m^2 + \dfrac{2}{3}m + \dfrac{1}{9}$

39. $x^2 - 1.0x + 0.25$

40. $y^2 - 1.4y + 0.49$

41. $2x^2 + 24x + 72$

42. $3y^2 - 48y + 192$

43. $16x^2 - 40x + 25$

44. $36y^2 - 60y + 25$

45. $49x^2 - 28xy + 4y^2$

46. $4z^2 - 12zw + 9w^2$

47. $64x^2 + 48xy + 9y^2$

48. $9t^2 + 24tr + 16r^2$

49. $50h^2 - 40hy + 8y^2$

50. $18x^2 + 48xy + 32y^2$

51. $4k^3 - 4k^2 + 9k$

52. $9r^3 + 6r^2 + 16r$

53. $25z^4 + 5z^3 + z^2$

54. In the polynomial $9y^2 + 14y + 25$, the first and last terms are perfect squares. Can the polynomial be factored? If it can, factor it. If it cannot, explain why it is not a perfect square trinomial.

Concept Check *Find the value of the indicated variable.*

55. Find a value of b so that $x^2 + bx + 25 = (x + 5)^2$.

56. For what value of c is $4m^2 - 12m + c = (2m - 3)^2$?

57. Find a so that $ay^2 - 12y + 4 = (3y - 2)^2$.

58. Find b so that $100a^2 + ba + 9 = (10a + 3)^2$.

Factor each binomial completely. Use your answers from Exercises 3 and 4 as necessary. See Examples 6 and 7.

59. $a^3 - 1$ **60.** $m^3 - 8$ **61.** $m^3 + 8$

62. $b^3 + 1$ **63.** $27x^3 - 64$ **64.** $64y^3 - 27$

65. $6p^3 + 6$ **66.** $81x^3 + 3$ **67.** $5x^3 + 40$

68. $128y^3 - 54$ **69.** $2x^3 - 16y^3$ **70.** $27w^3 - 216z^3$

71. $8p^3 + 729q^3$ **72.** $64x^3 + 125y^3$ **73.** $27a^3 + 64b^3$

74. $125m^3 - 8p^3$ **75.** $125t^3 + 8s^3$ **76.** $27r^3 + 1000s^3$

77. $8x^3 - 125y^6$ **78.** $27t^3 - 64s^6$ **79.** $27m^6 + 8n^3$

80. $1000r^6 + 27s^3$ **81.** $x^9 + y^9$ **82.** $x^9 - y^9$

The next problems are "brain busters." Extend the methods of factoring presented so far in this chapter to factor each polynomial completely.

83. $(m + n)^2 - (m - n)^2$ **84.** $(a - b)^3 - (a + b)^3$

85. $m^2 - p^2 + 2m + 2p$ **86.** $3r - 3k + 3r^2 - 3k^2$

PREVIEW EXERCISES

Solve each equation. See Sections 2.1 and 2.2.

87. $m - 4 = 0$ **88.** $3t + 2 = 0$ **89.** $4z - 9 = 0$

90. $2t + 10 = 0$ **91.** $9x - 6 = 0$ **92.** $7x = 0$

Summary Exercises on Factoring

The mixed exercises that follow are included to give you practice in selecting an appropriate method for factoring a particular polynomial. As you factor a polynomial, ask yourself these questions to decide on a suitable factoring technique.

Factoring a Polynomial

1. **Is there a common factor?** If so, factor it out.

2. **How many terms are in the polynomial?**

   Two terms: Check to see whether it is a difference of squares or a sum or difference of cubes. If so, factor as in **Section 6.4.**

   Three terms: Is it a perfect square trinomial? If the trinomial is not a perfect square, check to see whether the coefficient of the squared term is 1. If so, use the method of **Section 6.2.** If the coefficient of the squared term of the trinomial is not 1, use the general factoring methods of **Section 6.3.**

   Four terms: Try to factor the polynomial by grouping.

3. **Can any factors be factored further?** If so, factor them.

Factor each polynomial completely.

1. $a^2 - 4a - 12$

2. $a^2 + 17a + 72$

3. $6y^2 - 6y - 12$

4. $7y^6 + 14y^5 - 168y^4$

5. $6a + 12b + 18c$

6. $m^2 - 3mn - 4n^2$

7. $p^2 - 17p + 66$

8. $z^2 - 6z + 7z - 42$

9. $10z^2 - 7z - 6$

10. $2m^2 - 10m - 48$

11. $17x^3y^2 + 51xy$

12. $15y + 5$

13. $8a^5 - 8a^4 - 48a^3$

14. $8k^2 - 10k - 3$

15. $z^2 - 3za - 10a^2$

16. $50z^2 - 100$

17. $x^2 - 4x - 5x + 20$

18. $100n^2r^2 + 30nr^3 - 50n^2r$

19. $6n^2 - 19n + 10$

20. $9y^2 + 12y - 5$

21. $16x + 20$

22. $m^2 + 2m - 15$

23. $6y^2 - 5y - 4$

24. $m^2 - 81$

25. $6z^2 + 31z + 5$

26. $12x^2 + 47x - 4$

27. $4k^2 - 12k + 9$

28. $8p^2 + 23p - 3$

29. $54m^2 - 24z^2$

30. $8m^2 - 2m - 3$

31. $3k^2 + 4k - 4$

32. $45a^3b^5 - 60a^4b^2 + 75a^6b^4$

33. $14k^3 + 7k^2 - 70k$

34. $5 + r - 5s - rs$

35. $y^4 - 16$

36. $20y^5 - 30y^4$

37. $8m - 16m^2$

38. $k^2 - 16$

39. $z^3 - 8$

40. $y^2 - y - 56$

41. $k^2 + 9$

42. $27p^{10} - 45p^9 - 252p^8$

43. $32m^9 + 16m^5 + 24m^3$

44. $8m^3 + 125$

45. $16r^2 + 24rm + 9m^2$

46. $z^2 - 12z + 36$

47. $15h^2 + 11hg - 14g^2$

48. $5z^3 - 45z^2 + 70z$

49. $k^2 - 11k + 30$

50. $64p^2 - 100m^2$

51. $3k^3 - 12k^2 - 15k$

52. $y^2 - 4yk - 12k^2$

53. $1000p^3 + 27$

54. $64r^3 - 343$

55. $6 + 3m + 2p + mp$

56. $2m^2 + 7mn - 15n^2$

57. $16z^2 - 8z + 1$

58. $125m^4 - 400m^3n + 195m^2n^2$

59. $108m^2 - 36m + 3$

60. $100a^2 - 81y^2$

61. $x^2 - xy + y^2$

62. $4y^2 - 25$

63. $32z^3 + 56z^2 - 16z$

64. $10m^2 + 25m - 60$

65. $20 + 5m + 12n + 3mn$

66. $4 - 2q - 6p + 3pq$

67. $6a^2 + 10a - 4$

68. $36y^6 - 42y^5 - 120y^4$

69. $a^3 - b^3 + 2a - 2b$

70. $16k^2 - 48k + 36$

71. $64m^2 - 80mn + 25n^2$

72. $72y^3z^2 + 12y^2 - 24y^4z^2$

73. $8k^2 - 2kh - 3h^2$

74. $2a^2 - 7a - 30$

75. $2x^3 + 128$

76. $8a^3 - 27$

77. $10y^2 - 7yz - 6z^2$

78. $m^2 - 4m + 4$

79. $8a^2 + 23ab - 3b^2$

80. $a^4 - 625$

RELATING CONCEPTS (EXERCISES 81–88)

FOR INDIVIDUAL OR GROUP WORK

*A binomial may be both a difference of squares and a difference of cubes. One example of such a binomial is $x^6 - 1$. With the techniques of **Section 6.4**, one factoring method will give the completely factored form, while the other will not. **Work Exercises 81–88 in order** to determine the method to use if you have to make such a decision.*

81. Factor $x^6 - 1$ as the difference of squares.

82. The factored form obtained in Exercise 81 consists of a difference of cubes multiplied by a sum of cubes. Factor each binomial further.

83. Now start over and factor $x^6 - 1$ as the difference of cubes.

84. The factored form obtained in Exercise 83 consists of a binomial that is a difference of squares and a trinomial. Factor the binomial further.

85. Compare your results in Exercises 82 and 84. Which one of these is factored completely?

86. Verify that the trinomial in the factored form in Exercise 84 is the product of the two trinomials in the factored form in Exercise 82.

87. Use the results of Exercises 81–86 to complete the following statement: In general, if I must choose between factoring first with the method for the difference of squares or the method for the difference of cubes, I should choose the _____ method to eventually obtain the completely factored form.

88. Find the *completely* factored form of $x^6 - 729$ by using the knowledge you gained in Exercises 81–87.

6.5 Solving Quadratic Equations by Factoring

OBJECTIVES

1. Solve quadratic equations by factoring.
2. Solve other equations by factoring.

Galileo Galilei (1564–1642) developed theories to explain physical phenomena and set up experiments to test his ideas. According to legend, Galileo dropped objects of different weights from the Leaning Tower of Pisa to disprove the belief that heavier objects fall faster than lighter objects. He developed the formula

$$d = 16t^2$$

describing the motion of freely falling objects. In this formula, d is the distance in feet that an object falls (disregarding air resistance) in t seconds, regardless of weight.

The equation $d = 16t^2$ is a *quadratic equation,* the subject of this section. A quadratic equation contains a squared term and no terms of higher degree.

Quadratic Equation

A **quadratic equation** is an equation that can be written in the form

$$ax^2 + bx + c = 0,$$

where a, b, and c are real numbers, with $a \neq 0$.

The form $ax^2 + bx + c = 0$ is the **standard form** of a quadratic equation. For example,

$$x^2 + 5x + 6 = 0, \quad 2x^2 - 5x = 3, \quad \text{and} \quad x^2 = 4 \qquad \text{Quadratic equations}$$

are all quadratic equations, but only $x^2 + 5x + 6 = 0$ is in standard form.

Up to now, we have factored *expressions,* including many quadratic expressions of the form $ax^2 + bx + c$. In this section, we see how we can use factored quadratic expressions to solve quadratic *equations.*

OBJECTIVE 1 **Solve quadratic equations by factoring.** We use the **zero-factor property** to solve a quadratic equation by factoring.

Zero-Factor Property

If a and b are real numbers and if $ab = 0$, then $a = 0$ or $b = 0$.

That is, if the product of two numbers is 0, then at least one of the numbers must be 0. One number *must* be 0, but both *may* be 0.

EXAMPLE 1 **Using the Zero-Factor Property**

Solve each equation.

(a) $(x + 3)(2x - 1) = 0$

The product $(x + 3)(2x - 1)$ is equal to 0. By the zero-factor property, the only way that the product of these two factors can be 0 is if at least one of the factors equals 0. Therefore, either $x + 3 = 0$ or $2x - 1 = 0$. Solve each of these two linear equations as in **Chapter 2.**

$$\begin{array}{lll} x + 3 = 0 & \text{or} \quad 2x - 1 = 0 & \text{Zero-factor property} \\ x = -3 & 2x = 1 & \text{Isolate the variable term.} \\ & x = \dfrac{1}{2} & \text{Divide each side by 2.} \end{array}$$

The given equation, $(x + 3)(2x - 1) = 0$, has two solutions: -3 and $\frac{1}{2}$. Check these solutions by substituting -3 for x in the original equation, $(x + 3)(2x - 1) = 0$. Then start over and substitute $\frac{1}{2}$ for x.

If $x = -3$, then

$$(x + 3)(2x - 1) = 0$$

$$(-3 + 3)[2(-3) - 1] = 0 \quad ?$$

$$0(-7) = 0. \qquad \text{True}$$

If $x = \dfrac{1}{2}$, then

$$(x + 3)(2x - 1) = 0$$

$$\left(\dfrac{1}{2} + 3\right)\left(2 \cdot \dfrac{1}{2} - 1\right) = 0 \quad ?$$

$$\dfrac{7}{2}(1 - 1) = 0 \quad ?$$

$$\dfrac{7}{2} \cdot 0 = 0. \qquad \text{True}$$

Both -3 and $\frac{1}{2}$ result in true equations, so the solution set is $\left\{-3, \frac{1}{2}\right\}$.

(b)
$$y(3y - 4) = 0$$
$$y = 0 \quad \text{or} \quad 3y - 4 = 0 \qquad \text{Zero-factor property}$$
$$3y = 4$$

> Don't forget that 0 is a solution.

$$y = \frac{4}{3}$$

Check these solutions by substituting each one into the original equation. The solution set is $\left\{0, \frac{4}{3}\right\}$.

✔ **Now Try Exercises 13 and 15.**

▶ **NOTE** The word *or* as used in Example 1 means "one or the other or both."

In Example 1, each equation to be solved was given with the polynomial in factored form. If the polynomial in an equation is not already factored, first make sure that the equation is in standard form. Then factor.

EXAMPLE 2 Solving Quadratic Equations

Solve each equation.

(a) $x^2 - 5x = -6$

First, rewrite the equation in standard form by adding 6 to each side.

> Don't factor x out at this step.

$$x^2 - 5x = -6$$
$$x^2 - 5x + 6 = 0 \qquad \text{Add 6.}$$

Now factor $x^2 - 5x + 6$. Find two numbers whose product is 6 and whose sum is -5. These two numbers are -2 and -3, so the equation becomes

$$(x - 2)(x - 3) = 0. \qquad \text{Factor.}$$
$$x - 2 = 0 \quad \text{or} \quad x - 3 = 0 \qquad \text{Zero-factor property}$$
$$x = 2 \quad \text{or} \qquad x = 3 \qquad \text{Solve each equation.}$$

Check: If $x = 2$, then
$$x^2 - 5x = -6$$
$$2^2 - 5(2) = -6 \quad ?$$
$$4 - 10 = -6 \quad ?$$
$$-6 = -6. \qquad \text{True}$$

If $x = 3$, then
$$x^2 - 5x = -6$$
$$3^2 - 5(3) = -6 \quad ?$$
$$9 - 15 = -6 \quad ?$$
$$-6 = -6. \qquad \text{True}$$

Both solutions check, so the solution set is $\{2, 3\}$.

(b) $y^2 = y + 20$

Rewrite the equation in standard form.

$$y^2 = y + 20$$
$$y^2 - y - 20 = 0 \qquad \text{Subtract } y \text{ and 20.}$$
$$(y - 5)(y + 4) = 0 \qquad \text{Factor.}$$
$$y - 5 = 0 \quad \text{or} \quad y + 4 = 0 \qquad \text{Zero-factor property}$$
$$y = 5 \quad \text{or} \qquad y = -4 \qquad \text{Solve each equation.}$$

Check these solutions by substituting each one into the original equation. The solution set is $\{-4, 5\}$.

✔ **Now Try Exercises 31 and 35.**

In summary, follow these steps to solve quadratic equations by factoring.

Solving a Quadratic Equation by Factoring

Step 1   **Write the equation in standard form**—that is, with all terms on one side of the equals sign in descending powers of the variable and 0 on the other side.

Step 2   **Factor** completely.

Step 3   **Use the zero-factor property** to set each factor with a variable equal to 0, and solve the resulting equations.

Step 4   **Check** each solution in the original equation.

▶ **NOTE**   Not all quadratic equations can be solved by factoring. A more general method for solving such equations is given in **Chapter 9.**

EXAMPLE 3   **Solving a Quadratic Equation with a Common Factor**

Solve $4p^2 + 40 = 26p$.

$$4p^2 + 40 = 26p$$

$$4p^2 - 26p + 40 = 0 \qquad \text{Standard form}$$

$$2(2p^2 - 13p + 20) = 0 \qquad \text{Factor out 2.}$$

This 2 is *not* a solution of the equation.

$$2p^2 - 13p + 20 = 0 \qquad \text{Divide each side by 2.}$$

$$(2p - 5)(p - 4) = 0 \qquad \text{Factor.}$$

$$2p - 5 = 0 \quad \text{or} \quad p - 4 = 0 \qquad \text{Zero-factor property}$$

$$2p = 5 \qquad\qquad p = 4 \qquad \text{Solve each equation.}$$

$$p = \frac{5}{2}$$

Check that the solution set is $\left\{\frac{5}{2}, 4\right\}$ by substituting each solution into the original equation.

✔ **Now Try Exercise 41.**

▶ **CAUTION**   A common error is to include the common factor 2 as a solution in Example 3. *Only factors containing variables lead to solutions,* such as the factor y in the equation $y(3y - 4) = 0$ in Example 1(b).

EXAMPLE 4 Solving Quadratic Equations

Solve each equation.

(a) $16m^2 - 25 = 0$

Factor the left side of the equation as the difference of squares **(Section 6.4).**

$$(4m + 5)(4m - 5) = 0$$

$$4m + 5 = 0 \quad \text{or} \quad 4m - 5 = 0 \qquad \text{Zero-factor property}$$

$$4m = -5 \quad \text{or} \qquad 4m = 5 \qquad \text{Solve each equation.}$$

$$m = -\frac{5}{4} \quad \text{or} \qquad m = \frac{5}{4}$$

Check the two solutions, $-\frac{5}{4}$ and $\frac{5}{4}$, in the original equation. The solution set is $\left\{-\frac{5}{4}, \frac{5}{4}\right\}$.

(b)

$$y^2 = 2y$$

$$y^2 - 2y = 0 \qquad \text{Standard form}$$

> **Don't forget to set the variable factor *y* equal to 0.**

$$y(y - 2) = 0 \qquad \text{Factor.}$$

$$y = 0 \quad \text{or} \quad y - 2 = 0 \qquad \text{Zero-factor property}$$

$$y = 2 \qquad \text{Solve.}$$

The solution set is $\{0, 2\}$.

(c) $k(2k + 1) = 3$

Write the equation in standard form.

> **To be in standard form, 0 must be on the right side.**

$$k(2k + 1) = 3$$

$$2k^2 + k = 3 \qquad \text{Distributive property}$$

$$2k^2 + k - 3 = 0 \qquad \text{Subtract 3.}$$

$$(2k + 3)(k - 1) = 0 \qquad \text{Factor.}$$

$$2k + 3 = 0 \quad \text{or} \quad k - 1 = 0 \qquad \text{Zero-factor property}$$

$$2k = -3 \qquad\qquad k = 1$$

$$k = -\frac{3}{2}$$

The solution set is $\left\{-\frac{3}{2}, 1\right\}$.

✔ **Now Try Exercises 47, 51, and 55.**

▶ **CAUTION** In Example 4(b), it is tempting to begin by dividing both sides of the equation $y^2 = 2y$ by y to get $y = 2$. Note, however, that we do not get the other solution, 0, if we divide by a variable. (We may divide each side of an equation by a *nonzero* real number, however. For instance, in Example 3 we divided each side by 2.)

In Example 4(c), we could not use the zero-factor property to solve the equation $k(2k + 1) = 3$ in its given form because of the 3 on the right. *The zero-factor property applies only to a product that equals 0.*

OBJECTIVE 2 Solve other equations by factoring. We can also use the zero-factor property to solve equations that involve more than two factors with variables. (These equations are *not* quadratic equations. Why not?)

EXAMPLE 5 **Solving Equations with More than Two Variable Factors**

Solve each equation.

(a)
$$6z^3 - 6z = 0$$
$$6z(z^2 - 1) = 0 \qquad \text{Factor out } 6z.$$
$$6z(z + 1)(z - 1) = 0 \qquad \text{Factor } z^2 - 1.$$

By an extension of the zero-factor property, this product can equal 0 only if at least one of the factors is 0. Write and solve three equations, one for each factor with a variable.

$$6z = 0 \quad \text{or} \quad z + 1 = 0 \quad \text{or} \quad z - 1 = 0$$
$$z = 0 \quad \text{or} \quad z = -1 \quad \text{or} \quad z = 1$$

Check by substituting, in turn, 0, -1, and 1 into the original equation. The solution set is $\{-1, 0, 1\}$.

(b)
$$(3x - 1)(x^2 - 9x + 20) = 0$$
$$(3x - 1)(x - 5)(x - 4) = 0 \qquad \text{Factor } x^2 - 9x + 20.$$
$$3x - 1 = 0 \quad \text{or} \quad x - 5 = 0 \quad \text{or} \quad x - 4 = 0 \qquad \text{Zero-factor property}$$
$$x = \frac{1}{3} \quad \text{or} \qquad x = 5 \quad \text{or} \qquad x = 4$$

The solutions of the original equation are $\frac{1}{3}$, 4, and 5. *Check* each solution to verify that the solution set is $\left\{\frac{1}{3}, 4, 5\right\}$.

✔ **Now Try Exercises 61 and 65.**

▶ **CAUTION** In Example 5(b), it would be unproductive to begin by multiplying the two factors together. Keep in mind that the zero-factor property requires the *product* of two or more factors to equal 0. Always consider first whether an equation is given in an appropriate form for the zero-factor property to apply.

EXAMPLE 6 **Solving an Equation Requiring Multiplication before Factoring**

Solve $(3x + 1)x = (x + 1)^2 + 5$.

The zero-factor property requires the *product* of two or more factors to equal 0. To write this equation in the required form, we must first multiply on both sides and collect terms on one side.

$$(3x + 1)x = (x + 1)^2 + 5$$
$$3x^2 + x = x^2 + 2x + 1 + 5 \qquad \text{Multiply.}$$
$$3x^2 + x = x^2 + 2x + 6 \qquad \text{Combine like terms.}$$
$$2x^2 - x - 6 = 0 \qquad \text{Standard form}$$
$$(2x + 3)(x - 2) = 0 \qquad \text{Factor.}$$
$$2x + 3 = 0 \quad \text{or} \quad x - 2 = 0 \qquad \text{Zero-factor property}$$
$$x = -\frac{3}{2} \quad \text{or} \qquad x = 2$$

Check that the solution set is $\left\{-\frac{3}{2}, 2\right\}$.

✔ **Now Try Exercise 75.**

6.5 EXERCISES

Concept Check In Exercises 1–5, fill in the blank with the correct response.

1. A quadratic equation in x is an equation that can be put into the form _____ $= 0$.

2. The form $ax^2 + bx + c = 0$ is called _____ form.

3. If a quadratic equation is in standard form, to solve the equation we should begin by attempting to _____ the polynomial.

4. The equation $x^3 + x^2 + x = 0$ is not a quadratic equation, because _____.

5. If a quadratic equation $ax^2 + bx + c = 0$ has $c = 0$, then _____ *must* be a solution because _____ is a factor of the polynomial.

6. *Concept Check* Identify each equation as *linear* or *quadratic*.
   (a) $2x - 5 = 6$
   (b) $x^2 - 5 = -4$
   (c) $x^2 + 2x - 3 = 2x^2 - 2$
   (d) $5^2x + 2 = 0$

7. As shown in Example 5, the zero-factor property can be extended to more than two factors. For example, to solve $(x + 3)(2x - 7)(x - 4) = 0$, we would set each factor equal to 0 and solve three equations. Find the solution set of that equation.

8. *Concept Check* The number 9 is a *double solution* of the equation $(x - 9)^2 = 0$. Why is this so?

▨ 9. Students often become confused as to how to handle a constant, such as 2 in the equation $2x(3x - 4) = 0$. How would you explain to someone how to solve this equation and how to handle the constant 2?

▨ 10. Define *quadratic equation* in your own words, without using an algebraic expression. Then give examples.

Solve each equation and check your solutions. See Example 1.

11. $(x + 5)(x - 2) = 0$ 12. $(x - 1)(x + 8) = 0$ ⊙ 13. $(2m - 7)(m - 3) = 0$

14. $(6k + 5)(k + 4) = 0$ 15. $t(6t + 5) = 0$ 16. $w(4w + 1) = 0$

17. $2x(3x - 4) = 0$ 18. $6y(4y + 9) = 0$

19. $\left(x + \dfrac{1}{2}\right)\left(2x - \dfrac{1}{3}\right) = 0$ 20. $\left(a + \dfrac{2}{3}\right)\left(5a - \dfrac{1}{2}\right) = 0$

21. $(0.5z - 1)(2.5z + 2) = 0$ 22. $(0.25x + 1)(x - 0.5) = 0$

23. $(x - 9)(x - 9) = 0$ 24. $(2y + 1)(2y + 1) = 0$

25. *Concept Check* Look at this "solution."
    WHAT WENT WRONG?

    $$3x(5x - 4) = 0$$
    $$x = 3 \quad \text{or} \quad x = 0 \quad \text{or} \quad 5x - 4 = 0$$
    $$x = \frac{4}{5}$$

    The solution set is $\left\{3, 0, \frac{4}{5}\right\}$.

26. *Concept Check* Look at this "solution."
    WHAT WENT WRONG?

    $$x(7x - 1) = 0$$
    $$7x - 1 = 0 \qquad \text{Zero-factor property}$$
    $$x = \frac{1}{7}$$

    The solution set is $\left\{\frac{1}{7}\right\}$.

Solve each equation and check your solutions. See Examples 2–4.

27. $y^2 + 3y + 2 = 0$ 28. $p^2 + 8p + 7 = 0$ 29. $y^2 - 3y + 2 = 0$

30. $r^2 - 4r + 3 = 0$ ⊙ 31. $x^2 = 24 - 5x$ 32. $t^2 = 2t + 15$

33. $x^2 = 3 + 2x$ **34.** $m^2 = 4 + 3m$ **35.** $z^2 + 3z = -2$

36. $p^2 - 2p = 3$ **37.** $m^2 + 8m + 16 = 0$ **38.** $b^2 - 6b + 9 = 0$

39. $3x^2 + 5x - 2 = 0$ **40.** $6r^2 - r - 2 = 0$ ☻ **41.** $12p^2 = 8 - 10p$

42. $18x^2 = 12 + 15x$ **43.** $9s^2 + 12s = -4$ **44.** $36x^2 + 60x = -25$

45. $y^2 - 9 = 0$ **46.** $m^2 - 100 = 0$ **47.** $16k^2 - 49 = 0$

48. $4w^2 - 9 = 0$ **49.** $n^2 = 121$ **50.** $x^2 = 400$

Solve each equation and check your solutions. See Examples 4–6.

51. $x^2 = 7x$ **52.** $t^2 = 9t$ **53.** $6r^2 = 3r$

54. $10y^2 = -5y$ ☻ **55.** $x(x - 7) = -10$ **56.** $r(r - 5) = -6$

57. $3z(2z + 7) = 12$ **58.** $4b(2b + 3) = 36$

59. $2y(y + 13) = 136$ **60.** $t(3t - 20) = -12$

☻ **61.** $(2r + 5)(3r^2 - 16r + 5) = 0$ **62.** $(3m + 4)(6m^2 + m - 2) = 0$

63. $(2x + 7)(x^2 + 2x - 3) = 0$ **64.** $(x + 1)(6x^2 + x - 12) = 0$

65. $9y^3 - 49y = 0$ **66.** $16r^3 - 9r = 0$

67. $r^3 - 2r^2 - 8r = 0$ **68.** $x^3 - x^2 - 6x = 0$

69. $a^3 + a^2 - 20a = 0$ **70.** $y^3 - 6y^2 + 8y = 0$

71. $r^4 = 2r^3 + 15r^2$ **72.** $x^3 = 3x + 2x^2$

73. $3x(x + 1) = (2x + 3)(x + 1)$ **74.** $2k(k + 3) = (3k + 1)(k + 3)$

☻ **75.** $x^2 + (x + 1)^2 = (x + 2)^2$ **76.** $(x - 7)^2 + x^2 = (x + 1)^2$

77. $(2x)^2 = (2x + 4)^2 - (x + 5)^2$ **78.** $5 - (x - 1)^2 = (x - 2)^2$

79. $6p^2(p + 1) = 4(p + 1) - 5p(p + 1)$

80. $6x^2(2x + 3) - 5x(2x + 3) = 4(2x + 3)$

81. $(k + 3)^2 - (2k - 1)^2 = 0$

82. $(4y - 3)^3 - 9(4y - 3) = 0$

83. Galileo's formula describing the motion of freely falling objects, $d = 16t^2$, was given at the beginning of this section. The distance d in feet an object falls depends on the time t elapsed, in seconds. (This is an example of an important mathematical concept, the *function*.)

(a) Use Galileo's formula and complete the following table. (*Hint:* Substitute each given value into the formula and solve for the unknown value.)

t in seconds	0	1	2	3		
d in feet	0	16			256	576

✐ (b) When $t = 0$, $d = 0$. Explain this in the context of the problem.

✐ (c) When you substituted 256 for d and solved for t, you should have found two solutions: 4 and -4. Why doesn't -4 make sense as an answer?

✐ **84.** Explain why the quadratic equation $x^2 + x - 1 = 0$ cannot be solved by the method described in this section.

TECHNOLOGY INSIGHTS (EXERCISES 85–88)

*In **Section 3.2**, we showed how an equation in one variable can be solved with a graphing calculator by getting 0 on one side and then replacing 0 with y to get a corresponding equation in two variables. The x-values of the x-intercepts of the graph of the two-variable equation then give the solutions of the original equation.*

Use the calculator screens to determine the solution set of each quadratic equation. Verify your answers by substitution.

85. $x^2 + 0.4x - 0.05 = 0$

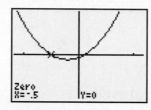 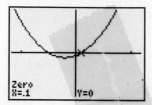

86. $2x^2 - 7.2x + 6.3 = 0$

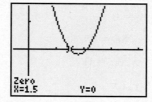 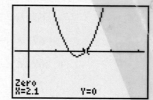

87. $2x^2 + 7.2x + 5.5 = 0$

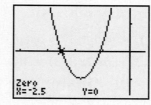 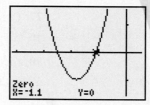

88. $4x^2 - x - 33 = 0$

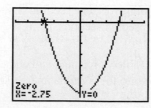 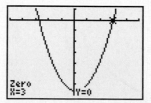

PREVIEW EXERCISES

*Solve each problem. See **Sections 2.4 and 2.5**.*

89. Florida has 9 more counties than California. Together, the two states have 125 counties. How many counties does each state have?

90. If a number is doubled and 6 is subtracted from this result, the answer is 3684. The unknown number is the year that Texas was admitted to the Union. What year was Texas admitted?

91. The length of the rectangle is 3 m more than its width. The perimeter of the rectangle is 34 m. Find the width of the rectangle.

x

$x + 3$

92. A rectangle has a length 4 m less than twice its width. The perimeter of the rectangle is 4 m more than five times its width. Find the width of the rectangle.

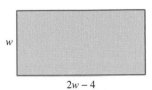

w

$2w - 4$

93. Twice the sum of two consecutive integers is 28 more than the second integer. Find the integers.

94. The sum of three consecutive integers is 10 less than four times the third integer. Find the integers.

95. The area of a triangle with base 12 in. is 48 in.^{2}. Find the height of the triangle.

6.6 Applications of Quadratic Equations

OBJECTIVES

1 Solve problems involving geometric figures.

2 Solve problems involving consecutive integers.

3 Solve problems by using the Pythagorean formula.

4 Solve problems by using given quadratic models.

We can now use factoring to solve quadratic equations that arise in application problems. We follow the same six problem-solving steps given in **Section 2.4.**

Solving an Applied Problem

Step 1 **Read** the problem carefully until you understand what is given and what is to be found.

Step 2 **Assign a variable** to represent the unknown value, using diagrams or tables as needed. Write down what the variable represents. If necessary, express any other unknown values in terms of the variable.

Step 3 **Write an equation,** using the variable expression(s).

Step 4 **Solve** the equation.

Step 5 **State the answer.** Does it seem reasonable?

Step 6 **Check** the answer in the words of the original problem.

OBJECTIVE 1 Solve problems involving geometric figures. Some of the applied problems in this section require one of the formulas given on the inside covers of the text.

EXAMPLE 1 Solving an Area Problem

Abe Biggs wants to plant a flower bed in a triangular area in a corner of his garden. One leg of the right-triangular flower bed will be 2 m shorter than the other leg, and he wants the bed to have an area of 24 m². See Figure 1. Find the lengths of the legs.

Step 1 **Read** the problem carefully. We need to find the lengths of the legs of a right triangle with area 24 m².

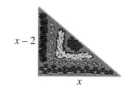
$x - 2$

x

FIGURE 1

Step 2 **Assign a variable.**

Let $x =$ the length of one leg.

Then $x - 2 =$ the length of the other leg.

Step 3 **Write an equation.** The area of a right triangle is given by the formula

$$\text{area} = \frac{1}{2} \times \text{base} \times \text{height} = \frac{1}{2}bh.$$

In a right triangle, the legs are the base and height, so we substitute 24 for the area, x for the base, and $x - 2$ for the height in the formula.

$$A = \frac{1}{2}bh$$

$$24 = \frac{1}{2}x(x - 2) \qquad \text{Let } A = 24, b = x, h = x - 2.$$

Step 4 **Solve.**

	$48 = x(x - 2)$	Multiply by 2.
	$48 = x^2 - 2x$	Distributive property
	$x^2 - 2x - 48 = 0$	Standard form
	$(x + 6)(x - 8) = 0$	Factor.
$x + 6 = 0$ or	$x - 8 = 0$	Zero-factor property
$x = -6$ or	$x = 8$	

Step 5 **State the answer.** The solutions are -6 and 8. Because a triangle cannot have a side of negative length, we discard the solution -6. Then the lengths of the legs will be 8 m and $8 - 2 = 6$ m.

Step 6 **Check.** The length of one leg is 2 m less than the length of the other leg, and the area is $\frac{1}{2}(8)(6) = 24$ m², as required.

✔ **Now Try Exercise 7.**

▶ CAUTION In solving applied problems, *always check solutions* against physical facts and discard any answers that are not appropriate.

OBJECTIVE 2 Solve problems involving consecutive integers. Recall from our work in **Section 2.4** that **consecutive integers** are integers that are next to each other on a number line, such as 5 and 6, or -11 and -10. **Consecutive odd integers** are *odd* integers that are next to each other, such as 5 and 7, or -13 and -11. **Consecutive even integers** are defined similarly; for example, 4 and 6 are consecutive even integers, as are -10 and -8. (In this book, we will list consecutive integers in increasing order when solving applications.)

> ▶ **PROBLEM-SOLVING HINT** In consecutive integer problems, if x represents the first integer, then, for
>
> | | |
> |---|---|
> | two consecutive integers, use | $x, \quad x + 1$; |
> | three consecutive integers, use | $x, \quad x + 1, \quad x + 2$; |
> | two consecutive even or odd integers, use | $x, \quad x + 2$; |
> | three consecutive even or odd integers, use | $x, \quad x + 2, \quad x + 4$. |

EXAMPLE 2 Solving a Consecutive Integer Problem

The product of the second and third of three consecutive integers is 2 more than 7 times the first integer. Find the integers.

Step 1 **Read** the problem. Note that the integers are consecutive.

Step 2 **Assign a variable.**

Let $x =$ the first integer.

Then $x + 1 =$ the second integer,

and $x + 2 =$ the third integer.

Step 3 **Write an equation.**

The product of the second and third is 2 more than 7 times the first.

$$(x + 1)(x + 2) = 7x + 2$$

Step 4 **Solve.**

$$x^2 + 3x + 2 = 7x + 2 \qquad \text{Multiply.}$$
$$x^2 - 4x = 0 \qquad \text{Standard form}$$
$$x(x - 4) = 0 \qquad \text{Factor.}$$
$$x = 0 \quad \text{or} \quad x = 4 \qquad \text{Zero-factor property}$$

Step 5 **State the answer.** The solutions 0 and 4 each lead to a correct answer:

$$0, 1, 2 \quad \text{or} \quad 4, 5, 6.$$

Step 6 **Check.** The product of the second and third integers must equal 2 more than 7 times the first. Since $1 \cdot 2 = 7 \cdot 0 + 2$ and $5 \cdot 6 = 7 \cdot 4 + 2$, both sets of consecutive integers satisfy the statement of the problem.

✔ **Now Try Exercise 19.**

OBJECTIVE 3 Solve problems by using the Pythagorean formula.

> **Pythagorean Formula**
>
> If a right triangle has longest side of length c and two other sides of lengths a and b, then
>
> $$a^2 + b^2 = c^2.$$
>
> Leg a Hypotenuse c 90°   Leg b
>
> The longest side, the **hypotenuse,** is opposite the right angle. The two shorter sides are the **legs** of the triangle.

EXAMPLE 3 Using the Pythagorean Formula

Ed and Mark leave their office, Ed traveling north and Mark traveling east. When Mark is 1 mi farther than Ed from the office, the distance between them is 2 mi more than Ed's distance from the office. Find their distances from the office and the distance between them.

Step 1 **Read** the problem again. There will be three answers to this problem.

Step 2 **Assign a variable.** Let x represent Ed's distance from the office, $x + 1$ represent Mark's distance from the office, and $x + 2$ represent the distance between them. Place these expressions on a right triangle, as in Figure 2.

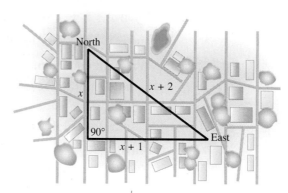

FIGURE 2

Step 3 **Write an equation.** Substitute into the Pythagorean formula.

$$a^2 + b^2 = c^2$$

$$x^2 + (x + 1)^2 = (x + 2)^2$$

> Be careful to substitute properly.

Step 4 **Solve.**

$$x^2 + x^2 + 2x + 1 = x^2 + 4x + 4$$

$$x^2 - 2x - 3 = 0 \qquad \text{Standard form}$$

$$(x - 3)(x + 1) = 0 \qquad \text{Factor.}$$

$$x - 3 = 0 \quad \text{or} \quad x + 1 = 0 \qquad \text{Zero-factor property}$$

$$x = 3 \quad \text{or} \qquad x = -1$$

Step 5   **State the answer.** Since -1 cannot represent a distance, 3 is the only possible answer. Ed's distance is 3 mi, Mark's distance is $3 + 1 = 4$ mi, and the distance between them is $3 + 2 = 5$ mi.

Step 6   **Check.** Since $3^2 + 4^2 = 5^2$, the answers are correct.

✔ **Now Try Exercise 25.**

> **CAUTION**   In solving a problem involving the Pythagorean formula, be sure that the expressions for the sides are properly placed.
>
> $$(\text{one leg})^2 + (\text{other leg})^2 = \text{hypotenuse}^2$$

OBJECTIVE ④ **Solve problems by using given quadratic models.**   In Examples 1–3, we wrote quadratic equations to model, or mathematically describe, various situations and then solved the equations. In the last two examples of this section, we are given the quadratic models and must use them to determine data.

EXAMPLE 4   **Finding the Height of a Ball**

A tennis player's serve travels 180 ft per sec (123 mph). If she hits the ball directly upward, the height h of the ball in feet at time t in seconds is modeled by the quadratic equation

$$h = -16t^2 + 180t + 6.$$

How long will it take for the ball to reach a height of 206 ft?

A height of 206 ft means that $h = 206$, so we substitute 206 for h in the equation.

$$h = -16t^2 + 180t + 6$$
$$206 = -16t^2 + 180t + 6 \qquad \text{Let } h = 206.$$

To solve the equation, we first write it in standard form. For convenience, we reverse the sides of the equation.

$$-16t^2 + 180t + 6 = 206$$
$$-16t^2 + 180t - 200 = 0 \qquad \text{Standard form}$$
$$4t^2 - 45t + 50 = 0 \qquad \text{Divide by } -4.$$
$$(4t - 5)(t - 10) = 0 \qquad \text{Factor.}$$
$$4t - 5 = 0 \quad \text{or} \quad t - 10 = 0 \qquad \text{Zero-factor property}$$
$$t = \frac{5}{4} \quad \text{or} \qquad t = 10$$

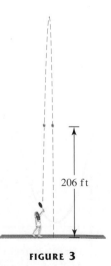

206 ft

FIGURE 3

Since we found two acceptable answers, the ball will be 206 ft above the ground twice (once on its way up and once on its way down): at $\frac{5}{4}$ sec and at 10 sec. See Figure 3.

✔ **Now Try Exercise 31.**

EXAMPLE 5   Modeling Increases in Spending on Hospital Services

The annual percent increase y in spending on hospital services in the years 1994–2001 can be modeled by the quadratic equation

$$y = 0.37x^2 - 4.1x + 12,$$

where $x = 4$ represents 1994, $x = 5$ represents 1995, and so on. (*Source:* Center for Studying Health System Change.)

(a) Use the model to find the annual percent increase, to the nearest tenth, in 1997.

Since $x = 4$ represents 1994, $x = 7$ represents 1997. Substitute 7 for x in the equation.

$$y = 0.37(7)^2 - 4.1(7) + 12 \qquad \text{Let } x = 7.$$
$$y = 1.4 \qquad\qquad\qquad\qquad \text{Round to the nearest tenth.}$$

Spending on hospital services increased about 1.4% in 1997.

(b) Repeat part (a) for 2001.

$$y = 0.37(11)^2 - 4.1(11) + 12 \qquad \text{For 2001, let } x = 11.$$
$$y = 11.7$$

In 2001, spending on hospital services increased about 11.7%.

(c) The model used in parts (a) and (b) was developed with the data in the table that follows. How do the results in parts (a) and (b) compare against the actual data from the table?

Year	Percent Increase
1994	1.8
1995	0.8
1996	0.5
1997	1.3
1998	3.4
1999	5.8
2000	7.1
2001	12.0

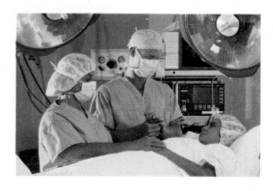

From the table, the actual data for 1997 is 1.3%. Our answer, 1.4%, is slightly high. For 2001, the actual data is 12.0%, so our answer of 11.7% in part (b) is a little low.

✔ **Now Try Exercise 33.**

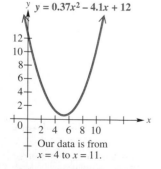

$y = 0.37x^2 - 4.1x + 12$

Our data is from $x = 4$ to $x = 11$.

FIGURE 4

▶ **NOTE**   A graph of the quadratic equation from Example 5 is shown in Figure 4. Notice that the basic shape of this graph follows the general pattern of the data in the table, which decreases from 1994 to 1996 and then increases from 1997 to 2001. We saw this type of graph, a *parabola,* in **Section 5.4.**

6.6 EXERCISES

● *Complete solution available on Video Lectures on CD/DVD*

Now Try Exercise

1. *Concept Check*   To review the six problem-solving steps first introduced in **Section 2.4,** complete each statement.

   Step 1: _____ the problem carefully until you understand what is given and what must be found.

   Step 2: Assign a _____ to represent the unknown value.

   Step 3: Write a(n) _____ using the variable expression(s).

   Step 4: _____ the equation.

   Step 5: State the _____.

   Step 6: _____ the answer in the words of the _____ problem.

2. A student solves an applied problem and gets 6 or −3 for the length of the side of a square. Which of these answers is reasonable? Explain.

In Exercises 3–6, a figure and a corresponding geometric formula are given. Using x as the variable, complete Steps 3–6 for each problem. (Refer to the steps in Exercise 1 as needed.)

3.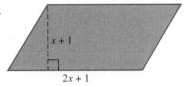

Area of a parallelogram: $A = bh$

The area of this parallelogram is 45 sq. units. Find its base and height.

4.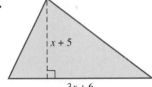

Area of a triangle: $A = \dfrac{1}{2}bh$

The area of this triangle is 60 sq. units. Find its base and height.

5.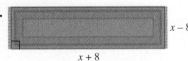

Area of a rectangular rug: $A = LW$

The area of this rug is 80 sq. units. Find its length and width.

6.

Volume of a rectangular Chinese box: $V = LWH$

The volume of this box is 192 cu. units. Find its length and width.

Solve each problem. Check your answers to be sure that they are reasonable. Refer to the formulas on the inside covers. See Example 1.

● 7. The length of a standard jewel case is 2 cm more than its width. The area of the rectangular top of the case is 168 cm². Find the length and width of the jewel case.

8. A hardcover book is 3 in. longer than it is wide. The area of the cover is 54 in.². Find the length and width of the book.

9. A 10-gal aquarium is 3 in. higher than it is wide. Its length is 21 in., and its volume is 2730 in.³. What are the height and width of the aquarium?

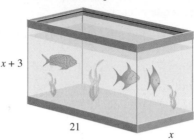

10. A toolbox is 2 ft high, and its width is 3 ft less than its length. If its volume is 80 ft³, find the length and width of the box.

11. The area of a triangle is 30 in.². The base of the triangle measures 2 in. more than twice the height of the triangle. Find the measures of the base and the height.

12. A certain triangle has its base equal in measure to its height. The area of the triangle is 72 m². Find the equal base and height measure.

13. The dimensions of an HPf1905 flat-panel monitor are such that its length is 3 in. more than its width. If the length were doubled and if the width were decreased by 1 in., the area would be increased by 150 in.². What are the length and width of the flat panel?

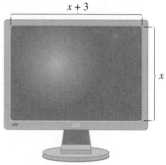

14. The keyboard that accompanies the monitor in Exercise 13 is 11 in. longer than it is wide. If the length were doubled and if 2 in. were added to the width, the area would be increased by 198 in.². What are the length and width of the keyboard? (*Source:* Author's computer.)

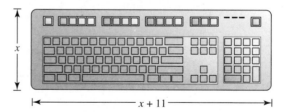

15. A square mirror has sides measuring 2 ft less than the sides of a square painting. If the difference between their areas is 32 ft², find the lengths of the sides of the mirror and the painting.

16. The sides of one square have length 3 m more than the sides of a second square. If the area of the larger square is subtracted from 4 times the area of the smaller square, the result is 36 m². What are the lengths of the sides of each square?

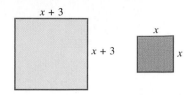

Solve each problem. See Example 2.

17. The product of the numbers on two consecutive volumes of research data is 420. Find the volume numbers. See the figure.

18. The product of the page numbers on two facing pages of a book is 600. Find the page numbers.

⊙ **19.** The product of the second and third of three consecutive integers is 2 more than 10 times the first integer. Find the integers.

20. The product of the first and third of three consecutive integers is 3 more than 3 times the second integer. Find the integers.

21. Find three consecutive odd integers such that 3 times the sum of all three is 18 more than the product of the first and second integers.

22. Find three consecutive odd integers such that the sum of all three is 42 less than the product of the second and third integers.

23. Find three consecutive even integers such that the sum of the squares of the first and second integers is equal to the square of the third integer.

24. Find three consecutive even integers such that the square of the sum of the first and second integers is equal to twice the third integer.

Solve each problem. See Example 3.

⊙ **25.** The hypotenuse of a right triangle is 1 cm longer than the longer leg. The shorter leg is 7 cm shorter than the longer leg. Find the length of the longer leg of the triangle.

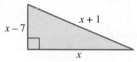

26. The longer leg of a right triangle is 1 m longer than the shorter leg. The hypotenuse is 1 m shorter than twice the shorter leg. Find the length of the shorter leg of the triangle.

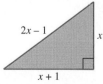

27. Tram works due north of home. Her husband Alan works due east. They leave for work at the same time. By the time Tram is 5 mi from home, the distance between them is 1 mi more than Alan's distance from home. How far from home is Alan?

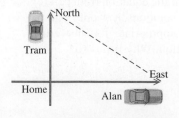

28. Two cars left an intersection at the same time. One traveled north. The other traveled 14 mi farther, but to the east. How far apart were they at that time if the distance between them was 4 mi more than the distance traveled east?

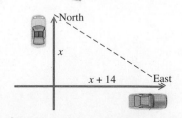

29. A ladder is leaning against a building. The distance from the bottom of the ladder to the building is 4 ft less than the length of the ladder. How high up the side of the building is the top of the ladder if that distance is 2 ft less than the length of the ladder?

30. A lot has the shape of a right triangle with one leg 2 m longer than the other. The hypotenuse is 2 m less than twice the length of the shorter leg. Find the length of the shorter leg.

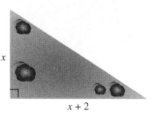

Solve each problem. See Examples 4 and 5.

31. An object projected from a height of 48 ft with an initial velocity of 32 ft per sec after t seconds has height

$$h = -16t^2 + 32t + 48.$$

(a) After how many seconds is the height 64 ft? (*Hint:* Let $h = 64$ and solve.)

(b) After how many seconds is the height 60 ft?

(c) After how many seconds does the object hit the ground? (*Hint:* When the object hits the ground, $h = 0$.)

(d) The quadratic equation from part (c) has two solutions, yet only one of them is appropriate for answering the question. Why is this so?

32. If an object is projected upward from ground level with an initial velocity of 64 ft per sec, its height h in feet t seconds later is

$$h = -16t^2 + 64t.$$

(a) After how many seconds is the height 48 ft?

(b) The object reaches its maximum height 2 sec after it is propelled. What is this maximum height?

(c) After how many seconds does the object hit the ground?

(d) Find the number of seconds after which the height is 60 ft.

(e) What is the physical interpretation of why part (d) has two answers?

(f) The quadratic equation from part (c) has two solutions, yet only one of them is appropriate for answering the question. Why is this so?

33. The table shows the number of cellular phone subscribers (in millions) in the United States.

Year	Cellular Phones (in millions)
1990	5
1992	11
1994	24
1996	44
1998	69
2000	109
2002	141
2004	182

Source: CTIA.

We used the preceding data to develop the quadratic equation

$$y = 0.740x^2 + 2.56x + 3.46,$$

which models the number y of cellular phone subscribers (in millions) in the year x, where $x = 0$ represents 1990, $x = 2$ represents 1992, and so on.

(a) Use the model to find the number of subscribers in 1996, to the nearest tenth. How does the result compare against the actual data in the table?

(b) What value of x corresponds to 2004?

(c) Use the model to find the number of cellular phone subscribers in 2004, to the nearest tenth. How does the result compare against the actual data in the table?

(d) Assuming that the trend in the data continues, use the quadratic equation to predict the number of cellular phone subscribers in 2006, to the nearest tenth.

RELATING CONCEPTS (EXERCISES 34–42)

FOR INDIVIDUAL OR GROUP WORK

The U.S. trade deficit represents the amount by which exports are less than imports. The deficit provides not only a sign of economic prosperity, but also a warning of potential decline. The data in the table shows the U.S. trade deficit for 1995 through 2000.

Year	Deficit (in billions of dollars)
1995	97.5
1996	104.3
1997	104.7
1998	164.3
1999	271.3
2000	378.7

Source: U.S. Department of Commerce.

*Use the data to **work Exercises 34–42 in order.***

34. How much did the trade deficit increase from 1999 to 2000? What percent increase is this (to the nearest percent)?

(continued)

35. The U.S. trade deficit might be approximated by the linear equation

$$y = 40.8x + 66.9,$$

where y is the deficit in billions of dollars. Here, $x = 0$ represents 1995, $x = 1$ represents 1996, and so on. Use this equation to approximate the trade deficits in 1997, 1999, and 2000.

36. How do your answers from Exercise 35 compare against the actual data in the table?

37. The trade deficit y (in billions of dollars) might also be approximated by the quadratic equation

$$y = 18.5x^2 - 33.4x + 104,$$

where $x = 0$ again represents 1995, $x = 1$ represents 1996, and so on. Use this equation to approximate the trade deficits in 1997, 1999, and 2000.

38. Compare your answers from Exercise 37 with the actual data in the table. Which equation, the linear one in Exercise 35 or the quadratic one in Exercise 37, models the data better?

39. We can also see graphically why the linear equation is not a very good model for the data. To do so, write the data from the table as a set of ordered pairs (x, y), where x represents the years since 1995 and y represents the trade deficit in billions of dollars.

40. Plot the ordered pairs from Exercise 39 on a graph. Recall from **Chapter 3** that a linear equation has a straight line for its graph. Do the ordered pairs you plotted lie in a linear pattern?

41. Assuming that the trend in the data continues, and since the quadratic equation modeled the data fairly well, use the quadratic equation to predict the trade deficit for the year 2002.

42. The actual trade deficit for 2002 was $417.9 billion.

(a) How does the actual deficit for 2002 compare with your prediction from Exercise 41?

(b) Should the quadratic equation be used to predict the U.S. trade deficit for years after 2000? Explain.

PREVIEW EXERCISES

*Write each fraction in lowest terms. See **Section 1.1**.*

43. $\dfrac{50}{72}$

44. $\dfrac{26}{156}$

45. $\dfrac{26}{13}$

46. $\dfrac{-18}{18}$

47. $\dfrac{48}{-27}$

48. $\dfrac{-35}{-21}$

Chapter **6** | **Group Activity**

FACTORING TRINOMIALS BY USING KEY NUMBERS AND FACTOR PAIRS

Objective   Use an organized approach to factoring by grouping.

To factor a trinomial by using the FOIL method, we must find the outer and inner coefficients that sum to give the coefficient of the middle term. Our approach begins with a **key number,** found by multiplying the coefficients of the first and last terms. For the trinomial $6x^2 - x - 2$, for instance, the key number is -12, since $6(-2) = -12$.

Step 1   Display the factors of -12 by entering $Y_1 = -12/X$ in a graphing calculator (Screen 1) and using an automatic table (Screen 2). Factors of -12 are automatically displayed in **factor pairs** as $1, -12$; $2, -6$; $3, -4$; $4, -3$; and $6, -2$ (Screen 3). You could scroll up or down to find other factors. Note that $5, -2.4$ and $7, -1.714$ are not factor pairs, since -2.4 and -1.714 are not integers.

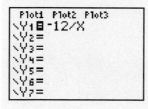

SCREEN 1

SCREEN 2

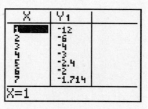

SCREEN 3

Step 2   Find the pair of factors that sum to the *middle* term coefficient, -1. We can let the calculator do this, too. Enter $Y_2 = X + -12/X$. In this case, X is one of the factors and $-12/X$ is the other, so Y_2 will give the sum (Screen 4). Look for -1 in the Y_2 column in Screen 5. (You may have to scroll up or down to find it.)

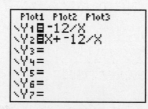

SCREEN 4

SCREEN 5

Step 3   Screen 5 shows that the coefficients of the outer and inner products are 3 and -4. Write $6x^2 - x - 2$ as $6x^2 + 3x - 4x - 2$. Using factoring by grouping,

$$(6x^2 + 3x) + (-4x - 2) = 3x(2x + 1) - 2(2x + 1)$$
$$= (2x + 1)(3x - 2).$$

(continued)

A. To factor each trinomial given in the column heads of the following table, first find the key number, and then use a calculator to help you find the coefficients of the outer and inner products of FOIL.

Trinomial	$3x^2 - 2x - 8$	$2x^2 - 11x + 15$	$10x^2 + 11x - 6$	$4x^2 + 5x + 3$
Key Number				
Outer, Inner Coefficients				
Factor by Grouping				(*Hint:* What does it mean if the middle term coefficient is *not* listed in the Y_2 column?)

B. Factor each trinomial by grouping.

Chapter **6** **SUMMARY** *View the Interactive Summary on the Pass the Test CD.*

KEY TERMS

6.1 factor
factored form
common factor
greatest common
factor (GCF)

6.2 prime polynomial
6.4 perfect square
trinomial

6.5 quadratic equation
standard form

6.6 hypotenuse
legs

TEST YOUR WORD POWER

See how well you have learned the vocabulary in this chapter. Answers, with examples, follow the Quick Review.

1. Factoring is
   A. a method of multiplying polynomials
   B. the process of writing a polynomial as a product
   C. the answer in a multiplication problem
   D. a way to add the terms of a polynomial.

2. A polynomial is in **factored form** when
   A. it is prime
   B. it is written as a sum

   C. the squared term has a coefficient of 1
   D. it is written as a product.

3. A **perfect square trinomial** is a trinomial
   A. that can be factored as the square of a binomial
   B. that cannot be factored
   C. that is multiplied by a binomial
   D. all of whose terms are perfect squares.

4. A **quadratic equation** is a polynomial equation of
   A. degree one
   B. degree two
   C. degree three
   D. degree four.

5. A **hypotenuse** is
   A. either of the two shorter sides of a triangle
   B. the shortest side of a triangle
   C. the side opposite the right angle in a triangle
   D. the longest side in any triangle.

QUICK REVIEW

Concepts	Examples

6.1 THE GREATEST COMMON FACTOR; FACTORING BY GROUPING

Finding the Greatest Common Factor (GCF)
1. Include the largest numerical factor of every term.
2. Include each variable that is a factor of every term raised to the least exponent that appears in a term.

Find the greatest common factor of
$$4x^2y, \qquad -6x^2y^3, \qquad 2xy^2.$$
$$4x^2y = 2^2 \cdot x^2 \cdot y$$
$$-6x^2y^3 = -1 \cdot 2 \cdot 3 \cdot x^2 \cdot y^3$$
$$2xy^2 = 2 \cdot x \cdot y^2$$

The greatest common factor is $2xy$.

Factoring by Grouping

Step 1 Group the terms.

Step 2 Factor out the greatest common factor in each group.

Step 3 Factor out a common binomial factor from the result of Step 2.

Step 4 If necessary, try a different grouping.

Factor by grouping.
$$3x^2 + 5x - 24xy - 40y = (3x^2 + 5x) + (-24xy - 40y)$$
$$= x(3x + 5) - 8y(3x + 5)$$
$$= (3x + 5)(x - 8y)$$

(continued)

Concepts	Examples

6.2 FACTORING TRINOMIALS

To factor $x^2 + bx + c$, find m and n such that $mn = c$ and $m + n = b$.

$$mn = c$$
$$\downarrow$$
$$x^2 + bx + c$$
$$\uparrow$$
$$m + n = b$$

Then $x^2 + bx + c = (x + m)(x + n)$.

Check by multiplying.

Factor $x^2 + 6x + 8$.

$$mn = 8$$
$$\downarrow$$
$$x^2 + 6x + 8$$
$$\uparrow$$
$$m + n = 6$$

$m = 2$ and $n = 4$

$$x^2 + 6x + 8 = (x + 2)(x + 4)$$
Check: $(x + 2)(x + 4) = x^2 + 4x + 2x + 8$
$$= x^2 + 6x + 8$$

6.3 MORE ON FACTORING TRINOMIALS

To factor $ax^2 + bx + c$,

By Grouping
Find m and n.

$$mn = ac$$
$$\downarrow \qquad \downarrow$$
$$ax^2 + bx + c$$
$$\uparrow$$
$$m + n = b$$

By Trial and Error
Use FOIL in reverse.

Factor $3x^2 + 14x - 5$.

$$-15$$

$mn = -15, m + n = 14$

$m = -1$ and $n = 15$ ← Choice of m and n

By trial and error or by grouping,
$$3x^2 + 14x - 5 = (3x - 1)(x + 5).$$

6.4 SPECIAL FACTORING TECHNIQUES

Difference of Squares
$$x^2 - y^2 = (x + y)(x - y)$$

Perfect Square Trinomials
$$x^2 + 2xy + y^2 = (x + y)^2$$
$$x^2 - 2xy + y^2 = (x - y)^2$$

Difference of Cubes
$$x^3 - y^3 = (x - y)(x^2 + xy + y^2)$$

Sum of Cubes
$$x^3 + y^3 = (x + y)(x^2 - xy + y^2)$$

Factor.

$$4x^2 - 9 = (2x + 3)(2x - 3)$$

$$9x^2 + 6x + 1 = (3x + 1)^2$$
$$4x^2 - 20x + 25 = (2x - 5)^2$$

$$m^3 - 8 = m^3 - 2^3 = (m - 2)(m^2 + 2m + 4)$$

$$z^3 + 27 = z^3 + 3^3 = (z + 3)(z^2 - 3z + 9)$$

6.5 SOLVING QUADRATIC EQUATIONS BY FACTORING

Zero-Factor Property

If a and b are real numbers and if $ab = 0$, then $a = 0$ or $b = 0$.

If $(x - 2)(x + 3) = 0$, then $x - 2 = 0$ or $x + 3 = 0$.

(continued)

Concepts	Examples

Solving a Quadratic Equation by Factoring

Solve $2x^2 = 7x + 15$.

Step 1 Write in standard form.

$$2x^2 - 7x - 15 = 0$$

Step 2 Factor.

$$(2x + 3)(x - 5) = 0$$

Step 3 Use the zero-factor property.

$$2x + 3 = 0 \qquad \text{or} \qquad x - 5 = 0$$
$$2x = -3 \qquad\qquad x = 5$$
$$x = -\frac{3}{2}$$

Step 4 Check.

Both solutions satisfy the original equation. The solution set is $\left\{-\frac{3}{2}, 5\right\}$.

6.6 APPLICATIONS OF QUADRATIC EQUATIONS

Pythagorean Formula

In a right triangle, the square of the hypotenuse equals the sum of the squares of the legs.

$$a^2 + b^2 = c^2$$

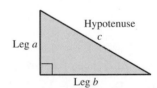

Hypotenuse
c

Leg *a*

Leg *b*

In a right triangle, one leg measures 2 ft longer than the other. The hypotenuse measures 4 ft longer than the shorter leg. Find the lengths of the three sides of the triangle.

Let $x = $ the length of the shorter leg. Then

$$x^2 + (x + 2)^2 = (x + 4)^2.$$

Verify that the solutions of this equation are -2 and 6. Discard -2 as a solution. Check that the sides are 6, $6 + 2 = 8$, and $6 + 4 = 10$ ft in length.

Answers to Test Your Word Power

1. B; *Example:*
$x^2 - 5x - 14 = (x - 7)(x + 2)$

2. D; *Example:* The factored form of
$x^2 - 5x - 14$ is $(x - 7)(x + 2)$.

3. A; *Example:* $a^2 + 2a + 1$ is a perfect square trinomial; its factored form is $(a + 1)^2$.

4. B; *Examples:*
$y^2 - 3y + 2 = 0, x^2 - 9 = 0,$
$2m^2 = 6m + 8$

5. C; *Example:* In Figure 2 of **Section 6.6,** the hypotenuse is the side labeled $x + 2$.

Chapter 6 REVIEW EXERCISES

[6.1] *Factor out the greatest common factor, or factor by grouping.*

1. $7t + 14$

2. $60z^3 + 30z$

3. $2xy - 8y + 3x - 12$

4. $6y^2 + 9y + 4xy + 6x$

[6.2] *Factor completely.*

5. $x^2 + 5x + 6$

6. $y^2 - 13y + 40$

7. $q^2 + 6q - 27$

8. $r^2 - r - 56$

9. $r^2 - 4rs - 96s^2$

10. $p^2 + 2pq - 120q^2$

11. $8p^3 - 24p^2 - 80p$ **12.** $3x^4 + 30x^3 + 48x^2$ **13.** $p^7 - p^6q - 2p^5q^2$

14. $3r^5 - 6r^4s - 45r^3s^2$ **15.** $9x^4y - 9x^3y - 54x^2y$ **16.** $2x^7 + 2x^6y - 12x^5y^2$

[6.3]

17. *Concept Check* To begin factoring $6r^2 - 5r - 6$, what are the possible first terms of the two binomial factors if we consider only positive integer coefficients?

18. *Concept Check* What is the first step you would use to factor $2z^3 + 9z^2 - 5z$?

Factor completely.

19. $2k^2 - 5k + 2$ **20.** $3r^2 + 11r - 4$ **21.** $6r^2 - 5r - 6$

22. $10z^2 - 3z - 1$ **23.** $8v^2 + 17v - 21$ **24.** $24x^5 - 20x^4 + 4x^3$

25. $-6x^2 + 3x + 30$ **26.** $10r^3s + 17r^2s^2 + 6rs^3$ **27.** $48x^4y + 4x^3y^2 - 4x^2y^3$

28. *Concept Check* On a quiz, a student factored $16x^2 - 24x + 5$ by grouping as follows:

$$16x^2 - 24x + 5 = 16x^2 - 4x - 20x + 5$$

$$= 4x(4x - 1) - 5(4x - 1). \quad \text{His answer}$$

He thought his answer was correct, since it checked by multiplication. **WHAT WENT WRONG?** Give the correct factored form.

[6.4]

29. *Concept Check* Which one of the following is the difference of squares?

A. $32x^2 - 1$ **B.** $4x^2y^2 - 25z^2$ **C.** $x^2 + 36$ **D.** $25y^3 - 1$

30. *Concept Check* Which one of the following is a perfect square trinomial?

A. $x^2 + x + 1$ **B.** $y^2 - 4y + 9$ **C.** $4x^2 + 10x + 25$ **D.** $x^2 - 20x + 100$

Factor completely.

31. $n^2 - 49$ **32.** $25b^2 - 121$ **33.** $49y^2 - 25w^2$

34. $144p^2 - 36q^2$ **35.** $x^2 + 100$ **36.** $r^2 - 12r + 36$

37. $9t^2 - 42t + 49$ **38.** $m^3 + 1000$ **39.** $125k^3 + 64x^3$

40. $343x^3 - 64$ **41.** $1000 - 27x^6$ **42.** $x^6 - y^6$

[6.5] *Solve each equation and check your solutions.*

43. $(4t + 3)(t - 1) = 0$ **44.** $(x + 7)(x - 4)(x + 3) = 0$

45. $x(2x - 5) = 0$ **46.** $z^2 + 4z + 3 = 0$

47. $m^2 - 5m + 4 = 0$ **48.** $x^2 = -15 + 8x$

49. $3z^2 - 11z - 20 = 0$ **50.** $81t^2 - 64 = 0$

51. $y^2 = 8y$ **52.** $n(n - 5) = 6$

53. $t^2 - 14t + 49 = 0$ **54.** $t^2 = 12(t - 3)$

55. $(5z + 2)(z^2 + 3z + 2) = 0$ **56.** $x^2 = 9$

[6.6] *Solve each problem.*

57. The length of a rug is 6 ft more than the width. The area is 40 ft². Find the length and width of the rug.

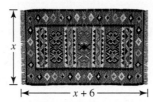

58. The surface area S of a box is given by

$$S = 2WH + 2WL + 2LH.$$

A treasure chest from a sunken galleon has the dimensions shown in the figure. Its surface area is 650 ft². Find its width.

59. The length of a rectangle is three times the width. If the width were increased by 3 m while the length remained the same, the new rectangle would have an area of 30 m². Find the length and width of the original rectangle.

60. The volume of a rectangular box is 120 m³. The width of the box is 4 m, and the height is 1 m less than the length. Find the length and height of the box.

61. The product of two consecutive integers is 29 more than their sum. What are the integers?

62. Two cars left an intersection at the same time. One traveled west, and the other traveled 14 mi less, but to the south. How far apart were they at that time, if the distance between them was 16 mi more than the distance traveled south?

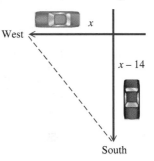

63. Annual revenue in billions of dollars for eBay is shown in the table.

Year	Annual Revenue (in billions of dollars)
2002	1.21
2003	2.17
2004	3.27

Source: eBay.

Using the data, we developed the quadratic equation

$$y = 0.07x^2 + 0.89x + 1.21$$

to model eBay revenues y in year x, where $x = 0$ represents 2002, $x = 1$ represents 2003, and so on. Because only three years of data were used to determine the model, we must be careful about using it to predict revenue for years beyond 2004.

(a) Use the model to predict annual revenue for eBay in 2005.

(b) The revenue for eBay for the first half of 2005 was $2.12 billion. Given this information, do you think your prediction in part (a) is reliable? Explain.

If an object is projected upward with an initial velocity of 128 *ft per sec, its height h after t seconds is* $h = 128t - 16t^2$. *Find the height of the object after each time listed.*

64. 1 sec   **65.** 2 sec   **66.** 4 sec

67. When does the object just described return to the ground?

MIXED REVIEW EXERCISES

68. *Concept Check*   Which of the following is *not* factored completely?

   A. $3(7t)$ **B.** $3x(7t + 4)$ **C.** $(3 + x)(7t + 4)$ **D.** $3(7t + 4) + x(7t + 4)$

69. Although $(2x + 8)(3x - 4) = 6x^2 + 16x - 32$ is a true statement, the polynomial is not factored completely. Explain why and give the completely factored form.

Factor completely.

70. $z^2 - 11zx + 10x^2$

71. $3k^2 + 11k + 10$

72. $15m^2 + 20m - 12mp - 16p$

73. $y^4 - 625$

74. $6m^3 - 21m^2 - 45m$

75. $24ab^3c^2 - 56a^2bc^3 + 72a^2b^2c$

76. $25a^2 + 15ab + 9b^2$

77. $12x^2yz^3 + 12xy^2z - 30x^3y^2z^4$

78. $2a^5 - 8a^4 - 24a^3$

79. $12r^2 + 18rq - 10r - 15q$

80. $1000a^3 + 27$

81. $49t^2 + 56t + 16$

Solve.

82. $t(t - 7) = 0$ **83.** $x^2 + 3x = 10$ **84.** $25x^2 + 20x + 4 = 0$

85. The numbers of alternative-fueled vehicles in use in the United States, in thousands, for the years 2001–2004 are given in the table.

ALTERNATIVE-FUELED VEHICLES

Year	Number (in thousands)
2001	456
2002	471
2003	510
2004	548*

* Projected
Source: Energy Information Administration, U.S. Dept. of Energy.

Using statistical methods, we constructed the quadratic equation

$$y = 5.5x^2 + 4.1x + 445$$

to model the number of vehicles y in year x. Here, we used $x = 1$ for 2001, $x = 2$ for 2002, and so on. Because only four years of data were used to determine the model, we must be particularly careful about using it to estimate for years before 2001 or after 2004.

   (a) What prediction for 2005 is given by the equation?

   (b) Why might the prediction for 2005 be unreliable?

86. A lot is in the shape of a right triangle. The hypotenuse is 3 m longer than the longer leg. The longer leg is 6 m longer than twice the length of the shorter leg. Find the lengths of the sides of the lot.

87. A pyramid has a rectangular base with a length that is 2 m more than its width. The height of the pyramid is 6 m, and its volume is 48 m³. Find the length and width of the base.

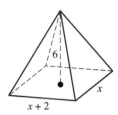

88. The product of the first and second of three consecutive integers is equal to 23 plus the third. Find the integers.

89. If an object is dropped, the distance d in feet it falls in t seconds (disregarding air resistance) is given by the quadratic equation

$$d = 16t^2.$$

Find the distance an object would fall in 4 sec.

90. Repeat Exercise 89 for 8 sec.

91. The floor plan for a house is a rectangle with length 7 m more than its width. The area is 170 m². Find the width and length of the house.

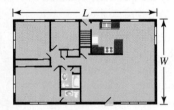

92. The triangular sail of a schooner has an area of 30 m². The height of the sail is 4 m more than the base. Find the base of the sail.

Chapter 6 TEST

View the complete solutions to all Chapter Test exercises on the Pass the Test CD.

1. *Concept Check* Which one of the following is the correct completely factored form of $2x^2 - 2x - 24$?

A. $(2x + 6)(x - 4)$ **B.** $(x + 3)(2x - 8)$
C. $2(x + 4)(x - 3)$ **D.** $2(x + 3)(x - 4)$

Factor each polynomial completely. If the polynomial is prime, say so.

2. $12x^2 - 30x$

3. $2m^3n^2 + 3m^3n - 5m^2n^2$

4. $2ax - 2bx + ay - by$

5. $x^2 - 5x - 24$

6. $2x^2 + x - 3$

7. $10z^2 - 17z + 3$

8. $t^2 + 2t + 3$

9. $x^2 + 36$

10. $12 - 6a + 2b - ab$

11. $9y^2 - 64$

12. $4x^2 - 28xy + 49y^2$

13. $-2x^2 - 4x - 2$

14. $6t^4 + 3t^3 - 108t^2$

15. $r^3 - 125$

16. $8k^3 + 64$

17. $x^4 - 81$

18. $81x^4 - 16y^4$

19. $9x^6y^4 + 12x^3y^2 + 4$

📝 **20.** Why is $(p + 3)(p + 3)$ not the correct factored form of $p^2 + 9$?

Solve each equation.

21. $2r^2 - 13r + 6 = 0$

22. $25x^2 - 4 = 0$

23. $x(x - 20) = -100$

24. $t^3 = 9t$

Solve each problem.

25. The length of a rectangular flower bed is 3 ft less than twice its width. The area of the bed is 54 ft². Find the dimensions of the flower bed.

26. Find two consecutive integers such that the square of the sum of the two integers is 11 more than the first integer.

27. A carpenter needs to cut a brace to support a wall stud, as shown in the figure. The brace should be 7 ft less than three times the length of the stud. If the brace will be anchored on the floor 15 ft away from the stud, how long should the brace be?

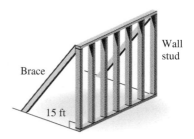

28. The number y of U.S. cable TV networks from 1984 through 2003 can be approximated by the quadratic equation

$$y = 1.06x^2 - 4.77x + 47.9,$$

where $x = 0$ represents 1984, $x = 1$ represents 1985, and so on. (*Source:* National Cable and Telecommunications Association; FCC annual report.) Use the model to estimate the number of cable TV channels in the year 2000. Round your answer to the nearest whole number.

Chapters 1–6 CUMULATIVE REVIEW EXERCISES

Solve each equation.

1. $3x + 2(x - 4) = 4(x - 2)$

2. $0.3x + 0.9x = 0.06$

3. $\dfrac{2}{3}m - \dfrac{1}{2}(m - 4) = 3$

4. Solve for P: $A = P + Prt$.

5. Find the measures of the marked angles.

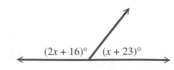

Solve each problem.

6. At the 2006 Winter Olympics in Torino, Italy, the top medal winner was Germany, which won a total of 29 medals. Germany won 1 more silver medal than gold and 5 more gold medals than bronze. Find the number of each type of medal won. (*Source:* www.infoplease.com.)

7. In December 2004, approximately 69.4 million people in the United States used broadband connections to access the Internet from home. This was a 36% increase from the same month the previous year. How many people in the United States used home broadband connections in December 2003? (*Source:* Nielsen Media Research.)

8. From a list of "everyday items" often taken for granted, adults were recently surveyed as to those items they wouldn't want to live without. Complete the results shown in the table if 500 adults were surveyed.

Item	Percent That Wouldn't Want to Live Without	Number That Wouldn't Want to Live Without
Toilet paper	69%	
Zipper	42%	
Frozen foods		190
Self-stick note pads		75

(Other items included tape, hair spray, panty hose, paper clips, and Velcro®.)
Source: Market Facts for Kleenex Cottonelle.

9. Fill in each blank with *positive* or *negative*. The point with coordinates (a, b) is in

   (a) quadrant II if a is _____ and b is _____.
   (b) quadrant III if a is _____ and b is _____.

Consider the equation $y = 12x + 3$. Find the following.

10. The x- and y-intercepts   11. The slope   12. The graph

13. The points on the graph show the number of U.S. radio stations in the years 1995–2002, along with the graph of a linear equation that models the data.

   ✏ (a) Use the ordered pairs shown on the graph to find the slope of the line to the nearest whole number. Interpret the slope.

   (b) Use the graph to estimate the number of radio stations in the year 2000. Write your answer as an ordered pair of the form (year, number of radio stations).

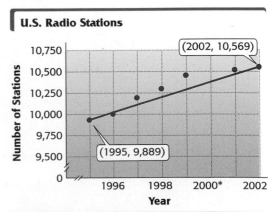

U.S. Radio Stations

(2002, 10,569)

(1995, 9,889)

Source: M Street Corporation.
* Data for 2000 unavailable.

14. The equation of the line from Exercise 13 defines a relation R.

    (a) Find the value of R in 2002.   **(b)** Is the relation a function?

Solve each system of equations.

15. $4x - y = -6$
    $2x + 3y = 4$

16. $5x + 3y = 10$
    $2x + \dfrac{6}{5}y = 5$

Evaluate each expression.

17. $\left(\dfrac{3}{4}\right)^{-2}$

18. $\left(\dfrac{4^{-3} \cdot 4^4}{4^5}\right)^{-1}$

Simplify each expression, and write the answer with only positive exponents. Assume that no denominators are 0.

19. $\dfrac{(p^2)^3 p^{-4}}{(p^{-3})^{-1} p}$

20. $\dfrac{(m^{-2})^3 m}{m^5 m^{-4}}$

Perform each indicated operation.

21. $(2k^2 + 4k) - (5k^2 - 2) - (k^2 + 8k - 6)$ **22.** $(9x + 6)(5x - 3)$

23. $(3p + 2)^2$ **24.** $\dfrac{8x^4 + 12x^3 - 6x^2 + 20x}{2x}$

25. To make a pound of honey, bees may travel 55,000 mi and visit more than 2,000,000 flowers. (*Source: Home & Garden.*) Write the two given numbers in scientific notation.

Factor completely.

26. $2a^2 + 7a - 4$ **27.** $10m^2 + 19m + 6$ **28.** $8t^2 + 10tv + 3v^2$

29. $4p^2 - 12p + 9$ **30.** $25r^2 - 81t^2$ **31.** $2pq + 6p^3q + 8p^2q$

Solve each equation.

32. $6m^2 + m - 2 = 0$ **33.** $8x^2 = 64x$

34. The length of the hypotenuse of a right triangle is twice the length of the shorter leg, plus 3 m. The longer leg is 7 m longer than the shorter leg. Find the lengths of the sides.

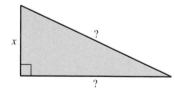

Rational Expressions and Applications

In the 1994 movie *Little Big League,* the young Billy Heywood inherits the Minnesota Twins baseball team and becomes its manager. He leads the team to a division championship, but before the biggest game of the year, he can't keep his mind on his job because a homework problem is giving him trouble.

If Joe can paint a house in 3 hours and Sam can paint the same house in 5 hours, how long does it take for them to do it together?

With the help of one of his players, Billy is able to solve the problem, and the team goes on to victory. In Example 3 of Section 7.7, we use *rational expressions,* the subject of this chapter, to solve Billy Heywood's problem.

7.1 The Fundamental Property of Rational Expressions

The quotient of two integers (with denominator not 0), such as $\frac{2}{3}$ or $-\frac{3}{4}$, is called a *rational number*. In the same way, the quotient of two polynomials with denominator not equal to 0 is called a *rational expression*.

> **Rational Expression**
>
> A **rational expression** is an expression of the form $\frac{P}{Q}$, where P and Q are polynomials, with $Q \neq 0$.

$$\frac{-6x}{x^3 + 8}, \quad \frac{9x}{y + 3}, \quad \text{and} \quad \frac{2m^3}{8} \qquad \text{Examples of rational expressions}$$

Our work with rational expressions requires much of what we learned in **Chapters 5 and 6** on polynomials and factoring, as well as the rules for fractions from **Section 1.1**.

OBJECTIVE 1 Find the numerical value of a rational expression. We use substitution to evaluate a rational expression for a given value of the variable.

EXAMPLE 1 Evaluating Rational Expressions

Find the numerical value of $\dfrac{3x + 6}{2x - 4}$ when **(a)** $x = 1$ and **(b)** $x = -2$.

(a) $\dfrac{3x + 6}{2x - 4} = \dfrac{3(1) + 6}{2(1) - 4}$ Let $x = 1$.

$= \dfrac{9}{-2} = -\dfrac{9}{2} \qquad \dfrac{a}{-b} = -\dfrac{a}{b}$

> Use parentheses around −2 to avoid errors.

(b) $\dfrac{3x + 6}{2x - 4} = \dfrac{3(-2) + 6}{2(-2) - 4}$ Let $x = -2$.

$= \dfrac{0}{-8} = 0 \qquad \dfrac{0}{a} = 0$

✔ **Now Try Exercise 3.**

OBJECTIVE 2 Find the values of the variable for which a rational expression is undefined. In the definition of a rational expression $\frac{P}{Q}$, Q cannot equal 0. *The denominator of a rational expression cannot equal 0 because division by 0 is undefined.*

For instance, in the rational expression

$$\frac{3x + 6}{2x - 4} \leftarrow \text{Denominator cannot equal 0.}$$

from Example 1, the variable x can take on any real number value except 2. If x is 2, then the denominator becomes $2(2) - 4 = 0$, making the expression undefined. Thus, x cannot equal 2. We indicate this restriction by writing $x \neq 2$.

▶ **NOTE** *The numerator of a rational expression may be any real number.* If the numerator equals 0 and the denominator does not equal 0, then the rational expression equals 0. See Example 1(b).

To determine the values for which a rational expression is undefined, use the following procedure.

Determining When a Rational Expression is Undefined

Step 1 Set the denominator of the rational expression equal to 0.

Step 2 Solve this equation.

Step 3 The solutions of the equation are the values that make the rational expression undefined.

EXAMPLE 2 Finding Values That Make Rational Expressions Undefined

Find any values of the variable for which each rational expression is undefined.

(a) $\dfrac{p + 5}{3p + 2}$

We must find any value of p that makes the *denominator* equal to 0, since division by 0 is undefined.

Step 1 Set the denominator equal to 0.
$$3p + 2 = 0$$

Step 2 Solve this equation.

$$3p = -2 \qquad \text{Subtract 2.}$$

$$p = -\frac{2}{3} \qquad \text{Divide by 3.}$$

Step 3 The given expression is undefined for $-\frac{2}{3}$, so $p \neq -\frac{2}{3}$.

(b) $\dfrac{8x^2 + 1}{x - 3}$

The denominator $x - 3 = 0$ when x is 3. The given expression is undefined for 3, so $x \neq 3$.

(c) $\dfrac{9m^2}{m^2 - 5m + 6}$

$$m^2 - 5m + 6 = 0 \qquad \text{Set the denominator equal to 0.}$$

$$(m - 2)(m - 3) = 0 \qquad \text{Factor.}$$

$$m - 2 = 0 \quad \text{or} \quad m - 3 = 0 \qquad \text{Zero-factor property}$$

$$m = 2 \quad \text{or} \qquad m = 3 \qquad \text{Solve for } m.$$

The given expression is undefined for 2 and 3, so $m \neq 2$ and $m \neq 3$.

(d) $\dfrac{2r}{r^2 + 1}$

This denominator will not equal 0 for any value of r, because r^2 is always greater than or equal to 0, and adding 1 makes the sum greater than 0. Thus, there are no values for which this rational expression is undefined.

✔ **Now Try Exercises 17, 19, and 23.**

OBJECTIVE 3 **Write rational expressions in lowest terms.** A fraction such as $\frac{2}{3}$ is said to be in *lowest terms.* How can "lowest terms" be defined? We use the idea of greatest common factor for this definition, which applies to all rational expressions.

Lowest Terms

A rational expression $\frac{P}{Q}$ ($Q \neq 0$) is in **lowest terms** if the greatest common factor of its numerator and denominator is 1.

The properties of rational numbers also apply to rational expressions. We use the **fundamental property of rational expressions** to write a rational expression in lowest terms.

Fundamental Property of Rational Expressions

If $\frac{P}{Q}$ ($Q \neq 0$) is a rational expression and if K represents any polynomial, where $K \neq 0$, then

$$\frac{PK}{QK} = \frac{P}{Q}.$$

This property is based on the identity property of multiplication, since

$$\frac{PK}{QK} = \frac{P}{Q} \cdot \frac{K}{K} = \frac{P}{Q} \cdot 1 = \frac{P}{Q}.$$

The next example shows how to write both a rational number and a rational expression in lowest terms. Notice the similarity in the procedures: In both cases, we factor and then divide out the greatest common factor.

EXAMPLE 3 **Writing in Lowest Terms**

Write each rational expression in lowest terms.

(a) $\dfrac{30}{72}$

Begin by factoring.

$$\frac{30}{72} = \frac{2 \cdot 3 \cdot 5}{2 \cdot 2 \cdot 2 \cdot 3 \cdot 3}$$

(b) $\dfrac{14k^2}{2k^3}$

Write k^2 as $k \cdot k$ and k^3 as $k \cdot k \cdot k$.

$$\frac{14k^2}{2k^3} = \frac{2 \cdot 7 \cdot k \cdot k}{2 \cdot k \cdot k \cdot k}$$

Group any factors common to the numerator and denominator.

$$\frac{30}{72} = \frac{5 \cdot (2 \cdot 3)}{2 \cdot 2 \cdot 3 \cdot (2 \cdot 3)} \qquad\qquad \frac{14k^2}{2k^3} = \frac{7(2 \cdot k \cdot k)}{k(2 \cdot k \cdot k)}$$

Use the fundamental property.

$$\frac{30}{72} = \frac{5}{2 \cdot 2 \cdot 3} = \frac{5}{12} \qquad\qquad \frac{14k^2}{2k^3} = \frac{7}{k}$$

✔ **Now Try Exercise 27.**

To write a rational expression in lowest terms, follow these steps.

> **Writing a Rational Expression in Lowest Terms**
>
> *Step 1*   **Factor** the numerator and denominator completely.
>
> *Step 2*   **Use the fundamental property** to divide out any common factors.

EXAMPLE 4   Writing in Lowest Terms

Write in lowest terms.

(a)
$$\frac{3x - 12}{5x - 20} = \frac{3(x - 4)}{5(x - 4)}$$ Factor both numerator and denominator. (Step 1)

x ≠ 4, since the denominator is 0 when x is 4.

$$= \frac{3}{5}$$ Divide out the common factor. (Step 2)

The given expression is equal to $\frac{3}{5}$ for all values of x, where $x \neq 4$.

(b)
$$\frac{2y^2 - 8}{2y + 4} = \frac{2(y^2 - 4)}{2(y + 2)}$$ Factor. (Step 1)

y ≠ −2, since the denominator is 0 for this value.

$$= \frac{2(y + 2)(y - 2)}{2(y + 2)}$$ Be sure to factor completely.

$$= y - 2$$ Fundamental property (Step 2)

(c)
$$\frac{m^2 + 2m - 8}{2m^2 - m - 6} = \frac{(m + 4)(m - 2)}{(2m + 3)(m - 2)}$$ Factor.

m ≠ −$\frac{3}{2}$ and m ≠ 2, since the denominator is 0 for these values.

$$= \frac{m + 4}{2m + 3}$$ Fundamental property

✔ **Now Try Exercises 33, 41, and 49.**

From now on, we write statements of equality of rational expressions with the understanding that they apply only to those real numbers that make neither denominator equal to 0.

▶ **C A U T I O N** *Rational expressions cannot be written in lowest terms until after the numerator and denominator have been factored. Only common factors can be divided out, not common terms.* For example,

$$\frac{6x + 9}{4x + 6} = \frac{3(2x + 3)}{2(2x + 3)} = \frac{3}{2} \qquad \frac{6 + x}{4x} \leftarrow \text{Numerator cannot be factored.}$$

Divide out the
common factor.

Already in lowest terms

EXAMPLE 5 **Writing in Lowest Terms (Factors Are Opposites)**

Write $\dfrac{x - y}{y - x}$ in lowest terms.

At first glance, there does not seem to be any way in which $x - y$ and $y - x$ can be factored to get a common factor. However, the denominator $y - x$ can be factored as

$$y - x = -1(-y + x) = -1(x - y). \qquad \text{Factor out } -1.$$

Be careful
with signs.

Now, use the fundamental property to simplify.

$$\frac{x - y}{y - x} = \frac{1(x - y)}{-1(x - y)} = \frac{1}{-1} = -1$$

✔ **Now Try Exercise 67.**

▶ **N O T E** Either the numerator or the denominator could have been factored in the first step in Example 5. Factor -1 from the numerator, and confirm that the result is the same.

In Example 5, notice that $y - x$ is the *opposite* (or additive inverse) of $x - y$. A general rule for this situation follows.

If the numerator and the denominator of a rational expression are opposites, as in $\dfrac{x - y}{y - x}$, then the rational expression is equal to -1.

Based on this result, the following are true:

Numerator and
denominator
are opposites.
$$\frac{q - 7}{7 - q} = -1 \quad \text{and} \quad \frac{-5a + 2b}{5a - 2b} = -1.$$

However, the following expression cannot be simplified further:

$$\frac{x - 2}{x + 2}. \quad \leftarrow \begin{array}{l} \text{Numerator and denominator} \\ \text{are } not \text{ opposites.} \end{array}$$

EXAMPLE 6   Writing in Lowest Terms (Factors Are Opposites)

Write each rational expression in lowest terms.

(a) $\dfrac{2 - m}{m - 2}$

Since $2 - m$ and $m - 2$ are opposites, this expression equals -1.

(b) $\dfrac{4x^2 - 9}{6 - 4x}$

$= \dfrac{(2x + 3)(2x - 3)}{2(3 - 2x)}$ Factor the numerator and denominator.

$= \dfrac{(2x + 3)(2x - 3)}{2(-1)(2x - 3)}$ Write $3 - 2x$ in the denominator as $-1(2x - 3)$.

$= \dfrac{2x + 3}{2(-1)}$ Fundamental property

$= \dfrac{2x + 3}{-2}$,   or $-\dfrac{2x + 3}{2}$ $\dfrac{a}{-b} = -\dfrac{a}{b}$

(c) $\dfrac{3 + r}{3 - r}$ $3 - r$ is *not* the opposite of $3 + r$.

This rational expression is already in lowest terms.

✔ **Now Try Exercises 69 and 73.**

OBJECTIVE 4   Recognize equivalent forms of rational expressions.   When working with rational expressions, it is important to be able to recognize equivalent forms of an expression. For example, the common fraction $-\frac{5}{6}$ can also be written $\frac{-5}{6}$ and $\frac{5}{-6}$.

Consider the final rational expression from Example 6(b),

$$-\dfrac{2x + 3}{2}.$$

The $-$ sign representing the factor -1 is in front of the expression, even with the fraction bar. The factor -1 may instead be placed in the numerator or in the denominator. Some other equivalent forms of this rational expression are

Use parentheses.

$$\dfrac{-(2x + 3)}{2} \qquad \text{and} \qquad \dfrac{2x + 3}{-2}.$$

Multiply *each* term in the numerator by -1.

By the distributive property,

$$\dfrac{-(2x + 3)}{2} \quad \text{can also be written} \quad \dfrac{-2x - 3}{2}.$$

▶ **CAUTION** $\frac{-2x + 3}{2}$ is *not* an equivalent form of $\frac{-(2x + 3)}{2}$. The sign preceding 3 in the numerator of $\frac{-2x + 3}{2}$ should be $-$ rather than $+$. **Be careful to apply the distributive property correctly.**

EXAMPLE 7 Writing Equivalent Forms of a Rational Expression

Write four equivalent forms of the rational expression

$$-\frac{3x + 2}{x - 6}.$$

If we apply the negative sign to the numerator, we obtain the equivalent forms

$$\frac{-(3x + 2)}{x - 6} \quad \text{and, by the distributive property,} \quad \frac{-3x - 2}{x - 6}.$$

If we apply the negative sign to the denominator, we obtain

$$\frac{3x + 2}{-(x - 6)} \quad \text{or, distributing once again,} \quad \frac{3x + 2}{-x + 6}.$$

> **Be careful with signs.**

✔ **Now Try Exercise 77.**

▶ **CAUTION** Recall that $-\frac{5}{6} \neq \frac{-5}{-6}$. Thus, in Example 7, it would be incorrect to distribute the negative sign to *both* the numerator *and* the denominator. (Doing this would actually lead to the *opposite* of the original expression.)

CONNECTIONS

In **Chapter 5,** we used long division to find the quotient of two polynomials. For example, we found $(2x^2 + 5x - 12) \div (2x - 3)$ as follows:

$$
\begin{array}{r}
x + 4 \\
2x - 3 \overline{\smash{\big)}\ 2x^2 + 5x - 12} \\
\underline{2x^2 - 3x} \\
8x - 12 \\
\underline{8x - 12} \\
0
\end{array}
$$

The quotient is $x + 4$. We get the same quotient by expressing the division problem as a rational expression (fraction) and writing this rational expression in lowest terms.

$$
\begin{aligned}
\frac{2x^2 + 5x - 12}{2x - 3} & \\
= \frac{(2x - 3)(x + 4)}{2x - 3} \quad & \text{Factor.} \\
= x + 4 \quad & \text{Fundamental property}
\end{aligned}
$$

For Discussion or Writing

What kind of division problem has a quotient that cannot be found by writing a fraction in lowest terms? Try using rational expressions to solve each division problem. Then use long division and compare.

1. $(3x^2 + 11x + 8) \div (x + 2)$ **2.** $(x^3 - 8) \div (x^2 + 2x + 4)$

7.1 EXERCISES

▨ Now Try Exercise

📝 **1.** Define *rational expression* in your own words, and give an example.

Find the numerical value of each rational expression when (a) $x = 2$ and (b) $x = -3$. See Example 1.

2. $\dfrac{5x - 2}{4x}$

⊕ **3.** $\dfrac{3x + 1}{5x}$

4. $\dfrac{2x^2 - 4x}{3x - 1}$

5. $\dfrac{x^2 - 4}{2x + 1}$

6. $\dfrac{(-3x)^2}{4x + 12}$

7. $\dfrac{(-2x)^3}{3x + 9}$

8. $\dfrac{5x + 2}{2x^2 + 11x + 12}$

9. $\dfrac{7 - 3x}{3x^2 - 7x + 2}$

10. *Concept Check* Fill in each blank with the correct response: The rational expression $\dfrac{x + 5}{x - 3}$ is undefined when x is _____, so $x \neq$ _____. This rational expression is equal to 0 when $x =$ _____.

📝 **11.** Why can't the denominator of a rational expression equal 0?

📝 **12.** If 2 is substituted for x in the rational expression $\dfrac{x - 2}{x^2 - 4}$, the result is $\dfrac{0}{0}$. An often-heard statement is "Any number divided by itself is 1." Does this mean that this expression is equal to 1 for $x = 2$? If not, explain.

Find any values of the variable for which each rational expression is undefined. Write answers with $\neq$. See Example 2.

13. $\dfrac{12}{5y}$

14. $\dfrac{-7}{3z}$

15. $\dfrac{x + 1}{x - 6}$

16. $\dfrac{m - 2}{m - 5}$

⊕ **17.** $\dfrac{4x^2}{3x + 5}$

18. $\dfrac{2x^3}{3x + 4}$

19. $\dfrac{5m + 2}{m^2 + m - 6}$

20. $\dfrac{2r - 5}{r^2 - 5r + 4}$

21. $\dfrac{x^2 + 3x}{4}$

22. $\dfrac{x^2 - 4x}{6}$

23. $\dfrac{3x - 1}{x^2 + 2}$

24. $\dfrac{4q + 2}{q^2 + 9}$

25. **(a)** Identify the two *terms* in the numerator and the two *terms* in the denominator of the rational expression $\dfrac{x^2 + 4x}{x + 4}$.

📝 **(b)** Describe the steps you would use to write the rational expression in part (a) in lowest terms. (*Hint:* It simplifies to x.)

26. *Concept Check* Only one of the following rational expressions can be simplified. Which one is it?

A. $\dfrac{x^2 + 2}{x^2}$ **B.** $\dfrac{x^2 + 2}{2}$ **C.** $\dfrac{x^2 + y^2}{y^2}$ **D.** $\dfrac{x^2 - 5x}{x}$

Write each rational expression in lowest terms. See Examples 3 and 4.

⊕ **27.** $\dfrac{18r^3}{6r}$

28. $\dfrac{27p^2}{3p}$

29. $\dfrac{4(y - 2)}{10(y - 2)}$

30. $\dfrac{15(m - 1)}{9(m - 1)}$

31. $\dfrac{(x + 1)(x - 1)}{(x + 1)^2}$

32. $\dfrac{(t + 5)(t - 3)}{(t - 1)(t + 5)}$

⊕ **33.** $\dfrac{7m + 14}{5m + 10}$

34. $\dfrac{8z - 24}{4z - 12}$

35. $\dfrac{6m - 18}{7m - 21}$

36. $\dfrac{5r + 20}{3r + 12}$

37. $\dfrac{m^2 - n^2}{m + n}$

38. $\dfrac{a^2 - b^2}{a - b}$

39. $\dfrac{2t + 6}{t^2 - 9}$

40. $\dfrac{5s - 25}{s^2 - 25}$

41. $\dfrac{12m^2 - 3}{8m - 4}$

42. $\dfrac{20p^2 - 45}{6p - 9}$

43. $\dfrac{3m^2 - 3m}{5m - 5}$

44. $\dfrac{6t^2 - 6t}{2t - 2}$

45. $\dfrac{9r^2 - 4s^2}{9r + 6s}$

46. $\dfrac{16x^2 - 9y^2}{12x - 9y}$

47. $\dfrac{5k^2 - 13k - 6}{5k + 2}$

48. $\dfrac{7t^2 - 31t - 20}{7t + 4}$

49. $\dfrac{x^2 + 2x - 15}{x^2 + 6x + 5}$

50. $\dfrac{y^2 - 5y - 14}{y^2 + y - 2}$

51. $\dfrac{2x^2 - 3x - 5}{2x^2 - 7x + 5}$

52. $\dfrac{3x^2 + 8x + 4}{3x^2 - 4x - 4}$

*These exercises involve factoring by grouping (**Section 6.1**) and factoring sums and differences of cubes (**Section 6.4**). Write each rational expression in lowest terms.*

53. $\dfrac{zw + 4z - 3w - 12}{zw + 4z + 5w + 20}$

54. $\dfrac{km + 4k + 4m + 16}{km + 4k + 5m + 20}$

55. $\dfrac{pr + qr + ps + qs}{pr + qr - ps - qs}$

56. $\dfrac{ac - ad + bc - bd}{ac - ad - bc + bd}$

57. $\dfrac{m^2 - n^2 - 4m - 4n}{2m - 2n - 8}$

58. $\dfrac{x^2y + y + x^2z + z}{xy + xz}$

59. $\dfrac{1 + p^3}{1 + p}$

60. $\dfrac{x^3 - 27}{x - 3}$

61. $\dfrac{b^3 - a^3}{a^2 - b^2}$

62. $\dfrac{k^3 + 8}{k^2 - 4}$

63. $\dfrac{z^3 + 27}{z^3 - 3z^2 + 9z}$

64. $\dfrac{1 - 8r^3}{8r^2 + 4r + 2}$

65. *Concept Check* Which two of the following rational expressions equal -1?

  A. $\dfrac{2x + 3}{2x - 3}$ **B.** $\dfrac{2x - 3}{3 - 2x}$ **C.** $\dfrac{2x + 3}{3 + 2x}$ **D.** $\dfrac{2x + 3}{-2x - 3}$

66. *Concept Check* Make the correct choice for the blank: $\dfrac{4 - r^2}{4 + r^2}$ _____ equal to -1.
  (is/is not)

Write each rational expression in lowest terms. See Examples 5 and 6.

67. $\dfrac{6 - t}{t - 6}$

68. $\dfrac{2 - k}{k - 2}$

69. $\dfrac{m^2 - 1}{1 - m}$

70. $\dfrac{a^2 - b^2}{b - a}$

71. $\dfrac{q^2 - 4q}{4q - q^2}$

72. $\dfrac{z^2 - 5z}{5z - z^2}$

73. $\dfrac{p + 6}{p - 6}$

74. $\dfrac{5 - x}{5 + x}$

75. *Concept Check* Only one of the following rational expressions is *not* equivalent to $\frac{x - 3}{4 - x}$. Which one is it?

  A. $\dfrac{3 - x}{x - 4}$ **B.** $\dfrac{x + 3}{4 + x}$ **C.** $-\dfrac{3 - x}{4 - x}$ **D.** $-\dfrac{x - 3}{x - 4}$

76. *Concept Check* Make the correct choice for the blank: $\frac{5 + 2x}{3 - x}$ and $\frac{-5 - 2x}{x - 3}$ _____ equivalent rational expressions.
  (are/are not)

Write four equivalent forms for each rational expression. See Example 7.

77. $-\dfrac{x + 4}{x - 3}$

78. $-\dfrac{x + 6}{x - 1}$

79. $-\dfrac{2x - 3}{x + 3}$

80. $-\dfrac{5x - 6}{x + 4}$

81. $-\dfrac{3x - 1}{5x - 6}$

82. $-\dfrac{2x - 9}{7x - 1}$

83. The area of the rectangle is represented by

$$x^4 + 10x^2 + 21.$$

What is the width? $\left(\textit{Hint:} \text{ Use } W = \frac{A}{L}.\right)$

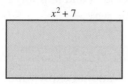

$x^2 + 7$

84. The volume of the box is represented by

$$(x^2 + 8x + 15)(x + 4).$$

Find the polynomial that represents the area of the bottom of the box.

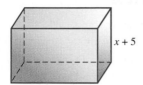

$x + 5$

Solve each problem.

85. The average number of vehicles waiting in line to enter a sports arena parking area is approximated by the rational expression

$$\frac{x^2}{2(1 - x)},$$

where x is a quantity between 0 and 1 known as the **traffic intensity.** (*Source:* Mannering, F., and W. Kilareski, *Principles of Highway Engineering and Traffic Control,* John Wiley and Sons, 1990.) To the nearest tenth, find the average number of vehicles waiting if the traffic intensity is

(a) 0.1   **(b)** 0.8   **(c)** 0.9.

(d) What happens to waiting time as traffic intensity increases?

86. The percent of deaths caused by smoking is modeled by the rational expression

$$\frac{x - 1}{x},$$

where x is the number of times a smoker is more likely than a nonsmoker to die of lung cancer. This is called the **incidence rate.** (*Source:* Walker, A., *Observation and Inference: An Introduction to the Methods of Epidemiology,* Epidemiology Resources Inc., 1991.) For example, $x = 10$ means that a smoker is 10 times more likely than a nonsmoker to die of lung cancer. Find the percent of deaths if the incidence rate is

(a) 5   **(b)** 10   **(c)** 20.

(d) Can the incidence rate equal 0? Explain.

PREVIEW EXERCISES

*Multiply or divide as indicated. Write each answer in lowest terms. See **Section 1.1.***

87. $\dfrac{2}{3} \cdot \dfrac{5}{6}$

88. $\dfrac{3}{7} \cdot \dfrac{2}{5}$

89. $\dfrac{6}{15} \cdot \dfrac{25}{3}$

90. $\dfrac{10}{8} \div \dfrac{7}{12}$

91. $\dfrac{10}{3} \div \dfrac{5}{6}$

92. $\dfrac{7}{12} \div \dfrac{15}{4}$

7.2 Multiplying and Dividing Rational Expressions

OBJECTIVE 1 Multiply rational expressions. The product of two fractions is found by multiplying the numerators and multiplying the denominators. Rational expressions are multiplied in the same way.

> **Multiplying Rational Expressions**
>
> The product of the rational expressions $\frac{P}{Q}$ and $\frac{R}{S}$ is
>
> $$\frac{P}{Q} \cdot \frac{R}{S} = \frac{PR}{QS}.$$
>
> That is, to multiply rational expressions, multiply the numerators and multiply the denominators.

In the next example, the parallel discussion with rational numbers and rational expressions lets you compare the steps.

EXAMPLE 1 Multiplying Rational Expressions

Multiply. Write each answer in lowest terms.

(a) $\dfrac{3}{10} \cdot \dfrac{5}{9}$ $\qquad$ **(b)** $\dfrac{6}{x} \cdot \dfrac{x^2}{12}$

Indicate the product of the numerators and the product of the denominators.

$$\frac{3}{10} \cdot \frac{5}{9} = \frac{3 \cdot 5}{10 \cdot 9} \qquad \frac{6}{x} \cdot \frac{x^2}{12} = \frac{6 \cdot x^2}{x \cdot 12}$$

Leave the products in factored form because common factors are needed to write the product in lowest terms. Factor the numerator and denominator to further identify any common factors. Then use the fundamental property to write each product in lowest terms.

$$\frac{3}{10} \cdot \frac{5}{9} = \frac{3 \cdot 5}{2 \cdot 5 \cdot 3 \cdot 3} = \frac{1}{6} \qquad \frac{6}{x} \cdot \frac{x^2}{12} = \frac{6 \cdot x \cdot x}{2 \cdot 6 \cdot x} = \frac{x}{2}$$

✔ **Now Try Exercise 3.**

> ▶ **NOTE** It is also possible to divide out common factors in the numerator and denominator *before* multiplying the rational expressions. For example,
>
> $$\frac{6}{5} \cdot \frac{35}{22} = \frac{2 \cdot 3}{5} \cdot \frac{5 \cdot 7}{2 \cdot 11} \qquad \text{Identify common factors.}$$
>
> $$= \frac{3}{1} \cdot \frac{7}{11} \qquad \text{Divide out common factors.}$$
>
> $$= \frac{21}{11}. \qquad \text{Multiply.}$$

EXAMPLE 2 Multiplying Rational Expressions

Multiply. Write the answer in lowest terms.

$$\frac{x + y}{2x} \cdot \frac{x^2}{(x + y)^2}$$

$$= \frac{(x + y)x^2}{2x(x + y)^2} \qquad \text{Multiply numerators.}$$
$$\text{Multiply denominators.}$$

$$= \frac{(x + y)x \cdot x}{2x(x + y)(x + y)} \qquad \text{Factor; identify common factors.}$$

$$= \frac{x}{2(x + y)} \qquad \text{Divide out common factors.}$$

The quotients $\frac{x}{x}$ and $\frac{(x + y)}{(x + y)}$ both equal 1, justifying the final product, $\frac{x}{2(x + y)}$.

✔ **Now Try Exercise 9.**

EXAMPLE 3 Multiplying Rational Expressions

Multiply. Write the answer in lowest terms.

$$\frac{x^2 + 3x}{x^2 - 3x - 4} \cdot \frac{x^2 - 5x + 4}{x^2 + 2x - 3}$$

$$= \frac{(x^2 + 3x)(x^2 - 5x + 4)}{(x^2 - 3x - 4)(x^2 + 2x - 3)} \qquad \text{Definition of multiplication}$$

$$= \frac{x(x + 3)(x - 4)(x - 1)}{(x - 4)(x + 1)(x + 3)(x - 1)} \qquad \text{Factor.}$$

$$= \frac{x}{x + 1} \qquad \text{Divide out common factors.}$$

The quotients $\frac{x + 3}{x + 3}, \frac{x - 4}{x - 4}$, and $\frac{x - 1}{x - 1}$ all equal 1, justifying the final product $\frac{x}{x + 1}$.

✔ **Now Try Exercise 43.**

OBJECTIVE 2 Divide rational expressions. Suppose you have $\frac{7}{8}$ gal of milk and want to find how many quarts you have. Since 1 qt is $\frac{1}{4}$ gal, you ask yourself, "How many $\frac{1}{4}$s are there in $\frac{7}{8}$?" This would be interpreted as

$$\frac{7}{8} \div \frac{1}{4} \quad \text{or} \quad \frac{\frac{7}{8}}{\frac{1}{4}}. \quad \leftarrow\text{The fraction bar means division.}$$

The fundamental property of rational expressions discussed earlier can be applied to rational number values of P, Q, and K. With $P = \frac{7}{8}$, $Q = \frac{1}{4}$, and $K = 4$ (K is the reciprocal of $Q = \frac{1}{4}$),

$$\frac{P}{Q} = \frac{P \cdot K}{Q \cdot K} = \frac{\frac{7}{8} \cdot 4}{\frac{1}{4} \cdot 4} = \frac{\frac{7}{8} \cdot 4}{1} = \frac{7}{8} \cdot \frac{4}{1}.$$

So, to divide $\frac{7}{8}$ by $\frac{1}{4}$, we multiply $\frac{7}{8}$ by the reciprocal of $\frac{1}{4}$, namely, 4. Since $\frac{7}{8}(4) = \frac{7}{2}$, there are $\frac{7}{2}$, or $3\frac{1}{2}$, qt in $\frac{7}{8}$ gal.

The preceding discussion illustrates dividing common fractions. Division of rational expressions is defined in the same way.

Dividing Rational Expressions

If $\frac{P}{Q}$ and $\frac{R}{S}$ are any two rational expressions with $\frac{R}{S} \neq 0$, then

$$\frac{P}{Q} \div \frac{R}{S} = \frac{P}{Q} \cdot \frac{S}{R} = \frac{PS}{QR}.$$

That is, to divide one rational expression by another rational expression, multiply the first rational expression by the reciprocal of the second rational expression.

The next example shows the division of two rational numbers and the division of two rational expressions.

EXAMPLE 4 Dividing Rational Expressions

Divide. Write each answer in lowest terms.

(a) $\dfrac{5}{8} \div \dfrac{7}{16}$

(b) $\dfrac{y}{y + 3} \div \dfrac{4y}{y + 5}$

Multiply the first expression by the reciprocal of the second.

$$\frac{5}{8} \div \frac{7}{16} = \frac{5}{8} \cdot \frac{16}{7} \quad \longleftarrow \begin{array}{l}\text{Reciprocal} \\ \text{of } \frac{7}{16}\end{array}$$

$$= \frac{5 \cdot 16}{8 \cdot 7}$$

$$= \frac{5 \cdot 8 \cdot 2}{8 \cdot 7}$$

$$= \frac{10}{7}$$

$$\frac{y}{y + 3} \div \frac{4y}{y + 5}$$

$$= \frac{y}{y + 3} \cdot \frac{y + 5}{4y} \quad \longleftarrow \begin{array}{l}\text{Reciprocal} \\ \text{of } \frac{4y}{y + 5}\end{array}$$

$$= \frac{y(y + 5)}{(y + 3)(4y)}$$

$$= \frac{y + 5}{4(y + 3)}$$

✔ **Now Try Exercise 19.**

EXAMPLE 5 Dividing Rational Expressions

Divide. Write the answer in lowest terms.

$$\frac{(3m)^2}{(2p)^3} \div \frac{6m^3}{16p^2}$$

$$= \frac{(3m)^2}{(2p)^3} \cdot \frac{16p^2}{6m^3} \qquad \text{Multiply by the reciprocal.}$$

$$\begin{array}{l}(3m)^2 = 3^2m^2; \\ (2p)^3 = 2^3p^3\end{array} \quad = \frac{9m^2}{8p^3} \cdot \frac{16p^2}{6m^3} \qquad \text{Power rule for exponents}$$

$$= \frac{9 \cdot 16m^2p^2}{8 \cdot 6p^3m^3} \qquad \text{Multiply numerators.} \\ \text{Multiply denominators.}$$

$$= \frac{3}{mp} \qquad \text{Lowest terms}$$

✔ **Now Try Exercise 17.**

EXAMPLE 6 Dividing Rational Expressions

Divide. Write the answer in lowest terms.

$$\frac{x^2 - 4}{(x + 3)(x - 2)} \div \frac{(x + 2)(x + 3)}{-2x}$$

$$= \frac{x^2 - 4}{(x + 3)(x - 2)} \cdot \frac{-2x}{(x + 2)(x + 3)} \qquad \text{Multiply by the reciprocal.}$$

$$= \frac{-2x(x^2 - 4)}{(x + 3)(x - 2)(x + 2)(x + 3)} \qquad \text{Multiply numerators.} \\ \text{Multiply denominators.}$$

$$= \frac{-2x(x + 2)(x - 2)}{(x + 3)(x - 2)(x + 2)(x + 3)} \qquad \text{Factor the numerator.}$$

$$= \frac{-2x}{(x + 3)^2}, \quad \text{or} \quad -\frac{2x}{(x + 3)^2} \qquad \text{Lowest terms; } \frac{-a}{b} = -\frac{a}{b}$$

✔ **Now Try Exercise 35.**

EXAMPLE 7 Dividing Rational Expressions (Factors Are Opposites)

Divide. Write the answer in lowest terms.

$$\frac{m^2 - 4}{m^2 - 1} \div \frac{2m^2 + 4m}{1 - m}$$

$$= \frac{m^2 - 4}{m^2 - 1} \cdot \frac{1 - m}{2m^2 + 4m} \qquad \text{Multiply by the reciprocal.}$$

$$= \frac{(m^2 - 4)(1 - m)}{(m^2 - 1)(2m^2 + 4m)} \qquad \text{Multiply numerators.} \\ \text{Multiply denominators.}$$

$$= \frac{(m + 2)(m - 2)(1 - m)}{(m + 1)(m - 1)(2m)(m + 2)} \qquad \text{Factor; } 1 - m \text{ and } m - 1 \text{ are opposites.}$$

$$= \frac{-1(m - 2)}{2m(m + 1)} \qquad \text{From \textbf{Section 7.1}, } \frac{1 - m}{m - 1} = -1.$$

$$= \frac{-m + 2}{2m(m + 1)}, \quad \text{or} \quad \frac{2 - m}{2m(m + 1)} \qquad \text{Distribute } -1 \text{ in the numerator;} \\ \text{rewrite } -m + 2 \text{ as } 2 - m.$$

✔ **Now Try Exercise 37.**

In summary, use the following steps to multiply or divide rational expressions.

> ### Multiplying or Dividing Rational Expressions
>
> *Step 1* **Note the operation.** If the operation is division, use the definition of division to rewrite it as multiplication.
>
> *Step 2* **Multiply** numerators and multiply denominators.
>
> *Step 3* **Factor** all numerators and denominators completely.
>
> *Step 4* **Write in lowest terms** using the fundamental property.
>
> *Note: Steps 2 and 3 may be interchanged based on personal preference.*

7.2 EXERCISES

Complete solution available on Video Lectures on CD/DVD

Now Try Exercise

1. Concept Check Match each multiplication problem in Column I with the correct product in Column II.

I		**II**
(a) $\dfrac{5x^3}{10x^4} \cdot \dfrac{10x^7}{2x}$		**A.** $\dfrac{2}{5x^5}$
(b) $\dfrac{10x^4}{5x^3} \cdot \dfrac{10x^7}{2x}$		**B.** $\dfrac{5x^5}{2}$
(c) $\dfrac{5x^3}{10x^4} \cdot \dfrac{2x}{10x^7}$		**C.** $\dfrac{1}{10x^7}$
(d) $\dfrac{10x^4}{5x^3} \cdot \dfrac{2x}{10x^7}$		**D.** $10x^7$

2. Concept Check Match each division problem in Column I with the correct quotient in Column II.

I		**II**
(a) $\dfrac{5x^3}{10x^4} \div \dfrac{10x^7}{2x}$		**A.** $\dfrac{5x^5}{2}$
(b) $\dfrac{10x^4}{5x^3} \div \dfrac{10x^7}{2x}$		**B.** $10x^7$
(c) $\dfrac{5x^3}{10x^4} \div \dfrac{2x}{10x^7}$		**C.** $\dfrac{2}{5x^5}$
(d) $\dfrac{10x^4}{5x^3} \div \dfrac{2x}{10x^7}$		**D.** $\dfrac{1}{10x^7}$

Multiply. Write each answer in lowest terms. See Examples 1 and 2.

3. $\dfrac{15a^2}{14} \cdot \dfrac{7}{5a}$

4. $\dfrac{27k^3}{9k} \cdot \dfrac{24}{9k^2}$

5. $\dfrac{12x^4}{18x^3} \cdot \dfrac{-8x^5}{4x^2}$

6. $\dfrac{12m^5}{-2m^2} \cdot \dfrac{6m^6}{28m^3}$

7. $\dfrac{2(c+d)}{3} \cdot \dfrac{18}{6(c+d)^2}$

8. $\dfrac{4(y-2)}{x} \cdot \dfrac{3x}{6(y-2)^2}$

9. $\dfrac{(x-y)^2}{2} \cdot \dfrac{24}{3(x-y)}$

10. $\dfrac{(a+b)^2}{5} \cdot \dfrac{30}{2(a+b)}$

11. $\dfrac{t-4}{8} \cdot \dfrac{4t^2}{t-4}$

12. $\dfrac{z+9}{12} \cdot \dfrac{3z^2}{z+9}$

13. $\dfrac{3x}{x+3} \cdot \dfrac{(x+3)^2}{6x^2}$

14. $\dfrac{(t-2)^2}{4t^2} \cdot \dfrac{2t}{t-2}$

Divide. Write each answer in lowest terms. See Examples 4 and 5.

15. $\dfrac{9z^4}{3z^5} \div \dfrac{3z^2}{5z^3}$

16. $\dfrac{35q^8}{9q^5} \div \dfrac{25q^6}{10q^5}$

17. $\dfrac{4t^4}{2t^5} \div \dfrac{(2t)^3}{-6}$

18. $\dfrac{-12a^6}{3a^2} \div \dfrac{(2a)^3}{27a}$

19. $\dfrac{3}{2y-6} \div \dfrac{6}{y-3}$

20. $\dfrac{4m+16}{10} \div \dfrac{3m+12}{18}$

21. $\dfrac{7t+7}{-6} \div \dfrac{4t+4}{15}$

22. $\dfrac{8z-16}{-20} \div \dfrac{3z-6}{40}$

23. $\dfrac{2x}{x-1} \div \dfrac{x^2}{x+2}$

24. $\dfrac{y^2}{y+1} \div \dfrac{3y}{y-3}$

25. $\dfrac{(x-3)^2}{6x} \div \dfrac{x-3}{x^2}$

26. $\dfrac{2a}{a+4} \div \dfrac{a^2}{(a+4)^2}$

✍ **27.** Use an example to explain how to multiply rational expressions.

✍ **28.** Use an example to explain how to divide rational expressions.

Multiply or divide. Write each answer in lowest terms. See Examples 3, 6, and 7.

29. $\dfrac{5x - 15}{3x + 9} \cdot \dfrac{4x + 12}{6x - 18}$

30. $\dfrac{8r + 16}{24r - 24} \cdot \dfrac{6r - 6}{3r + 6}$

31. $\dfrac{2 - t}{8} \div \dfrac{t - 2}{6}$

32. $\dfrac{4}{m - 2} \div \dfrac{16}{2 - m}$

33. $\dfrac{27 - 3z}{4} \cdot \dfrac{12}{2z - 18}$

34. $\dfrac{5 - x}{5 + x} \cdot \dfrac{x + 5}{x - 5}$

35. $\dfrac{p^2 + 4p - 5}{p^2 + 7p + 10} \div \dfrac{p - 1}{p + 4}$

36. $\dfrac{z^2 - 3z + 2}{z^2 + 4z + 3} \div \dfrac{z - 1}{z + 1}$

37. $\dfrac{m^2 - 4}{16 - 8m} \div \dfrac{m + 2}{8}$

38. $\dfrac{2}{3 - x} \div \dfrac{2x + 6}{x^2 - 9}$

39. $\dfrac{2x^2 - 7x + 3}{x - 3} \cdot \dfrac{x + 2}{x - 1}$

40. $\dfrac{3x^2 - 5x - 2}{x - 2} \cdot \dfrac{x - 3}{x + 1}$

41. $\dfrac{2k^2 - k - 1}{2k^2 + 5k + 3} \div \dfrac{4k^2 - 1}{2k^2 + k - 3}$

42. $\dfrac{2m^2 - 5m - 12}{m^2 + m - 20} \div \dfrac{4m^2 - 9}{m^2 + 4m - 5}$

43. $\dfrac{2k^2 + 3k - 2}{6k^2 - 7k + 2} \cdot \dfrac{4k^2 - 5k + 1}{k^2 + k - 2}$

44. $\dfrac{2m^2 - 5m - 12}{m^2 - 10m + 24} \div \dfrac{4m^2 - 9}{m^2 - 9m + 18}$

45. $\dfrac{m^2 + 2mp - 3p^2}{m^2 - 3mp + 2p^2} \div \dfrac{m^2 + 4mp + 3p^2}{m^2 + 2mp - 8p^2}$

46. $\dfrac{r^2 + rs - 12s^2}{r^2 - rs - 20s^2} \div \dfrac{r^2 - 2rs - 3s^2}{r^2 + rs - 30s^2}$

47. $\dfrac{m^2 + 3m + 2}{m^2 + 5m + 4} \cdot \dfrac{m^2 + 10m + 24}{m^2 + 5m + 6}$

48. $\dfrac{z^2 - z - 6}{z^2 - 2z - 8} \cdot \dfrac{z^2 + 7z + 12}{z^2 - 9}$

49. $\dfrac{y^2 + y - 2}{y^2 + 3y - 4} \div \dfrac{y + 2}{y + 3}$

50. $\dfrac{r^2 + r - 6}{r^2 + 4r - 12} \div \dfrac{r + 3}{r - 1}$

51. $\dfrac{2m^2 + 7m + 3}{m^2 - 9} \cdot \dfrac{m^2 - 3m}{2m^2 + 11m + 5}$

52. $\dfrac{m^2 + 2mp - 3p^2}{m^2 - 3mp + 2p^2} \div \dfrac{m^2 + 4mp + 3p^2}{m^2 + 2mp - 8p^2}$

53. $\dfrac{r^2 + rs - 12s^2}{r^2 - rs - 20s^2} \div \dfrac{r^2 - 2rs - 3s^2}{r^2 + rs - 30s^2}$

54. $\dfrac{(x + 1)^3(x + 4)}{x^2 + 5x + 4} \div \dfrac{x^2 + 2x + 1}{x^2 + 3x + 2}$

55. $\dfrac{(q - 3)^4(q + 2)}{q^2 + 3q + 2} \div \dfrac{q^2 - 6q + 9}{q^2 + 4q + 4}$

56. $\dfrac{(x + 4)^3(x - 3)}{x^2 - 9} \div \dfrac{x^2 + 8x + 16}{x^2 + 6x + 9}$

*These exercises involve grouping symbols (**Section 1.2**), factoring by grouping (**Section 6.1**), and factoring sums and differences of cubes (**Section 6.4**). Multiply or divide as indicated. Write each answer in lowest terms.*

57. $\dfrac{x + 5}{x + 10} \div \left(\dfrac{x^2 + 10x + 25}{x^2 + 10x} \cdot \dfrac{10x}{x^2 + 15x + 50} \right)$

58. $\dfrac{m - 8}{m - 4} \div \left(\dfrac{m^2 - 12m + 32}{8m} \cdot \dfrac{m^2 - 8m}{m^2 - 8m + 16} \right)$

59. $\dfrac{3a - 3b - a^2 + b^2}{4a^2 - 4ab + b^2} \cdot \dfrac{4a^2 - b^2}{2a^2 - ab - b^2}$

60. $\dfrac{4r^2 - t^2 + 10r - 5t}{2r^2 + rt + 5r} \cdot \dfrac{4r^3 + 4r^2t + rt^2}{2r + t}$

61. $\dfrac{-x^3 - y^3}{x^2 - 2xy + y^2} \div \dfrac{3y^2 - 3xy}{x^2 - y^2}$

62. $\dfrac{b^3 - 8a^3}{4a^3 + 4a^2b + ab^2} \div \dfrac{4a^2 + 2ab + b^2}{-a^3 - ab^3}$

63. If the rational expression $\frac{5x^2y^3}{2pq}$ represents the area of a rectangle and $\frac{2xy}{p}$ represents the length, what rational expression represents the width?

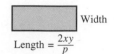
Width

Length $= \dfrac{2xy}{p}$

The area is $\dfrac{5x^2y^3}{2pq}$.

64. *Concept Check*   If you are given the problem

$$\frac{4y + 12}{2y - 10} \div \frac{?}{y^2 - y - 20} = \frac{2(y + 4)}{y - 3},$$

what must be the polynomial that is represented by the question mark?

> **PREVIEW EXERCISES**
>
> *Write the prime factored form of each number. See **Section 1.1.***
>
> **65.** 18   **66.** 48   **67.** 108   **68.** 60
>
> *Find the greatest common factor of each group of terms. See **Section 6.1.***
>
> **69.** $24m, 18m^2, 6$ **70.** $14x^2, 28x, 7$ **71.** $84q^3, 90q^6$ **72.** $54b^3, 36b^4$

7.3 Least Common Denominators

OBJECTIVES

1. Find the least common denominator for a group of fractions.

2. Rewrite rational expressions with given denominators.

OBJECTIVE 1 **Find the least common denominator for a group of fractions.**   Adding or subtracting rational expressions (to be discussed in **Section 7.4**) often requires a **least common denominator (LCD),** the simplest expression that is divisible by all of the denominators in all of the expressions. For example, the least common denominator for the fractions $\frac{2}{9}$ and $\frac{5}{12}$ is 36, because 36 is the smallest positive number divisible by both 9 and 12.

We can often find least common denominators by inspection. For example, the LCD for $\frac{1}{6}$ and $\frac{2}{3m}$ is $6m$. In other cases, we find the LCD by a procedure similar to that used in **Section 6.1** for finding the greatest common factor.

> **Finding the Least Common Denominator (LCD)**
>
> *Step 1*   **Factor** each denominator into prime factors.
>
> *Step 2*   **List each different denominator factor** the *greatest* number of times it appears in any of the denominators.
>
> *Step 3*   **Multiply** the denominator factors from Step 2 to get the LCD.

When each denominator is factored into prime factors, every prime factor must be a factor of the least common denominator.

In Example 1, we find the LCD for both numerical and algebraic denominators.

EXAMPLE 1 Finding the LCD

Find the LCD for each pair of fractions.

(a) $\dfrac{1}{24}, \dfrac{7}{15}$ **(b)** $\dfrac{1}{8x}, \dfrac{3}{10x}$

Step 1 Write each denominator in factored form with numerical coefficients in prime factored form.

$$24 = 2 \cdot 2 \cdot 2 \cdot 3 = 2^3 \cdot 3 \qquad\qquad 8x = 2 \cdot 2 \cdot 2 \cdot x = 2^3 \cdot x$$
$$15 = 3 \cdot 5 \qquad\qquad\qquad\qquad\qquad 10x = 2 \cdot 5 \cdot x$$

Step 2 Find the LCD by taking each different factor the *greatest* number of times it appears as a factor in any of the denominators.

The factor 2 appears three times in one product and not at all in the other, so the greatest number of times 2 appears is three. The greatest number of times both 3 and 5 appear is one.

Here, 2 appears three times in one product and once in the other, so the greatest number of times 2 appears is three. The greatest number of times 5 appears is one, and the greatest number of times x appears in either product is one.

Step 3   LCD $= 2 \cdot 2 \cdot 2 \cdot 3 \cdot 5$ LCD $= 2 \cdot 2 \cdot 2 \cdot 5 \cdot x$
$$= 2^3 \cdot 3 \cdot 5 \qquad\qquad\qquad\qquad = 2^3 \cdot 5 \cdot x$$
$$= 120 \qquad\qquad\qquad\qquad\qquad\;\, = 40x$$

✔ **Now Try Exercises 5 and 11.**

EXAMPLE 2 Finding the LCD

Find the LCD for $\dfrac{5}{6r^2}$ and $\dfrac{3}{4r^3}$.

Step 1 Factor each denominator.
$$6r^2 = 2 \cdot 3 \cdot r^2$$
$$4r^3 = 2 \cdot 2 \cdot r^3 = 2^2 \cdot r^3$$

Step 2 The greatest number of times 2 appears is two, the greatest number of times 3 appears is one, and the greatest number of times r appears is three; therefore,

Step 3   LCD $= 2^2 \cdot 3 \cdot r^3 = 12r^3$.

✔ **Now Try Exercise 13.**

▶ **CAUTION** When finding the LCD, use each factor the *greatest* number of times it appears in any *single* denominator, not the *total* number of times it appears. For instance, the greatest number of times r appears as a factor in one denominator in Example 2 is 3, *not* 5.

EXAMPLE 3 Finding the LCD

Find the LCD for the fractions in each list.

(a) $\dfrac{6}{5m}, \dfrac{4}{m^2 - 3m}$

$$\left. \begin{array}{l} 5m = 5 \cdot m \\ m^2 - 3m = m(m - 3) \end{array} \right\} \text{ Factor each denominator.}$$

Use each different factor the greatest number of times it appears. The greatest number of times 5 appears is one, the greatest number of times m appears is one, and the greatest number of times $m - 3$ appears is one; therefore,

$$\text{LCD} = 5 \cdot m \cdot (m - 3) = 5m(m - 3).$$

> Be sure to include m as a factor in the LCD.

Because m is not a *factor* of $m - 3$, both m and $m - 3$ must appear in the LCD.

(b) $\dfrac{1}{r^2 - 4r - 5}, \dfrac{3}{r^2 - r - 20}, \dfrac{1}{r^2 - 10r + 25}$

$$\left. \begin{array}{l} r^2 - 4r - 5 = (r - 5)(r + 1) \\ r^2 - r - 20 = (r - 5)(r + 4) \\ r^2 - 10r + 25 = (r - 5)^2 \end{array} \right\} \text{ Factor each denominator.}$$

Use each different factor the greatest number of times it appears as a factor.

$$\text{LCD} = (r - 5)^2(r + 1)(r + 4)$$

(c) $\dfrac{1}{q - 5}, \dfrac{3}{5 - q}$

The expressions $q - 5$ and $5 - q$ are opposites of each other because

$$-(q - 5) = -q + 5 = 5 - q.$$

Therefore, either $q - 5$ or $5 - q$ can be used as the LCD.

✔ **Now Try Exercises 19, 31, and 41.**

OBJECTIVE 2 Rewrite rational expressions with given denominators. Once the LCD has been found, the next step in preparing to add or subtract two rational expressions is to use the fundamental property to write equivalent rational expressions.

Writing a Rational Expression with a Specified Denominator

Step 1 **Factor** both denominators.

Step 2 **Decide what factor(s) the denominator must be multiplied by** in order to equal the specified denominator.

Step 3 **Multiply** the rational expression by that factor divided by itself. (That is, multiply by 1.)

EXAMPLE 4 Writing Rational Expressions with Given Denominators

Rewrite each rational expression with the indicated denominator.

(a) $\dfrac{3}{8} = \dfrac{?}{40}$

(b) $\dfrac{9k}{25} = \dfrac{?}{50k}$

Step 1 For each example, first factor the denominator on the right. Then compare the denominator on the left with the one on the right to decide what factors are missing. (It may sometimes be necessary to factor both denominators.)

$$\dfrac{3}{8} = \dfrac{?}{5 \cdot 8} \qquad\qquad \dfrac{9k}{25} = \dfrac{?}{25 \cdot 2k}$$

Step 2 A factor of 5 is missing.

Factors of 2 and k are missing.

Step 3 Multiply $\frac{3}{8}$ by $\frac{5}{5}$.

Multiply $\frac{9k}{25}$ by $\frac{2k}{2k}$.

$$\dfrac{3}{8} = \dfrac{3}{8} \cdot \dfrac{5}{5} = \dfrac{15}{40} \qquad\qquad \dfrac{9k}{25} = \dfrac{9k}{25} \cdot \dfrac{2k}{2k} = \dfrac{18k^2}{50k}$$

$$\dfrac{5}{5} = 1 \xrightarrow{\qquad} \qquad\qquad \dfrac{2k}{2k} = 1 \xrightarrow{\qquad}$$

✔ **Now Try Exercises 51 and 53.**

EXAMPLE 5 Writing Rational Expressions with Given Denominators

Rewrite each rational expression with the indicated denominator.

(a) $\dfrac{8}{3x + 1} = \dfrac{?}{12x + 4}$

$$\dfrac{8}{3x + 1} = \dfrac{?}{4(3x + 1)} \xleftarrow{\qquad} \text{Factor the denominator on the right.}$$

The missing factor is 4, so multiply the fraction on the left by $\frac{4}{4}$.

$$\dfrac{8}{3x + 1} \cdot \dfrac{4}{4} = \dfrac{32}{12x + 4} \qquad \text{Fundamental property}$$

(b) $\dfrac{12p}{p^2 + 8p} = \dfrac{?}{p^3 + 4p^2 - 32p}$

Factor $p^2 + 8p$ as $p(p + 8)$. Compare $p(p + 8)$ with the denominator on the right, which factors as $p(p + 8)(p - 4)$. The factor $p - 4$ is missing, so multiply $\frac{12p}{p(p + 8)}$ by $\frac{p - 4}{p - 4}$.

$$\dfrac{12p}{p^2 + 8p} = \dfrac{12p}{p(p + 8)} \cdot \dfrac{p - 4}{p - 4} \qquad \text{Fundamental property}$$

$$= \dfrac{12p(p - 4)}{p(p + 8)(p - 4)} \qquad \begin{array}{l}\text{Multiply numerators.}\\ \text{Multiply denominators.}\end{array}$$

$$= \dfrac{12p^2 - 48p}{p^3 + 4p^2 - 32p} \qquad \text{Multiply the factors.}$$

✔ **Now Try Exercises 57 and 61.**

▶ **NOTE** In **Section 7.4,** we add and subtract rational expressions, a process that sometimes requires the steps illustrated in Examples 4 and 5. While it is beneficial to leave the denominator in factored form, we multiplied the factors in the denominator in Example 5 to give the answer in the same form as the original problem.

7.3 EXERCISES

Concept Check *Choose the correct response in Exercises 1–4.*

1. Suppose that the greatest common factor of a and b is 1. Then the least common denominator for $\frac{1}{a}$ and $\frac{1}{b}$ is

    A. a **B.** b **C.** ab **D.** 1.

2. If a is a factor of b, then the least common denominator for $\frac{1}{a}$ and $\frac{1}{b}$ is

    A. a **B.** b **C.** ab **D.** 1.

3. The least common denominator for $\frac{11}{20}$ and $\frac{1}{2}$ is

    A. 40   **B.** 2   **C.** 20   **D.** none of these.

4. Suppose that we wish to write the fraction $\frac{1}{(x-4)^2(y-3)}$ with denominator $(x-4)^3(y-3)^2$. Then we must multiply both the numerator and the denominator by

    A. $(x-4)(y-3)$ **B.** $(x-4)^2$ **C.** $x-4$ **D.** $(x-4)^2(y-3)$.

Find the LCD for the fractions in each list. See Examples 1–3.

◉ **5.** $\dfrac{7}{15}, \dfrac{21}{20}$

6. $\dfrac{9}{10}, \dfrac{12}{25}$

7. $\dfrac{17}{100}, \dfrac{23}{120}, \dfrac{43}{180}$

8. $\dfrac{17}{250}, \dfrac{-21}{300}, \dfrac{127}{360}$

9. $\dfrac{9}{x^2}, \dfrac{8}{x^5}$

10. $\dfrac{12}{m^7}, \dfrac{13}{m^8}$

11. $\dfrac{-2}{5p}, \dfrac{13}{6p}$

12. $\dfrac{14}{15k}, \dfrac{9}{4k}$

◉ **13.** $\dfrac{17}{15y^2}, \dfrac{55}{36y^4}$

14. $\dfrac{4}{25m^3}, \dfrac{-9}{10m^4}$

15. $\dfrac{5}{21r^3}, \dfrac{7}{12r^5}$

16. $\dfrac{9}{35t^2}, \dfrac{5}{49t^6}$

17. $\dfrac{13}{5a^2b^3}, \dfrac{29}{15a^5b}$

18. $\dfrac{-7}{3r^4s^5}, \dfrac{-22}{9r^6s^8}$

◉ **19.** $\dfrac{7}{6p}, \dfrac{15}{4p-8}$

20. $\dfrac{7}{8k}, \dfrac{-23}{12k-24}$

21. $\dfrac{9}{28m^2}, \dfrac{3}{12m-20}$

22. $\dfrac{14}{27a^3}, \dfrac{8}{9a-45}$

23. $\dfrac{7}{5b-10}, \dfrac{11}{6b-12}$

24. $\dfrac{3}{7x^2+21x}, \dfrac{1}{5x^2+15x}$

25. $\dfrac{37}{6r-12}, \dfrac{25}{9r-18}$

26. $\dfrac{-14}{5p-30}, \dfrac{5}{6p-36}$

27. $\dfrac{5}{12p+60}, \dfrac{17}{p^2+5p}, \dfrac{16}{p^2+10p+25}$

28. $\dfrac{13}{r^2+7r}, \dfrac{-3}{5r+35}, \dfrac{-7}{r^2+14r+49}$

29. $\dfrac{3}{8y+16}, \dfrac{22}{y^2+3y+2}$

30. $\dfrac{-2}{9m-18}, \dfrac{-9}{m^2-7m+10}$

31. $\dfrac{5}{c-d}, \dfrac{8}{d-c}$

32. $\dfrac{4}{y-x}, \dfrac{7}{x-y}$

33. $\dfrac{12}{m-3}, \dfrac{-4}{3-m}$

34. $\dfrac{-17}{8-a}, \dfrac{2}{a-8}$

35. $\dfrac{29}{p-q}, \dfrac{18}{q-p}$

36. $\dfrac{16}{z-x}, \dfrac{8}{x-z}$

37. $\dfrac{3}{k^2+5k}, \dfrac{2}{k^2+3k-10}$

38. $\dfrac{1}{z^2-4z}, \dfrac{4}{z^2-3z-4}$

39. $\dfrac{6}{a^2+6a}, \dfrac{-5}{a^2+3a-18}$

40. $\dfrac{8}{y^2-5y}, \dfrac{-2}{y^2-2y-15}$

41. $\dfrac{5}{p^2+8p+15}, \dfrac{3}{p^2-3p-18}, \dfrac{2}{p^2-p-30}$

42. $\dfrac{10}{y^2-10y+21}, \dfrac{2}{y^2-2y-3}, \dfrac{5}{y^2-6y-7}$

43. $\dfrac{-5}{k^2+2k-35}, \dfrac{-8}{k^2+3k-40}, \dfrac{9}{k^2-2k-15}$

44. $\dfrac{19}{z^2+4z-12}, \dfrac{16}{z^2+z-30}, \dfrac{6}{z^2+2z-24}$

RELATING CONCEPTS (EXERCISES 45–50)

FOR INDIVIDUAL OR GROUP WORK

Work Exercises 45–50 in order.

45. Suppose that you want to write $\frac{3}{4}$ as an equivalent fraction with denominator 28. By what number must you multiply both the numerator and the denominator?

46. If you write $\frac{3}{4}$ as an equivalent fraction with denominator 28, by what number are you actually multiplying the fraction?

47. What property of multiplication is being used when we write a common fraction as an equivalent one with a larger denominator? (See **Section 1.7.**)

48. Suppose that you want to write $\frac{2x+5}{x-4}$ as an equivalent fraction with denominator $7x-28$. By what number must you multiply both the numerator and the denominator?

49. If you write $\frac{2x+5}{x-4}$ as an equivalent fraction with denominator $7x-28$, by what number are you actually multiplying the fraction?

50. Repeat Exercise 47, changing "a common" to "an algebraic."

Rewrite each rational expression with the indicated denominator. See Examples 4 and 5.

51. $\dfrac{4}{11} = \dfrac{?}{55}$

52. $\dfrac{6}{7} = \dfrac{?}{42}$

53. $\dfrac{-5}{k} = \dfrac{?}{9k}$

54. $\dfrac{-3}{q} = \dfrac{?}{6q}$

55. $\dfrac{15m^2}{8k} = \dfrac{?}{32k^4}$

56. $\dfrac{5t^2}{3y} = \dfrac{?}{9y^2}$

57. $\dfrac{19z}{2z - 6} = \dfrac{?}{6z - 18}$

58. $\dfrac{2r}{5r - 5} = \dfrac{?}{15r - 15}$

59. $\dfrac{-2a}{9a - 18} = \dfrac{?}{18a - 36}$

60. $\dfrac{-5y}{6y + 18} = \dfrac{?}{24y + 72}$

61. $\dfrac{6}{k^2 - 4k} = \dfrac{?}{k(k - 4)(k + 1)}$

62. $\dfrac{15}{m^2 - 9m} = \dfrac{?}{m(m - 9)(m + 8)}$

63. $\dfrac{36r}{r^2 - r - 6} = \dfrac{?}{(r - 3)(r + 2)(r + 1)}$

64. $\dfrac{4m}{m^2 - 8m + 15} = \dfrac{?}{(m - 5)(m - 3)(m + 2)}$

65. $\dfrac{a + 2b}{2a^2 + ab - b^2} = \dfrac{?}{2a^3b + a^2b^2 - ab^3}$

66. $\dfrac{m - 4}{6m^2 + 7m - 3} = \dfrac{?}{12m^3 + 14m^2 - 6m}$

67. $\dfrac{4r - t}{r^2 + rt + t^2} = \dfrac{?}{t^3 - r^3}$

68. $\dfrac{3x - 1}{x^2 + 2x + 4} = \dfrac{?}{x^3 - 8}$

69. $\dfrac{2(z - y)}{y^2 + yz + z^2} = \dfrac{?}{y^4 - z^3y}$

70. $\dfrac{2p + 3q}{p^2 + 2pq + q^2} = \dfrac{?}{(p + q)(p^3 + q^3)}$

PREVIEW EXERCISES

*Add or subtract as indicated. Write each answer in lowest terms. See **Section 1.1**.*

71. $\dfrac{3}{4} + \dfrac{7}{4}$

72. $\dfrac{2}{5} + \dfrac{9}{5}$

73. $\dfrac{1}{2} + \dfrac{7}{8}$

74. $\dfrac{2}{3} + \dfrac{8}{27}$

75. $\dfrac{7}{5} - \dfrac{3}{4}$

76. $\dfrac{11}{6} - \dfrac{2}{5}$

77. $\dfrac{4}{3} - \dfrac{1}{4}$

78. $\dfrac{7}{8} - \dfrac{10}{3}$

7.4 Adding and Subtracting Rational Expressions

OBJECTIVES

1 Add rational expressions having the same denominator.

2 Add rational expressions having different denominators.

3 Subtract rational expressions.

To add and subtract rational expressions, we find least common denominators and write equivalent fractions with the LCD.

OBJECTIVE 1 Add rational expressions having the same denominator. We find the sum of two such rational expressions with the same procedure that we used in **Section 1.1** for adding two fractions having the same denominator.

> **Adding Rational Expressions Having the Same Denominator**
>
> If $\frac{P}{Q}$ and $\frac{R}{Q}$ ($Q \neq 0$) are rational expressions, then
>
> $$\frac{P}{Q} + \frac{R}{Q} = \frac{P+R}{Q}.$$
>
> That is, to add rational expressions with the same denominator, add the numerators and keep the same denominator.

EXAMPLE 1 Adding Rational Expressions with the Same Denominator

Add. Write each answer in lowest terms.

(a) $\dfrac{4}{9} + \dfrac{2}{9}$ **(b)** $\dfrac{3x}{x+1} + \dfrac{3}{x+1}$

The denominators are the same, so the sum is found by adding the two numerators and keeping the same (common) denominator.

$$\frac{4}{9} + \frac{2}{9}$$
$$= \frac{4+2}{9}$$
$$= \frac{6}{9}$$
$$= \frac{2 \cdot 3}{3 \cdot 3}$$
$$= \frac{2}{3}$$

$$\frac{3x}{x+1} + \frac{3}{x+1}$$
$$= \frac{3x+3}{x+1}$$
$$= \frac{3(x+1)}{x+1}$$
$$= 3$$

✔ **Now Try Exercises 9 and 19.**

OBJECTIVE 2 Add rational expressions having different denominators. We use the following steps, which are the same as those we used in **Section 1.1** to add fractions having different denominators.

> **Adding Rational Expressions Having Different Denominators**
>
> *Step 1* **Find the least common denominator (LCD).**
>
> *Step 2* **Rewrite each rational expression** as an equivalent rational expression with the LCD as the denominator.
>
> *Step 3* **Add** the numerators to get the numerator of the sum. The LCD is the denominator of the sum.
>
> *Step 4* **Write in lowest terms** using the fundamental property.

EXAMPLE 2 Adding Rational Expressions with Different Denominators

Add. Write each answer in lowest terms.

(a) $\dfrac{1}{12} + \dfrac{7}{15}$　　　　　　　　　**(b)** $\dfrac{2}{3y} + \dfrac{1}{4y}$

Step 1 First find the LCD, using the methods of the previous section.

$$12 = 2 \cdot 2 \cdot 3 = 2^2 \cdot 3 \qquad\qquad 3y = 3 \cdot y$$
$$15 = 3 \cdot 5 \qquad\qquad 4y = 2 \cdot 2 \cdot y = 2^2 \cdot y$$
$$\text{LCD} = 2^2 \cdot 3 \cdot 5 = 60 \qquad\qquad \text{LCD} = 2^2 \cdot 3 \cdot y = 12y$$

Step 2 Now rewrite each rational expression as a fraction with the LCD (60 and 12*y*, respectively) as the denominator.

$$\frac{1}{12} + \frac{7}{15} = \frac{1(5)}{12(5)} + \frac{7(4)}{15(4)} \qquad\qquad \frac{2}{3y} + \frac{1}{4y} = \frac{2(4)}{3y(4)} + \frac{1(3)}{4y(3)}$$
$$= \frac{5}{60} + \frac{28}{60} \qquad\qquad\qquad = \frac{8}{12y} + \frac{3}{12y}$$

Step 3 Since the fractions now have common denominators, add the numerators.

Step 4 Write in lowest terms if necessary.

$$\frac{5}{60} + \frac{28}{60} = \frac{5 + 28}{60} \qquad\qquad \frac{8}{12y} + \frac{3}{12y} = \frac{8 + 3}{12y}$$
$$= \frac{33}{60} = \frac{11}{20} \qquad\qquad\qquad = \frac{11}{12y}$$

✔ **Now Try Exercises 25 and 31.**

EXAMPLE 3 Adding Rational Expressions

Add. Write the answer in lowest terms.

$$\frac{2x}{x^2 - 1} + \frac{-1}{x + 1}$$

Step 1 Since the denominators are different, find the LCD.

$$\left. \begin{array}{l} x^2 - 1 = (x + 1)(x - 1) \\ x + 1 \text{ is prime.} \end{array} \right\} \quad \text{The LCD is } (x + 1)(x - 1).$$

Step 2 Rewrite each rational expression with the LCD as the denominator.

$$\frac{2x}{x^2 - 1} + \frac{-1}{x + 1}$$ LCD $= (x + 1)(x - 1)$

$$= \frac{2x}{(x + 1)(x - 1)} + \frac{-1(x - 1)}{(x + 1)(x - 1)}$$ Multiply the second fraction by $\frac{x - 1}{x - 1}$.

$$= \frac{2x}{(x + 1)(x - 1)} + \frac{-x + 1}{(x + 1)(x - 1)}$$ Distributive property

Step 3 $$= \frac{2x - x + 1}{(x + 1)(x - 1)}$$ Add numerators; keep the same denominator.

$$= \frac{x + 1}{(x + 1)(x - 1)}$$ Combine like terms.

Step 4 $$= \frac{1(x + 1)}{(x + 1)(x - 1)}$$ Identity property of multiplication

Remember to write 1 in the numerator.

$$= \frac{1}{x - 1}$$ Divide out common factors.

✔ **Now Try Exercise 41.**

EXAMPLE 4 Adding Rational Expressions

Add. Write the answer in lowest terms.

$$\frac{2x}{x^2 + 5x + 6} + \frac{x + 1}{x^2 + 2x - 3}$$

$$= \frac{2x}{(x + 2)(x + 3)} + \frac{x + 1}{(x + 3)(x - 1)}$$ Factor the denominators.

$$= \frac{2x(x - 1)}{(x + 2)(x + 3)(x - 1)} + \frac{(x + 1)(x + 2)}{(x + 2)(x + 3)(x - 1)}$$ The LCD is $(x + 2)(x + 3)(x - 1)$.

$$= \frac{2x(x - 1) + (x + 1)(x + 2)}{(x + 2)(x + 3)(x - 1)}$$ Add numerators; keep the same denominator.

$$= \frac{2x^2 - 2x + x^2 + 3x + 2}{(x + 2)(x + 3)(x - 1)}$$ Multiply.

$$= \frac{3x^2 + x + 2}{(x + 2)(x + 3)(x - 1)}$$ Combine like terms.

The numerator cannot be factored here, so the expression is in lowest terms.

✔ **Now Try Exercise 43.**

▶ **NOTE** If the final expression in Example 4 could be written in lower terms, the numerator would have a factor of $x + 2$, $x + 3$, or $x - 1$. Therefore, it is only necessary to check for possible factored forms of the numerator that would contain one of these binomials.

EXAMPLE 5 Adding Rational Expressions with Denominators That Are Opposites

Add. Write the answer in lowest terms.

$$\frac{y}{y-2} + \frac{8}{2-y}$$ The denominators are opposites.

$$= \frac{y}{y-2} + \frac{8(-1)}{(2-y)(-1)}$$ Multiply $\frac{8}{2-y}$ by $\frac{-1}{-1}$ to get a common denominator.

$$= \frac{y}{y-2} + \frac{-8}{-2+y}$$ Distributive property

$$= \frac{y}{y-2} + \frac{-8}{y-2}$$ Rewrite $-2 + y$ as $y - 2$.

$$= \frac{y-8}{y-2}$$ Add numerators; keep the same denominator.

If we had chosen $2 - y$ as the common denominator, the final answer would be $\frac{8-y}{2-y}$, which is equivalent to $\frac{y-8}{y-2}$.

✔ **Now Try Exercise 51.**

OBJECTIVE 3 Subtract rational expressions. To subtract rational expressions having the same denominator, use the following rule.

Subtracting Rational Expressions Having the Same Denominator

If $\frac{P}{Q}$ and $\frac{R}{Q}$ ($Q \neq 0$) are rational expressions, then

$$\frac{P}{Q} - \frac{R}{Q} = \frac{P-R}{Q}.$$

That is, to subtract rational expressions with the same denominator, subtract the numerators and keep the same denominator.

EXAMPLE 6 Subtracting Rational Expressions with the Same Denominator

Subtract. Write the answer in lowest terms.

$$\frac{2m}{m-1} - \frac{m+3}{m-1}$$ *Use parentheses to avoid errors.*

$$= \frac{2m - (m+3)}{m-1}$$ Subtract numerators; keep the same denominator.

Be careful with signs. $$= \frac{2m - m - 3}{m-1}$$ Distributive property

$$= \frac{m-3}{m-1}$$ Combine like terms.

✔ **Now Try Exercise 15.**

▶ **CAUTION** Sign errors often occur in subtraction problems like the one in Example 6. The numerator of the fraction being subtracted must be treated as a single quantity. *Be sure to use parentheses after the subtraction sign.*

We subtract rational expressions having different denominators using a procedure similar to the one used to add rational expressions having different denominators.

EXAMPLE 7 Subtracting Rational Expressions with Different Denominators

Subtract. Write the answer in lowest terms.

$$\frac{9}{x-2} - \frac{3}{x}$$ The LCD is $x(x-2)$.

$$= \frac{9x}{x(x-2)} - \frac{3(x-2)}{x(x-2)}$$ Write each expression with the LCD.

(Be careful with signs.) $$= \frac{9x - 3(x-2)}{x(x-2)}$$ Subtract numerators; keep the same denominator.

$$= \frac{9x - 3x + 6}{x(x-2)}$$ Distributive property

$$= \frac{6x + 6}{x(x-2)}$$ Combine like terms.

$$= \frac{6(x+1)}{x(x-2)}$$ Factor the numerator.

✔ **Now Try Exercise 39.**

▶ **NOTE** We factored the final numerator in Example 7 to get $\frac{6(x+1)}{x(x-2)}$; however, the fundamental property does not apply, since there are no common factors that allow us to write the answer in lower terms.

EXAMPLE 8 Subtracting Rational Expressions with Denominators That Are Opposites

Subtract. Write the answer in lowest terms.

$$\frac{3x}{x-5} - \frac{2x-25}{5-x}$$ The denominators are opposites.

$$= \frac{3x}{x-5} - \frac{(2x-25)(-1)}{(5-x)(-1)}$$ Multiply $\frac{2x-25}{5-x}$ by $\frac{-1}{-1}$ to get a common denominator.

(Subtract the *entire* numerator. Use parentheses to show this.) $$= \frac{3x}{x-5} - \frac{-2x+25}{x-5}$$ $(5-x)(-1) = -5 + x = x - 5$

(Be careful with signs.) $$= \frac{3x - (-2x+25)}{x-5}$$ Subtract numerators.

$$= \frac{3x + 2x - 25}{x-5}$$ Distributive property

$$= \frac{5x - 25}{x - 5} \qquad \text{Combine like terms.}$$

$$= \frac{5(x - 5)}{x - 5} \qquad \text{Factor.}$$

$$= 5 \qquad \text{Divide out the common factor.}$$

✔ **Now Try Exercise 53.**

EXAMPLE 9 Subtracting Rational Expressions

Subtract. Write the answer in lowest terms.

$$\frac{6x}{x^2 - 2x + 1} - \frac{1}{x^2 - 1}$$

$$= \frac{6x}{(x - 1)(x - 1)} - \frac{1}{(x - 1)(x + 1)} \qquad \begin{array}{l}\text{Factor the denominators.}\\ \text{LCD} = (x - 1)(x - 1)(x + 1)\end{array}$$

$$= \frac{6x(x + 1)}{(x - 1)(x - 1)(x + 1)} - \frac{1(x - 1)}{(x - 1)(x - 1)(x + 1)} \qquad \begin{array}{l}\text{Fundamental}\\ \text{property}\end{array}$$

Use $(x - 1)$ twice in the LCD.

$$= \frac{6x(x + 1) - 1(x - 1)}{(x - 1)(x - 1)(x + 1)} \qquad \text{Subtract numerators.}$$

$$= \frac{6x^2 + 6x - x + 1}{(x - 1)(x - 1)(x + 1)} \qquad \text{Distributive property}$$

$$= \frac{6x^2 + 5x + 1}{(x - 1)(x - 1)(x + 1)} \qquad \text{Combine like terms.}$$

$$= \frac{(2x + 1)(3x + 1)}{(x - 1)^2(x + 1)} \qquad \text{Factor the numerator.}$$

✔ **Now Try Exercise 63.**

7.4 EXERCISES

● *Complete solution available on Video Lectures on CD/DVD*

Now Try Exercise

Concept Check *Match each expression in Column I with the correct sum or difference in Column II.*

I

1. $\dfrac{x}{x + 6} + \dfrac{6}{x + 6}$

2. $\dfrac{2x}{x - 6} - \dfrac{12}{x - 6}$

3. $\dfrac{6}{x - 6} - \dfrac{x}{x - 6}$

4. $\dfrac{6}{x + 6} - \dfrac{x}{x + 6}$

5. $\dfrac{x}{x + 6} - \dfrac{6}{x + 6}$

6. $\dfrac{1}{x} + \dfrac{1}{6}$

7. $\dfrac{1}{6} - \dfrac{1}{x}$

8. $\dfrac{1}{6x} - \dfrac{1}{6x}$

II

A. 2

B. $\dfrac{x - 6}{x + 6}$

C. -1

D. $\dfrac{6 + x}{6x}$

E. 1

F. 0

G. $\dfrac{x - 6}{6x}$

H. $\dfrac{6 - x}{x + 6}$

Note: When adding and subtracting rational expressions, several different equivalent forms of the answer often exist. If your answer does not look exactly like the one given in the back of the book, check to see whether you have written an equivalent form.

Add or subtract. Write each answer in lowest terms. See Examples 1 and 6.

9. $\dfrac{4}{m} + \dfrac{7}{m}$

10. $\dfrac{5}{p} + \dfrac{11}{p}$

11. $\dfrac{5}{y+4} - \dfrac{1}{y+4}$

12. $\dfrac{4}{y+3} - \dfrac{1}{y+3}$

13. $\dfrac{x}{x+y} + \dfrac{y}{x+y}$

14. $\dfrac{a}{a+b} + \dfrac{b}{a+b}$

15. $\dfrac{5m}{m+1} - \dfrac{1+4m}{m+1}$

16. $\dfrac{4x}{x+2} - \dfrac{2+3x}{x+2}$

17. $\dfrac{a+b}{2} - \dfrac{a-b}{2}$

18. $\dfrac{x-y}{2} - \dfrac{x+y}{2}$

19. $\dfrac{x^2}{x+5} + \dfrac{5x}{x+5}$

20. $\dfrac{t^2}{t-3} + \dfrac{-3t}{t-3}$

21. $\dfrac{y^2-3y}{y+3} + \dfrac{-18}{y+3}$

22. $\dfrac{r^2-8r}{r-5} + \dfrac{15}{r-5}$

23. Explain with an example how to add or subtract rational expressions with the same denominator.

24. Explain with an example how to add or subtract rational expressions with different denominators.

Add or subtract. Write each answer in lowest terms. See Examples 2, 3, 4, and 7.

25. $\dfrac{z}{5} + \dfrac{1}{3}$

26. $\dfrac{p}{8} + \dfrac{3}{5}$

27. $\dfrac{5}{7} - \dfrac{r}{2}$

28. $\dfrac{10}{9} - \dfrac{z}{3}$

29. $-\dfrac{3}{4} - \dfrac{1}{2x}$

30. $-\dfrac{5}{8} - \dfrac{3}{2a}$

31. $\dfrac{6}{5x} + \dfrac{9}{2x}$

32. $\dfrac{3}{2x} + \dfrac{4}{7x}$

33. $\dfrac{x+1}{6} + \dfrac{3x+3}{9}$

34. $\dfrac{2x-6}{4} + \dfrac{x+5}{6}$

35. $\dfrac{x+3}{3x} + \dfrac{2x+2}{4x}$

36. $\dfrac{x+2}{5x} + \dfrac{6x+3}{3x}$

37. $\dfrac{7}{3p^2} - \dfrac{2}{p}$

38. $\dfrac{12}{5m^2} - \dfrac{5}{m}$

39. $\dfrac{1}{k+4} - \dfrac{2}{k}$

40. $\dfrac{3}{m+1} - \dfrac{4}{m}$

41. $\dfrac{x}{x-2} + \dfrac{-8}{x^2-4}$

42. $\dfrac{2x}{x-1} + \dfrac{-4}{x^2-1}$

43. $\dfrac{4m}{m^2+3m+2} + \dfrac{2m-1}{m^2+6m+5}$

44. $\dfrac{a}{a^2+3a-4} + \dfrac{4a}{a^2+7a+12}$

45. $\dfrac{4y}{y^2-1} - \dfrac{5}{y^2+2y+1}$

46. $\dfrac{2x}{x^2-16} - \dfrac{3}{x^2+8x+16}$

47. $\dfrac{t}{t+2} + \dfrac{5-t}{t} - \dfrac{4}{t^2+2t}$

48. $\dfrac{2p}{p-3} + \dfrac{2+p}{p} - \dfrac{-6}{p^2-3p}$

49. *Concept Check* What are the two possible LCDs that could be used for the sum $\dfrac{10}{m-2} + \dfrac{5}{2-m}$?

50. *Concept Check* If one form of the correct answer to a sum or difference of rational expressions is $\dfrac{4}{k-3}$, what would an alternative form of the answer be if the denominator is $3-k$?

Add or subtract. Write each answer in lowest terms. See Examples 5 and 8.

51. $\dfrac{4}{x-5} + \dfrac{6}{5-x}$

52. $\dfrac{10}{m-2} + \dfrac{5}{2-m}$

53. $\dfrac{-1}{1-y} - \dfrac{4y-3}{y-1}$

54. $\dfrac{-4}{p-3} - \dfrac{p+1}{3-p}$

55. $\dfrac{2}{x-y^2} + \dfrac{7}{y^2-x}$

56. $\dfrac{-8}{p-q^2} + \dfrac{3}{q^2-p}$

57. $\dfrac{x}{5x-3y} - \dfrac{y}{3y-5x}$

58. $\dfrac{t}{8t-9s} - \dfrac{s}{9s-8t}$

59. $\dfrac{3}{4p-5} + \dfrac{9}{5-4p}$

60. $\dfrac{8}{3-7y} - \dfrac{2}{7y-3}$

*In these subtraction problems, the rational expression that follows the subtraction sign has a numerator with more than one term. **Be careful with signs** and find each difference. See Example 9.*

61. $\dfrac{2m}{m-n} - \dfrac{5m+n}{2m-2n}$

62. $\dfrac{5p}{p-q} - \dfrac{3p+1}{4p-4q}$

63. $\dfrac{5}{x^2-9} - \dfrac{x+2}{x^2+4x+3}$

64. $\dfrac{1}{a^2-1} - \dfrac{a-1}{a^2+3a-4}$

65. $\dfrac{2q+1}{3q^2+10q-8} - \dfrac{3q+5}{2q^2+5q-12}$

66. $\dfrac{4y-1}{2y^2+5y-3} - \dfrac{y+3}{6y^2+y-2}$

Perform each indicated operation. See Examples 1–9.

67. $\dfrac{4}{r^2-r} + \dfrac{6}{r^2+2r} - \dfrac{1}{r^2+r-2}$

68. $\dfrac{6}{k^2+3k} - \dfrac{1}{k^2-k} + \dfrac{2}{k^2+2k-3}$

69. $\dfrac{x+3y}{x^2+2xy+y^2} + \dfrac{x-y}{x^2+4xy+3y^2}$

70. $\dfrac{m}{m^2-1} + \dfrac{m-1}{m^2+2m+1}$

71. $\dfrac{r+y}{18r^2+9ry-2y^2} + \dfrac{3r-y}{36r^2-y^2}$

72. $\dfrac{2x-z}{2x^2+xz-10z^2} \cdot \dfrac{x+z}{x^2-4z^2}$

73. Refer to the rectangle in the figure.

  (a) Find an expression that represents its perimeter. Give the simplified form.

  (b) Find an expression that represents its area. Give the simplified form.

$$\dfrac{3k+1}{10}$$

$$\dfrac{5}{6k+2}$$

74. Refer to the triangle in the figure. Find an expression that represents its perimeter.

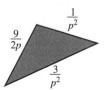

A ***concours d'elegance*** *is a competition in which a maximum of* 100 *points is awarded to a car based on its general attractiveness. The rational expression*

$$\frac{1010}{49(101 - x)} - \frac{10}{49}$$

approximates the cost, in thousands of dollars, of restoring a car so that it will win x points.
    Use this information to work Exercises 75 and 76.

75. Simplify the given expression by performing the indicated subtraction.

76. Use the simplified expression from Exercise 75 to determine how much it would cost to win 95 points.

PREVIEW EXERCISES

*Perform the indicated operations, using the order of operations as necessary. See **Section 1.1.***

77. $\dfrac{\dfrac{5}{6} + \dfrac{7}{6}}{\dfrac{2}{3} - \dfrac{1}{3}}$
   78. $\dfrac{\dfrac{3}{8} - \dfrac{5}{8}}{\dfrac{1}{4} + \dfrac{7}{4}}$
   79. $\dfrac{\dfrac{3}{2} - \dfrac{5}{4}}{\dfrac{7}{4} + \dfrac{1}{3}}$
   80. $\dfrac{\dfrac{5}{7} - \dfrac{3}{14}}{\dfrac{5}{3} + \dfrac{1}{2}}$

7.5 Complex Fractions

OBJECTIVES

1 Simplify a complex fraction by writing it as a division problem (Method 1).

2 Simplify a complex fraction by multiplying by the least common denominator (Method 2).

The quotient of two mixed numbers in arithmetic, such as $2\frac{1}{2} \div 3\frac{1}{4}$, can be written as a fraction:

$$2\frac{1}{2} \div 3\frac{1}{4} = \frac{2\frac{1}{2}}{3\frac{1}{4}} = \frac{2 + \frac{1}{2}}{3 + \frac{1}{4}}.$$

The last expression is the quotient of expressions that involve fractions. In algebra, some rational expressions also have fractions in the numerator, or denominator, or both.

Complex Fraction

A rational expression with one or more fractions in the numerator, or denominator, or both is called a **complex fraction.**

Examples of complex fractions include

$$\frac{2 + \frac{1}{2}}{3 + \frac{1}{4}}, \quad \frac{\frac{3x^2 - 5x}{6x^2}}{2x - \frac{1}{x}}, \quad \text{and} \quad \frac{3 + x}{5 - \frac{2}{x}}. \qquad \text{Complex fractions}$$

The parts of a complex fraction are named as follows.

$$\left.\begin{array}{c}\dfrac{2}{p} - \dfrac{1}{q} \\[2mm] \dfrac{3}{p} + \dfrac{5}{q}\end{array}\right\}$$ ←Numerator of complex fraction
←Main fraction bar
←Denominator of complex fraction

OBJECTIVE ❶ Simplify a complex fraction by writing it as a division problem (Method 1).
Since the main fraction bar represents division in a complex fraction, one method of simplifying a complex fraction involves division.

Method 1

To simplify a complex fraction,

Step 1 Write both the numerator and denominator as single fractions.

Step 2 Change the complex fraction to a division problem.

Step 3 Perform the indicated division.

Once again, in this section the first example shows complex fractions from both arithmetic and algebra.

EXAMPLE 1 Simplifying Complex Fractions (Method 1)

Simplify each complex fraction.

(a) $\dfrac{\dfrac{2}{3} + \dfrac{5}{9}}{\dfrac{1}{4} + \dfrac{1}{12}}$

(b) $\dfrac{6 + \dfrac{3}{x}}{\dfrac{x}{4} + \dfrac{1}{8}}$

Step 1 First, write each numerator as a single fraction.

$$\frac{2}{3} + \frac{5}{9} = \frac{2(3)}{3(3)} + \frac{5}{9}$$
$$= \frac{6}{9} + \frac{5}{9} = \frac{11}{9}$$

$$6 + \frac{3}{x} = \frac{6}{1} + \frac{3}{x}$$
$$= \frac{6x}{x} + \frac{3}{x} = \frac{6x + 3}{x}$$

Do the same thing with each denominator.

$$\frac{1}{4} + \frac{1}{12} = \frac{1(3)}{4(3)} + \frac{1}{12}$$
$$= \frac{3}{12} + \frac{1}{12} = \frac{4}{12}$$

$$\frac{x}{4} + \frac{1}{8} = \frac{x(2)}{4(2)} + \frac{1}{8}$$
$$= \frac{2x}{8} + \frac{1}{8} = \frac{2x + 1}{8}$$

Step 2 The original complex fraction can now be written as follows:

$$\frac{\dfrac{11}{9}}{\dfrac{4}{12}} = \frac{11}{9} \div \frac{4}{12}.$$

$$\frac{\dfrac{6x + 3}{x}}{\dfrac{2x + 1}{8}} = \frac{6x + 3}{x} \div \frac{2x + 1}{8}.$$

Step 3 Now use the rule for division and the fundamental property.

Multiply by the reciprocal.

$$\frac{11}{9} \div \frac{4}{12} = \frac{11}{9} \cdot \frac{12}{4}$$

$$= \frac{11 \cdot 3 \cdot 4}{3 \cdot 3 \cdot 4}$$

$$= \frac{11}{3}$$

Multiply by the reciprocal.

$$\frac{6x+3}{x} \div \frac{2x+1}{8} = \frac{6x+3}{x} \cdot \frac{8}{2x+1}$$

$$= \frac{3(2x+1)}{x} \cdot \frac{8}{2x+1}$$

$$= \frac{24}{x}$$

✔ **Now Try Exercises 1 and 15.**

EXAMPLE 2 Simplifying a Complex Fraction (Method 1)

Simplify the complex fraction.

$$\frac{\dfrac{xp}{q^3}}{\dfrac{p^2}{qx^2}} = \frac{xp}{q^3} \div \frac{p^2}{qx^2} = \frac{xp}{q^3} \cdot \frac{qx^2}{p^2} = \frac{x^3}{q^2 p} \qquad \text{Multiply by the reciprocal.}$$

✔ **Now Try Exercise 9.**

EXAMPLE 3 Simplifying a Complex Fraction (Method 1)

Simplify the complex fraction.

$$\frac{\dfrac{3}{x+2} - 4}{\dfrac{2}{x+2} + 1} = \frac{\dfrac{3}{x+2} - \dfrac{4(x+2)}{x+2}}{\dfrac{2}{x+2} + \dfrac{1(x+2)}{x+2}} \qquad \text{Write both second terms with a denominator of } x+2.$$

$$= \frac{\dfrac{3 - 4(x+2)}{x+2}}{\dfrac{2 + 1(x+2)}{x+2}} \qquad \begin{array}{l}\text{Subtract in the numerator.}\\[1.5em]\text{Add in the denominator.}\end{array}$$

Be careful with signs.

$$= \frac{\dfrac{3 - 4x - 8}{x+2}}{\dfrac{2 + x + 2}{x+2}} \qquad \text{Distributive property}$$

$$= \frac{\dfrac{-5 - 4x}{x+2}}{\dfrac{4 + x}{x+2}} \qquad \text{Combine like terms.}$$

$$= \frac{-5 - 4x}{x+2} \cdot \frac{x+2}{4+x} \qquad \text{Multiply by the reciprocal.}$$

$$= \frac{-5 - 4x}{4 + x} \qquad \text{Divide out the common factor.}$$

✔ **Now Try Exercise 33.**

OBJECTIVE 2 Simplify a complex fraction by multiplying by the least common denominator (Method 2). Since any expression can be multiplied by a form of 1 to get an equivalent expression, we can multiply both the numerator and the denominator of a complex fraction by the same nonzero expression to get an equivalent rational expression. If we choose the expression to be the LCD of all the fractions within the complex fraction, the complex fraction will be simplified. This is Method 2.

Method 2

To simplify a complex fraction,

Step 1 Find the LCD of all fractions within the complex fraction.

Step 2 Multiply both the numerator and the denominator of the complex fraction by this LCD using the distributive property as necessary. Write in lowest terms.

EXAMPLE 4 Simplifying Complex Fractions (Method 2)

Simplify each complex fraction.

(a) $\dfrac{\dfrac{2}{3} + \dfrac{5}{9}}{\dfrac{1}{4} + \dfrac{1}{12}}$

(b) $\dfrac{6 + \dfrac{3}{x}}{\dfrac{x}{4} + \dfrac{1}{8}}$

Step 1 Find the LCD for all denominators in the complex fraction.

The LCD for 3, 9, 4, and 12 is 36. | The LCD for x, 4, and 8 is $8x$.

Step 2 Multiply numerator and denominator of the complex fraction by the LCD.

$$\dfrac{\dfrac{2}{3} + \dfrac{5}{9}}{\dfrac{1}{4} + \dfrac{1}{12}} = \dfrac{36\left(\dfrac{2}{3} + \dfrac{5}{9}\right)}{36\left(\dfrac{1}{4} + \dfrac{1}{12}\right)}$$

$$\dfrac{6 + \dfrac{3}{x}}{\dfrac{x}{4} + \dfrac{1}{8}} = \dfrac{8x\left(6 + \dfrac{3}{x}\right)}{8x\left(\dfrac{x}{4} + \dfrac{1}{8}\right)}$$

Multiply each term by 36.
$$= \dfrac{36\left(\dfrac{2}{3}\right) + 36\left(\dfrac{5}{9}\right)}{36\left(\dfrac{1}{4}\right) + 36\left(\dfrac{1}{12}\right)}$$

Multiply each term by 8x.
$$= \dfrac{8x(6) + 8x\left(\dfrac{3}{x}\right)}{8x\left(\dfrac{x}{4}\right) + 8x\left(\dfrac{1}{8}\right)}$$ Distributive property

$$= \dfrac{24 + 20}{9 + 3}$$

$$= \dfrac{48x + 24}{2x^2 + x}$$ Multiply.

$$= \dfrac{44}{12} = \dfrac{4 \cdot 11}{4 \cdot 3}$$

$$= \dfrac{24(2x + 1)}{x(2x + 1)}$$ Factor.

$$= \dfrac{11}{3}$$

$$= \dfrac{24}{x}$$ Divide out the common factor.

✔ **Now Try Exercises 2 and 19.**

EXAMPLE 5 Simplifying a Complex Fraction (Method 2)

Simplify the complex fraction.

$$\frac{\dfrac{3}{5m} - \dfrac{2}{m^2}}{\dfrac{9}{2m} + \dfrac{3}{4m^2}} = \frac{20m^2\left(\dfrac{3}{5m} - \dfrac{2}{m^2}\right)}{20m^2\left(\dfrac{9}{2m} + \dfrac{3}{4m^2}\right)}$$

The LCD for $5m$, m^2, $2m$, and $4m^2$ is $20m^2$.

$$= \frac{20m^2\left(\dfrac{3}{5m}\right) - 20m^2\left(\dfrac{2}{m^2}\right)}{20m^2\left(\dfrac{9}{2m}\right) + 20m^2\left(\dfrac{3}{4m^2}\right)}$$

Distributive property

$$= \frac{12m - 40}{90m + 15}$$

Multiply and simplify.

✔ **Now Try Exercise 25.**

Either method can be used to simplify a complex fraction. Some students choose one method and stick with it to eliminate confusion. Others use Method 1 for problems like Example 2, which is the quotient of two fractions, and Method 2 for problems like Examples 1, 3, 4, and 5, which have sums or differences in the numerators, or denominators, or both.

EXAMPLE 6 Deciding on a Method and Simplifying Complex Fractions

Simplify each complex fraction.

(a) $$\frac{\dfrac{1}{y} + \dfrac{2}{y+2}}{\dfrac{4}{y} - \dfrac{3}{y+2}} = \frac{\left(\dfrac{1}{y} + \dfrac{2}{y+2}\right) \cdot y(y+2)}{\left(\dfrac{4}{y} - \dfrac{3}{y+2}\right) \cdot y(y+2)}$$

Use Method 2, since there are sums and differences in the numerator and denominator. Multiply by the LCD, $y(y+2)$.

$$= \frac{\left(\dfrac{1}{y}\right)y(y+2) + \left(\dfrac{2}{y+2}\right)y(y+2)}{\left(\dfrac{4}{y}\right)y(y+2) - \left(\dfrac{3}{y+2}\right)y(y+2)}$$

Distributive property

$$= \frac{1(y+2) + 2y}{4(y+2) - 3y}$$

Fundamental property

$$= \frac{y + 2 + 2y}{4y + 8 - 3y}$$

Distributive property

$$= \frac{3y + 2}{y + 8}$$

Combine like terms.

Be careful not to use $y + 2$ as the LCD. Because y appears in two denominators, it must be a factor in the LCD.

(b) $\dfrac{\dfrac{x+2}{x-3}}{\dfrac{x^2-4}{x^2-9}} = \dfrac{x+2}{x-3} \div \dfrac{x^2-4}{x^2-9}$ This is a quotient of two rational expressions. Use Method 1.

$= \dfrac{x+2}{x-3} \cdot \dfrac{x^2-9}{x^2-4}$ Definition of division

$= \dfrac{x+2}{x-3} \cdot \dfrac{(x+3)(x-3)}{(x+2)(x-2)}$ Factor.

$= \dfrac{x+3}{x-2}$ Divide out the common factors.

✔ **Now Try Exercises 29 and 31.**

7.5 EXERCISES

Note: In many problems involving complex fractions, several different equivalent forms of the answer exist. If your answer does not look exactly like the one given in the back of the book, check to see whether you have written an equivalent form.

◉ **1. Concept Check** Consider the complex fraction $\dfrac{\frac{1}{2} - \frac{1}{3}}{\frac{5}{6} - \frac{1}{12}}$. Answer each part, outlining Method 1 for simplifying this complex fraction.

(a) To combine the terms in the numerator, we must find the LCD of $\frac{1}{2}$ and $\frac{1}{3}$. What is this LCD? Determine the simplified form of the numerator of the complex fraction.

(b) To combine the terms in the denominator, we must find the LCD of $\frac{5}{6}$ and $\frac{1}{12}$. What is this LCD? Determine the simplified form of the denominator of the complex fraction.

(c) Now use the results from parts (a) and (b) to write the complex fraction as a division problem using the symbol $\div$.

(d) Perform the operation from part (c) to obtain the final simplification.

◉ **2. Concept Check** Consider the complex fraction given in Exercise 1: $\dfrac{\frac{1}{2} - \frac{1}{3}}{\frac{5}{6} - \frac{1}{12}}$. Answer each part, outlining Method 2 for simplifying this complex fraction.

(a) We must determine the LCD of all the fractions within the complex fraction. What is this LCD?

(b) Multiply every term in the complex fraction by the LCD found in part (a), but do not yet combine the terms in the numerator and the denominator.

(c) Combine the terms from part (b) to obtain the simplified form of the complex fraction.

✎ **3.** Which complex fraction is equivalent to $\dfrac{3 - \frac{1}{2}}{2 - \frac{1}{4}}$? Answer this question without showing any work, and explain your reasoning.

A. $\dfrac{3 + \frac{1}{2}}{2 + \frac{1}{4}}$ **B.** $\dfrac{-3 + \frac{1}{2}}{2 - \frac{1}{4}}$ **C.** $\dfrac{-3 - \frac{1}{2}}{-2 - \frac{1}{4}}$ **D.** $\dfrac{-3 + \frac{1}{2}}{-2 + \frac{1}{4}}$

✏ **4.** Only one of these choices is equal to $\dfrac{\frac{1}{2} + \frac{1}{4}}{\frac{1}{3} + \frac{1}{12}}$. Which one is it? Answer this question without showing any work, and explain your reasoning.

A. $\dfrac{9}{5}$ **B.** $-\dfrac{9}{5}$ **C.** $-\dfrac{5}{9}$ **D.** -12

✏ **5.** Describe Method 1 for simplifying complex fractions. Illustrate with the example $\dfrac{\frac{1}{2}}{\frac{2}{3}}$.

✏ **6.** Describe Method 2 for simplifying complex fractions. Illustrate with the example $\dfrac{\frac{1}{2}}{\frac{2}{3}}$.

Simplify each complex fraction. Use either method. See Examples 1–6.

7. $\dfrac{-\dfrac{4}{3}}{\dfrac{2}{9}}$

8. $\dfrac{-\dfrac{5}{6}}{\dfrac{5}{4}}$

9. $\dfrac{\dfrac{x}{y^2}}{\dfrac{x^2}{y}}$

10. $\dfrac{\dfrac{p^4}{r}}{\dfrac{p^2}{r^2}}$

11. $\dfrac{\dfrac{4a^4b^3}{3a}}{\dfrac{2ab^4}{b^2}}$

12. $\dfrac{\dfrac{2r^4t^2}{3t}}{\dfrac{5r^2t^5}{3r}}$

13. $\dfrac{\dfrac{m+2}{3}}{\dfrac{m-4}{m}}$

14. $\dfrac{\dfrac{q-5}{q}}{\dfrac{q+5}{3}}$

15. $\dfrac{\dfrac{2}{x} - 3}{\dfrac{2-3x}{2}}$

16. $\dfrac{6 + \dfrac{2}{r}}{\dfrac{3r+1}{4}}$

17. $\dfrac{\dfrac{1}{x} + x}{\dfrac{x^2+1}{8}}$

18. $\dfrac{\dfrac{3}{m} - m}{\dfrac{3-m^2}{4}}$

19. $\dfrac{a - \dfrac{5}{a}}{a + \dfrac{1}{a}}$

20. $\dfrac{q + \dfrac{1}{q}}{q + \dfrac{4}{q}}$

21. $\dfrac{\dfrac{5}{8} + \dfrac{2}{3}}{\dfrac{7}{3} - \dfrac{1}{4}}$

22. $\dfrac{\dfrac{6}{5} - \dfrac{1}{9}}{\dfrac{2}{5} + \dfrac{5}{3}}$

23. $\dfrac{\dfrac{1}{x^2} + \dfrac{1}{y^2}}{\dfrac{1}{x} - \dfrac{1}{y}}$

24. $\dfrac{\dfrac{1}{a^2} - \dfrac{1}{b^2}}{\dfrac{1}{a} - \dfrac{1}{b}}$

25. $\dfrac{\dfrac{2}{p^2} - \dfrac{3}{5p}}{\dfrac{4}{p} + \dfrac{1}{4p}}$

26. $\dfrac{\dfrac{2}{m^2} - \dfrac{3}{m}}{\dfrac{2}{5m^2} + \dfrac{1}{3m}}$

27. $\dfrac{\dfrac{5}{x^2y} - \dfrac{2}{xy^2}}{\dfrac{3}{x^2y^2} + \dfrac{4}{xy}}$

28. $\dfrac{\dfrac{1}{m^3p} + \dfrac{2}{mp^2}}{\dfrac{4}{mp} + \dfrac{1}{m^2p}}$

29. $\dfrac{\dfrac{1}{4} - \dfrac{1}{a^2}}{\dfrac{1}{2} + \dfrac{1}{a}}$

30. $\dfrac{\dfrac{1}{9} - \dfrac{1}{m^2}}{\dfrac{1}{3} + \dfrac{1}{m}}$

31. $\dfrac{\dfrac{1}{z+5}}{\dfrac{4}{z^2-25}}$

32. $\dfrac{\dfrac{1}{a+1}}{\dfrac{2}{a^2-1}}$

33. $\dfrac{\dfrac{1}{m+1} - 1}{\dfrac{1}{m+1} + 1}$

34. $\dfrac{\dfrac{2}{x-1}+2}{\dfrac{2}{x-1}-2}$

35. $\dfrac{\dfrac{1}{m-1}+\dfrac{2}{m+2}}{\dfrac{2}{m+2}-\dfrac{1}{m-3}}$

36. $\dfrac{\dfrac{5}{r+3}-\dfrac{1}{r-1}}{\dfrac{2}{r+2}+\dfrac{3}{r+3}}$

37. *Concept Check* In a fraction, what operation does the fraction bar represent?

38. *Concept Check* What property of real numbers justifies Method 2 of simplifying complex fractions?

RELATING CONCEPTS (EXERCISES 39–42)

FOR INDIVIDUAL OR GROUP WORK

To find the average of two numbers, we add them and divide by 2. Suppose that we wish to find the average of $\frac{3}{8}$ and $\frac{5}{6}$. **Work Exercises 39–42 in order,** *to see how a complex fraction occurs in a problem like this.*

39. Write in symbols: the sum of $\frac{3}{8}$ and $\frac{5}{6}$, divided by 2. Your result should be a complex fraction.

40. Use Method 1 to simplify the complex fraction from Exercise 39.

41. Use Method 2 to simplify the complex fraction from Exercise 39.

42. Your answers in Exercises 40 and 41 should be the same. Which method did you prefer? Why?

The next exercises, which involve expressions called **continued fractions,** *are "brain busters." Simplify by starting at "the bottom" and working upward.*

43. $1+\dfrac{1}{1+\dfrac{1}{1+1}}$

44. $5+\dfrac{5}{5+\dfrac{5}{5+5}}$

45. $7-\dfrac{3}{5+\dfrac{2}{4-2}}$

46. $3-\dfrac{2}{4+\dfrac{2}{4-2}}$

47. $r+\dfrac{r}{4-\dfrac{2}{6+2}}$

48. $\dfrac{2q}{7}-\dfrac{q}{6+\dfrac{8}{4+4}}$

PREVIEW EXERCISES

*Simplify. See **Section 1.8.***

49. $9\left(\dfrac{4x}{3}+\dfrac{2}{9}\right)$

50. $8\left(\dfrac{3r}{4}+\dfrac{9}{8}\right)$

51. $-12\left(\dfrac{11p^2}{3}-\dfrac{9p}{4}\right)$

52. $6\left(\dfrac{5z^2}{2}-\dfrac{8z}{3}\right)$

*Solve each equation. See **Sections 2.3 and 6.5.***

53. $3x+5=7x+3$

54. $9z+2=7z+6$

55. $6(z-3)+5=8z-3$

56. $k^2+3k-4=0$

7.6 Solving Equations with Rational Expressions

OBJECTIVE 1 Distinguish between operations with rational expressions and equations with terms that are rational expressions. Before solving equations with rational expressions, you must understand the difference between sums and differences of terms with rational coefficients, or rational *expressions,* and *equations* with terms that are rational expressions. *Sums and differences are expressions to simplify. Equations are solved.*

EXAMPLE 1 Distinguishing between Expressions and Equations

Identify each of the following as an *expression* or an *equation.* Then simplify the expression or solve the equation.

(a) $\dfrac{3}{4}x - \dfrac{2}{3}x$

This difference of two terms represents an expression. There is no equals sign.

$$\dfrac{3}{4}x - \dfrac{2}{3}x \qquad \text{This is an expression to simplify.}$$

$$= \dfrac{9}{12}x - \dfrac{8}{12}x \qquad \text{The LCD is 12. Write each coefficient with this LCD.}$$

$$= \dfrac{1}{12}x \qquad \text{Combine like terms, using the distributive property: } \tfrac{9}{12}x - \tfrac{8}{12}x = \left(\tfrac{9}{12} - \tfrac{8}{12}\right)x.$$

(b) $\qquad \dfrac{3}{4}x - \dfrac{2}{3}x = \dfrac{1}{2} \qquad$ This is an equation to solve.

$$12\left(\dfrac{3}{4}x - \dfrac{2}{3}x\right) = 12\left(\dfrac{1}{2}\right) \qquad \begin{array}{l}\text{Use the multiplication property of equality}\\\text{to clear fractions; multiply by the LCD, 12.}\end{array}$$

Multiply *each* term by 12.

$$12\left(\dfrac{3}{4}x\right) - 12\left(\dfrac{2}{3}x\right) = 12\left(\dfrac{1}{2}\right) \qquad \text{Distributive property}$$

$$9x - 8x = 6 \qquad \text{Multiply.}$$

$$x = 6 \qquad \text{Combine like terms.}$$

Check: $\qquad \dfrac{3}{4}x - \dfrac{2}{3}x = \dfrac{1}{2} \qquad$ Original equation

$$\dfrac{3}{4}(6) - \dfrac{2}{3}(6) = \dfrac{1}{2} \quad ? \qquad \text{Let } x = 6.$$

$$\dfrac{9}{2} - 4 = \dfrac{1}{2} \quad ? \qquad \text{Multiply.}$$

$$\dfrac{1}{2} = \dfrac{1}{2} \qquad \text{True}$$

Since a true statement results, {6} is the solution set of the equation.

✔ Now Try Exercises 1 and 3.

The ideas of Example 1 can be summarized as follows.

Uses of the LCD

When adding or subtracting rational expressions, keep the LCD throughout the simplification.

When solving an equation, multiply each side by the LCD so that denominators are eliminated.

OBJECTIVE 2 **Solve equations with rational expressions.**   When an equation involves fractions, as in Example 1(b), we use the multiplication property of equality to clear the fractions. Choose as multiplier the LCD of all denominators in the fractions of the equation.

EXAMPLE 2 **Solving an Equation with Rational Expressions**

Solve $\dfrac{p}{2} - \dfrac{p-1}{3} = 1$.

$$6\left(\frac{p}{2} - \frac{p-1}{3}\right) = 6(1) \qquad \text{Multiply each side by the LCD, 6.}$$

$$6\left(\frac{p}{2}\right) - 6\left(\frac{p-1}{3}\right) = 6(1) \qquad \text{Distributive property}$$

$$3p - 2(p-1) = 6 \qquad \text{Use parentheses around } p-1 \text{ to avoid errors.}$$

$$3p - 2(p) - 2(-1) = 6 \qquad \text{Distributive property}$$

Be careful with signs.

$$3p - 2p + 2 = 6 \qquad \text{Multiply.}$$

$$p + 2 = 6 \qquad \text{Combine like terms.}$$

$$p = 4 \qquad \text{Subtract 2.}$$

Check to see that {4} is the solution set by replacing p with 4 in the original equation.

✔ **Now Try Exercise 37.**

▶ **CAUTION**   Note that the use of the LCD here is different from its use in **Section 7.5.** Here, we use the multiplication property of equality to multiply each side of an *equation* by the LCD. Earlier, we used the fundamental property to multiply a *fraction* by another fraction that had the LCD as both its numerator and denominator. Be careful not to confuse these two methods.

Recall from **Section 7.1** that the denominator of a rational expression cannot equal 0, since division by 0 is undefined. *Therefore, when solving an equation with rational expressions that have variables in the denominator, the solution cannot be a number that makes the denominator equal 0.*

EXAMPLE 3 Solving an Equation with Rational Expressions

Solve $\dfrac{x}{x-2} = \dfrac{2}{x-2} + 2$. Check the proposed solution.

$$\dfrac{x}{x-2} = \dfrac{2}{x-2} + 2 \qquad \begin{array}{l} x \neq 2 \text{, since both denominators equal 0} \\ \text{if } x \text{ is 2.} \end{array}$$

$$(x-2)\left(\dfrac{x}{x-2}\right) = (x-2)\left(\dfrac{2}{x-2} + 2\right) \qquad \begin{array}{l}\text{Multiply each side by} \\ \text{the LCD, } x-2.\end{array}$$

$$(x-2)\left(\dfrac{x}{x-2}\right) = (x-2)\left(\dfrac{2}{x-2}\right) + (x-2)(2) \qquad \text{Distributive property}$$

$$x = 2 + 2x - 4 \qquad \text{Simplify.}$$

$$x = -2 + 2x \qquad \text{Combine like terms.}$$

$$-x = -2 \qquad \text{Subtract } 2x.$$

$$x = 2 \qquad \text{Multiply by } -1.$$

As noted, x cannot equal 2, since replacing x with 2 in the original equation causes the denominators to equal 0.

Check: $\qquad \dfrac{x}{x-2} = \dfrac{2}{x-2} + 2 \qquad$ Original equation

$$\dfrac{2}{2-2} = \dfrac{2}{2-2} + 2 \quad ? \qquad \text{Let } x = 2.$$

Division by 0 is undefined. $\qquad \dfrac{2}{0} = \dfrac{2}{0} + 2 \qquad ?$

Thus, 2 must be rejected as a solution, and the solution set is $\emptyset$.

✔ **Now Try Exercise 31.**

While it is always a good idea to check solutions to guard against arithmetic and algebraic errors, *it is essential to check proposed solutions when variables appear in denominators in the original equation.* Some students like to determine which numbers cannot be solutions *before* solving the equation, as we did in Example 3.

The steps used to solve an equation with rational expressions follow.

Solving an Equation with Rational Expressions

Step 1 **Multiply each side of the equation by the LCD** to clear the equation of fractions.

Step 2 **Solve** the resulting equation.

Step 3 **Check** each proposed solution by substituting it into the original equation. Reject any that cause a denominator to equal 0.

EXAMPLE 4 Solving an Equation with Rational Expressions

Solve $\dfrac{2}{x^2 - x} = \dfrac{1}{x^2 - 1}$. Check the proposed solution.

Step 1 Factor the denominators to find the LCD.

$$\frac{2}{x(x - 1)} = \frac{1}{(x + 1)(x - 1)} \qquad \text{The LCD is } x(x + 1)(x - 1).$$

Notice that $x \neq 0, -1$, or 1; otherwise a denominator will equal 0. Thus, 0, -1, and 1 cannot be solutions of this equation.

$$x(x + 1)(x - 1)\frac{2}{x(x - 1)} = x(x + 1)(x - 1)\frac{1}{(x + 1)(x - 1)} \qquad \begin{array}{l}\text{Multiply by}\\\text{the LCD.}\end{array}$$

Step 2
$$\begin{aligned}
2(x + 1) &= x && \text{Divide out common factors.}\\
2x + 2 &= x && \text{Distributive property}\\
x + 2 &= 0 && \text{Subtract } x.\\
x &= -2 && \text{Subtract 2.}
\end{aligned}$$

Step 3 The proposed solution is -2, which does not make any denominator equal 0.

Check:
$$\frac{2}{x^2 - x} = \frac{1}{x^2 - 1} \qquad \text{Original equation}$$

$$\frac{2}{(-2)^2 - (-2)} = \frac{1}{(-2)^2 - 1} \qquad ? \quad \text{Let } x = -2.$$

$$\frac{2}{4 + 2} = \frac{1}{4 - 1} \qquad ?$$

$$\frac{1}{3} = \frac{1}{3} \qquad \text{True}$$

Thus, the solution set is $\{-2\}$.

✔ **Now Try Exercise 43.**

EXAMPLE 5 Solving an Equation with Rational Expressions

Solve $\dfrac{2m}{m^2 - 4} + \dfrac{1}{m - 2} = \dfrac{2}{m + 2}$.

Factor the first denominator on the left to find the LCD.

$$\frac{2m}{(m + 2)(m - 2)} + \frac{1}{m - 2} = \frac{2}{m + 2} \qquad \text{The LCD is } (m + 2)(m - 2).$$

Here $m \neq -2$ or 2, since a denominator is 0 for these values. Thus, -2 and 2 cannot be solutions of this equation.

$$(m + 2)(m - 2)\left(\frac{2m}{(m + 2)(m - 2)} + \frac{1}{m - 2}\right) \qquad \text{Multiply by the LCD.}$$

$$= (m + 2)(m - 2)\frac{2}{m + 2}$$

$$(m + 2)(m - 2)\frac{2m}{(m + 2)(m - 2)} + (m + 2)(m - 2)\frac{1}{m - 2}$$

$$= (m + 2)(m - 2)\frac{2}{m + 2} \qquad \text{Distributive property}$$

$$2m + m + 2 = 2(m - 2) \qquad \text{Divide out common factors.}$$

$$3m + 2 = 2m - 4 \qquad \text{Combine like terms; distributive property}$$

$$m = -6 \qquad \text{Subtract } 2m; \text{ subtract 2.}$$

A check verifies that $\{-6\}$ is the solution set.

✔ **Now Try Exercise 53.**

EXAMPLE 6 Solving an Equation with Rational Expressions

Solve $\dfrac{1}{x - 1} + \dfrac{1}{2} = \dfrac{2}{x^2 - 1}$.

Factor the denominator on the right.

x ≠ 1, −1 or a denominator is 0.

$$\frac{1}{x - 1} + \frac{1}{2} = \frac{2}{(x + 1)(x - 1)} \qquad \text{The LCD is } 2(x + 1)(x - 1).$$

$$2(x + 1)(x - 1)\left(\frac{1}{x - 1} + \frac{1}{2}\right) = 2(x + 1)(x - 1)\frac{2}{(x + 1)(x - 1)}$$

Multiply by the LCD.

$$2(x + 1)(x - 1)\frac{1}{x - 1} + 2(x + 1)(x - 1)\frac{1}{2} = 2(x + 1)(x - 1)\frac{2}{(x + 1)(x - 1)}$$

Distributive property

$$2(x + 1) + (x + 1)(x - 1) = 4 \qquad \text{Simplify.}$$

$$2x + 2 + x^2 - 1 = 4 \qquad \text{Distributive property}$$

$$x^2 + 2x + 1 = 4 \qquad \text{Combine like terms.}$$

$$x^2 + 2x - 3 = 0 \qquad \text{Subtract 4.}$$

$$(x + 3)(x - 1) = 0 \qquad \text{Factor.}$$

$$x + 3 = 0 \quad \text{or} \quad x - 1 = 0 \qquad \text{Zero-factor property}$$

$$x = -3 \quad \text{or} \quad x = 1 \qquad \text{Solve for } x.$$

Proposed solutions are -3 and 1. However, 1 makes an original denominator equal 0, so 1 is not a solution. Check that -3 is a solution.

Check:

$$\frac{1}{x - 1} + \frac{1}{2} = \frac{2}{x^2 - 1} \qquad \text{Original equation}$$

$$\frac{1}{-3 - 1} + \frac{1}{2} = \frac{2}{(-3)^2 - 1} \quad ? \qquad \text{Let } x = -3.$$

$$\frac{1}{-4} + \frac{1}{2} = \frac{2}{9 - 1} \quad ? \qquad \text{Simplify.}$$

$$\frac{1}{4} = \frac{1}{4} \qquad \text{True}$$

The solution set is $\{-3\}$.

✔ **Now Try Exercise 63.**

EXAMPLE 7 Solving an Equation with Rational Expressions

Solve $\dfrac{1}{k^2 + 4k + 3} + \dfrac{1}{2k + 2} = \dfrac{3}{4k + 12}$.

Factor each denominator.

$$\frac{1}{(k + 1)(k + 3)} + \frac{1}{2(k + 1)} = \frac{3}{4(k + 3)} \qquad \text{The LCD is } 4(k + 1)(k + 3).$$

$k \neq -1, -3$

$$4(k + 1)(k + 3)\left(\frac{1}{(k + 1)(k + 3)} + \frac{1}{2(k + 1)}\right)$$

$$= 4(k + 1)(k + 3)\frac{3}{4(k + 3)} \qquad \text{Multiply by the LCD.}$$

$$4(k + 1)(k + 3)\frac{1}{(k + 1)(k + 3)} + 2 \cdot 2(k + 1)(k + 3)\frac{1}{2(k + 1)}$$

$$= 4(k + 1)(k + 3)\frac{3}{4(k + 3)} \qquad \text{Distributive property}$$

Do *not* add
4 + 2 here.

$$4 + 2(k + 3) = 3(k + 1) \qquad \text{Simplify.}$$
$$4 + 2k + 6 = 3k + 3 \qquad \text{Distributive property}$$
$$2k + 10 = 3k + 3 \qquad \text{Combine like terms.}$$
$$7 = k \qquad \text{Subtract } 2k \text{ and } 3.$$

The proposed solution, 7, does not make an original denominator equal 0. A check shows that the algebra is correct, so {7} is the solution set.

✔ **Now Try Exercise 65.**

OBJECTIVE 3 Solve a formula for a specified variable. Solving a formula for a specified variable was first discussed in **Section 2.5.** *Remember to treat the variable for which you are solving as if it were the only variable, and all others as if they were constants.*

EXAMPLE 8 Solving for a Specified Variable

Solve each formula for the specified variable.

(a) $F = \dfrac{k}{d - D}$ for d

Our goal is to get d alone on one side of the equation.

$$F = \frac{k}{d - D} \qquad \text{Given equation}$$

$$F(d - D) = \frac{k}{d - D}(d - D) \qquad \begin{array}{l}\text{Multiply by } d - D \text{ to}\\ \text{clear the fraction.}\end{array}$$

$$F(d - D) = k \qquad \text{Simplify.}$$

$$Fd - FD = k \qquad \text{Distributive property}$$

$$Fd = k + FD \qquad \text{Add } FD.$$

$$d = \frac{k + FD}{F}, \quad \text{or} \quad d = \frac{k}{F} + D \qquad \text{Divide by } F.$$

(b) $\dfrac{1}{a} = \dfrac{1}{b} + \dfrac{1}{c}$ for c

Our goal is to get c alone on one side of the equation.

$$\dfrac{1}{a} = \dfrac{1}{b} + \dfrac{1}{c}$$ Given equation

$$abc\left(\dfrac{1}{a}\right) = abc\left(\dfrac{1}{b} + \dfrac{1}{c}\right)$$ Multiply by the LCD, abc.

$$abc\left(\dfrac{1}{a}\right) = abc\left(\dfrac{1}{b}\right) + abc\left(\dfrac{1}{c}\right)$$ Distributive property

> We need both terms with c on the *same* side.

$$bc = ac + ab$$ Simplify.

$$bc - ac = ab$$ Subtract ac.

$$c(b - a) = ab$$ Factor out c.

$$c = \dfrac{ab}{b - a}$$ Divide by $b - a$.

Solving differently would lead to

$$c = \dfrac{-ab}{a - b},$$

an equivalent form of the correct answer.

✔ **Now Try Exercises 77 and 83.**

▶ **CAUTION** Students often have trouble in the step that involves factoring out the variable for which they are solving. In Example 8(b), we had to factor out c on the left side so that we could divide both sides by $b - a$.
When solving an equation for a specified variable, be sure that the specified variable appears alone on only one side of the equals sign in the final equation.

7.6 EXERCISES

Complete solution available on Video Lectures on CD/DVD

Now Try Exercise

Identify each of the following as an expression *or an* equation. *Then simplify the expression or solve the equation. See Example 1.*

1. $\dfrac{7}{8}x + \dfrac{1}{5}x$

2. $\dfrac{4}{7}x + \dfrac{3}{5}x$

3. $\dfrac{7}{8}x + \dfrac{1}{5}x = 1$

4. $\dfrac{4}{7}x + \dfrac{3}{5}x = 1$

5. $\dfrac{3}{5}x - \dfrac{7}{10}x$

6. $\dfrac{2}{3}x - \dfrac{7}{4}x$

7. $\dfrac{3}{5}x - \dfrac{7}{10}x = 1$

8. $\dfrac{2}{3}x - \dfrac{7}{4}x = -13$

When solving an equation with variables in denominators, we must determine the values that cause these denominators to equal 0, so that we can reject these values if they appear as possible solutions. Find all values for which at least one denominator is equal to 0. Write answers using $\neq$. Do not solve. See Examples 3–7.

9. $\dfrac{3}{x+2} - \dfrac{5}{x} = 1$

10. $\dfrac{7}{x} + \dfrac{9}{x-3} = 5$

11. $\dfrac{-1}{(x+3)(x-4)} = \dfrac{1}{2x+1}$

12. $\dfrac{8}{(x-8)(x+2)} = \dfrac{7}{3x-10}$

13. $\dfrac{4}{x^2+8x-9} + \dfrac{1}{x^2-4} = 0$

14. $\dfrac{-3}{x^2+9x-10} - \dfrac{12}{x^2-16} = 0$

15. What is wrong with the following problem? "Solve $\dfrac{2}{3x} + \dfrac{1}{5x}$."

16. Explain how the LCD is used in a different way when adding and subtracting rational expressions as compared to solving equations with rational expressions.

Solve each equation, and check your solutions. See Examples 1(b), 2, and 3.

17. $\dfrac{5}{m} - \dfrac{3}{m} = 8$

18. $\dfrac{4}{y} + \dfrac{1}{y} = 2$

19. $\dfrac{5}{y} + 4 = \dfrac{2}{y}$

20. $\dfrac{11}{q} = 3 - \dfrac{1}{q}$

21. $\dfrac{3x}{5} - 6 = x$

22. $\dfrac{5t}{4} + t = 9$

23. $\dfrac{4m}{7} + m = 11$

24. $a - \dfrac{3a}{2} = 1$

25. $\dfrac{z-1}{4} = \dfrac{z+3}{3}$

26. $\dfrac{r-5}{2} = \dfrac{r+2}{3}$

27. $\dfrac{3p+6}{8} = \dfrac{3p-3}{16}$

28. $\dfrac{2z+1}{5} = \dfrac{7z+5}{15}$

29. $\dfrac{2x+3}{x} = \dfrac{3}{2}$

30. $\dfrac{5-2x}{x} = \dfrac{1}{4}$

31. $\dfrac{k}{k-4} - 5 = \dfrac{4}{k-4}$

32. $\dfrac{-5}{a+5} = \dfrac{a}{a+5} + 2$

33. $\dfrac{q+2}{3} + \dfrac{q-5}{5} = \dfrac{7}{3}$

34. $\dfrac{t}{6} + \dfrac{4}{3} = \dfrac{t-2}{3}$

35. $\dfrac{x}{2} = \dfrac{5}{4} + \dfrac{x-1}{4}$

36. $\dfrac{8p}{5} = \dfrac{3p-4}{2} + \dfrac{5}{2}$

Solve each equation, and check your solutions. **Be careful with signs.** *See Example 2.*

37. $\dfrac{a+7}{8} - \dfrac{a-2}{3} = \dfrac{4}{3}$

38. $\dfrac{x+3}{7} - \dfrac{x+2}{6} = \dfrac{1}{6}$

39. $\dfrac{p}{2} - \dfrac{p-1}{4} = \dfrac{5}{4}$

40. $\dfrac{r}{6} - \dfrac{r-2}{3} = -\dfrac{4}{3}$

41. $\dfrac{3x}{5} - \dfrac{x-5}{7} = 3$

42. $\dfrac{8k}{5} - \dfrac{3k-4}{2} = \dfrac{5}{2}$

Solve each equation, and check your solutions. See Examples 3–7.

43. $\dfrac{4}{x^2-3x} = \dfrac{1}{x^2-9}$

44. $\dfrac{2}{t^2-4} = \dfrac{3}{t^2-2t}$

45. $\dfrac{2}{m} = \dfrac{m}{5m+12}$

46. $\dfrac{x}{4-x} = \dfrac{2}{x}$

47. $\dfrac{-2}{z+5} + \dfrac{3}{z-5} = \dfrac{20}{z^2-25}$

48. $\dfrac{3}{r+3} - \dfrac{2}{r-3} = \dfrac{-12}{r^2-9}$

49. $\dfrac{3}{x-1} + \dfrac{2}{4x-4} = \dfrac{7}{4}$

50. $\dfrac{2}{p+3} + \dfrac{3}{8} = \dfrac{5}{4p+12}$

51. $\dfrac{x}{3x+3} = \dfrac{2x-3}{x+1} - \dfrac{2x}{3x+3}$

52. $\dfrac{2k+3}{k+1} - \dfrac{3k}{2k+2} = \dfrac{-2k}{2k+2}$

53. $\dfrac{2p}{p^2-1} = \dfrac{2}{p+1} - \dfrac{1}{p-1}$

54. $\dfrac{2x}{x^2-16} - \dfrac{2}{x-4} = \dfrac{4}{x+4}$

55. $\dfrac{5x}{14x+3} = \dfrac{1}{x}$

56. $\dfrac{m}{8m+3} = \dfrac{1}{3m}$

57. $\dfrac{2}{x-1} - \dfrac{2}{3} = \dfrac{-1}{x+1}$

58. $\dfrac{5}{p-2} = 7 - \dfrac{10}{p+2}$

59. $\dfrac{x}{2x+2} = \dfrac{-2x}{4x+4} + \dfrac{2x-3}{x+1}$

60. $\dfrac{5t+1}{3t+3} = \dfrac{5t-5}{5t+5} + \dfrac{3t-1}{t+1}$

61. $\dfrac{8x+3}{x} = 3x$

62. $\dfrac{2}{x} = \dfrac{x}{5x-12}$

63. $\dfrac{1}{x+4} + \dfrac{x}{x-4} = \dfrac{-8}{x^2-16}$

64. $\dfrac{x}{x-3} + \dfrac{4}{x+3} = \dfrac{18}{x^2-9}$

65. $\dfrac{4}{3x+6} - \dfrac{3}{x+3} = \dfrac{8}{x^2+5x+6}$

66. $\dfrac{-13}{t^2+6t+8} + \dfrac{4}{t+2} = \dfrac{3}{2t+8}$

67. $\dfrac{3x}{x^2+5x+6} = \dfrac{5x}{x^2+2x-3} - \dfrac{2}{x^2+x-2}$

68. $\dfrac{m}{m^2+m-2} + \dfrac{m}{m^2-1} = \dfrac{m}{m^2+3m+2}$

69. $\dfrac{x+4}{x^2-3x+2} - \dfrac{5}{x^2-4x+3} = \dfrac{x-4}{x^2-5x+6}$

70. $\dfrac{3}{r^2+r-2} - \dfrac{1}{r^2-1} = \dfrac{7}{2(r^2+3r+2)}$

71. *Concept Check* If you are solving a formula for the letter k, and your steps lead to the equation $kr - mr = km$, what would be your next step?

72. *Concept Check* If you are solving a formula for the letter k, and your steps lead to the equation $kr - km = mr$, what would be your next step?

Solve each formula for the specified variable. See Example 8.

73. $m = \dfrac{kF}{a}$ for F

74. $I = \dfrac{kE}{R}$ for E

75. $m = \dfrac{kF}{a}$ for a

76. $I = \dfrac{kE}{R}$ for R

77. $I = \dfrac{E}{R+r}$ for R

78. $I = \dfrac{E}{R+r}$ for r

79. $h = \dfrac{2A}{B+b}$ for A

80. $d = \dfrac{2S}{n(a+L)}$ for S

81. $d = \dfrac{2S}{n(a+L)}$ for a

82. $h = \dfrac{2A}{B + b}$ for B **83.** $\dfrac{1}{x} = \dfrac{1}{y} - \dfrac{1}{z}$ for y **84.** $\dfrac{3}{k} = \dfrac{1}{p} + \dfrac{1}{q}$ for q

85. $\dfrac{2}{r} + \dfrac{3}{s} + \dfrac{1}{t} = 1$ for t **86.** $\dfrac{5}{p} + \dfrac{2}{q} + \dfrac{3}{r} = 1$ for r

87. $9x + \dfrac{3}{z} = \dfrac{5}{y}$ for z **88.** $-3t - \dfrac{4}{p} = \dfrac{6}{s}$ for p

---

PREVIEW EXERCISES

Write a mathematical expression for each exercise. See Section 2.7.

89. Andrew drives from Philadelphia to Pittsburgh, a distance of 288 mi, in t hours. Find his rate in miles per hour.

90. Tyler drives for 10 hr, traveling from City A to City B, a distance of d kilometers. Find his rate in kilometers per hour.

91. Jack flies his small plane from St. Louis to Chicago, a distance of 289 mi, at z miles per hour. Find his time in hours.

92. Joshua can do a job in x hours. What portion of the job is done in 1 hr?

Summary Exercises on Rational Expressions and Equations

Students often confuse *simplifying expressions* with *solving equations*. We review the four operations to simplify the rational expressions $\frac{1}{x}$ and $\frac{1}{x-2}$ as follows.

Add: $\dfrac{1}{x} + \dfrac{1}{x-2}$

$= \dfrac{1(x-2)}{x(x-2)} + \dfrac{x(1)}{x(x-2)}$ Write with a common denominator.

$= \dfrac{x - 2 + x}{x(x-2)}$ Add numerators; keep the same denominator.

$= \dfrac{2x - 2}{x(x-2)}$ Combine like terms.

Subtract: $\dfrac{1}{x} - \dfrac{1}{x-2}$

$= \dfrac{1(x-2)}{x(x-2)} - \dfrac{x(1)}{x(x-2)}$ Write with a common denominator.

$= \dfrac{x - 2 - x}{x(x-2)}$ Subtract numerators; keep the same denominator.

$= \dfrac{-2}{x(x-2)}$ Combine like terms.

Multiply: $\dfrac{1}{x} \cdot \dfrac{1}{x-2}$

$= \dfrac{1}{x(x-2)}$ Multiply numerators and multiply denominators.

Divide: $\dfrac{1}{x} \div \dfrac{1}{x-2}$

$= \dfrac{1}{x} \cdot \dfrac{x-2}{1} = \dfrac{x-2}{x}$ Multiply by the reciprocal of the second fraction.

By contrast, consider the *equation*

$$\frac{1}{x} + \frac{1}{x-2} = \frac{3}{4}.$$

> $x \neq 0$ and $x \neq 2$ since a denominator is 0 for these values.

$4x(x-2)\dfrac{1}{x} + 4x(x-2)\dfrac{1}{x-2} = 4x(x-2)\dfrac{3}{4}$ Multiply each side by the LCD, $4x(x-2)$, to clear fractions.

$4(x-2) + 4x = 3x(x-2)$ Simplify.

$4x - 8 + 4x = 3x^2 - 6x$ Distributive property

$3x^2 - 14x + 8 = 0$ Get 0 on one side.

$(3x - 2)(x - 4) = 0$ Factor.

$3x - 2 = 0$ or $x - 4 = 0$ Zero-factor property

$x = \dfrac{2}{3}$ or $x = 4$ Solve for x.

Both $\frac{2}{3}$ and 4 are solutions, since neither makes a denominator equal 0. Check to confirm that the solution set is $\left\{\frac{2}{3}, 4\right\}$.

Points to Remember when Working with Rational Expressions and Equations

1. When simplifying rational expressions, the fundamental property is applied only after numerators and denominators have been *factored*.

2. When adding and subtracting rational expressions, the common denominator must be kept throughout the problem and in the final result.

3. Always check to see if the answer is in lowest terms; if it is not, use the fundamental property.

4. When solving equations with rational expressions, the LCD is used to clear the equation of fractions. Multiply each side by the LCD. (Notice how this use differs from that of the LCD in Point 2.)

5. When solving equations with rational expressions, reject any proposed solution that causes an original denominator to equal 0.

For each exercise, indicate "expression" if an expression is to be simplified or "equation" if an equation is to be solved. Then simplify the expression or solve the equation.

1. $\dfrac{4}{p} + \dfrac{6}{p}$

2. $\dfrac{x^3y^2}{x^2y^4} \cdot \dfrac{y^5}{x^4}$

3. $\dfrac{1}{x^2 + x - 2} \div \dfrac{4x^2}{2x - 2}$

4. $\dfrac{8}{m - 5} = 2$

5. $\dfrac{2y^2 + y - 6}{2y^2 - 9y + 9} \cdot \dfrac{y^2 - 2y - 3}{y^2 - 1}$

6. $\dfrac{2}{k^2 - 4k} + \dfrac{3}{k^2 - 16}$

7. $\dfrac{x - 4}{5} = \dfrac{x + 3}{6}$

8. $\dfrac{3t^2 - t}{6t^2 + 15t} \div \dfrac{6t^2 + t - 1}{2t^2 - 5t - 25}$

9. $\dfrac{4}{p + 2} + \dfrac{1}{3p + 6}$

10. $\dfrac{1}{x} + \dfrac{1}{x - 3} = -\dfrac{5}{4}$

11. $\dfrac{3}{t - 1} + \dfrac{1}{t} = \dfrac{7}{2}$

12. $\dfrac{6}{y} - \dfrac{2}{3y}$

13. $\dfrac{5}{4z} - \dfrac{2}{3z}$

14. $\dfrac{k + 2}{3} = \dfrac{2k - 1}{5}$

15. $\dfrac{1}{m^2 + 5m + 6} + \dfrac{2}{m^2 + 4m + 3}$

16. $\dfrac{2k^2 - 3k}{20k^2 - 5k} \div \dfrac{2k^2 - 5k + 3}{4k^2 + 11k - 3}$

17. $\dfrac{2}{x + 1} + \dfrac{5}{x - 1} = \dfrac{10}{x^2 - 1}$

18. $\dfrac{3}{x + 3} + \dfrac{4}{x + 6} = \dfrac{9}{x^2 + 9x + 18}$

19. $\dfrac{4t^2 - t}{6t^2 + 10t} \div \dfrac{8t^2 + 2t - 1}{3t^2 + 11t + 10}$

20. $\dfrac{x}{x - 2} + \dfrac{3}{x + 2} = \dfrac{8}{x^2 - 4}$

7.7 Applications of Rational Expressions

OBJECTIVES

1 Solve problems about numbers.

2 Solve problems about distance, rate, and time.

3 Solve problems about work.

In **Section 7.6,** we solved equations with rational expressions; now we can solve applications that involve this type of equation. The six-step problem-solving method of **Section 2.4** still applies.

OBJECTIVE 1 Solve problems about numbers.

EXAMPLE 1 Solving a Problem about an Unknown Number

If the same number is added to both the numerator and the denominator of the fraction $\frac{2}{5}$, the result is equivalent to $\frac{2}{3}$. Find the number.

Step 1 **Read** the problem carefully. We are trying to find a number.

Step 2 **Assign a variable.** Here, we let $x =$ the number added to the numerator and the denominator.

Step 3 **Write an equation.** The fraction $\frac{2 + x}{5 + x}$ represents the result of adding the same number to both the numerator and the denominator. Since this result is equivalent to $\frac{2}{3}$, the equation is

$$\frac{2 + x}{5 + x} = \frac{2}{3}.$$

Step 4 **Solve** this equation.

$$3(5 + x)\frac{2 + x}{5 + x} = 3(5 + x)\frac{2}{3} \qquad \text{Multiply by the LCD, } 3(5 + x).$$

$$3(2 + x) = 2(5 + x)$$

$$6 + 3x = 10 + 2x \qquad \text{Distributive property}$$

$$x = 4 \qquad \text{Subtract } 2x; \text{ subtract } 6.$$

Step 5 **State the answer.** The number is 4.

Step 6 **Check** the solution in the words of the original problem. If 4 is added to both the numerator and the denominator of $\frac{2}{5}$, the result is $\frac{6}{9} = \frac{2}{3}$, as required.

✔ **Now Try Exercise 3.**

OBJECTIVE 2 **Solve problems about distance, rate, and time.** Recall from **Chapter 2** the following formulas relating distance, rate, and time.

Distance, Rate, and Time Relationship

$$d = rt \qquad r = \frac{d}{t} \qquad t = \frac{d}{r}$$

You may wish to refer to Example 5 in **Section 2.7** to review the basic use of these formulas. We continue our work with motion problems here.

EXAMPLE 2 **Solving a Problem about Distance, Rate, and Time**

The Tickfaw River has a current of 3 mph. A motorboat takes as long to go 12 mi downstream as to go 8 mi upstream. What is the speed of the boat in still water?

Step 1 **Read** the problem again. We must find the speed of the boat in still water.

Step 2 **Assign a variable.** Let $x =$ the speed of the boat in still water.

Because the current pushes the boat when the boat is going downstream, the speed of the boat downstream will be the *sum* of the speed of the boat and the speed of the current, $x + 3$ mph. Because the current slows down the boat when the boat is going upstream, the boat's speed going upstream is given by the *difference* between the speed of the boat and the speed of the current, $x - 3$ mph. See Figure 1 on the next page.

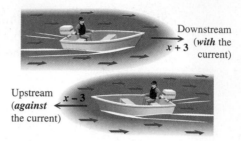

FIGURE 1

This information is summarized in the following table.

	d	r	t
Downstream	12	x + 3	
Upstream	8	x − 3	

Fill in the times by using the formula $t = \frac{d}{r}$.

The time downstream is the distance divided by the rate, or

$$t = \frac{d}{r} = \frac{12}{x + 3}, \quad \text{Time downstream}$$

and the time upstream is that distance divided by that rate, or

$$t = \frac{d}{r} = \frac{8}{x - 3}. \quad \text{Time upstream}$$

The completed table follows.

	d	r	t
Downstream	12	x + 3	$\frac{12}{x + 3}$
Upstream	8	x − 3	$\frac{8}{x - 3}$

Times are equal.

Step 3 **Write an equation.** Since the time downstream equals the time upstream, the two times from the table must be equal, giving the equation

$$\frac{12}{x + 3} = \frac{8}{x - 3}.$$

Step 4 **Solve.**

$$(x + 3)(x - 3)\frac{12}{x + 3} = (x + 3)(x - 3)\frac{8}{x - 3} \qquad \text{Multiply by the LCD,} \\ (x + 3)(x - 3).$$

$$12(x - 3) = 8(x + 3)$$

$$12x - 36 = 8x + 24 \qquad \text{Distributive property}$$

$$4x = 60 \qquad \text{Subtract } 8x; \text{ add } 36.$$

$$x = 15 \qquad \text{Divide by 4.}$$

Step 5 **State the answer.** The speed of the boat in still water is 15 mph.

Step 6 **Check.** First we find the speed of the boat going downstream, which is 15 + 3 = 18 mph. Traveling 12 mi would take

$$t = \frac{d}{r} = \frac{12}{18} = \frac{2}{3} \text{ hr.}$$

Next, the speed of the boat going upstream is 15 − 3 = 12 mph, and traveling 8 mi would take

$$t = \frac{d}{r} = \frac{8}{12} = \frac{2}{3} \text{ hr.}$$

The time upstream equals the time downstream, as required.

✔ **Now Try Exercise 21.**

OBJECTIVE 3 **Solve problems about work.** Suppose that you can mow your lawn in 4 hr. Then after 1 hr, you will have mowed $\frac{1}{4}$ of the lawn. After 2 hr, you will have mowed $\frac{2}{4}$, or $\frac{1}{2}$, of the lawn, and so on. This idea is generalized as follows.

Rate of Work

If a job can be completed in *t* units of time, then the rate of work is

$$\frac{1}{t} \text{ job per unit of time.}$$

▶ **PROBLEM-SOLVING HINT** Recall that the formula *d* = *rt* says that distance traveled is equal to rate of travel multiplied by time traveled. Similarly, the fractional part of a job accomplished is equal to the rate of work multiplied by the time worked. In the lawn-mowing example, after 3 hr, the fractional part of the job done is

$$\underbrace{\frac{1}{4}}_{\substack{\text{Rate of} \\ \text{work}}} \cdot \underbrace{3}_{\substack{\text{Time} \\ \text{worked}}} = \underbrace{\frac{3}{4}}_{\substack{\text{Fractional part} \\ \text{of job done}}}.$$

After 4 hr, $\frac{1}{4}(4) = 1$ whole job has been done.

EXAMPLE 3 **Solving a Problem about Work Rates**

"If Joe can paint a house in 3 hr and Sam can paint the same house in 5 hr, how long does it take for them to do it together?" (*Source*: The movie *Little Big League.*)

Step 1 **Read** the problem again. We are looking for time working together.

Step 2 **Assign a variable.** Let $x =$ the number of hours it takes Joe and Sam to paint the house, working together.

Certainly, x will be less than 3, since Joe alone can complete the job in 3 hr. We begin by making a table. Based on the preceding discussion, Joe's rate alone is $\frac{1}{3}$ job per hour, and Sam's rate is $\frac{1}{5}$ job per hour.

	Rate	Time Working Together	Fractional Part of the Job Done when Working Together
Joe	$\frac{1}{3}$	x	$\frac{1}{3}x$
Sam	$\frac{1}{5}$	x	$\frac{1}{5}x$

Sum is 1 whole job.

Step 3 **Write an equation.** Together, Joe and Sam complete 1 whole job, so we add their individual fractional parts and set the sum equal to 1.

$$\underbrace{\text{Fractional part}}_{\text{done by Joe}} + \underbrace{\text{Fractional part}}_{\text{done by Sam}} = \underbrace{1 \text{ whole job.}}$$

$$\frac{1}{3}x \quad + \quad \frac{1}{5}x \quad = \quad 1$$

Step 4 **Solve.**
$$15\left(\frac{1}{3}x + \frac{1}{5}x\right) = 15(1) \qquad \text{Multiply by the LCD, 15.}$$

$$15\left(\frac{1}{3}x\right) + 15\left(\frac{1}{5}x\right) = 15(1) \qquad \text{Distributive property}$$

$$5x + 3x = 15$$

$$8x = 15 \qquad \text{Combine like terms.}$$

$$x = \frac{15}{8} \qquad \text{Divide by 8.}$$

Step 5 **State the answer.** Working together, Joe and Sam can paint the house in $\frac{15}{8}$ hr, or $1\frac{7}{8}$ hr.

Step 6 **Check** to be sure the answer is correct.

✔ **Now Try Exercise 35.**

▶ **NOTE** An alternative approach in work problems is to consider the part of the job that can be done in 1 hr. For instance, in Example 3 Joe can do the entire job in 3 hr and Sam can do it in 5 hr. Thus, their work rates, as we saw in Example 3, are $\frac{1}{3}$ and $\frac{1}{5}$, respectively. Since it takes them x hours to complete the job working together, in 1 hr they can paint $\frac{1}{x}$ of the house. The amount

(continued)

painted by Joe in 1 hr plus the amount painted by Sam in 1 hr must equal the amount they can do together. This relationship leads to the equation

Amount by Sam

$$\underset{\text{Amount by Joe} \rightarrow}{\frac{1}{3}} + \frac{1}{5} = \frac{1}{x}. \leftarrow \text{Amount together}$$

Compare this equation with the one in Example 3. Multiplying each side by $15x$ leads to

$$5x + 3x = 15,$$

the same equation found in the third line of Step 4 in the example. The same solution results.

7.7 EXERCISES

Concept Check   Use Steps 2 and 3 of the six-step method to set up the equation you would use to solve each problem. (Remember that Step 1 is to read the problem carefully.) Do not actually solve the equation. See Example 1.

1. The numerator of the fraction $\frac{5}{6}$ is increased by an amount so that the value of the resulting fraction is equivalent to $\frac{13}{3}$. By what amount was the numerator increased?

   (a) Let $x =$ _____. (*Step 2*)

   (b) Write an expression for "the numerator of the fraction $\frac{5}{6}$ is increased by an amount."
   (c) Set up an equation to solve the problem. (*Step 3*)

2. If the same number is added to the numerator and subtracted from the denominator of $\frac{23}{12}$, the resulting fraction is equivalent to $\frac{3}{2}$. What is the number?

   (a) Let $x =$ _____. (*Step 2*)

   (b) Write an expression for "a number is added to the numerator of $\frac{23}{12}$." Then write an expression for "the same number is subtracted from the denominator of $\frac{23}{12}$."
   (c) Set up an equation to solve the problem. (*Step 3*)

Solve each problem. See Example 1.

3. In a certain fraction, the denominator is 6 more than the numerator. If 3 is added to both the numerator and the denominator, the resulting fraction is equivalent to $\frac{5}{7}$. What was the original fraction (*not* written in lowest terms)?

4. In a certain fraction, the denominator is 4 less than the numerator. If 3 is added to both the numerator and the denominator, the resulting fraction is equivalent to $\frac{3}{2}$. What was the original fraction?

5. The numerator of a certain fraction is four times the denominator. If 6 is added to both the numerator and the denominator, the resulting fraction is equivalent to 2. What was the original fraction (*not* written in lowest terms)?

6. The denominator of a certain fraction is three times the numerator. If 2 is added to the numerator and subtracted from the denominator, the resulting fraction is equivalent to 1. What was the original fraction (*not* written in lowest terms)?

7. One-third of a number is 2 more than one-sixth of the same number. What is the number?

8. One-sixth of a number is 5 more than the same number. What is the number?

9. A quantity, $\frac{2}{3}$ of it, $\frac{1}{2}$ of it, and $\frac{1}{7}$ of it, added together, equals 33. What is the quantity? (*Source:* Rhind Mathematical Papyrus.)

10. A quantity, $\frac{3}{4}$ of it, $\frac{1}{2}$ of it, and $\frac{1}{3}$ of it, added together, equals 93. What is the quantity? (*Source:* Rhind Mathematical Papyrus.)

Solve each problem. See Example 5 in **Section 2.7** *(pages 154 and 155).*

11. In the 2006 Winter Olympics, Svetlana Zhurova of Russia won the 500-m speed skating event for women. Her rate was 6.530 m per sec. What was her time (to the nearest hundredth of a second)? (*Source:* www.espn.com)

12. In the 2004 Summer Olympics, Jody Henry of Australia won the women's 100-m freestyle swimming event. Her rate was 1.854 m per sec. What was her time (to the nearest hundredth of a second)? (*Source: World Almanac and Book of Facts 2006.*)

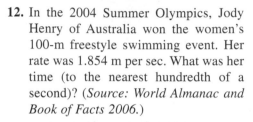

13. Meseret Defar of Ethiopia won the women's 5000-m race in the 2004 Olympics with a time of 14.761 min. What was her rate (to three decimal places)? (*Source: World Almanac and Book of Facts 2006.*)

14. The winner of the women's 1500-m run in the 2004 Olympics was Kelly Holmes of Great Britain with a time of 3.965 min. What was her rate (to three decimal places)? (*Source: World Almanac and Book of Facts 2006.*)

15. The winner of the 2005 Daytona 500 (mile) race was Jeff Gordon, who drove his Chevrolet to victory with a rate of 135.173 mph. What was his time (to the nearest thousandth of an hour)? (*Source: World Almanac and Book of Facts 2006.*)

16. In 2005, Dan Wheldon drove his Dallara-Honda to victory in the Indianapolis 500 (mile) race. His rate was 157.603 mph. What was his time (to the nearest thousandth of an hour)? (*Source: World Almanac and Book of Facts 2006.*)

Concept Check *Solve each problem.*

17. Suppose Stephanie walks D miles at R mph in the same time that Wally walks d miles at r mph. Give an equation relating D, R, d, and r.

18. If a migrating hawk travels m mph in still air, what is its rate when it flies into a steady headwind of 5 mph? What is its rate with a tailwind of 5 mph?

Set up the equation you would use to solve each problem. Do not actually solve the equation. See Example 2.

19. Mitch Levy flew his airplane 500 mi against the wind in the same time it took him to fly 600 mi with the wind. If the speed of the wind was 10 mph, what was the speed of his plane in still air? (Let x = speed of the plane in still air.)

	d	r	t
Against the Wind	500	x − 10	
With the Wind	600	x + 10	

20. Jane Saska can row 4 mph in still water. She takes as long to row 8 mi upstream as 24 mi downstream. How fast is the current? (Let x = speed of the current.)

	d	r	t
Upstream	8	4 − x	
Downstream	24	4 + x	

Solve each problem. See Example 2.

21. A boat can go 20 mi against a current in the same time that it can go 60 mi with the current. The current is 4 mph. Find the speed of the boat in still water.

22. Vince Grosso can fly his plane 200 mi against the wind in the same time it takes him to fly 300 mi with the wind. The wind blows at 30 mph. Find the speed of his plane in still air.

23. The sanderling is a small shorebird about 6.5 in. long, with a thin, dark bill and a wide, white wing stripe. If a sanderling can fly 30 mi with the wind in the same time it can fly 18 mi against the wind when the wind speed is 8 mph, what is the speed of the bird in still air? (*Source:* U.S. Geological Survey.)

24. Airplanes usually fly faster from west to east than from east to west because the prevailing winds go from west to east. The air distance between Chicago and London is about 4000 mi, while the air distance between New York and London is about 3500 mi. If a jet can fly eastbound from Chicago to London in the same time it can fly westbound from London to New York in a 35-mph wind, what is the speed of the plane in still air? (*Source: Encyclopaedia Britannica.*)

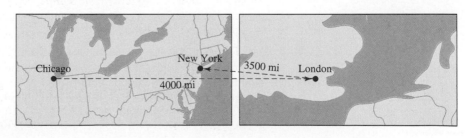

25. An airplane maintaining a constant airspeed takes as long to go 450 mi with the wind as it does to go 375 mi against the wind. If the wind is blowing at 15 mph, what is the speed of the plane in still air?

26. A river has a current of 4 km per hr. Find the speed of Lynn McTernan's boat in still water if it goes 40 km downstream in the same time that it takes to go 24 km upstream.

27. Connie McNair's boat goes 12 mph. Find the rate of the current of the river if she can go 6 mi upstream in the same amount of time she can go 10 mi downstream.

28. Howie Sorkin can travel 8 mi upstream in the same time it takes him to go 12 mi downstream. His boat goes 15 mph in still water. What is the rate of the current?

29. The distance from Seattle, Washington, to Victoria, British Columbia, is about 148 mi by ferry. It takes about 4 hr less to travel by the same ferry from Victoria to Vancouver, British Columbia, a distance of about 74 mi. What is the average speed of the ferry?

30. Driving from Tulsa to Detroit, Jeff averaged 50 mph. He figured that if he had averaged 60 mph, his driving time would have decreased 3 hr. How far is it from Tulsa to Detroit?

Concept Check   *Solve each problem.*

31. If it takes Elayn 10 hr to do a job, what is her rate?

32. If it takes Clay 12 hr to do a job, how much of the job does he do in 8 hr?

In Exercises 33 and 34, set up the equation you would use to solve each problem. Do not actually solve the equation. See Example 3.

33. Working alone, Jorge can paint a room in 8 hr. Caterina can paint the same room working alone in 6 hr. How long will it take them if they work together? (Let t represent the time they work together.)

	r	t	w
Jorge		t	
Caterina		t	

34. Edwin Bedford can tune up his Chevy in 2 hr working alone. His son, Beau, can do the job in 3 hr working alone. How long would it take them if they worked together? (Let t represent the time they work together.)

	r	t	w
Edwin		t	
Beau		t	

Solve each problem. See Example 3.

35. Ms. Tseng, a high school mathematics teacher, gave a test on perimeter, area, and volume to her geometry classes. Working alone, it would take her 4 hr to grade the tests. Her student teacher, Jonah Schmidt, would take 6 hr to grade the same tests. How long would it take them to grade these tests if they work together?

36. Zachary and Samuel are brothers who share a bedroom. By himself, Zachary can completely mess up their room in 20 min, while it would take Samuel only 12 min to do the same thing. How long would it take them to mess up the room together?

37. A pump can pump the water out of a flooded basement in 10 hr. A smaller pump takes 12 hr. How long would it take to pump the water from the basement with both pumps?

38. Lou Viggiano's copier can do a printing job in 7 hr. Bo Wojtowicz's copier can do the same job in 12 hr. How long would it take to do the job with both copiers?

39. An experienced employee can enter tax data into a computer twice as fast as a new employee. Working together, it takes the employees 2 hr. How long would it take the experienced employee working alone?

40. One roofer can put a new roof on a house three times faster than another. Working together, they can roof a house in 4 days. How long would it take the faster roofer working alone?

41. One pipe can fill a swimming pool in 6 hr, and another pipe can do it in 9 hr. How long will it take the two pipes working together to fill the pool $\frac{3}{4}$ full?

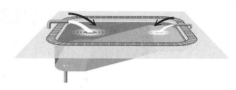

42. An inlet pipe can fill a swimming pool in 9 hr, and an outlet pipe can empty the pool in 12 hr. Through an error, both pipes are left open. How long will it take to fill the pool?

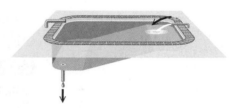

The next two problems are "brain busters."

43. A cold-water faucet can fill a sink in 12 min, and a hot-water faucet can fill it in 15 min. The drain can empty the sink in 25 min. If both faucets are on and the drain is open, how long will it take to fill the sink?

44. Refer to Exercise 42. Assume that the error was discovered after both pipes had been running for 3 hr and the outlet pipe was then closed. How much more time would then be required to fill the pool? (*Hint:* Consider how much of the job had been done when the error was discovered.)

PREVIEW EXERCISES

*Solve each equation for k. See **Section 2.2**.*

45. $200 = 15k$ **46.** $16 = 9k$ **47.** $180 = \dfrac{k}{20}$ **48.** $92 = \dfrac{k}{3}$

*Solve each formula for k. See **Section 2.5**.*

49. $y = kx$ **50.** $y = kx^2$ **51.** $y = \dfrac{k}{x}$ **52.** $y = \dfrac{k}{x^2}$

7.8 Variation

OBJECTIVE 1 **Solve direct variation problems.** Suppose that gasoline costs \$3.00 per gal. Then 1 gal costs \$3.00, 2 gal costs 2(\$3.00) = \$6.00, 3 gal costs 3(\$3.00) = \$9.00, and so on. Each time, the total cost is obtained by multiplying the number of gallons by the price per gallon. In general, if k equals the price per gallon and x equals the number of gallons, then the total cost y is equal to kx. Thus,

As the *number of gallons increases,* the *total cost increases.*

The reverse is also true:

As the *number of gallons decreases,* the *total cost decreases.*

The preceding discussion presents an example of *variation.* As in the gasoline example, *two variables vary directly if one is a constant multiple of the other.*

Direct Variation

y **varies directly as** *x* if there exists a constant k such that
$$y = kx.$$

Also, y is said to be *proportional to x.* The constant k in the equation for direct variation is a numerical value, such as 3.00 in the gasoline price discussion. This value is called the **constant of variation.**

EXAMPLE 1 **Using Direct Variation**

Suppose y varies directly as x, and $y = 20$ when $x = 4$. Find y when $x = 9$.

Since y varies directly as x, there is a constant k such that $y = kx$. We know that $y = 20$ when $x = 4$. Substituting these values into $y = kx$ and solving for k gives

$$y = kx \qquad \text{Equation for direct variation}$$
$$20 = k \cdot 4 \qquad \text{Substitute the given values.}$$
$$k = 5. \quad \longleftarrow \text{Constant of variation}$$

Since $y = kx$ and $k = 5$,

$$y = 5x. \qquad \text{Let } k = 5.$$

Therefore, when $x = 9$, $\quad y = 5x = 5 \cdot 9 = 45.$ $\qquad$ Let $x = 9$.

Thus, $y = 45$ when $x = 9$.

✔ **Now Try Exercise 19.**

Solving a Variation Problem

Step 1 Write the variation equation.

Step 2 Substitute the appropriate given values and solve for k.

Step 3 Rewrite the variation equation with the value of k from Step 2.

Step 4 Substitute the remaining values, solve for the unknown, and find the required answer.

The direct variation equation $y = kx$ is a linear equation. However, other kinds of variation involve other types of equations. For example, one variable can be proportional to a power of another variable.

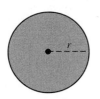

$A = \pi r^2$

FIGURE 2

> ### Direct Variation as a Power
>
> **y varies directly as the nth power of x** if there exists a real number k such that
> $$y = kx^n.$$

An example of direct variation as a power is the formula for the area of a circle, $A = \pi r^2$. Here, π is the constant of variation, and the area varies directly as the square of the radius. See Figure 2.

EXAMPLE 2   Solving a Direct Variation Problem

The distance a body falls from rest varies directly as the square of the time it falls (disregarding air resistance). If a sky diver falls 64 ft in 2 sec, how far will she fall in 8 sec?

Step 1   If d represents the distance the sky diver falls and t the time it takes to fall, then d is a function of t, and, for some constant k,
$$d = kt^2.$$

Step 2   To find the value of k, use the fact that the sky diver falls 64 ft in 2 sec.

$d = kt^2$	Variation equation
$64 = k(2)^2$	Let $d = 64$ and $t = 2$.
$64 = 4k$	Apply the exponent.
$k = 16$	Divide by 4.

Step 3   Using 16 for k, the variation equation becomes
$$d = 16t^2.$$

Step 4   Now let $t = 8$ to find the number of feet the sky diver will fall in 8 sec.
$$d = 16(8)^2 = 16 \cdot 64 = 1024 \qquad \text{Let } t = 8.$$

The sky diver will fall 1024 ft in 8 sec.

✔ **Now Try Exercise 41.**

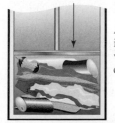

As pressure increases, volume decreases.

FIGURE 3

OBJECTIVE ② Solve inverse variation problems.   In direct variation, where $k > 0$, as x increases, y increases. Similarly, as x decreases, y decreases. Another type of variation is *inverse variation*. With inverse variation, where $k > 0$,

As one variable *increases,* the other variable *decreases.*

For example, in a closed space, volume decreases as pressure increases, as illustrated by a trash compactor. See Figure 3. As the compactor presses down, the pressure on the trash increases; in turn, the trash occupies a smaller space.

Inverse Variation

y varies inversely as x if there exists a real number k such that

$$y = \frac{k}{x}.$$

Also, **y varies inversely as the nth power of x** if there exists a real number k such that

$$y = \frac{k}{x^n}.$$

Another example of inverse variation comes from the distance formula.

$$d = rt \qquad \text{Distance formula}$$

$$t = \frac{d}{r} \qquad \text{Divide each side by } r.$$

In the form $t = \frac{d}{r}$, t (time) varies inversely as r (rate or speed), with d (distance) serving as the constant of variation. For example, if the distance between Chicago and Des Moines is 300 mi, then

$$t = \frac{300}{r}$$

and the values of r and t might be any of the following:

$$
\left.
\begin{array}{l}
r = 50,\, t = 6 \\
r = 60,\, t = 5 \\
r = 75,\, t = 4
\end{array}
\right\}
\begin{array}{l}
\text{As } r \text{ increases,} \\
t \text{ decreases.}
\end{array}
\qquad
\left.
\begin{array}{l}
r = 30,\, t = 10 \\
r = 25,\, t = 12 \\
r = 20,\, t = 15.
\end{array}
\right\}
\begin{array}{l}
\text{As } r \text{ decreases,} \\
t \text{ increases.}
\end{array}
$$

If we *increase* the rate (speed) at which we drive, the time *decreases*. If we *decrease* the rate (speed) at which we drive, the time *increases*.

EXAMPLE 3 Using Inverse Variation

Suppose y varies inversely as x, and $y = 3$ when $x = 8$. Find y when $x = 6$.

Since y varies inversely as x, there is a constant k such that $y = \frac{k}{x}$. We know that $y = 3$ when $x = 8$, so we can find k.

$$y = \frac{k}{x} \qquad \text{Equation for inverse variation}$$

$$3 = \frac{k}{8} \qquad \text{Substitute the given values.}$$

$$k = 24 \qquad \text{Multiply by 8; rewrite } 24 = k \text{ as } k = 24.$$

Since $y = \frac{24}{x}$, we let $x = 6$ and find y.

$$y = \frac{24}{x} = \frac{24}{6} = 4$$

Therefore, when $x = 6$, $y = 4$.

✔ Now Try Exercise 23.

EXAMPLE 4 Using Inverse Variation

In the manufacturing of a certain medical syringe, the cost of producing the syringe varies inversely as the number produced. If 10,000 syringes are produced, the cost is $2 per syringe. Find the cost per syringe to produce 25,000 syringes.

$$\text{Let} \quad x = \text{the number of syringes produced,}$$
$$\text{and} \quad c = \text{the cost per unit.}$$

Here, as production increases, cost decreases, and as production decreases, cost increases. Since c varies inversely as x, there is a constant k such that

$$c = \frac{k}{x} \qquad \text{Equation for inverse variation}$$

$$2 = \frac{k}{10,000} \qquad \text{Substitute the given values.}$$

$$k = 20,000. \qquad \text{Multiply by 10,000; rewrite.}$$

Since $c = \frac{k}{x}$,

$$c = \frac{20,000}{25,000} = 0.80. \qquad \text{Let } k = 20,000 \text{ and } x = 25,000.$$

The cost per syringe to make 25,000 syringes is $0.80.

✔ **Now Try Exercise 37.**

7.8 EXERCISES

Concept Check *Use personal experience or intuition to determine whether the situation suggests direct or inverse variation.* *

1. The number of different lottery tickets you buy and your probability of winning that lottery

2. The rate and the distance traveled by a pickup truck in 3 hr

3. The amount of pressure put on the accelerator of a car and the speed of the car

4. The surface area of a balloon and its diameter

5. The number of days until the end of the baseball season and the number of home runs that Alex Rodriguez has

6. The amount of gasoline you pump and the amount you will pay

7. The number of days from now until December 25 and the magnitude of the frenzy of Christmas shopping

8. Your age and the probability that you believe in Santa Claus

_____

*The authors thank Linda Kodama of Kapiolani Community College for suggesting the inclusion of exercises of this type.

Concept Check *Determine whether each equation represents* direct *or* inverse *variation.*

9. $y = \dfrac{3}{x}$ **10.** $y = \dfrac{8}{x}$ **11.** $y = 10x^2$ **12.** $y = 2x^3$

13. $y = 50x$ **14.** $y = 100x$ **15.** $y = \dfrac{12}{x^2}$ **16.** $y = \dfrac{10}{x^3}$

17. *Concept Check* Fill in each blank with the correct response.

    (a) If the constant of variation is positive and y varies directly as x, then as x increases, y _____.
            (increases/decreases)

    (b) If the constant of variation is positive and y varies inversely as x, then as x increases, y _____.
            (increases/decreases)

18. Bill Veeck was the owner of several major league baseball teams in the 1950s and 1960s. He was known to often sit in the stands and enjoy games with his paying customers. Here is a quote attributed to him:

> *I have discovered in 20 years of moving around a ballpark, that the knowledge of the game is usually in inverse proportion to the price of the seats.*

Explain in your own words the meaning of his statement. (To prove his point, Veeck once allowed the fans to vote on managerial decisions.)

Solve each problem involving direct or inverse variation. See Examples 1 and 3.

19. If x varies directly as y, and $x = 27$ when $y = 6$, find x when $y = 2$.

20. If z varies directly as x, and $z = 30$ when $x = 8$, find z when $x = 4$.

21. If d varies directly as t, and $d = 150$ when $t = 3$, find d when $t = 5$.

22. If d varies directly as r, and $d = 200$ when $r = 40$, find d when $r = 60$.

23. If x varies inversely as y, and $x = 3$ when $y = 8$, find y when $x = 4$.

24. If z varies inversely as x, and $z = 50$ when $x = 2$, find z when $x = 25$.

25. If p varies inversely as q, and $p = 7$ when $q = 6$, find p when $q = 2$.

26. If m varies inversely as r, and $m = 12$ when $r = 8$, find m when $r = 16$.

27. If m varies inversely as p^2, and $m = 20$ when $p = 2$, find m when $p = 5$.

28. If a varies inversely as b^2, and $a = 48$ when $b = 4$, find a when $b = 7$.

29. If p varies inversely as q^2, and $p = 4$ when $q = \frac{1}{2}$, find p when $q = \frac{3}{2}$.

30. If z varies inversely as x^2, and $z = 9$ when $x = \frac{2}{3}$, find z when $x = \frac{5}{4}$.

Solve each variation problem. See Examples 1–4.

31. The interest on an investment varies directly as the rate of interest. If the interest is \$48 when the interest rate is 5%, find the interest when the rate is 4.2%.

32. For a given base, the area of a triangle varies directly as its height. Find the area of a triangle with a height of 6 in., if the area is 10 in.² when the height is 4 in.

33. Hooke's law for an elastic spring states that the distance a spring stretches varies directly with the force applied. If a force of 75 lb stretches a certain spring 16 in., how much will a force of 200 lb stretch the spring?

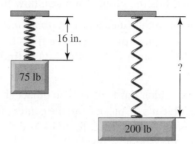

34. The pressure exerted by water at a given point varies directly with the depth of the point beneath the surface of the water. Water exerts 4.34 lb per in.² for every 10 ft traveled below the water's surface. What is the pressure exerted on a scuba diver at 20 ft?

35. Over a specified distance, speed varies inversely with time. If a Dodge Viper on a test track goes a certain distance in one-half minute at 160 mph, what speed is needed to go the same distance in three-fourths minute?

36. For a constant area, the length of a rectangle varies inversely as the width. The length of a rectangle is 27 ft when the width is 10 ft. Find the width of a rectangle with the same area if the length is 18 ft.

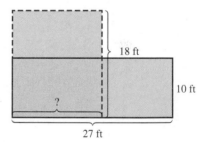

37. The current in a simple electrical circuit varies inversely as the resistance. If the current is 20 amps when the resistance is 5 ohms, find the current when the resistance is 8 ohms.

38. If the temperature is constant, the pressure of a gas in a container varies inversely as the volume of the container. If the pressure is 10 lb per ft² in a container with volume 3 ft³, what is the pressure in a container with volume 1.5 ft³?

39. The force required to compress a spring varies directly as the change in the length of the spring. If a force of 12 lb is required to compress a certain spring 3 in., how much force is required to compress the spring 5 in.?

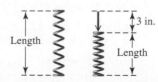

40. In the inversion of raw sugar, the rate of change of the amount of raw sugar varies directly as the amount of raw sugar remaining. The rate is 200 kg per hr when there are 800 kg left. What is the rate of change per hour when only 100 kg are left?

41. The area of a circle varies directly as the square of its radius. A circle with radius 3 in. has area 28.278 in.^{2}. What is the area of a circle with radius 4.1 in. (to the nearest thousandth)?

42. For a body falling freely from rest (disregarding air resistance), the distance the body falls varies directly as the square of the time. If an object is dropped from the top of a tower 400 ft high and hits the ground in 5 sec, how far did it fall in the first 3 sec?

43. The amount of light (measured in *footcandles*) produced by a light source varies inversely as the square of the distance from the source. If the amount of light produced 4 ft from a light source is 75 footcandles, find the amount of light produced 9 ft from the same source.

44. The force with which Earth attracts an object above Earth's surface varies inversely as the square of the object's distance from the center of Earth. If an object 4000 mi from the center of Earth is attracted with a force of 160 lb, find the force of attraction on an object 6000 mi from the center of Earth.

TECHNOLOGY INSIGHTS (EXERCISES 45–48)

In each table, Y_1 either varies directly or varies inversely as X. Tell which type of variation is illustrated.

45.

X	Y₁
0	0
1	2.5
2	5
3	7.5
4	10
5	12.5
6	15

X=0

46.

X	Y₁
0	0
1	3.2
2	6.4
3	9.6
4	12.8
5	16
6	19.2

X=0

47.

X	Y₁
1	24
2	12
3	8
4	6
5	4.8
6	4
7	3.4286

X=1

48.

X	Y₁
1	48
2	24
3	16
4	12
5	9.6
6	8
7	6.8571

X=1

PREVIEW EXERCISES

Find each power. See Sections 1.2 and 5.1.

49. 8^2 **50.** $(-4)^2$ **51.** -12^2 **52.** 1.5^2

Find the value of $a^2 + b^2$ for the given values of a and b. See Section 1.3.

53. $a = 5, b = 12$ **54.** $a = 8, b = 15$

Chapter 7 Group Activity

BUYING A CAR

Objective Use a complex fraction to calculate monthly car payments.

You are shopping for a midsize car and have put aside a certain amount of money each month for a car payment. Your instructor will assign this amount to you. You have narrowed your choices to the cars listed in the table.

Year/Make/Model	Base Retail Price	Miles per Gallon (city)	Miles per Gallon (highway)
2006 Ford Fusion	$17,145	21	29
2006 Honda Accord Hybrid	30,990	25	34
2006 Pontiac Grand Prix	21,330	19	28
2006 Lexus ES 330	32,300	21	29
2006 Chevrolet Malibu	16,365	21	30
2006 Infiniti G35	31,200	17	24

Source: www.edmunds.com; www.automobiles.honda.com

As a group, work through the following steps to determine which car you can afford to buy.

A. Decide which cars you think are within your budget.

B. Select one of the cars you identified in part A. Have each group member calculate the monthly payment for this car using a different financing option. Use the following formula, where P is the principal, r is the interest rate, and m is the number of monthly payments, along with the financing options table:

Financing Options	
Time (in years)	Interest Rate
3	6.0%
4	6.5%
5	7.0%

$$\text{Monthly Payment} = \frac{\dfrac{Pr}{12}}{1 - \left(\dfrac{12}{12 + r}\right)^{m}}.$$

C. Have each group member determine the amount of money paid in interest over the duration of the loan for the group's financing option.

D. Consider fuel expenses.

  1. Assume that you will travel an average of 75 mi in the city and 400 mi on the highway each week. How many gallons of gas will you need to buy each month?

  2. Given typical prices for gas in your area at this time, how much money will you need to have available for buying gas?

E. Repeat parts B–D as necessary until your group can reach a consensus on the car it will buy and the financing option it will use. Write a paragraph to explain the choices.

KEY TERMS

7.1 rational expression
lowest terms

7.3 least common
denominator (LCD)

7.5 complex fraction
7.8 direct variation

constant of variation
inverse variation

TEST YOUR WORD POWER

See how well you have learned the vocabulary in this chapter. Answers, with examples, follow the Quick Review.

1. A **rational expression** is
   A. an algebraic expression made up of a term or the sum of a finite number of terms with real coefficients and whole number exponents
   B. a polynomial equation of degree 2
   C. an expression with one or more fractions in the numerator, or denominator, or both

   D. the quotient of two polynomials with denominator not 0.

2. A **complex fraction** is
   A. an algebraic expression made up of a term or the sum of a finite number of terms with real coefficients and whole number exponents

   B. a polynomial equation of degree 2
   C. a rational expression with one or more fractions in the numerator, or denominator, or both
   D. the quotient of two polynomials with denominator not 0.

QUICK REVIEW

Concepts	Examples

7.1 THE FUNDAMENTAL PROPERTY OF RATIONAL EXPRESSIONS

To find the value(s) for which a rational expression is undefined, set the denominator equal to 0 and solve the equation.

Find the values for which the expression $\dfrac{x-4}{x^2-16}$ is undefined.

$$x^2 - 16 = 0$$
$$(x-4)(x+4) = 0 \quad \text{Factor.}$$
$$x - 4 = 0 \quad \text{or} \quad x + 4 = 0 \quad \text{Zero-factor property}$$
$$x = 4 \quad \text{or} \quad x = -4$$

The rational expression is undefined for 4 and -4, so $x \neq 4$ and $x \neq -4$.

Writing a Rational Expression in Lowest Terms

Step 1 Factor the numerator and denominator.

Step 2 Use the fundamental property to divide out common factors.

There are often several different equivalent forms of a rational expression.

Write $\dfrac{x^2-1}{(x-1)^2}$ in lowest terms.

$$\frac{x^2-1}{(x-1)^2} = \frac{(x-1)(x+1)}{(x-1)(x-1)}$$
$$= \frac{x+1}{x-1}$$

Give four equivalent forms of $-\dfrac{x-1}{x+2}$.

Distribute the $-$ sign in the numerator to get $\frac{-(x-1)}{x+2}$ or $\frac{-x+1}{x+2}$; do so in the denominator to get $\frac{x-1}{-(x+2)}$ or $\frac{x-1}{-x-2}$.

(continued)

Concepts	Examples

7.2 MULTIPLYING AND DIVIDING RATIONAL EXPRESSIONS

Multiplying or Dividing Rational Expressions

Step 1 Note the operation. If the operation is division, use the definition of division to rewrite as multiplication.

Step 2 Multiply numerators and multiply denominators.

Step 3 Factor numerators and denominators completely.

Step 4 Write in lowest terms, using the fundamental property.

Note: Steps 2 and 3 may be interchanged based on personal preference.

Multiply. $\dfrac{3x + 9}{x - 5} \cdot \dfrac{x^2 - 3x - 10}{x^2 - 9}$

$$= \frac{(3x + 9)(x^2 - 3x - 10)}{(x - 5)(x^2 - 9)}$$

$$= \frac{3(x + 3)(x - 5)(x + 2)}{(x - 5)(x + 3)(x - 3)}$$

$$= \frac{3(x + 2)}{x - 3}$$

Divide. $\dfrac{2x + 1}{x + 5} \div \dfrac{6x^2 - x - 2}{x^2 - 25}$

$$= \frac{2x + 1}{x + 5} \cdot \frac{x^2 - 25}{6x^2 - x - 2} \qquad \text{Multiply by the reciprocal.}$$

$$= \frac{(2x + 1)(x^2 - 25)}{(x + 5)(6x^2 - x - 2)}$$

$$= \frac{(2x + 1)(x + 5)(x - 5)}{(x + 5)(2x + 1)(3x - 2)}$$

$$= \frac{x - 5}{3x - 2}$$

7.3 LEAST COMMON DENOMINATORS

Finding the LCD

Step 1 Factor each denominator into prime factors.

Step 2 List each different factor the greatest number of times it appears.

Step 3 Multiply the factors from Step 2 to get the LCD.

Find the LCD for $\dfrac{3}{k^2 - 8k + 16}$ and $\dfrac{1}{4k^2 - 16k}$.

$$\left. \begin{aligned} k^2 - 8k + 16 &= (k - 4)^2 \\ 4k^2 - 16k &= 4k(k - 4) \end{aligned} \right\} \quad \begin{aligned} &\text{Factor each} \\ &\text{denominator.} \end{aligned}$$

$$\begin{aligned} \text{LCD} &= (k - 4)^2 \cdot 4 \cdot k \\ &= 4k(k - 4)^2 \end{aligned}$$

Writing a Rational Expression with a Specified Denominator

Step 1 Factor both denominators.

Step 2 Decide what factor(s) the denominator must be multiplied by in order to equal the specified denominator.

Step 3 Multiply the rational expression by that factor divided by itself. (That is, multiply by 1.)

Find the numerator: $\dfrac{5}{2z^2 - 6z} = \dfrac{?}{4z^3 - 12z^2}$.

$$\frac{5}{2z(z - 3)} = \frac{?}{4z^2(z - 3)}$$

$2z(z - 3)$ must be multiplied by $2z$ in order to obtain $4z^2(z - 3)$.

$$\frac{5}{2z(z - 3)} \cdot \frac{2z}{2z} = \frac{10z}{4z^2(z - 3)} = \frac{10z}{4z^3 - 12z^2}$$

(continued)

Concepts	Examples

7.4 **ADDING AND SUBTRACTING RATIONAL EXPRESSIONS**

Adding Rational Expressions

Step 1 Find the LCD.

Step 2 Rewrite each rational expression with the LCD as denominator.

Step 3 Add the numerators to get the numerator of the sum. The LCD is the denominator of the sum.

Step 4 Write in lowest terms.

Add. $\dfrac{2}{3m+6} + \dfrac{m}{m^2-4}$

$\left.\begin{array}{l} 3m+6 = 3(m+2) \\ m^2-4 = (m+2)(m-2) \end{array}\right\}$ The LCD is $3(m+2)(m-2)$.

$= \dfrac{2(m-2)}{3(m+2)(m-2)} + \dfrac{3m}{3(m+2)(m-2)}$

$= \dfrac{2m-4+3m}{3(m+2)(m-2)}$

$= \dfrac{5m-4}{3(m+2)(m-2)}$

Subtracting Rational Expressions

Follow the same steps as for addition, but subtract in Step 3.

Subtract. $\dfrac{6}{k+4} - \dfrac{2}{k}$ The LCD is $k(k+4)$.

$= \dfrac{6k}{(k+4)k} - \dfrac{2(k+4)}{k(k+4)}$

$= \dfrac{6k-2(k+4)}{k(k+4)}$

$= \dfrac{6k-2k-8}{k(k+4)}$ *Be careful with signs when subtracting the numerators.*

$= \dfrac{4k-8}{k(k+4)},$ or $\dfrac{4(k-2)}{k(k+4)}$

7.5 **COMPLEX FRACTIONS**

Simplifying Complex Fractions

Method 1 Simplify the numerator and denominator separately. Then divide the simplified numerator by the simplified denominator.

Simplify.

Method 1 $\dfrac{\dfrac{1}{a}-a}{1-a} = \dfrac{\dfrac{1}{a}-\dfrac{a^2}{a}}{1-a} = \dfrac{\dfrac{1-a^2}{a}}{1-a}$

$= \dfrac{1-a^2}{a} \div (1-a)$

$= \dfrac{1-a^2}{a} \cdot \dfrac{1}{1-a}$ Multiply by the reciprocal.

$= \dfrac{(1-a)(1+a)}{a(1-a)} = \dfrac{1+a}{a}$

Method 2 Multiply the numerator and denominator of the complex fraction by the LCD of all the denominators in the complex fraction. Write in lowest terms.

Method 2 $\dfrac{\dfrac{1}{a}-a}{1-a} = \dfrac{\left(\dfrac{1}{a}-a\right)a}{(1-a)a} = \dfrac{\dfrac{a}{a}-a^2}{(1-a)a}$

$= \dfrac{1-a^2}{(1-a)a} = \dfrac{(1+a)(1-a)}{(1-a)a}$

$= \dfrac{1+a}{a}$

(continued)

Concepts	Examples

7.6 SOLVING EQUATIONS WITH RATIONAL EXPRESSIONS

Solving Equations with Rational Expressions

Step 1 Multiply each side of the equation by the LCD to clear the equation of fractions.

Step 2 Solve the resulting equation.

Solve $\quad x \neq 1 \quad \dfrac{2}{x-1} + \dfrac{3}{4} = \dfrac{5}{x-1}$. The LCD is $4(x-1)$.

$$4(x-1)\left(\frac{2}{x-1} + \frac{3}{4}\right) = 4(x-1)\left(\frac{5}{x-1}\right)$$

$$4(x-1)\left(\frac{2}{x-1}\right) + 4(x-1)\left(\frac{3}{4}\right) = 4(x-1)\left(\frac{5}{x-1}\right)$$

$$8 + 3(x-1) = 20$$
$$8 + 3x - 3 = 20$$
$$3x = 15$$
$$x = 5$$

Step 3 Check each proposed solution.

The proposed solution, 5, checks. The solution set is $\{5\}$.

7.7 APPLICATIONS OF RATIONAL EXPRESSIONS

Solving Problems about Distance, Rate, and Time

Use the six-step method.

Step 1 Read the problem carefully.

Step 2 Assign a variable. Use a table to identify distance, rate, and time. Solve $d = rt$ for the unknown quantity in the table.

On a trip from Sacramento to Monterey, Marge traveled at an average speed of 60 mph. The return trip, at an average speed of 64 mph, took $\frac{1}{4}$ hr less. How far did she travel between the two cities?

Let x = the unknown distance.

	d	r	$t = \dfrac{d}{r}$
Going	x	60	$\dfrac{x}{60}$
Returning	x	64	$\dfrac{x}{64}$

Step 3 Write an equation. From the wording in the problem, decide the relationship between the quantities. Use those expressions to write an equation.

Since the time for the return trip was $\frac{1}{4}$ hr less, the time going equals the time returning plus $\frac{1}{4}$.

$$\frac{x}{60} = \frac{x}{64} + \frac{1}{4}$$

Step 4 Solve the equation.

$$16x = 15x + 240 \qquad \text{Multiply by the LCD, 960.}$$
$$x = 240 \qquad \text{Subtract } 15x.$$

Step 5 State the answer.

She traveled 240 mi.

Step 6 Check the solution.

The trip there took $\frac{240}{60} = 4$ hr, while the return trip took $\frac{240}{64} = 3\frac{3}{4}$ hr, which is $\frac{1}{4}$ hr less time. The solution checks.

Solving Problems about Work

Step 1 Read the problem carefully.

Step 2 Assign a variable. State what the variable represents. Put the information from the problem into a table. If a job is done in t units of time, the rate is $\frac{1}{t}$.

It takes the regular mail carrier 6 hr to cover her route. A substitute takes 8 hr to cover the same route. How long would it take them to cover the route together?

Let x = the number of hours required to cover the route together.

(continued)

Concepts	Examples

The rate of the regular carrier is $\frac{1}{6}$ job per hr; the rate of the substitute is $\frac{1}{8}$ job per hr. Multiply rate by time to get the fractional part of the job done.

	Rate	Time	Part of the Job Done
Regular	$\frac{1}{6}$	x	$\frac{1}{6}x$
Substitute	$\frac{1}{8}$	x	$\frac{1}{8}x$

Step 3 Write an equation. The sum of the fractional parts should equal 1 (whole job).

Step 4 Solve the equation.

$$\frac{1}{6}x + \frac{1}{8}x = 1$$

$$24\left(\frac{1}{6}x + \frac{1}{8}x\right) = 24(1) \qquad \text{The LCD is 24.}$$

$$4x + 3x = 24 \qquad \text{Distributive property}$$

$$7x = 24 \qquad \text{Combine like terms.}$$

$$x = \frac{24}{7} \qquad \text{Divide by 7.}$$

Steps 5 and 6 State the answer and check the solution.

It would take them $\frac{24}{7}$, or $3\frac{3}{7}$, hr to cover the route together. The solution checks because $\frac{1}{6}\left(\frac{24}{7}\right) + \frac{1}{8}\left(\frac{24}{7}\right) = 1$.

7.8 VARIATION

Solving Variation Problems

Step 1 Write the variation equation. Use

$y = kx$ or $y = kx^n$ Direct variation

$y = \dfrac{k}{x}$ or $y = \dfrac{k}{x^n}$. Inverse variation

Step 2 Find k by substituting the appropriate given values of x and y into the equation.

Step 3 Rewrite the variation equation with the value of k from Step 2.

Step 4 Substitute the remaining values, and solve for the unknown.

If y varies inversely as x, and $y = 4$ when $x = 9$, find y when $x = 6$.

The equation for inverse variation is

$$y = \frac{k}{x}.$$

$$4 = \frac{k}{9} \qquad \text{Substitute given values.}$$

$$k = 36 \qquad \text{Multiply by 9; rewrite.}$$

$$y = \frac{36}{x} \qquad k = 36$$

$$y = \frac{36}{6} = 6 \qquad \text{Let } x = 6.$$

Answers to Test Your Word Power

1. D; *Examples*: $-\dfrac{3}{4y}$, $\dfrac{5x^3}{x+2}$, $\dfrac{a+3}{a^2 - 4a - 5}$ **2.** C; *Examples*: $\dfrac{\frac{2}{3}}{\frac{4}{7}}$, $\dfrac{x - \frac{1}{y}}{x + \frac{1}{y}}$, $\dfrac{\frac{2}{a+1}}{a^2 - 1}$

| Chapter | **7** | **REVIEW EXERCISES** |

[7.1] *Find the numerical value of each rational expression when* **(a)** $x = -2$ *and* **(b)** $x = 4$.

1. $\dfrac{4x - 3}{5x + 2}$

2. $\dfrac{3x}{x^2 - 4}$

Find any values of the variable for which each rational expression is undefined. Write answers with $\neq$.

3. $\dfrac{4}{x - 3}$

4. $\dfrac{y + 3}{2y}$

5. $\dfrac{2k + 1}{3k^2 + 17k + 10}$

✎ **6.** How would you determine the values of the variable for which a rational expression is undefined?

Write each rational expression in lowest terms.

7. $\dfrac{5a^3b^3}{15a^4b^2}$

8. $\dfrac{m - 4}{4 - m}$

9. $\dfrac{4x^2 - 9}{6 - 4x}$

10. $\dfrac{4p^2 + 8pq - 5q^2}{10p^2 - 3pq - q^2}$

Write four equivalent forms for each rational expression.

11. $-\dfrac{4x - 9}{2x + 3}$

12. $-\dfrac{8 - 3x}{3 - 6x}$

[7.2] *Multiply or divide, and write each answer in lowest terms.*

13. $\dfrac{18p^3}{6} \cdot \dfrac{24}{p^4}$

14. $\dfrac{8x^2}{12x^5} \cdot \dfrac{6x^4}{2x}$

15. $\dfrac{x - 3}{4} \cdot \dfrac{5}{2x - 6}$

16. $\dfrac{2r + 3}{r - 4} \cdot \dfrac{r^2 - 16}{6r + 9}$

17. $\dfrac{6a^2 + 7a - 3}{2a^2 - a - 6} \div \dfrac{a + 5}{a - 2}$

18. $\dfrac{y^2 - 6y + 8}{y^2 + 3y - 18} \div \dfrac{y - 4}{y + 6}$

19. $\dfrac{2p^2 + 13p + 20}{p^2 + p - 12} \cdot \dfrac{p^2 + 2p - 15}{2p^2 + 7p + 5}$

20. $\dfrac{3z^2 + 5z - 2}{9z^2 - 1} \cdot \dfrac{9z^2 + 6z + 1}{z^2 + 5z + 6}$

[7.3] *Find the least common denominator for the fractions in each list.*

21. $\dfrac{4}{9y}, \dfrac{7}{12y^2}, \dfrac{5}{27y^4}$

22. $\dfrac{3}{x^2 + 4x + 3}, \dfrac{5}{x^2 + 5x + 4}$

Rewrite each rational expression with the given denominator.

23. $\dfrac{3}{2a^3} = \dfrac{?}{10a^4}$

24. $\dfrac{9}{x - 3} = \dfrac{?}{18 - 6x}$

25. $\dfrac{-3y}{2y - 10} = \dfrac{?}{50 - 10y}$

26. $\dfrac{4b}{b^2 + 2b - 3} = \dfrac{?}{(b + 3)(b - 1)(b + 2)}$

[7.4] *Add or subtract, and write each answer in lowest terms.*

27. $\dfrac{10}{x} + \dfrac{5}{x}$

28. $\dfrac{6}{3p} - \dfrac{12}{3p}$

29. $\dfrac{9}{k} - \dfrac{5}{k - 5}$

30. $\dfrac{4}{y} + \dfrac{7}{7+y}$

31. $\dfrac{m}{3} - \dfrac{2+5m}{6}$

32. $\dfrac{12}{x^2} - \dfrac{3}{4x}$

33. $\dfrac{5}{a-2b} + \dfrac{2}{a+2b}$

34. $\dfrac{4}{k^2-9} - \dfrac{k+3}{3k-9}$

35. $\dfrac{8}{z^2+6z} - \dfrac{3}{z^2+4z-12}$

36. $\dfrac{11}{2p-p^2} - \dfrac{2}{p^2-5p+6}$

[7.5]

37. Simplify the complex fraction $\dfrac{\dfrac{a^4}{b^2}}{\dfrac{a^3}{b}}$ by

    (a) Method 1 as described in **Section 7.5.**

    (b) Method 2 as described in **Section 7.5.**

    (c) Explain which method you prefer, and why.

Simplify each complex fraction.

38. $\dfrac{\dfrac{2}{3} - \dfrac{1}{6}}{\dfrac{1}{4} + \dfrac{2}{5}}$

39. $\dfrac{\dfrac{y-3}{y}}{\dfrac{y+3}{4y}}$

40. $\dfrac{\dfrac{1}{p} - \dfrac{1}{q}}{\dfrac{1}{q-p}}$

41. $\dfrac{x + \dfrac{1}{w}}{x - \dfrac{1}{w}}$

42. $\dfrac{\dfrac{1}{r+t} - 1}{\dfrac{1}{r+t} + 1}$

[7.6]

43. *Concept Check* Before even beginning to solve, how do you know that 2 cannot be a solution of the equation found in Exercise 46?

Solve each equation, and check your solutions.

44. $\dfrac{4-z}{z} + \dfrac{3}{2} = \dfrac{-4}{z}$

45. $\dfrac{3x-1}{x-2} = \dfrac{5}{x-2} + 1$

46. $\dfrac{3}{m-2} + \dfrac{1}{m-1} = \dfrac{7}{m^2-3m+2}$

Solve each formula for the specified variable.

47. $m = \dfrac{Ry}{t}$ for t

48. $x = \dfrac{3y-5}{4}$ for y

49. $p^2 = \dfrac{4}{3m-q}$ for m

[7.7] *Solve each problem.*

50. In a certain fraction, the denominator is 5 less than the numerator. If 5 is added to both the numerator and the denominator, the resulting fraction is equivalent to $\frac{5}{4}$. Find the original fraction (*not* written in lowest terms).

51. The denominator of a certain fraction is six times the numerator. If 3 is added to the numerator and subtracted from the denominator, the resulting fraction is equivalent to $\frac{2}{5}$. Find the original fraction (*not* written in lowest terms).

52. A plane flies 350 mi with the wind in the same time that it can fly 310 mi against the wind. The plane has a speed of 165 mph in still air. Find the speed of the wind.

53. Dennis Manieri can plant his garden in 5 hr working alone. A friend can do the same job in 8 hr. How long would it take them if they worked together?

54. The head gardener can mow the lawns in the city park twice as fast as his assistant. Working together, they can complete the job in $1\frac{1}{3}$ hr. How long would it take the head gardener working alone?

[7.8] *Solve each problem.*

55. *Concept Check* The longer the term of your subscription to *ESPN: The Magazine,* the less you will have to pay per year. Is this an example of direct or inverse variation?

56. If a parallelogram has a fixed area, the height varies inversely as the base. A parallelogram has a height of 8 cm and a base of 12 cm. Find the height if the base is changed to 24 cm.

57. If y varies directly as x, and $x = 12$ when $y = 5$, find x when $y = 3$.

MIXED REVIEW EXERCISES

Perform each indicated operation.

58. $\dfrac{4}{m-1} - \dfrac{3}{m+1}$

59. $\dfrac{8p^5}{5} \div \dfrac{2p^3}{10}$

60. $\dfrac{r-3}{8} \div \dfrac{3r-9}{4}$

61. $\dfrac{\dfrac{5}{x} - 1}{\dfrac{5-x}{3x}}$

62. $\dfrac{4}{z^2 - 2z + 1} - \dfrac{3}{z^2 - 1}$

Solve.

63. $a = \dfrac{v - w}{t}$ for v

64. $\dfrac{2}{z} - \dfrac{z}{z+3} = \dfrac{1}{z+3}$

65. With spraying equipment, Lizette Foley can paint the woodwork in a small house in 8 hr. Laura Hillerbrand needs 14 hr to complete the same job painting by hand. If Lizette and Laura work together, how long will it take them to paint the woodwork?

66. Rob Fusco flew his plane 400 km with the wind in the same time it took him to go 200 km against the wind. The speed of the wind is 50 km per hr. Find the speed of the plane in still air.

67. In a rectangle of constant area, the length and the width vary inversely. When the length is 24, the width is 2. What is the width when the length is 12?

68. If w varies inversely as z, and $w = 16$ when $z = 3$, find w when $z = 2$.

RELATING CONCEPTS   (EXERCISES 69–78)

FOR INDIVIDUAL OR GROUP WORK

In these exercises, we summarize the various concepts involving rational expressions.
Work Exercises 69–78 in order.
   Let *P*, *Q*, and *R* be rational expressions defined as follows:

$$P = \frac{6}{x + 3} \qquad Q = \frac{5}{x + 1} \qquad R = \frac{4x}{x^2 + 4x + 3}.$$

69. Find the value or values for which the expression is undefined.

   (a) *P*   **(b)** *Q*   **(c)** *R*

70. Find and express $(P \cdot Q) \div R$ in lowest terms.

71. Why is $(P \cdot Q) \div R$ not defined if $x = 0$?

72. Find the LCD for *P*, *Q*, and *R*.

73. Perform the operations and express $P + Q - R$ in lowest terms.

74. Simplify the complex fraction $\frac{P + Q}{R}$.

75. Solve the equation $P + Q = R$.

76. How does your answer to Exercise 69 help you work Exercise 75?

77. Suppose that a car travels 6 mi in $x + 3$ min. Explain why *P* represents the rate of the car (in miles per minute).

78. For what value or values of *x* is $R = \frac{40}{77}$?

Chapter **7** TEST

View the complete solutions to all Chapter Test exercises on the Pass the Test CD.

1. Find the numerical value of $\dfrac{6r + 1}{2r^2 - 3r - 20}$ when **(a)** $r = -2$ and **(b)** $r = 4$.

2. Find any values for which $\dfrac{3x - 1}{x^2 - 2x - 8}$ is undefined. Write your answer with $\neq$.

3. Write four rational expressions equivalent to $-\dfrac{6x - 5}{2x + 3}$.

Write each rational expression in lowest terms.

4. $\dfrac{-15x^6 y^4}{5x^4 y}$

5. $\dfrac{6a^2 + a - 2}{2a^2 - 3a + 1}$

Multiply or divide. Write each answer in lowest terms.

6. $\dfrac{5(d - 2)}{9} \div \dfrac{3(d - 2)}{5}$

7. $\dfrac{6k^2 - k - 2}{8k^2 + 10k + 3} \cdot \dfrac{4k^2 + 7k + 3}{3k^2 + 5k + 2}$

8. $\dfrac{4a^2 + 9a + 2}{3a^2 + 11a + 10} \div \dfrac{4a^2 + 17a + 4}{3a^2 + 2a - 5}$

Find the least common denominator for the fractions in each list.

9. $\dfrac{-3}{10p^2}, \dfrac{21}{25p^3}, \dfrac{-7}{30p^5}$

10. $\dfrac{r + 1}{2r^2 + 7r + 6}, \dfrac{-2r + 1}{2r^2 - 7r - 15}$

Rewrite each rational expression with the given denominator.

11. $\dfrac{15}{4p} = \dfrac{?}{64p^3}$

12. $\dfrac{3}{6m - 12} = \dfrac{?}{42m - 84}$

Add or subtract. Write each answer in lowest terms.

13. $\dfrac{4x + 2}{x + 5} + \dfrac{-2x + 8}{x + 5}$

14. $\dfrac{-4}{y + 2} + \dfrac{6}{5y + 10}$

15. $\dfrac{x + 1}{3 - x} + \dfrac{x^2}{x - 3}$

16. $\dfrac{3}{2m^2 - 9m - 5} - \dfrac{m + 1}{2m^2 - m - 1}$

Simplify each complex fraction.

17. $\dfrac{\dfrac{2p}{k^2}}{\dfrac{3p^2}{k^3}}$

18. $\dfrac{\dfrac{1}{x + 3} - 1}{1 + \dfrac{1}{x + 3}}$

Solve.

19. $\dfrac{2x}{x - 3} + \dfrac{1}{x + 3} = \dfrac{-6}{x^2 - 9}$

20. $F = \dfrac{k}{d - D}$ for D

Solve each problem.

21. A boat goes 7 mph in still water. It takes as long to go 20 mi upstream as 50 mi downstream. Find the speed of the current.

22. Abdalla Elusta can paint a room in his house, working alone, in 5 hr. His neighbor can do the job in 4 hr. How long will it take them to paint the room if they work together?

23. If x varies directly as y, and $x = 12$ when $y = 4$, find x when $y = 9$.

24. Under certain conditions, the length of time that it takes for fruit to ripen during the growing season varies inversely as the average maximum temperature during the season. If it takes 25 days for fruit to ripen with an average maximum temperature of 80°, find the number of days it would take at 75°. Round your answer to the nearest whole number.

Chapters 1–7 CUMULATIVE REVIEW EXERCISES

1. Use the order of operations to evaluate $3 + 4\left(\frac{1}{2} - \frac{3}{4}\right)$.

Solve.

2. $3(2y - 5) = 2 + 5y$

3. $A = \frac{1}{2}bh$ for b

4. $\frac{2 + m}{2 - m} = \frac{3}{4}$

5. $5y \le 6y + 8$

6. $5m - 9 > 2m + 3$

7. For the graph of $4x + 3y = -12$,

    (a) what is the *x*-intercept?   **(b)** what is the *y*-intercept?

Sketch each graph.

8. $y = -3x + 2$

9. $y = -x^2 + 1$

Solve each system.

10. $4x - y = -7$
    $5x + 2y = 1$

11. $5x + 2y = 7$
    $10x + 4y = 12$

Simplify each expression. Write with only positive exponents.

12. $\dfrac{(2x^3)^{-1} \cdot x}{2^3 x^5}$

13. $\dfrac{(m^{-2})^3 m}{m^5 m^{-4}}$

14. $\dfrac{2p^3 q^4}{8p^5 q^3}$

Perform each indicated operation.

15. $(2k^2 + 3k) - (k^2 + k - 1)$

16. $8x^2 y^2(9x^4 y^5)$

17. $(2a - b)^2$

18. $(y^2 + 3y + 5)(3y - 1)$

19. $\dfrac{12p^3 + 2p^2 - 12p + 4}{2p - 2}$

20. A computer can do one operation in 1.4×10^{-7} sec. How long would it take for the computer to do one trillion (10^{12}) operations?

Factor completely.

21. $8t^2 + 10tv + 3v^2$

22. $8r^2 - 9rs + 12s^2$

23. $16x^4 - 1$

Solve each equation.

24. $r^2 = 2r + 15$

25. $(r - 5)(2r + 1)(3r - 2) = 0$

Solve each problem.

26. One number is 4 more than another. The product of the numbers is 2 less than the smaller number. Find the smaller number.

27. The length of a rectangle is 2 m less than twice the width. The area is 60 m². Find the width of the rectangle.

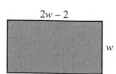

28. One of the following is equal to 1 for *all* real numbers. Which one is it?

A. $\dfrac{k^2 + 2}{k^2 + 2}$ **B.** $\dfrac{4 - m}{4 - m}$ **C.** $\dfrac{2x + 9}{2x + 9}$ **D.** $\dfrac{x^2 - 1}{x^2 - 1}$

29. Which one of the following rational expressions is *not* equivalent to $\dfrac{4 - 3x}{7}$?

A. $-\dfrac{-4 + 3x}{7}$ **B.** $-\dfrac{4 - 3x}{-7}$ **C.** $\dfrac{-4 + 3x}{-7}$ **D.** $\dfrac{-(3x + 4)}{7}$

Perform each operation and write the answer in lowest terms.

30. $\dfrac{5}{q} - \dfrac{1}{q}$

31. $\dfrac{3}{7} + \dfrac{4}{r}$

32. $\dfrac{4}{5q - 20} - \dfrac{1}{3q - 12}$

33. $\dfrac{2}{k^2 + k} - \dfrac{3}{k^2 - k}$

34. $\dfrac{7z^2 + 49z + 70}{16z^2 + 72z - 40} \div \dfrac{3z + 6}{4z^2 - 1}$

35. $\dfrac{\dfrac{4}{a} + \dfrac{5}{2a}}{\dfrac{7}{6a} - \dfrac{1}{5a}}$

36. When solving the equation $\dfrac{1}{x - 4} = \dfrac{3}{2x}$, which values of x cannot be proposed solutions? Write your answer with $\neq$.

Solve each equation. Check your solutions.

37. $\dfrac{r + 2}{5} = \dfrac{r - 3}{3}$

38. $\dfrac{1}{x} = \dfrac{1}{x + 1} + \dfrac{1}{2}$

Solve each problem.

39. Jody Harris can weed the yard in 3 hr. Frances Keiper can weed the same yard in 2 hr. How long would it take them if they worked together?

40. The circumference of a circle varies directly as its radius. A circle with circumference 9.42 in. has radius approximately 1.5 in. Find the circumference of a circle with radius 5.25 in.

$C = 9.42$ in.

Roots and Radicals

The London Eye opened on New Year's Eve in 1999. This unique Ferris wheel features 32 observation capsules and has a diameter of 135 m. Located on the bank of the Thames River, it faces the Houses of Parliament and is the fourth-tallest structure in London. (*Source:* www.londoneye.com)

The formula

$$\text{sight distance} = 111.7\sqrt{\text{height of structure in kilometers}}$$

can be used to determine how far one can see (in kilometers) from the top of a structure on a clear day. (*Source: A Sourcebook of Applications of School Mathematics,* NCTM, 1980.) In Exercise 77 of Section 8.6, we use this formula to determine the truth of the claim that passengers on the London Eye can see Windsor Castle, 25 mi away.

8.1 Evaluating Roots

OBJECTIVES

1. Find square roots.
2. Decide whether a given root is rational, irrational, or not a real number.
3. Find decimal approximations for irrational square roots.
4. Use the Pythagorean formula.
5. Use the distance formula.
6. Find cube, fourth, and other roots.

In **Section 1.2,** we discussed the idea of the *square* of a number. Recall that squaring a number means multiplying the number by itself.

$$\text{If}\quad a = 8, \qquad \text{then}\quad a^2 = 8 \cdot 8 = 64.$$
$$\text{If}\quad a = -4, \qquad \text{then}\quad a^2 = (-4)(-4) = 16.$$
$$\text{If}\quad a = -\frac{1}{2}, \quad \text{then}\quad a^2 = \left(-\frac{1}{2}\right)\left(-\frac{1}{2}\right) = \frac{1}{4}.$$

In this chapter, we consider the opposite process.

$$\text{If}\quad a^2 = 49, \qquad \text{then}\quad a = ?.$$
$$\text{If}\quad a^2 = 100, \quad \text{then}\quad a = ?.$$
$$\text{If}\quad a^2 = 25, \qquad \text{then}\quad a = ?.$$

OBJECTIVE ❶ Find square roots. To find a in the preceding three statements, we must find a number that, when multiplied by itself, results in the given number. The number a is called a **square root** of the number a^2.

EXAMPLE 1 Finding All Square Roots of a Number

Find all square roots of 49.

To find a square root of 49, think of a number that, when multiplied by itself, gives 49. One square root is 7, because $7 \cdot 7 = 49$. Another square root of 49 is -7, because $(-7)(-7) = 49$. The number 49 has two square roots: 7 and -7; one is positive, and one is negative.

✔ **Now Try Exercise 7.**

The **positive** or **principal square root** of a number is written with the symbol $\sqrt{}$. For example, the positive square root of 121 is 11, written

$$\sqrt{121} = 11.$$

The symbol $-\sqrt{}$ is used for the **negative square root** of a number. For example, the negative square root of 121 is -11, written

$$-\sqrt{121} = -11.$$

The symbol $\sqrt{}$, called a **radical sign,** always represents the positive square root $\big(\text{except that } \sqrt{0} = 0\big)$. The number inside the radical sign is called the **radicand,** and the entire expression—radical sign and radicand—is called a **radical.**

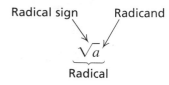

Radical sign   Radicand

$\sqrt{a}$

Radical

An algebraic expression containing a radical is called a **radical expression.**

Radicals have a long mathematical history. The radical sign $\sqrt{}$ has been used since 16-century Germany and was probably derived from the letter R. The radical

Early radical symbol

symbol in the margin on the preceding page comes from the Latin word *radix,* for *root.* It was first used by Leonardo of Pisa (Fibonacci) in 1220.

We summarize our discussion of square roots as follows.

Square Roots of *a*

If *a* is a positive real number, then

$$\sqrt{a} \text{ is the positive or principal square root of } a,$$

and $-\sqrt{a}$ is the negative square root of *a.*

For nonnegative *a,*

$$\sqrt{a} \cdot \sqrt{a} = \left(\sqrt{a}\right)^2 = a \quad \text{and} \quad -\sqrt{a} \cdot \left(-\sqrt{a}\right) = \left(-\sqrt{a}\right)^2 = a.$$

Also, $\sqrt{0} = 0.$

EXAMPLE 2 Finding Square Roots

Find each square root.

(a) $\sqrt{144}$

The radical $\sqrt{144}$ represents the positive or principal square root of 144. Think of a positive number whose square is 144.

$$12^2 = 144, \quad \text{so} \quad \sqrt{144} = 12.$$

(b) $-\sqrt{1024}$

This symbol represents the negative square root of 1024. A calculator with a square root key can be used to find $\sqrt{1024} = 32$. Then, $-\sqrt{1024} = -32$.

(c) $\sqrt{\dfrac{4}{9}} = \dfrac{2}{3}$ **(d)** $-\sqrt{\dfrac{16}{49}} = -\dfrac{4}{7}$ **(e)** $\sqrt{0.81} = 0.9$

> ✔ **Now Try Exercises 19, 21, and 23.**

As shown in the preceding definition, when the square root of a positive real number is squared, the result is that positive real number. $\left(\text{Also, } \left(\sqrt{0}\right)^2 = 0.\right)$

EXAMPLE 3 Squaring Radical Expressions

Find the *square* of each radical expression.

(a) $\sqrt{13}$

$$\left(\sqrt{13}\right)^2 = 13 \qquad \text{Definition of square root}$$

(b) $-\sqrt{29}$

$$\left(-\sqrt{29}\right)^2 = 29 \qquad \text{The square of a } negative \text{ number is positive.}$$

(c) $\sqrt{p^2 + 1}$

$$\left(\sqrt{p^2 + 1}\right)^2 = p^2 + 1$$

> ✔ **Now Try Exercises 27, 29, and 33.**

OBJECTIVE ② **Decide whether a given root is rational, irrational, or not a real number.**
All numbers with square roots that are rational are called **perfect squares.**

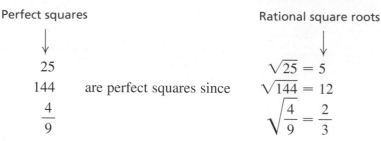

Perfect squares   Rational square roots

$$25$$
$$144 \quad \text{are perfect squares since} \quad \sqrt{144} = 12$$
$$\frac{4}{9}$$

$$\sqrt{25} = 5$$
$$\sqrt{144} = 12$$
$$\sqrt{\frac{4}{9}} = \frac{2}{3}$$

A number that is not a perfect square has a square root that is not a rational number. For example, $\sqrt{5}$ is not a rational number because it cannot be written as the ratio of two integers. Its decimal equivalent (or approximation) neither terminates nor repeats. However, $\sqrt{5}$ is a real number and corresponds to a point on the number line. As mentioned in **Section 1.4,** a real number that is not rational is called an *irrational number.* The number $\sqrt{5}$ is irrational. Many square roots of integers are irrational.

If a is a positive real number that is not a perfect square, then $\sqrt{a}$ is irrational.

Not every number has a real number square root. For example, there is no real number that can be squared to get -36. (The square of a real number can never be negative.) Because of this, $\sqrt{-36}$ *is not a real number.*

If a is a negative real number, then $\sqrt{a}$ is not a real number.

▶ **CAUTION** Be careful not to confuse $\sqrt{-36}$ and $-\sqrt{36}$. $\sqrt{-36}$ is not a real number, since there is no real number that can be squared to get -36. However, $-\sqrt{36}$ is the negative square root of 36, which is -6.

EXAMPLE 4 Identifying Types of Square Roots

Tell whether each square root is *rational, irrational,* or *not a real number.*

(a) $\sqrt{17}$

Because 17 is not a perfect square, $\sqrt{17}$ is irrational.

(b) $\sqrt{64}$

The number 64 is a perfect square, 8^2, so $\sqrt{64} = 8$, a rational number.

(c) $\sqrt{-25}$

There is no real number whose square is -25. Therefore, $\sqrt{-25}$ is not a real number.

✔ **Now Try Exercises 39, 41, and 47.**

▶ **NOTE** Not all irrational numbers are square roots of integers. For example, π (approximately 3.14159) is an irrational number that is not a square root of any integer.

OBJECTIVE 3 Find decimal approximations for irrational square roots. Even if a number is irrational, a decimal that approximates the number can be found using a calculator. For example, if we use a calculator to find $\sqrt{10}$, the display might show 3.16227766, which is an *approximation* of $\sqrt{10}$, not an exact rational value. See **Appendix A** for instructions on using the square root key on a calculator.

EXAMPLE 5 Approximating Irrational Square Roots

Find a decimal approximation for each square root. Round answers to the nearest thousandth.

(a) $\sqrt{11}$

Using the square root key on a calculator gives $3.31662479 \approx 3.317$, where $\approx$ means "is approximately equal to."

(b) $\sqrt{39} \approx 6.245$ Use a calculator.   **(c)** $-\sqrt{740} \approx -27.203$

✔ **Now Try Exercises 45 and 49.**

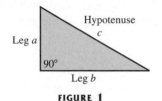

Hypotenuse
c

Leg a

$90°$

Leg b

FIGURE 1

OBJECTIVE 4 Use the Pythagorean formula. Many applications of square roots require the use of the Pythagorean formula. Recall from **Section 6.6** that if c is the length of the hypotenuse of a right triangle, and a and b are the lengths of the two legs, then

$$a^2 + b^2 = c^2.$$

See Figure 1.

In the next example, we use the fact that if $k > 0$, then the positive solution of the equation $x^2 = k$ is $\sqrt{k}$. (See page 614.)

EXAMPLE 6 Using the Pythagorean Formula

Find the length of the unknown side of each right triangle with sides a, b, and c, where c is the hypotenuse.

(a) $a = 3, b = 4$

Use the Pythagorean formula to find c^2 first.

$$a^2 + b^2 = c^2$$
$$3^2 + 4^2 = c^2 \quad \text{Let } a = 3 \text{ and } b = 4.$$
$$9 + 16 = c^2 \quad \text{Square.}$$
$$25 = c^2 \quad \text{Add.}$$

Since the length of a side of a triangle must be a positive number, find the positive square root of 25 to get c.

$$c = \sqrt{25} = 5$$

(b) $c = 9, b = 5$

Substitute the given values into the Pythagorean formula. Then solve for a^2.

$$a^2 + b^2 = c^2$$
$$a^2 + 5^2 = 9^2 \qquad \text{Let } c = 9 \text{ and } b = 5.$$
$$a^2 + 25 = 81 \qquad \text{Square.}$$
$$a^2 = 56 \qquad \text{Subtract 25.}$$

Use a calculator to find $a = \sqrt{56} \approx 7.483$.

✔ **Now Try Exercises 61 and 63.**

▶ **CAUTION** Be careful not to make the common mistake of thinking that $\sqrt{a^2 + b^2}$ equals $a + b$. As Example 6(a) shows, $\sqrt{9 + 16} = \sqrt{25} = 5$. However, $\sqrt{9} + \sqrt{16} = 3 + 4 = 7$. Since $5 \neq 7$, in general,

$$\sqrt{a^2 + b^2} \neq a + b.$$

EXAMPLE 7 Using the Pythagorean Formula to Solve an Application

A ladder 10 ft long leans against a wall. The foot of the ladder is 6 ft from the base of the wall. How high up the wall does the top of the ladder rest?

Step 1 **Read** the problem again.

Step 2 **Assign a variable.** As shown in Figure 2, a right triangle is formed with the ladder as the hypotenuse. Let a represent the height of the top of the ladder, measured straight down to the ground.

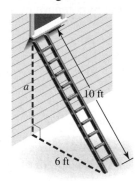

FIGURE 2

Step 3 **Write an equation** using the Pythagorean formula.

Substitute carefully.

$$c^2 = a^2 + b^2$$
$$10^2 = a^2 + 6^2 \qquad \text{Let } c = 10 \text{ and } b = 6.$$

Step 4 **Solve.**
$$100 = a^2 + 36 \qquad \text{Square.}$$
$$64 = a^2 \qquad \text{Subtract 36.}$$
$$\sqrt{64} = a$$
$$a = 8 \qquad \sqrt{64} = 8$$

We choose the positive square root of 64 because a represents a length.

Step 5 **State the answer.** The top of the ladder rests 8 ft up the wall.

Step 6 **Check.** From Figure 2, we see that we must have

$$8^2 + 6^2 = 10^2 \quad ?$$
$$64 + 36 = 100. \qquad \text{True}$$

The check confirms that the top of the ladder rests 8 ft up the wall.

✔ **Now Try Exercise 69.**

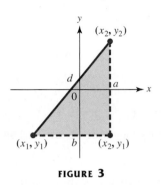

FIGURE 3

OBJECTIVE 5 **Use the distance formula.** Consider Figure 3. The distance between the points (x_2, y_2) and (x_2, y_1) is $a = y_2 - y_1$, and the distance between the points (x_1, y_1) and (x_2, y_1) is $b = x_2 - x_1$. From the Pythagorean formula,

$$d^2 = (x_2 - x_1)^2 + (y_2 - y_1)^2.$$

Taking the square root of each side, we get the **distance formula.**

Distance Formula

The distance between the points (x_1, y_1) and (x_2, y_2) is

$$d = \sqrt{(x_2 - x_1)^2 + (y_2 - y_1)^2}.$$

EXAMPLE 8 **Using the Distance Formula**

Find the distance between $(-3, 4)$ and $(2, 5)$.

$$d = \sqrt{(x_2 - x_1)^2 + (y_2 - y_1)^2} \qquad \text{Distance formula}$$
$$= \sqrt{(2 - (-3))^2 + (5 - 4)^2} \qquad \text{Let } (x_1, y_1) = (-3, 4)$$
$$\text{and } (x_2, y_2) = (2, 5).$$
$$= \sqrt{5^2 + 1^2}$$
$$= \sqrt{26}$$

Start with the x-value and the y-value of the same point.

✔ **Now Try Exercise 89.**

OBJECTIVE 6 **Find cube, fourth, and other roots.** Finding the square root of a number is the inverse (reverse) of squaring a number. In a similar way, there are inverses to finding the cube of a number or to finding the fourth or greater power of a number. These inverses are, respectively, the **cube root,** written $\sqrt[3]{a}$, and the **fourth root,** written $\sqrt[4]{a}$. Similar symbols are used for other roots. In general, we have the following.

$\sqrt[n]{a}$

The *n*th root of *a* is written $\sqrt[n]{a}$.

In $\sqrt[n]{a}$, the number n is the **index** or **order** of the radical.

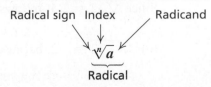

Radical sign   Index   Radicand

$\sqrt[n]{a}$

Radical

We could write $\sqrt[2]{a}$ instead of $\sqrt{a}$, but the simpler symbol $\sqrt{a}$ is customary, since the square root is the most commonly used root.

When working with cube roots or fourth roots, it is helpful to memorize the first few *perfect cubes* ($2^3 = 8$, $3^3 = 27$, and so on) and the first few *perfect fourth powers* ($2^4 = 16$, $3^4 = 81$, and so on).

EXAMPLE 9 Finding Cube Roots

Find each cube root.

(a) $\sqrt[3]{8}$

Look for a number that can be cubed to give 8. Because $2^3 = 8$, $\sqrt[3]{8} = 2$.

(b) $\sqrt[3]{-8} = -2$ because $(-2)^3 = -8$.

(c) $-\sqrt[3]{216} = -6$ because $\sqrt[3]{216} = 6$; thus, $-\sqrt[3]{216} = -6$.

> ✔ **Now Try Exercises 95, 97, and 99.**

Notice in Example 9(b) that we can find the cube root of a negative number. (Contrast this with the square root of a negative number, which is not real.) In fact, the cube root of a positive number is positive, and the cube root of a negative number is negative. ***There is only one real number cube root for each real number.***

When a radical has an ***even index*** (square root, fourth root, and so on), ***the radicand must be nonnegative*** to yield a real number root. Also,

$$\sqrt{a}, \sqrt[4]{a}, \sqrt[6]{a}, \text{ and so on are positive (principal) roots;}$$

$$-\sqrt{a}, -\sqrt[4]{a}, -\sqrt[6]{a}, \text{ and so on are negative roots.}$$

EXAMPLE 10 Finding Other Roots

Find each root.

(a) $\sqrt[4]{16} = 2$ because 2 is positive and $2^4 = 16$.

(b) $-\sqrt[4]{16}$

From part (a), $\sqrt[4]{16} = 2$, so the negative root is $-\sqrt[4]{16} = -2$.

(c) $\sqrt[4]{-16}$

For a real number fourth root, the radicand must be nonnegative. There is no real number that equals $\sqrt[4]{-16}$.

(d) $-\sqrt[5]{32}$

First find $\sqrt[5]{32}$. Because 2 is the number whose fifth power is 32, $\sqrt[5]{32} = 2$. Since $\sqrt[5]{32} = 2$, it follows that

$$-\sqrt[5]{32} = -2.$$

(e) $\sqrt[5]{-32} = -2$, because $(-2)^5 = -32$.

> ✔ **Now Try Exercises 101, 105, and 107.**

CONNECTIONS

Although Pythagoras may have written the first proof of the Pythagorean relationship, there is evidence that the Babylonians knew the concept quite well. The figure on the left illustrates the formula with a tile pattern. In the figure, the side of the square along the hypotenuse measures 5 units, while the sides along the legs measure 3 and 4 units. If we let $a = 3$, $b = 4$, and $c = 5$, the Pythagorean formula is satisfied.

$$a^2 + b^2 = c^2$$
$$3^2 + 4^2 = 5^2 \quad ?$$
$$25 = 25 \qquad \text{True}$$

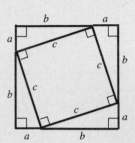

For Discussion or Writing

The diagram on the right can be used to verify the Pythagorean formula. To do so, express the area of the figure in two ways: first as the area of the large square, and then as the sum of the areas of the smaller square and the four right triangles. Finally, set the areas equal and simplify the equation.

8.1 EXERCISES

Concept Check *Decide whether each statement is* true *or* false. *If false, tell why.*

1. Every positive number has two real square roots.

2. A negative number has negative real square roots.

3. Every nonnegative number has two real square roots.

4. The positive square root of a positive number is its principal square root.

5. The cube root of every real number has the same sign as the number itself.

6. Every positive number has three real cube roots.

Find all square roots of each number. See Example 1.

⊙ **7.** 9   **8.** 16   **9.** 64   **10.** 100   **11.** 144

12. 225   **13.** $\dfrac{25}{196}$ **14.** $\dfrac{81}{400}$ **15.** 900   **16.** 1600

Find each square root. See Examples 2 and 4(c).

17. $\sqrt{1}$ **18.** $\sqrt{4}$ ⊙ **19.** $\sqrt{49}$ **20.** $\sqrt{81}$ **21.** $-\sqrt{121}$

22. $-\sqrt{196}$ **23.** $-\sqrt{\dfrac{144}{121}}$ **24.** $-\sqrt{\dfrac{49}{36}}$ **25.** $\sqrt{-121}$ **26.** $\sqrt{-64}$

Find the square of each radical expression. See Example 3.

⊙ **27.** $\sqrt{19}$ **28.** $\sqrt{59}$ **29.** $-\sqrt{19}$ **30.** $-\sqrt{99}$

31. $\sqrt{\dfrac{2}{3}}$ **32.** $\sqrt{\dfrac{5}{7}}$ **33.** $\sqrt{3x^2 + 4}$ **34.** $\sqrt{9y^2 + 3}$

Concept Check *What must be true about the variable a for each statement in Exercises 35–38 to be true?*

35. $\sqrt{a}$ represents a positive number.

36. $-\sqrt{a}$ represents a negative number.

37. $\sqrt{a}$ is not a real number.

38. $-\sqrt{a}$ is not a real number.

Determine whether each number is rational, irrational, *or* not a real number. *If a number is rational, give its exact value. If a number is irrational, give a decimal approximation to the nearest thousandth. Use a calculator as necessary. See Examples 4 and 5.*

39. $\sqrt{25}$ **40.** $\sqrt{169}$ **41.** $\sqrt{29}$ **42.** $\sqrt{33}$

43. $-\sqrt{64}$ **44.** $-\sqrt{81}$ **45.** $-\sqrt{300}$ **46.** $-\sqrt{500}$

47. $\sqrt{-29}$ **48.** $\sqrt{-47}$ **49.** $\sqrt{1200}$ **50.** $\sqrt{1500}$

Concept Check *Without using a calculator, determine between which two consecutive integers each square root lies. For example, $\sqrt{75}$ is between 8 and 9, because $\sqrt{64} = 8$, $\sqrt{81} = 9$, and $64 < 75 < 81$.*

51. $\sqrt{94}$ **52.** $\sqrt{43}$ **53.** $\sqrt{51}$ **54.** $\sqrt{30}$

55. $-\sqrt{40}$ **56.** $-\sqrt{63}$ **57.** $\sqrt{23.2}$ **58.** $\sqrt{10.3}$

Work Exercises 59 and 60 without using a calculator.

59. Choose the best estimate for the length and width (in meters) of this rectangle.

A. 11 by 6   **B.** 11 by 7
C. 10 by 7   **D.** 10 by 6

60. Choose the best estimate for the base and height (in feet) of this triangle.

A. $b = 8, h = 5$ **B.** $b = 8, h = 4$
C. $b = 9, h = 5$ **D.** $b = 9, h = 4$

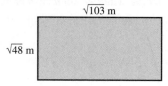

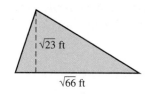

Find the length of the unknown side of each right triangle with sides a, b, and c, where c is the hypotenuse. See Figure 1 and Example 6. Give any decimal approximations to the nearest thousandth.

61. $a = 8, b = 15$ **62.** $a = 24, b = 10$ **63.** $a = 6, c = 9$

64. $b = 12, c = 17$ **65.** $a = 11, b = 4$ **66.** $a = 13, b = 9$

Use the Pythagorean formula to solve each problem. See Example 7.

67. The diagonal of a rectangle measures 25 cm. The width of the rectangle is 7 cm. Find the length of the rectangle.

68. The length of a rectangle is 40 m, and the width is 9 m. Find the measure of the diagonal of the rectangle.

69. Tyler is flying a kite on 100 ft of string. How high is it above his hand (vertically) if the horizontal distance between Tyler and the kite is 60 ft?

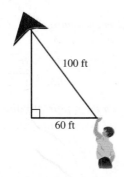

100 ft

60 ft

70. A guy wire is attached to the mast of a shortwave transmitting antenna at a point 96 ft above ground level. If the wire is staked to the ground 72 ft from the base of the mast, how long is the wire?

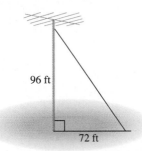

96 ft

72 ft

71. A surveyor measured the distances shown in the figure. Find the distance across the lake between points R and S.

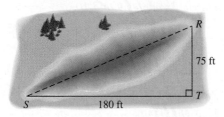

R

75 ft

S 180 ft T

72. A boat is being pulled toward a dock with a rope attached at water level. When the boat is 24 ft from the dock, 30 ft of rope is extended. What is the height of the dock above the water?

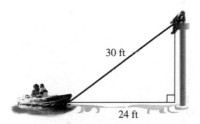

30 ft

24 ft

73. Following Hurricane Katrina, thousands of pine trees in southeastern Louisiana formed right triangles as shown in the photo. Suppose that, for a small such tree, the vertical distance from the base of the broken tree to the point of the break is 4.5 ft. The length of the broken part is 12 ft. How far along the ground (to the nearest tenth) is it from the base of the tree to the point where the broken part touches the ground?

74. One of the authors of this text recently purchased a new rear-projection Toshiba 51H84 television. A television set is "sized" according to the diagonal measurement of the viewing screen. The author purchased a 51-in. TV, so the TV measures 51 in. from one corner of the viewing screen diagonally to the other corner. The viewing screen is 44.5 in. wide. Find the height of the viewing screen (to the nearest tenth).

75. What is the value of x (to the nearest thousandth) in the figure?

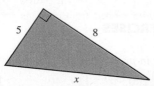

5

8

x

76. What is the value of y (to the nearest thousandth) in the figure?

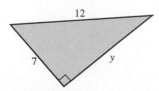

12

7

y

77. *Concept Check*   Use specific values for *a* and *b* different from those given in the "Caution" following Example 6 to show that $\sqrt{a^2 + b^2} \ne a + b$.

78. *Concept Check*   Why would the values $a = 0$ and $b = 1$ *not* be satisfactory in Exercise 77?

Concept Check   Find each square root. Express the answer as a whole number or as a square root. Do not use a calculator.

79. $\sqrt{3^2 + 4^2}$ **80.** $\sqrt{6^2 + 8^2}$ **81.** $\sqrt{8^2 + 15^2}$

82. $\sqrt{5^2 + 12^2}$ **83.** $\sqrt{(-2)^2 + 3^2}$ **84.** $\sqrt{(-1)^2 + 5^2}$

Find the distance between each pair of points. Express the answer as a whole number or as a square root. Do not use a calculator. See Example 8.

85. (5, 7) and (1, 4)   **86.** (8, 13) and (3, 1)

87. (2, 9) and (−3, −3)   **88.** (4, 6) and (−4, −9)

89. (−1, −2) and (−3, 1)   **90.** (−3, −6) and (−4, 0)

91. $\left(-\dfrac{1}{4}, \dfrac{2}{3}\right)$ and $\left(\dfrac{3}{4}, -\dfrac{1}{3}\right)$ **92.** $\left(\dfrac{2}{5}, \dfrac{3}{2}\right)$ and $\left(-\dfrac{8}{5}, \dfrac{1}{2}\right)$

Find each root. See Examples 9 and 10.

93. $\sqrt[3]{1}$ **94.** $\sqrt[3]{27}$ **95.** $\sqrt[3]{125}$ **96.** $\sqrt[3]{1000}$

97. $\sqrt[3]{-27}$ **98.** $\sqrt[3]{-64}$ **99.** $-\sqrt[3]{8}$ **100.** $-\sqrt[3]{343}$

101. $\sqrt[4]{625}$ **102.** $\sqrt[4]{10,000}$ **103.** $\sqrt[4]{-1}$ **104.** $\sqrt[4]{-625}$

105. $-\sqrt[4]{81}$ **106.** $-\sqrt[4]{256}$ **107.** $\sqrt[5]{-1024}$ **108.** $\sqrt[5]{-100,000}$

Use a calculator with a cube root key to find each root. Round to the nearest thousandth.

109. $\sqrt[3]{12}$ **110.** $\sqrt[3]{74}$ **111.** $\sqrt[3]{130.6}$

112. $\sqrt[3]{251.8}$ **113.** $\sqrt[3]{-87}$ **114.** $\sqrt[3]{-95}$

115. One of the many proofs of the Pythagorean formula was given in the Connections box in this section. Another one, attributed to the Hindu mathematician Bhāskara, is based on the figures shown here. The figure on the left is made up of the same square and triangles as the figure on the right. Use this information to derive the Pythagorean formula.

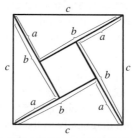 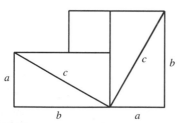

116. Investigate the derivation of the Pythagorean formula that was given by President James Garfield, and write a report on it.

PREVIEW EXERCISES

*Write each number in prime factored form. See **Section 6.1**.*

117. 72   **118.** 100   **119.** 40   **120.** 242   **121.** 23   **122.** 41

8.2 Multiplying, Dividing, and Simplifying Radicals

OBJECTIVE 1 Multiply square root radicals. In this section, we develop several rules for finding products and quotients of radicals. To illustrate the rule for products, notice that

$$\sqrt{4} \cdot \sqrt{9} = 2 \cdot 3 = 6 \quad \text{and} \quad \sqrt{4 \cdot 9} = \sqrt{36} = 6,$$

showing that
$$\sqrt{4} \cdot \sqrt{9} = \sqrt{4 \cdot 9}.$$

This result is a particular case of the more general product rule for radicals.

Product Rule for Radicals

For nonnegative real numbers a and b,

$$\sqrt{a} \cdot \sqrt{b} = \sqrt{a \cdot b} \quad \text{and} \quad \sqrt{a \cdot b} = \sqrt{a} \cdot \sqrt{b}.$$

That is, the product of two square roots is the square root of the product, and the square root of a product is the product of the square roots.

▶ **CAUTION** In general, $\sqrt{x + y} \neq \sqrt{x} + \sqrt{y}$. To see why this is so, let $x = 16$ and $y = 9$.

$$\sqrt{16 + 9} = \sqrt{25} = 5, \quad \text{but} \quad \sqrt{16} + \sqrt{9} = 4 + 3 = 7.$$

EXAMPLE 1 Using the Product Rule to Multiply Radicals

Use the product rule for radicals to find each product. Assume that $a \geq 0$.

(a) $\sqrt{2} \cdot \sqrt{3} = \sqrt{2 \cdot 3} = \sqrt{6}$ Product rule

(b) $\sqrt{7} \cdot \sqrt{5} = \sqrt{35}$ Product rule

(c) $\sqrt{11} \cdot \sqrt{a} = \sqrt{11a}$ **(d)** $\sqrt{31} \cdot \sqrt{31} = 31$

☑ **Now Try Exercises 7, 15, and 17.**

OBJECTIVE 2 Simplify radicals by using the product rule. *A square root radical is simplified when no perfect square factor remains under the radical sign.* We accomplish this by using the product rule in the form $\sqrt{a \cdot b} = \sqrt{a} \cdot \sqrt{b}$.

EXAMPLE 2 Using the Product Rule to Simplify Radicals

Simplify each radical.

(a) $\sqrt{20}$

     Because 20 has a perfect square factor of 4, we can write

$$\sqrt{20} = \sqrt{4 \cdot 5} \qquad \text{4 is a perfect square.}$$
$$= \sqrt{4} \cdot \sqrt{5} \qquad \text{Product rule}$$
$$= 2\sqrt{5}. \qquad \sqrt{4} = 2$$

Thus, $\sqrt{20} = 2\sqrt{5}$. Because 5 has no perfect square factor (other than 1), $2\sqrt{5}$ is called the **simplified form** of $\sqrt{20}$. Note that $2\sqrt{5}$ represents a product whose factors are 2 and $\sqrt{5}$.

We could also factor 20 into prime factors and look for pairs of like factors. Each pair of like factors produces one factor outside the radical in the simplified form.

$$\sqrt{20} = \sqrt{2 \cdot 2 \cdot 5} = 2\sqrt{5}$$

(b) $\sqrt{72}$

Look for the *greatest* perfect square factor of 72. This number is 36, so

$$\sqrt{72} = \sqrt{36 \cdot 2} \qquad \text{36 is a perfect square.}$$
$$= \sqrt{36} \cdot \sqrt{2} \qquad \text{Product rule}$$
$$= 6\sqrt{2}. \qquad \sqrt{36} = 6$$

We could also factor 72 into its prime factors and look for pairs of like factors.

$$\sqrt{72} = \sqrt{2 \cdot 2 \cdot 2 \cdot 3 \cdot 3} = 2 \cdot 3 \cdot \sqrt{2} = 6\sqrt{2}$$

In either case, we obtain $6\sqrt{2}$ as the simplified form of $\sqrt{72}$. However, our work is simpler if we begin with the greatest perfect square factor.

(c) $\sqrt{300} = \sqrt{100 \cdot 3} \qquad \text{100 is a perfect square.}$
$$= \sqrt{100} \cdot \sqrt{3} \qquad \text{Product rule}$$
$$= 10\sqrt{3} \qquad \sqrt{100} = 10$$

(d) $\sqrt{15}$

The number 15 has no perfect square factors (except 1), so $\sqrt{15}$ cannot be simplified further.

> ✔ **Now Try Exercises 23, 29, and 33.**

EXAMPLE 3 **Multiplying and Simplifying Radicals**

Find each product and simplify.

(a) $\sqrt{9} \cdot \sqrt{75} = 3\sqrt{75} \qquad \sqrt{9} = 3$
$$= 3\sqrt{25 \cdot 3} \qquad \text{25 is a perfect square.}$$
$$= 3\sqrt{25} \cdot \sqrt{3} \qquad \text{Product rule}$$
$$= 3 \cdot 5\sqrt{3} \qquad \sqrt{25} = 5$$
$$= 15\sqrt{3} \qquad \text{Multiply.}$$

Notice that we could have used the product rule to get $\sqrt{9} \cdot \sqrt{75} = \sqrt{675}$ and then simplified. However, the product rule as used here allows us to obtain the final answer without using a large number like 675.

(b) $\sqrt{8} \cdot \sqrt{12} = \sqrt{8 \cdot 12} \qquad \text{Product rule}$
$$= \sqrt{4 \cdot 2 \cdot 4 \cdot 3} \qquad \text{Factor; 4 is a perfect square.}$$
$$= \sqrt{4} \cdot \sqrt{4} \cdot \sqrt{2 \cdot 3} \qquad \text{Product rule}$$
$$= 2 \cdot 2 \cdot \sqrt{6} \qquad \sqrt{4} = 2$$
$$= 4\sqrt{6} \qquad \text{Multiply.}$$

> ✔ **Now Try Exercises 43 and 47.**

> ▶ **NOTE** We could also simplify the product in Example 3(b) as follows.
>
> $$\sqrt{8} \cdot \sqrt{12} = \sqrt{4 \cdot 2} \cdot \sqrt{4 \cdot 3}$$
> $$= 2\sqrt{2} \cdot 2\sqrt{3} \qquad \sqrt{4} = 2$$
> $$= 2 \cdot 2 \cdot \sqrt{2} \cdot \sqrt{3} \qquad \text{Commutative property}$$
> $$= 4\sqrt{6} \qquad\qquad \text{Multiply.}$$
>
> There is often more than one way to find such a product.

OBJECTIVE 3 **Simplify radicals by using the quotient rule.** The quotient rule for radicals is similar to the product rule. It, too, can be used either way.

Quotient Rule for Radicals

If a and b are nonnegative real numbers and $b \neq 0$, then

$$\sqrt{\frac{a}{b}} = \frac{\sqrt{a}}{\sqrt{b}} \quad \text{and} \quad \frac{\sqrt{a}}{\sqrt{b}} = \sqrt{\frac{a}{b}}.$$

That is, the square root of a quotient is the quotient of the square roots, and the quotient of two square roots is the square root of the quotient.

EXAMPLE 4 **Using the Quotient Rule to Simplify Radicals**

Simplify each radical.

(a) $\sqrt{\dfrac{25}{9}} = \dfrac{\sqrt{25}}{\sqrt{9}} = \dfrac{5}{3}$ Quotient rule

(b) $\dfrac{\sqrt{288}}{\sqrt{2}} = \sqrt{\dfrac{288}{2}} = \sqrt{144} = 12$ **(c)** $\sqrt{\dfrac{3}{4}} = \dfrac{\sqrt{3}}{\sqrt{4}} = \dfrac{\sqrt{3}}{2}$

✔ **Now Try Exercises 51, 53, and 57.**

EXAMPLE 5 **Using the Quotient Rule to Divide Radicals**

Simplify $\dfrac{27\sqrt{15}}{9\sqrt{3}}$.

Use multiplication of fractions and the quotient rule as follows.

$$\frac{27\sqrt{15}}{9\sqrt{3}} = \frac{27}{9} \cdot \frac{\sqrt{15}}{\sqrt{3}} = \frac{27}{9} \cdot \sqrt{\frac{15}{3}} = 3\sqrt{5}$$

✔ **Now Try Exercise 59.**

Some problems require both the product and quotient rules.

EXAMPLE 6 Using Both the Product and Quotient Rules

Simplify $\sqrt{\dfrac{3}{5}} \cdot \sqrt{\dfrac{1}{5}}$.

$$\sqrt{\dfrac{3}{5}} \cdot \sqrt{\dfrac{1}{5}} = \sqrt{\dfrac{3}{5} \cdot \dfrac{1}{5}} \qquad \text{Product rule}$$

$$= \sqrt{\dfrac{3}{25}} \qquad \text{Multiply fractions.}$$

$$= \dfrac{\sqrt{3}}{\sqrt{25}} \qquad \text{Quotient rule}$$

$$= \dfrac{\sqrt{3}}{5} \qquad \sqrt{25} = 5$$

✔ **Now Try Exercise 61.**

OBJECTIVE ④ Simplify radicals involving variables. Radicals can also involve variables. An example is $\sqrt{x^2}$. Simplifying such radicals can get a little tricky. If x represents a nonnegative number, then $\sqrt{x^2} = x$. If x represents a negative number, then $\sqrt{x^2} = -x$, the *opposite* of x (which is positive). For example,

$$\sqrt{5^2} = 5, \quad \text{but} \quad \sqrt{(-5)^2} = \sqrt{25} = 5, \quad \text{the opposite of } -5.$$

This means that the square root of a squared number is always nonnegative. We can use absolute value to express this.

$\sqrt{a^2}$

For any real number a, $\qquad \sqrt{a^2} = |a|.$

The product and quotient rules apply when variables appear under the radical sign, as long as the variables represent only *nonnegative* real numbers. ***To avoid negative radicands, variables under radical signs are assumed to be nonnegative in this text.*** Therefore, absolute value bars are not necessary, since, for $x \geq 0$, $|x| = x$.

EXAMPLE 7 Simplifying Radicals Involving Variables

Simplify each radical. Remember, we assume that all variables represent nonnegative real numbers.

(a) $\sqrt{x^4} = x^2$, since $(x^2)^2 = x^4$.

(b) $\sqrt{25m^6} = \sqrt{25} \cdot \sqrt{m^6} \qquad$ Product rule
$\qquad\quad = 5m^3 \qquad\qquad\qquad (m^3)^2 = m^6$

(c) $\sqrt{8p^{10}} = \sqrt{4 \cdot 2 \cdot p^{10}} \qquad\quad$ 4 is a perfect square.
$\qquad\quad = \sqrt{4} \cdot \sqrt{2} \cdot \sqrt{p^{10}} \qquad$ Product rule
$\qquad\quad = 2 \cdot \sqrt{2} \cdot p^5 \qquad\qquad (p^5)^2 = p^{10}$
$\qquad\quad = 2p^5\sqrt{2}$

(d) $\sqrt{r^9} = \sqrt{r^8 \cdot r}$

$= \sqrt{r^8} \cdot \sqrt{r}$ Product rule

$= r^4\sqrt{r}$ $(r^4)^2 = r^8$

(e) $\sqrt{\dfrac{5}{x^2}} = \dfrac{\sqrt{5}}{\sqrt{x^2}}$ Quotient rule

$= \dfrac{\sqrt{5}}{x}$ $x \neq 0$

✔ **Now Try Exercises 63, 69, 71, 75, and 85.**

▶ **NOTE** A quick way to find the square root of a variable raised to an even power is to divide the exponent by the index, 2. For example,

$$\sqrt{x^6} = x^3 \quad \text{and} \quad \sqrt{x^{10}} = x^5.$$

$6 \div 2 = 3$ $10 \div 2 = 5$

OBJECTIVE 5 **Simplify other roots.** The product and quotient rules for radicals also work for other roots. To simplify cube roots, look for factors that are *perfect cubes*. A **perfect cube** is a number with a rational cube root. For example, $\sqrt[3]{64} = 4$, and because 4 is a rational number, 64 is a perfect cube. Other roots are handled in a similar manner.

Properties of Radicals

For all real numbers for which the indicated roots exist,

$$\sqrt[n]{a} \cdot \sqrt[n]{b} = \sqrt[n]{ab} \quad \text{and} \quad \frac{\sqrt[n]{a}}{\sqrt[n]{b}} = \sqrt[n]{\frac{a}{b}} \quad (b \neq 0).$$

EXAMPLE 8 **Simplifying Other Roots**

Simplify each radical.

(a) $\sqrt[3]{32} = \sqrt[3]{8 \cdot 4}$ 8 is a perfect cube.

Remember to write the root index 3 in each radical.

$= \sqrt[3]{8} \cdot \sqrt[3]{4}$ Product rule

$= 2\sqrt[3]{4}$

(b) $\sqrt[4]{32} = \sqrt[4]{16 \cdot 2}$ 16 is a perfect fourth power.

Remember to write the root index 4 in each radical.

$= \sqrt[4]{16} \cdot \sqrt[4]{2}$ Product rule

$= 2\sqrt[4]{2}$

(c) $\sqrt[3]{\dfrac{27}{125}} = \dfrac{\sqrt[3]{27}}{\sqrt[3]{125}} = \dfrac{3}{5}$ Quotient rule

✔ **Now Try Exercises 91, 97, and 99.**

Other roots of radicals involving variables can also be simplified. To simplify cube roots with variables, use the fact that for any real number a,

$$\sqrt[3]{a^3} = a.$$

This is true whether a is positive or negative.

EXAMPLE 9 Simplifying Cube Roots Involving Variables

Simplify each radical.

(a) $\sqrt[3]{m^6} = m^2 \qquad (m^2)^3 = m^6$

(b) $\sqrt[3]{27x^{12}} = \sqrt[3]{27} \cdot \sqrt[3]{x^{12}} \qquad$ Product rule

$\qquad\qquad = 3x^4 \qquad\qquad\quad 3^3 = 27; (x^4)^3 = x^{12}$

(c) $\sqrt[3]{32a^4} = \sqrt[3]{8a^3 \cdot 4a} \qquad 8a^3$ is a perfect cube.

$\qquad\qquad = \sqrt[3]{8a^3} \cdot \sqrt[3]{4a} \qquad$ Product rule

$\qquad\qquad = 2a\sqrt[3]{4a} \qquad\quad (2a)^3 = 8a^3$

(d) $\sqrt[3]{\dfrac{y^3}{125}} = \dfrac{\sqrt[3]{y^3}}{\sqrt[3]{125}} \qquad$ Quotient rule

$\qquad\qquad = \dfrac{y}{5}$

✔ **Now Try Exercises 105, 107, 111, and 113.**

8.2 EXERCISES

⊕ *Complete solution available on Video Lectures on CD/DVD*

Now Try Exercise

Concept Check *Decide whether each statement is* true *or* false. *If false, show why.*

1. $\sqrt{9} \cdot \sqrt{16} = \sqrt{9 \cdot 16}$

2. $\sqrt{9 + 16} = \sqrt{9} + \sqrt{16}$

3. $\sqrt{0.5} = \sqrt{\dfrac{1}{2}}$

4. For nonnegative real numbers x and y, $\sqrt{xy} = \sqrt{x} \cdot \sqrt{y}$.

5. $\sqrt{(-6)^2} = -6$

6. $\sqrt[3]{(-6)^3} = -6$

Find each product. See Example 1.

⊕ **7.** $\sqrt{3} \cdot \sqrt{5}$

8. $\sqrt{3} \cdot \sqrt{7}$

9. $\sqrt{2} \cdot \sqrt{11}$

10. $\sqrt{2} \cdot \sqrt{15}$

11. $\sqrt{6} \cdot \sqrt{7}$

12. $\sqrt{5} \cdot \sqrt{6}$

13. $\sqrt{3} \cdot \sqrt{27}$

14. $\sqrt{2} \cdot \sqrt{8}$

15. $\sqrt{13} \cdot \sqrt{13}$

16. $\sqrt{17} \cdot \sqrt{17}$

17. $\sqrt{13} \cdot \sqrt{r}, r \geq 0$

18. $\sqrt{19} \cdot \sqrt{k}, k \geq 0$

✎ **19.** Explain why $\sqrt{x} \cdot \sqrt{x} = x$ for all $x \geq 0$.

✎ **20.** Explain why $\sqrt[3]{x} \cdot \sqrt[3]{x} \cdot \sqrt[3]{x} = x$ for all x.

21. Which one of the following radicals is simplified? See Example 2.

   A. $\sqrt{47}$ **B.** $\sqrt{45}$ **C.** $\sqrt{48}$ **D.** $\sqrt{44}$

22. *Concept Check*   If p is a prime number, is $\sqrt{p}$ in simplified form?

Simplify each radical. See Example 2.

23. $\sqrt{45}$ **24.** $\sqrt{27}$ **25.** $\sqrt{24}$ **26.** $\sqrt{44}$

27. $\sqrt{90}$ **28.** $\sqrt{56}$ **29.** $\sqrt{75}$ **30.** $\sqrt{18}$

31. $\sqrt{125}$ **32.** $\sqrt{80}$ **33.** $\sqrt{145}$ **34.** $\sqrt{110}$

35. $\sqrt{160}$ **36.** $\sqrt{128}$ **37.** $-\sqrt{700}$ **38.** $-\sqrt{600}$

39. $3\sqrt{27}$ **40.** $9\sqrt{8}$ **41.** $5\sqrt{50}$ **42.** $6\sqrt{40}$

Find each product and simplify. See Example 3.

43. $\sqrt{9} \cdot \sqrt{32}$ **44.** $\sqrt{9} \cdot \sqrt{50}$ **45.** $\sqrt{12} \cdot \sqrt{48}$

46. $\sqrt{50} \cdot \sqrt{72}$ **47.** $\sqrt{12} \cdot \sqrt{30}$ **48.** $\sqrt{30} \cdot \sqrt{24}$

49. Simplify the product $\sqrt{8} \cdot \sqrt{32}$ in two ways. First, multiply 8 by 32 and simplify the square root of this product. Second, simplify $\sqrt{8}$, simplify $\sqrt{32}$, and then multiply. How do the answers compare? Make a conjecture (an educated guess) about whether the correct answer can always be obtained using either method when simplifying a product such as this.

50. Simplify the radical $\sqrt{288}$ in two ways. First, factor 288 as $144 \cdot 2$ and then simplify. Second, factor 288 as $48 \cdot 6$ and then simplify. How do the answers compare? Make a conjecture concerning the quickest way to simplify such a radical.

Simplify each radical expression. See Examples 4–6.

51. $\sqrt{\dfrac{16}{225}}$ **52.** $\sqrt{\dfrac{9}{100}}$ **53.** $\sqrt{\dfrac{7}{16}}$ **54.** $\sqrt{\dfrac{13}{25}}$

55. $\sqrt{\dfrac{4}{50}}$ **56.** $\sqrt{\dfrac{14}{72}}$ **57.** $\dfrac{\sqrt{75}}{\sqrt{3}}$ **58.** $\dfrac{\sqrt{200}}{\sqrt{2}}$

59. $\dfrac{30\sqrt{10}}{5\sqrt{2}}$ **60.** $\dfrac{50\sqrt{20}}{2\sqrt{10}}$ **61.** $\sqrt{\dfrac{5}{2}} \cdot \sqrt{\dfrac{125}{8}}$ **62.** $\sqrt{\dfrac{8}{3}} \cdot \sqrt{\dfrac{512}{27}}$

Simplify each radical. Assume that all variables represent nonnegative real numbers. See Example 7.

63. $\sqrt{m^2}$ **64.** $\sqrt{k^2}$ **65.** $\sqrt{y^4}$ **66.** $\sqrt{s^4}$

67. $\sqrt{36z^2}$ **68.** $\sqrt{49n^2}$ **69.** $\sqrt{400x^6}$ **70.** $\sqrt{900y^8}$

71. $\sqrt{18x^8}$ **72.** $\sqrt{20r^{10}}$ **73.** $\sqrt{45c^{14}}$ **74.** $\sqrt{50d^{20}}$

75. $\sqrt{z^5}$ **76.** $\sqrt{y^3}$ **77.** $\sqrt{a^{13}}$ **78.** $\sqrt{p^{17}}$

79. $\sqrt{64x^7}$ **80.** $\sqrt{25t^{11}}$ **81.** $\sqrt{x^6y^{12}}$ **82.** $\sqrt{a^8b^{10}}$

83. $\sqrt{81m^4n^2}$ **84.** $\sqrt{100c^4d^6}$ **85.** $\sqrt{\dfrac{7}{x^{10}}}, x \neq 0$ **86.** $\sqrt{\dfrac{14}{z^{12}}}, z \neq 0$

87. $\sqrt{\dfrac{y^4}{100}}$ **88.** $\sqrt{\dfrac{w^8}{144}}$ **89.** $\sqrt{\dfrac{x^4y^6}{169}}$ **90.** $\sqrt{\dfrac{w^8z^{10}}{400}}$

Simplify each radical. See Example 8.

91. $\sqrt[3]{40}$ **92.** $\sqrt[3]{48}$ **93.** $\sqrt[3]{54}$ **94.** $\sqrt[3]{135}$

95. $\sqrt[3]{128}$ **96.** $\sqrt[3]{192}$ **97.** $\sqrt[4]{80}$ **98.** $\sqrt[4]{243}$

99. $\sqrt[3]{\dfrac{8}{27}}$ **100.** $\sqrt[3]{\dfrac{64}{125}}$ **101.** $\sqrt[3]{-\dfrac{216}{125}}$ **102.** $\sqrt[3]{-\dfrac{1}{64}}$

Simplify each radical. See Example 9.

103. $\sqrt[3]{p^3}$ **104.** $\sqrt[3]{w^3}$ **105.** $\sqrt[3]{x^9}$ **106.** $\sqrt[3]{y^{18}}$

107. $\sqrt[3]{64z^6}$ **108.** $\sqrt[3]{125a^{15}}$ **109.** $\sqrt[3]{343a^9b^3}$ **110.** $\sqrt[3]{216m^3n^6}$

111. $\sqrt[3]{16t^5}$ **112.** $\sqrt[3]{24x^4}$ **113.** $\sqrt[3]{\dfrac{m^{12}}{8}}$ **114.** $\sqrt[3]{\dfrac{n^9}{27}}$

115. In Example 2(a), we showed *algebraically* that $\sqrt{20}$ is equal to $2\sqrt{5}$. To give *numerical support* to this result, use a calculator to do the following:

     (a) Use your calculator to find a decimal approximation for $\sqrt{20}$. Record as many digits as the calculator shows.

     (b) Use your calculator to find a decimal approximation for $\sqrt{5}$, and then multiply the result by 2. Record as many digits as the calculator shows.

     (c) Your results in parts (a) and (b) should be the same. A mathematician would not accept this numerical exercise as *proof* that $\sqrt{20}$ is equal to $2\sqrt{5}$. Explain why.

116. *Concept Check*   On your calculator, multiply the approximations for $\sqrt{3}$ and $\sqrt{5}$. Now predict what your calculator will show when you find an approximation for $\sqrt{15}$. What rule stated in this section justifies your answer?

117. When we multiply two radicals with variables under the radical sign, such as $\sqrt{a} \cdot \sqrt{b} = \sqrt{ab}$, why is it important to know that both a and b represent nonnegative numbers?

118. Is it necessary to restrict k to a nonnegative number to say that $\sqrt[3]{k} \cdot \sqrt[3]{k} \cdot \sqrt[3]{k} = k$? Why or why not?

The volume of a cube is found with the formula $V = s^3$, where s is the length of an edge of the cube. Use this information in Exercises 119 and 120.

119. A container in the shape of a cube has a volume of 216 cm³. What is the depth of the container?

120. A cube-shaped box must be constructed to contain 128 ft³. What should the dimensions (height, width, and length) of the box be?

The volume of a sphere is found with the formula $V = \frac{4}{3}\pi r^3$, where r is the length of the radius of the sphere. Use this information in Exercises 121 and 122.

121. A ball in the shape of a sphere has a volume of 288π in.³. What is the radius of the ball?

122. Suppose that the volume of the ball described in Exercise 121 is multiplied by 8. How is the radius affected?

Work Exercises 123 and 124 without using a calculator.

123. Choose the best estimate for the area (in square inches) of this rectangle.

 A. 45   **B.** 72   **C.** 80   **D.** 90

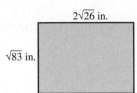

$2\sqrt{26}$ in.

$\sqrt{83}$ in.

124. Choose the best estimate for the area (in square feet) of this triangle.

 A. 20   **B.** 40   **C.** 60   **D.** 80

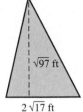

$\sqrt{97}$ ft

$2\sqrt{17}$ ft

PREVIEW EXERCISES

*Combine like terms. See **Section 1.8.***

125. $4x + 7 - 9x + 12$

126. $9x^2 + 3x^2 - 2x + 4x - 8 + 1$

127. $2xy + 3x^2y - 9xy + 8x^2y$

128. $x + 3y + 12z$

8.3 Adding and Subtracting Radicals

OBJECTIVES

1 Add and subtract radicals.

2 Simplify radical sums and differences.

3 Simplify more complicated radical expressions.

OBJECTIVE 1 Add and subtract radicals. We add or subtract radicals by using the distributive property. For example,

$$8\sqrt{3} + 6\sqrt{3}$$
$$= (8 + 6)\sqrt{3}$$
$$= 14\sqrt{3}.$$

$$2\sqrt{11} - 7\sqrt{11}$$
$$= (2 - 7)\sqrt{11}$$
$$= -5\sqrt{11}.$$

Only **like radicals**—those which are multiples of the *same root* of the *same number*—can be combined in this way. In the preceding examples, $8\sqrt{3}$ and $6\sqrt{3}$ are like radicals, as are $2\sqrt{11}$ and $-7\sqrt{11}$. By contrast, examples of *unlike radicals* are

$$2\sqrt{5} \quad \text{and} \quad 2\sqrt{3}, \quad \text{Radicands are different.}$$

as well as

$$2\sqrt{3} \quad \text{and} \quad 2\sqrt[3]{3}. \quad \text{Indexes are different.}$$

EXAMPLE 1 Adding and Subtracting Like Radicals

Add or subtract, as indicated.

(a) $3\sqrt{6} + 5\sqrt{6}$
$$= (3 + 5)\sqrt{6}$$
$$= 8\sqrt{6}$$

(b) $5\sqrt{10} - 7\sqrt{10}$
$$= (5 - 7)\sqrt{10}$$
$$= -2\sqrt{10}$$

(c) $\sqrt{7} + 2\sqrt{7}$

$= 1\sqrt{7} + 2\sqrt{7}$

$= (1 + 2)\sqrt{7}$

$= 3\sqrt{7}$

(d) $\sqrt{5} + \sqrt{5}$

$= 1\sqrt{5} + 1\sqrt{5}$

$= (1 + 1)\sqrt{5}$

$= 2\sqrt{5}$

(e) $\sqrt{3} + \sqrt{7}$ cannot be added by the distributive property.

✔ **Now Try Exercises 1, 3, 5, 7, and 11.**

OBJECTIVE 2 Simplify radical sums and differences. Sometimes, one or more radical expressions in a sum or difference must be simplified. Then, any like radicals that result can be added or subtracted.

EXAMPLE 2 Adding and Subtracting Radicals That Must Be Simplified

Add or subtract, as indicated.

(a) $3\sqrt{2} + \sqrt{8}$

$= 3\sqrt{2} + \sqrt{4 \cdot 2}$ Factor.

$= 3\sqrt{2} + \sqrt{4} \cdot \sqrt{2}$ Product rule

$= 3\sqrt{2} + 2\sqrt{2}$ $\sqrt{4} = 2$

$= 5\sqrt{2}$ Add like radicals.

(b)

$\sqrt{18} - \sqrt{27} = \sqrt{9 \cdot 2} - \sqrt{9 \cdot 3}$ Factor.

$= \sqrt{9} \cdot \sqrt{2} - \sqrt{9} \cdot \sqrt{3}$ Product rule

These terms cannot be combined.

$= 3\sqrt{2} - 3\sqrt{3}$ $\sqrt{9} = 3$

Because $\sqrt{2}$ and $\sqrt{3}$ are unlike radicals, this difference cannot be simplified further.

(c) $2\sqrt{12} + 3\sqrt{75}$

$= 2\left(\sqrt{4} \cdot \sqrt{3}\right) + 3\left(\sqrt{25} \cdot \sqrt{3}\right)$ Product rule

$= 2\left(2\sqrt{3}\right) + 3\left(5\sqrt{3}\right)$ $\sqrt{4} = 2; \sqrt{25} = 5$

$= 4\sqrt{3} + 15\sqrt{3}$ Multiply.

$= 19\sqrt{3}$ Add like radicals.

(d) $3\sqrt[3]{16} + 5\sqrt[3]{2}$

$= 3\left(\sqrt[3]{8} \cdot \sqrt[3]{2}\right) + 5\sqrt[3]{2}$ Product rule

$= 3\left(2\sqrt[3]{2}\right) + 5\sqrt[3]{2}$ $\sqrt[3]{8} = 2$

$= 6\sqrt[3]{2} + 5\sqrt[3]{2}$ Multiply.

$= 11\sqrt[3]{2}$ Add like radicals.

✔ **Now Try Exercises 9, 13, 15, and 33.**

OBJECTIVE 3 **Simplify more complicated radical expressions.** When simplifying more complicated radical expressions, recall the rules for the order of operations from **Section 1.2.**

EXAMPLE 3 **Simplifying Radical Expressions**

Simplify each radical expression. As before, assume that all variables represent nonnegative real numbers.

(a) $\sqrt{5} \cdot \sqrt{15} + 4\sqrt{3}$

$$= \sqrt{5 \cdot 15} + 4\sqrt{3} \qquad \text{Product rule}$$

$$= \sqrt{75} + 4\sqrt{3} \qquad \text{Multiply.}$$

$$= \sqrt{25 \cdot 3} + 4\sqrt{3} \qquad \text{25 is a perfect square.}$$

$$= \sqrt{25} \cdot \sqrt{3} + 4\sqrt{3} \qquad \text{Product rule}$$

$$= 5\sqrt{3} + 4\sqrt{3} \qquad \sqrt{25} = 5$$

$$= 9\sqrt{3} \qquad \text{Add like radicals.}$$

(b) $\sqrt{12k} + \sqrt{27k}$

$$= \sqrt{4 \cdot 3k} + \sqrt{9 \cdot 3k} \qquad \text{Factor.}$$

$$= \sqrt{4} \cdot \sqrt{3k} + \sqrt{9} \cdot \sqrt{3k} \qquad \text{Product rule}$$

$$= 2\sqrt{3k} + 3\sqrt{3k} \qquad \sqrt{4} = 2; \sqrt{9} = 3$$

$$= 5\sqrt{3k} \qquad \text{Add like radicals.}$$

(c) $3x\sqrt{50} + \sqrt{2x^2}$

$$= 3x\sqrt{25 \cdot 2} + \sqrt{x^2 \cdot 2} \qquad \text{Factor.}$$

$$= 3x\sqrt{25} \cdot \sqrt{2} + \sqrt{x^2} \cdot \sqrt{2} \qquad \text{Product rule}$$

$$= 3x \cdot 5\sqrt{2} + x\sqrt{2} \qquad \sqrt{25} = 5; \sqrt{x^2} = x$$

$$= 15x\sqrt{2} + x\sqrt{2} \qquad \text{Multiply.}$$

$$= 16x\sqrt{2} \qquad \text{Add like radicals.}$$

(d) $2\sqrt[3]{32m^3} - \sqrt[3]{108m^3}$

$$= 2\sqrt[3]{(8m^3)4} - \sqrt[3]{(27m^3)4} \qquad \text{Factor.}$$

$$= 2(2m)\sqrt[3]{4} - 3m\sqrt[3]{4} \qquad \sqrt[3]{8m^3} = 2m; \sqrt[3]{27m^3} = 3m$$

$$= 4m\sqrt[3]{4} - 3m\sqrt[3]{4} \qquad \text{Multiply.}$$

$$= m\sqrt[3]{4} \qquad \text{Subtract like radicals.}$$

✔ **Now Try Exercises 27, 41, 51, and 59.**

▶ **CAUTION** *A sum or difference of radicals can be simplified only if the radicals are like radicals.* Thus, $\sqrt{5} + 3\sqrt{5} = 4\sqrt{5}$, but $\sqrt{5} + 5\sqrt{3}$ cannot be simplified further. Also, $2\sqrt{3} + 5\sqrt[3]{3}$ cannot be simplified further.

8.3 EXERCISES

Add or subtract wherever possible. See Examples 1, 2, and 3(a).

1. $2\sqrt{3} + 5\sqrt{3}$ **2.** $6\sqrt{5} + 8\sqrt{5}$ **3.** $4\sqrt{7} - 9\sqrt{7}$ **4.** $6\sqrt{2} - 8\sqrt{2}$

5. $\sqrt{6} + \sqrt{6}$ **6.** $\sqrt{11} + \sqrt{11}$ **7.** $\sqrt{17} + 2\sqrt{17}$ **8.** $3\sqrt{19} + \sqrt{19}$

9. $5\sqrt{3} + \sqrt{12}$ **10.** $3\sqrt{23} - \sqrt{23}$ **11.** $\sqrt{6} + \sqrt{7}$ **12.** $\sqrt{14} + \sqrt{17}$

13. $-\sqrt{12} + \sqrt{75}$ **14.** $2\sqrt{27} - \sqrt{300}$

15. $2\sqrt{50} - 5\sqrt{72}$ **16.** $6\sqrt{18} - 4\sqrt{32}$

17. $5\sqrt{7} - 2\sqrt{28} + 6\sqrt{63}$ **18.** $3\sqrt{11} + 5\sqrt{44} - 3\sqrt{99}$

19. $9\sqrt{24} - 2\sqrt{54} + 3\sqrt{20}$ **20.** $2\sqrt{8} - 5\sqrt{32} + 2\sqrt{48}$

21. $5\sqrt{72} - 3\sqrt{48} - 4\sqrt{128}$ **22.** $4\sqrt{50} + 3\sqrt{12} + 5\sqrt{45}$

23. $\dfrac{1}{4}\sqrt{288} + \dfrac{1}{6}\sqrt{72}$ **24.** $\dfrac{2}{3}\sqrt{27} + \dfrac{3}{4}\sqrt{48}$ **25.** $\dfrac{3}{5}\sqrt{75} - \dfrac{2}{3}\sqrt{45}$

26. $\dfrac{5}{8}\sqrt{128} - \dfrac{3}{4}\sqrt{160}$ **27.** $\sqrt{3} \cdot \sqrt{7} + 2\sqrt{21}$ **28.** $\sqrt{13} \cdot \sqrt{2} + 3\sqrt{26}$

29. $\sqrt{6} \cdot \sqrt{2} + 3\sqrt{3}$ **30.** $4\sqrt{15} \cdot \sqrt{3} - 2\sqrt{5}$ **31.** $4\sqrt[3]{16} - 3\sqrt[3]{54}$

32. $5\sqrt[3]{128} + 3\sqrt[3]{250}$ **33.** $3\sqrt[3]{24} + 6\sqrt[3]{81}$ **34.** $2\sqrt[4]{48} - \sqrt[4]{243}$

35. $5\sqrt[4]{32} + 2\sqrt[4]{32} \cdot \sqrt[4]{4}$ **36.** $8\sqrt[3]{48} + 10\sqrt[3]{3} \cdot \sqrt[3]{18}$

37. *Concept Check* The distributive property, which says that $a(b + c) = ab + ac$ and $ba + ca = (b + c)a$, provides the justification for adding and subtracting like radicals. While we usually skip the step that indicates this property, we could not make the statement $2\sqrt{3} + 4\sqrt{3} = 6\sqrt{3}$ without it. Write an equation showing how the distributive property is actually used in this statement.

38. Refer to Example 1(e), and explain why $\sqrt{3} + \sqrt{7}$ cannot be further simplified. Confirm, by using calculator approximations, that $\sqrt{3} + \sqrt{7}$ is *not* equal to $\sqrt{10}$.

Perform each indicated operation. Assume that all variables represent nonnegative real numbers. See Example 3.

39. $\sqrt{32x} - \sqrt{18x}$ **40.** $\sqrt{125t} - \sqrt{80t}$ **41.** $\sqrt{27r} + \sqrt{48r}$

42. $\sqrt{6x^2} + x\sqrt{54}$ **43.** $\sqrt{75x^2} + x\sqrt{300}$ **44.** $\sqrt{20y^2} - 3y\sqrt{5}$

45. $3\sqrt{8x^2} - 4x\sqrt{2}$ **46.** $\sqrt{2b^2} + 3b\sqrt{18}$ **47.** $5\sqrt{75p^2} - 4\sqrt{27p^2}$

48. $3\sqrt{32k} + 6\sqrt{8k}$ **49.** $2\sqrt{125x^2z} + 8x\sqrt{80z}$ **50.** $4p\sqrt{63m} + 6\sqrt{28mp^2}$

51. $3k\sqrt{24k^2h^2} + 9h\sqrt{54k^3}$ **52.** $6r\sqrt{27r^2s} + 3r^2\sqrt{3s}$ **53.** $6\sqrt[3]{8p^2} - 2\sqrt[3]{27p^2}$

54. $5\sqrt[3]{27x^2} + 8\sqrt[3]{8x^2}$ **55.** $5\sqrt[4]{m^3} + 8\sqrt[4]{16m^3}$ **56.** $5\sqrt[4]{m^4} + 3\sqrt[4]{81m^4}$

57. $2\sqrt[4]{p^5} - 5p\sqrt[4]{16p}$ **58.** $8k\sqrt[3]{54k} + 6\sqrt[3]{16k^4}$

59. $-5\sqrt[3]{256z^4} - 2z\sqrt[3]{32z}$ **60.** $10\sqrt[3]{4m^4} - 3m\sqrt[3]{32m}$

61. $2\sqrt[4]{6k^7} - k\sqrt[4]{96k^3}$ **62.** $\dfrac{3}{2}\sqrt[3]{16a^4b^5} - ab\sqrt[3]{54ab^2}$

63. In the directions for Exercises 39–62, we made the assumption that all variables represent nonnegative real numbers. However, in Exercises 53, 58, 59, 60, and 62, variables actually *may* represent negative numbers. Explain why this is so.

64. Find the perimeter of each figure.

(a)

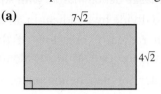

(b)

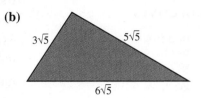

Perform the indicated operations, following the rules for the order of operations we have used throughout the book. Express all answers in simplest form.

65. $\sqrt{(-3 - 6)^2 + (2 - 4)^2}$

66. $\sqrt{(-9 - 3)^2 + (3 - 8)^2}$

67. $\sqrt{(2 - (-2))^2 + (-1 - 2)^2}$

68. $\sqrt{(3 - 1)^2 + (2 - (-1))^2}$

69. $\sqrt{(-5)^2 - 4(1)(-6)}$

70. $\sqrt{(-7)^2 - 4(1)(-8)}$

71. $\sqrt{(-10)^2 - 4(3)(-8)}$

72. $\sqrt{(-2)^2 - 4(5)(-3)}$

73. $\sqrt{(-4)^2 - 4(2)(1)}$

74. $\sqrt{(-6)^2 - 4(1)(-3)}$

RELATING CONCEPTS (EXERCISES 75–78)

FOR INDIVIDUAL OR GROUP WORK

Adding and subtracting like radicals is no different from adding and subtracting like terms. **Work Exercises 75–78 in order.**

75. Combine like terms: $5x^2y + 3x^2y - 14x^2y$.

76. Combine like terms:

$$5(p - 2q)^2(a + b) + 3(p - 2q)^2(a + b) - 14(p - 2q)^2(a + b).$$

77. Combine like radicals: $5a^2\sqrt{xy} + 3a^2\sqrt{xy} - 14a^2\sqrt{xy}$.

78. Compare your answers in Exercises 75–77. How are they alike? How are they different?

PREVIEW EXERCISES

*Perform each operation. See **Section 8.1.***

79. $(\sqrt{6})^2$

80. $(\sqrt{25})^2$

81. $\sqrt[3]{2} \cdot \sqrt[3]{4}$

*Simplify each radical. See **Section 8.2.***

82. $\sqrt{288}$

83. $\sqrt{7500}$

84. $\sqrt{x^2y^6}$, $x \geq 0, y \geq 0$

8.4 Rationalizing the Denominator

OBJECTIVE 1 Rationalize denominators with square roots. Although calculators now make it fairly easy to divide by a radical in an expression such as $\frac{1}{\sqrt{2}}$, it is sometimes easier to work with radical expressions if the denominators do not contain any radicals. For example, the radical in the denominator of $\frac{1}{\sqrt{2}}$ can be eliminated by multiplying the numerator and denominator by $\sqrt{2}$, since $\sqrt{2} \cdot \sqrt{2} = \sqrt{4} = 2$.

$$\frac{1}{\sqrt{2}} = \frac{1 \cdot \sqrt{2}}{\sqrt{2} \cdot \sqrt{2}} = \frac{\sqrt{2}}{2} \qquad \text{Multiply by } \tfrac{\sqrt{2}}{\sqrt{2}} = 1.$$

This process of changing the denominator from a radical (an irrational number) to a rational number is called **rationalizing the denominator.** *The value of the radical expression is not changed; only the form is changed, because the expression has been multiplied by 1 in the form* $\frac{\sqrt{2}}{\sqrt{2}}$.

EXAMPLE 1 Rationalizing Denominators

Rationalize each denominator.

(a)
$$\frac{9}{\sqrt{6}} = \frac{9 \cdot \sqrt{6}}{\sqrt{6} \cdot \sqrt{6}} \qquad \text{Multiply by } \tfrac{\sqrt{6}}{\sqrt{6}} = 1.$$
$$= \frac{9\sqrt{6}}{6} \qquad \sqrt{6} \cdot \sqrt{6} = \sqrt{36} = 6$$
$$= \frac{3\sqrt{6}}{2} \qquad \text{Lowest terms}$$

(b) $\dfrac{12}{\sqrt{8}}$

The denominator could be rationalized here by multiplying by $\sqrt{8}$ in both numerator and denominator. However, the result can be found more directly by first simplifying the denominator.

$$\sqrt{8} = \sqrt{4} \cdot \sqrt{2} = 2\sqrt{2}$$

Then multiply the numerator and denominator by $\sqrt{2}$.

$$\frac{12}{\sqrt{8}} = \frac{12}{2\sqrt{2}} \qquad \sqrt{8} = 2\sqrt{2}$$
$$= \frac{12 \cdot \sqrt{2}}{2\sqrt{2} \cdot \sqrt{2}} \qquad \text{Multiply by } \tfrac{\sqrt{2}}{\sqrt{2}}.$$
$$= \frac{12 \cdot \sqrt{2}}{2 \cdot 2} \qquad \sqrt{2} \cdot \sqrt{2} = \sqrt{4} = 2$$
$$= \frac{12\sqrt{2}}{4} \qquad \text{Multiply.}$$
$$= 3\sqrt{2} \qquad \text{Lowest terms}$$

✔ **Now Try Exercises 5 and 13.**

▶ **N O T E** In Example 1(b), we could also have rationalized the original denominator, $\sqrt{8}$, by multiplying by $\sqrt{2}$, since $\sqrt{8} \cdot \sqrt{2} = \sqrt{16} = 4$.

$$\frac{12}{\sqrt{8}} = \frac{12 \cdot \sqrt{2}}{\sqrt{8} \cdot \sqrt{2}} = \frac{12\sqrt{2}}{\sqrt{16}} = \frac{12\sqrt{2}}{4} = 3\sqrt{2}$$

When simplifying radicals, be aware that there are often several approaches which yield the same correct answer.

OBJECTIVE 2 **Write radicals in simplified form.** A radical is considered to be in simplified form if the following three conditions are met.

Conditions for Simplified Form of a Radical

1. The radicand contains no factor (except 1) that is a perfect square (in dealing with square roots), a perfect cube (in dealing with cube roots), and so on.

2. The radicand has no fractions.

3. No denominator contains a radical.

In the examples that follow, we simplify radicals according to these conditions.

EXAMPLE 2 **Simplifying a Radical**

Simplify $\sqrt{\dfrac{27}{5}}$.

This form of the radical violates condition 2. To begin, use the quotient rule.

$$\sqrt{\frac{27}{5}} = \frac{\sqrt{27}}{\sqrt{5}} \qquad \text{Quotient rule}$$

$$= \frac{\sqrt{27} \cdot \sqrt{5}}{\sqrt{5} \cdot \sqrt{5}} \qquad \text{Rationalize the denominator.}$$

$$= \frac{\sqrt{27} \cdot \sqrt{5}}{5} \qquad \sqrt{5} \cdot \sqrt{5} = 5$$

$$= \frac{\sqrt{9 \cdot 3} \cdot \sqrt{5}}{5} \qquad \text{Factor.}$$

$$= \frac{\sqrt{9} \cdot \sqrt{3} \cdot \sqrt{5}}{5} \qquad \text{Product rule}$$

$$= \frac{3 \cdot \sqrt{3} \cdot \sqrt{5}}{5} \qquad \sqrt{9} = 3$$

$$= \frac{3\sqrt{15}}{5} \qquad \text{Product rule}$$

✔ **Now Try Exercise 19.**

EXAMPLE 3 Simplifying a Product of Radicals

Simplify $\sqrt{\dfrac{5}{8}} \cdot \sqrt{\dfrac{1}{6}}$.

$$\sqrt{\dfrac{5}{8}} \cdot \sqrt{\dfrac{1}{6}} = \sqrt{\dfrac{5}{8} \cdot \dfrac{1}{6}} \qquad \text{Product rule}$$

$$= \sqrt{\dfrac{5}{48}} \qquad \text{Multiply fractions.}$$

$$= \dfrac{\sqrt{5}}{\sqrt{48}} \qquad \text{Quotient rule}$$

$$= \dfrac{\sqrt{5}}{\sqrt{16} \cdot \sqrt{3}} \qquad \text{Product rule}$$

$$= \dfrac{\sqrt{5}}{4\sqrt{3}} \qquad \sqrt{16} = 4$$

$$= \dfrac{\sqrt{5} \cdot \sqrt{3}}{4\sqrt{3} \cdot \sqrt{3}} \qquad \text{Rationalize the denominator.}$$

$$= \dfrac{\sqrt{15}}{4 \cdot 3} \qquad \text{Product rule; } \sqrt{3} \cdot \sqrt{3} = 3$$

$$= \dfrac{\sqrt{15}}{12} \qquad \text{Multiply.}$$

✔ **Now Try Exercise 45.**

EXAMPLE 4 Simplifying a Quotient of Radicals

Simplify $\dfrac{\sqrt{4x}}{\sqrt{y}}$. Assume that x and y are positive real numbers.

$$\dfrac{\sqrt{4x}}{\sqrt{y}} = \dfrac{\sqrt{4x} \cdot \sqrt{y}}{\sqrt{y} \cdot \sqrt{y}} = \dfrac{\sqrt{4xy}}{y} = \dfrac{2\sqrt{xy}}{y}$$

✔ **Now Try Exercise 59.**

EXAMPLE 5 Simplifying a Radical Quotient

Simplify $\sqrt{\dfrac{2x^2y}{3}}$. Assume that x and y are nonnegative real numbers.

$$\sqrt{\dfrac{2x^2y}{3}} = \dfrac{\sqrt{2x^2y}}{\sqrt{3}} \qquad \text{Quotient rule}$$

$$= \dfrac{\sqrt{2x^2y} \cdot \sqrt{3}}{\sqrt{3} \cdot \sqrt{3}} \qquad \text{Rationalize the denominator.}$$

$$= \frac{\sqrt{6x^2 y}}{3} \qquad \text{Product rule}$$

$$= \frac{\sqrt{x^2}\sqrt{6y}}{3} \qquad \text{Product rule}$$

$$= \frac{x\sqrt{6y}}{3} \qquad \sqrt{x^2} = x, \text{ since } x \geq 0.$$

✔ **Now Try Exercise 67.**

OBJECTIVE 3 **Rationalize denominators with cube roots.** To rationalize a denominator with a cube root, we change the radicand in the denominator to a perfect cube.

EXAMPLE 6 **Rationalizing Denominators with Cube Roots**

Rationalize each denominator.

(a) $\sqrt[3]{\dfrac{3}{2}}$

First write the expression as a quotient of radicals. Then multiply numerator and denominator by a sufficient number of factors of 2 to make the denominator a perfect cube. This will eliminate the radical in the denominator. Here, multiply by $\sqrt[3]{2 \cdot 2} = \sqrt[3]{2^2}$.

$$\sqrt[3]{\frac{3}{2}} = \frac{\sqrt[3]{3}}{\sqrt[3]{2}} = \frac{\sqrt[3]{3} \cdot \sqrt[3]{2^2}}{\sqrt[3]{2} \cdot \sqrt[3]{2^2}} = \frac{\sqrt[3]{3 \cdot 2^2}}{\sqrt[3]{2^3}} = \frac{\sqrt[3]{12}}{2} \qquad \sqrt[3]{2^3} = \sqrt[3]{8} = 2$$

Be careful not to multiply by $\sqrt[3]{2}$ here.

└── Denominator is a perfect cube.

(b) $\dfrac{\sqrt[3]{3}}{\sqrt[3]{4}}$

Since $\sqrt[3]{4} \cdot \sqrt[3]{2} = \sqrt[3]{2^2} \cdot \sqrt[3]{2} = \sqrt[3]{2^3} = 2$, multiply numerator and denominator by $\sqrt[3]{2}$.

$$\frac{\sqrt[3]{3}}{\sqrt[3]{4}} = \frac{\sqrt[3]{3} \cdot \sqrt[3]{2}}{\sqrt[3]{2^2} \cdot \sqrt[3]{2}} = \frac{\sqrt[3]{6}}{\sqrt[3]{2^3}} = \frac{\sqrt[3]{6}}{2}$$

Be careful not to multiply by $\sqrt[3]{2^2}$ here.

(c) $\dfrac{\sqrt[3]{2}}{\sqrt[3]{3x^2}} \quad (x \neq 0)$

Multiply numerator and denominator by enough factors of 3 and of x to get a perfect cube in the denominator. Here, multiply by $\sqrt[3]{3^2 x}$ (that is, $\sqrt[3]{9x}$), since $\sqrt[3]{3x^2} \cdot \sqrt[3]{3^2 x} = \sqrt[3]{(3x)^3} = 3x$.

$$\frac{\sqrt[3]{2}}{\sqrt[3]{3x^2}} = \frac{\sqrt[3]{2} \cdot \sqrt[3]{3^2 x}}{\sqrt[3]{3x^2} \cdot \sqrt[3]{3^2 x}} = \frac{\sqrt[3]{18x}}{\sqrt[3]{(3x)^3}} = \frac{\sqrt[3]{18x}}{3x}$$

Be careful not to multiply by $\sqrt[3]{3x^2}$ here.

└── Denominator is a perfect cube.

✔ **Now Try Exercises 77, 79, and 83.**

▶ **CAUTION** A common error in a problem like the one in Example 6(a) is to multiply by $\sqrt[3]{2}$ instead of $\sqrt[3]{2^2}$. Doing this would give a denominator of $\sqrt[3]{2} \cdot \sqrt[3]{2} = \sqrt[3]{4}$. Because 4 is not a perfect cube, the denominator is still not rationalized.

8.4 EXERCISES

🌐 *Complete solution available on Video Lectures on CD/DVD*

▨ *Now Try Exercise*

Rationalize each denominator. See Examples 1 and 2.

1. $\dfrac{6}{\sqrt{5}}$ **2.** $\dfrac{3}{\sqrt{2}}$ **3.** $\dfrac{5}{\sqrt{5}}$ **4.** $\dfrac{15}{\sqrt{15}}$

🌐 **5.** $\dfrac{4}{\sqrt{6}}$ **6.** $\dfrac{15}{\sqrt{10}}$ **7.** $\dfrac{8\sqrt{3}}{\sqrt{5}}$ **8.** $\dfrac{9\sqrt{6}}{\sqrt{5}}$

9. $\dfrac{12\sqrt{10}}{8\sqrt{3}}$ **10.** $\dfrac{9\sqrt{15}}{6\sqrt{2}}$ **11.** $\dfrac{8}{\sqrt{27}}$ **12.** $\dfrac{12}{\sqrt{18}}$

13. $\dfrac{6}{\sqrt{200}}$ **14.** $\dfrac{10}{\sqrt{300}}$ **15.** $\dfrac{12}{\sqrt{72}}$ **16.** $\dfrac{21}{\sqrt{45}}$

17. $\dfrac{\sqrt{10}}{\sqrt{5}}$ **18.** $\dfrac{\sqrt{6}}{\sqrt{3}}$ 🌐 **19.** $\sqrt{\dfrac{40}{3}}$ **20.** $\sqrt{\dfrac{5}{8}}$

21. $\sqrt{\dfrac{1}{32}}$ **22.** $\sqrt{\dfrac{1}{8}}$ **23.** $\sqrt{\dfrac{9}{5}}$ **24.** $\sqrt{\dfrac{16}{7}}$

25. $\dfrac{-3}{\sqrt{50}}$ **26.** $\dfrac{-5}{\sqrt{75}}$ **27.** $\dfrac{63}{\sqrt{45}}$ **28.** $\dfrac{27}{\sqrt{32}}$ **29.** $\dfrac{\sqrt{8}}{\sqrt{24}}$

30. $\dfrac{\sqrt{5}}{\sqrt{10}}$ **31.** $-\sqrt{\dfrac{1}{5}}$ **32.** $-\sqrt{\dfrac{1}{6}}$ **33.** $\sqrt{\dfrac{13}{5}}$ **34.** $\sqrt{\dfrac{17}{11}}$

35. *Concept Check* To rationalize the denominator of an expression such as $\frac{4}{\sqrt{3}}$, we multiply both the numerator and denominator by $\sqrt{3}$. By what number are we actually multiplying the given expression, and what property of real numbers justifies the fact that our result is equal to the given expression?

36. In Example 1(a), we showed algebraically that $\frac{9}{\sqrt{6}} = \frac{3\sqrt{6}}{2}$. Support this result numerically by finding the decimal approximation of $\frac{9}{\sqrt{6}}$ on your calculator and then finding the decimal approximation of $\frac{3\sqrt{6}}{2}$. What do you notice?

Simplify. See Example 3.

37. $\sqrt{\dfrac{7}{13}} \cdot \sqrt{\dfrac{13}{3}}$ **38.** $\sqrt{\dfrac{19}{20}} \cdot \sqrt{\dfrac{20}{3}}$ **39.** $\sqrt{\dfrac{21}{7}} \cdot \sqrt{\dfrac{21}{8}}$ **40.** $\sqrt{\dfrac{5}{8}} \cdot \sqrt{\dfrac{5}{6}}$

41. $\sqrt{\dfrac{1}{12}} \cdot \sqrt{\dfrac{1}{3}}$ **42.** $\sqrt{\dfrac{1}{8}} \cdot \sqrt{\dfrac{1}{2}}$ **43.** $\sqrt{\dfrac{2}{9}} \cdot \sqrt{\dfrac{9}{2}}$ **44.** $\sqrt{\dfrac{4}{3}} \cdot \sqrt{\dfrac{3}{4}}$

45. $\sqrt{\dfrac{3}{4}} \cdot \sqrt{\dfrac{1}{5}}$ **46.** $\sqrt{\dfrac{1}{10}} \cdot \sqrt{\dfrac{10}{3}}$ **47.** $\sqrt{\dfrac{17}{3}} \cdot \sqrt{\dfrac{17}{6}}$ **48.** $\sqrt{\dfrac{1}{11}} \cdot \sqrt{\dfrac{33}{16}}$

49. $\sqrt{\dfrac{2}{5}} \cdot \sqrt{\dfrac{3}{10}}$ **50.** $\sqrt{\dfrac{9}{8}} \cdot \sqrt{\dfrac{7}{16}}$ **51.** $\sqrt{\dfrac{16}{27}} \cdot \sqrt{\dfrac{1}{9}}$ **52.** $\sqrt{\dfrac{5}{8}} \cdot \sqrt{\dfrac{5}{6}}$

Simplify each radical. Assume that all variables represent positive real numbers. See Examples 4 and 5.

53. $\sqrt{\dfrac{6}{p}}$ **54.** $\sqrt{\dfrac{5}{x}}$ **55.** $\sqrt{\dfrac{3}{y}}$ **56.** $\sqrt{\dfrac{9}{k}}$

57. $\sqrt{\dfrac{16}{m}}$ **58.** $\sqrt{\dfrac{2z^2}{x}}$ **59.** $\dfrac{\sqrt{3p^2}}{\sqrt{q}}$ **60.** $\dfrac{\sqrt{5a^3}}{\sqrt{b}}$

61. $\dfrac{\sqrt{7x^3}}{\sqrt{y}}$ **62.** $\dfrac{\sqrt{4r^3}}{\sqrt{s}}$ **63.** $\sqrt{\dfrac{6p^3}{3m}}$ **64.** $\sqrt{\dfrac{a^3b}{6}}$

65. $\sqrt{\dfrac{x^2}{4y}}$ **66.** $\sqrt{\dfrac{m^2n}{2}}$ **67.** $\sqrt{\dfrac{9a^2r}{5}}$ **68.** $\sqrt{\dfrac{2x^2z^4}{3y}}$

69. *Concept Check* Which one of the following would be an appropriate choice for multi-plying the numerator and the denominator of $\dfrac{\sqrt[3]{2}}{\sqrt[3]{5}}$ in order to rationalize the denominator?

A. $\sqrt[3]{5}$ **B.** $\sqrt[3]{25}$ **C.** $\sqrt[3]{2}$ **D.** $\sqrt[3]{3}$

70. *Concept Check* In Example 6(b), we multiplied the numerator and denominator of $\dfrac{\sqrt[3]{3}}{\sqrt[3]{4}}$ by $\sqrt[3]{2}$ to rationalize the denominator. Suppose we had chosen to multiply by $\sqrt[3]{16}$ instead. Would we have obtained the correct answer after all simplifications were done?

Rationalize each denominator. Assume that variables in denominators are nonzero. See Example 6.

71. $\sqrt[3]{\dfrac{1}{2}}$ **72.** $\sqrt[3]{\dfrac{1}{4}}$ **73.** $\sqrt[3]{\dfrac{1}{32}}$ **74.** $\sqrt[3]{\dfrac{1}{5}}$

75. $\sqrt[3]{\dfrac{1}{11}}$ **76.** $\sqrt[3]{\dfrac{3}{2}}$ **77.** $\sqrt[3]{\dfrac{2}{5}}$ **78.** $\sqrt[3]{\dfrac{4}{9}}$

79. $\dfrac{\sqrt[3]{4}}{\sqrt[3]{7}}$ **80.** $\dfrac{\sqrt[3]{5}}{\sqrt[3]{10}}$ **81.** $\sqrt[3]{\dfrac{3}{4y^2}}$ **82.** $\sqrt[3]{\dfrac{3}{25x^2}}$

83. $\dfrac{\sqrt[3]{7m}}{\sqrt[3]{36n}}$ **84.** $\dfrac{\sqrt[3]{11p}}{\sqrt[3]{49q}}$ **85.** $\sqrt[4]{\dfrac{1}{8}}$ **86.** $\sqrt[4]{\dfrac{1}{27}}$

PREVIEW EXERCISES

Find each product. See Sections 5.5 and 5.6.

87. $(4x + 7)(8x - 3)$ **88.** $ab(3a^2b - 2ab^2 + 7)$ **89.** $(6x - 1)(6x + 1)$

90. $(r + 7)(r - 7)$ **91.** $(p + q)(a - m)$ **92.** $(3w - 8)^2$

8.5 More Simplifying and Operations with Radicals

OBJECTIVES

1 Simplify products of radical expressions.

2 Use conjugates to rationalize denominators of radical expressions.

3 Write radical expressions with quotients in lowest terms.

The conditions for which a radical is in simplest form were listed in the previous section. A set of guidelines to use when you are simplifying radical expressions follows.

Simplifying Radical Expressions

1. If a radical represents a rational number, use that rational number in place of the radical.

   Examples: $\sqrt{49} = 7, \quad \sqrt{\dfrac{169}{9}} = \dfrac{13}{3}$

2. If a radical expression contains products of radicals, use the product rule for radicals, $\sqrt[n]{a} \cdot \sqrt[n]{b} = \sqrt[n]{ab}$, to get a single radical.

   Examples: $\sqrt{5} \cdot \sqrt{x} = \sqrt{5x}, \qquad \sqrt[3]{3} \cdot \sqrt[3]{2} = \sqrt[3]{6}$

3. If a radicand of a square root radical has a factor that is a perfect square, express the radical as the product of the positive square root of the perfect square and the remaining radical factor. A similar statement applies to higher roots.

   Examples: $\sqrt{20} = \sqrt{4 \cdot 5} = \sqrt{4} \cdot \sqrt{5} = 2\sqrt{5}$

   $\sqrt[3]{16} = \sqrt[3]{8 \cdot 2} = \sqrt[3]{8} \cdot \sqrt[3]{2} = 2\sqrt[3]{2}$

4. If a radical expression contains sums or differences of radicals, use the distributive property to combine like radicals.

   Examples: $3\sqrt{2} + 4\sqrt{2}$ can be combined to get $7\sqrt{2}$.

   $3\sqrt{2} + 4\sqrt{3}$ cannot be simplified further.

5. Rationalize any denominator containing a radical.

   Examples: $\dfrac{5}{\sqrt{3}} = \dfrac{5 \cdot \sqrt{3}}{\sqrt{3} \cdot \sqrt{3}} = \dfrac{5\sqrt{3}}{3}$

   $\sqrt{\dfrac{3}{2}} = \dfrac{\sqrt{3}}{\sqrt{2}} = \dfrac{\sqrt{3} \cdot \sqrt{2}}{\sqrt{2} \cdot \sqrt{2}} = \dfrac{\sqrt{6}}{2}$

   $\sqrt[3]{\dfrac{1}{4}} = \dfrac{\sqrt[3]{1}}{\sqrt[3]{4}} = \dfrac{\sqrt[3]{1} \cdot \sqrt[3]{2}}{\sqrt[3]{4} \cdot \sqrt[3]{2}} = \dfrac{\sqrt[3]{2}}{\sqrt[3]{8}} = \dfrac{\sqrt[3]{2}}{2}$

OBJECTIVE 1 Simplify products of radical expressions. Use the preceding guidelines.

EXAMPLE 1 Multiplying Radical Expressions

Find each product and simplify.

(a) $\sqrt{5}\left(\sqrt{8} - \sqrt{32}\right)$

Start by simplifying $\sqrt{8}$ and $\sqrt{32}$.

$$\sqrt{8} = 2\sqrt{2} \quad \text{and} \quad \sqrt{32} = 4\sqrt{2}.$$

Now simplify inside the parentheses.

$$\sqrt{5}(\sqrt{8} - \sqrt{32})$$
$$= \sqrt{5}(2\sqrt{2} - 4\sqrt{2}) \qquad \sqrt{8} = 2\sqrt{2};\ \sqrt{32} = 4\sqrt{2}$$
$$= \sqrt{5}(-2\sqrt{2}) \qquad \text{Subtract like radicals.}$$
$$= -2\sqrt{5 \cdot 2} \qquad \text{Product rule}$$
$$= -2\sqrt{10} \qquad \text{Multiply.}$$

(b) $(\sqrt{3} + 2\sqrt{5})(\sqrt{3} - 4\sqrt{5})$

We can find the products of sums of radicals in the same way that we found the product of binomials in **Section 5.5,** using the FOIL method.

$$(\sqrt{3} + 2\sqrt{5})(\sqrt{3} - 4\sqrt{5})$$
$$= \underbrace{\sqrt{3}(\sqrt{3})}_{\text{First}} + \underbrace{\sqrt{3}(-4\sqrt{5})}_{\text{Outer}} + \underbrace{2\sqrt{5}(\sqrt{3})}_{\text{Inner}} + \underbrace{2\sqrt{5}(-4\sqrt{5})}_{\text{Last}}$$
$$= 3 - 4\sqrt{15} + 2\sqrt{15} - 8 \cdot 5 \qquad \text{Product rule}$$
$$= 3 - 2\sqrt{15} - 40 \qquad \text{Add like radicals.}$$
$$= -37 - 2\sqrt{15} \qquad \text{Combine terms.}$$

> This does *not* equal $-39\sqrt{15}$.

(c) $(\sqrt{3} + \sqrt{21})(\sqrt{3} - \sqrt{7})$
$$= \sqrt{3}(\sqrt{3}) + \sqrt{3}(-\sqrt{7}) + \sqrt{21}(\sqrt{3}) + \sqrt{21}(-\sqrt{7}) \qquad \text{FOIL}$$
$$= 3 - \sqrt{21} + \sqrt{63} - \sqrt{147} \qquad \text{Product rule}$$
$$= 3 - \sqrt{21} + \sqrt{9} \cdot \sqrt{7} - \sqrt{49} \cdot \sqrt{3} \qquad \text{9 and 49 are perfect squares.}$$
$$= 3 - \sqrt{21} + 3\sqrt{7} - 7\sqrt{3} \qquad \sqrt{9} = 3;\ \sqrt{49} = 7$$

Since there are no like radicals, no terms can be combined.

> ✔ **Now Try Exercises 5, 11, and 27.**

The special products of binomials discussed in **Section 5.6** can be applied to radicals. Example 2 uses the rules for the square of a binomial:

$$(x + y)^2 = x^2 + 2xy + y^2 \quad \text{and} \quad (x - y)^2 = x^2 - 2xy + y^2.$$

EXAMPLE 2 Using Special Products with Radicals

Find each product. Assume that $x \geq 0$.

(a) $(\sqrt{10} - 7)^2$

Follow the second pattern for squaring a binomial. Let $x = \sqrt{10}$ and $y = 7$.
$$(\sqrt{10} - 7)^2 = (\sqrt{10})^2 - 2(\sqrt{10})(7) + 7^2$$
$$= 10 - 14\sqrt{10} + 49 \qquad (\sqrt{10})^2 = 10;\ 7^2 = 49$$
$$= 59 - 14\sqrt{10} \qquad \text{Combine terms.}$$

> Do *not* try to combine further here.

(b) $(2\sqrt{3} + 4)^2 = (2\sqrt{3})^2 + 2(2\sqrt{3})(4) + 4^2$ $x = 2\sqrt{3}; y = 4$

$= 12 + 16\sqrt{3} + 16$ $(2\sqrt{3})^2 = 4 \cdot 3 = 12$

$= 28 + 16\sqrt{3}$ Do *not* try to combine further here.

(c) $(5 - \sqrt{x})^2 = 5^2 - 2(5)(\sqrt{x}) + (\sqrt{x})^2$

$= 25 - 10\sqrt{x} + x$

✔ **Now Try Exercises 15, 17, and 37.**

▶ **CAUTION**   Be careful! In Examples 2(a) and (b),

$$59 - 14\sqrt{10} \neq 45\sqrt{10} \quad \text{and} \quad 28 + 16\sqrt{3} \neq 44\sqrt{3}.$$

Only like radicals can be combined.

Example 3 uses the rule for the product of the sum and difference of two terms,

$$(x + y)(x - y) = x^2 - y^2.$$

EXAMPLE 3   **Using a Special Product with Radicals**

Find each product. Assume that $x \geq 0$.

(a) $(4 + \sqrt{3})(4 - \sqrt{3})$

Use the pattern given above. Let $x = 4$ and $y = \sqrt{3}$.

$(4 + \sqrt{3})(4 - \sqrt{3}) = 4^2 - (\sqrt{3})^2$

$= 16 - 3$ $4^2 = 16; (\sqrt{3})^2 = 3$

$= 13$

(b) $(\sqrt{x} - \sqrt{6})(\sqrt{x} + \sqrt{6}) = (\sqrt{x})^2 - (\sqrt{6})^2$

$= x - 6$ $(\sqrt{x})^2 = x; (\sqrt{6})^2 = 6$

✔ **Now Try Exercises 21 and 41.**

OBJECTIVE 2   **Use conjugates to rationalize denominators of radical expressions.**
Notice that the results in Example 3 do not contain radicals. The pairs of expressions being multiplied, $4 + \sqrt{3}$ and $4 - \sqrt{3}$, and $\sqrt{x} - \sqrt{6}$ and $\sqrt{x} + \sqrt{6}$, are called **conjugates** of each other. Conjugates can be used to rationalize the denominators in more complicated quotients, such as

$$\frac{2}{4 - \sqrt{3}}.$$

By Example 3(a), if this denominator, $4 - \sqrt{3}$, is multiplied by $4 + \sqrt{3}$, then the product, $(4 - \sqrt{3})(4 + \sqrt{3})$, is the rational number 13. Multiplying the numerator and denominator of the quotient by $4 + \sqrt{3}$ gives

$$\frac{2}{4 - \sqrt{3}} = \frac{2(4 + \sqrt{3})}{(4 - \sqrt{3})(4 + \sqrt{3})} = \frac{2(4 + \sqrt{3})}{13}.$$

The denominator has now been rationalized; it contains no radicals.

Using Conjugates to Simplify a Radical Expression

To simplify a radical expression with two terms in the denominator, where at least one of those terms is a square root radical, multiply numerator and denominator by the conjugate of the denominator.

EXAMPLE 4 Using Conjugates to Rationalize Denominators

Simplify by rationalizing each denominator. Assume that $x \geq 0$.

(a) $\dfrac{5}{3 + \sqrt{5}} = \dfrac{5(3 - \sqrt{5})}{(3 + \sqrt{5})(3 - \sqrt{5})}$ Multiply by the conjugate.

$= \dfrac{5(3 - \sqrt{5})}{3^2 - (\sqrt{5})^2}$ $(x + y)(x - y) = x^2 - y^2$

$= \dfrac{5(3 - \sqrt{5})}{9 - 5}$ $3^2 = 9; (\sqrt{5})^2 = 5$

$= \dfrac{5(3 - \sqrt{5})}{4}$ Subtract.

(b) $\dfrac{6 + \sqrt{2}}{\sqrt{2} - 5} = \dfrac{(6 + \sqrt{2})(\sqrt{2} + 5)}{(\sqrt{2} - 5)(\sqrt{2} + 5)}$ Multiply by the conjugate.

$= \dfrac{6\sqrt{2} + 30 + 2 + 5\sqrt{2}}{2 - 25}$ FOIL; $(x + y)(x - y) = x^2 - y^2$

$= \dfrac{11\sqrt{2} + 32}{-23}$ Combine like terms.

$= \dfrac{-11\sqrt{2} - 32}{23}$ $\dfrac{x}{-y} = \dfrac{-x}{y}$

(c) $\dfrac{4}{3 + \sqrt{x}} = \dfrac{4(3 - \sqrt{x})}{(3 + \sqrt{x})(3 - \sqrt{x})}$ Multiply by the conjugate.

$= \dfrac{4(3 - \sqrt{x})}{9 - x}$ $3^2 = 9; (\sqrt{x})^2 = x$

✔ **Now Try Exercises 47, 55, and 65.**

OBJECTIVE 3 Write radical expressions with quotients in lowest terms.

EXAMPLE 5 Writing a Radical Quotient in Lowest Terms

Write $\dfrac{3\sqrt{3} + 9}{12}$ in lowest terms.

Factor the numerator and denominator, and then use the fundamental property from **Section 7.1** to divide out the common factor.

$$\dfrac{3\sqrt{3} + 9}{12} = \dfrac{3(\sqrt{3} + 3)}{3(4)} = 1 \cdot \dfrac{\sqrt{3} + 3}{4} = \dfrac{\sqrt{3} + 3}{4}$$

Don't simplify yet!　　Now, simplify.

✔ **Now Try Exercise 69.**

▶ **CAUTION** An expression like the one in Example 5 can be simplified only by factoring a common factor from the denominator and *each* term of the numerator. For example,

$$\frac{4 + 8\sqrt{5}}{4} \neq 1 + 8\sqrt{5}.$$

First factor to get $\dfrac{4 + 8\sqrt{5}}{4} = \dfrac{4(1 + 2\sqrt{5})}{4} = 1 + 2\sqrt{5}.$

8.5 EXERCISES

🌐 *Complete solution available on Video Lectures on CD/DVD*

▨ *Now Try Exercise*

In this exercise set, we assume that variables are such that no negative numbers appear as radicals in square roots and such that no denominators are zero.

In Exercises 1–4, perform the operations mentally, and write the answers without doing intermediate steps.

1. $\sqrt{49} + \sqrt{36}$ **2.** $\sqrt{100} - \sqrt{81}$ **3.** $\sqrt{2} \cdot \sqrt{8}$ **4.** $\sqrt{8} \cdot \sqrt{8}$

Simplify each expression. Use the five guidelines given in this section. See Examples 1–3.

🌐 **5.** $\sqrt{5}(\sqrt{3} - \sqrt{7})$ **6.** $\sqrt{7}(\sqrt{10} + \sqrt{3})$ **7.** $2\sqrt{5}(\sqrt{2} + 3\sqrt{5})$

8. $3\sqrt{7}(2\sqrt{7} + 4\sqrt{5})$ **9.** $3\sqrt{14} \cdot \sqrt{2} - \sqrt{28}$ **10.** $7\sqrt{6} \cdot \sqrt{3} - 2\sqrt{18}$

11. $(2\sqrt{6} + 3)(3\sqrt{6} + 7)$ **12.** $(4\sqrt{5} - 2)(2\sqrt{5} - 4)$

13. $(5\sqrt{7} - 2\sqrt{3})(3\sqrt{7} + 4\sqrt{3})$ **14.** $(2\sqrt{10} + 5\sqrt{2})(3\sqrt{10} - 3\sqrt{2})$

🌐 **15.** $(8 - \sqrt{7})^2$ **16.** $(6 - \sqrt{11})^2$

17. $(2\sqrt{7} + 3)^2$ **18.** $(4\sqrt{5} + 5)^2$

19. $(\sqrt{6} + 1)^2$ **20.** $(\sqrt{7} + 2)^2$

🌐 **21.** $(5 - \sqrt{2})(5 + \sqrt{2})$ **22.** $(3 - \sqrt{5})(3 + \sqrt{5})$

23. $(\sqrt{8} - \sqrt{7})(\sqrt{8} + \sqrt{7})$ **24.** $(\sqrt{12} - \sqrt{11})(\sqrt{12} + \sqrt{11})$

25. $(\sqrt{78} - \sqrt{76})(\sqrt{78} + \sqrt{76})$ **26.** $(\sqrt{85} - \sqrt{82})(\sqrt{85} + \sqrt{82})$

27. $(\sqrt{2} + \sqrt{3})(\sqrt{6} - \sqrt{2})$ **28.** $(\sqrt{3} + \sqrt{5})(\sqrt{15} - \sqrt{5})$

29. $(\sqrt{10} - \sqrt{5})(\sqrt{5} + \sqrt{20})$ **30.** $(\sqrt{6} - \sqrt{3})(\sqrt{3} + \sqrt{18})$

31. $(\sqrt{5} + \sqrt{30})(\sqrt{6} + \sqrt{3})$ **32.** $(\sqrt{10} - \sqrt{20})(\sqrt{2} - \sqrt{5})$

33. $(5\sqrt{7} - 2\sqrt{3})^2$ **34.** $(8\sqrt{2} - 3\sqrt{3})^2$

✐ **35.** In Example 1(b), the original expression simplifies to $-37 - 2\sqrt{15}$. Students often try to further simplify expressions like this by combining the -37 and the -2 to get $-39\sqrt{15}$, which is incorrect. Explain why.

36. Find each product mentally.

    (a) $(\sqrt{x} + \sqrt{y})(\sqrt{x} - \sqrt{y})$ **(b)** $(\sqrt{28} - \sqrt{14})(\sqrt{28} + \sqrt{14})$

Simplify each radical expression. See Examples 1–3.

37. $\left(7 + \sqrt{x}\right)^2$

38. $\left(12 - \sqrt{r}\right)^2$

39. $\left(3\sqrt{t} + \sqrt{7}\right)\left(2\sqrt{t} - \sqrt{14}\right)$

40. $\left(2\sqrt{z} - \sqrt{3}\right)\left(\sqrt{z} - \sqrt{5}\right)$

41. $\left(\sqrt{3m} + \sqrt{2n}\right)\left(\sqrt{3m} - \sqrt{2n}\right)$

42. $\left(\sqrt{4p} - \sqrt{3k}\right)\left(\sqrt{4p} + \sqrt{3k}\right)$

43. *Concept Check* Determine the expression by which you should multiply the numerator and denominator to rationalize each denominator.

(a) $\dfrac{1}{\sqrt{5} + \sqrt{3}}$ (b) $\dfrac{3}{\sqrt{6} - \sqrt{5}}$

44. *Concept Check* If you try to rationalize the denominator of $\dfrac{2}{4 + \sqrt{3}}$ by multiplying the numerator and denominator by $4 + \sqrt{3}$, what problem arises? What should you multiply by?

Rationalize each denominator. Write quotients in lowest terms. See Example 4.

45. $\dfrac{1}{2 + \sqrt{5}}$

46. $\dfrac{1}{4 + \sqrt{15}}$

47. $\dfrac{7}{2 - \sqrt{11}}$

48. $\dfrac{38}{5 - \sqrt{6}}$

49. $\dfrac{\sqrt{12}}{\sqrt{3} + 1}$

50. $\dfrac{\sqrt{18}}{\sqrt{2} - 1}$

51. $\dfrac{2\sqrt{3}}{\sqrt{3} + 5}$

52. $\dfrac{\sqrt{12}}{2 - \sqrt{10}}$

53. $\dfrac{\sqrt{2} + 3}{\sqrt{3} - 1}$

54. $\dfrac{\sqrt{5} + 2}{2 - \sqrt{3}}$

55. $\dfrac{6 - \sqrt{5}}{\sqrt{2} + 2}$

56. $\dfrac{3 + \sqrt{2}}{\sqrt{2} + 1}$

57. $\dfrac{2\sqrt{6} + 1}{\sqrt{2} + 5}$

58. $\dfrac{3\sqrt{2} - 4}{\sqrt{3} + 2}$

59. $\dfrac{\sqrt{7} + \sqrt{2}}{\sqrt{3} - \sqrt{2}}$

60. $\dfrac{\sqrt{6} + \sqrt{5}}{\sqrt{3} + \sqrt{5}}$

61. $\dfrac{\sqrt{5}}{\sqrt{2} + \sqrt{3}}$

62. $\dfrac{\sqrt{3}}{\sqrt{2} + \sqrt{3}}$

63. $\dfrac{\sqrt{108}}{3 + 3\sqrt{3}}$

64. $\dfrac{9\sqrt{8}}{6\sqrt{2} - 6}$

65. $\dfrac{8}{4 - \sqrt{x}}$

66. $\dfrac{12}{6 + \sqrt{y}}$

67. $\dfrac{1}{\sqrt{x} + \sqrt{y}}$

68. $\dfrac{2}{\sqrt{x} - \sqrt{y}}$

Write each quotient in lowest terms. See Example 5.

69. $\dfrac{5\sqrt{7} - 10}{5}$

70. $\dfrac{6\sqrt{5} - 9}{3}$

71. $\dfrac{2\sqrt{3} + 10}{8}$

72. $\dfrac{4\sqrt{6} + 6}{10}$

73. $\dfrac{12 - 2\sqrt{10}}{4}$

74. $\dfrac{9 - 6\sqrt{2}}{12}$

75. $\dfrac{16 + 8\sqrt{2}}{24}$

76. $\dfrac{25 + 5\sqrt{3}}{10}$

The next problems are "brain busters." Perform each operation and express the answer in simplest form.

77. $\sqrt[3]{4}\left(\sqrt[3]{2} - 3\right)$

78. $\sqrt[3]{5}\left(4\sqrt[3]{5} - \sqrt[3]{25}\right)$

79. $2\sqrt[4]{2}\left(3\sqrt[4]{8} + 5\sqrt[4]{4}\right)$

80. $6\sqrt[4]{9}\left(2\sqrt[4]{9} - \sqrt[4]{27}\right)$

81. $\left(\sqrt[3]{2} - 1\right)\left(\sqrt[3]{4} + 3\right)$

82. $\left(\sqrt[3]{9} + 5\right)\left(\sqrt[3]{3} - 4\right)$

83. $\left(\sqrt[3]{5} - \sqrt[3]{4}\right)\left(\sqrt[3]{25} + \sqrt[3]{20} + \sqrt[3]{16}\right)$

84. $\left(\sqrt[3]{4} + \sqrt[3]{2}\right)\left(\sqrt[3]{16} - \sqrt[3]{8} + \sqrt[3]{4}\right)$

Solve each problem.

85. The radius of the circular top or bottom of a tin can with surface area S and height h is given by

$$r = \dfrac{-h + \sqrt{h^2 + 0.64S}}{2}.$$

What radius should be used to make a can with height 12 in. and surface area 400 in.^{2}?

86. If an investment of P dollars grows to A dollars in 2 yr, the annual rate of return on the investment is given by

$$r = \frac{\sqrt{A} - \sqrt{P}}{\sqrt{P}}.$$

First rationalize the denominator and then find the annual rate of return (as a percent) if $50,000 increases to $58,320.

In Exercises 87 and 88, (a) give the answer as a simplified radical and (b) use a calculator to give the answer correct to the nearest thousandth.

87. The period p of a pendulum is the time it takes for it to swing from one extreme to the other and back again. The value of p in seconds is given by

$$p = k \cdot \sqrt{\frac{L}{g}},$$

where L is the length of the pendulum, g is the acceleration due to gravity, and k is a constant. Find the period when $k = 6, L = 9$ ft, and $g = 32$ ft per sec^2.

88. The velocity v of a meteor approaching Earth is given by

$$v = \frac{k}{\sqrt{d}}$$

kilometers per second, where d is the distance of the meteor from the center of Earth and k is a constant. What is the velocity of a meteor that is 6000 km away from the center of Earth if $k = 450$?

RELATING CONCEPTS (EXERCISES 89–94)

FOR INDIVIDUAL OR GROUP WORK

Work Exercises 89–94 in order, to see why a common student error is indeed an error.

89. Use the distributive property to write $6(5 + 3x)$ as a sum.

90. Your answer in Exercise 89 should be $30 + 18x$. Why can we not combine these two terms to get $48x$?

91. Repeat Exercise 14 from earlier in this exercise set.

92. Your answer in Exercise 91 should be $30 + 18\sqrt{5}$. Many students will, in error, try to combine these terms to get $48\sqrt{5}$. Why is this wrong?

93. Write the expression similar to $30 + 18x$ that simplifies to $48x$. Then write the expression similar to $30 + 18\sqrt{5}$ that simplifies to $48\sqrt{5}$.

94. Write a short explanation of the similarities between combining like terms and combining like radicals.

> ### PREVIEW EXERCISES
>
> *Solve each equation. See **Section 6.5.***
>
> **95.** $(2x - 1)(4x - 3) = 0$
>
> **96.** $(5x + 6)^2 = 0$
>
> **97.** $x^2 + 4x + 3 = 0$
>
> **98.** $x^2 - 6x + 9 = 0$
>
> **99.** $x(x + 2) = 3$
>
> **100.** $x(x + 4) = 21$

Summary Exercises on Operations with Radicals

Perform all indicated operations and express each answer in simplest form. Assume that all variables represent positive numbers.

1. $5\sqrt{10} - 8\sqrt{10}$

2. $\sqrt{5}(\sqrt{5} - \sqrt{3})$

3. $(1 + \sqrt{3})(2 - \sqrt{6})$

4. $\sqrt{98} - \sqrt{72} + \sqrt{50}$

5. $(3\sqrt{5} - 2\sqrt{7})^2$

6. $\dfrac{3}{\sqrt{6}}$

7. $\dfrac{1 + \sqrt{2}}{1 - \sqrt{2}}$

8. $\dfrac{8}{\sqrt{7} - \sqrt{5}}$

9. $(\sqrt{3} + 6)(\sqrt{3} - 6)$

10. $\dfrac{1}{\sqrt{t} + \sqrt{3}}$

11. $\sqrt[3]{8x^3y^5z^6}$

12. $\dfrac{12}{\sqrt[3]{9}}$

13. $\dfrac{5}{\sqrt{6} - 1}$

14. $\sqrt{\dfrac{2}{3x}}$

15. $\dfrac{6\sqrt{3}}{5\sqrt{12}}$

16. $\dfrac{8\sqrt{50}}{2\sqrt{25}}$

17. $\dfrac{-4}{\sqrt[3]{4}}$

18. $\dfrac{\sqrt{6} - \sqrt{5}}{\sqrt{6} + \sqrt{5}}$

19. $\sqrt{75x} - \sqrt{12x}$

20. $(5 + 3\sqrt{3})^2$

21. $\sqrt[3]{\dfrac{16}{81}}$

22. $(\sqrt{107} - \sqrt{106})(\sqrt{107} + \sqrt{106})$

23. $x\sqrt[4]{x^5} - 3\sqrt[4]{x^9} + x^2\sqrt[4]{x}$

24. $\sqrt[3]{16t^2} - \sqrt[3]{54t^2} + \sqrt[3]{128t^2}$

25. $(1 + \sqrt[3]{3})(1 - \sqrt[3]{3} + \sqrt[3]{9})$

Students often have trouble distinguishing between the following two types of problems:

Simplifying a Radical Involving a Square Root	Solving an Equation by Using Square Roots
Exercise: Simplify $\sqrt{25}$.	*Exercise:* Solve $x^2 = 25$.
Answer: 5	*Answer:* $\{-5, 5\}$
In this situation, $\sqrt{25}$ represents the positive square root of 25, namely, 5.	In this situation, $x^2 = 25$ has either of two solutions: the negative square root of 25 or the positive square root of 25—that is, -5 or 5. (See Exercise 36.)

Use the preceding information to work Exercises 26–35.

26. (a) Simplify $\sqrt{36}$.
 (b) Solve $x^2 = 36$.

27. (a) Simplify $\sqrt{81}$.
 (b) Solve $x^2 = 81$.

28. (a) Solve $x^2 = 4$.
(b) Simplify $-\sqrt{4}$.

29. (a) Solve $x^2 = 9$.
(b) Simplify $-\sqrt{9}$.

30. (a) Solve $x^2 = \dfrac{1}{4}$.

(b) Simplify $\sqrt{\dfrac{1}{4}}$.

31. (a) Solve $x^2 = \dfrac{1}{49}$.

(b) Simplify $\sqrt{\dfrac{1}{49}}$.

32. (a) Simplify $-\sqrt{\dfrac{16}{25}}$.

(b) Solve $x^2 = \dfrac{16}{25}$.

33. (a) Simplify $-\sqrt{\dfrac{49}{100}}$.

(b) Solve $x^2 = \dfrac{49}{100}$.

34. (a) Solve $x^2 = 0.04$.
(b) Simplify $\sqrt{0.04}$.

35. (a) Solve $x^2 = 0.16$.
(b) Simplify $\sqrt{0.16}$.

36. Use the zero-factor property **(Section 6.5)** to show that the solution set of $x^2 = 25$ is $\{-5, 5\}$.

8.6 Solving Equations with Radicals

OBJECTIVES

1 Solve radical equations having square root radicals.

2 Identify equations with no solutions.

3 Solve equations by squaring a binomial.

4 Solve radical equations having cube root radicals.

A **radical equation** is an equation having a variable in the radicand, such as

$$\sqrt{x + 1} = 3 \quad \text{or} \quad 3\sqrt{x} = \sqrt{8x + 9}.$$

OBJECTIVE 1 **Solve radical equations having square root radicals.** To solve radical equations having square root radicals, we need a new property, called the *squaring property*.

Squaring Property of Equality

If each side of a given equation is squared, then all solutions of the original equation are *among* the solutions of the squared equation.

▶ **CAUTION** Be very careful with the squaring property: Using this property can give a new equation with *more* solutions than the original equation has. For example, starting with the equation $x = 4$ and squaring both sides gives

$$x^2 = 4^2 \quad \text{or} \quad x^2 = 16.$$

This last equation, $x^2 = 16$, has either of *two* solutions, 4 or -4, while the original equation, $x = 4$, has only *one* solution, 4. Because of this possibility, checking is more than just a guard against algebraic errors when solving an equation with radicals. It is an essential part of the solution process. *All proposed solutions from the squared equation must be checked in the original equation.*

EXAMPLE 1 Using the Squaring Property of Equality

Solve $\sqrt{x + 1} = 3$.

Use the squaring property of equality to square both sides of the equation.

$$\left(\sqrt{x + 1}\right)^2 = 3^2$$

$$x + 1 = 9 \qquad\qquad \left(\sqrt{x + 1}\right)^2 = x + 1$$

$$x = 8 \qquad\qquad \text{Subtract 1.}$$

Check:

$$\sqrt{x + 1} = 3 \qquad\qquad \text{Original equation}$$

$$\sqrt{8 + 1} = 3 \quad ? \quad \text{Let } x = 8.$$

> A check is essential.

$$\sqrt{9} = 3 \quad ?$$

$$3 = 3 \qquad\qquad \text{True}$$

Because this statement is true, $\{8\}$ is the solution set of $\sqrt{x + 1} = 3$. In this case, the equation obtained by squaring had just one solution, which also satisfied the original equation.

✔ **Now Try Exercise 3.**

EXAMPLE 2 Using the Squaring Property with a Radical on Each Side

Solve $3\sqrt{x} = \sqrt{x + 8}$.

$$3\sqrt{x} = \sqrt{x + 8}$$

$$\left(3\sqrt{x}\right)^2 = \left(\sqrt{x + 8}\right)^2 \qquad\qquad \text{Squaring property}$$

$$3^2\left(\sqrt{x}\right)^2 = \left(\sqrt{x + 8}\right)^2 \qquad\qquad (ab)^2 = a^2b^2$$

$$9x = x + 8 \qquad\qquad \left(\sqrt{x}\right)^2 = x; \left(\sqrt{x + 8}\right)^2 = x + 8$$

$$8x = 8 \qquad\qquad \text{Subtract } x.$$

$$x = 1 \qquad\qquad \text{Divide by 8.}$$

Check:

$$3\sqrt{x} = \sqrt{x + 8} \qquad\qquad \text{Original equation}$$

$$3\sqrt{1} = \sqrt{1 + 8} \quad ? \quad \text{Let } x = 1.$$

$$3(1) = \sqrt{9} \quad ?$$

$$3 = 3 \qquad\qquad \text{True}$$

The solution set of $3\sqrt{x} = \sqrt{x + 8}$ is $\{1\}$.

✔ **Now Try Exercise 15.**

▶ **CAUTION** Do not write the final result obtained in the check in the solution set. In Example 2, the solution set is $\{1\}$, *not* $\{3\}$.

OBJECTIVE 2 Identify equations with no solutions. Not all radical equations have solutions, as shown in Examples 3 and 4.

EXAMPLE 3 Using the Squaring Property when One Side Is Negative

Solve $\sqrt{x} = -3$.

$$\sqrt{x} = -3$$

$$\left(\sqrt{x}\right)^2 = (-3)^2 \qquad \text{Squaring property}$$

$$x = 9$$

Check:

$$\sqrt{x} = -3 \qquad \text{Original equation}$$

$$\sqrt{9} = -3 \qquad ? \quad \text{Let } x = 9.$$

$$3 = -3 \qquad \text{False}$$

Because the statement $3 = -3$ is false, the number 9 is not a solution of the given equation and is said to be an **extraneous solution;** it must be discarded. In fact, $\sqrt{x} = -3$ has no solution. The solution set is $\emptyset$.

> ✔ **Now Try Exercise 11.**

▶ **NOTE** Because $\sqrt{x}$ represents the *principal* or *nonnegative* square root of x in Example 3, we might have seen immediately that there is no solution.

Use the following steps when solving an equation with radicals.

Solving a Radical Equation

Step 1 **Isolate a radical.** Arrange the terms so that a radical is isolated on one side of the equation.

Step 2 **Square both sides.**

Step 3 **Combine like terms.**

Step 4 **Repeat Steps 1–3** if there is still a term with a radical.

Step 5 **Solve the equation.** Find all proposed solutions.

Step 6 **Check** all proposed solutions in the original equation.

EXAMPLE 4 Using the Squaring Property with a Quadratic Expression

Solve $x = \sqrt{x^2 + 5x + 10}$.

Step 1 The radical is already isolated on the right side of the equation.

Step 2 Square both sides.

$$x^2 = \left(\sqrt{x^2 + 5x + 10}\right)^2 \qquad \text{Squaring property}$$

$$x^2 = x^2 + 5x + 10 \qquad \left(\sqrt{x^2 + 5x + 10}\right)^2 = x^2 + 5x + 10$$

Step 3 $\quad 0 = 5x + 10 \qquad \text{Subtract } x^2.$

Step 4 This step is not needed.

Step 5 $\quad -10 = 5x \qquad \text{Subtract 10.}$

$$-2 = x \qquad \text{Divide by 5.}$$

Step 6 *Check:*

$$x = \sqrt{x^2 + 5x + 10}$$

$$-2 = \sqrt{(-2)^2 + 5(-2) + 10} \quad ? \quad \text{Let } x = -2.$$

$$-2 = \sqrt{4 - 10 + 10} \quad ? \quad \text{Multiply.}$$

$$-2 = 2 \quad \text{False}$$

The principal square root of a quantity *cannot* be negative.

Since substituting -2 for x leads to a false result, the equation has no solution, and the solution set is $\emptyset$.

✔ **Now Try Exercise 21.**

OBJECTIVE 3 Solve equations by squaring a binomial. The next examples use the following rules from **Section 5.6:**

$$(x + y)^2 = x^2 + 2xy + y^2 \quad \text{and} \quad (x - y)^2 = x^2 - 2xy + y^2.$$

By the second pattern, for example,

$$(x - 3)^2 = x^2 - 2x(3) + 3^2$$
$$= x^2 - 6x + 9.$$

EXAMPLE 5 Using the Squaring Property when One Side Has Two Terms

Solve $\sqrt{2x - 3} = x - 3$.

Square each side, using the preceding result to square the binomial on the right side of the equation.

$$\left(\sqrt{2x - 3}\right)^2 = (x - 3)^2$$

$$2x - 3 = x^2 - 6x + 9$$

Remember the middle term when squaring.

This equation is quadratic because of the x^2-term. To solve it, as shown in **Section 6.5,** we must write the equation in standard form.

$$x^2 - 8x + 12 = 0 \quad \text{Subtract } 2x \text{ and add } 3.$$

$$(x - 6)(x - 2) = 0 \quad \text{Factor.}$$

$$x - 6 = 0 \quad \text{or} \quad x - 2 = 0 \quad \text{Zero-factor property}$$

$$x = 6 \quad \text{or} \quad x = 2 \quad \text{Solve.}$$

Check both of these proposed solutions in the original equation.

If $x = 6$, then

$$\sqrt{2x - 3} = x - 3$$

$$\sqrt{2(6) - 3} = 6 - 3 \quad ?$$

$$\sqrt{12 - 3} = 3 \quad ?$$

$$\sqrt{9} = 3 \quad ?$$

$$3 = 3. \quad \text{True}$$

If $x = 2$, then

$$\sqrt{2x - 3} = x - 3$$

$$\sqrt{2(2) - 3} = 2 - 3 \quad ?$$

$$\sqrt{4 - 3} = -1 \quad ?$$

$$\sqrt{1} = -1 \quad ?$$

$$1 = -1. \quad \text{False}$$

Only 6 is a solution of the equation, so the solution set is $\{6\}$.

✔ **Now Try Exercise 35.**

Remember to use Step 1 and isolate the radical before squaring both sides. For example, suppose we want to solve $\sqrt{9x} - 1 = 2x$. If we skip Step 1 and square both sides, we get

$$\left(\sqrt{9x} - 1\right)^2 = (2x)^2$$
$$9x - 2\sqrt{9x} + 1 = 4x^2,$$

a more complicated equation that still contains a radical. This is why we should begin by isolating the radical on one side of the equation, as shown in Example 6.

EXAMPLE 6 **Rewriting an Equation before Using the Squaring Property**

Solve $\sqrt{9x} - 1 = 2x$.

$$\sqrt{9x} = 2x + 1 \qquad \text{Add 1 to isolate the radical.}$$
$$\left(\sqrt{9x}\right)^2 = (2x + 1)^2 \qquad \text{Square both sides.}$$
$$9x = 4x^2 + 4x + 1 \qquad \text{Remember the middle term when squaring.}$$

$$4x^2 - 5x + 1 = 0 \qquad \text{Subtract } 9x; \text{ standard form}$$
$$(4x - 1)(x - 1) = 0 \qquad \text{Factor.}$$
$$4x - 1 = 0 \quad \text{or} \quad x - 1 = 0 \qquad \text{Zero-factor property}$$
$$x = \frac{1}{4} \quad \text{or} \qquad x = 1 \qquad \text{Solve.}$$

Check:

If $x = \dfrac{1}{4}$, then

$$\sqrt{9x} - 1 = 2x$$

$$\sqrt{9\left(\frac{1}{4}\right)} - 1 = 2\left(\frac{1}{4}\right) \quad ?$$

$$\frac{1}{2} = \frac{1}{2}. \qquad \text{True}$$

If $x = 1$, then

$$\sqrt{9x} - 1 = 2x$$

$$\sqrt{9(1)} - 1 = 2(1) \quad ?$$

$$2 = 2. \qquad \text{True}$$

Both solutions check, so the solution set is $\left\{\frac{1}{4}, 1\right\}$.

✔ **Now Try Exercise 49.**

▶ **CAUTION** Errors often occur when both sides of an equation are squared. For instance, in Example 6, when both sides of

$$\sqrt{9x} = 2x + 1$$

are squared, the *entire* binomial $2x + 1$ must be squared to get $4x^2 + 4x + 1$. It is incorrect to square the $2x$ and the 1 separately to get $4x^2 + 1$.

Some radical equations require squaring twice, as in the next example.

EXAMPLE 7 Using the Squaring Property Twice

Solve $\sqrt{21 + x} = 3 + \sqrt{x}$.

$$\sqrt{21 + x} = 3 + \sqrt{x}$$

$$\left(\sqrt{21 + x}\right)^2 = \left(3 + \sqrt{x}\right)^2 \qquad \text{Square both sides.}$$

$$21 + x = 9 + 6\sqrt{x} + x$$

$$12 = 6\sqrt{x} \qquad \text{Subtract 9; subtract } x.$$

$$2 = \sqrt{x} \qquad \text{Divide by 6.}$$

$$2^2 = \left(\sqrt{x}\right)^2 \qquad \text{Square both sides again.}$$

$$4 = x$$

Check: If $x = 4$, then

$$\sqrt{21 + x} = 3 + \sqrt{x} \qquad \text{Original equation}$$

$$\sqrt{21 + 4} = 3 + \sqrt{4} \qquad ?$$

$$\sqrt{25} = 3 + 2 \qquad ?$$

$$5 = 5. \qquad \text{True}$$

The solution set is $\{4\}$.

✔ **Now Try Exercise 55.**

OBJECTIVE 4 Solve radical equations having cube root radicals. We can extend the concept of raising both sides of an equation to a power in order to solve radical equations with cube roots.

EXAMPLE 8 Solving Equations with Cube Root Radicals

Solve each equation.

(a) $\sqrt[3]{5x} = \sqrt[3]{3x + 1}$

$$\left(\sqrt[3]{5x}\right)^3 = \left(\sqrt[3]{3x + 1}\right)^3 \qquad \text{Cube each side.}$$

$$5x = 3x + 1$$

$$2x = 1 \qquad \text{Subtract } 3x.$$

$$x = \frac{1}{2} \qquad \text{Divide by 2.}$$

Check: $\qquad \sqrt[3]{5x} = \sqrt[3]{3x + 1} \qquad \text{Original equation}$

$$\sqrt[3]{5\left(\frac{1}{2}\right)} = \sqrt[3]{3\left(\frac{1}{2}\right) + 1} \qquad ? \qquad \text{Let } x = \frac{1}{2}.$$

$$\sqrt[3]{\frac{5}{2}} = \sqrt[3]{\frac{5}{2}} \qquad \text{True}$$

The solution set is $\left\{\frac{1}{2}\right\}$.

(b)
$$\sqrt[3]{x^2} = \sqrt[3]{26x + 27}$$
$$(\sqrt[3]{x^2})^3 = (\sqrt[3]{26x + 27})^3 \qquad \text{Cube each side.}$$
$$x^2 = 26x + 27$$
$$x^2 - 26x - 27 = 0 \qquad \text{Standard form}$$
$$(x + 1)(x - 27) = 0 \qquad \text{Factor.}$$
$$x + 1 = 0 \quad \text{or} \quad x - 27 = 0 \qquad \text{Zero-factor property}$$
$$x = -1 \quad \text{or} \qquad x = 27$$

Check:

If $x = -1$, then
$$\sqrt[3]{x^2} = \sqrt[3]{26x + 27}$$
$$\sqrt[3]{(-1)^2} = \sqrt[3]{26(-1) + 27} \quad ?$$
$$\sqrt[3]{1} = \sqrt[3]{1} \quad ?$$
$$1 = 1. \qquad \text{True}$$

If $x = 27$, then
$$\sqrt[3]{x^2} = \sqrt[3]{26x + 27}$$
$$\sqrt[3]{(27)^2} = \sqrt[3]{26(27) + 27} \quad ?$$
$$\sqrt[3]{729} = \sqrt[3]{729} \quad ?$$
$$9 = 9. \qquad \text{True}$$

Both proposed solutions check, so the solution set is $\{-1, 27\}$.

✔ **Now Try Exercises 61 and 63.**

8.6 EXERCISES

● *Complete solution available on Video Lectures on CD/DVD*

▨ *Now Try Exercise*

Solve each equation. See Examples 1–4.

1. $\sqrt{x} = 7$ **2.** $\sqrt{k} = 10$ ● **3.** $\sqrt{x + 2} = 3$ **4.** $\sqrt{x + 7} = 5$

5. $\sqrt{r - 4} = 9$ **6.** $\sqrt{k - 12} = 3$ **7.** $\sqrt{4 - t} = 7$ **8.** $\sqrt{9 - s} = 5$

9. $\sqrt{2t + 3} = 0$ **10.** $\sqrt{5x - 4} = 0$ ● **11.** $\sqrt{t} = -5$ **12.** $\sqrt{p} = -8$

13. $\sqrt{w - 4} = 7$ **14.** $\sqrt{t + 3} = 10$ ● **15.** $\sqrt{10x - 8} = 3\sqrt{x}$

16. $\sqrt{17t - 4} = 4\sqrt{t}$ **17.** $5\sqrt{x} = \sqrt{10x + 15}$ **18.** $4\sqrt{x} = \sqrt{20x - 16}$

19. $\sqrt{3x - 5} = \sqrt{2x + 1}$ **20.** $\sqrt{5x + 2} = \sqrt{3x + 8}$ ● **21.** $k = \sqrt{k^2 - 5k - 15}$

22. $x = \sqrt{x^2 - 2x - 6}$ **23.** $7x = \sqrt{49x^2 + 2x - 10}$ **24.** $6x = \sqrt{36x^2 + 5x - 5}$

25. $\sqrt{2x + 2} = \sqrt{3x - 5}$ **26.** $\sqrt{x + 2} = \sqrt{2x - 5}$ **27.** $\sqrt{5x - 5} = \sqrt{4x + 1}$

28. $\sqrt{3m + 3} = \sqrt{5m - 1}$ **29.** $\sqrt{3x - 8} = -2$ **30.** $\sqrt{6t + 4} = -3$

31. *Concept Check* The first step in solving the equation $\sqrt{2x + 1} = x - 7$ is to square each side of the equation. Errors often occur in solving equations such as this one when the right side of the equation is squared incorrectly. What is the square of the right side?

▨ **32.** *Concept Check* Consider the following "solution." **WHAT WENT WRONG?** Give the correct solution set.

$$-\sqrt{x - 1} = -4$$
$$-(x - 1) = 16 \qquad \text{Square both sides.}$$
$$-x + 1 = 16 \qquad \text{Distributive property}$$
$$-x = 15 \qquad \text{Subtract 1.}$$
$$x = -15 \qquad \text{Multiply by } -1.$$

Solution set: $\{-15\}$

Solve each equation. See Examples 5 and 6.

33. $\sqrt{5x + 11} = x + 3$ **34.** $\sqrt{5x + 1} = x + 1$ **35.** $\sqrt{2x + 1} = x - 7$

36. $\sqrt{3x + 10} = 2x - 5$ **37.** $\sqrt{x + 2} - 2 = x$ **38.** $\sqrt{x + 1} - 1 = x$

39. $\sqrt{12x + 12} + 10 = 2x$ **40.** $\sqrt{4x + 5} + 5 = 2x$

41. $\sqrt{6x + 7} - 1 = x + 1$ **42.** $\sqrt{8x + 8} - 1 = 2x + 1$

43. $2\sqrt{x + 7} = x - 1$ **44.** $3\sqrt{x + 13} = x + 9$

45. $\sqrt{2x} + 4 = x$ **46.** $\sqrt{3x} + 6 = x$

47. $\sqrt{x} + 9 = x + 3$ **48.** $\sqrt{x} + 3 = x - 9$

49. $3\sqrt{x - 2} = x - 2$ **50.** $2\sqrt{x + 4} = x + 1$

51. Explain how you can tell that the equation $\sqrt{x} = -8$ has no real number solution without performing any algebraic steps.

52. *Concept Check* Consider the following "solution." *WHAT WENT WRONG?* Give the correct solution set.

$$\sqrt{3x + 4} + \sqrt{x + 5} = 7$$
$$3x + 4 + x + 5 = 49 \qquad \text{Square each side.}$$
$$4x + 9 = 49 \qquad \text{Combine terms.}$$
$$4x = 40 \qquad \text{Subtract 9.}$$
$$x = 10 \qquad \text{Divide by 4.}$$

Solve each equation. See Example 7.

53. $\sqrt{3x + 3} + \sqrt{x + 2} = 5$ **54.** $\sqrt{2x + 1} + \sqrt{x + 4} = 3$

55. $\sqrt{x + 6} = \sqrt{x + 72}$ **56.** $\sqrt{x - 4} = \sqrt{x - 32}$

57. $\sqrt{3x + 4} - \sqrt{2x - 4} = 2$ **58.** $\sqrt{1 - x} + \sqrt{x + 9} = 4$

59. $\sqrt{2x + 11} + \sqrt{x + 6} = 2$ **60.** $\sqrt{x + 9} + \sqrt{x + 16} = 7$

Solve each equation. (Hint: In Exercises 67 and 68, extend the concepts to fourth root radicals.) See Example 8.

61. $\sqrt[3]{2x} = \sqrt[3]{5x + 2}$ **62.** $\sqrt[3]{4x + 3} = \sqrt[3]{2x - 1}$ **63.** $\sqrt[3]{x^2} = \sqrt[3]{8 + 7x}$

64. $\sqrt[3]{x^2} = \sqrt[3]{8 - 7x}$ **65.** $\sqrt[3]{3x^2 - 9x + 8} = \sqrt[3]{x}$ **66.** $\sqrt[3]{5x^2 - 6x + 2} = \sqrt[3]{x}$

67. $\sqrt[4]{x^2 + 24x} = 3$ **68.** $\sqrt[4]{x^2 + 6x} = 2$

Solve each problem.

69. The square root of the sum of a number and 4 is 5. Find the number.

70. A certain number is the same as the square root of the product of 8 and the number. Find the number.

71. Three times the square root of 2 equals the square root of the sum of some number and 10. Find the number.

72. The negative square root of a number equals that number decreased by 2. Find the number.

Solve each problem. Give answers to the nearest tenth.

73. Police sometimes use the following procedure to estimate the speed at which a car was traveling at the time of an accident: A police officer drives the car involved in the accident under conditions similar to those during which the accident took place and then skids to a stop. If the car is driven at 30 mph, then the speed at the time of the accident is given by

$$s = 30\sqrt{\frac{a}{p}},$$

where *a* is the length of the skid marks left at the time of the accident and *p* is the length of the skid marks in the police test. Find *s* for the following values of *a* and *p*.

(a) $a = 862$ ft; $p = 156$ ft
(b) $a = 382$ ft; $p = 96$ ft
(c) $a = 84$ ft; $p = 26$ ft

74. A formula for calculating the distance *d* one can see from an airplane to the horizon on a clear day is

$$d = 1.22\sqrt{x},$$

where *x* is the altitude of the plane in feet and *d* is given in miles. How far can one see to the horizon in a plane flying at the following altitudes?

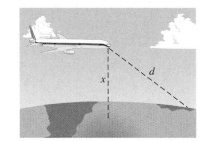

(a) 15,000 ft
(b) 18,000 ft
(c) 24,000 ft

75. A surveyor wants to find the height of a building. At a point 110.0 ft from the base of the building, he sights to the top of the building and finds the distance to be 193.0 ft. How high is the building?

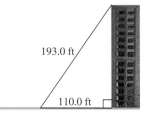

193.0 ft

110.0 ft

76. Two towns are separated by dense woods. To go from Town B to Town A, it is necessary to travel due west for 19.0 mi and then turn due north and travel for 14.0 mi. How far apart are the towns?

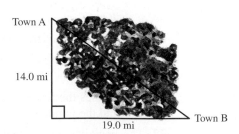

Town A

14.0 mi

Town B

19.0 mi

On a clear day, the maximum distance in kilometers that you can see from a tall building is given by the formula

$$\text{sight distance} = 111.7 \sqrt{\text{height of building in kilometers}}.$$

(Source: A Sourcebook of Applications of School Mathematics, NCTM, 1980.)

Use the conversion equations 1 ft ≈ 0.3048 m *and* 1 km ≈ 0.621371 mi *as necessary to solve each problem. Round your answers to the nearest mile.*

77. As mentioned in the chapter opener, the London Eye is a unique form of a Ferris wheel that features 32 observation capsules. It is the fourth-tallest structure in London, with a diameter of 135 m. (*Source:* www.londoneye.com) Does the formula justify the claim that on a clear day passengers on the London Eye can see Windsor Castle, 25 mi away?

78. The Empire State Building opened in 1931 on 5th Avenue in New York City. The building is 1250 ft high. (The antenna reaches to 1454 ft.) The observation deck, located on the 102nd floor, is at a height of 1050 ft. (*Source:* www.esbnyc.com) How far could you see on a clear day from the observation deck?

79. The twin Petronas Towers in Kuala Lumpur, Malaysia, were built in 1998. Both towers are 1483 ft high (including the spires). (*Source: World Almanac and Book of Facts 2000.*) How far would one of the builders have been able to see on a clear day from the top of a spire?

80. The Khufu Pyramid in Giza (also known as the Cheops Pyramid) was built in about 2566 B.C. to a height, at that time, of 482 ft. It is now only about 450 ft high. How far would one of the original builders of the pyramid have been able to see from the top of the pyramid?

RELATING CONCEPTS (EXERCISES 81–86)

FOR INDIVIDUAL OR GROUP WORK

The most common formula for the area of a triangle is $A = \frac{1}{2}bh$, where b is the length of the base and h is the height. What if the height is not known? What if we know only the lengths of the sides? Another formula, known as **Heron's formula,** allows us to calculate the area of a triangle if we know the lengths of the sides a, b, and c. First, let s equal the **semiperimeter,** which is one-half the perimeter.

$$s = \frac{1}{2}(a + b + c)$$

The area A is given by the formula
$$A = \sqrt{s(s - a)(s - b)(s - c)}.$$

For example, the familiar 3–4–5 right triangle has area $A = \frac{1}{2}(3)(4) = 6$ square units, calculated with the familiar formula. From Heron's formula, $s = \frac{1}{2}(3 + 4 + 5) = 6$. Therefore,
$$A = \sqrt{6(6 - 3)(6 - 4)(6 - 5)} = \sqrt{36} = 6.$$

The area is 6 square units, as expected.

Consider the following figure, and *work Exercises 81–86 in order.*

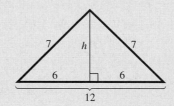

81. The lengths of the sides of the entire triangle are 7, 7, and 12. Find the semi-perimeter s.

82. Now use Heron's formula to find the area of the entire triangle. Write it as a simplified radical.

83. Find the value of h by using the Pythagorean formula.

84. Find the area of each of the congruent right triangles forming the entire triangle by using the formula $A = \frac{1}{2}bh$.

85. Double your result from Exercise 84 to determine the area of the entire triangle.

86. How do your answers in Exercises 82 and 85 compare? (*Note:* They should be equal, since the area of the entire triangle is unique.)

PREVIEW EXERCISES

Simplify each expression. Write the answer in exponential form with only positive exponents. See Sections 5.1 and 5.2.

87. $(5^2)^3$

88. $3^{-4} \cdot 3^{-1}$

89. $\dfrac{a^{-2}a^3}{a^4}$

90. $(2x^3)^{-1}$

91. $\left(\dfrac{p}{3}\right)^{-2}$

92. $\left(\dfrac{2y^3}{y^{-1}}\right)^{-2}$

93. $\dfrac{(c^3)^2c^4}{(c^{-1})^3}$

94. $\dfrac{(m^2)^4m^{-1}}{(m^3)^{-1}}$

8.7 Using Rational Numbers as Exponents

OBJECTIVES

1 Define and use expressions of the form $a^{1/n}$.

2 Define and use expressions of the form $a^{m/n}$.

3 Apply the rules for exponents using rational exponents.

4 Use rational exponents to simplify radicals.

OBJECTIVE 1 Define and use expressions of the form $a^{1/n}$. We now consider how an expression such as $5^{1/2}$ should be defined. We want to define $5^{1/2}$ so that all the rules for exponents developed earlier in this book still hold. Then we should define $5^{1/2}$ so that

$$5^{1/2} \cdot 5^{1/2} = 5^{1/2+1/2} = 5^1 = 5.$$

This agrees with the product rule for exponents from **Section 5.1.** By definition,

$$\left(\sqrt{5}\right)\left(\sqrt{5}\right) = 5.$$

Since both $5^{1/2} \cdot 5^{1/2}$ and $\sqrt{5} \cdot \sqrt{5}$ equal 5,

$$5^{1/2} \text{ should equal } \sqrt{5}.$$

Similarly,

$$5^{1/3} \cdot 5^{1/3} \cdot 5^{1/3} = 5^{1/3+1/3+1/3} = 5^{3/3} = 5,$$

and

$$\sqrt[3]{5} \cdot \sqrt[3]{5} \cdot \sqrt[3]{5} = \sqrt[3]{5^3} = 5,$$

so

$$5^{1/3} \text{ should equal } \sqrt[3]{5}.$$

These examples suggest the following definition.

$a^{1/n}$

If a is a nonnegative number and n is a positive integer, then

$$a^{1/n} = \sqrt[n]{a}.$$

Notice that the denominator of the rational exponent is the index of the radical.

EXAMPLE 1 Using the Definition of $a^{1/n}$

Simplify by first writing in radical form.

(a) $16^{1/2}$

By the definition of $a^{1/n}$, $\quad 16^{1/2} = \sqrt{16} = 4.$

(b) $27^{1/3} = \sqrt[3]{27} = 3$ **(c)** $216^{1/3} = \sqrt[3]{216} = 6$ **(d)** $64^{1/6} = \sqrt[6]{64} = 2$

The denominator is the index.

✔ **Now Try Exercises 5 and 7.**

OBJECTIVE 2 Define and use expressions of the form $a^{m/n}$. Now we can define a more general exponential expression, such as $16^{3/4}$. By the power rule, $(a^m)^n = a^{mn}$, so

$$16^{3/4} = (16^{1/4})^3 = \left(\sqrt[4]{16}\right)^3 = 2^3 = 8.$$

However, $16^{3/4}$ can also be written as

$$16^{3/4} = (16^3)^{1/4} = (4096)^{1/4} = \sqrt[4]{4096} = 8.$$

— Same answer

Either way, the answer is the same. As the example suggests, taking the root first involves smaller numbers and is often easier. This example suggests the following definition for $a^{m/n}$.

$a^{m/n}$

If a is a nonnegative number and m and n are integers with $n > 0$, then
$$a^{m/n} = (a^{1/n})^m = \left(\sqrt[n]{a}\right)^m.$$

EXAMPLE 2 Using the Definition of $a^{m/n}$

Evaluate.

(a) $9^{3/2} = (9^{1/2})^3 = 3^3 = 27$

(b) $64^{2/3} = (64^{1/3})^2 = 4^2 = 16$

(c) $-32^{4/5} = -(32^{1/5})^4 = -2^4 = -16$

The base is 32, not -32.

Be careful with signs here.

✔ **Now Try Exercises 13, 15, and 21.**

Earlier, a^{-n} was defined as

$$a^{-n} = \frac{1}{a^n}$$

for nonzero numbers a and integers n. This same result applies to negative rational exponents.

$a^{-m/n}$

If a is a positive number and m and n are integers with $n > 0$, then
$$a^{-m/n} = \frac{1}{a^{m/n}}.$$

EXAMPLE 3 Using the Definition of $a^{-m/n}$

Evaluate.

(a) $32^{-3/5} = \frac{1}{32^{3/5}} = \frac{1}{(32^{1/5})^3} = \frac{1}{2^3} = \frac{1}{8}$

(b) $27^{-4/3} = \frac{1}{27^{4/3}} = \frac{1}{(27^{1/3})^4} = \frac{1}{3^4} = \frac{1}{81}$

Think:
$32^{1/5} = \sqrt[5]{32} = 2$.

This is *not* the same as $-27^{4/3}$.

✔ **Now Try Exercises 25 and 27.**

▶ **CAUTION** In Example 3(b), a common mistake is to write $27^{-4/3}$ as $-27^{3/4}$. *This is incorrect.* The negative exponent does not indicate a negative number. Also, the negative exponent indicates to use the reciprocal of the *base*, not the reciprocal of the *exponent*.

OBJECTIVE 3 Apply the rules for exponents using rational exponents. All the rules for exponents given earlier still hold when the exponents are fractions.

EXAMPLE 4 Using the Rules for Exponents with Fractional Exponents

Simplify. Write each answer in exponential form with only positive exponents.

Keep the same base.

(a) $3^{2/3} \cdot 3^{5/3} = 3^{2/3+5/3} = 3^{7/3}$

Keep the same base.

(b) $\dfrac{5^{1/4}}{5^{3/4}} = 5^{1/4-3/4} = 5^{-2/4} = 5^{-1/2} = \dfrac{1}{5^{1/2}}$

(c) $(9^{1/4})^2 = 9^{2(1/4)} = 9^{2/4} = 9^{1/2} = \sqrt{9} = 3$

(d) $\left(\dfrac{9}{4}\right)^{5/2} = \dfrac{9^{5/2}}{4^{5/2}} = \dfrac{(9^{1/2})^5}{(4^{1/2})^5} = \dfrac{(\sqrt{9})^5}{(\sqrt{4})^5} = \dfrac{3^5}{2^5}$

Use parentheses to avoid errors.

(e) $\dfrac{2^{1/2} \cdot 2^{-1}}{2^{-3/2}} = \dfrac{2^{1/2+(-1)}}{2^{-3/2}} = \dfrac{2^{-1/2}}{2^{-3/2}} = 2^{-1/2-(-3/2)} = 2^{2/2} = 2^1 = 2$

✔ **Now Try Exercises 31, 35, 39, 43, and 47.**

EXAMPLE 5 Using Fractional Exponents with Variables

Simplify. Write each answer in exponential form with only positive exponents. Assume that all variables represent positive numbers.

(a) $m^{1/5} \cdot m^{3/5} = m^{1/5+3/5} = m^{4/5}$

(b) $\dfrac{p^{5/3}}{p^{4/3}} = p^{5/3-4/3} = p^{1/3}$

(c) $(x^2 y^{1/2})^4 = (x^2)^4 (y^{1/2})^4 = x^8 y^2$

(d) $\left(\dfrac{z^{1/4}}{w^{1/3}}\right)^5 = \dfrac{(z^{1/4})^5}{(w^{1/3})^5} = \dfrac{z^{5/4}}{w^{5/3}}$

(e) $\dfrac{k^{2/3} \cdot k^{-1/3}}{k^{5/3}} = k^{2/3+(-1/3)-5/3} = k^{-4/3} = \dfrac{1}{k^{4/3}}$

Use parentheses to avoid errors.

✔ **Now Try Exercises 49, 51, 53, 55, and 57.**

▶ **CAUTION** Errors often occur in problems like those in Examples 4 and 5 because students try to convert the expressions to radicals. Remember that the *rules of exponents* apply here.

OBJECTIVE 4 Use rational exponents to simplify radicals. Sometimes it is easier to simplify a radical by first writing it in exponential form.

EXAMPLE 6 Simplifying Radicals by Using Rational Exponents

Simplify each radical by first writing it in exponential form.

(a) $\sqrt[6]{9^3} = (9^3)^{1/6} = 9^{3/6} = 9^{1/2} = \sqrt{9} = 3$

(b) $\left(\sqrt[4]{m}\right)^2 = (m^{1/4})^2 = m^{2/4} = m^{1/2} = \sqrt{m}, \quad m \geq 0.$

✔ **Now Try Exercises 59 and 65.**

8.7 EXERCISES

Concept Check *Decide which one of the four choices is* not *equal to the given expression.*

1. $49^{1/2}$

   A. -7 **B.** 7 **C.** $\sqrt{49}$ **D.** $49^{0.5}$

2. $81^{1/2}$

   A. 9 **B.** $\sqrt{81}$ **C.** $81^{0.5}$ **D.** $\dfrac{81}{2}$

3. $-64^{1/3}$

   A. $-\sqrt{16}$ **B.** -4 **C.** 4 **D.** $-\sqrt[3]{64}$

4. $-125^{1/3}$

   A. $-\sqrt{25}$ **B.** -5 **C.** $-\sqrt[3]{125}$ **D.** 5

Simplify by first writing in radical form. See Examples 1–3.

5. $25^{1/2}$ **6.** $121^{1/2}$ **7.** $64^{1/3}$ **8.** $125^{1/3}$ **9.** $16^{1/4}$

10. $81^{1/4}$ **11.** $32^{1/5}$ **12.** $243^{1/5}$ **13.** $4^{3/2}$ **14.** $9^{5/2}$

15. $27^{2/3}$ **16.** $8^{5/3}$ **17.** $16^{3/4}$ **18.** $64^{5/3}$ **19.** $32^{2/5}$

20. $144^{3/2}$ **21.** $-8^{2/3}$ **22.** $-27^{5/3}$ **23.** $-64^{1/3}$ **24.** $-125^{5/3}$

25. $49^{-3/2}$ **26.** $9^{-5/2}$ **27.** $216^{-2/3}$ **28.** $32^{-4/5}$ **29.** $-16^{-5/4}$

30. $-81^{-3/4}$

Simplify. Write answers in exponential form with only positive exponents. Assume that all variables represent positive numbers. See Examples 4 and 5.

31. $2^{1/3} \cdot 2^{7/3}$ **32.** $5^{2/3} \cdot 5^{5/3}$ **33.** $6^{1/4} \cdot 6^{-3/4}$ **34.** $12^{2/5} \cdot 12^{-1/5}$

35. $\dfrac{15^{3/4}}{15^{5/4}}$ **36.** $\dfrac{7^{3/5}}{7^{-1/5}}$ **37.** $\dfrac{11^{-2/7}}{11^{-3/7}}$ **38.** $\dfrac{4^{-2/3}}{4^{1/3}}$

39. $(8^{3/2})^2$ **40.** $(5^{2/5})^{10}$ **41.** $(6^{1/3})^{3/2}$ **42.** $(7^{2/5})^{5/3}$

43. $\left(\dfrac{25}{4}\right)^{3/2}$ **44.** $\left(\dfrac{8}{27}\right)^{2/3}$ **45.** $\dfrac{2^{2/5} \cdot 2^{-3/5}}{2^{7/5}}$ **46.** $\dfrac{3^{-3/4} \cdot 3^{5/4}}{3^{-1/4}}$

47. $\dfrac{6^{-2/9}}{6^{1/9} \cdot 6^{-5/9}}$ **48.** $\dfrac{8^{6/7}}{8^{2/7} \cdot 8^{-1/7}}$ **49.** $x^{2/5} \cdot x^{7/5}$ **50.** $y^{2/3} \cdot y^{5/3}$

51. $\dfrac{r^{4/9}}{r^{3/9}}$ **52.** $\dfrac{s^{5/6}}{s^{4/6}}$ **53.** $(m^3 n^{1/4})^{2/3}$ **54.** $(p^4 q^{1/2})^{4/3}$

55. $\left(\dfrac{a^{2/3}}{b^{1/4}}\right)^6$ **56.** $\left(\dfrac{t^{3/7}}{s^{1/3}}\right)^{21}$ **57.** $\dfrac{m^{3/4} \cdot m^{-1/4}}{m^{1/3}}$ **58.** $\dfrac{q^{5/6} \cdot q^{-1/6}}{q^{1/3}}$

Simplify each radical by first writing it in exponential form. Give the answer as an integer or a radical in simplest form. Assume that all variables represent nonnegative numbers. See Example 6.

59. $\sqrt[6]{4^3}$ **60.** $\sqrt[9]{8^3}$ **61.** $\sqrt[8]{16^2}$ **62.** $\sqrt[9]{27^3}$

63. $\sqrt[4]{a^2}$ **64.** $\sqrt[9]{b^3}$ **65.** $\sqrt[6]{k^4}$ **66.** $\sqrt[8]{m^4}$

Use the exponential key on your calculator to find the following roots. For example, to find $\sqrt[5]{32}$, *enter 32 and then raise it to the 1/5 power. (The exponent 1/5 may be entered as 0.2 if you wish.) Refer to* **Appendix A** *if necessary. If the root is irrational, round it to the nearest thousandth.*

67. $\sqrt[6]{64}$ **68.** $\sqrt[5]{243}$ **69.** $\sqrt[7]{84}$ **70.** $\sqrt[9]{16}$

Solve each problem.

71. The formula

$$d = 1.22\sqrt{x}$$

has sometimes been used to calculate the distance in miles one can see from an airplane to the horizon on a clear day. Here, x is in kilometers.

(a) Write the formula using a rational exponent.
(b) Find d to the nearest hundredth if the altitude x is 30,000 ft.

72. A biologist has shown that the number of different plant species S on a Galápagos island is related to the area of the island, A (in square miles), by

$$S = 28.6A^{1/3}.$$

How many plant species would exist on such an island with the following areas?

(a) 8 mi^2 **(b)** 27,000 mi^2

73. Explain in your own words why $7^{1/2}$ is defined as $\sqrt{7}$.

74. Explain in your own words why $7^{1/3}$ is defined as $\sqrt[3]{7}$.

PREVIEW EXERCISES

Find the real square roots of each number. Simplify where possible. See **Section 8.1.**

75. 121   **76.** 625   **77.** 0.49   **78.** 0.81

79. $\dfrac{1}{4}$ **80.** $\dfrac{1}{9}$ **81.** $\dfrac{4}{25}$ **82.** $\dfrac{16}{49}$

Find and simplify the positive square root of each number. See **Section 8.2.**

83. 236   **84.** 160   **85.** 147   **86.** 320

Chapter **8** **Group Activity**

COMPARING TELEVISION SIZES

Objective Use the Pythagorean formula to find the dimensions of different television sets.

Television sets are identified by the diagonal measurement of the viewing screen. For example, a 19-in. TV measures 19 in. from one corner of the viewing screen diagonally to the other corner.

19 in.

A. The table gives some common TV sizes, as well as their corresponding widths. Use the Pythagorean formula,

$$a^2 + b^2 = c^2,$$

to find the heights of the viewing screens. Round up to the next whole number. (*Hint:* The TV size is the hypotenuse.)

TV Set	TV Size	Width in Inches	Height in Inches
A	13 in.	10	
B	19 in.	15	
C	25 in.	21	
D	27 in.	22	
E	32 in.	25	
F	36 in.	30	

Source: Data based on information from a Sears catalog.

B. The dimensions in the table correspond only to television *viewing screens*. Each television set also has an outer border of $1\frac{1}{2}$ in., as well as an additional 4 in. for a control panel at the bottom of the set. Also, it is recommended that a minimum of 2 in. of space be allowed above the set for ventilation.

   Consider the following entertainment centers, with dimensions for television space as given. Find the largest television set from the table in part A that will fit into each entertainment center.

   1. 27 in. by 29 in.   **2.** 28 in. by 25 in.   **3.** 20 in. by 20 in.

Chapter **8** **SUMMARY** *View the Interactive Summary on the Pass the Test CD.*

KEY TERMS

8.1 square root
principal square root
radicand
radical
radical expression

perfect square
cube root
fourth root
index (order)

8.2 perfect cube
8.3 like radicals
8.4 rationalizing the
denominator

8.5 conjugate
8.6 radical equation
squaring property
extraneous solution

NEW SYMBOLS

$\sqrt{}$ radical sign

$\approx$ is approximately
equal to

$\sqrt[3]{a}$ cube root of a
$\sqrt[n]{a}$ nth root of a

$a^{1/m}$ mth root of a

TEST YOUR WORD POWER

See how well you have learned the vocabulary in this chapter. Answers, with examples, follow the Quick Review.

1. The **square root** of a number is
   A. the number raised to the second power
   B. the number under a radical sign
   C. a number that, when multiplied by itself, gives the original number
   D. the inverse of the number.

2. A **radical** is
   A. a symbol that indicates the nth root
   B. an algebraic expression containing a square root
   C. the positive nth root of a number
   D. a radical sign and the number or expression under it.

3. The **principal root** of a positive number with even index n is
   A. the positive nth root of the number
   B. the negative nth root of the number
   C. the square root of the number
   D. the cube root of the number.

4. **Like radicals** are
   A. radicals in simplest form
   B. algebraic expressions containing radicals
   C. multiples of the same root of the same number
   D. radicals with the same index.

5. **Rationalizing the denominator** is the process of
   A. eliminating fractions from a radical expression
   B. changing the denominator of a fraction from a radical to a rational number
   C. clearing a radical expression of radicals
   D. multiplying radical expressions.

6. The **conjugate** of $a + b$ is
   A. $a - b$
   B. $a \cdot b$
   C. $a \div b$
   D. $(a + b)^2$.

QUICK REVIEW

Concepts	Examples

8.1 EVALUATING ROOTS

If a is a positive real number, then

$\sqrt{a}$ is the positive square root of a;

$-\sqrt{a}$ is the negative square root of a; $\sqrt{0} = 0$.

If a is a negative real number, then $\sqrt{a}$ is not a real number.

$$\sqrt{49} = 7$$
$$-\sqrt{81} = -9$$
$$\sqrt{-25} \text{ is not a real number.}$$

(continued)

Concepts	Examples

If a is a positive rational number, then $\sqrt{a}$ is rational if a is a perfect square and $\sqrt{a}$ is irrational if a is not a perfect square.

$\sqrt{\dfrac{4}{9}}, \sqrt{16}$ are rational. $\qquad \sqrt{\dfrac{2}{3}}, \sqrt{21}$ are irrational.

Each real number has exactly one real cube root.

$$\sqrt[3]{27} = 3 \qquad \sqrt[3]{-8} = -2$$

8.2 MULTIPLYING, DIVIDING, AND SIMPLIFYING RADICALS

Product Rule for Radicals

For nonnegative real numbers a and b,

$$\sqrt{a} \cdot \sqrt{b} = \sqrt{ab} \quad \text{and} \quad \sqrt{a \cdot b} = \sqrt{a} \cdot \sqrt{b}.$$

$$\sqrt{5} \cdot \sqrt{7} = \sqrt{35}$$
$$\sqrt{8} \cdot \sqrt{2} = \sqrt{16} = 4$$
$$\sqrt{48} = \sqrt{16 \cdot 3} = \sqrt{16} \cdot \sqrt{3} = 4\sqrt{3}$$

Quotient Rule for Radicals

If a and b are nonnegative real numbers and b is not 0, then

$$\frac{\sqrt{a}}{\sqrt{b}} = \sqrt{\frac{a}{b}} \quad \text{and} \quad \sqrt{\frac{a}{b}} = \frac{\sqrt{a}}{\sqrt{b}}.$$

$$\frac{\sqrt{8}}{\sqrt{2}} = \sqrt{\frac{8}{2}} = \sqrt{4} = 2 \qquad \sqrt{\frac{25}{64}} = \frac{\sqrt{25}}{\sqrt{64}} = \frac{5}{8}$$

If all indicated roots are real, then

$$\sqrt[n]{a} \cdot \sqrt[n]{b} = \sqrt[n]{ab} \quad \text{and} \quad \frac{\sqrt[n]{a}}{\sqrt[n]{b}} = \sqrt[n]{\frac{a}{b}} \ (b \neq 0).$$

$$\sqrt[3]{5} \cdot \sqrt[3]{3} = \sqrt[3]{15} \qquad \frac{\sqrt[4]{12}}{\sqrt[4]{4}} = \sqrt[4]{\frac{12}{4}} = \sqrt[4]{3}$$

8.3 ADDING AND SUBTRACTING RADICALS

Add and subtract like radicals by using the distributive property. *Only like radicals can be combined in this way.*

$$2\sqrt{5} + 4\sqrt{5} = (2 + 4)\sqrt{5} \qquad \sqrt{8} + \sqrt{32} = 2\sqrt{2} + 4\sqrt{2}$$
$$= 6\sqrt{5} \qquad\qquad\qquad\qquad = 6\sqrt{2}$$

8.4 RATIONALIZING THE DENOMINATOR

The denominator of a radical can be rationalized by multiplying both the numerator and denominator by a number that will eliminate the radical from the denominator.

$$\frac{2}{\sqrt{3}} = \frac{2 \cdot \sqrt{3}}{\sqrt{3} \cdot \sqrt{3}} = \frac{2\sqrt{3}}{3}$$

$$\sqrt[3]{\frac{5}{6}} = \frac{\sqrt[3]{5} \cdot \sqrt[3]{6^2}}{\sqrt[3]{6} \cdot \sqrt[3]{6^2}} = \frac{\sqrt[3]{180}}{6}$$

8.5 MORE SIMPLIFYING AND OPERATIONS WITH RADICALS

When appropriate, use the rules for adding and multiplying polynomials to simplify radical expressions.

$$\sqrt{6}(\sqrt{5} - \sqrt{7}) = \sqrt{30} - \sqrt{42}$$
$$(\sqrt{5} - \sqrt{3})(\sqrt{5} + \sqrt{3}) = 5 - 3 = 2$$

Any denominators with radicals should be rationalized.

$$\frac{3}{\sqrt{6}} = \frac{3\sqrt{6}}{6} = \frac{\sqrt{6}}{2}$$

If a radical expression contains two terms in the denominator and at least one of those terms is a square root radical, multiply both numerator and denominator by the conjugate of the denominator.

$$\frac{6}{\sqrt{7} - \sqrt{2}} = \frac{6}{\sqrt{7} - \sqrt{2}} \cdot \frac{\sqrt{7} + \sqrt{2}}{\sqrt{7} + \sqrt{2}}$$

$$= \frac{6(\sqrt{7} + \sqrt{2})}{7 - 2} \qquad \text{Multiply fractions.}$$

$$= \frac{6(\sqrt{7} + \sqrt{2})}{5} \qquad \text{Subtract.}$$

(continued)

Concepts	Examples

8.6 SOLVING EQUATIONS WITH RADICALS

Solving a Radical Equation

Step 1 Isolate a radical.

Step 2 Square both sides. (By the squaring property of equality, all solutions of the original equation are *among* the solutions of the squared equation.)

Step 3 Combine like terms.

Step 4 If there is still a term with a radical, repeat Steps 1–3.

Step 5 Solve the equation for proposed solutions.

Step 6 Check all proposed solutions from Step 5 in the original equation.

Solve $\sqrt{2x - 3} + x = 3$.

$$\sqrt{2x - 3} = 3 - x \qquad \text{Isolate the radical.}$$

$$\left(\sqrt{2x - 3}\right)^2 = (3 - x)^2 \qquad \text{Square both sides.}$$

$$2x - 3 = 9 - 6x + x^2$$

Remember the middle term.

$$0 = x^2 - 8x + 12 \qquad \text{Standard form}$$

$$0 = (x - 2)(x - 6) \qquad \text{Factor.}$$

$$x - 2 = 0 \quad \text{or} \quad x - 6 = 0 \qquad \text{Zero-factor property}$$

$$x = 2 \quad \text{or} \quad x = 6 \qquad \text{Solve.}$$

Verify that 2 is the only solution (6 is extraneous). The solution set is $\{2\}$.

8.7 USING RATIONAL NUMBERS AS EXPONENTS

Assume that $a \geq 0$, m and n are integers, and $n > 0$. Then

$$a^{1/n} = \sqrt[n]{a}$$

$$a^{m/n} = \sqrt[n]{a^m} = \left(\sqrt[n]{a}\right)^m$$

$$a^{-m/n} = \frac{1}{a^{m/n}} \quad (a \neq 0)$$

$$8^{1/3} = \sqrt[3]{8} = 2$$

$$(81)^{3/4} = \sqrt[4]{81^3} = \left(\sqrt[4]{81}\right)^3 = 3^3 = 27$$

$$36^{-3/2} = \frac{1}{36^{3/2}} = \frac{1}{(36^{1/2})^3} = \frac{1}{6^3} = \frac{1}{216}$$

Answers to Test Your Word Power

1. C; *Examples:* 6 is a square root of 36, since $6^2 = 6 \cdot 6 = 36$; -6 is also a square root of 36.

2. D; *Examples:* $\sqrt{144}$, $\sqrt{4xy^2}$, $\sqrt{4 + t^2}$

3. A; *Examples:* $\sqrt{36} = 6$, $\sqrt[4]{81} = 3$, $\sqrt[6]{64} = 2$

4. C; *Examples:* $\sqrt{7}$ and $3\sqrt{7}$ are like radicals; so are $2\sqrt[3]{6k}$ and $5\sqrt[3]{6k}$.

5. B; *Example:* To rationalize the denominator of $\dfrac{5}{\sqrt{3} + 1}$, multiply numerator and denominator by $\sqrt{3} - 1$ to get $\dfrac{5(\sqrt{3} - 1)}{2}$.

6. A; *Example:* The conjugate of $\sqrt{3} + 1$ is $\sqrt{3} - 1$.

Chapter **8** REVIEW EXERCISES

[8.1] *Find all square roots of each number.*

1. 49   **2.** 81   **3.** 196   **4.** 121   **5.** 225   **6.** 729

Find each indicated root. If the root is not a real number, say so.

7. $\sqrt{16}$ **8.** $-\sqrt{36}$ **9.** $\sqrt[3]{1000}$ **10.** $\sqrt[4]{81}$

11. $\sqrt{-8100}$ **12.** $-\sqrt{4225}$ **13.** $\sqrt{\dfrac{49}{36}}$ **14.** $\sqrt{\dfrac{100}{81}}$

15. *Concept Check* If $\sqrt{a}$ is not a real number, then what kind of number must a be?

16. Find the value of x in the figure.

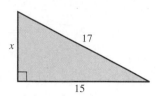

17. An HP f1905 computer monitor has viewing screen dimensions as shown in the figure. Find the diagonal measure of the viewing screen to the nearest tenth. (*Source:* Author's computer.)

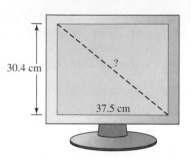

Determine whether each number is rational, irrational, *or* not a real number. *If the number is rational, give its exact value. If the number is irrational, give a decimal approximation rounded to the nearest thousandth.*

18. $\sqrt{111}$

19. $-\sqrt{25}$

20. $\sqrt{-4}$

[8.2] *Simplify each expression.*

21. $\sqrt{5} \cdot \sqrt{15}$

22. $-\sqrt{27}$

23. $\sqrt{160}$

24. $\sqrt[3]{-1331}$

25. $\sqrt[3]{1728}$

26. $\sqrt{12} \cdot \sqrt{27}$

27. $\sqrt{32} \cdot \sqrt{48}$

28. $\sqrt{50} \cdot \sqrt{125}$

Use the product rule, the quotient rule, or both to simplify each expression.

29. $-\sqrt{\dfrac{121}{400}}$

30. $\sqrt{\dfrac{3}{49}}$

31. $\sqrt{\dfrac{7}{169}}$

32. $\sqrt{\dfrac{1}{6}} \cdot \sqrt{\dfrac{5}{6}}$

33. $\sqrt{\dfrac{2}{5}} \cdot \sqrt{\dfrac{2}{45}}$

34. $\dfrac{3\sqrt{10}}{\sqrt{5}}$

35. $\dfrac{24\sqrt{12}}{6\sqrt{3}}$

36. $\dfrac{8\sqrt{150}}{4\sqrt{75}}$

Simplify each expression. Assume that all variables represent nonnegative real numbers.

37. $\sqrt{p} \cdot \sqrt{p}$

38. $\sqrt{k} \cdot \sqrt{m}$

39. $\sqrt{r^{18}}$

40. $\sqrt{x^{10}y^{16}}$

41. $\sqrt{a^{15}b^{21}}$

42. $\sqrt{121x^6y^{10}}$

43. Use a calculator to find approximations for $\sqrt{0.5}$ and $\frac{\sqrt{2}}{2}$. Based on your results, do you think that these two expressions represent the same number? If so, verify it algebraically.

[8.3] *Simplify, and combine terms where possible.*

44. $3\sqrt{2} + 6\sqrt{2}$

45. $3\sqrt{75} + 2\sqrt{27}$

46. $4\sqrt{12} + \sqrt{48}$

47. $4\sqrt{24} - 3\sqrt{54} + \sqrt{6}$

48. $2\sqrt{7} - 4\sqrt{28} + 3\sqrt{63}$

49. $\dfrac{2}{5}\sqrt{75} + \dfrac{3}{4}\sqrt{160}$

50. $\dfrac{1}{3}\sqrt{18} + \dfrac{1}{4}\sqrt{32}$

51. $\sqrt{15} \cdot \sqrt{2} + 5\sqrt{30}$

Simplify each expression. Assume that all variables represent nonnegative real numbers.

52. $\sqrt{4x} + \sqrt{36x} - \sqrt{9x}$

53. $\sqrt{16p} + 3\sqrt{p} - \sqrt{49p}$

54. $\sqrt{20m^2} - m\sqrt{45}$

55. $3k\sqrt{8k^2n} + 5k^2\sqrt{2n}$

[8.4] *Perform each indicated operation and write answers in simplest form. Assume that all variables represent positive real numbers.*

56. $\dfrac{8\sqrt{2}}{\sqrt{5}}$

57. $\dfrac{5}{\sqrt{5}}$

58. $\dfrac{12}{\sqrt{24}}$

59. $\dfrac{\sqrt{2}}{\sqrt{15}}$

60. $\sqrt{\dfrac{2}{5}}$

61. $\sqrt{\dfrac{5}{14}} \cdot \sqrt{28}$

62. $\sqrt{\dfrac{2}{7}} \cdot \sqrt{\dfrac{1}{3}}$

63. $\sqrt{\dfrac{r^2}{16x}}$

64. $\sqrt[3]{\dfrac{1}{3}}$

65. $\sqrt[3]{\dfrac{2}{7}}$

66. The radius r of a cone in terms of its volume V is given by the formula

$$r = \sqrt{\dfrac{3V}{\pi h}}.$$

Write this radical in simplified form.

[8.5] *Simplify each expression.*

67. $-\sqrt{3}\left(\sqrt{5} + \sqrt{27}\right)$

68. $3\sqrt{2}\left(\sqrt{3} + 2\sqrt{2}\right)$

69. $\left(2\sqrt{3} - 4\right)\left(5\sqrt{3} + 2\right)$

70. $\left(5\sqrt{7} + 2\right)^2$

71. $\left(\sqrt{5} - \sqrt{7}\right)\left(\sqrt{5} + \sqrt{7}\right)$

72. $\left(2\sqrt{3} + 5\right)\left(2\sqrt{3} - 5\right)$

Rationalize each denominator.

73. $\dfrac{1}{2 + \sqrt{5}}$

74. $\dfrac{2}{\sqrt{2} - 3}$

75. $\dfrac{\sqrt{8}}{\sqrt{2} + 6}$

76. $\dfrac{\sqrt{3}}{1 + \sqrt{3}}$

77. $\dfrac{\sqrt{5} - 1}{\sqrt{2} + 3}$

78. $\dfrac{2 + \sqrt{6}}{\sqrt{3} - 1}$

Write each quotient in lowest terms.

79. $\dfrac{15 + 10\sqrt{6}}{15}$

80. $\dfrac{3 + 9\sqrt{7}}{12}$

81. $\dfrac{6 + \sqrt{192}}{2}$

[8.6] *Solve each equation.*

82. $\sqrt{m} - 5 = 0$

83. $\sqrt{p} + 4 = 0$

84. $\sqrt{k + 1} = 7$

85. $\sqrt{5m + 4} = 3\sqrt{m}$

86. $\sqrt{2p + 3} = \sqrt{5p - 3}$

87. $\sqrt{4x + 1} = x - 1$

88. $\sqrt{-2k - 4} = k + 2$

89. $\sqrt{2 - x} + 3 = x + 7$

90. $\sqrt[3]{x + 4} = \sqrt[3]{16 - 2x}$

91. $\sqrt{x + 4} - \sqrt{x - 4} = 2$

92. $\sqrt{5x + 6} + \sqrt{3x + 4} = 2$

[8.7] *Simplify each expression. Assume that all variables represent positive real numbers.*

93. $81^{1/2}$

94. $-125^{1/3}$

95. $7^{2/3} \cdot 7^{7/3}$

96. $\dfrac{13^{4/5}}{13^{-3/5}}$

97. $\dfrac{x^{1/4} \cdot x^{5/4}}{x^{3/4}}$

98. $\sqrt[8]{49^4}$

MIXED REVIEW EXERCISES

Simplify each expression. Assume that all variables represent positive real numbers.

99. $64^{2/3}$

100. $2\sqrt{27} + 3\sqrt{75} - \sqrt{300}$

101. $\dfrac{1}{5 + \sqrt{2}}$

102. $\sqrt{\dfrac{1}{3}} \cdot \sqrt{\dfrac{24}{5}}$

103. $\sqrt{50y^2}$

104. $\sqrt[3]{-125}$

105. $-\sqrt{5}\left(\sqrt{2} + \sqrt{75}\right)$

106. $\sqrt{\dfrac{16r^3}{3s}}$

107. $\dfrac{12 + 6\sqrt{13}}{12}$

108. $-\sqrt{162} + \sqrt{8}$

109. $\left(\sqrt{5} - \sqrt{2}\right)^2$

110. $\left(6\sqrt{7} + 2\right)\left(4\sqrt{7} - 1\right)$

111. $-\sqrt{121}$

112. $\dfrac{x^{8/3}}{x^{2/3}}$

Solve.

113. $\sqrt{x + 2} = x - 4$

114. $\sqrt{k} + 3 = 0$

115. $\sqrt{1 + 3t} - t = -3$

116. The *fall speed,* in miles per hour, of a vehicle running off the road into a ditch is given by the formula

$$S = \dfrac{2.74D}{\sqrt{h}},$$

where D is the horizontal distance traveled from the level surface to the bottom of the ditch and h is the height (or depth) of the ditch. What is the fall speed (to the nearest tenth) of a vehicle that traveled 32 ft horizontally into a ditch 5 ft deep?

Chapter 8 TEST

On this test, assume that all variables represent positive real numbers.

1. Find all square roots of 196.

2. Consider $\sqrt{142}$.

    (a) Determine whether it is rational or irrational.

    (b) Find a decimal approximation to the nearest thousandth.

3. Explain why $\sqrt{-5}$ is not a real number.

Simplify where possible.

4. $-\sqrt{27}$

5. $\sqrt{\dfrac{128}{25}}$

6. $\sqrt[3]{32}$

7. $\dfrac{20\sqrt{18}}{5\sqrt{3}}$

8. $3\sqrt{28} + \sqrt{63}$

9. $3\sqrt{27x} - 4\sqrt{48x} + 2\sqrt{3x}$

10. $\sqrt[3]{32x^2y^3}$

11. $\left(6 - \sqrt{5}\right)\left(6 + \sqrt{5}\right)$

12. $\left(2 - \sqrt{7}\right)\left(3\sqrt{2} + 1\right)$

13. $\left(\sqrt{5} + \sqrt{6}\right)^2$

14. $\sqrt[3]{16x^4} - 2\sqrt[3]{128x^4}$

15. $\sqrt[3]{\dfrac{2}{3}}$

Solve each problem.

16. Find the measure of the unknown leg of this right triangle.

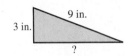

9 in.

3 in.

?

(a) Give its length in simplified radical form.
(b) Give the length to the nearest thousandth.

17. In electronics, the impedance Z of an alternating series circuit is given by the formula

$$Z = \sqrt{R^2 + X^2},$$

where R is the resistance and X is the reactance, both in ohms. Find the value of the impedance Z if $R = 40$ ohms and $X = 30$ ohms. (*Source:* Cooke, Nelson M., and Orleans, Joseph B., *Mathematics Essential to Electricity and Radio*, McGraw-Hill, 1943.)

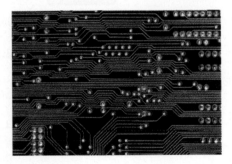

18. The radius r of a sphere in terms of its surface area S is given by the formula

$$r = \sqrt{\dfrac{S}{4\pi}}.$$

Write this radical in simplified form.

Rationalize each denominator.

19. $\dfrac{5\sqrt{2}}{\sqrt{7}}$

20. $\sqrt{\dfrac{2}{3x}}$

21. $\dfrac{-2}{\sqrt[3]{4}}$

22. $\dfrac{-3}{4 - \sqrt{3}}$

Write each expression in lowest terms.

23. $\dfrac{2 + \sqrt{8}}{4}$

24. $\dfrac{\sqrt{12} + 3\sqrt{128}}{6}$

Solve each equation.

25. $\sqrt{2x + 6} + 4 = 2$

26. $\sqrt{x + 1} = 5 - x$

27. $3\sqrt{x} - 1 = 2x$

28. $\sqrt{2x + 9} + \sqrt{x + 5} = 2$

Simplify each expression.

29. $8^{4/3}$

30. $-125^{2/3}$

31. $5^{3/4} \cdot 5^{1/4}$

32. $\dfrac{\left(3^{1/4}\right)^3}{3^{7/4}}$

✎ **33.** *Concept Check*   Consider the following "solution." *WHAT WENT WRONG?* Give the correct solution set.

$$\sqrt{2x + 1} + 5 = 0$$
$$\sqrt{2x + 1} = -5 \qquad \text{Subtract 5.}$$
$$2x + 1 = 25 \qquad \text{Square both sides.}$$
$$2x = 24 \qquad \text{Subtract 1.}$$
$$x = 12 \qquad \text{Divide by 2.}$$

The solution set is $\{12\}$.

Chapters 1–8 CUMULATIVE REVIEW EXERCISES

Simplify each expression.

1. $3(6 + 7) + 6 \cdot 4 - 3^2$

2. $\dfrac{3(6 + 7) + 3}{2(4) - 1}$

3. $|-6| - |-3|$

Solve each equation or inequality.

4. $5(k - 4) - k = k - 11$

5. $-\dfrac{3}{4}y \le 12$

6. $5z + 3 - 4 > 2z + 9 + z$

7. U.S. production of corn for grain reached record levels in 2004, when 1.7 billion more bushels of corn were produced than in 2003. Total production for the two years was 21.9 billion bushels. How much corn was produced for grain in each of these years? (*Source:* U.S. Department of Agriculture.)

Graph.

8. $-4x + 5y = -20$

9. $x = 2$

10. $2x - 5y > 10$

11. The graph on the next page shows a linear equation that models total federal spending on each of the two major-party political conventions, in millions of dollars.

✎ **(a)** Use the ordered pairs shown on the graph to find the slope of the line to the nearest hundredth. Interpret the slope.

(b) Use the slope from part (a) and the ordered pair (2004, 15.0) to find the equation of the line that models the data. Write the equation in slope–intercept form.

(c) Use the equation from part (b) to project federal spending for each convention for 2008. Round your answer to the nearest tenth.

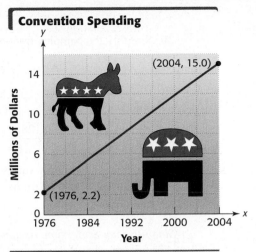

Convention Spending

Source: Federal Election Commission.

Solve each system of equations.

12. $4x - y = 19$
$3x + 2y = -5$

13. $2x - y = 6$
$3y = 6x - 18$

14. Des Moines and Chicago are 345 mi apart. Two cars start from these cities, traveling toward each other. They meet after 3 hr. The car from Chicago has an average speed 7 mph faster than the other car. Find the average speed of each car. (*Source: State Farm Road Atlas.*)

Simplify and write each expression without negative exponents. Assume that variables represent positive real numbers.

15. $(3x^6)(2x^2y)^2$

16. $\left(\dfrac{3^2 y^{-2}}{2^{-1} y^3}\right)^{-3}$

17. Subtract $7x^3 - 8x^2 + 4$ from $10x^3 + 3x^2 - 9$.

18. Divide: $\dfrac{8t^3 - 4t^2 - 14t + 15}{2t + 3}$.

Factor each polynomial completely.

19. $m^2 + 12m + 32$

20. $25t^4 - 36$

21. $12a^2 + 4ab - 5b^2$

22. $81z^2 + 72z + 16$

Solve each quadratic equation.

23. $x^2 - 7x = -12$

24. $(x + 4)(x - 1) = -6$

25. For what real number(s) is the expression $\dfrac{3}{x^2 + 5x - 14}$ undefined?

Perform each indicated operation. Express answers in lowest terms.

26. $\dfrac{x^2 - 3x - 4}{x^2 + 3x} \cdot \dfrac{x^2 + 2x - 3}{x^2 - 5x + 4}$

27. $\dfrac{t^2 + 4t - 5}{t + 5} \div \dfrac{t - 1}{t^2 + 8t + 15}$

28. $\dfrac{y}{y^2 - 1} + \dfrac{y}{y + 1}$

29. $\dfrac{2}{x + 3} - \dfrac{4}{x - 1}$

30. Simplify the complex fraction: $\dfrac{\dfrac{2}{3} + \dfrac{1}{2}}{\dfrac{1}{9} - \dfrac{1}{6}}$.

Solve each equation.

31. $\dfrac{x}{x + 8} - \dfrac{3}{x - 8} = \dfrac{128}{x^2 - 64}$

32. $A = \dfrac{B + CD}{BC + D}$ for B

33. The speed of a pulley varies inversely as its diameter. One kind of pulley, with diameter 9 in., turns at 450 revolutions per min. Find the speed of a similar pulley with diameter 10 in.

Simplify each expression. Assume that all variables represent nonnegative real numbers.

34. $\sqrt{27} - 2\sqrt{12} + 6\sqrt{75}$

35. $\dfrac{2}{\sqrt{3} + \sqrt{5}}$

36. $\sqrt{200x^2y^5}$

37. $\left(3\sqrt{2} + 1\right)\left(4\sqrt{2} - 3\right)$

38. Solve $\sqrt{x} + 2 = x - 10$.

Evaluate.

39. $16^{5/4}$

40. $\dfrac{8^{-7/3}}{8^{-1/3}}$

Quadratic Equations

In this chapter, we develop methods of solving quadratic equations. The graphs of such equations in two variables, called *parabolas,* have many applications. For example, the Parkes radio telescope, pictured here, has a *parabolic* dish shape with diameter 210 ft and depth 32 ft. (*Source:* Mar, J., and Liebowitz, H., *Structure Technology for Large Radio and Radar Telescope Systems,* The MIT Press, 1969.)

In Section 9.5, we use a graph to model the parabolic shape of the Parkes radio telescope and find a quadratic equation that represents the graph.

9.1 Solving Quadratic Equations by the Square Root Property

In **Section 6.5,** we solved quadratic equations by factoring. However, since not all quadratic equations can easily be solved by factoring, we must develop other methods. In this chapter, we do just that.

Recall that a **quadratic equation** is an equation that can be written in the form

$$ax^2 + bx + c = 0 \qquad \text{Standard form}$$

for real numbers a, b, and c, with $a \neq 0$. As seen in **Section 6.5,** we can solve the quadratic equation $x^2 + 4x + 3 = 0$ by factoring, using the zero-factor property.

$$x^2 + 4x + 3 = 0$$

$$(x + 3)(x + 1) = 0 \qquad \text{Factor.}$$

$$x + 3 = 0 \qquad \text{or} \qquad x + 1 = 0 \qquad \text{Zero-factor property}$$

$$x = -3 \qquad \text{or} \qquad x = -1 \qquad \text{Solve each equation.}$$

The solution set is $\{-3, -1\}$.

OBJECTIVE 1 Solve equations of the form $x^2 = k$, where $k > 0$. We can solve equations such as $x^2 = 9$ by factoring as follows.

$$x^2 = 9$$

$$x^2 - 9 = 0 \qquad \text{Subtract 9.}$$

$$(x + 3)(x - 3) = 0 \qquad \text{Factor.}$$

$$x + 3 = 0 \qquad \text{or} \qquad x - 3 = 0 \qquad \text{Zero-factor property}$$

$$x = -3 \qquad \text{or} \qquad x = 3 \qquad \text{Solve each equation.}$$

The solution set is $\{-3, 3\}$.

We might also have solved $x^2 = 9$ by noticing that x must be a number whose square is 9. Thus, $x = \sqrt{9} = 3$ or $x = -\sqrt{9} = -3$. This is generalized as the **square root property.**

Square Root Property

If k is a positive number and if $x^2 = k$, then

$$x = \sqrt{k} \qquad \text{or} \qquad x = -\sqrt{k}.$$

The solution set is $\{-\sqrt{k}, \sqrt{k}\}$, which can be written $\{\pm\sqrt{k}\}$. ($\pm$ is read "positive or negative" or "plus or minus.")

▶ **NOTE** When we solve an equation, we must find *all* values of the variable that satisfy the equation. Therefore, we want both the positive and negative square roots of k.

EXAMPLE 1 Solving Quadratic Equations of the Form $x^2 = k$

Solve each equation. Write radicals in simplified form.

(a) $x^2 = 16$

By the square root property, if $x^2 = 16$, then

$$x = \sqrt{16} = 4 \quad \text{or} \quad x = -\sqrt{16} = -4.$$

Check each solution by substituting it for x in the original equation. The solution set is

$$\{-4, 4\}, \quad \text{or} \quad \{\pm 4\}.$$

This notation indicates *two* solutions, one positive and one negative.

(b) $z^2 = 5$

The solutions are $z = \sqrt{5}$ or $z = -\sqrt{5}$, so the solution set is $\{\pm\sqrt{5}\}$.

(c)
$$5m^2 - 40 = 0$$
$$5m^2 = 40 \qquad \text{Add 40.}$$
$$m^2 = 8 \qquad \text{Divide by 5.}$$

Don't stop here. Simplify the radicals.

$$m = \sqrt{8} \quad \text{or} \quad m = -\sqrt{8} \qquad \text{Square root property}$$
$$m = 2\sqrt{2} \quad \text{or} \quad m = -2\sqrt{2} \qquad \sqrt{8} = \sqrt{4} \cdot \sqrt{2} = 2\sqrt{2}$$

The solution set is $\{\pm 2\sqrt{2}\}$.

(d) $p^2 = -4$

Since -4 is a negative number and since the square of a real number cannot be negative, there is no real number solution of this equation. (To use the square root property, k must be positive.) The solution set is $\emptyset$.

(e)
$$3x^2 + 5 = 11$$
$$3x^2 = 6 \qquad \text{Subtract 5.}$$
$$x^2 = 2 \qquad \text{Divide by 3.}$$
$$x = \sqrt{2} \quad \text{or} \quad x = -\sqrt{2} \qquad \text{Square root property}$$

The solution set is $\{\pm\sqrt{2}\}$.

✔ **Now Try Exercises 5, 7, 11, and 21.**

OBJECTIVE 2 Solve equations of the form $(ax + b)^2 = k$, where $k > 0$. In each equation in Example 1, the exponent 2 appeared with a single variable as its base. We can extend the square root property to solve equations in which the base is a binomial.

EXAMPLE 2 Solving Quadratic Equations of the Form $(x + b)^2 = k$

Solve each equation.

(a) $(x - 3)^2 = 16$

Apply the square root property, using $x - 3$ as the base.

$$(x - 3)^2 = 16$$

$x - 3 = \sqrt{16}$ or $x - 3 = -\sqrt{16}$ Square root property

$x - 3 = 4$ or $x - 3 = -4$ $\sqrt{16} = 4$

$x = 7$ or $x = -1$ Add 3.

Check each answer in the original equation.

$(x - 3)^2 = 16$	$(x - 3)^2 = 16$
$(7 - 3)^2 = 16$? Let $x = 7$.	$(-1 - 3)^2 = 16$? Let $x = -1$.
$4^2 = 16$?	$(-4)^2 = 16$?
$16 = 16$ True	$16 = 16$ True

The solution set is $\{-1, 7\}$.

(b) $(x - 1)^2 = 6$

$x - 1 = \sqrt{6}$ or $x - 1 = -\sqrt{6}$ Square root property

$x = 1 + \sqrt{6}$ or $x = 1 - \sqrt{6}$ Add 1.

Check: $\left(1 + \sqrt{6} - 1\right)^2 = \left(\sqrt{6}\right)^2 = 6;$

$\left(1 - \sqrt{6} - 1\right)^2 = \left(-\sqrt{6}\right)^2 = 6.$

The solution set is $\left\{1 + \sqrt{6}, 1 - \sqrt{6}\right\}$, or $\left\{1 \pm \sqrt{6}\right\}$.

✔ **Now Try Exercises 29 and 33.**

EXAMPLE 3   Solving a Quadratic Equation of the Form $(ax + b)^2 = k$

Solve $(3r - 2)^2 = 27$.

$3r - 2 = \sqrt{27}$ or $3r - 2 = -\sqrt{27}$ Square root property

$3r - 2 = 3\sqrt{3}$ or $3r - 2 = -3\sqrt{3}$ $\sqrt{27} = \sqrt{9} \cdot \sqrt{3} = 3\sqrt{3}$

$3r = 2 + 3\sqrt{3}$ or $3r = 2 - 3\sqrt{3}$ Add 2.

$r = \dfrac{2 + 3\sqrt{3}}{3}$ or $r = \dfrac{2 - 3\sqrt{3}}{3}$ Divide by 3.

The solution set is $\left\{\dfrac{2 \pm 3\sqrt{3}}{3}\right\}$.

✔ **Now Try Exercise 41.**

▶ **CAUTION**   The solutions in Example 3 are fractions that cannot be simplified, since 3 is *not* a common factor in the numerator.

EXAMPLE 4   Recognizing a Quadratic Equation with No Real Solutions

Solve $(x + 3)^2 = -9$.

Because the square root of -9 is not a real number, the solution set is $\emptyset$.

✔ **Now Try Exercise 31.**

OBJECTIVE 3 Use formulas involving squared variables.

EXAMPLE 5 Finding the Length of a Bass

The formula

$$w = \frac{L^2 g}{1200}$$

is used to approximate the weight of a bass, in pounds, given its length L and its girth g, both measured in inches. Approximate the length of a bass weighing 2.20 lb and having girth 10 in. (*Source: Sacramento Bee,* November 29, 2000.)

$w = \dfrac{L^2 g}{1200}$	Given formula
$2.20 = \dfrac{L^2 \cdot 10}{1200}$	$w = 2.20,\ g = 10$
$2640 = 10L^2$	Multiply by 1200.
$L^2 = 264$	Divide by 10.
$L = \sqrt{264}$ or $L = -\sqrt{264}$	Square root property

A calculator shows that $\sqrt{264} \approx 16.25$, so the length of the bass is a little more than 16 in. (We discard the negative solution $-\sqrt{264} \approx -16.25$, since L represents length.)

✔ **Now Try Exercise 55.**

9.1 EXERCISES

⊕ *Complete solution available on Video Lectures on CD/DVD*

▢ *Now Try Exercise*

Concept Check *Match each equation in Column I with the correct description of its solution in Column II.*

I	II
1. $x^2 = 10$	**A.** No real number solutions
2. $x^2 = -4$	**B.** Two integer solutions
3. $x^2 = \dfrac{9}{16}$	**C.** Two irrational solutions
4. $x^2 = 9$	**D.** Two rational solutions that are not integers

Solve each equation by using the square root property. Simplify all radicals. See Example 1.

⊕ **5.** $x^2 = 81$ **6.** $z^2 = 121$ **7.** $k^2 = 14$ **8.** $m^2 = 22$

9. $t^2 = 48$ **10.** $x^2 = 54$ **11.** $z^2 = -100$ **12.** $m^2 = -64$

13. $x^2 = \dfrac{25}{4}$ **14.** $m^2 = \dfrac{36}{121}$ **15.** $z^2 = 2.25$ **16.** $w^2 = 56.25$

17. $r^2 - 3 = 0$ **18.** $x^2 - 13 = 0$ **19.** $7k^2 = 4$ **20.** $3p^2 = 10$

21. $3n^2 - 72 = 0$ **22.** $5z^2 - 200 = 0$ **23.** $5a^2 + 4 = 8$ **24.** $4p^2 - 3 = 7$

25. $2t^2 + 7 = 61$ **26.** $3x^2 - 8 = 64$

27. *Concept Check* When a student was asked to solve $x^2 = 81$, she wrote $\{9\}$ as her answer. Her teacher did not give her full credit. The student argued that because $9^2 = 81$, her answer had to be correct. ***WHAT WENT WRONG?*** Give the correct solution set.

✏ **28.** Explain the square root property for solving equations, and illustrate with an example.

Solve each equation by using the square root property. Simplify all radicals. See Examples 2–4.

🌐 **29.** $(x - 3)^2 = 25$ **30.** $(k - 7)^2 = 16$ 🌐 **31.** $(z + 5)^2 = -13$

32. $(m + 2)^2 = -17$ **33.** $(x - 8)^2 = 27$ **34.** $(p - 5)^2 = 40$

35. $(3k + 2)^2 = 49$ **36.** $(5t + 3)^2 = 36$ **37.** $(4x - 3)^2 = 9$

38. $(7z - 5)^2 = 25$ **39.** $(5 - 2x)^2 = 30$ **40.** $(3 - 2a)^2 = 70$

🌐 **41.** $(3k + 1)^2 = 18$ **42.** $(5z + 6)^2 = 75$ **43.** $\left(\dfrac{1}{2}x + 5\right)^2 = 12$

44. $\left(\dfrac{1}{3}m + 4\right)^2 = 27$ **45.** $(4k - 1)^2 - 48 = 0$ **46.** $(2s - 5)^2 - 180 = 0$

✏ **47.** Johnny solved the equation in Exercise 39 and wrote his answer as $\left\{\dfrac{5 + \sqrt{30}}{2}, \dfrac{5 - \sqrt{30}}{2}\right\}$. Linda solved the same equation and wrote her answer as $\left\{\dfrac{-5 + \sqrt{30}}{-2}, \dfrac{-5 - \sqrt{30}}{-2}\right\}$. The teacher gave them both full credit. Explain why both students were correct, although their answers look different.

✏ **48.** In the solutions $\dfrac{2 \pm 3\sqrt{3}}{3}$ found in Example 3 of this section, why is it not valid to simplify the answers by dividing out the 3's in the numerator and denominator?

Use a calculator with a square root key to solve each equation. Round your answers to the nearest hundredth.

49. $(k + 2.14)^2 = 5.46$ **50.** $(r - 3.91)^2 = 9.28$

51. $(2.11p + 3.42)^2 = 9.58$ **52.** $(1.71m - 6.20)^2 = 5.41$

Solve each problem. See Example 5.

53. One expert at marksmanship can hold a silver dollar at forehead level, drop it, draw his gun, and shoot the coin as it passes waist level. The distance traveled by a falling object is given by

$$d = 16t^2,$$

where d is the distance (in feet) the object falls in t seconds. If the coin falls about 4 ft, use the formula to estimate the time that elapses between the dropping of the coin and the shot.

54. The illumination produced by a light source depends on the distance from the source. For a particular light source, this relationship can be expressed as

$$I = \frac{4050}{d^2},$$

where I is the amount of illumination in footcandles and d is the distance from the light source (in feet). How far from the source is the illumination equal to 50 footcandles?

55. The area A of a circle with radius r is given by the formula

$$A = \pi r^2.$$

If a circle has area 81π in.^{2}, what is its radius?

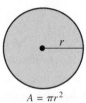

$A = \pi r^2$

56. The surface area S of a sphere with radius r is given by the formula

$$S = 4\pi r^2.$$

If a sphere has surface area 36π ft^2, what is its radius?

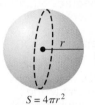

$S = 4\pi r^2$

The amount A that P dollars invested at an annual rate of interest r will grow to in 2 yr is

$$A = P(1 + r)^2.$$

57. At what interest rate will \$100 grow to \$110.25 in 2 yr?

58. At what interest rate will \$500 grow to \$572.45 in 2 yr?

PREVIEW EXERCISES

Simplify all radicals, and combine like terms. Express fractions in lowest terms. See **Sections 8.3 and 8.4.**

59. $\dfrac{4}{5} + \sqrt{\dfrac{48}{25}}$

60. $12 + \sqrt{\dfrac{2}{3}}$

61. $\dfrac{6 + \sqrt{24}}{8}$

62. $\dfrac{12 - \sqrt{27}}{9}$

Factor each perfect square trinomial. See **Section 6.4.**

63. $x^2 - 10x + 25$

64. $z^2 + 4z + 4$

65. $x^2 - 7x + \dfrac{49}{4}$

66. $z^2 + z + \dfrac{1}{4}$

9.2 Solving Quadratic Equations by Completing the Square

OBJECTIVE 1 Solve quadratic equations by completing the square when the coefficient of the squared term is 1. The methods we have studied so far are not enough to solve the equation

$$x^2 + 6x + 7 = 0.$$

If we could write the equation in the form $(x + 3)^2$ equals a constant, we could solve it with the square root property discussed in **Section 9.1.** To do that, we need to have a perfect square trinomial on one side of the equation.

Recall from **Section 6.4** that a perfect square trinomial has the form

$$x^2 + 2kx + k^2 \qquad \text{or} \qquad x^2 - 2kx + k^2,$$

where k represents a number.

EXAMPLE 1 Creating Perfect Square Trinomials

Complete each trinomial so that it is a perfect square. Then factor the trinomial.

(a) $x^2 + 8x + $ _____

The perfect square trinomial will have the form $x^2 + 2kx + k^2$. Thus, the middle term, $8x$, must equal $2kx$.

$$8x = 2kx \leftarrow\text{Solve this equation for } k.$$
$$4 = k \qquad \text{Divide each side by } 2x.$$

Therefore, $k = 4$ and $k^2 = 4^2 = 16$. The required perfect square trinomial is

$$x^2 + 8x + 16, \quad \text{which factors as} \quad (x + 4)^2.$$

(b) $x^2 - 18x + $ _____

Here the perfect square trinomial will have the form $x^2 - 2kx + k^2$. The middle term, $-18x$, must equal $-2kx$.

$$-18x = -2kx \leftarrow \text{Solve this equation for } k.$$
$$9 = k \qquad \text{Divide each side by } -2x.$$

Thus, $k = 9$ and $k^2 = 9^2 = 81$. The required perfect square trinomial is

$$x^2 - 18x + 81, \quad \text{which factors as} \quad (x - 9)^2.$$

✔ **Now Try Exercises 1 and 3.**

EXAMPLE 2 Rewriting an Equation to Use the Square Root Property

Solve $x^2 + 6x + 7 = 0$.

$$x^2 + 6x = -7 \qquad \text{Subtract 7 from each side.}$$

To solve this equation with the square root property, the quantity on the left side, $x^2 + 6x$, must be written as a perfect square trinomial in the form $x^2 + 2kx + k^2$.

$$x^2 + 6x + \text{_____}$$

Here, $2kx = 6x$, so $k = 3$ and $k^2 = 9$. The required perfect square trinomial is

$$x^2 + 6x + 9, \quad \text{which factors as} \quad (x + 3)^2.$$

Therefore, if we add 9 to each side of $x^2 + 6x = -7$, the equation will have a perfect square trinomial on the left side, as needed.

$$x^2 + 6x + 9 = -7 + 9 \qquad \text{Add 9.}$$
$$(x + 3)^2 = 2 \qquad \text{Factor; add.}$$

Now use the square root property to complete the solution.

$$x + 3 = \sqrt{2} \qquad \text{or} \qquad x + 3 = -\sqrt{2}$$
$$x = -3 + \sqrt{2} \qquad \text{or} \qquad x = -3 - \sqrt{2}$$

Check by substituting $-3 + \sqrt{2}$ and $-3 - \sqrt{2}$ for x in the original equation. The solution set is $\left\{-3 \pm \sqrt{2}\right\}$.

✔ **Now Try Exercise 13.**

The process of changing the form of the equation in Example 2 from

$$x^2 + 6x + 7 = 0 \qquad \text{to} \qquad (x + 3)^2 = 2$$

is called **completing the square.** Completing the square changes only the form of the equation. To see this, multiply out the left side of $(x + 3)^2 = 2$ and combine terms. Then subtract 2 from both sides to see that the result is $x^2 + 6x + 7 = 0$.

Look again at the original equation in Example 2.

$$x^2 + 6x + 7 = 0$$

If we take half the coefficient of x, which is 6 here, and square it, we get 9.

$$\frac{1}{2} \cdot 6 = 3 \qquad \text{and} \qquad 3^2 = 9$$

Coefficient of x Quantity added to each side

To complete the square in Example 2, we added 9 to each side.

EXAMPLE 3 **Completing the Square to Solve a Quadratic Equation**

Solve $x^2 - 8x = 5$.

To complete the square on $x^2 - 8x$, take half the coefficient of x and square it.

$$\frac{1}{2}(-8) = -4 \qquad \text{and} \qquad (-4)^2 = 16$$

Coefficient of x

Add the result, 16, to each side of the equation.

$$x^2 - 8x = 5 \qquad\qquad \text{Given equation}$$
$$x^2 - 8x + 16 = 5 + 16 \qquad \text{Add 16.}$$
$$(x - 4)^2 = 21 \qquad\qquad \text{Factor on the left; add on the right.}$$
$$x - 4 = \pm\sqrt{21} \qquad\qquad \text{Square root property}$$
$$x = 4 \pm \sqrt{21} \qquad\qquad \text{Add 4.}$$

A check indicates that the solution set is $\left\{4 \pm \sqrt{21}\right\}$.

✔ **Now Try Exercise 15.**

OBJECTIVE ❷ Solve quadratic equations by completing the square when the coefficient of the squared term is not 1. Up to this point, the quadratic equations we have solved by completing the square have had 1 as the coefficient of the second-degree term. If a quadratic equation has the form

$$ax^2 + bx + c = 0, \quad \text{where } a \neq 1,$$

then to obtain 1 as the coefficient of x^2, we first divide each side of the equation by a.

The steps used to solve a quadratic equation $ax^2 + bx + c = 0$ by completing the square are summarized here.

Solving a Quadratic Equation by Completing the Square

Step 1 **Be sure the second-degree term has coefficient 1.** If the coefficient of the second-degree term is 1, go to Step 2. If it is not 1, but some other nonzero number a, divide each side of the equation by a.

Step 2 **Write in correct form.** Make sure that all variable terms are on one side of the equation and that all constant terms are on the other.

Step 3 **Complete the square.** Take half the coefficient of the first-degree term, and square it. Add the square to both sides of the equation. Factor the variable side and combine terms on the other side.

Step 4 **Solve** the equation by using the square root property.

EXAMPLE 4 **Solving a Quadratic Equation by Completing the Square**

Solve $4z^2 + 24z - 13 = 0$.

Step 1 **Be sure the second-degree term has coefficient 1.** Divide each side by 4 so that the coefficient of z^2 is 1.

$$4z^2 + 24z - 13 = 0$$

$$z^2 + 6z - \frac{13}{4} = 0 \qquad \text{Divide by 4.}$$

Step 2 **Write in correct form.** Add $\frac{13}{4}$ to each side to get the variable terms on the left and the constant on the right.

$$z^2 + 6z = \frac{13}{4} \qquad \text{Add } \tfrac{13}{4}.$$

Step 3 **Complete the square.** To complete the square, take half the coefficient of z and square the result: $\left[\frac{1}{2}(6)\right]^2 = 3^2 = 9$.

$$z^2 + 6z + 9 = \frac{13}{4} + 9 \qquad \text{Be sure to add 9 to } each \text{ side.}$$

$$(z + 3)^2 = \frac{49}{4} \qquad \text{Factor on the left; add on the right.}$$

Step 4 Solve. Use the square root property.

$$z + 3 = \frac{7}{2} \qquad \text{or} \qquad z + 3 = -\frac{7}{2} \qquad \text{Square root property;}$$
$$\pm\sqrt{\frac{49}{4}} = \pm\frac{7}{2}$$

$$z = -3 + \frac{7}{2} \qquad \text{or} \qquad z = -3 - \frac{7}{2} \qquad \text{Subtract 3.}$$

$$z = \frac{1}{2} \qquad \text{or} \qquad z = -\frac{13}{2}$$

Check by substituting each solution into the original equation. The solution set is $\left\{-\frac{13}{2}, \frac{1}{2}\right\}$. (We give the solutions in the order in which they appear on a number line.)

✔ **Now Try Exercise 19.**

EXAMPLE 5 Solving a Quadratic Equation by Completing the Square

Solve $4p^2 + 8p + 5 = 0$.

$$4p^2 + 8p + 5 = 0$$

$$p^2 + 2p + \frac{5}{4} = 0 \qquad \text{Divide by 4.}$$

The coefficient of the squared term must be 1.

$$p^2 + 2p = -\frac{5}{4} \qquad \text{Add } -\frac{5}{4} \text{ to each side.}$$

The coefficient of p is 2. Take half of 2, square the result, and add it to both sides: $\left[\frac{1}{2}(2)\right]^2 = 1^2 = 1$.

$$p^2 + 2p + 1 = -\frac{5}{4} + 1 \qquad \text{Add 1.}$$

$$(p + 1)^2 = -\frac{1}{4} \qquad \text{Factor on the left; add on the right.}$$

We cannot use the square root property to solve this equation, because the square root of $-\frac{1}{4}$ is not a real number. The solution set is $\emptyset$.

✔ **Now Try Exercise 21.**

OBJECTIVE 3 Simplify an equation before solving.

EXAMPLE 6 Simplifying an Equation before Completing the Square

Solve $(x + 3)(x - 1) = 2$.

$$(x + 3)(x - 1) = 2$$
$$x^2 + 2x - 3 = 2 \qquad \text{Multiply by using the FOIL method.}$$
$$x^2 + 2x = 5 \qquad \text{Add 3.}$$
$$x^2 + 2x + 1 = 5 + 1 \qquad \text{Add } \left[\frac{1}{2}(2)\right]^2 = 1.$$
$$(x + 1)^2 = 6 \qquad \text{Factor on the left; add on the right.}$$
$$x + 1 = \sqrt{6} \qquad \text{or} \qquad x + 1 = -\sqrt{6} \qquad \text{Square root property}$$
$$x = -1 + \sqrt{6} \qquad \text{or} \qquad x = -1 - \sqrt{6} \qquad \text{Subtract 1.}$$

The solution set is $\left\{-1 \pm \sqrt{6}\right\}$.

✔ **Now Try Exercise 29.**

▶ **NOTE** The solutions $-1 \pm \sqrt{6}$ given in Example 6 are *exact*. In applications, decimal solutions are more appropriate. Using the square root key of a calculator yields $\sqrt{6} \approx 2.449$. Approximating the two solutions gives

$$x \approx 1.449 \quad \text{and} \quad x \approx -3.449.$$

OBJECTIVE 4 **Solve applied problems that require quadratic equations.** There are many practical applications of quadratic equations. The next example illustrates an application from physics.

EXAMPLE 7 **Solving a Velocity Problem**

If a ball is projected into the air from ground level with an initial velocity of 64 ft per sec, its altitude (height) s in feet in t seconds is given by the formula

$$s = -16t^2 + 64t.$$

At what times will the ball be 48 ft above the ground?

Since s represents the height, we let $s = 48$ in the formula and solve this equation for the time t by completing the square.

$48 = -16t^2 + 64t$	Let $s = 48$.
$-3 = t^2 - 4t$	Divide by -16.
$t^2 - 4t = -3$	Reverse the sides.
$t^2 - 4t + 4 = -3 + 4$	Add $[\frac{1}{2}(-4)]^2 = 4$.
$(t - 2)^2 = 1$	Factor; add.
$t - 2 = 1 \quad$ or $\quad t - 2 = -1$	Square root property
$t = 3 \quad$ or $\quad t = 1$	Add 2.

The ball reaches a height of 48 ft twice, once on the way up and again on the way down. It takes 1 sec to reach 48 ft on the way up, and then after 3 sec, the ball reaches 48 ft again on the way down.

✔ **Now Try Exercise 37.**

9.2 EXERCISES

Complete each trinomial so that it is a perfect square. Then factor the trinomial. See Example 1.

⊕ **1.** $x^2 + 10x +$ _____ **2.** $x^2 + 16x +$ _____ **3.** $z^2 - 20z +$ _____

4. $a^2 - 32a +$ _____ **5.** $x^2 + 2x +$ _____ **6.** $m^2 - 2m +$ _____

7. $p^2 - 5p +$ _____ **8.** $x^2 + 3x +$ _____

9. *Concept Check* Which step is an appropriate way to begin solving the quadratic equation $2x^2 - 4x = 9$ by completing the square?

　　A. Add 4 to each side of the equation. 　　**B.** Factor the left side as $2x(x - 2)$.
　　C. Factor the left side as $x(2x - 4)$. 　　**D.** Divide each side by 2.

10. *Concept Check* In Example 3 of **Section 6.5,** we solved the quadratic equation $4p^2 - 26p + 40 = 0$ by factoring. If we were to solve by completing the square, would we get the same solution set, $\left\{\frac{5}{2}, 4\right\}$?

Solve each equation by completing the square. See Examples 2 and 3.

11. $x^2 - 4x = -3$ **12.** $p^2 - 2p = 8$ **13.** $x^2 + 2x - 5 = 0$

14. $r^2 + 4r + 1 = 0$ **15.** $x^2 - 8x = -4$ **16.** $m^2 - 4m = 14$

17. $z^2 + 6z + 9 = 0$ **18.** $k^2 - 8k + 16 = 0$

Solve each equation by completing the square. See Examples 4–6.

19. $4x^2 + 4x = 3$ **20.** $9x^2 + 3x = 2$ **21.** $2p^2 - 2p + 3 = 0$

22. $3q^2 - 3q + 4 = 0$ **23.** $3a^2 - 9a + 5 = 0$ **24.** $6b^2 - 8b - 3 = 0$

25. $3k^2 + 7k = 4$ **26.** $2k^2 + 5k = 1$ **27.** $(x + 3)(x - 1) = 5$

28. $(z - 8)(z + 2) = 24$ **29.** $(r - 3)(r - 5) = 2$ **30.** $(k - 1)(k - 7) = 1$

31. $-x^2 + 2x = -5$ **32.** $-x^2 + 4x = 1$

*Solve each equation by completing the square. Give **(a)** exact solutions and **(b)** solutions rounded to the nearest thousandth.*

33. $3r^2 - 2 = 6r + 3$ **34.** $4p + 3 = 2p^2 + 2p$

35. $(x + 1)(x + 3) = 2$ **36.** $(x - 3)(x + 1) = 1$

Solve each problem. See Example 7.

37. If an object is projected upward on the surface of Mars from ground level with an initial velocity of 104 ft per sec, its altitude (height) s in feet in t seconds is given by the formula $s = -13t^2 + 104t$. At what times will the object be 195 ft above the ground?

38. After how many seconds will the object in Exercise 37 return to the surface? (*Hint:* When it returns to the surface, $s = 0$.)

39. If an object is projected upward from ground level on Earth with an initial velocity of 96 ft per sec, its altitude (height) s in feet in t seconds is given by the formula $s = -16t^2 + 96t$. At what times will the object be at a height of 80 ft? (*Hint:* Let $s = 80$.)

40. At what times will the object described in Exercise 39 be at a height of 100 ft? Round your answers to the nearest tenth.

41. A farmer has a rectangular cattle pen with perimeter 350 ft and area 7500 ft². What are the dimensions of the pen? (*Hint:* Use the figure to set up the equation.)

42. The base of a triangle measures 1 m more than three times the height of the triangle. The area of the triangle is 15 m². Find the lengths of the base and the height.

43. Two cars travel at right angles to each other from an intersection until they are 17 mi apart. At that point, one car has gone 7 mi farther than the other. How far did the slower car travel? (*Hint:* Use the Pythagorean formula.)

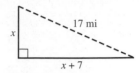

44. Two painters are painting a house in a development of new homes. One of the painters takes 2 hr longer to paint a house working alone than the other painter takes. When they do the job together, they can complete it in 4.8 hr. How long would it take the faster painter alone to paint the house? (Give your answer to the nearest tenth.)

RELATING CONCEPTS (EXERCISES 45–48)

FOR INDIVIDUAL OR GROUP WORK

We have discussed "completing the square" so far only in an algebraic sense. However, this procedure can literally be applied to a geometric figure so that it becomes a square. For example, to complete the square for $x^2 + 8x$, begin with a square having a side of length x. Add four rectangles of width 1 to the right side and to the bottom, as shown in the figure on the left. To "complete the square," fill in the bottom right corner with 16 squares of area 1, as shown in the figure on the right.

Work Exercises 45–48 in order.

45. What is the area of the original square?

46. What is the area of the figure after the 8 rectangles are added?

47. What is the area of the figure after the 16 small squares are added?

48. At what point did we "complete the square"?

PREVIEW EXERCISES

Write each quotient in lowest terms. Simplify the radicals. See **Section 8.5.**

49. $\dfrac{8 - 6\sqrt{3}}{6}$ **50.** $\dfrac{4 + \sqrt{28}}{2}$ **51.** $\dfrac{6 - \sqrt{45}}{6}$ **52.** $\dfrac{8 + \sqrt{32}}{4}$

Evaluate the expression $\sqrt{b^2 - 4ac}$ for the given values of a, b, and c. Simplify the radicals. See **Sections 1.3, 8.1, and 8.2.**

53. $a = 1, b = 2, c = -4$ **54.** $a = 9, b = 30, c = 25$

55. $a = 4, b = -28, c = 49$ **56.** $a = 1, b = -4, c = -5$

9.3 Solving Quadratic Equations by the Quadratic Formula

OBJECTIVES

1. Identify the values of a, b, and c in a quadratic equation.

2. Use the quadratic formula to solve quadratic equations.

3. Solve quadratic equations with only one solution.

4. Solve quadratic equations with fractions.

We can solve any quadratic equation by completing the square, but the method is tedious. In this section, we complete the square on the general quadratic equation

$$ax^2 + bx + c = 0 \quad (a \neq 0)$$

to get the *quadratic formula*, which gives the solutions of any quadratic equation.

OBJECTIVE 1 Identify the values of a, b, and c in a quadratic equation. To solve a quadratic equation with the quadratic formula, we must first identify the values of a, b, and c in the standard form of the quadratic equation.

EXAMPLE 1 Determining Values of a, b, and c in Quadratic Equations

Identify the values of a, b, and c in each quadratic equation $ax^2 + bx + c = 0$.

$$\overset{a}{\downarrow} \quad \overset{b}{\downarrow} \quad \overset{c}{\downarrow}$$

(a) $2x^2 + 3x - 5 = 0$

Here, $a = 2$, $b = 3$, and $c = -5$.

(b) $-x^2 + 2 = 6x$

First write the equation in standard form $ax^2 + bx + c = 0$.

$$-x^2 + 2 = 6x$$
$$-x^2 - 6x + 2 = 0 \qquad \text{Subtract 6x.}$$

Here, $a = -1$, $b = -6$, and $c = 2$. (Notice that the coefficient of x^2 is understood to be -1.)

(c) $5x^2 - 12 = 0$

The x-term is missing, so write the equation as

$$5x^2 + 0x - 12 = 0.$$

Then $a = 5$, $b = 0$, and $c = -12$.

(d)

> The equation is not in standard form.

$$(2x - 7)(x + 4) = -23$$
$$2x^2 + x - 28 = -23 \qquad \text{Use the FOIL method.}$$
$$2x^2 + x - 5 = 0 \qquad \text{Add 23.}$$

Now, identify the required values: $a = 2$, $b = 1$, and $c = -5$.

✔ **Now Try Exercises 1, 7, and 9.**

OBJECTIVE 2 Use the quadratic formula to solve quadratic equations. To develop the quadratic formula, we follow the steps given in **Section 9.2** for completing the square on $ax^2 + bx + c = 0$. For comparison, we also show the corresponding steps for solving $2x^2 + x - 5 = 0$ (from Example 1(d)).

Step 1 Transform so that the coefficient of the second-degree term is equal to 1.

$$2x^2 + x - 5 = 0 \qquad\qquad ax^2 + bx + c = 0 \quad (a > 0)$$

$$x^2 + \frac{1}{2}x - \frac{5}{2} = 0 \quad \text{Divide by 2.} \qquad x^2 + \frac{b}{a}x + \frac{c}{a} = 0 \quad \text{Divide by } a.$$

Step 2 Write the equation so that the variable terms with x are alone on the left side.

$$x^2 + \frac{1}{2}x = \frac{5}{2} \quad \text{Add } \tfrac{5}{2}. \qquad x^2 + \frac{b}{a}x = -\frac{c}{a} \quad \text{Subtract } \tfrac{c}{a}.$$

Step 3 Add the square of half the coefficient of x to both sides, factor the left side, and combine terms on the right.

$$x^2 + \frac{1}{2}x + \frac{1}{16} = \frac{5}{2} + \frac{1}{16} \quad \text{Add } \tfrac{1}{16}. \qquad x^2 + \frac{b}{a}x + \frac{b^2}{4a^2} = -\frac{c}{a} + \frac{b^2}{4a^2}$$

$$\text{Add } \frac{b^2}{4a^2}.$$

$$\left(x + \frac{1}{4}\right)^2 = \frac{41}{16} \quad \begin{array}{l}\text{Factor;}\\ \text{add on}\\ \text{right.}\end{array} \qquad \left(x + \frac{b}{2a}\right)^2 = \frac{b^2 - 4ac}{4a^2}$$

$$\text{Factor; add on right.}$$

Step 4 Use the square root property to complete the solution.

$$x + \frac{1}{4} = \pm\sqrt{\frac{41}{16}} \qquad\qquad x + \frac{b}{2a} = \pm\sqrt{\frac{b^2 - 4ac}{4a^2}}$$

$$x + \frac{1}{4} = \pm\frac{\sqrt{41}}{4} \qquad\qquad x + \frac{b}{2a} = \pm\frac{\sqrt{b^2 - 4ac}}{2a}$$

$$x = -\frac{1}{4} \pm \frac{\sqrt{41}}{4} \qquad\qquad x = -\frac{b}{2a} \pm \frac{\sqrt{b^2 - 4ac}}{2a}$$

$$x = \frac{-1 \pm \sqrt{41}}{4} \qquad\qquad x = \frac{-b \pm \sqrt{b^2 - 4ac}}{2a}$$

The final result in the column on the right is called the **quadratic formula.** (It is also valid for $a < 0$.) ***It is a key formula that should be memorized.*** Notice that there are two values: one for the $+$ sign and one for the $-$ sign.

Quadratic Formula

The solutions of the quadratic equation $ax^2 + bx + c = 0$, $a \neq 0$, are

$$x = \frac{-b + \sqrt{b^2 - 4ac}}{2a} \quad \text{and} \quad x = \frac{-b - \sqrt{b^2 - 4ac}}{2a}$$

or, in compact form, $\qquad x = \dfrac{-b \pm \sqrt{b^2 - 4ac}}{2a}.$

▶ **CAUTION** Notice that the fraction bar is under $-b$ as well as the radical. *When using this formula, be sure to find the values of $-b \pm \sqrt{b^2 - 4ac}$ first. Then divide those results by the value of $2a$.*

EXAMPLE 2 Solving a Quadratic Equation by the Quadratic Formula

Solve $2x^2 - 7x - 9 = 0$.

In this equation, $a = 2$, $b = -7$, and $c = -9$.

$$x = \frac{-b \pm \sqrt{b^2 - 4ac}}{2a} \qquad \text{Quadratic formula}$$

Be sure to write $-b$ in the numerator.

$$x = \frac{-(-7) \pm \sqrt{(-7)^2 - 4(2)(-9)}}{2(2)} \qquad \begin{array}{l}\text{Substitute } a = 2,\\ b = -7, \text{ and } c = -9.\end{array}$$

$$x = \frac{7 \pm \sqrt{49 + 72}}{4}$$

$$x = \frac{7 \pm \sqrt{121}}{4}$$

$$x = \frac{7 \pm 11}{4} \qquad \sqrt{121} = 11$$

Find the two solutions by first using the plus sign and then using the minus sign:

$$x = \frac{7 + 11}{4} = \frac{18}{4} = \frac{9}{2} \qquad \text{or} \qquad x = \frac{7 - 11}{4} = \frac{-4}{4} = -1.$$

Check each solution in the original equation. The solution set is $\left\{-1, \frac{9}{2}\right\}$.

✔ **Now Try Exercise 19.**

EXAMPLE 3 Rewriting a Quadratic Equation before Solving

Solve $x^2 = 2x + 1$.

Write the given equation in standard form (with 0 on one side).

$$x^2 - 2x - 1 = 0 \qquad \text{Subtract } 2x; \text{ subtract 1.}$$

Then $a = 1$, $b = -2$, and $c = -1$.

$$x = \frac{-b \pm \sqrt{b^2 - 4ac}}{2a} \qquad \text{Quadratic formula}$$

$$x = \frac{-(-2) \pm \sqrt{(-2)^2 - 4(1)(-1)}}{2(1)} \qquad \begin{array}{l}\text{Substitute } a = 1,\\ b = -2, \text{ and } c = -1.\end{array}$$

Be careful substituting the negative values.

$$x = \frac{2 \pm \sqrt{4 + 4}}{2}$$

$$x = \frac{2 \pm \sqrt{8}}{2}$$

$$x = \frac{2 \pm 2\sqrt{2}}{2} \qquad \sqrt{8} = \sqrt{4} \cdot \sqrt{2} = 2\sqrt{2}$$

Factor first. Then divide out the common factor.

$$x = \frac{2(1 \pm \sqrt{2})}{2} \qquad \text{Factor to write in lowest terms.}$$

$$x = 1 \pm \sqrt{2} \qquad \text{Divide out the common factor.}$$

The solution set is $\left\{1 \pm \sqrt{2}\right\}$.

✔ **Now Try Exercise 21.**

OBJECTIVE 3 Solve quadratic equations with only one solution. In the quadratic formula, the quantity under the radical, $b^2 - 4ac$, is called the **discriminant.** When the discriminant equals 0, the equation has just one rational number solution, and the trinomial $ax^2 + bx + c$ is a perfect square.

EXAMPLE 4 Solving a Quadratic Equation with Only One Solution

Solve $4x^2 + 25 = 20x$.

$$4x^2 - 20x + 25 = 0 \qquad \text{Subtract } 20x.$$

Here, $a = 4$, $b = -20$, and $c = 25$. By the quadratic formula,

$$x = \frac{-(-20) \pm \sqrt{(-20)^2 - 4(4)(25)}}{2(4)} = \frac{20 \pm 0}{8} = \frac{5}{2}.$$

In this case, the discriminant $b^2 - 4ac = 0$, and the trinomial $4x^2 - 20x + 25$ is a perfect square, $(2x - 5)^2$. There is just one solution in the solution set $\left\{\frac{5}{2}\right\}$.

✔ **Now Try Exercise 17.**

▶ **NOTE** The single solution of the equation in Example 4 is a rational number. If all solutions of a quadratic equation are rational, the equation can be solved by factoring as well.

OBJECTIVE 4 Solve quadratic equations with fractions.

EXAMPLE 5 Solving a Quadratic Equation with Fractions

Solve $\dfrac{1}{10}t^2 = \dfrac{2}{5}t - \dfrac{1}{2}$.

$$10\left(\frac{1}{10}t^2\right) = 10\left(\frac{2}{5}t - \frac{1}{2}\right) \qquad \begin{array}{l}\text{Clear fractions; multiply}\\ \text{by the LCD, 10.}\end{array}$$

$$t^2 = 10\left(\frac{2}{5}t\right) - 10\left(\frac{1}{2}\right) \qquad \text{Distributive property}$$

$$t^2 = 4t - 5$$

$$t^2 - 4t + 5 = 0 \qquad \text{Standard form}$$

Identify $a = 1$, $b = -4$, and $c = 5$. By the quadratic formula,

$$t = \frac{-(-4) \pm \sqrt{(-4)^2 - 4(1)(5)}}{2(1)} \qquad \begin{array}{l}\text{Substitute into the}\\ \text{quadratic formula.}\end{array}$$

$$t = \frac{4 \pm \sqrt{16 - 20}}{2} \qquad \text{Perform the operations.}$$

$$t = \frac{4 \pm \sqrt{-4}}{2}.$$

Because $\sqrt{-4}$ does not represent a real number, the solution set is $\varnothing$.

✔ **Now Try Exercise 49.**

9.3 EXERCISES

Complete solution available on Video Lectures on CD/DVD

Now Try Exercise

If necessary, write each equation in standard form $ax^2 + bx + c = 0$. Then identify the values of a, b, and c. Do not actually solve the equation. See Example 1.

1. $3x^2 + 4x - 8 = 0$ **2.** $9x^2 + 2x - 3 = 0$ **3.** $-8x^2 - 2x - 3 = 0$

4. $-2x^2 + 3x - 8 = 0$ **5.** $3x^2 = 4x + 2$ **6.** $5x^2 = 3x - 6$

7. $3x^2 = -7x$ **8.** $9x^2 = 8x$

9. $(x - 3)(x + 4) = 0$ **10.** $(x + 6)^2 = 3$

11. $9(x - 1)(x + 2) = 8$ **12.** $(3x - 1)(2x + 5) = x(x - 1)$

13. Why is the restriction $a \neq 0$ necessary in the definition of a quadratic equation?

14. To solve the quadratic equation $-2x^2 - 4x + 3 = 0$, we might choose to use $a = -2$, $b = -4$, and $c = 3$. Or we might instead decide to first multiply both sides by -1, obtaining the equation $2x^2 + 4x - 3 = 0$, and then use $a = 2$, $b = 4$, and $c = -3$. Show that in either case we obtain the same solution set.

15. A student writes the quadratic formula as $x = -b \pm \dfrac{\sqrt{b^2 - 4ac}}{2a}$. Is this correct? If not, explain the error, and give the correct formula.

16. Another student writes the quadratic formula as $x = -b \pm \sqrt{\dfrac{b^2 - 4ac}{2a}}$. Is this correct? If not, explain the error, and give the correct formula.

Use the quadratic formula to solve each equation. Simplify all radicals, and write all answers in lowest terms. See Examples 2–4.

17. $p^2 - 4p + 4 = 0$ **18.** $9x^2 + 6x + 1 = 0$ **19.** $k^2 + 12k - 13 = 0$

20. $r^2 - 8r - 9 = 0$ **21.** $2x^2 + 12x = -5$ **22.** $5m^2 + m = 1$

23. $2x^2 = 5 + 3x$ **24.** $2z^2 = 30 + 7z$ **25.** $6x^2 + 6x = 0$

26. $4n^2 - 12n = 0$ **27.** $7x^2 = 12x$ **28.** $9r^2 = 11r$

29. $x^2 - 24 = 0$ **30.** $z^2 - 96 = 0$ **31.** $25x^2 - 4 = 0$

32. $16x^2 - 9 = 0$ **33.** $3x^2 - 2x + 5 = 10x + 1$ **34.** $4x^2 - x + 4 = x + 7$

35. $-2x^2 = -3x + 2$ **36.** $-x^2 = -5x + 20$ **37.** $2x^2 + x + 5 = 0$

38. $3x^2 + 2x + 8 = 0$ **39.** $(x + 3)(x + 2) = 15$ **40.** $(2x + 1)(x + 1) = 7$

Use the quadratic formula to solve each equation. **(a)** *Give solutions in exact form, and* **(b)** *use a calculator to give solutions correct to the nearest thousandth.*

41. $2x^2 + 2x = 5$ **42.** $5x^2 = 3 - x$ **43.** $x^2 = 1 + x$ **44.** $x^2 = 2 + 4x$

Use the quadratic formula to solve each equation. See Example 5.

45. $\dfrac{3}{2}k^2 - k - \dfrac{4}{3} = 0$ **46.** $\dfrac{2}{5}x^2 - \dfrac{3}{5}x - 1 = 0$ **47.** $\dfrac{1}{2}x^2 + \dfrac{1}{6}x = 1$

48. $\dfrac{2}{3}z^2 - \dfrac{4}{9}z = \dfrac{1}{3}$ **49.** $\dfrac{3}{8}x^2 - x + \dfrac{17}{24} = 0$ **50.** $\dfrac{1}{3}x^2 + \dfrac{8}{9}x + \dfrac{7}{9} = 0$

51. $0.5x^2 = x + 0.5$ **52.** $0.25x^2 = -1.5x - 1$

53. $0.6x - 0.4x^2 = -1$ **54.** $0.5m^2 = 0.5m - 1$

Solve each problem.

55. Solve the formula $S = 2\pi rh + \pi r^2$ for r by writing it in the form $ar^2 + br + c = 0$ and then using the quadratic formula.

56. Solve the formula $V = \pi r^2 h + \pi R^2 h$ for r, using the method described in Exercise 55.

57. A frog is sitting on a stump 3 ft above the ground. He hops off the stump and lands on the ground 4 ft away. During his leap, his height h with respect to the ground is given by

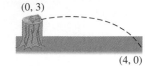

(0, 3)

(4, 0)

$$h = -0.5x^2 + 1.25x + 3,$$

where x is the distance in feet from the base of the stump and h is in feet. How far was the frog from the base of the stump when he was 1.25 ft above the ground?

58. An astronaut on the moon throws a baseball upward. The altitude (height) h of the ball, in feet, x seconds after he throws it, is given by the equation

$$h = -2.7x^2 + 30x + 6.5.$$

At what times is the ball 12 ft above the moon's surface?

59. A rule for estimating the number of board feet of lumber that can be cut from a log depends on the diameter of the log. To find the diameter d required to get 9 board ft of lumber, we use the equation

$$\left(\frac{d - 4}{4}\right)^2 = 9.$$

Solve this equation for d. Are both answers reasonable?

60. An old Babylonian problem asks for the length of the side of a square, given that the area of the square minus the length of a side is 870. Find the length of the side. (*Source:* Eves, Howard, *An Introduction to the History of Mathematics,* Sixth Edition, Saunders College Publishing, 1990.)

RELATING CONCEPTS (EXERCISES 61–66)

FOR INDIVIDUAL OR GROUP WORK

*In **Chapter 6,** we presented methods for factoring trinomials. There is a way to determine beforehand whether a trinomial of the form $ax^2 + bx + c$ can be factored. The key is the discriminant, $b^2 - 4ac$.* **Work Exercises 61–66 in order.**

61. Each of the following trinomials is factorable. Find the discriminant for each one.

    (a) $18x^2 - 9x - 2$ **(b)** $5x^2 + 7x - 6$
    (c) $48x^2 + 14x + 1$ **(d)** $x^2 - 5x - 24$

62. What do you notice about the discriminants you found in Exercise 61?

63. Factor each of the trinomials in Exercise 61.

64. Each of the following trinomials is not factorable by any of the methods of **Chapter 6.** Find the discriminant for each one.

    (a) $2x^2 + x - 5$ **(b)** $2x^2 + x + 5$ **(c)** $x^2 + 6x + 6$ **(d)** $3x^2 + 2x - 9$

65. Are any of the discriminants you found in Exercise 64 perfect squares?

66. Make a conjecture (an educated guess) concerning when a trinomial of the form $ax^2 + bx + c$ is factorable. Then use your conjecture to determine whether each trinomial is factorable. (Do not actually factor.)

    (a) $42x^2 + 117x + 66$ **(b)** $99x^2 + 186x - 24$ **(c)** $58x^2 + 184x + 27$

PREVIEW EXERCISES

*Perform the indicated operations. See **Sections 5.4, 5.5, and 5.6.***

67. $(4 + 6z) + (-9 + 2z)$ **68.** $(10 - 3t) - (5 - 7t)$ **69.** $4 - (6 - 3k)$

70. $7x(3 - 4x)$ **71.** $(4 + 3r)(6 - 5r)$ **72.** $(5 + 2x)(5 - 2x)$

Summary Exercises on Quadratic Equations

Four algebraic methods have now been introduced for solving quadratic equations written in the form $ax^2 + bx + c = 0$.

Method	Advantages	Disadvantages
1. Factoring	It is usually the fastest method.	Not all equations can be solved by factoring. Some factorable polynomials are difficult to factor.
2. Square root property	It is the simplest method for solving equations of the form $(ax + b)^2 = $ a number.	Few equations are given in this form.
3. Completing the square	It can always be used. (Also, the procedure is useful in other areas of mathematics.)	It requires more steps than other methods.
4. Quadratic formula	It can always be used.	It is more difficult than factoring because of the $\sqrt{b^2 - 4ac}$ expression.

Solve each quadratic equation by the method of your choice.

1. $s^2 = 36$

2. $x^2 + 3x = -1$

3. $x^2 - \dfrac{100}{81} = 0$

4. $81t^2 = 49$

5. $z^2 - 4z + 3 = 0$

6. $w^2 + 3w + 2 = 0$

7. $z(z - 9) = -20$

8. $x^2 + 3x - 2 = 0$

9. $(3k - 2)^2 = 9$

10. $(2s - 1)^2 = 10$

11. $(x + 6)^2 = 121$

12. $(5k + 1)^2 = 36$

13. $(3r - 7)^2 = 24$

14. $(7p - 1)^2 = 32$

15. $(5x - 8)^2 = -6$

16. $2t^2 + 1 = t$

17. $-2x^2 = -3x - 2$

18. $-2x^2 + x = -1$

19. $8z^2 = 15 + 2z$

20. $3k^2 = 3 - 8k$

21. $0.1x^2 - 0.2x = 0.1$

22. $0.3x^2 + 0.5x = -0.1$

23. $5z^2 - 22z = -8$

24. $z(z + 6) + 4 = 0$

25. $(x + 2)(x + 1) = 10$

26. $16x^2 + 40x + 25 = 0$

27. $4x^2 = -1 + 5x$

28. $2p^2 = 2p + 1$

29. $3m(3m + 4) = 7$

30. $5x - 1 + 4x^2 = 0$

31. $\dfrac{r^2}{2} + \dfrac{7r}{4} + \dfrac{11}{8} = 0$

32. $t(15t + 58) = -48$

33. $9k^2 = 16(3k + 4)$

34. $\frac{1}{5}x^2 + x + 1 = 0$ **35.** $x^2 - x + 3 = 0$ **36.** $4m^2 - 11m + 8 = -2$

37. $-3x^2 + 4x = -4$ **38.** $z^2 - \frac{5}{12}z = \frac{1}{6}$ **39.** $5k^2 + 19k = 2k + 12$

40. $\frac{1}{2}n^2 - n = \frac{15}{2}$ **41.** $k^2 - \frac{4}{15} = -\frac{4}{15}k$ **42.** $(x + 2)(x - 4) = 16$

43. *Concept Check* How many real solutions are there for a quadratic equation that has a negative number as its radicand in the quadratic formula?

9.4 Complex Numbers

OBJECTIVES

1 Write complex numbers as multiples of i.

2 Add and subtract complex numbers.

3 Multiply complex numbers.

4 Write complex number quotients in standard form.

5 Solve quadratic equations with complex number solutions.

Some quadratic equations have no real number solutions. For example, the numbers

$$\frac{4 \pm \sqrt{-4}}{2},$$

which occurred in the solution of Example 5 in **Section 9.3,** are *not* real numbers, because -4 appears as the radicand. To ensure that every quadratic equation has a solution, we need a new set of numbers that includes the real numbers. This new set of numbers is defined with a new number i, called the **imaginary unit,** such that

$$i = \sqrt{-1} \quad \text{and} \quad i^2 = -1.$$

OBJECTIVE 1 Write complex numbers as multiples of i. We can write numbers such as $\sqrt{-4}$, $\sqrt{-5}$, and $\sqrt{-8}$ as multiples of i, using the properties of i to define any square root of a negative number as follows.

$\sqrt{-b}$

For any positive real number b, $\sqrt{-b} = i\sqrt{b}.$

EXAMPLE 1 Simplifying Square Roots of Negative Numbers

Write each number as a multiple of i.

(a) $\sqrt{-4} = i\sqrt{4} = i \cdot 2 = 2i$ **(b)** $\sqrt{-5} = i\sqrt{5}$

(c) $\sqrt{-8} = i\sqrt{8} = i \cdot \sqrt{4} \cdot \sqrt{2} = i \cdot 2 \cdot \sqrt{2} = 2i\sqrt{2}$

✔ **Now Try Exercises 1 and 3.**

▶ **CAUTION** It is easy to mistake $\sqrt{2}i$ for $\sqrt{2i}$, with the i under the radical. For this reason, it is customary to write the factor i first when it is multiplied by a radical. For example, we usually write $i\sqrt{2}$ rather than $\sqrt{2}i$.

Numbers that are nonzero multiples of i are *pure imaginary numbers*. The *complex numbers* include all real numbers and all imaginary numbers.

Complex Number

A **complex number** is a number of the form $a + bi,$ where a and b are real numbers. If $a = 0$ and $b \neq 0$, then the number bi is a **pure imaginary number.**

For example, the real number 2 is a complex number, since it can be written as $2 + 0i$. Also, the pure imaginary number $3i = 0 + 3i$ is a complex number. Other complex numbers are

$$3 - 2i, \qquad 1 + i\sqrt{2}, \qquad \text{and} \qquad -5 + 4i.$$

In the complex number $a + bi$, a is called the **real part** and b is called the **imaginary part.***

A complex number written in the form $a + bi$ (or $a + ib$) is in **standard form.** Figure 1 shows the relationships among the various types of numbers discussed in this book. (Compare this figure with Figure 8 in **Section 1.4.**)

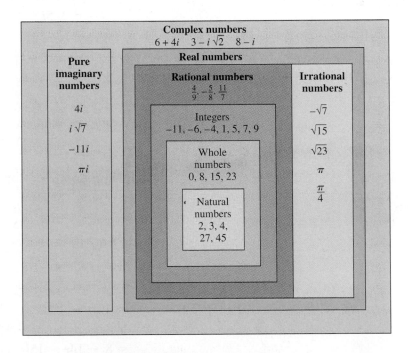

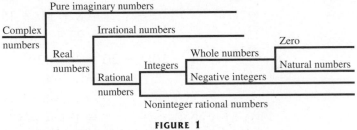

FIGURE 1

---

*Some texts refer to bi as the imaginary part.

OBJECTIVE 2 Add and subtract complex numbers. Adding and subtracting complex numbers is similar to adding and subtracting binomials.

> ### Adding and Subtracting Complex Numbers
>
> To add complex numbers, add their real parts and add their imaginary parts.
> To subtract complex numbers, add the additive inverse (or opposite).

EXAMPLE 2 Adding and Subtracting Complex Numbers

Add or subtract.

(a) $(2 - 6i) + (7 + 4i) = (2 + 7) + (-6 + 4)i$ Add real parts;
add imaginary parts.

$= 9 - 2i$ Standard form

(b) $3i + (-2 - i) = -2 + (3 - 1)i$ $-i = -1i$

$= -2 + 2i$

(c) $(2 + 6i) - (-4 + i) = (2 + 6i) + (4 - i)$ Add the additive inverse.

$= (2 + 4) + (6 - 1)i$ Properties of real numbers

$= 6 + 5i$

(d) $(-1 + 2i) - 4 = (-1 - 4) + 2i$

$= -5 + 2i$

✔ **Now Try Exercises 9 and 11.**

OBJECTIVE 3 Multiply complex numbers. We multiply complex numbers as we do polynomials. Since $i^2 = -1$ by definition, whenever i^2 appears, we replace it with -1.

EXAMPLE 3 Multiplying Complex Numbers

Find each product.

(a) $3i(2 - 5i) = 6i - 15i^2$ Distributive property

$= 6i - 15(-1)$ $i^2 = -1$

$= 6i + 15$

$= 15 + 6i$ Standard form

(b) $(4 - 3i)(2 + 5i) = 4(2) + 4(5i) + (-3i)2 + (-3i)5i$ Use the FOIL method.

$= 8 + 20i - 6i - 15i^2$

$= 8 + 14i - 15(-1)$

$= 8 + 14i + 15$

$= 23 + 14i$

(c) $(1 + 2i)(1 - 2i) = 1^2 - (2i)^2$ $(x + y)(x - y) = x^2 - y^2$

$= 1 - 4i^2$ $(xy)^2 = x^2y^2$

$= 1 - 4(-1)$

$= 1 + 4 = 5$

✔ **Now Try Exercises 15 and 19.**

OBJECTIVE 4 Write complex number quotients in standard form. The quotient of two complex numbers is expressed in standard form by changing the denominator into a real number. For example, to write

$$\frac{8 + i}{1 + 2i}$$

in standard form, the denominator must be a real number. As seen in Example 3(c), the product $(1 + 2i)(1 - 2i)$ is 5, a real number. This suggests multiplying the numerator and denominator of the given quotient by $1 - 2i$ as follows.

$$\frac{8 + i}{1 + 2i} = \frac{8 + i}{1 + 2i} \cdot \frac{1 - 2i}{1 - 2i} \qquad \frac{1 - 2i}{1 - 2i} = 1$$

$$= \frac{8 - 16i + i - 2i^2}{1 - 4i^2} \qquad \text{Multiply.}$$

$$= \frac{8 - 16i + i - 2(-1)}{1 - 4(-1)} \qquad \boxed{\text{Use parentheses around } -1 \text{ to avoid errors.}}$$

$$= \frac{10 - 15i}{5} \qquad \text{Combine terms.}$$

$\boxed{\text{Factor first. Then divide out the common factor.}}$ $\qquad = \frac{5(2 - 3i)}{5} \qquad \text{Factor.}$

$$= 2 - 3i \qquad \text{Divide out the common factor.}$$

We used a similar method to rationalize some radical expressions in **Chapter 8.** The complex numbers $1 + 2i$ and $1 - 2i$ are *conjugates*. That is, the **conjugate** of the complex number $a + bi$ is $a - bi$. Multiplying the complex number $a + bi$ by its conjugate $a - bi$ gives the real number $a^2 + b^2$.

Product of Conjugates

$$(a + bi)(a - bi) = a^2 + b^2$$

That is, the product of a complex number and its conjugate is the sum of the squares of the real and imaginary parts.

EXAMPLE 4 Dividing Complex Numbers

Write each quotient in standard form.

(a) $\dfrac{-4 + i}{2 - i} = \dfrac{-4 + i}{2 - i} \cdot \dfrac{2 + i}{2 + i}$ $\qquad$ Multiply numerator and denominator by the conjugate of the denominator.

$$= \frac{-8 - 4i + 2i + i^2}{4 - i^2} \qquad \text{Multiply.}$$

$$= \frac{-8 - 4i + 2i - 1}{4 - (-1)} \qquad \boxed{\text{Be careful with signs.}}$$

$$= \frac{-9 - 2i}{5} \qquad \text{Combine terms.}$$

$$= -\frac{9}{5} - \frac{2}{5}i \qquad \text{Standard form}$$

(b) $\dfrac{3 + i}{-i} = \dfrac{3 + i}{-i} \cdot \dfrac{i}{i}$ The conjugate of $0 - i$ is $0 + i$, or i.

$$= \dfrac{3i + i^2}{-i^2}$$

$$= \dfrac{-1 + 3i}{-(-1)}$$ $i^2 = -1$; commutative property

Be careful with signs.

$$= -1 + 3i$$

✔ **Now Try Exercise 21.**

OBJECTIVE 5 **Solve quadratic equations with complex number solutions.** Quadratic equations that have no real solutions do have complex solutions.

EXAMPLE 5 Solving a Quadratic Equation with Complex Solutions (Square Root Property)

Solve $(x + 3)^2 = -25$.

$$(x + 3)^2 = -25$$

$x + 3 = \sqrt{-25}$ or $x + 3 = -\sqrt{-25}$ Square root property

$x + 3 = 5i$ or $x + 3 = -5i$ $\sqrt{-25} = 5i$

$x = -3 + 5i$ or $x = -3 - 5i$ Add -3.

The solution set is $\{-3 \pm 5i\}$.

✔ **Now Try Exercise 27.**

EXAMPLE 6 Solving a Quadratic Equation with Complex Solutions (Quadratic Formula)

Solve $2p^2 = 4p - 5$.

Write the equation in standard form as $2p^2 - 4p + 5 = 0$.

$$p = \dfrac{-b \pm \sqrt{b^2 - 4ac}}{2a}$$ Quadratic formula

$$p = \dfrac{-(-4) \pm \sqrt{(-4)^2 - 4(2)(5)}}{2(2)}$$ Substitute $a = 2$, $b = -4$, and $c = 5$.

$$p = \dfrac{4 \pm \sqrt{16 - 40}}{4}$$

$$p = \dfrac{4 \pm \sqrt{-24}}{4}$$

$$p = \dfrac{4 \pm 2i\sqrt{6}}{4}$$ $\sqrt{-24} = i\sqrt{24} = i \cdot \sqrt{4} \cdot \sqrt{6}$ $= i \cdot 2 \cdot \sqrt{6} = 2i\sqrt{6}$

Factor first. Then divide out the common factor.

$$p = \dfrac{2(2 \pm i\sqrt{6})}{2(2)}$$ Factor out 2.

$$p = \frac{2 \pm i\sqrt{6}}{2} \qquad \text{Divide out the common factor.}$$

$$p = \frac{2}{2} \pm \frac{i\sqrt{6}}{2} \qquad \text{Separate into real and imaginary parts.}$$

$$p = 1 \pm \frac{\sqrt{6}}{2}i \qquad \text{Standard form}$$

The solution set is $\left\{ 1 \pm \frac{\sqrt{6}}{2}i \right\}$.

✔ **Now Try Exercise 35.**

9.4 EXERCISES

◉ *Complete solution available on Video Lectures on CD/DVD*

    Now Try Exercise

Write each number as a multiple of i. See Example 1.

◉ **1.** $\sqrt{-9}$ **2.** $\sqrt{-36}$ **3.** $\sqrt{-20}$ **4.** $\sqrt{-27}$

5. $\sqrt{-18}$ **6.** $\sqrt{-50}$ **7.** $\sqrt{-125}$ **8.** $\sqrt{-98}$

Add or subtract as indicated. See Example 2.

◉ **9.** $(2 + 8i) + (3 - 5i)$ **10.** $(4 + 5i) + (7 - 2i)$

11. $(8 - 3i) - (2 + 6i)$ **12.** $(1 + i) - (3 - 2i)$

13. $(3 - 4i) + (6 - i) - (3 + 2i)$ **14.** $(5 + 8i) - (4 + 2i) + (3 - i)$

Find each product. See Example 3.

◉ **15.** $(3 + 2i)(4 - i)$ **16.** $(9 - 2i)(3 + i)$ **17.** $(5 - 4i)(3 - 2i)$

18. $(10 + 6i)(8 - 4i)$ **19.** $(3 + 6i)(3 - 6i)$ **20.** $(11 - 2i)(11 + 2i)$

Write each quotient in standard form. See Example 4.

◉ **21.** $\dfrac{17 + i}{5 + 2i}$ **22.** $\dfrac{21 + i}{4 + i}$ **23.** $\dfrac{40}{2 + 6i}$

24. $\dfrac{13}{3 + 2i}$ **25.** $\dfrac{i}{4 - 3i}$ **26.** $\dfrac{-i}{1 + 2i}$

Solve each quadratic equation for complex solutions by the square root property. Write solutions in standard form. See Example 5.

◉ **27.** $(a + 1)^2 = -4$ **28.** $(p - 5)^2 = -36$ **29.** $(k - 3)^2 = -5$

30. $(x + 6)^2 = -12$ **31.** $(3x + 2)^2 = -18$ **32.** $(4z - 1)^2 = -20$

Solve each quadratic equation for complex solutions by the quadratic formula. Write solutions in standard form. See Example 6.

33. $m^2 - 2m + 2 = 0$ **34.** $x^2 + x + 3 = 0$ ◉ **35.** $2r^2 + 3r + 5 = 0$

36. $3q^2 = 2q - 3$ **37.** $p^2 - 3p + 4 = 0$ **38.** $2x^2 = -x - 3$

39. $5x^2 + 3 = 2x$ **40.** $6x^2 + 2x + 1 = 0$ **41.** $2m^2 + 7 = -2m$

42. $4z^2 + 2z + 3 = 0$ **43.** $r^2 + 3 = r$ **44.** $4q^2 - 2q + 3 = 0$

Exercises 45 and 46 refer to quadratic equations having real number coefficients.

✎ **45.** Suppose you are solving a quadratic equation by the quadratic formula. How can you tell, before completing the solution, whether the equation will have solutions that are not real numbers?

46. *Concept Check* Refer to Examples 5 and 6, and complete the following: If a quadratic equation has solutions that are nonreal complex numbers, they are _____ of each other.

Concept Check *Answer* true *or* false *to each statement. If false, say why.*

47. Every real number is a complex number.

48. Every pure imaginary number is a complex number.

49. Every complex number is a real number.

50. Some complex numbers are pure imaginary numbers.

PREVIEW EXERCISES

Graph each linear equation. See **Section 3.2.**

51. $2x - 3y = 6$ **52.** $y = 4x - 3$ **53.** $3x + 5y = 15$

Evaluate each expression if $x = 3$. See **Section 1.3.**

54. $x^2 - 8$ **55.** $2x^2 - x + 1$ **56.** $(x - 1)^2$

9.5 More on Graphing Quadratic Equations; Quadratic Functions

OBJECTIVES

1 Graph quadratic equations of the form $y = ax^2 + bx + c$ $(a \neq 0)$.

2 Use a graph to determine the number of real solutions of a quadratic equation.

3 Use a quadratic function to solve an application.

In **Section 5.4,** we graphed the quadratic equation $y = x^2$. By plotting points, we obtained the graph of a **parabola,** shown here in Figure 2.

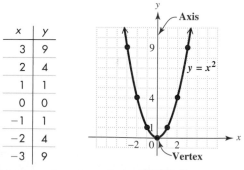

x	y
3	9
2	4
1	1
0	0
−1	1
−2	4
−3	9

FIGURE 2

Recall that the lowest point on this graph is called the **vertex** of the parabola. (If the parabola opens downward, the vertex is the highest point.) The vertical line through the vertex is called the **axis,** or **axis of symmetry.** The two halves of the parabola are mirror images of each other across this axis.

We have also seen that other quadratic equations, such as $y = -x^2 + 3$ and $y = (x + 2)^2$, have parabolas as their graphs. See Figures 3 and 4.

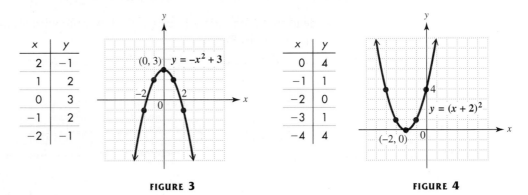

x	y
2	-1
1	2
0	3
-1	2
-2	-1

x	y
0	4
-1	1
-2	0
-3	1
-4	4

FIGURE 3

FIGURE 4

OBJECTIVE 1 Graph quadratic equations of the form $y = ax^2 + bx + c$ ($a \neq 0$). Every equation of the form

$$y = ax^2 + bx + c,$$

with $a \neq 0$, has a graph that is a parabola. As the preceding graphs suggest, the vertex is the most important point to locate when graphing a quadratic equation.

EXAMPLE 1 Graphing a Parabola by Finding the Vertex and Intercepts

Graph $y = x^2 - 2x - 3$.

We must find the vertex of the graph. *Because of its symmetry, if a parabola has two x-intercepts, the x-value of the vertex is exactly halfway between them.* See Figure 3, for example. Therefore, we begin by finding the x-intercepts. Let $y = 0$ in the equation and solve for x.

$$0 = x^2 - 2x - 3$$
$$x^2 - 2x - 3 = 0 \qquad \text{Reverse sides.}$$
$$(x + 1)(x - 3) = 0 \qquad \text{Factor.}$$
$$x + 1 = 0 \quad \text{or} \quad x - 3 = 0 \qquad \text{Zero-factor property}$$
$$x = -1 \quad \text{or} \quad x = 3$$

There are two x-intercepts: $(-1, 0)$ and $(3, 0)$.

Since the x-value of the vertex is halfway between the x-values of the two x-intercepts, it is half their sum.

$$x = \frac{1}{2}(-1 + 3) = 1 \longleftarrow x\text{-value of the vertex}$$

Find the corresponding y-value by substituting 1 for x in the equation.

$$y = 1^2 - 2(1) - 3 = -4 \longleftarrow y\text{-value of the vertex}$$

The vertex is $(1, -4)$. The axis is the line $x = 1$.

To find the y-intercept, substitute $x = 0$ in the equation.

$$y = 0^2 - 2(0) - 3 = -3$$

The y-intercept is $(0, -3)$.

Plot the three intercepts and the vertex. Find additional ordered pairs as needed. For example, if $x = 2$, then

$$y = 2^2 - 2(2) - 3 = -3,$$

leading to the ordered pair $(2, -3)$. A table with all these ordered pairs is shown with the graph in Figure 5.

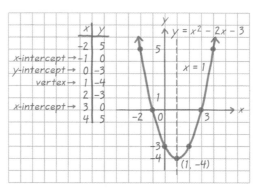

FIGURE 5

✔ **Now Try Exercise 7.**

We can generalize from Example 1. The x-coordinates of the x-intercepts for the equation $y = ax^2 + bx + c$, by the quadratic formula, are

$$x = \frac{-b + \sqrt{b^2 - 4ac}}{2a} \qquad \text{and} \qquad x = \frac{-b - \sqrt{b^2 - 4ac}}{2a}.$$

Thus, the x-value of the vertex is

$$x = \frac{1}{2}\left(\frac{-b + \sqrt{b^2 - 4ac}}{2a} + \frac{-b - \sqrt{b^2 - 4ac}}{2a} \right)$$

$$x = \frac{1}{2}\left(\frac{-b + \sqrt{b^2 - 4ac} - b - \sqrt{b^2 - 4ac}}{2a} \right)$$

$$x = \frac{1}{2}\left(\frac{-2b}{2a} \right)$$

$$x = -\frac{b}{2a}.$$

For the equation in Example 1, $y = x^2 - 2x - 3$, $a = 1$, and $b = -2$. Thus, the x-value of the vertex is

$$x = -\frac{b}{2a} = -\frac{-2}{2(1)} = 1,$$

which is the same x-value for the vertex we found in Example 1. (The x-value of the vertex is $x = -\frac{b}{2a}$ even if the graph has no x-intercepts.)

A procedure for graphing quadratic equations follows.

Graphing the Parabola $y = ax^2 + bx + c$

Step 1 **Find the vertex.** Let $x = -\frac{b}{2a}$, and find the corresponding y-value by substituting for x in the equation.

Step 2 **Find the y-intercept.**

Step 3 **Find the x-intercepts** (if they exist).

Step 4 **Plot** the intercepts and the vertex.

Step 5 **Find and plot additional ordered pairs** near the vertex and intercepts as needed, using symmetry about the axis of the parabola.

EXAMPLE 2 Graphing a Parabola

Graph $y = x^2 - 4x + 1$.

Step 1 Find the vertex. The x-value of the vertex is

$$x = -\frac{b}{2a} = -\frac{-4}{2(1)} = 2. \qquad a = 1, b = -4$$

The y-value of the vertex is

$$y = 2^2 - 4(2) + 1 = -3,$$

so the vertex is $(2, -3)$. The axis is the line $x = 2$.

Step 2 Now find the y-intercept. Let $x = 0$ in $y = x^2 - 4x + 1$.

$$y = 0^2 - 4(0) + 1 = 1$$

The y-intercept is $(0, 1)$.

Step 3 Let $y = 0$ to get the x-intercepts. The equation becomes $0 = x^2 - 4x + 1$, which cannot be solved by factoring. Use the quadratic formula to solve for x.

$$x = \frac{-(-4) \pm \sqrt{(-4)^2 - 4(1)(1)}}{2(1)} \qquad \text{Let } a = 1, b = -4, c = 1 \text{ in the quadratic formula.}$$

$$x = \frac{4 \pm \sqrt{12}}{2}$$

$$x = \frac{4 \pm 2\sqrt{3}}{2} \qquad\qquad \sqrt{12} = \sqrt{4} \cdot \sqrt{3} = 2\sqrt{3}$$

Factor first. Then divide out the common factor.

$$x = \frac{2(2 \pm \sqrt{3})}{2} \qquad\qquad \text{Factor.}$$

$$x = 2 \pm \sqrt{3} \qquad\qquad \text{Divide out 2.}$$

Use a calculator to find that the x-intercepts are $(3.7, 0)$ and $(0.3, 0)$ to the nearest tenth.

Steps 4 Plot the intercepts, vertex, and the additional points shown in the table.
and 5 Connect these points with a smooth curve. The graph is shown in Figure 6.

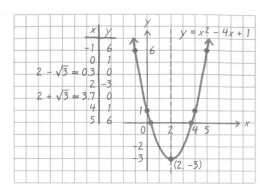

FIGURE 6

✔ **Now Try Exercise 9.**

OBJECTIVE ❷ **Use a graph to determine the number of real solutions of a quadratic equation.** Using the vertical line test **(Section 3.6),** we see that the graph of an equation of the form

$$y = ax^2 + bx + c$$

is the graph of a function. A function defined by an equation of the form

$$f(x) = ax^2 + bx + c \quad (a \neq 0)$$

is called a **quadratic function.** The **domain** (possible x-values) of a quadratic function is the set of all real numbers, or $(-\infty, \infty)$; the **range** (the resulting y-values) can be determined after the function is graphed. In Example 2, the domain is $(-\infty, \infty)$, and from Figure 6, we see that the range is $[-3, \infty)$.

In Example 2, we found that the x-intercepts of the graph of $y = x^2 - 4x + 1$ (where $y = 0$) are

$$2 - \sqrt{3} \approx 0.3 \quad \text{and} \quad 2 + \sqrt{3} \approx 3.7.$$

This means that $2 - \sqrt{3} \approx 0.3$ and $2 + \sqrt{3} \approx 3.7$ are also the solutions of the equation $0 = x^2 - 4x + 1$.

---

x-Intercepts of the Graph of a Quadratic Function

The real number solutions of a quadratic equation $ax^2 + bx + c = 0$ are the x-values of the x-intercepts of the graph of the corresponding quadratic function defined by $f(x) = ax^2 + bx + c$.

The fact that the graph of a quadratic function can intersect the x-axis in two, one, or no points justifies why some quadratic equations have two, some have one, and some have no real solutions.

EXAMPLE 3 Determining the Number of Real Solutions from Graphs

Decide from the graphs in Figures 7–9 how many real number solutions the corresponding equation $f(x) = 0$ has. Give the solution set.

(a) Figure 7 shows the graph of $f(x) = x^2 - 3$. The corresponding equation, $x^2 - 3 = 0$, has two real solutions, $\sqrt{3}$ and $-\sqrt{3}$, which correspond to the x-intercepts. The solution set is $\left\{\pm\sqrt{3}\right\}$.

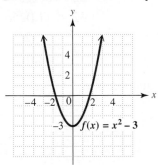

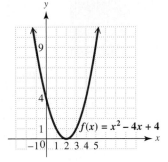

FIGURE 7   **FIGURE 8**

(b) Figure 8 shows the graph of $f(x) = x^2 - 4x + 4$. The corresponding equation, $x^2 - 4x + 4 = 0$, has one real solution, 2, which is the x-value of the x-intercept of the graph. The solution set is $\{2\}$.

(c) Figure 9 shows the graph of $f(x) = x^2 + 2$. The equation $x^2 + 2 = 0$ has no real solutions, since there are no x-intercepts. The solution set over the domain of real numbers is $\emptyset$. (The equation *does* have two pure imaginary solutions: $i\sqrt{2}$ and $-i\sqrt{2}$.)

✔ **Now Try Exercises 13, 15, and 17.**

OBJECTIVE 3 Use a quadratic function to solve an application.

EXAMPLE 4 Finding the Equation of a Parabolic Satellite Dish

The Parkes radio telescope has a parabolic dish shape with diameter 210 ft and depth 32 ft. Figure 10(a) shows a diagram of such a dish, and Figure 10(b) shows how a cross section of the dish can be modeled by a graph, with the vertex of the parabola at the origin of a coordinate system. (*Source:* Mar, J., and H. Liebowitz, *Structure Technology for Large Radio and Radar Telescope Systems,* The MIT Press, 1969.)

Find the equation of this graph.

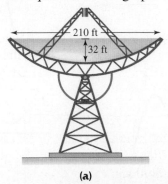

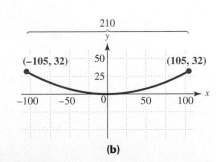

(a)   (b)

FIGURE 10

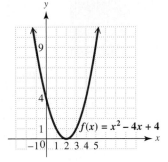

FIGURE 9

Because the vertex is at the origin, the equation will be of the form

$$y = ax^2.$$

As shown in Figure 10(b), one point on the graph has coordinates (105, 32). Letting $x = 105$ and $y = 32$, we can solve for a.

$y = ax^2$	General equation
$32 = a(105)^2$	Substitute for x and y.
$32 = 11{,}025a$	$105^2 = 11{,}025$
$a = \dfrac{32}{11{,}025}$	Divide by 11,025.

Thus, the equation is $y = \frac{32}{11{,}025}x^2$.

✔ **Now Try Exercise 35.**

9.5 EXERCISES

🌐 *Now Try Exercise*

🖉 **1.** In your own words, explain what is meant by the vertex of a parabola.

🖉 **2.** In your own words, explain what is meant by the line of symmetry of a parabola that opens upward or downward.

Give the coordinates of the vertex and sketch the graph of each equation. See Examples 1 and 2.

3. $y = x^2 - 6$ **4.** $y = -x^2 + 2$ **5.** $y = (x + 3)^2$

6. $y = (x - 4)^2$ 🌐 **7.** $y = x^2 + 2x + 3$ **8.** $y = x^2 - 4x + 3$

🌐 **9.** $y = x^2 - 8x + 16$ **10.** $y = x^2 + 6x + 9$ **11.** $y = -x^2 + 6x - 5$

12. $y = -x^2 - 4x - 3$

Decide from each graph how many real solutions $f(x) = 0$ has. Then give the solution set (of real solutions). See Example 3.

🌐 **13.**
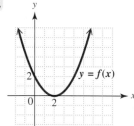

14.
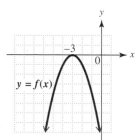

15.
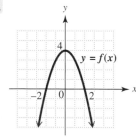

16.
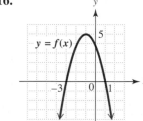

17.
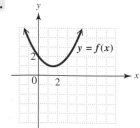

18.
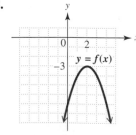

19. *Concept Check* Based on your work in Exercises 3–12, what seems to be the direction in which the parabola $y = ax^2 + bx + c$ opens if $a > 0$? if $a < 0$?

20. *Concept Check* How many real solutions does a quadratic equation have if its corresponding graph has **(a)** no x-intercepts, **(b)** one x-intercept, **(c)** two x-intercepts? See Examples 1–3.

TECHNOLOGY INSIGHTS (EXERCISES 21–24)

*The connection between the solutions of an equation and the x-intercepts of its graph enables us to solve quadratic equations with a graphing calculator. With the equation in the form $ax^2 + bx + c = 0$, enter $ax^2 + bx + c$ as Y_1, and then direct the calculator to find the x-intercepts of the graph. (These are also referred to as **zeros** of the function.)*

For example, to solve $x^2 - 5x - 6 = 0$ graphically, refer to the three screens shown here. The displays at the bottoms of the lower two screens show the two solutions: -1 and 6.

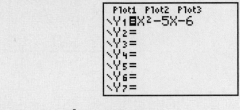

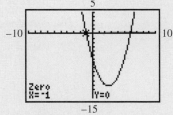 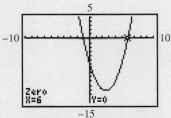

Determine the solution set of each quadratic equation by observing the corresponding screens. Then use the method of your choice to verify your answers by solving the quadratic equation.

21. $x^2 - x - 6 = 0$

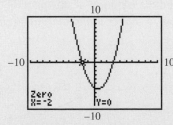 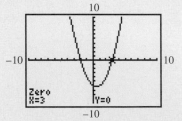

22. $x^2 + 6x + 5 = 0$

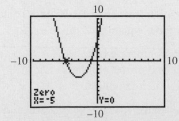 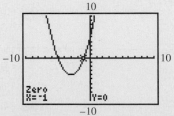

(continued)

23. $2x^2 - x - 3 = 0$

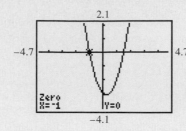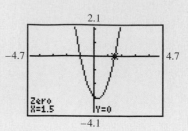

24. $4x^2 - 11x - 3 = 0$

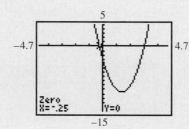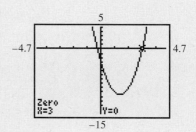

Find the domain and range of each function graphed in the indicated exercise.

25. Exercise 13 **26.** Exercise 14 **27.** Exercise 15

28. Exercise 16 **29.** Exercise 17 **30.** Exercise 18

Given $f(x) = 2x^2 - 5x + 3$, find each of the following.

31. $f(0)$ **32.** $f(1)$ **33.** $f(-2)$ **34.** $f(-1)$

Solve each problem. See Example 4.

35. The U.S. Naval Research Laboratory designed a giant radio telescope that had a diameter of 300 ft and a maximum depth of 44 ft. The graph depicts a cross section of that telescope. Find the equation of this parabola. (*Source:* Mar, J., and H. Liebowitz, *Structure Technology for Large Radio and Radar Telescope Systems,* The MIT Press, 1969.)

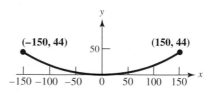

36. Suppose the telescope in Exercise 35 had a diameter of 400 ft and a maximum depth of 50 ft. Find the equation of this parabola.

37. Find two numbers whose sum is 80 and whose product is a maximum. (*Hint:* Let x represent one of the numbers. Then $80 - x$ represents the other. A quadratic function represents their product.)

38. Find two numbers whose sum is 300 and whose product is a maximum.

RELATING CONCEPTS (EXERCISES 39–44)

FOR INDIVIDUAL OR GROUP WORK

We can use a graphing calculator to illustrate how the graph of $y = x^2$ can be transformed through arithmetic operations. **Work Exercises 39–44 in order.**

39. In the standard viewing window of your calculator, graph the following one at a time, leaving the previous graphs on the screen as you move along.

$$Y_1 = x^2 \qquad Y_2 = 2x^2 \qquad Y_3 = 3x^2 \qquad Y_4 = 4x^2$$

Describe the effect the successive coefficients have on the parabola.

40. Repeat Exercise 39 for the following.

$$Y_1 = x^2 \qquad Y_2 = \frac{1}{2}x^2 \qquad Y_3 = \frac{1}{4}x^2 \qquad Y_4 = \frac{1}{8}x^2$$

41. In the standard viewing window of your calculator, graph the following pair of parabolas on the same screen.

$$Y_1 = x^2 \qquad Y_2 = -x^2$$

Describe how the graph of Y_2 can be obtained from the graph of Y_1.

42. In the standard viewing window of your calculator, graph the following parabolas on the same screen.

$$Y_1 = -x^2 \qquad Y_2 = -2x^2 \qquad Y_3 = -3x^2 \qquad Y_4 = -4x^2$$

Make a conjecture about what happens when the coefficient of x^2 is negative.

43. In the standard viewing window of your calculator, graph the following one at a time, leaving the previous graphs on the screen as you move along.

$$Y_1 = x^2 \qquad Y_2 = x^2 + 3 \qquad Y_3 = x^2 - 6$$

Describe the effect that adding or subtracting a constant has on the parabola.

44. Repeat Exercise 43 for the following.

$$Y_1 = x^2 \qquad Y_2 = (x + 3)^2 \qquad Y_3 = (x - 6)^2$$

Chapter 9 | Group Activity

HOW IS A RADIO TELESCOPE LIKE A WOK?

Objective Determine appropriate domains and ranges when graphing parabolas on a graphing calculator.

In Example 4 of **Section 9.5,** we explored the parabolic shape of a radio telescope. This activity uses that example and another to show the importance of setting appropriate domains and ranges on a graphing calculator. One student should write the answers to the questions, while the other one does the graphing on a calculator. When you start part B, switch tasks.

A. In Example 4 of **Section 9.5,** an equation was found to represent the shape of the Parkes radio telescope. Using this equation, graph the parabola on a graphing calculator.

 1. Use a standard viewing window—that is, Xmin = −10, Xmax = 10, Ymin = −10, and Ymax = 10. Describe your graph. Does its shape look similar to the picture of the telescope in **Section 9.5?**

 2. Change the domain and range settings to Xmin = −50, Xmax = 50, Ymin = −5, and Ymax = 50. You may want to change Xscl and Yscl to 5. How does the graph look now?

 3. Continue to adjust the domain and range settings until the calculator graph is close in shape to the picture of the telescope in **Section 9.5.** What are your settings?

B. A wok used in Asian cooking has a parabolic shape. Just as the shape of a parabola focuses radio waves from space, a wok focuses heat and oil for cooking.

 1. Find an equation that would model the cross section of a wok with a diameter of 14 in. and a depth of 4 in.

 2. Graph the equation you found in Exercise 1, using a standard viewing window on a calculator. Describe the graph of the equation. Describe the differences between this graph and the graph in part A, Exercise 1.

 3. Adjust the settings for domain and range until the graph looks similar to the cross section of a wok.

 4. How are a wok and a radio telescope alike?

 5. How is the range (or domain) of a function different from the range (or domain) setting on a graphing calculator?

Chapter **9** **SUMMARY** *View the Interactive Summary on the Pass the Test CD.*

KEY TERMS

9.1 quadratic equation
9.2 completing the square
9.3 quadratic formula
    discriminant

9.4 complex number
    pure imaginary number
    real part
    imaginary part

standard form (of a
   complex number)
conjugate (of a
   complex number)

9.5 parabola
    vertex
    axis (of symmetry)
    quadratic function

NEW SYMBOLS

$\pm$ positive or negative
(plus or minus)

i imaginary unit

TEST YOUR WORD POWER

See how well you have learned the vocabulary in this chapter. Answers, with examples, follow the Quick Review.

1. A **quadratic equation** is an equation that can be written in the form
   A. $Ax + By = C$
   B. $ax^2 + bx + c = 0$
   C. $Ax + B = 0$
   D. $y = mx + b$.

2. A **complex number** is
   A. a real number that includes a complex fraction
   B. a nonzero multiple of i
   C. a number of the form $a + bi$, where a and b are real numbers
   D. the square root of -1.

3. A **pure imaginary number** is
   A. a complex number $a + bi$, where $a = 0, b \neq 0$
   B. a number that does not exist
   C. a complex number $a + bi$, where $b = 0$
   D. any real number.

4. A **parabola** is the graph of
   A. any equation in two variables
   B. a linear equation
   C. an equation of degree three
   D. a quadratic equation in two variables.

5. The **vertex** of a parabola is
   A. the point where the graph intersects the y-axis

   B. the point where the graph intersects the x-axis
   C. the lowest point on a parabola that opens up or the highest point on a parabola that opens down
   D. the origin.

6. The **axis** of a vertical parabola is
   A. either the x-axis or the y-axis
   B. the vertical line through the vertex
   C. the horizontal line through the vertex
   D. the x-axis.

QUICK REVIEW

Concepts	Examples

9.1 SOLVING QUADRATIC EQUATIONS BY THE SQUARE ROOT PROPERTY

Square Root Property
If k is positive and if $x^2 = k$, then
$$x = \sqrt{k} \quad \text{or} \quad x = -\sqrt{k}.$$
The solution set, $\{-\sqrt{k}, \sqrt{k}\}$, can be written $\{\pm\sqrt{k}\}$.

Solve $(2x + 1)^2 = 5$.

$2x + 1 = \sqrt{5}$	or	$2x + 1 = -\sqrt{5}$
$2x = -1 + \sqrt{5}$	or	$2x = -1 - \sqrt{5}$
$x = \dfrac{-1 + \sqrt{5}}{2}$	or	$x = \dfrac{-1 - \sqrt{5}}{2}$

Solution set: $\left\{ \dfrac{-1 \pm \sqrt{5}}{2} \right\}$

(continued)

Concepts	Examples

9.2 SOLVING QUADRATIC EQUATIONS BY COMPLETING THE SQUARE

Solving a Quadratic Equation by Completing the Square

Step 1 If the coefficient of the second-degree term is 1, go to Step 2. If it is not 1, divide each side of the equation by this coefficient.

Step 2 Make sure that all variable terms are on one side of the equation and all constant terms are on the other.

Step 3 Take half the coefficient of x, square it, and add the square to each side of the equation. Factor the variable side and combine terms on the other side.

Step 4 Use the square root property to solve the equation.

Solve $2x^2 + 4x - 1 = 0$.

$$x^2 + 2x - \frac{1}{2} = 0 \qquad \text{Divide by 2.}$$

$$x^2 + 2x = \frac{1}{2} \qquad \text{Add } \tfrac{1}{2}.$$

$$x^2 + 2x + 1 = \frac{1}{2} + 1 \qquad \text{Add } \left[\tfrac{1}{2}(2)\right]^2 = 1.$$

$$(x + 1)^2 = \frac{3}{2} \qquad \text{Factor; add.}$$

$$x + 1 = \sqrt{\frac{3}{2}} = \frac{\sqrt{6}}{2} \quad \text{or} \quad x + 1 = -\sqrt{\frac{3}{2}} = -\frac{\sqrt{6}}{2}$$

$$x = -1 + \frac{\sqrt{6}}{2} \quad \text{or} \quad x = -1 - \frac{\sqrt{6}}{2}$$

$$x = \frac{-2 + \sqrt{6}}{2} \quad \text{or} \quad x = \frac{-2 - \sqrt{6}}{2}$$

Solution set: $\left\{ \dfrac{-2 \pm \sqrt{6}}{2} \right\}$

9.3 SOLVING QUADRATIC EQUATIONS BY THE QUADRATIC FORMULA

Quadratic Formula
The solutions of $ax^2 + bx + c = 0$, $a \neq 0$, are

$$x = \frac{-b \pm \sqrt{b^2 - 4ac}}{2a}.$$

The discriminant of the quadratic equation is $b^2 - 4ac$.

Solve $3x^2 - 4x - 2 = 0$.

$$x = \frac{-(-4) \pm \sqrt{(-4)^2 - 4(3)(-2)}}{2(3)} \qquad a = 3, b = -4, c = -2$$

$$x = \frac{4 \pm \sqrt{16 + 24}}{6}$$

$$x = \frac{4 \pm \sqrt{40}}{6} = \frac{4 \pm 2\sqrt{10}}{6} \qquad \text{Discriminant: 40}$$

$$x = \frac{2(2 \pm \sqrt{10})}{2(3)} = \frac{2 \pm \sqrt{10}}{3} \qquad \begin{array}{l}\text{Factor; then}\\ \text{divide out 2.}\end{array}$$

Solution set: $\left\{ \dfrac{2 \pm \sqrt{10}}{3} \right\}$

9.4 COMPLEX NUMBERS

The imaginary unit is i, where

$$i = \sqrt{-1} \quad \text{and} \quad i^2 = -1.$$

For the positive number b, $\sqrt{-b} = i\sqrt{b}$.

$$\sqrt{-19} = i\sqrt{19}$$

(continued)

Concepts	Examples
Addition Add complex numbers by adding the real parts and adding the imaginary parts.	Add: $(3 + 6i) + (-9 + 2i) = (3 - 9) + (6 + 2)i$ $= -6 + 8i$
Subtraction To subtract complex numbers, change the number following the subtraction sign to its negative and add.	Subtract: $(5 + 4i) - (2 - 4i) = (5 + 4i) + (-2 + 4i)$ $= (5 - 2) + (4 + 4)i$ $= 3 + 8i$
Multiplication Multiply complex numbers in the same way polynomials are multiplied. Replace i^2 with -1.	Multiply: $(7 + i)(3 - 4i)$ $= 7(3) + 7(-4i) + i(3) + i(-4i)$ FOIL method $= 21 - 28i + 3i - 4i^2$ $= 21 - 25i - 4(-1)$ $i^2 = -1$ $= 21 - 25i + 4$ $= 25 - 25i$
Division Divide complex numbers by multiplying the numerator and the denominator by the conjugate of the denominator.	Divide: $\dfrac{2}{6 + i} = \dfrac{2}{6 + i} \cdot \dfrac{6 - i}{6 - i}$ $= \dfrac{2(6 - i)}{36 - i^2}$ $= \dfrac{12 - 2i}{36 - (-1)}$ $i^2 = -1$ $= \dfrac{12 - 2i}{37}$ $= \dfrac{12}{37} - \dfrac{2}{37}i$ Standard form
Complex Solutions A quadratic equation may have nonreal complex solutions. This occurs when the discriminant is negative. The quadratic formula will give complex solutions in such cases.	Solve for all complex solutions of $x^2 + x + 1 = 0$. $x = \dfrac{-1 \pm \sqrt{1^2 - 4(1)(1)}}{2(1)}$ $a = 1, b = 1, c = 1$ $x = \dfrac{-1 \pm \sqrt{1 - 4}}{2}$ $x = \dfrac{-1 \pm \sqrt{-3}}{2}$ $x = \dfrac{-1 \pm i\sqrt{3}}{2}$ $\sqrt{-b} = i\sqrt{b}$ Solution set: $\left\{ -\dfrac{1}{2} \pm \dfrac{\sqrt{3}}{2}i \right\}$

(continued)

Concepts	Examples

9.5 **MORE ON GRAPHING QUADRATIC EQUATIONS; QUADRATIC FUNCTIONS**

To graph $y = ax^2 + bx + c,$

Step 1 Find the vertex: $x = -\frac{b}{2a}$; find y by substituting this value for x in the equation.

Graph $y = 2x^2 - 5x - 3$.

$$x = -\frac{b}{2a} = -\frac{-5}{2(2)} = \frac{5}{4}$$

$$y = 2\left(\frac{5}{4}\right)^2 - 5\left(\frac{5}{4}\right) - 3$$

$$= 2\left(\frac{25}{16}\right) - \frac{25}{4} - 3$$

$$= \frac{25}{8} - \frac{50}{8} - \frac{24}{8} = -\frac{49}{8}$$

The vertex is $\left(\frac{5}{4}, -\frac{49}{8}\right)$.

Step 2 Find the y-intercept.

$$y = 2(0)^2 - 5(0) - 3 = -3$$

The y-intercept is $(0, -3)$.

Step 3 Find the x-intercepts (if they exist).

$$0 = 2x^2 - 5x - 3$$
$$0 = (2x + 1)(x - 3)$$

$$2x + 1 = 0 \qquad \text{or} \qquad x - 3 = 0$$
$$2x = -1 \qquad \text{or} \qquad x = 3$$
$$x = -\frac{1}{2} \qquad \text{or} \qquad x = 3$$

The x-intercepts are $\left(-\frac{1}{2}, 0\right)$ and $(3, 0)$.

Step 4 Plot the intercepts and the vertex.

Step 5 Find and plot additional ordered pairs near the vertex and intercepts as needed.

x	y
$-\frac{1}{2}$	0
0	-3
$\frac{5}{4}$	$-\frac{49}{8}$
2	-5
3	0

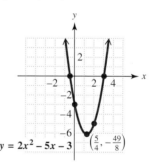

$y = 2x^2 - 5x - 3$ $\left(\frac{5}{4}, -\frac{49}{8}\right)$

Answers to Test Your Word Power

1. B; *Examples:* $z^2 + 6z + 9 = 0$, $y^2 - 2y = 8$, $(x + 3)(x - 1) = 5$

2. C; *Examples:* -5 (or $-5 + 0i$), $7i$ (or $0 + 7i$), $\sqrt{2} - 4i$

3. A; *Examples:* $2i, -13i, i\sqrt{6}$

4. D; *Examples:* See Figures 2–9 in Section 9.5.

5. C; *Example:* The graph of $y = (x + 3)^2$ has vertex $(-3, 0)$, which is the lowest point on the graph.

6. B; *Example:* The axis of the graph of $y = (x + 3)^2$ is the line $x = -3$.

Chapter 9 REVIEW EXERCISES

[9.1] *Solve each equation by using the square root property. Give only real number solutions. Express all radicals in simplest form.*

1. $z^2 = 144$ **2.** $x^2 = 37$ **3.** $m^2 = 128$ **4.** $(k + 2)^2 = 25$

5. $(r - 3)^2 = 10$ **6.** $(2p + 1)^2 = 14$ **7.** $(3k + 2)^2 = -3$ **8.** $(5x + 3)^2 = 0$

[9.2] *Solve each equation by completing the square. Give only real number solutions.*

9. $m^2 + 6m + 5 = 0$ **10.** $p^2 + 4p = 7$

11. $-x^2 + 5 = 2x$ **12.** $2z^2 - 3 = -8z$

13. $5k^2 - 3k - 2 = 0$ **14.** $(4x + 1)(x - 1) = -7$

Solve each problem.

15. If an object is projected upward on Earth from a height of 50 ft, with an initial velocity of 32 ft per sec, then its altitude (height) after t seconds is given by $h = -16t^2 + 32t + 50$, where h is in feet. At what times will the object be at a height of 30 ft?

16. Find the lengths of the three sides of the right triangle shown.

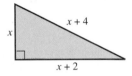

17. *Concept Check* What must be added to $x^2 + 3x$ to make it a perfect square?

[9.3]

18. Consider the equation $x^2 - 9 = 0$.

    (a) Solve the equation by factoring.

    (b) Solve the equation by the square root property.

    (c) Solve the equation by the quadratic formula.

    (d) Compare your answers. If a quadratic equation can be solved by both factoring and the quadratic formula, should you always get the same results? Explain.

Solve each equation by using the quadratic formula. Give only real number solutions.

19. $x^2 - 2x - 4 = 0$ **20.** $3k^2 + 2k = -3$ **21.** $2p^2 + 8 = 4p + 11$

22. $-4x^2 + 7 = 2x$ **23.** $\frac{1}{4}p^2 = 2 - \frac{3}{4}p$ **24.** $3x^2 - x - 2 = 0$

25. *Concept Check* Why is this *not* the statement of the quadratic formula for $ax^2 + bx + c = 0$?

$$x = -b \pm \frac{\sqrt{b^2 - 4ac}}{2a}$$

[9.4] *Perform each indicated operation.*

26. $(3 + 5i) + (2 - 6i)$ **27.** $(-2 - 8i) - (4 - 3i)$ **28.** $(6 - 2i)(3 + i)$

29. $(2 + 3i)(2 - 3i)$ **30.** $\frac{1 + i}{1 - i}$ **31.** $\frac{5 + 6i}{2 + 3i}$

32. ***Concept Check*** What is the conjugate of the real number a?

📝 **33.** Is it possible to multiply a complex number by its conjugate and get a product that is not a real number? Explain.

Find the complex solutions of each quadratic equation.

34. $(m + 2)^2 = -3$ **35.** $(3p - 2)^2 = -8$ **36.** $3k^2 = 2k - 1$

37. $h^2 + 3h = -8$ **38.** $4q^2 + 2 = 3q$ **39.** $9z^2 + 2z + 1 = 0$

[9.5] *Identify the vertex and sketch the graph of each equation.*

40. $y = -3x^2$ **41.** $y = -x^2 + 5$ **42.** $y = (x + 4)^2$

43. $y = x^2 - 2x + 1$ **44.** $y = -x^2 + 2x + 3$ **45.** $y = x^2 + 4x + 2$

Decide from the graph how many real number solutions the equation $f(x) = 0$ has. Determine the solution set (of real solutions) for $f(x) = 0$ from the graph.

46.

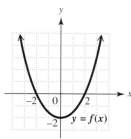

47.

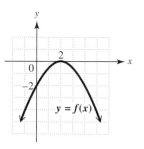

48.

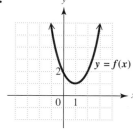

49. Give the domain and range of the graphs in Exercises 46–48.

50. Suppose that a telescope has a diameter of 200 ft and a maximum depth of 30 ft. Find the equation for a cross section of the parabolic dish. (*Hint:* Refer to Example 4 and Exercise 35 in **Section 9.5.**)

MIXED REVIEW EXERCISES

Solve by any method. Give only real number solutions.

51. $(2t - 1)(t + 1) = 54$ **52.** $(2p + 1)^2 = 100$ **53.** $(k + 2)(k - 1) = 3$

54. $6t^2 + 7t - 3 = 0$ **55.** $2x^2 + 3x + 2 = x^2 - 2x$ **56.** $x^2 + 2x + 5 = 7$

57. $m^2 - 4m + 10 = 0$ **58.** $k^2 - 9k + 10 = 0$ **59.** $(3x + 5)^2 = 0$

60. $\dfrac{1}{2}r^2 = \dfrac{7}{2} - r$ **61.** $x^2 + 4x = 1$ **62.** $7x^2 - 8 = 5x^2 + 8$

63. The owners of Cole's Baseball Cards have found that the price p, in dollars, of a particular Cal Ripken card depends on the demand d, in hundreds, for the card, according to the function defined by $p = -(d - 6)^2 + 10$. What demand produces a price of \$6 for the card?

64. Find the vertex of the parabola from Exercise 63. Give the corresponding demand and price.

Chapter **9** **TEST**

*Items marked * require knowledge of complex numbers.*

Solve by using the square root property.

1. $x^2 = 39$ **2.** $(z + 3)^2 = 64$ **3.** $(4x + 3)^2 = 24$

Solve by completing the square.

4. $x^2 - 4x = 6$ **5.** $2x^2 + 12x - 3 = 0$

Solve by the quadratic formula.

6. $5x^2 + 2x = 0$ **7.** $2x^2 + 5x - 3 = 0$

8. $3w^2 + 2 = 6w$ ***9.** $4x^2 + 8x + 11 = 0$

10. $t^2 - \dfrac{5}{3}t + \dfrac{1}{3} = 0$

Solve by the method of your choice.

11. $p^2 - 2p - 1 = 0$ **12.** $(2x + 1)^2 = 18$

13. $(x - 5)(2x - 1) = 1$ **14.** $t^2 + 25 = 10t$

Solve each problem.

15. If an object is projected vertically into the air from ground level on Earth with an initial velocity of 64 ft per sec, its altitude (height) s in feet after t seconds is given by the formula

$$s = -16t^2 + 64t.$$

At what time will the object be at a height of 64 ft?

16. Find the lengths of the three sides of the right triangle.

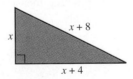

*Perform each indicated operation.

17. $(3 + i) + (-2 + 3i) - (6 - i)$ **18.** $(6 + 5i)(-2 + i)$

19. $(3 - 8i)(3 + 8i)$ **20.** $\dfrac{15 - 5i}{7 + i}$

Identify the vertex and sketch the graph of each equation.

21. $y = x^2 - 6x + 9$ **22.** $y = -x^2 - 2x - 4$ **23.** $f(x) = x^2 + 6x + 7$

24. Refer to the equation in Exercise 23.

    (a) Determine the number of real solutions of $x^2 + 6x + 7 = 0$ by looking at the graph.

(b) Use the quadratic formula to find the exact values of the real solutions. Give the solution set.

(c) Use a calculator to find approximations for the solutions. Round your answers to the nearest thousandth.

25. Find two numbers whose sum is 400 and whose product is a maximum.

Chapters **1–9** **CUMULATIVE REVIEW EXERCISES**

Note: This cumulative review exercise set can be considered a final examination for the course.

Perform each indicated operation.

1. $\dfrac{-4 \cdot 3^2 + 2 \cdot 3}{2 - 4 \cdot 1}$

2. $|-3| - |1 - 6|$

3. $-9 - (-8)(2) + 6 - (6 + 2)$

4. $-4r + 14 + 3r - 7$

5. $13k - 4k + k - 14k + 2k$

6. $5(4m - 2) - (m + 7)$

Solve each equation.

7. $x - 5 = 13$

8. $3k - 9k - 8k + 6 = -64$

9. $\dfrac{3}{5}t - \dfrac{1}{10} = \dfrac{3}{2}$

10. $2(m - 1) - 6(3 - m) = -4$

Solve each problem.

11. Find the measures of the marked angles.

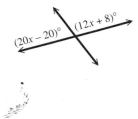

12. The perimeter of a basketball court is 288 ft. The width of the court is 44 ft less than the length. What are the dimensions of the court?

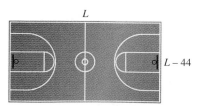

13. Solve the formula $P = 2L + 2W$ for L.

Solve each inequality and graph the solution set.

14. $-8m < 16$

15. $-9p + 2(8 - p) - 6 \geq 4p - 50$

Graph each equation or inequality.

16. $2x + 3y = 6$　　　　　**17.** $y = 3$　　　　　**18.** $2x - 5y < 10$

19. Find the slope of the line through $(-1, 4)$ and $(5, 2)$.

20. Write an equation of a line with slope 2 and y-intercept $(0, 3)$. Give the equation in the form $Ax + By = C$.

Solve each system of equations.

21.　$2x + y = -4$
　　　$-3x + 2y = 13$

22.　$3x - 5y = 8$
　　　$-6x + 10y = 16$

23. In February 2006, you could buy three AT&T Trimline® corded phones and two GE corded speakerphones for $98.95. You could also buy two of the AT&T phones and three of the GE phones for $115.95. Find the price for a single phone of each model. (*Source:* www.circuitcity.com)

24. Graph the solution set of the system of inequalities.

$$2x + y \le 4$$
$$x - y > 2$$

Simplify each expression. Write answers with positive exponents.

25. $(3^2 \cdot x^{-4})^{-1}$　　　　　**26.** $\left(\dfrac{b^{-3}c^4}{b^5c^3}\right)^{-2}$　　　　　**27.** $\left(\dfrac{5}{3}\right)^{-3}$

Perform each indicated operation.

28. $(5x^5 - 9x^4 + 8x^2) - (9x^2 + 8x^4 - 3x^5)$　　**29.** $(2x - 5)(x^3 + 3x^2 - 2x - 4)$

30. $\dfrac{3x^3 + 10x^2 - 7x + 4}{x + 4}$

31. (a) The number of possible hands in contract bridge is about 6,350,000,000. Write this number in scientific notation.

　　(b) The body of a 150-lb person contains about 2.3×10^{-4} lb of copper. Write this number without using exponents.

Factor.

32. $16x^3 - 48x^2y$　　　　　**33.** $2a^2 - 5a - 3$

34. $16x^4 - 1$　　　　　**35.** $25m^2 - 20m + 4$

Solve.

36. $x^2 + 3x - 54 = 0$

37. The length of a rectangle is 2.5 times its width. The area is 1000 m². Find the length.

Simplify each expression as much as possible.

38. $\dfrac{2}{a - 3} \div \dfrac{5}{2a - 6}$　　　　　**39.** $\dfrac{1}{k} - \dfrac{2}{k - 1}$

40. $\dfrac{2}{a^2 - 4} + \dfrac{3}{a^2 - 4a + 4}$　　　　　**41.** $\dfrac{\dfrac{1}{a} + \dfrac{1}{b}}{\dfrac{1}{a} - \dfrac{1}{b}}$

Solve.

42. $\dfrac{1}{x+3} + \dfrac{1}{x} = \dfrac{7}{10}$

43. Jake's boat goes 12 mph in still water. Find the speed of the current of the river if he can go 6 mi upstream in the same time that he can go 10 mi downstream.

Simplify each expression as much as possible.

44. $\sqrt{100}$

45. $\dfrac{6\sqrt{6}}{\sqrt{5}}$

46. $\sqrt[3]{\dfrac{7}{16}}$

47. $3\sqrt{5} - 2\sqrt{20} + \sqrt{125}$

48. $\sqrt[3]{16a^3b^4} - \sqrt[3]{54a^3b^4}$

49. Solve $\sqrt{x+2} = x - 4$.

50. Simplify.

(a) $8^{2/3}$ (b) $-16^{1/4}$

Solve each quadratic equation, using the method indicated. Give only real solutions.

51. $(3x+2)^2 = 12$ (square root property)

52. $-x^2 + 5 = 2x$ (completing the square)

53. $2x(x-2) - 3 = 0$ (quadratic formula)

54. $(4x+1)(x-1) = -3$ (any method)

Solve each problem. (Items marked * require knowledge of complex numbers.)

55. In a right triangle, the lengths of the sides are consecutive integers. Use the Pythagorean formula to find these lengths.

*56. Perform the indicated operations. Give answers in standard form.

(a) $(-9 + 3i) + (4 + 2i) - (-5 - 3i)$ (b) $\dfrac{-17 - i}{-3 + i}$

*57. Find the complex solutions of $2x^2 + 2x = -9$.

58. Graph the quadratic function defined by $f(x) = -x^2 - 2x + 1$, and identify the vertex. Give the domain and range.

Appendix A

An Introduction to Calculators

There is little doubt that the appearance of handheld calculators more than three decades ago and the later development of scientific and graphing calculators have changed the methods of learning and studying mathematics forever. For example, computations with tables of logarithms and slide rules made up an important part of mathematics courses prior to 1970. Today, with the widespread availability of calculators, these topics are studied only for their historical significance.

Calculators come in a large array of different types, sizes, and prices. *For the course for which this textbook is intended, the most appropriate type is the scientific calculator,* which costs $10–$20.

In this introduction, we explain some of the features of scientific and graphing calculators. However, remember that calculators vary among manufacturers and models and that, while the methods explained here apply to many of them, they may not apply to your specific calculator. *This introduction is only a guide and is not intended to take the place of your owner's manual.* Always refer to the manual whenever you need an explanation of how to perform a particular operation.

Scientific Calculators

Scientific calculators are capable of much more than the typical four-function calculator that you might use for balancing your checkbook. Most scientific calculators use *algebraic logic.* (Models sold by Texas Instruments, Sharp, Casio, and Radio Shack, for example, use algebraic logic.) A notable exception is Hewlett-Packard, a company whose calculators use *Reverse Polish Notation* (RPN). In this introduction, we explain the use of calculators with algebraic logic.

Arithmetic Operations   To perform an operation of arithmetic, simply enter the first number, press the operation key $\boxed{+}$, $\boxed{-}$, $\boxed{\times}$, or $\boxed{\div}$, enter the second number, and then press the $\boxed{=}$ key. For example, to add 4 and 3, use the following keystrokes.

Change Sign Key   The key marked $\boxed{+/-}$ allows you to change the sign of a display. This is particularly useful when you wish to enter a negative number. For example, to enter -3, use the following keystrokes.

$$\boxed{3} \quad \boxed{+/-} \quad \boxed{-3}$$

Memory Key   Scientific calculators can hold a number in memory for later use. The label of the memory key varies among models; two of these are $\boxed{M}$ and $\boxed{STO}$. The $\boxed{M+}$ and $\boxed{M-}$ keys allow you to add to or subtract from the value currently in memory. The memory recall key, labeled $\boxed{MR}$, $\boxed{RM}$, or $\boxed{RCL}$, allows you to retrieve the value stored in memory.

Suppose that you wish to store the number 5 in memory. Enter 5, and then press the key for memory. You can then perform other calculations. When you need to retrieve the 5, press the key for memory recall.

If a calculator has a constant memory feature, the value in memory will be retained even after the power is turned off. Some advanced calculators have more than one memory. Read the owner's manual for your model to see exactly how memory is activated.

Clearing/Clear Entry Keys   The key $\boxed{C}$ or $\boxed{CE}$ allows you to clear the display or clear the last entry entered into the display. In some models, pressing the $\boxed{C}$ key once will clear the last entry, while pressing it twice will clear the entire operation in progress.

Second Function Key   This key, usually marked $\boxed{2nd}$, is used in conjunction with another key to activate a function that is printed *above* an operation key (and not on the key itself). For example, suppose you wish to find the square of a number, and the squaring function (explained in more detail later) is printed above another key. You would need to press $\boxed{2nd}$ before the desired squaring function can be activated.

Square Root Key   Pressing $\boxed{\sqrt{}}$ or $\boxed{\sqrt{x}}$ will give the square root (or an approximation of the square root) of the number in the display. On some scientific calculators, the square root key is pressed *before* entering the number, while other calculators use the opposite order. Experiment with your calculator to see which method it uses. For example, to find the square root of 36, use the following keystrokes.

$$\boxed{\sqrt{}} \;\; \boxed{3} \;\; \boxed{6} \;\; \boxed{6} \qquad \text{or} \qquad \boxed{3} \;\; \boxed{6} \;\; \boxed{\sqrt{}} \;\; \boxed{6}$$

The square root of 2 is an example of an irrational number (**Chapter 8**). The calculator will give an approximation of its value, since the decimal for $\sqrt{2}$ never terminates and never repeats. The number of digits shown will vary among models. To find an approximation for $\sqrt{2}$, use the following keystrokes.

$$\boxed{\sqrt{}} \;\; \boxed{2} \;\; \boxed{1.4142136} \qquad \text{or} \qquad \boxed{2} \;\; \boxed{\sqrt{}} \;\; \boxed{1.4142136}$$

An approximation for $\sqrt{2}$

Squaring Key   The $\boxed{x^2}$ key allows you to square the entry in the display. For example, to square 35.7, use the following keystrokes.

$$\boxed{3} \;\; \boxed{5} \;\; \boxed{.} \;\; \boxed{7} \;\; \boxed{x^2} \;\; \boxed{1274.49}$$

The squaring key and the square root key are often found together, with one of them being a second function (that is, activated by the second function key previously described).

Reciprocal Key   The key marked $\boxed{1/x}$ is the reciprocal key. (When two numbers have a product of 1, they are called *reciprocals*. See **Chapter 1.**) Suppose that you wish to find the reciprocal of 5. Use the following keystrokes.

$$\boxed{5} \quad \boxed{1/x} \quad \boxed{\qquad 0.2}$$

Inverse Key   Some calculators have an inverse key, marked $\boxed{\text{INV}}$. Inverse operations are operations that "undo" each other. For example, the operations of squaring and taking the square root are inverse operations. The use of the $\boxed{\text{INV}}$ key varies among different models of calculators, so read your owner's manual carefully.

Exponential Key   The key marked $\boxed{x^y}$ or $\boxed{y^x}$ allows you to raise a number to a power. For example, if you wish to raise 4 to the fifth power (that is, find 4^5, as explained in **Chapter 1**), use the following keystrokes.

$$\boxed{4} \quad \boxed{x^y} \quad \boxed{5} \quad \boxed{=} \quad \boxed{\qquad 1024}$$

Root Key   Some calculators have a key specifically marked $\boxed{\sqrt[x]{x}}$ or $\boxed{\sqrt[x]{y}}$; with others, the operation of taking roots is accomplished by using the inverse key in conjunction with the exponential key. Suppose, for example, your calculator is of the latter type and you wish to find the fifth root of 1024. Use the following keystrokes.

$$\boxed{1} \quad \boxed{0} \quad \boxed{2} \quad \boxed{4} \quad \boxed{\text{INV}} \quad \boxed{x^y} \quad \boxed{5} \quad \boxed{=} \quad \boxed{\qquad 4}$$

Notice how this "undoes" the operation explained in the discussion of the exponential key.

Pi Key   The number π is an important number in mathematics. It occurs, for example, in the area and circumference formulas for a circle. One popular model gives the following display when the $\boxed{\pi}$ key is pressed. (Because π is irrational, the display shows only an approximation.)

$$\boxed{3.1415927} \qquad \text{An approximation for } \pi$$

Methods of Display   When decimal approximations are shown on scientific calculators, they are either *truncated* or *rounded*. To see how a particular model is programmed, evaluate 1/18 as an example. If the display shows 0.0555555 (last digit 5), the calculator truncates the display. If the display shows 0.0555556 (last digit 6), the calculator rounds the display.

When very large or very small numbers are obtained as answers, scientific calculators often express these numbers in scientific notation (**Chapter 5**). For example, if you multiply 6,265,804 by 8,980,591, the display might look like this:

$$\boxed{5.6270623 \ 13}$$

The 13 at the far right means that the number on the left is multiplied by 10^{13}. This means that the decimal point must be moved 13 places to the right if the answer is to be expressed in its usual form. Even then, the value obtained will only be an approximation: 56,270,623,000,000.

Graphing Calculators

While you are not expected to have a graphing calculator to study from this book, we include the following as background information and reference should your course or future courses require the use of graphing calculators.

Basic Features   In addition to possessing the typical keys found on scientific calculators, graphing calculators have keys that can be used to create graphs, make tables, analyze data, and change settings. One of the major differences between graphing and scientific calculators is that a graphing calculator has a larger viewing screen with graphing capabilities. The following screens illustrate the graphs of $Y = X$ and $Y = X^2$.

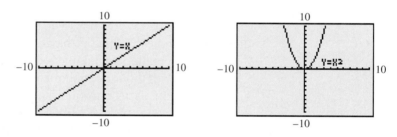

If you look closely at the screens, you will see that the graphs appear to be jagged rather than smooth. The reason for this is that graphing calculators have much lower resolution than computer screens. Because of this, graphs generated by graphing calculators must be interpreted carefully.

Editing Input   The screen of a graphing calculator can display several lines of text at a time. This feature allows you to view both previous and current expressions. If an incorrect expression is entered, an error message is displayed. The erroneous expression can be viewed and corrected by using various editing keys, much like a word-processing program. You do not need to enter the entire expression again. Many graphing calculators can also recall past expressions for editing or updating. The screen on the left shows how two expressions are evaluated. The final line is entered incorrectly, and the resulting error message is shown in the screen on the right.

Order of Operations   Arithmetic operations on graphing calculators are usually entered as they are written in mathematical expressions. For example, to evaluate $\sqrt{36}$ you would first press the square root key and then enter 36. See the left screen below. The order of operations on a graphing calculator is also important, and current models assist the user by inserting parentheses when typical errors might occur. The open parenthesis that follows the square root symbol is automatically entered by the calculator so that an expression such as $\sqrt{2 \times 8}$ will not be calculated incorrectly as $\sqrt{2} \times 8$. Compare the two entries and their results in the screen on the right.

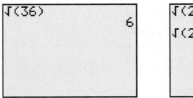
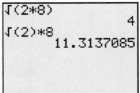

Viewing Windows   The viewing window for a graphing calculator is similar to the viewfinder in a camera. A camera usually cannot take a photograph of an entire view of a scene. The camera must be centered on some object and can capture only a portion of the available scenery. A camera with a zoom lens can photograph different views of the same scene by zooming in and out. Graphing calculators have similar capabilities. The xy-coordinate plane is infinite. The calculator screen can show only a finite, rectangular region in the plane, and it must be specified before the graph can be drawn. This is done by setting both minimum and maximum values for the x- and y-axes. The scale (distance between tick marks) is usually specified as well. Determining an appropriate viewing window for a graph is often a challenge, and many times it will take a few attempts before a satisfactory window is found.

The screen on the left shows a standard viewing window, and the graph of $Y = 2X + 1$ is shown on the right. Using a different window would give a different view of the line.

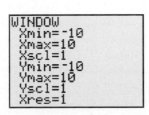
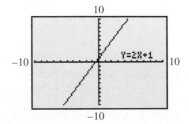

Locating Points on a Graph: Tracing and Tables   Graphing calculators allow you to trace along the graph of an equation and display the coordinates of points on the graph. For example, the screen on the left at the top of the next page indicates that the point $(2, 5)$ lies on the graph of $Y = 2X + 1$. Tables for equations can also be displayed. The screen on the right on the next page shows a partial table for this same equation. Note the middle of the screen, which indicates that when $X = 2$, $Y = 5$.

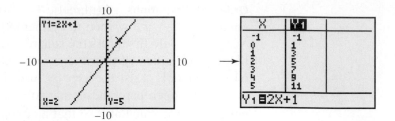

Additional Features   There are many features of graphing calculators that go far beyond the scope of this book. These calculators can be programmed, much like computers. Many of them can solve equations at the stroke of a key, analyze statistical data, and perform symbolic algebraic manipulations. Calculators also provide the opportunity to ask "What if . . . ?" more easily. Values in algebraic expressions can be altered and conjectures tested quickly.

Final Comments   Despite the power of today's calculators, they cannot replace human thought. *In the entire problem-solving process, your brain is the most important component.* Calculators are only tools, and like any tool, they must be used appropriately in order to enhance our ability to understand mathematics. Mathematical insight may often be the quickest and easiest way to solve a problem; a calculator may be neither needed nor appropriate. By applying mathematical concepts, you can make the decision whether to use a calculator.

Review of Decimals and Percents

OBJECTIVE ❶ Add and subtract decimals. A **decimal** is a number written with a decimal point, such as 4.2.

EXAMPLE 1 Adding and Subtracting Decimals

Add or subtract as indicated.

(a) $6.92 + 14.8 + 3.217$

Place the numbers in a column with decimal points lined up and add. If you like, attach 0's to make all the numbers the same length; this is a good way to avoid errors.

$$
\begin{array}{r}
6.92 \\
14.8 \\
+\ 3.217 \\
\hline
24.937
\end{array}
\quad \text{or} \quad
\begin{array}{r}
6.920 \\
14.800 \\
+\ 3.217 \\
\hline
24.937
\end{array}
\quad \text{Attach zeros.}
$$

Be sure to line up decimal points.

(b) $47.6 - 32.509$

Write the numbers in a column, attaching 0's to 47.6.

$$
\begin{array}{r}
47.6 \\
-32.509
\end{array}
\quad \text{becomes} \quad
\begin{array}{r}
47.600 \\
-32.509 \\
\hline
15.091
\end{array}
$$

(c) $3 - 0.253$

$$
\begin{array}{r}
3.000 \\
-0.253 \\
\hline
2.747
\end{array}
\quad \text{Attach 0's to 3 as needed.}
$$

✔ **Now Try Exercises 1 and 5.**

OBJECTIVE ❷ Multiply and divide decimals.

EXAMPLE 2 Multiplying Decimals

Multiply.

(a) 29.3×4.52

Multiply as if the numbers were whole numbers. To find the number of decimal places in the answer, add the numbers of decimal places in the factors.

$$
\begin{array}{r}
29.3 \\
\times\ 4.52 \\
\hline
5\ 86 \\
14\ 6\ 5 \\
117\ 2 \\
\hline
132.4\ 36
\end{array}
$$

29.3	1 decimal place in first factor
× 4.52	2 decimal places in second factor
	1 + 2 = 3
132.4 36	3 decimal places in answer

(b) 7.003×55.8

$$
\begin{array}{r}
7.003 \\
\times\ 55.8 \\
\hline
5\ 602\ 4 \\
35\ 015 \\
350\ 15 \\
\hline
390.767\ 4
\end{array}
$$

7.003	3 decimal places
× 55.8	1 decimal place
	3 + 1 = 4
390.767 4	4 decimal places in answer

✔ **Now Try Exercises 11 and 15.**

EXAMPLE 3 **Dividing Decimals**

Divide: $279.45 \div 24.3$.

    Move the decimal point in 24.3 one place to the right, to get the whole number 243. Move the decimal point the same number of places in 279.45.

$$243.\overline{)2794.5} \qquad \text{Move one decimal place to the right.}$$

Bring the decimal point straight up and divide as with whole numbers.

$$
\begin{array}{r}
11.5 \\
243.\overline{)2794.5} \\
243 \\
\hline
364 \\
243 \\
\hline
121\ 5 \\
121\ 5 \\
\hline
0
\end{array}
\qquad \text{Move the decimal point straight up.}
$$

✔ **Now Try Exercise 19.**

OBJECTIVE 3 **Convert percents to decimals and decimals to percents.** The word **percent** means "per one hundred." Percent is written with the sign %. One percent means "one per one hundred."

$$1\% = 0.01 \qquad \text{or} \qquad 1\% = \frac{1}{100}$$

EXAMPLE 4 **Converting between Decimals and Percents**

(a) Write 75% as a decimal.

$$75\% = 75 \cdot 1\% = 75 \cdot 0.01 = 0.75$$

The fraction form $1\% = \frac{1}{100}$ can also be used to convert 75% to a decimal.

$$75\% = 75 \cdot 1\% = 75 \cdot \frac{1}{100} = \frac{75}{100} = 0.75$$

(b) Write 2.63 as a percent.

$$2.63 = 263 \cdot 0.01 = 263 \cdot 1\% = 263\%$$

> ✔ **Now Try Exercises 23 and 33.**

OBJECTIVE 4 Find percentages by multiplying. A part of a whole is called a **percentage.** For example, since 50% represents $\frac{50}{100} = \frac{1}{2}$ of a whole, 50% of 800 is half of 800, or 400. Multiply to find percentages, as in the next example.

EXAMPLE 5 Finding Percentages

Find each percentage.

(a) 15% of 600

The word *of* indicates multiplication here.

$$15\% \cdot 600 = 0.15 \cdot 600 = 90$$

(b) 125% of 80

$$125\% \cdot 80 = 1.25 \cdot 80 = 100$$

(c) What percent of 52 is 7.8?

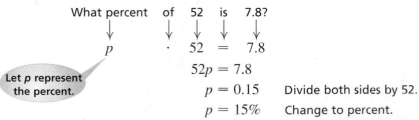

$$52p = 7.8$$
$$p = 0.15 \qquad \text{Divide both sides by 52.}$$
$$p = 15\% \qquad \text{Change to percent.}$$

> ✔ **Now Try Exercises 41 and 53.**

EXAMPLE 6 Using Percent in a Consumer Problem

A DVD movie with a regular price of $18 is on sale at 22% off. Find the amount of the discount.

The discount is 22% of $18. Use the fact that *of* indicates multiplication.

$$22\% \cdot 18 = 0.22 \cdot 18 = 3.96, \quad \text{so the discount is } \$3.96.$$

> ✔ **Now Try Exercise 57.**

APPENDIX B EXERCISES

Now Try Exercise

Perform each indicated operation. See Examples 1–3.

1. $14.23 + 9.81 + 74.63 + 18.715$ **2.** $89.416 + 21.32 + 478.91 + 298.213$

3. $19.74 - 6.53$ **4.** $27.96 - 8.39$ **5.** $219 - 68.51$ **6.** $283 - 12.42$

7. 48.96
    37.421
    + 9.72

8. 9.71
    4.8
    3.6
    5.2
    +8.17

9. 8.6
    −3.751

10. 27.8
    −13.582

11. 39.6×4.2 **12.** 18.7×2.3 **13.** 42.1×3.9 **14.** 19.63×4.08

15. 0.042×32 **16.** 571×2.9 **17.** $24.84 \div 6$ **18.** $32.84 \div 4$

19. $7.6266 \div 3.42$ **20.** $14.9202 \div 2.43$ **21.** $2496 \div 0.52$ **22.** $0.56984 \div 0.034$

Convert each percent to a decimal. See Example 4(a).

23. 53%   **24.** 38%   **25.** 129%   **26.** 174%

27. 96%   **28.** 11%   **29.** 0.9%   **30.** 0.1%

Convert each decimal to a percent. See Example 4(b).

31. 0.80   **32.** 0.75   **33.** 0.007   **34.** 1.4

35. 0.67   **36.** 0.003   **37.** 0.125   **38.** 0.983

Respond to each statement or question. Round your answer to the nearest hundredth if appropriate. See Example 5.

39. What is 14% of 780?

40. Find 12% of 350.

41. Find 22% of 1086.

42. What is 20% of 1500?

43. 4 is what percent of 80?

44. 1300 is what percent of 2000?

45. What percent of 5820 is 6402?

46. What percent of 75 is 90?

47. 121 is what percent of 484?

48. What percent of 3200 is 64?

49. Find 118% of 125.8.

50. Find 3% of 128.

51. What is 91.72% of 8546.95?

52. Find 12.741% of 58.902.

53. What percent of 198.72 is 14.68?

54. 586.3 is what percent of 765.4?

Solve each problem. See Example 6.

55. A retailer has $23,000 invested in her business. She finds that she is earning 12% per year on this investment. How much money is she earning per year?

56. Jeff Cole recently bought a duplex for $644,000. He expects to earn 8% per year on the purchase price. How many dollars per year will he earn?

57. For a recent tour of the eastern United States, a travel agent figured that the trip totaled 2300 mi, with 35% of the trip by air. How many miles of the trip were by air?

58. Parish National Bank pays 3.2% interest per year. What is the annual interest on an account of $3000?

59. An ad for steel-belted radial tires promises 15% better mileage than with other tires. Ken Grace's car now goes 420 mi on a tank of gas. If he switched to the new tires, how many extra miles could he drive on a tank of gas?

60. A home worth $250,000 is located in an area where home prices are increasing at a rate of 6% per year. By how much would the value of this home increase in 1 year?

61. A family of four with a monthly income of $2000 spends 90% of its earnings and saves the rest. Find the *annual* savings of this family.

Appendix C

Sets

OBJECTIVE 1 List the elements of a set. A **set** is a collection of things. The objects in a set are called the **elements** of the set. A set is represented by listing its elements between **braces,** { }.* The order in which the elements of a set are listed is unimportant.

EXAMPLE 1 Listing the Elements of Sets

Represent each set by listing its elements.

(a) The set of states in the United States that border the Pacific Ocean is

{California, Oregon, Washington, Hawaii, Alaska}.

(b) The set of all counting numbers less than $6 = \{1, 2, 3, 4, 5\}$.

✔ **Now Try Exercises 1 and 3.**

OBJECTIVE 2 Learn the vocabulary and symbols used to discuss sets. Capital letters are used to name sets. To state that 5 is an element of

$$S = \{1, 2, 3, 4, 5\},$$

write $5 \in S$. The statement $6 \notin S$ means that 6 is not an element of S.

A set with no elements is called the **empty set** or the **null set.** The symbol $\emptyset$ or { } is used for the empty set. If we let A be the set of all negative natural numbers, then A is the empty set.

$$A = \emptyset \quad \text{or} \quad A = \{ \ \}$$

▶ **CAUTION** Do not make the common error of writing the empty set as $\{\emptyset\}$.

✔ **Now Try Exercise 5.**

In any discussion of sets, there is some set that includes all the elements under consideration. This set is called the **universal set** for that situation. For example, if the discussion is about presidents of the United States, then the set of all presidents of the United States is the universal set. The universal set is denoted U.

OBJECTIVE 3 Decide whether a set is finite or infinite. In Example 1, there are five elements in the set in part (a) and five in part (b). If the number of elements in a set is either 0 or a counting number, then the set is **finite.** By contrast, the set of natural

---

*Some people refer to this convention as *roster notation*.

numbers is an **infinite** set, because there is no final natural number. We can list the elements of the set of natural numbers as

$$N = \{1, 2, 3, 4, \ldots\},$$

where the three dots indicate that the set continues indefinitely. Not all infinite sets can be listed in this way. For example, there is no way to list the elements in the set of all real numbers between 1 and 2.

EXAMPLE 2 Distinguishing between Finite and Infinite Sets

List the elements of each set if possible. Decide whether each set is finite or infinite.

(a) The set of all integers
One way to list the elements is $\{\ldots, -2, -1, 0, 1, 2, \ldots\}$. The set is infinite.

(b) The set of all natural numbers between 0 and 5
$\{1, 2, 3, 4\}$ The set is finite.

(c) The set of all irrational numbers
This is an infinite set whose elements cannot be listed.

✔ **Now Try Exercise 11.**

Two sets are equal if they have exactly the same elements. Thus, the set of natural numbers and the set of positive integers are equal sets. Also, the sets

$$\{1, 2, 4, 7\} \quad \text{and} \quad \{4, 2, 7, 1\}$$

are equal. The order of the elements does not make a difference.

OBJECTIVE 4 Decide whether a given set is a subset of another set.   If all elements of a set A are also elements of another set B, then we say that A is a **subset** of B, written $A \subseteq B$. We use the symbol $A \nsubseteq B$ to mean that A is not a subset of B.

EXAMPLE 3 Using Subset Notation

Let $A = \{1, 2, 3, 4\}$, $B = \{1, 4\}$, and $C = \{1\}$. Then

$$B \subseteq A, \quad C \subseteq A, \quad \text{and} \quad C \subseteq B,$$

but

$$A \nsubseteq B, \quad A \nsubseteq C, \quad \text{and} \quad B \nsubseteq C.$$

✔ **Now Try Exercises 21 and 25.**

The empty set is defined to be a subset of any set. Thus, the set $M = \{a, b\}$ has four subsets: $\{a, b\}, \{a\}, \{b\}$, and $\emptyset$. How many subsets does $N = \{a, b, c\}$ have? There is one subset with three elements: $\{a, b, c\}$. There are three subsets with two elements:

$$\{a, b\}, \quad \{a, c\}, \quad \text{and} \quad \{b, c\}.$$

There are three subsets with one element:

$$\{a\}, \quad \{b\}, \quad \text{and} \quad \{c\}.$$

There is one subset with no elements: $\emptyset$. Thus, set N has eight subsets.

The following generalization can be made.

Number of Subsets of a Set

A set with n elements has 2^n subsets.

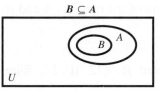

$B \subseteq A$

FIGURE 1

To illustrate the relationships between sets, **Venn diagrams** are often used. A rectangle represents the universal set, U. The sets under discussion are represented by regions within the rectangle. The Venn diagram in Figure 1 shows that $B \subseteq A$.

OBJECTIVE 5 **Find the complement of a set.**   For every set A, there is a set A', the **complement** of A, that contains all the elements of U that are not in A. The shaded region in the Venn diagram in Figure 2 represents A'.

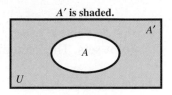

A' is shaded.

FIGURE 2

EXAMPLE 4 Determining Complements of a Set

Given $U = \{a, b, c, d, e, f, g\}$, $A = \{a, b, c\}$, $B = \{a, d, f, g\}$, and $C = \{d, e\}$, it follows that

$$A' = \{d, e, f, g\}, \quad B' = \{b, c, e\}, \quad \text{and} \quad C' = \{a, b, c, f, g\}.$$

✔ **Now Try Exercises 45 and 47.**

OBJECTIVE 6 **Find the union and the intersection of two sets.**   The **union** of two sets A and B, written $A \cup B$, is the set of all elements of A together with all elements of B. Thus, for the sets in Example 4,

$$A \cup B = \{a, b, c, d, f, g\} \quad \text{and} \quad A \cup C = \{a, b, c, d, e\}.$$

In Figure 3, the shaded region is the union of sets A and B.

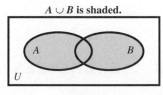

$A \cup B$ is shaded.

FIGURE 3

EXAMPLE 5 Finding the Union of Two Sets

If $M = \{2, 5, 7\}$ and $N = \{1, 2, 3, 4, 5\}$, then

$$M \cup N = \{1, 2, 3, 4, 5, 7\}.$$

✔ **Now Try Exercise 55.**

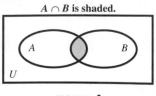

$A \cap B$ is shaded.

FIGURE 4

The **intersection** of two sets A and B, written $A \cap B$, is the set of all elements that belong to both A and B. For example, if

$$A = \{\text{José, Ellen, Marge, Kevin}\}$$

and

$$B = \{\text{José, Patrick, Ellen, Sue}\},$$

then

$$A \cap B = \{\text{José, Ellen}\}.$$

The shaded region in Figure 4 represents the intersection of the two sets A and B.

EXAMPLE 6 Finding the Intersection of Two Sets

Suppose that $P = \{3, 9, 27\}$, $Q = \{2, 3, 10, 18, 27, 28\}$, and $R = \{2, 10, 28\}$. Find
(a) $P \cap Q$, **(b)** $Q \cap R$, and **(c)** $P \cap R$.

(a) $P \cap Q = \{3, 27\}$ **(b)** $Q \cap R = \{2, 10, 28\} = R$ **(c)** $P \cap R = \emptyset$

✔ **Now Try Exercises 49 and 51.**

Sets like P and R in Example 6 that have no elements in common are called **disjoint sets.** The Venn diagram in Figure 5 shows a pair of disjoint sets.

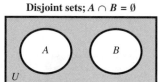

Disjoint sets; $A \cap B = \emptyset$

FIGURE 5

EXAMPLE 7 Using Set Operations

Let $U = \{2, 5, 7, 10, 14, 20\}$, $A = \{2, 10, 14, 20\}$, $B = \{5, 7\}$, and $C = \{2, 5, 7\}$.
Find **(a)** $A \cup B$, **(b)** $A \cap B$, **(c)** $B \cup C$, **(d)** $B \cap C$, and **(e)** A'.

(a) $A \cup B = \{2, 5, 7, 10, 14, 20\} = U$ **(b)** $A \cap B = \emptyset$

(c) $B \cup C = \{2, 5, 7\} = C$ **(d)** $B \cap C = \{5, 7\} = B$

(e) $A' = \{5, 7\} = B$

✔ **Now Try Exercises 53 and 57.**

APPENDIX C EXERCISES

Now Try Exercise

List the elements of each set. See Examples 1 and 2.

1. The set of all natural numbers less than 8

2. The set of all integers between 4 and 10

3. The set of seasons

4. The set of months of the year

5. The set of women presidents of the United States before 2006

6. The set of all living humans who are more than 200 years old

7. The set of letters of the alphabet between K and M

8. The set of letters of the alphabet between D and H

9. The set of positive even integers

10. The set of all multiples of 5

11. Which of the sets described in Exercises 1–10 are infinite sets?

12. Which of the sets described in Exercises 1–10 are finite sets?

Concept Check   *Tell whether each statement is* true *or* false.

13. $5 \in \{1, 2, 5, 8\}$ 14. $6 \in \{1, 2, 3, 4, 5\}$

15. $2 \in \{1, 3, 5, 7, 9\}$ 16. $1 \in \{6, 2, 5, 1\}$

17. $7 \notin \{2, 4, 6, 8\}$ 18. $7 \notin \{1, 3, 5, 7\}$

19. $\{2, 4, 9, 12, 13\} = \{13, 12, 9, 4, 2\}$ 20. $\{7, 11, 4\} = \{7, 11, 4, 0\}$

Let

$$A = \{1, 3, 4, 5, 7, 8\}, \quad B = \{2, 4, 6, 8\},$$
$$C = \{1, 3, 5, 7\}, \quad D = \{1, 2, 3\},$$
$$E = \{3, 7\}, \quad \text{and} \quad U = \{1, 2, 3, 4, 5, 6, 7, 8, 9, 10\}.$$

Tell whether each statement is true *or* false. *See Examples 3, 5, 6, and 7.*

21. $A \subseteq U$ 22. $D \subseteq A$ 23. $\emptyset \subseteq A$ 24. $\{1, 2\} \subseteq D$ 25. $C \subseteq A$

26. $A \subseteq C$ 27. $D \subseteq B$ 28. $E \subseteq C$ 29. $D \nsubseteq E$ 30. $E \nsubseteq A$

31. There are exactly 4 subsets of E.   32. There are exactly 8 subsets of D.

33. There are exactly 12 subsets of C.   34. There are exactly 16 subsets of B.

35. $\{4, 6, 8, 12\} \cap \{6, 8, 14, 17\} = \{6, 8\}$ 36. $\{2, 5, 9\} \cap \{1, 2, 3, 4, 5\} = \{2, 5\}$

37. $\{3, 1, 0\} \cap \{0, 2, 4\} = \{0\}$ 38. $\{4, 2, 1\} \cap \{1, 2, 3, 4\} = \{1, 2, 3\}$

39. $\{3, 9, 12\} \cap \emptyset = \{3, 9, 12\}$ 40. $\{3, 9, 12\} \cup \emptyset = \emptyset$

41. $\{3, 5, 7, 9\} \cup \{4, 6, 8\} = \emptyset$ 42. $\{1, 2, 3\} \cup \{1, 2, 3\} = \{1, 2, 3\}$

43. $\{4, 9, 11, 7, 3\} \cup \{1, 2, 3, 4, 5\} = \{1, 2, 3, 4, 5, 7, 9, 11\}$

44. $\{5, 10, 15, 20\} \cup \{5, 15, 30\} = \{5, 15\}$

Let

$$U = \{a, b, c, d, e, f, g, h\}, \quad A = \{a, b, c, d, e, f\},$$
$$B = \{a, c, e\}, \quad C = \{a, f\},$$
and
$$D = \{d\}.$$

List the elements in each set. See Examples 4–7.

45. A' 46. B' 47. C' 48. D'

49. $A \cap B$ 50. $B \cap A$ 51. $A \cap D$ 52. $B \cap D$

53. $B \cap C$ 54. $A \cup B$ 55. $B \cup D$ 56. $B \cup C$

57. $C \cup B$ 58. $C \cup D$ 59. $A \cap \emptyset$ 60. $B \cup \emptyset$

61. Name every pair of disjoint sets among sets A–D in the instruction line for Exercises 45–60.

Appendix D

Mean, Median, and Mode

OBJECTIVES

1 Find the mean of a list of numbers.

2 Find a weighted mean.

3 Find the median.

4 Find the mode.

OBJECTIVE 1 Find the mean of a list of numbers. Making sense of a long list of numbers can be difficult. So when we analyze data, one of the first things to look for is a *measure of central tendency*—a single number that we can use to represent the entire list of numbers. One such measure is the *average*, or **mean.** The mean can be found with the following formula.

Finding the Mean (Average)

$$\text{mean} = \frac{\text{sum of all values}}{\text{number of values}}$$

EXAMPLE 1 Finding the Mean (Average)

David had test scores of 84, 90, 95, 98, and 88. Find his mean (average) score.

Use the formula for finding the mean. Add up all the test scores and then divide the sum by the number of tests.

$$\text{mean} = \frac{84 + 90 + 95 + 98 + 88}{5} \quad \begin{matrix} \leftarrow \text{Sum of test scores} \\ \leftarrow \text{Number of tests} \end{matrix}$$

$$\text{mean} = \frac{455}{5}$$

$$\text{mean} = 91 \quad \text{Divide.}$$

David has a mean (average) score of 91.

✔ **Now Try Exercise 1.**

EXAMPLE 2 Applying the Mean (Average)

The sales of photo albums at Sarah's Card Shop for each day last week were $86, $103, $118, $117, $126, $158, and $149. Find the mean daily sales of photo albums.

To find the mean, add all the daily sales amounts and then divide the sum by the number of days (7).

$$\text{mean} = \frac{\$86 + \$103 + \$118 + \$117 + \$126 + \$158 + \$149}{7} \quad \begin{matrix} \leftarrow \text{Sum of sales} \\ \leftarrow \text{Number of days} \end{matrix}$$

$$\text{mean} = \frac{\$857}{7}$$

$$\text{mean} \approx \$122.43 \quad \text{Nearest cent}$$

✔ **Now Try Exercise 7.**

OBJECTIVE 2 Find a weighted mean. Some items in a list of data might appear more than once. In this case, we find a **weighted mean,** in which each value is "weighted" by multiplying it by the number of times it occurs.

EXAMPLE 3 Finding a Weighted Mean

The table shows the amount of contribution and the number of times the amount was given (frequency) to a food pantry. Find the weighted mean.

Contribution Value	Frequency
$ 3	4 ← 4 people each contributed $3.
$ 5	2
$ 7	1
$ 8	5
$ 9	3
$10	2
$12	1
$13	2

In most cases, the same amount was given by more than one person. For example, $3 was given by four people and $8 was given by five people. Other amounts, such as $12, were given by only one person.

To find the mean, multiply each contribution value by its frequency. Then add the products. Next, add the numbers in the *frequency* column to find the total number of values—that is, the total number of people who contributed money.

Value	Frequency	Product
$ 3	4	($3 · 4) = $12
$ 5	2	($5 · 2) = $10
$ 7	1	($7 · 1) = $ 7
$ 8	5	($8 · 5) = $40
$ 9	3	($9 · 3) = $27
$10	2	($10 · 2) = $20
$12	1	($12 · 1) = $12
$13	2	($13 · 2) = $26
Totals	20	$154

Finally, divide the totals.

$$\text{mean} = \frac{\$154}{20} = \$7.70$$

The mean contribution to the food pantry was $7.70.

✔ **Now Try Exercise 11.**

A common use of the weighted mean is to find a student's *grade point average (GPA),* as shown in the next example.

EXAMPLE 4   Applying the Weighted Mean

Find last-semester's GPA for a student who earned the grades listed in the table. Assume that A = 4, B = 3, C = 2, D = 1, and F = 0. The number of credits determines how many times the grade is counted (the frequency).

Course	Credits	Grade	Credits · Grade
Mathematics	4	A (= 4)	4 · 4 = 16
Speech	3	C (= 2)	3 · 2 = 6
English	3	B (= 3)	3 · 3 = 9
Computer science	2	A (= 4)	2 · 4 = 8
Theater	2	D (= 1)	2 · 1 = 2
Totals	14		41

It is common to round grade point averages to the nearest hundredth, so

$$\text{GPA} = \frac{41}{14} \approx 2.93.$$

✔ **Now Try Exercise 15.**

OBJECTIVE 3 Find the median.   Because it can be affected by extremely high or low numbers, the mean is often a poor indicator of central tendency for a list of numbers. Accordingly, in such cases, another measure of central tendency, called the *median,* can be used. The **median** divides a group of numbers in half; half the numbers lie above the median, and half lie below the median.

To find the median, list the numbers *in order* from *least* to *greatest.* If the list contains an *odd* number of items, the median is the *middle number.*

EXAMPLE 5   Finding the Median

Find the median for this list of prices.

$$\$7, \$23, \$15, \$6, \$18, \$12, \$24$$

First arrange the numbers in numerical order from least to greatest.

Least → 6, 7, 12, 15, 18, 23, 24   ← Greatest

Next, find the middle number in the list.

$$\underbrace{6, 7, 12,}\ 15,\ \underbrace{18, 23, 24}$$

Three are below.   ↓   Three are above.
Middle number

The median price is $15.

✔ **Now Try Exercise 19.**

If a list contains an *even* number of items, there is no single middle number. In this case, the median is defined as the mean (average) of the *middle two* numbers.

EXAMPLE 6 **Finding the Median**

Find the median for this list of ages, in years.

$$74, 7, 15, 13, 25, 28, 47, 59, 32, 68$$

First arrange the numbers in numerical order from least to greatest. Then, because the list has an even number of ages, find the middle *two* numbers.

Least → 7, 13, 15, 25, 28, 32, 47, 59, 68, 74 ← Greatest

Middle two numbers

The median age is the mean of the middle two numbers.

$$\text{median} = \frac{28 + 32}{2} = \frac{60}{2} = 30 \text{ yr}$$

✔ **Now Try Exercise 21.**

OBJECTIVE 4 Find the mode. Another statistical measure is the **mode,** which is the number that occurs *most often* in a list of numbers. For example, if the test scores for 10 students were

$$74, 81, 39, 74, 82, 80, 100, 92, 74, \text{ and } 85,$$

then the mode is 74. Three students earned a score of 74, so 74 appears more times on the list than any other score. It is *not* necessary to place the numbers in numerical order when looking for the mode, although that may help you find it more easily.

A list can have two modes; such a list is sometimes called **bimodal.** If no number occurs more frequently than any other number in a list, the list has *no mode*.

EXAMPLE 7 **Finding the Mode**

Find the mode for each list of numbers.

(a) 51, 32, 49, 51, 49, 90, 49, 60, 17, 60

The number 49 occurs three times, which is more often than any other number. Therefore, 49 is the mode.

(b) 482, 485, 483, 485, 487, 487, 489, 486

Because both 485 and 487 occur twice, each is a mode. This list is binomial.

(c) 10,708; 11,519; 10,972; 12,546; 13,905; 12,182

No number occurs more than once. This list has no mode.

✔ **Now Try Exercises 25, 27, and 29.**

Measures of Central Tendency

The **mean** is the sum of all the values, divided by the number of values. It is the mathematical *average*.

The **median** is the middle number in a group of values that are listed from least to greatest. It divides a group of numbers in half.

The **mode** is the value that occurs most often in a group of values.

APPENDIX D EXERCISES

Now Try Exercise

Find the mean for each list of numbers. Round answers to the nearest tenth when applicable. See Example 1.

1. Ages of infants at the child-care center (in months): 4, 9, 6, 4, 7, 10, 9

2. Monthly electric bills: $53, $77, $38, $29, $49, $48

3. Final exam scores: 92, 51, 59, 86, 68, 73, 49, 80

4. Quiz scores: 18, 25, 21, 8, 16, 13, 23, 19

5. Annual salaries: $31,900; $32,850; $34,930; $39,712; $38,340; $60,000

6. Numbers of people attending baseball games: 27,500; 18,250; 17,357; 14,298; 33,110

Solve each problem. See Examples 2 and 3.

7. The Athletic Shoe Store sold shoes at the following prices: $75.52, $36.15, $58.24, $21.86, $47.68, $106.57, $82.72, $52.14, $28.60, $72.92. Find the mean shoe sales amount.

8. In one evening, a waitress collected the following checks from her dinner customers: $30.10, $42.80, $91.60, $51.20, $88.30, $21.90, $43.70, $51.20. Find the mean dinner check amount.

9. The table shows the face value (policy amount) of life insurance policies sold and the number of policies sold for each amount by the New World Life Company during one week. Find the weighted mean amount for the policies sold.

10. Detroit Metro-Sales Company prepared the following table showing the gasoline mileage obtained by each of the cars in the company's automobile fleet. Find the weighted mean to determine the number of miles per gallon for the fleet of cars.

Policy Amount	Number of Policies Sold
$ 10,000	6
$ 20,000	24
$ 25,000	12
$ 30,000	8
$ 50,000	5
$100,000	3
$250,000	2

Miles per Gallon	Number of Autos
15	5
20	6
24	10
30	14
32	5
35	6
40	4

Find each weighted mean. Round answers to the nearest tenth when applicable. See Example 3.

11.

Quiz Scores	Frequency
3	4
5	2
6	5
8	5
9	2

12.

Credits per Student	Frequency
9	3
12	5
13	2
15	6
18	1

13.

Hours Worked	Frequency
12	4
13	2
15	5
19	3
22	1
23	5

14.

Students per Class	Frequency
25	1
26	2
29	5
30	4
32	3
33	5

Find the GPA for students earning the grades shown in the table. Assume that A = 4, B = 3, C = 2, D = 1, *and* F = 0. *Round answers to the nearest hundredth. See Example 4.*

15.

Course	Credits	Grade
Biology	4	B
Biology lab	2	A
Mathematics	5	C
Health	1	F
Psychology	3	B

16.

Course	Credits	Grade
Chemistry	3	A
English	3	B
Mathematics	4	B
Theater	2	C
Astronomy	3	C

17. Look again at the grades in Exercise 15. Find the student's GPA in each of these situations.

   (a) The student earned a B instead of an F in the one-credit class.
   (b) The student earned a B instead of a C in the five-credit class.
   (c) Both (a) and (b) happened.

18. List the credits for the courses you're taking at this time. List the lowest grade you think you will earn in each class and find your GPA. Then list the highest grade you think you will earn in each class and find your GPA.

Find the median for each list of numbers. See Examples 5 and 6.

19. Number of e-mail messages received: 9, 12, 14, 15, 23, 24, 28

20. Deliveries by a newspaper distributor: 99, 108, 109, 123, 126, 129, 146, 168, 170

21. Students enrolled in algebra each semester: 328, 549, 420, 592, 715, 483

22. Number of cars in the parking lot each day: 520, 523, 513, 1283, 338, 509, 290, 420

23. Number of computer service calls taken each day: 51, 48, 96, 40, 47, 23, 95, 56, 34, 48

24. Number of gallons of paint sold per week: 1072, 1068, 1093, 1042, 1056, 205, 1009, 1081

Find the mode(s) for each list of numbers. Indicate whether a list is bimodal or has no mode. See Example 7.

25. Number of samples taken each hour: 3, 8, 5, 1, 7, 6, 8, 4, 5, 8

26. Monthly water bills: $21, $32, $46, $32, $49, $32, $49, $25, $32

27. Ages of retirees (in years): 74, 68, 68, 68, 75, 75, 74, 74, 70, 77

28. Patients admitted to the hospital each week: 30, 19, 25, 78, 36, 20, 45, 85, 38

29. Number of boxes of candy sold by each child: 5, 9, 17, 3, 2, 8, 19, 1, 4, 20, 10, 6

30. Weights of soccer players (in pounds): 158, 161, 165, 162, 165, 157, 163, 162

Appendix E

Solving Quadratic Inequalities

OBJECTIVE 1 Solve quadratic inequalities and graph their solutions. A **quadratic inequality** is an inequality that involves a second-degree polynomial. Examples of quadratic inequalities are

$$2x^2 + 3x - 5 < 0, \qquad x^2 \leq 4, \qquad \text{and} \qquad x^2 + 5x + 6 > 0.$$

EXAMPLE 1 Solving a Quadratic Inequality Including Endpoints

Solve $x^2 - 3x - 10 \leq 0$.

To begin, we find the solutions of the corresponding quadratic *equation*.

$$x^2 - 3x - 10 = 0$$

$$(x - 5)(x + 2) = 0 \qquad \text{Factor.}$$

$$x - 5 = 0 \quad \text{or} \quad x + 2 = 0 \qquad \text{Zero-factor property}$$

$$x = 5 \quad \text{or} \qquad x = -2 \qquad \text{Solve each equation.}$$

Since 5 and -2 are the only values that satisfy $x^2 - 3x - 10 = 0$, all other values of x will make $x^2 - 3x - 10$ either less than 0 (< 0) or greater than 0 (> 0). The values $x = 5$ and $x = -2$ determine three regions (intervals) on the number line, as shown in Figure 1. Region A includes all numbers less than -2, Region B includes all numbers between -2 and 5, and Region C includes all numbers greater than 5.

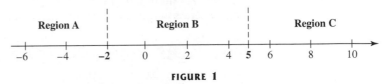

FIGURE 1

For all values of x in a given region, $x^2 - 3x - 10$ will have the same sign (either positive or negative). We test one value of x from each region to see which regions satisfy $x^2 - 3x - 10 \leq 0$. First, are the points in Region A part of the solution set? As a trial value, we choose any number less than -2, say, -6.

$$x^2 - 3x - 10 \leq 0 \qquad \text{Original inequality}$$

$$(-6)^2 - 3(-6) - 10 \leq 0 \quad ? \qquad \text{Let } x = -6.$$

$$36 + 18 - 10 \leq 0 \quad ? \qquad \text{Simplify.}$$

$$44 \leq 0 \qquad \text{False}$$

Since $44 \leq 0$ is false, the points in Region A do not belong to the solution set of the inequality.

683

What about Region B? We try the value $x = 0$ in the original inequality.

$$0^2 - 3(0) - 10 \leq 0 \qquad ? \qquad \text{Let } x = 0.$$
$$-10 \leq 0 \qquad\qquad \text{True}$$

Since $-10 \leq 0$ is true, the points in Region B do belong to the solution set. Try $x = 6$ to check Region C.

$$6^2 - 3(6) - 10 \leq 0 \qquad ? \qquad \text{Let } x = 6.$$
$$36 - 18 - 10 \leq 0 \qquad ? \qquad \text{Simplify.}$$
$$8 \leq 0 \qquad\qquad \text{False}$$

Since $8 \leq 0$ is false, the points in Region C do not belong to the solution set.

The points in Region B are the only ones that satisfy $x^2 - 3x - 10 < 0$. As shown in Figure 2, the solution set includes the points in Region B together with the endpoints -2 and 5, since the original inequality uses $\leq$. The solution set is written as the interval $[-2, 5]$.

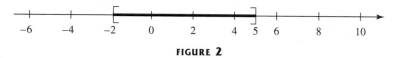

FIGURE 2

✔ **Now Try Exercise 11.**

To summarize, we use the following steps to solve a quadratic inequality.

Solving a Quadratic Inequality

Step 1 **Write the inequality as an equation and solve it.**

Step 2 **Determine intervals.** Use the solutions of the equation to divide a number line into intervals.

Step 3 **Test each interval.** Substitute a number from each interval into the original inequality to determine the intervals that satisfy the inequality.

Step 4 **Consider the endpoints separately.** The solutions from Step 1 are included in the solution set only if the inequality is $\leq$ or $\geq$.

EXAMPLE 2 Solving a Quadratic Inequality Excluding Endpoints

Solve $-x^2 - 5x - 6 < 0$.

It will be easier to factor if we multiply both sides by -1. We must also remember to reverse the direction of the inequality symbol:

> Reverse the inequality symbol when multiplying by a negative number.

$$x^2 + 5x + 6 > 0.$$

Step 1 Factoring $x^2 + 5x + 6$ in the corresponding equation $x^2 + 5x + 6 = 0$, we get $(x + 2)(x + 3) = 0$. The solutions of the equation are -2 and -3.

Step 2 These numbers determine three regions on a number line. See Figure 3. This time, these points will not belong to the solution set, because the inequality symbol *does not* include equality as well.

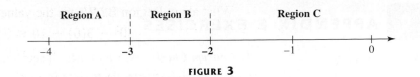

FIGURE 3

Step 3 To decide whether the points in Region A belong to the solution set, we select any number in Region A, such as -4, and substitute.

$$-x^2 - 5x - 6 < 0 \qquad \text{Original inequality}$$
$$-(-4)^2 - 5(-4) - 6 < 0 \quad ? \quad \text{Let } x = -4.$$
$$-16 + 20 - 6 < 0 \quad ? \quad \text{Simplify.}$$
$$-2 < 0 \qquad \text{True}$$

Since $-2 < 0$ is true, all the points in Region A belong to the solution set of the inequality.

For Region B, we choose a number between -3 and -2, say, $-2\frac{1}{2}$, or $-\frac{5}{2}$.

$$-\left(-\frac{5}{2}\right)^2 - 5\left(-\frac{5}{2}\right) - 6 < 0 \quad ? \quad \text{Let } x = -\frac{5}{2}.$$

$$-\frac{25}{4} + \frac{25}{2} - 6 < 0 \quad ? \quad \text{Simplify.}$$

$$\frac{1}{4} < 0 \qquad \text{False}$$

Since $\frac{1}{4} < 0$ is false, no point in Region B belongs to the solution set.

For Region C, we try the number 0.

$$-0^2 - 5(0) - 6 < 0 \quad ? \quad \text{Let } x = 0.$$
$$-6 < 0 \qquad \text{True}$$

Since $-6 < 0$ is true, the points in Region C belong to the solution set.

Step 4 The solutions include all values of x less than -3, together with all values of x greater than -2, as shown in the graph in Figure 4. Using inequalities, we may write the solutions as $x < -3$ *or* $x > -2$. Any number that satisfies *either* of these inequalities will be a solution, because *or* means "either one or the other or both." In interval notation, we write the solution set with the *union symbol* $\cup$ for *or*:

$$(-\infty, -3) \cup (-2, \infty).$$

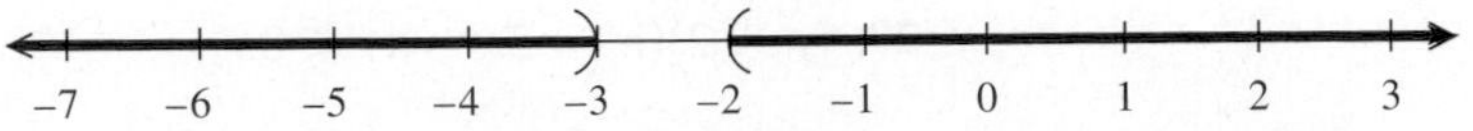

FIGURE 4

✔ **Now Try Exercise 9.**

CAUTION There is no shortcut way to write the solutions $x < -3$ or $x > -2$.

APPENDIX E EXERCISES

Now Try Exercise

1. *Concept Check* Which of these inequalities is not a quadratic inequality?

   A. $5x^2 - x + 7 < 0$ **B.** $x^2 + 4 > 0$ **C.** $3x + 5 \geq 0$ **D.** $(x - 1)(x + 2) \leq 0$

2. *Concept Check* To solve a quadratic inequality, we use the _____ of the corresponding equation to determine _____ on the number line.

Concept Check *Answer* true *or* false, *depending on whether the given value of x satisfies or does not satisfy the inequality.*

3. $(2x + 1)(x - 5) \geq 0$

   (a) $x = -\dfrac{1}{2}$ **(b)** $x = 5$ **(c)** $x = 0$ **(d)** $x = -6$

4. $(x - 6)(3x + 4) < 0$

   (a) $x = -3$ **(b)** $x = 6$ **(c)** $x = 0$ **(d)** $x = -\dfrac{4}{3}$

Solve each inequality and graph the solution set. See Examples 1 and 2.

5. $(a + 3)(a - 3) < 0$	**6.** $(b - 2)(b + 2) > 0$
7. $(a + 6)(a - 7) \geq 0$	**8.** $(z - 5)(z - 4) \leq 0$
9. $m^2 + 5m + 6 > 0$	**10.** $x^2 - 3x + 2 < 0$
11. $z^2 - 4z - 5 \leq 0$	**12.** $3p^2 - 5p - 2 \leq 0$
13. $5m^2 + 3m - 2 < 0$	**14.** $2k^2 + 7k - 4 > 0$
15. $6r^2 - 5r < 4$	**16.** $6r^2 + 7r > 3$
17. $q^2 - 7q < -6$	**18.** $2k^2 - 7k \leq 15$
19. $6m^2 + m - 1 > 0$	**20.** $30r^2 + 3r - 6 \leq 0$
21. $12p^2 + 11p + 2 < 0$	**22.** $a^2 - 16 < 0$
23. $9m^2 - 36 > 0$	**24.** $r^2 - 100 \geq 0$
25. $r^2 > 16$	**26.** $m^2 \geq 25$

Each given inequality is not quadratic, but it may be solved in a manner similar to the way a quadratic inequality is solved. Solve and graph each inequality. (Hint: Because these inequalities correspond to equations with three solutions, they determine four regions on a number line.)

27. $(a + 2)(3a - 1)(a - 4) \geq 0$	**28.** $(2p - 7)(p - 1)(p + 3) \leq 0$
29. $(r - 2)(r^2 - 3r - 4) < 0$	**30.** $(m + 5)(m^2 - m - 6) > 0$

Answers to Selected Exercises

In this section we provide the answers that we think most students will obtain when they work the exercises using the methods explained in the text. If your answer does not look exactly like the one given here, it is not necessarily wrong. In many cases, there are equivalent forms of the answer that are correct. For example, if the answer section shows $\frac{3}{4}$ and your answer is 0.75, you have obtained the right answer, but written it in a different (yet equivalent) form. Unless the directions specify otherwise, 0.75 is just as valid an answer as $\frac{3}{4}$.

In general, if your answer does not agree with the one given in the text, see whether it can be transformed into the other form. If it can, then it is the correct answer. If you still have doubts, talk with your instructor. You might also want to obtain a copy of the *Student's Solutions Manual* that goes with this book. Your college bookstore either has this manual or can order it for you.

CHAPTER 1   The Real Number System

Section 1.1   (pages 11–15)

1. true   **3.** false; the fraction $\frac{17}{51}$ is written in lowest terms as $\frac{1}{3}$.   **5.** false; *product* refers to multiplication, so the product of 8 and 2 is 16.   **7.** prime   **9.** composite; $2 \cdot 2 \cdot 2 \cdot 2 \cdot 2 \cdot 2$ **11.** composite; $2 \cdot 7 \cdot 13 \cdot 19$ **13.** neither
15. composite; $2 \cdot 3 \cdot 5$ **17.** composite; $2 \cdot 2 \cdot 5 \cdot 5 \cdot 5$ **19.** composite; $2 \cdot 2 \cdot 31$ **21.** prime   **23.** $\frac{1}{2}$ **25.** $\frac{5}{6}$
27. $\frac{1}{5}$ **29.** $\frac{6}{5}$ **31.** C   **33.** $\frac{24}{35}$ **35.** $\frac{6}{25}$ **37.** $\frac{6}{5}$, or $1\frac{1}{5}$ **39.** $\frac{65}{12}$, or $5\frac{5}{12}$ **41.** $\frac{38}{5}$, or $7\frac{3}{5}$ **43.** $\frac{10}{3}$, or $3\frac{1}{3}$
45. 12   **47.** $\frac{1}{16}$ **49.** $\frac{35}{24}$, or $1\frac{11}{24}$ **51.** $\frac{84}{47}$, or $1\frac{37}{47}$ **53.** To multiply two fractions, multiply their numerators to get the numerator of the product and multiply their denominators to get the denominator of the product. For example,
$\frac{2}{3} \cdot \frac{8}{5} = \frac{2 \cdot 8}{3 \cdot 5} = \frac{16}{15}$. To divide two fractions, replace the divisor with its reciprocal and then multiply. For example,
$\frac{2}{5} \div \frac{7}{9} = \frac{2}{5} \cdot \frac{9}{7} = \frac{2 \cdot 9}{5 \cdot 7} = \frac{18}{35}$.   **55.** $\frac{2}{3}$ **57.** $\frac{8}{9}$ **59.** $\frac{43}{8}$, or $5\frac{3}{8}$ **61.** $\frac{101}{20}$, or $5\frac{1}{20}$ **63.** $\frac{2}{3}$ **65.** $\frac{17}{36}$
67. $\frac{67}{20}$, or $3\frac{7}{20}$ **69.** $\frac{11}{12}$ **71.** 6 cups   **73.** $1\frac{1}{8}$ in.   **75.** $\frac{9}{16}$ in.   **77.** $618\frac{3}{4}$ ft   **79.** $5\frac{5}{24}$ in.   **81.** 8 cakes (There will be some sugar left over.)   **83.** $16\frac{5}{8}$ yd   **85.** $3\frac{3}{8}$ in.   **87.** $\frac{7}{100}$ **89.** more than $4\frac{19}{25}$ million   **91.** (a) $\frac{1}{2}$ (b) $\frac{1}{4}$ (c) $\frac{1}{3}$
(d) $\frac{1}{6}$

Section 1.2   (pages 21–24)

1. false; $4 + 3(8 - 2) = 4 + 3 \cdot 6 = 4 + 18 = 22$. The common error leading to 42 is adding 4 to 3 and then multiplying by 6. One must follow the order of operations.   **3.** false; the correct interpretation is $4 = 16 - 12$.   **5.** 49   **7.** 144   **9.** 64

11. 1000   **13.** 81   **15.** 1024   **17.** $\dfrac{16}{81}$ **19.** 0.064   **21.** Write the base as a factor the number of times indicated by the exponent. For example, $6^3 = 6 \cdot 6 \cdot 6 = 216$.   **23.** 32   **25.** 58   **27.** 22.2   **29.** $\dfrac{49}{30}$, or $1\dfrac{19}{30}$ **31.** 12   **33.** 13

35. 26   **37.** 4   **39.** 42   **41.** 5   **43.** 95   **45.** 90   **47.** 14   **49.** 9   **51.** Begin by squaring 2. Then subtract 1 to get a result of $4 - 1 = 3$ within the parentheses. Next, raise 3 to the third power to get $3^3 = 27$. Multiply this result by 3 to obtain 81. Finally, add this result to 4 to get 85, the final answer.   **53.** $16 \le 16$; true   **55.** $61 \le 60$; false   **57.** $0 \ge 0$; true   **59.** $45 \ge 46$; false   **61.** $66 > 72$; false   **63.** $2 \ge 3$; false   **65.** $3 \ge 3$; true   **67.** $15 = 5 + 10$ **69.** $9 > 5 - 4$

71. $16 \ne 19$ **73.** $\dfrac{1}{2} \le \dfrac{2}{4}$ **75.** Seven is less than nineteen; true   **77.** Three is not equal to six; true   **79.** Eight is greater than or equal to eleven; false   **81.** Answers will vary. One example is $5 + 3 \ge 2 \cdot 2$.   **83.** $30 > 5$ **85.** $1.3 \le 2.5$ **87.** is younger than   **89. (a)** $14.7 - 40 \cdot 0.13$ **(b)** 9.5 **(c)** 8.075; walking (5 mph)   **91.** Answers will vary.   **93.** 1998, 1999, 2000   **95.** $3 \cdot (6 + 4) \cdot 2 = 60$ **97.** $10 - (7 - 3) = 6$ **99.** $(8 + 2)^2 = 100$

Section 1.3   (pages 29–30)

1. 10   **3.** $12 + x$; 21   **5.** no   **7.** $2x^3 = 2 \cdot x \cdot x \cdot x$, while $2x \cdot 2x \cdot 2x = (2x)^3$.   **9.** The exponent 2 applies only to its base, which is x.   **11.** Answers will vary. Two such pairs are $x = 0$, $y = 6$ and $x = 1$, $y = 4$. To find a pair, choose one number, susbstitute it for a variable, and then calculate the value for the other variable.   **13. (a)** 13 **(b)** 15   **15. (a)** 20 **(b)** 30   **17. (a)** 64 **(b)** 144   **19. (a)** $\dfrac{5}{3}$ **(b)** $\dfrac{7}{3}$ **21. (a)** $\dfrac{7}{8}$ **(b)** $\dfrac{13}{12}$ **23. (a)** 52 **(b)** 114   **25. (a)** 25.836 **(b)** 38.754

27. (a) 24 **(b)** 28   **29. (a)** 12 **(b)** 33   **31. (a)** 6 **(b)** $\dfrac{9}{5}$ **33. (a)** $\dfrac{4}{3}$ **(b)** $\dfrac{13}{6}$ **35. (a)** $\dfrac{2}{7}$ **(b)** $\dfrac{16}{27}$ **37. (a)** 12 **(b)** 55

39. (a) 1 **(b)** $\dfrac{28}{17}$ **41. (a)** 3.684 **(b)** 8.841   **43.** $12x$ **45.** $x + 7$ **47.** $x - 2$ **49.** $7 - x$ **51.** $x - 6$ **53.** $\dfrac{12}{x}$

55. $6(x - 4)$ **57.** "Please excuse me, but I would like to point out that one *solves* an equation but *simplifies* an expression. You might change 'Solve' to 'Simplify'."   **59.** yes   **61.** no   **63.** yes   **65.** yes   **67.** yes   **69.** $x + 8 = 18$; 10

71. $16 - \dfrac{3}{4}x = 13$; 4   **73.** $2x + 1 = 5$; 2   **75.** $3x = 2x + 8$; 8   **77.** expression   **79.** equation   **81.** equation

83. 64.9 yr   **85.** 72.8 yr

Section 1.4   (pages 38–40)

1. 2,845,000   **3.** -2809 **5.** -2.4; 5.2   **7.** 52.59   **9.** 4   **11.** 0   **13.** One example is $\sqrt{12}$. There are others.

15. true   **17.** true   **19. (a)** 3, 7 **(b)** 0, 3, 7 **(c)** $-9, 0, 3, 7$ **(d)** $-9, -1\dfrac{1}{4}, -\dfrac{3}{5}, 0, 3, 5.9, 7$ **(e)** $-\sqrt{7}, \sqrt{5}$ **(f)** All are real numbers.   **21.** The *natural numbers* are the numbers with which we count. An example is 1. The *whole numbers* are the natural numbers with 0 also included. An example is 0. The *integers* are the whole numbers and their negatives. An example is -1. The *rational numbers* are the numbers that can be represented by a quotient of integers, such as $\dfrac{1}{2}$. The *irrational numbers,* such as $\sqrt{2}$, cannot be represented as a quotient of integers. The *real numbers* include all positive numbers, negative numbers, and zero. All the numbers just discussed are real.   **23.**   **25.**   **27.**

29. (a) A **(b)** A **(c)** B **(d)** B   **31. (a)** 4 **(b)** 4   **33. (a)** -6 **(b)** 6   **35.** 6   **37.** $-\dfrac{2}{3}$ **39.** 3   **41.** -12 **43.** -8

45. 3　**47.** $|-3.5|$, or 3.5　**49.** $-|-6|$, or -6　**51.** $|5-3|$, or 2　**53.** true　**55.** true　**57.** true　**59.** false

61. true　**63.** false　**65.** petroleum refineries, 2002 to 2003　**67.** construction machinery manufacturing, 2002 to 2003

In Exercises 69–73, answers will vary.

69. $\dfrac{1}{2}, \dfrac{5}{8}, 1\dfrac{3}{4}$　**71.** $-3\dfrac{1}{2}, -\dfrac{2}{3}, \dfrac{3}{7}$　**73.** $\sqrt{5}, \pi, -\sqrt{3}$　**75.** This is not true. The absolute value of 0 is 0, and 0 is not

positive. A more accurate way of describing absolute value is to say that *absolute value is never negative*, or *absolute value is always nonnegative.*

Section 1.5　(pages 48–53)

1. negative　**3.** negative　**5.** To add two numbers with the same sign, add their

absolute values and keep the same sign for the sum. For example, $3 + 4 = 7$ and $-3 + (-4) = -7$. To add two numbers with different signs, subtract the smaller absolute value from the larger absolute value, and use the sign of the number with the larger absolute value. For example, $6 + (-4) = 2$ and $(-6) + 4 = -2$.　**7.** -8　**9.** -12　**11.** 2　**13.** -2　**15.** 8.9

17. 12　**19.** 5　**21.** 2　**23.** -9　**25.** 0　**27.** $\dfrac{1}{2}$　**29.** $-\dfrac{19}{24}$　**31.** $-\dfrac{3}{4}$　**33.** -7.7　**35.** -8　**37.** 0　**39.** -20

41. -3　**43.** -4　**45.** -8　**47.** -14　**49.** 9　**51.** -4　**53.** 4　**55.** $\dfrac{3}{4}$　**57.** $-\dfrac{11}{8}$, or $-1\dfrac{3}{8}$　**59.** $\dfrac{15}{8}$, or $1\dfrac{7}{8}$

61. 11.6　**63.** -9.9　**65.** 10　**67.** -5　**69.** 11　**71.** -10　**73.** 22　**75.** -2　**77.** -6　**79.** -12

81. -5.90617　**83.** $-5 + 12 + 6$; 13　**85.** $[-19 + (-4)] + 14$; -9　**87.** $[-4 + (-10)] + 12$; -2

89. $\left[\dfrac{5}{7} + \left(-\dfrac{9}{7}\right)\right] + \dfrac{2}{7}$; $-\dfrac{2}{7}$　**91.** $4 - (-8)$; 12　**93.** $-2 - 8$; -10　**95.** $[9 + (-4)] - 7$; -2

97. $[8 - (-5)] - 12$; 1　**99.** -3.6 (billion dollars)　**101.** 28.2 (billion dollars)　**103.** 50,395 ft　**105.** 1345 ft

107. 136 ft　**109.** -12　**111.** $-56°$F　**113.** $-69°$F　**115.** -184 m　**117. (a)** 11.3% **(b)** Americans spent more money than they earned, which means they had to dip into savings or increase borrowing.　**119.** $2169　**121.** 17

123. $1045.55　**125.** $323.83　**127.** positive　**129.** positive

Section 1.6　(pages 62–65)

1. greater than 0　**3.** less than 0　**5.** greater than 0　**7.** equal to 0　**9.** undefined; 0; Examples include $\dfrac{1}{0}$, which is

undefined, and $\dfrac{0}{1}$, which equals 0.　**11.** -12　**13.** 12　**15.** 120　**17.** -33　**19.** -2.38　**21.** $\dfrac{5}{12}$　**23.** $-\dfrac{1}{6}$　**25.** 6

27. $-32, -16, -8, -4, -2, -1, 1, 2, 4, 8, 16, 32$　**29.** $-40, -20, -10, -8, -5, -4, -2, -1, 1, 2, 4, 5, 8, 10, 20, 40$

31. $-31, -1, 1, 31$　**33.** 3　**35.** -5　**37.** 7　**39.** -6　**41.** $\dfrac{32}{3}$, or $10\dfrac{2}{3}$　**43.** -4　**45.** 0　**47.** undefined

49. -11　**51.** -2　**53.** 35　**55.** 6　**57.** -18　**59.** 67　**61.** -8　**63.** 3　**65.** 7　**67.** 4　**69.** -1　**71.** 4

73. -3　**75.** negative　**77.** 47　**79.** 72　**81.** $-\dfrac{78}{25}$　**83.** 0　**85.** -23　**87.** 2　**89.** $9 + (-9)(2)$; -9

91. $-4 - 2(-1)(6)$; 8　**93.** $(1.5)(-3.2) - 9$; -13.8　**95.** $12[9 - (-8)]$; 204　**97.** $\dfrac{-12}{-5 + (-1)}$; 2

99. $\dfrac{15 + (-3)}{4(-3)}$; -1　**101.** $\dfrac{2}{3}[8 - (-1)]$; 6　**103.** $0.20(-5 \cdot 6)$; -6　**105.** $\left(\dfrac{1}{2} + \dfrac{5}{8}\right)\left(\dfrac{3}{5} - \dfrac{1}{3}\right)$; $\dfrac{3}{10}$

107. $\dfrac{-\dfrac{1}{2}\left(\dfrac{3}{4}\right)}{-\dfrac{2}{3}}; \dfrac{9}{16}$ **109.** $\dfrac{x}{3} = -3; -9$ **111.** $x - 6 = 4; 10$ **113.** $x + 5 = -5; -10$ **115.** $8\dfrac{2}{5}$

117. 2   **119.** 0   **121. (a)** 6 is divisible by 2. **(b)** 9 is not divisible by 2.   **123. (a)** 64 is divisible by 4. **(b)** 35 is not divisible by 4.   **125. (a)** 2 is divisible by 2 and $1 + 5 + 2 + 4 + 8 + 2 + 2 = 24$ is divisible by 3. **(b)** Although 0 is divisible by 2, $2 + 8 + 7 + 3 + 5 + 9 + 0 = 34$ is not divisible by 3.   **127. (a)** $4 + 1 + 1 + 4 + 1 + 0 + 7 = 18$ is divisible by 9. **(b)** $2 + 2 + 8 + 7 + 3 + 2 + 1 = 25$ is not divisible by 9.

Summary Exercises on Operations with Real Numbers   (page 66)

1. -16 **2.** 4   **3.** 0   **4.** -24 **5.** -17 **6.** 76   **7.** -18 **8.** 90   **9.** 38   **10.** 4   **11.** -5 **12.** 5

13. $-\dfrac{7}{2}$, or $-3\dfrac{1}{2}$ **14.** 4   **15.** 13   **16.** $\dfrac{5}{4}$, or $1\dfrac{1}{4}$ **17.** 9   **18.** $\dfrac{37}{10}$, or $3\dfrac{7}{10}$ **19.** 0   **20.** 25   **21.** 14

22. undefined   **23.** -4 **24.** $\dfrac{6}{5}$, or $1\dfrac{1}{5}$ **25.** -1 **26.** $\dfrac{52}{37}$, or $1\dfrac{15}{37}$ **27.** $\dfrac{17}{16}$, or $1\dfrac{1}{16}$ **28.** $-\dfrac{2}{3}$ **29.** 3.33

30. 1.02   **31.** -13 **32.** 0   **33.** 24   **34.** -7 **35.** 37   **36.** -3 **37.** -1 **38.** $\dfrac{1}{2}$ **39.** $-\dfrac{5}{13}$ **40.** 5

41. $-\dfrac{8}{27}$ **42.** 4

Section 1.7   (pages 74–76)

1. -12; commutative property   **3.** 3; commutative property   **5.** 7; associative property   **7.** 8; associative property
9. (a) B **(b)** F **(c)** C **(d)** I **(e)** B **(f)** D, F **(g)** B **(h)** A **(i)** G **(j)** H   **11.** commutative property   **13.** associative property
15. associative property   **17.** inverse property   **19.** inverse property   **21.** identity property   **23.** commutative property
25. distributive property   **27.** identity property   **29.** distributive property   **31.** The identity properties allow us to perform an operation so that the result is the number we started with. The inverse properties allow us to perform an operation that gives an identity element as a result.   **33.** identity property   **35.** 150   **37.** 2010   **39.** 400   **41.** 1400   **43.** 11
45. 0   **47.** -0.38 **49.** 1   **51.** Subtraction is not associative.   **53.** The expression following the first equals sign should be $-3(4) - 3(-6)$. The student forgot that 6 should be preceded by a $-$ sign. The correct work is $-3(4 - 6) = -3(4) - 3(-6) = -12 + 18 = 6$.   **55.** 85   **57.** $4t + 12$ **59.** $-8r - 24$ **61.** $-5y + 20$
63. $-16y - 20z$ **65.** $8(z + w)$ **67.** $7(2v + 5r)$ **69.** $24r + 32s - 40y$ **71.** $-24x - 9y - 12z$ **73.** $5(x + 3)$
75. $-4t - 3m$ **77.** $5c + 4d$ **79.** $q - 5r + 8s$ **81.** Answers will vary; for example, "putting on your socks" and "putting on your shoes."   **83.** 0   **84.** $-3(5) + (-3)(-5)$ **85.** -15 **86.** We must interpret $(-3)(-5)$ as 15, since it is the additive inverse of -15.   **87. (a)** no **(b)** distributive property

Section 1.8   (pages 80–82)

1. C   **3.** A   **5.** $4r + 11$ **7.** $5 + 2x - 6y$ **9.** $-7 + 3p$ **11.** $2 - 3x$ **13.** -12 **15.** 5   **17.** 1   **19.** -1
21. $\dfrac{1}{5}$ **23.** like   **25.** unlike   **27.** like   **29.** unlike   **31.** The student made a sign error when applying the distributive property: $7x - 2(3 - 2x) = 7x - 2(3) - 2(-2x)$. The correct answer is $11x - 6$.   **33.** $17y$ **35.** $-6a$ **37.** $13b$
39. $7k + 15$ **41.** $-4y$ **43.** $2x + 6$ **45.** $14 - 7m$ **47.** $-17 + x$ **49.** $23x$ **51.** $-\dfrac{1}{3}t - \dfrac{28}{3}$ **53.** $9y^2$

55. $-14p^3 + 5p^2$ **57.** $8x + 15$ **59.** $5x + 15$ **61.** $-4y + 22$ **63.** $-\dfrac{3}{2}y + 16$ **65.** $-16y + 63$ **67.** $4r + 15$

69. $12k - 5$ **71.** $-2k - 3$ **73.** $4k - 7$ **75.** $-23.7y - 12.6$ **77.** $(x + 3) + 5x$; $6x + 3$ **79.** $(13 + 6x) - (-7x)$; $13 + 13x$ **81.** $2(3x + 4) - (-4 + 6x)$; 12 **83.** Wording will vary. One example is "the difference between 9 times a number and the sum of the number and 2." **85.** $1000 + 5x$ (dollars) **86.** $750 + 3y$ (dollars) **87.** $1000 + 5x + 750 + 3y$ (dollars) **88.** $1750 + 5x + 3y$ (dollars)

Chapter 1 Review Exercises (pages 88–92)

1. $\dfrac{3}{4}$ **3.** $\dfrac{9}{40}$ **5.** 625 **7.** 0.0004 **9.** 27 **11.** 39 **13.** true **15.** false **17.** $5 + 2 \neq 10$ **19.** 30 **21.** 14

23. $x + 6$ **25.** $6x - 9$ **27.** yes **29.** $2x - 6 = 10$; 8 **31.** [number line graph with points at $-\frac{1}{2}$ and 2.5, marks at -4, -2, 0, 2, 4] **33.** rational numbers, real numbers

35. -10 **37.** $-\dfrac{3}{4}$ **39.** true **41.** true **43.** (a) 9 (b) 9 **45.** (a) -6 (b) 6 **47.** 12 **49.** -19 **51.** -6

53. -17 **55.** -21.8 **57.** -10 **59.** -11 **61.** 7 **63.** 10.31 **65.** 2 **67.** $(-31 + 12) + 19$; 0

69. $-4 - (-6)$; 2 **71.** -2 **73.** \$26.25 **75.** $-\$29$ **77.** 38 **79.** 36 **81.** $\dfrac{1}{2}$ **83.** -20 **85.** -24 **87.** 4

89. $-\dfrac{3}{4}$ **91.** -1 **93.** 1 **95.** -18 **97.** 125 **99.** $-4(5) - 9$; -29 **101.** $\dfrac{12}{8 + (-4)}$; 3 **103.** $8x = -24$; -3

105. 32 **107.** identity property **109.** inverse property **111.** associative property **113.** distributive property **115.** $7(y + 2)$ **117.** $3(2s + 5y)$ **119.** $25 - (5 - 2) = 22$ and $(25 - 5) - 2 = 18$. Because different groupings lead to different results, we conclude that in general subtraction is not associative. **121.** $11m$ **123.** $16p^2 + 2p$

125. $-2m + 29$ **127.** $-2(3x) - 7x$; $-13x$ **129.** 16 **131.** $\dfrac{8}{3}$, or $2\dfrac{2}{3}$ **133.** 2 **135.** $-\dfrac{3}{2}$, or $-1\dfrac{1}{2}$

137. $-\dfrac{28}{15}$, or $-1\dfrac{13}{15}$ **139.** $8x^2 - 21y^2$ **141.** Dividing 0 *by* a nonzero number gives a quotient of 0. However, dividing a number *by* 0 is undefined. **143.** $5(x + 7)$; $5x + 35$

Chapter 1 Test (pages 93–94)

[1.1] **1.** $\dfrac{7}{11}$ **2.** $\dfrac{241}{120}$, or $2\dfrac{1}{120}$ **3.** $\dfrac{19}{18}$, or $1\dfrac{1}{18}$ **4.** (a) 492 million (b) 861 million [1.2] **5.** true

[1.4] **6.** [number line graph with marks at -4, -2, 0, 2, 4] **7.** rational numbers, real numbers **8.** If -8 and -1 are both graphed on a number line, we see that the point for -8 is to the *left* of the point for -1. This indicates $-8 < -1$. [1.6] **9.** $\dfrac{-6}{2 + (-8)}$; 1

[1.1, 1.4–1.6] **10.** 4 **11.** $-\dfrac{17}{6}$, or $-2\dfrac{5}{6}$ **12.** 2 **13.** 6 **14.** 108 **15.** 3 **16.** $\dfrac{30}{7}$, or $4\dfrac{2}{7}$

[1.3, 1.5, 1.6] **17.** 6 **18.** 4 [1.4–1.6] **19.** -70 **20.** 3 **21.** 7000 m **22.** 15 **23.** (a) -1.86 (million students) (b) -1.25 (million students) (c) 1.59 (million students) (d) 0.83 (million students) [1.7] **24.** B **25.** D **26.** E **27.** A **28.** C **29.** distributive property **30.** (a) -18 (b) -18 (c) The distributive property assures us that the answers must be the same, because $a(b + c) = ab + ac$ for all a, b, c. [1.8] **31.** $21x$ **32.** $15x - 3$

CHAPTER 2   Linear Equations and Inequalities in One Variable

Section 2.1   (pages 100–102)

1. A and C   **3.** The addition property of equality says that the same number (or expression) added to each side of an equation results in an equivalent equation. *Example:* $-x$ can be added to each side of $2x + 3 = x - 5$ to get the equivalent equation $x + 3 = -5$.   **5.** $\{12\}$ **7.** $\{31\}$ **9.** $\{-3\}$ **11.** $\{4\}$ **13.** $\{-9\}$ **15.** $\{-10\}$ **17.** $\{-13\}$ **19.** $\{10\}$ **21.** $\{6.3\}$ **23.** $\{-16.9\}$ **25.** $\{-3\}$ **27.** $\{0\}$ **29.** $\{2\}$ **31.** $\{-6\}$ **33.** $\{-2\}$ **35.** $\{0\}$ **37.** $\{0\}$ **39.** $\{-2\}$ **41.** $\{-7\}$ **43.** $\{-30\}$ **45.** A sample answer might be "A linear equation in one variable is an equation that can be written using only one variable term with the variable to the first power."   **47.** $\{13\}$ **49.** $\{-4\}$ **51.** $\{0\}$ **53.** $\left\{\dfrac{7}{15}\right\}$ **55.** $\{7\}$ **57.** $\{-4\}$ **59.** $\{13\}$ **61.** $\{29\}$ **63.** $\{18\}$ **65.** $\{12\}$ **67.** Answers will vary. One example is $x - 6 = -8$.   **69.** $3x = 2x + 17; \{17\}$ **71.** $7x - 6x = -9; \{-9\}$ **73.** 1   **75.** x **77.** r

Section 2.2   (pages 106–108)

1. The multiplication property of equality says that the same nonzero number (or expression) multiplied on each side of the equation results in an equivalent equation. *Example:* Multiplying each side of $7x = 4$ by $\dfrac{1}{7}$ gives the equivalent equation $x = \dfrac{4}{7}$.

3. C   **5.** To get x alone on the left side, divide each side by 4, the coefficient of x.   **7.** $\dfrac{3}{2}$ **9.** 10   **11.** $-\dfrac{2}{9}$ **13.** -1 **15.** 6   **17.** -4 **19.** 0.12   **21.** -1 **23.** $\{6\}$ **25.** $\left\{\dfrac{15}{2}\right\}$ **27.** $\{-5\}$ **29.** $\{-4\}$ **31.** $\left\{-\dfrac{18}{5}\right\}$ **33.** $\{12\}$ **35.** $\{0\}$ **37.** $\{40\}$ **39.** $\{-12.2\}$ **41.** $\{-48\}$ **43.** $\{72\}$ **45.** $\{-35\}$ **47.** $\{14\}$ **49.** $\{18\}$ **51.** $\left\{-\dfrac{27}{35}\right\}$ **53.** $\{-12\}$ **55.** $\left\{\dfrac{3}{4}\right\}$ **57.** $\{-30\}$ **59.** $\{3\}$ **61.** $\{-5\}$ **63.** $\{7\}$ **65.** $\{0\}$ **67.** $\left\{-\dfrac{3}{5}\right\}$ **69.** $\{18\}$ **71.** Answers will vary. One example is $\dfrac{3}{2}x = -6$.   **73.** $4x = 6; \left\{\dfrac{3}{2}\right\}$ **75.** $\dfrac{x}{-5} = 2; \{-10\}$ **77.** $-3m - 5$ **79.** $-8 + 5p$ **81.** $\{5\}$

Section 2.3   (pages 115–117)

1. *Step 1:* Clear parentheses and combine like terms, as needed. *Step 2:* Use the addition property to get all variable terms on one side of the equation and all numbers on the other. Then combine like terms. *Step 3:* Use the multiplication property to get the equation in the form $x = $ a number. *Step 4:* Check the solution. Examples will vary.   **3.** D   **5.** $\{-1\}$ **7.** $\{5\}$ **9.** $\left\{\dfrac{4}{3}\right\}$ **11.** $\left\{-\dfrac{5}{3}\right\}$ **13.** $\{5\}$ **15.** $\varnothing$ **17.** $\{$all real numbers$\}$ **19.** $\{1\}$ **21.** $\varnothing$ **23.** $\{5\}$ **25.** $\{0\}$ **27.** $\{18\}$ **29.** $\{120\}$ **31.** $\{6\}$ **33.** $\{15,000\}$ **35.** $\{8\}$ **37.** $\{0\}$ **39.** $\{4\}$ **41.** $\{20\}$ **43.** $\{$all real numbers$\}$ **45.** $\varnothing$ **47.** $11 - q$ **49.** $\dfrac{9}{k}$ **51.** $x + 7$ **53.** $65 - h$ **55.** $a + 12; a - 2$ **57.** $25r$ **59.** $\dfrac{t}{5}$ **61.** $3b + 2d$ **63.** $-6 + x$ **65.** $x - 9$ **67.** $\dfrac{-6}{x}$ **69.** $12(x - 9)$

Summary Exercises on Solving Linear Equations   (page 117)

1. $\{-5\}$ **2.** $\{4\}$ **3.** $\{-5.1\}$ **4.** $\{12\}$ **5.** $\{-25\}$ **6.** $\{-6\}$ **7.** $\{-3\}$ **8.** $\{-16\}$ **9.** $\{7\}$ **10.** $\left\{-\dfrac{96}{5}\right\}$

11. $\{5\}$ **12.** $\{23.7\}$ **13.** {all real numbers}   **14.** $\{1\}$ **15.** $\{-6\}$ **16.** $\emptyset$ **17.** $\{6\}$ **18.** $\{3\}$ **19.** $\emptyset$ **20.** $\left\{\dfrac{7}{3}\right\}$

21. $\{25\}$ **22.** $\{-10.8\}$ **23.** $\{3\}$ **24.** $\{7\}$ **25.** $\{2\}$ **26.** {all real numbers}   **27.** $\{-2\}$ **28.** $\{70\}$ **29.** $\left\{\dfrac{14}{17}\right\}$

30. $\left\{-\dfrac{5}{2}\right\}$

Connections   **(page 125)**   Polya's Step 1 corresponds to our Steps 1 and 2. Polya's Step 2 corresponds to our Step 3. Polya's Step 3 corresponds to our Steps 4 and 5. Polya's Step 4 corresponds to our Step 6. Trial and error, or guessing and checking, fits into Polya's Step 2: Devise a plan.

Section 2.4   (pages 125–131)

1. D; there cannot be a fractional number of cars.   **3.** A; distance cannot be negative.   **5.** 3   **7.** 6   **9.** -3 **11.** California: 59 screens; New York: 48 screens   **13.** Democrats: 44; Republicans: 55   **15.** Bruce Springsteen and the E Street Band: \$115.9 million; Céline Dion: \$80.5 million   **17.** wins: 62; losses: 20   **19.** 1950 Denver nickel: \$8.00; 1945 Philadelphia nickel: \$7.00   **21.** onions: 81.3 kg; grilled steak: 536.3 kg   **23.** 168 DVDs   **25.** whole wheat: 25.6 oz; rye: 6.4 oz   **27.** American: 18; United: 11; Southwest: 26   **29.** shortest piece: 15 in.; middle piece: 20 in.; longest piece: 24 in.   **31.** 36 million mi   **33.** A and B: $40°$; C: $100°$ **35.** $k - m$ **37.** no   **39.** $x - 1$ **41.** $18°$ **43.** $39°$ **45.** $50°$ **47.** 68, 69   **49.** 10, 12   **51.** 101, 102   **53.** 10, 11   **55.** 18   **57.** 15, 17, 19   **59.** 2002: \$6.54 billion; 2003: \$6.67 billion; 2004: \$6.77 billion   **61.** 24   **63.** 20   **65.** 81

Section 2.5   (pages 137–141)

1. (a) The perimeter of a plane geometric figure is the distance around the figure.   **(b)** The area of a plane geometric figure is the measure of the surface covered or enclosed by the figure.   **3.** four   **5.** area   **7.** perimeter   **9.** area   **11.** area   **13.** $P = 26$ **15.** $A = 64$ **17.** $b = 4$ **19.** $t = 5.6$ **21.** $I = 1575$ **23.** $B = 14$ **25.** $r = 2.6$ **27.** $r = 10$ **29.** $A = 50.24$ **31.** $r = 6$ **33.** $V = 150$ **35.** $V = 52$ **37.** $V = 7234.56$ **39.** length: 18 in.; width: 9 in.   **41.** length: 14 m; width: 4 m   **43.** shortest: 5 in.; medium: 7 in.; longest: 8 in.   **45.** two equal sides: 7 m; third side: 10 m   **47.** about 154,000 ft^2 **49.** perimeter: 5.4 m; area: 1.8 m^2 **51.** 10 ft   **53.** 23,800.10 ft^2 **55.** length: 36 in.; volume: 11,664 in.^{3} **57.** $48°, 132°$ **59.** $51°, 51°$ **61.** $105°, 105°$ **63.** $t = \dfrac{d}{r}$ **65.** $b = \dfrac{A}{h}$ **67.** $d = \dfrac{C}{\pi}$ **69.** $H = \dfrac{V}{LW}$ **71.** $r = \dfrac{I}{pt}$ **73.** $h = \dfrac{2A}{b}$ **75.** $h = \dfrac{3V}{\pi r^2}$ **77.** $b = P - a - c$ **79.** $W = \dfrac{P - 2L}{2}$ **81.** $m = \dfrac{y - b}{x}$ **83.** $y = \dfrac{C - Ax}{B}$ **85.** $r = \dfrac{M - C}{C}$, or $r = \dfrac{M}{C} - 1$ **87.** $\{2\}$ **89.** $\{5000\}$ **91.** $\{28\}$ **93.** $\left\{-\dfrac{1}{12}\right\}$

Section 2.6   (pages 146–149)

1. (a) C   **(b)** D   **(c)** B   **(d)** A   **3.** $\dfrac{4}{3}$ **5.** $\dfrac{4}{3}$ **7.** $\dfrac{15}{2}$ **9.** $\dfrac{1}{5}$ **11.** $\dfrac{5}{6}$ **13.** 10 lb size; \$0.439   **15.** 32 oz size; \$0.093   **17.** 128 oz size; \$0.044   **19.** 36 oz size; \$0.049   **21.** 263 oz size; \$0.076   **23.** true   **25.** false   **27.** true   **29.** $\{35\}$ **31.** $\{7\}$ **33.** $\left\{\dfrac{45}{2}\right\}$ **35.** $\{2\}$ **37.** $\{-1\}$ **39.** $\{5\}$ **41.** $\left\{-\dfrac{31}{5}\right\}$ **43.** $\{-28\}$ **45.** \$30.00   **47.** \$8.75

49. $67.50　**51.** $48.90　**53.** 4 ft　**55.** 2.7 in.　**57.** 2.0 in.　**59.** $2\frac{5}{8}$ cups　**61.** $363.84　**63.** 50,000 fish

65. $x = 4$　**67.** $x = 1; y = 4$　**69. (a)**

(b) 54 ft　**71.** $144　**73.** $165　**75.** 4%　**77.** 10 yr

79. {6}　**81.** {4}

Section 2.7　(pages 157–163)

1. 45 L　**3.** $750　**5.** $17.50　**7.** A　**9. (a)** 520,000 **(b)** 960,000 **(c)** 240,000　**11.** 5.5%　**13.** D　**15.** 160 L

17. $13\frac{1}{3}$ L　**19.** 4 L　**21.** $7\frac{1}{2}$ gal　**23.** 20 mL　**25.** $2100 at 5%; $900 at 4%　**27.** $2500 at 6%; $13,500 at 5%

29. 10 nickels　**31.** 39-cent stamps: 28; 24-cent stamps: 17　**33.** Arabian Mocha: 7 lb; Colombian Decaf: 3.5 lb　**35.** A

37. 530 mi　**39.** 3.173 hr　**41.** 8.08 m per sec　**43.** 8.40 m per sec　**45.** 5 hr　**47.** $1\frac{3}{4}$ hr　**49.** $7\frac{1}{2}$ hr

51. eastbound: 300 mph; westbound: 450 mph　**53.** 40 mph; 60 mph　**55.** Bob: 7 yr old; Kevin: 21 yr old
57. width: 3 ft; length: 9 ft　**59.** $650　**61.** 0　**63.** 6　**65.** yes　**67.** no

Section 2.8　(pages 172–175)

1. Use a parenthesis if the symbol is $<$ or $>$. Use a square bracket if the symbol is $\leq$ or $\geq$.　**3.** $x > -4$　**5.** $x \leq 4$

7. $(-\infty, 4]$ ⟷|⟶　**9.** $(-\infty, -3)$ ⟷)⟶　**11.** $(4, \infty)$ ⟶(⟶

13. $[8, 10]$ ⊢[⊣⟶　**15.** $(0, 10]$ ⊢(⊣⟶　**17.** $(-4, 3)$ ⊢(⟶　**19.** It would imply

that the false statement $3 < -2$ is actually true.　**21.** $[1, \infty)$ ⟶[⟶　**23.** $[5, \infty)$ ⟶[⟶

25. $(-\infty, -11)$ ⟷)⟶　**27.** It must be reversed when one is multiplying or dividing by a negative number.

29. Divide by -5 and reverse the inequality symbol to get $x < -4$.　**31.** $(-\infty, 6)$ ⟷)⟶

33. $[-10, \infty)$ ⊢[⟶　**35.** $(-\infty, -3)$ ⟷)⟶　**37.** $(-\infty, 0]$ ⟷]⟶

39. $(20, \infty)$ ⊢(⟶　**41.** $[-3, \infty)$ ⊢[⟶　**43.** $[-5, \infty)$ ⊢[⟶

45. $(-\infty, 1)$ ⟷)⟶　**47.** $(-\infty, 0]$ ⟷]⟶　**49.** $[4, \infty)$ ⊢[⟶

51. $(-\infty, 32)$ ⟷)⟶　**53.** $\left[\frac{5}{12}, \infty\right)$ ⊢[⟶　**55.** $(-21, \infty)$ ⊢(⟶

57. $-1 < x < 2$　**59.** $-1 < x \leq 2$　**61.** $[-1, 6]$ ⊢[⟶]⟶　**63.** $\left(-\frac{11}{6}, -\frac{2}{3}\right)$ ⊢(⟶)⟶

65. $(1, 3)$ ⊢(⟶)⟶　**67.** $[-26, 6]$ ⊢[⟶]⟶　**69.** $[-3, 6]$ ⊢[⟶]⟶

71. $\left[-\dfrac{24}{5}, 0\right]$ **73.** $\{4\}$ **74.** $(4, \infty)$

75. $(-\infty, 4)$ **76.** The graph would be all the real numbers.   **77.** 83 or greater   **79.** all numbers greater than 16   **81.** It is never less than $-13°$ Fahrenheit.   **83.** 32 or greater   **85.** 15 min   **87.** 10.5 gal   **89.** $R = 5x - 100$ **91.** $P = R - C = (5x - 100) - (125 + 4x) = x - 225; \ x > 225$ **93. (a)** -7 **(b)** 23   **95. (a)** -14 **(b)** 22   **97. (a)** $-\dfrac{14}{5}$ **(b)** $-\dfrac{2}{5}$

Chapter 2 Review Exercises (pages 181–185)

1. $\{6\}$ **3.** $\{7\}$ **5.** $\{11\}$ **7.** $\{5\}$ **9.** $\{5\}$ **11.** $\left\{\dfrac{64}{5}\right\}$ **13.** {all real numbers}   **15.** {all real numbers}   **17.** $\emptyset$

19. Democrats: 44; Republicans: 57   **21.** Kegon Falls: 330 ft; Rhaiadr Falls: 240 ft   **23.** $80°$ **25.** $h = 11$

27. $r = 4.75$ **29.** $h = \dfrac{A}{b}$ **31.** $135°; 45°$ **33.** 2 cm   **35.** $42.2°; 92.8°$ **37.** $\dfrac{3}{2}$ **39.** $\dfrac{3}{4}$ **41.** $\left\{\dfrac{7}{2}\right\}$ **43.** $6\dfrac{2}{3}$ lb

45. 375 km   **47.** 25.5 oz size   **49.** approximately 55.2%   **51.** $5000 at 5%; $5000 at 6%   **53.** 13 hr

55. $[-4, \infty)$ **57.** $[-5, 6)$ **59.** $[-3, \infty)$

61. $[3, \infty)$ **63.** $(-\infty, -5)$ **65.** $\left[-2, \dfrac{3}{2}\right]$ **67.** 88 or more

69. $\{7\}$ **71.** $(-\infty, 2)$ **73.** $\{70\}$ **75.** $\emptyset$ **77.** Since $-(8 + 4x) = -1(8 + 4x)$, the first step is to distribute -1 over *both* terms in the parentheses, to get $3 - 8 - 4x$ on the left side of the equation. The student got $3 - 8 + 4x$ instead. The correct solution set is $\{-2\}$.   **79.** Golden Gate Bridge: 4200 ft; Brooklyn Bridge: 1596 ft   **81.** 8 qt   **83.** 44 m

Chapter 2 Test (pages 186–187)

[2.1–2.3] **1.** $\{-6\}$ **2.** $\{21\}$ **3.** $\emptyset$ **4.** $\{30\}$ **5.** {all real numbers}   [2.4] **6.** wins: 100; losses: 62   **7.** Hawaii: 4021 mi^2; Maui: 728 mi^2; Kauai: 551 mi^2 **8.** $50°$ [2.5] **9. (a)** $W = \dfrac{P - 2L}{2}$ **(b)** 18   **10.** $75°, 75°$ [2.6] **11.** $\{6\}$

12. $\{-29\}$ **13.** 8 slices for $2.19   **14.** 2300 mi   [2.7] **15.** $8000 at 3%; $14,000 at 4.5%   **16.** 4 hr

[2.8] **17.** $(-\infty, 4]$ **18.** $(-2, 6]$ **19.** 83 or more   **20.** When an inequality is multiplied or divided by a negative number, the direction of the inequality symbol must be reversed.

Chapters 1–2 Cumulative Review Exercises (pages 187–188)

[1.1] **1.** $\dfrac{3}{4}$ **2.** $\dfrac{37}{60}$ **3.** $\dfrac{48}{5}$ [1.2] **4.** $\dfrac{1}{2}x - 18$ **5.** $\dfrac{6}{x + 12} = 2$ [1.4] **6.** true   [1.5–1.6] **7.** 11   **8.** -8 **9.** 28

[1.3] **10.** $-\dfrac{19}{3}$ [1.7] **11.** distributive property   **12.** commutative property   [1.8] **13.** $2k - 11$ [2.1–2.3] **14.** $\{-1\}$

15. $\{-1\}$ **16.** $\{-12\}$ [2.6] **17.** $\{26\}$ [2.5] **18.** $y = \dfrac{24 - 3x}{4}$ **19.** $n = \dfrac{A - P}{iP}$

[2.8] **20.** $(-\infty, 1]$ 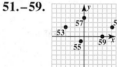 **21.** $(-1, 2]$ **[2.4] 22.** 4 cm; 9 cm; 27 cm

[2.5] **23.** 12.42 cm   [2.6] **24.** $\frac{25}{6}$, or $4\frac{1}{6}$ cups   [2.7] **25.** 40 mph; 60 mph

---

CHAPTER 3   Linear Equations and Inequalities in Two Variables; Functions

Section 3.1   (pages 199–203)

1. Ohio (OH): about 8 billion eggs; Iowa (IA): about 10 billion eggs   **3.** North Carolina (NC); about 2.5 billion eggs
5. between 1998 and 1999, 1999 and 2000, and 2003 and 2004   **7.** 2003: 6.0%; 2004: 5.5%; decline: 0.5%   **9.** does; do not
11. II   **13.** 3   **15.** A linear equation in one variable can be written in the form $Ax + B = C$, where $A \neq 0$. Examples are
$2x + 5 = 0$, $3x + 6 = 2$, and $x = -5$. A linear equation in two variables can be written in the form $Ax + By = C$, where A and
B cannot both equal 0. Examples are $2x + 3y = 8$, $3x = 5y$, and $x - y = 0$.   **17.** yes   **19.** yes   **21.** no   **23.** yes
25. yes   **27.** no   **29.** No, the ordered pair $(3, 4)$ represents the point 3 units to the right of the origin and 4 units up from the
x-axis. The ordered pair $(4, 3)$ represents the point 4 units to the right of the origin and 3 units up from the x-axis.   **31.** 17
33. -5 **35.** -1 **37.** -7 **39.** b **41.** 8; 6; 3; $(0, 8)$; $(6, 0)$; $(3, 4)$ **43.** -9; 4; 9; $(-9, 0)$; $(0, 4)$; $(9, 8)$
45. 12; 12; 12; $(12, 3)$; $(12, 8)$; $(12, 0)$ **47.** -10; -10; -10; $(4, -10)$; $(0, -10)$; $(-4, -10)$ **49.** -2; -2; -2; $(9, -2)$;
$(2, -2)$; $(0, -2)$

51.–59.   **61.** negative; negative   **63.** positive; negative   **65.** -3; 6; -2; 4

67. -3; 4; -6; $-\frac{4}{3}$ **69.** -4; -4; -4; -4 **71.** The points in each graph appear to lie on a
straight line.

73. (a) $(5, 45)$ **(b)** $(6, 50)$ **75. (a)** $(1996, 53.3)$, $(1997, 52.8)$, $(1998, 52.1)$, $(1999, 51.6)$, $(2000, 51.2)$, $(2001, 50.9)$
(b) $(2002, 51.0)$ means that in 2002, the graduation rate for 4-yr college students within 5 yr was 51.0%.
(c)   **(d)** The points appear to be approximated by a straight line. Graduation rates for 4-yr college students
within 5 yr are decreasing.

77. (a) 170; 153; 136; 119 **(b)** $(20, 170)$, $(40, 153)$, $(60, 136)$, $(80, 119)$ **(c)** 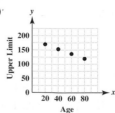   The points lie in a linear
pattern.

79. $\{-2\}$ **81.** $\{13\}$ **83.** $y = -x + 6$ **85.** $y = \dfrac{12 - 2x}{3}$

Connections **(page 212)** **1.** $3x + 4 - 2x - 7 - 4x - 3 = 0$ **2.** $5x - 15 - 3(x - 2) = 0$

Section 3.2 (pages 212–216)

1. 5; 5; 3 **3.** 1; 3; −1 **5.** −6; −2; −5 **7.** C **9.** D **11.** B

In Exercises 13 and 15, descriptions may vary. **13.** The graph of this equation is a line with *x*-intercept $(-3, 0)$ and *y*-intercept $(0, 9)$. **15.** The graph of this equation is a horizontal line with *y*-intercept $(0, -2)$. **17.** $(12, 0); (0, -8)$ **19.** $(0, 0); (0, 0)$ **21.** $(4, 0); (0, -10)$ **23.** $(2, 0); (0, 4)$ **25.** $(6, 0); (0, -2)$ **27.** $(4, 0);$ none **29.** none; $(0, 2.5)$ **31.** $y = 0; x = 0$

33. **35.** **37.** **39.** **41.** **43.**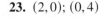

45. **47.** **49.** **51.** **53.** **55.**

57. Find two ordered pairs that satisfy the equation. Plot the corresponding points on a coordinate system. Draw a straight line through the two points. As a check, find a third ordered pair and verify that it lies on the line you drew.

59. **(a)** $(20, 170), (40, 153), (60, 136), (80, 119)$ **(b)** **(c)** 160 **(d)** 162 **(e)** They are quite close.

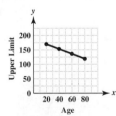

61. No. To go beyond the given data at either end assumes that the graph continues in the same way, which may not be true.

63. **(a)** 121 lb; 143 lb; 176 lb **(b)** **(c)** 68 in.; 68 in. **65.** **(a)** 1997: $67.5 billion; 2000: $73.5 billion; 2002: $77.5 billion **(b)** 1997: $67 billion; 2000: $74 billion; 2002: $78 billion **(c)** The values are very close.

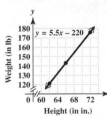

67. **(a)** **(b)** 3000 units **(c)** $10 **69.** 8; yes **71.** 7; yes **73.** $\dfrac{2}{3}$ **75.** $\dfrac{1}{2}$ **77.** $t = \dfrac{d}{r}$

79. $c = P - a - b$

Section 3.3 (pages 224–229)

1. Rise is the vertical change between two different points on a line. Run is the horizontal change between two different points on a line. **3.** 4 **5.** $-\dfrac{1}{2}$ **7.** 0 **9.** (a) C (b) A (c) D (d) B

In Exercises 11 and 13, sketches will vary. **11.** The line must fall from left to right. **13.** The line must be vertical.
15. The slope of a line is the ratio (or quotient) of the rise (the change in y) to the run (the change in x). **17.** (a) negative

(b) zero **19.** (a) positive (b) negative **21.** (a) zero (b) negative **23.** $\dfrac{8}{27}$ **25.** Because he found the difference

$3 - 5 = -2$ in the numerator, he should have subtracted in the same order in the denominator to get $-1 - 2 = -3$. The correct

slope is $\dfrac{-2}{-3} = \dfrac{2}{3}$. **27.** $\dfrac{5}{4}$ **29.** $\dfrac{3}{2}$ **31.** 0 **33.** -3 **35.** undefined **37.** $\dfrac{1}{4}$ **39.** $-\dfrac{1}{2}$ **41.** 5 **43.** $\dfrac{1}{4}$

45. $\dfrac{3}{2}$ **47.** $\dfrac{3}{2}$ **49.** 0 **51.** undefined **53.** $-3; \dfrac{1}{3}$ **55.** A **57.** $-\dfrac{2}{5}; -\dfrac{2}{5};$ parallel **59.** $\dfrac{8}{9}; -\dfrac{4}{3};$ neither

61. $\dfrac{3}{2}; -\dfrac{2}{3};$ perpendicular **63.** $5; \dfrac{1}{5};$ neither **65.** 232 thousand, or 232,000 **66.** positive; increased

67. 232,000 students **68.** -1.26 **69.** negative; decreased **70.** 1.26 students per computer **71.** The change for each

year is 0.1 billion (or 100,000,000) ft², so the graph is a straight line. **73.** 0.4 **75.** $(0, 4)$ **77.** $y = -\dfrac{2}{5}x + 3$

79. $y = \dfrac{10}{3}x - 10$ **81.** $y = 2x - 16$ **83.** $y = -\dfrac{1}{2}x - \dfrac{21}{10}$

Section 3.4 (pages 237–241)

1. E **3.** B **5.** (a) C (b) B (c) A (d) D **7.** $y = 3x - 3$ **9.** $y = -x + 3$ **11.** $y = 4x - 3$ **13.** $y = 3$
15. $x = 0$ **17.** **19.** **21.** **23.** **25.**

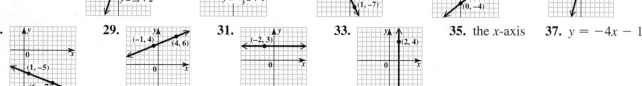

27. **29.** **31.** **33.** **35.** the x-axis **37.** $y = -4x - 1$

39. $y = \dfrac{2}{3}x + \dfrac{19}{3}$ **41.** $y = \dfrac{3}{4}x + 4$ **43.** $y = \dfrac{5}{2}x - 4$ **45.** $y = x$ (There are other forms as well.) **47.** $y = x + 6$

49. $y = -\dfrac{3}{5}x - \dfrac{11}{5}$ **51.** $y = \dfrac{1}{2}x + 2$ **53.** $y = -\dfrac{1}{3}x + \dfrac{22}{9}$ **55.** $(0, 32), (100, 212)$ **56.** $\dfrac{9}{5}$

57. $F - 32 = \dfrac{9}{5}(C - 0)$ **58.** $F = \dfrac{9}{5}C + 32$ **59.** $C = \dfrac{5}{9}(F - 32)$ **60.** $86°$ **61.** $10°$ **62.** $-40°$

63. $y = \dfrac{3}{4}x - \dfrac{9}{2}$ **65.** $y = -2x - 3$ **67.** (a) \$400 (b) \$0.25 (c) $y = 0.25x + 400$ (d) \$425 (e) 1500

69. (a) (1, 13,785), (3, 15,518), (5, 17,377), (7, 18,950), (9, 21,235)   **(b)** yes

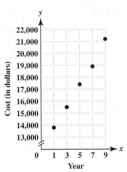

(c) $y = 952.8x + 12,659.6$ or $y = 952.8x + 12,659.8$ (depending on the point used)   **(d)** \$24,093

71. $y = -3x + 6$ **73.** $Y_1 = \dfrac{3}{4}X + 1$ **75.** $(-3, \infty)$ ———⟨——→ -3 **77.** $[5, \infty)$ ———[——→ 5

79. $(-\infty, 4]$ ←——]—→ 4

Summary Exercises on Linear Equations and Graphs   (page 242)

1.   **2.**   **3.**   **4.**   **5.**   **6.**

7.   **8.**   **9.**   **10.**   **11.**   **12.**

13.   **14.**   **15.**   **16.**   **17.**   **18.**

19. (a) B **(b)** D **(c)** A **(d)** C   **20.** A, B   **21.** $y = -3x - 6$ **22.** $y = \dfrac{3}{2}x + 12$ **23.** $y = -4x - 3$

24. $y = \dfrac{3}{5}x$ **25.** $x = 0$ **26.** $y = x - 3$ **27.** $y = \dfrac{2}{3}x$ **28.** $y = -2x - 4$ **29.** $y = x - 5$ **30.** $y = 0$

31. $y = \dfrac{5}{3}x + 5$ **32.** $y = -5x - 8$

Connections   (page 247)   1. Answers will vary.   **2. (a)** $(5, \infty)$ **(b)** $(-\infty, 5)$ **(c)** $\left[3, \dfrac{11}{3}\right)$

Section 3.5   (pages 247–249)

1. false   **3.** true   **5.** >   **7.** ≤   **9.** <   **11.** 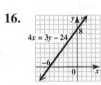   **13.**   **15.**

17. Use a dashed line if the symbol is $<$ or $>$. Use a solid line if the symbol is $\leq$ or $\geq$.

19. **21.** **23.** **25.** **27.** 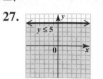 **29.**

31. Every point in quadrant IV has a positive x-value and a negative y-value. Substituting into $y > x$ would imply that a negative number is greater than a positive number, which is always false. Thus, the graph of $y > x$ cannot lie in quadrant IV. **33.** A
35. C **37. (a)** 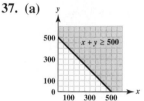 **(b)** $(500, 0)$, $(200, 400)$; other answers are possible. **39.** 5 **41.** 85 **43.** 16

Section 3.6 (pages 256–259)

1. 3; 3; $(1, 3)$ **3.** 5; 5; $(3, 5)$ **5.** The graph consists of the four points $(0, 2)$, $(1, 3)$, $(2, 4)$, and $(3, 5)$.
7. (a) domain: $\{-4, -2, 0\}$; range: $\{3, 1, 5, -8\}$ **(b)** not a function **9. (a)** domain: $\{A, B, C, D, E\}$; range: $\{2, 3, 6, 4\}$
(b) function **11. (a)** domain: $\{-4, -2, 0, 2, 3\}$; range: $\{-2, 0, 1, 2, 3\}$ **(b)** not a function **13.** function
15. not a function **17.** function **19.** function **21.** not a function **23.** domain: $(-\infty, \infty)$; range: $(-\infty, \infty)$
25. domain: $(-\infty, \infty)$; range: $[2, \infty)$ **27.** domain: $[0, \infty)$; range: $[0, \infty)$ **29.** $(2, 4)$ **30.** $(-1, -4)$ **31.** $\dfrac{8}{3}$

32. $f(x) = \dfrac{8}{3}x - \dfrac{4}{3}$ **33. (a)** 11 **(b)** 3 **(c)** -9 **35. (a)** 4 **(b)** 2 **(c)** 14 **37. (a)** 2 **(b)** 0 **(c)** 3 **39. (a)** 4 **(b)** 2

41. $\{(1970, 9.6), (1980, 14.1), (1990, 19.8), (2000, 28.4)\}$; yes **43.** $g(1980) = 14.1$; $g(1990) = 19.8$
45. For the year 2002, the function indicates 30.3 million foreign-born residents in the United States.
46. (a) domain: $\{1987, 1993, 2000, 2003\}$; range: $\{10.2, 11.6, 14.4, 15.1\}$; yes **(b)** 14.4; 2003

47. $y = 0.30625x - 598.32$ **48.** 1993: 12.0%; 2000: 14.2% **49.** $y = 0.4x - 785.6$ **50.** 1987: 9.2%; 2003: 15.6%;
the equation from Exercise 48 gives better approximations. The results in Exercise 48 vary by 0.4% and 0.2% from the data.
The results here give answers that vary by 1.0% and 0.5%. **51.** 4 **53.** 1 **55.** 1 **57.** $y = x + 1$
59. **61.** 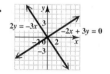 **63.**

Chapter 3 Review Exercises   (pages 265–268)

1. (a) $1.50 **(b)** about $0.40 **(c)** from 2000 to 2001 and 2001 to 2002 **(d)** from 1999 to 2000; about $0.35 per gal

3. $2; \dfrac{3}{2}; \dfrac{14}{3}$ **5.** 7; 7; 7   **7.** no   **9.** I   **11.** none   *Graph for Exercises 9 and 11*   **13.** I or III

15. $\left(\dfrac{8}{3}, 0\right); (0, 4)$ 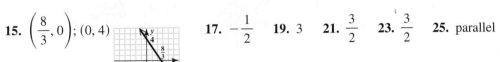 **17.** $-\dfrac{1}{2}$ **19.** 3   **21.** $\dfrac{3}{2}$ **23.** $\dfrac{3}{2}$ **25.** parallel   **27.** neither

29. $y = -\dfrac{1}{2}x + 4$ **31.** $y = \dfrac{2}{3}x + \dfrac{14}{3}$ **33.** $y = -\dfrac{1}{4}x + \dfrac{3}{2}$ **35.** $x = \dfrac{1}{3}$; it is not possible to express this equation in

the form $y = mx + b$.   **37.**   **39.** not a function; domain: $\{-2, 0, 2\}$; range: $\{4, 8, 5, 3\}$ **41.** not a function

43. function   **45. (a)** 8 **(b)** -1 **47. (a)** 5 **(b)** 2   **49.** C, D   **51.** D   **53.** B   **55.** $(0, 0); (0, 0); -\dfrac{1}{3}$

57. $y = -\dfrac{1}{4}x - \dfrac{5}{4}$ **59.** $y = -\dfrac{4}{7}x - \dfrac{23}{7}$ **61.**   **62.** 3.0%

63. Because the graph falls from left to right, the slope is negative.   **64.** $(1997, 44.2), (2002, 41.2)$

65. $y = -0.6x + 1242.4$ **66.** -0.6; yes   **67.** 43.6, 43.0, 42.4, 41.8   **68.** 40.6%; No. The equation is based on data only from 1997 through 2002.

Chapter 3 Test   (pages 268–270)

[3.1] **1.** $-6, -10, -5$ **2.** no   [3.2] **3.** To find the x-intercept, let $y = 0$, and to find the y-intercept, let $x = 0$.

4. x-intercept: $(2, 0)$; y-intercept: $(0, 6)$ **5.** x-intercept: $(0, 0)$; y-intercept: $(0, 0)$

6. x-intercept: $(-3, 0)$; y-intercept: none   **7.** x-intercept: none; y-intercept: $(0, 1)$

8. x-intercept: $(4, 0)$; y-intercept: $(0, -4)$ [3.3] **9.** $-\dfrac{8}{3}$ **10.** -2 **11.** undefined **12.** $\dfrac{5}{2}$ **13.** 0

[3.4] **14.** $y = 2x + 6$ **15.** $y = \dfrac{5}{2}x - 4$ **16.** $y = -9x + 12$ [3.5] **17.** **18.**

[3.1–3.3] **19.** The slope is positive, since food and drink sales are increasing. **20.** $(0, 43)$, $(30, 376)$; 11.1

21. 1990: \$265 billion; 1995: \$320.5 billion **22.** In 2000, food and drink sales were \$376 billion.

[3.6] **23. (a)** not a function **(b)** function; domain: $\{0, 1, 2\}$; range: $\{2\}$ **24.** not a function **25.** 1

Chapters 1–3 Cumulative Review Exercises (pages 270–271)

[1.1] **1.** $\dfrac{301}{40}$, or $7\dfrac{21}{40}$ **2.** 6 [1.5] **3.** 7 [1.6] **4.** $\dfrac{73}{18}$, or $4\dfrac{1}{18}$ [1.2–1.6] **5.** true **6.** -43 [1.7] **7.** distributive

property [1.8] **8.** $-p + 2$ [2.5] **9.** $h = \dfrac{3V}{\pi r^2}$ [2.3] **10.** $\{-1\}$ **11.** $\{2\}$ [2.6] **12.** $\{-13\}$

[2.8] **13.** $(-2.6, \infty)$ ⟶ (at -2.6) **14.** $(0, \infty)$ ⟶ (at 0) **15.** $(-\infty, -4]$ ⟵ (at -4)

[2.4, 2.5] **16.** high school diploma: \$21,577; bachelor's degree: \$40,013 **17.** 13 mi [3.1] **18. (a)** 89.33; 82.21; 77.15 **(b)** In 1980, the winning time was 85.27 sec. **19. (a)** \$7000 **(b)** \$10,000 **(c)** about \$30,000 [3.2] **20.** $(-4, 0)$; $(0, 3)$

[3.3] **21.** $\dfrac{3}{4}$ [3.2] **22.** [3.3] **23.** perpendicular [3.4] **24.** $y = 3x - 11$ **25.** $y = 4$

CHAPTER 4 Systems of Linear Equations and Inequalities

Section 4.1 (pages 279–283)

1. no **3.** yes **5.** yes **7.** no **9.** yes **11. (a)** B **(b)** C **(c)** D **(d)** A **13.** D; The ordered-pair solution must be on the y-axis, with $y < 0$. **15.** $\{(4, 2)\}$ **17.** $\{(0, 4)\}$ **19.** $\{(4, -1)\}$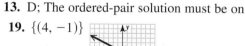

In Exercises 21–29, we do not show the graphs. **21.** $\{(1, 3)\}$ **23.** $\{(0, 2)\}$ **25.** $\emptyset$ (inconsistent system)
27. $\{(x, y) \mid 3x + y = 5\}$ (dependent equations) **29.** $\{(4, -3)\}$ **31.** If the coordinates of the point of intersection are not integers, the solution will be difficult to determine from a graph. **33. (a)** neither **(b)** intersecting lines **(c)** one solution
35. (a) dependent **(b)** one line **(c)** infinite number of solutions **37. (a)** inconsistent **(b)** parallel lines **(c)** no solution
39. (a) neither **(b)** intersecting lines **(c)** one solution **41. (a)** neither **(b)** intersecting lines **(c)** one solution **43.** 2001; about 800 newspapers **45.** 30 **47. (a)** 1997–2002 **(b)** 2001 **(c)** 2002 **(d)** $(1998, 30)$ (The y-value is approximate.)
(e) During the period 1997–2004, debit card use went from least popular to most popular of the three methods depicted.

49. B   **51.** A   **53.** $\{(-1, 5)\}$ **55.** $\{(0.25, -0.5)\}$ **57.** $y = -3x + 4$ **59.** $y = \dfrac{9}{2}x - 2$ **61.** $\{2\}$ **63.** $\{3\}$

65. $\left\{\dfrac{4}{5}\right\}$

Section 4.2   (pages 289–291)

1. No, it is not correct. The solution set is $\{(3, 0)\}$.   **3.** $\{(3, 9)\}$ **5.** $\{(7, 3)\}$ **7.** $\{(0, 5)\}$ **9.** $\{(-4, 8)\}$ **11.** $\{(3, -2)\}$

13. $\{(x, y) \mid 3x - y = 5\}$ **15.** $\left\{\left(\dfrac{1}{4}, -\dfrac{1}{2}\right)\right\}$ **17.** $\emptyset$ **19.** $\{(x, y) \mid 2x - y = -12\}$ **21.** **(a)** A false statement, such as

$0 = 3$, occurs. **(b)** A true statement, such as $0 = 0$, occurs.   **23.** $\{(2, 6)\}$ **25.** $\{(2, -4)\}$ **27.** $\{(-2, 1)\}$

29. $\left\{\left(13, -\dfrac{7}{5}\right)\right\}$ **31.** $\{(x, y) \mid x + 2y = 48\}$ **33.** To find the total cost, multiply the number of bicycles (x) by the cost per

bicycle ($400), and add the fixed cost ($5000). Thus, $y_1 = 400x + 5000$ gives this total cost (in dollars).   **34.** $y_2 = 600x$

35. $y_1 = 400x + 5000$, $y_2 = 600x$; solution set: $\{(25, 15{,}000)\}$ **36.** 25; 15,000; 15,000

37. $\{(2, 4)\}$ 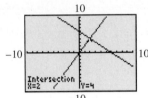   **39.** $\{(1, 5)\}$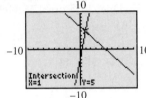

41. $\{(5, -3)\}$; the equations to input are $Y_1 = \dfrac{5 - 4X}{5}$ and $Y_2 = \dfrac{1 - 2X}{3}$.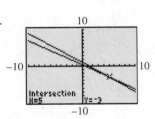

43. Adjust the viewing window so that the point appears.   **45.** $16x$ **47.** $10y$ **49.** $4x$ **51.** -2

Section 4.3   (pages 296–298)

1. false; multiply by -3.   **3.** true   **5.** $\{(4, 6)\}$ **7.** $\{(-1, -3)\}$ **9.** $\{(-2, 3)\}$ **11.** $\emptyset$ **13.** $\{(x, y) \mid x + 3y = 6\}$

15. $\left\{\left(\dfrac{1}{2}, 4\right)\right\}$ **17.** $\{(3, -6)\}$ **19.** $\{(0, 4)\}$ **21.** $\{(0, 0)\}$ **23.** $\{(7, 4)\}$ **25.** $\{(0, 3)\}$ **27.** $\{(3, 0)\}$ **29.** $\{(11, 15)\}$

31. $\{(x, y) \mid x - 3y = -4\}$ **33.** $\emptyset$ **35.** $\{(-3, 2)\}$ **37.** $\left\{\left(-\dfrac{5}{7}, -\dfrac{2}{7}\right)\right\}$ **39.** $\left\{\left(\dfrac{1}{8}, -\dfrac{5}{6}\right)\right\}$ **41.** $\{(0, 0)\}$

43. $1339 = 1996a + b$ **44.** $1536 = 2004a + b$ **45.** $1996a + b = 1339$, $2004a + b = 1536$; solution set:

$\{(24.625, -47{,}812.5)\}$ **46.** $y = 24.625x - 47{,}812.5$ **47.** 1486.8 (million); this is quite a bit less than the actual figure.

48. Since the data do not lie in a perfectly straight line, the quantity obtained from an equation determined in this way will

probably be "off" a bit. We cannot put too much faith in models such as this one, because not all sets of data points are linear in

nature.   **49.** goals: 29; assists: 77   **51.** gold: 37; silver: 34; bronze: 37   **53.** 8 ft   **55.** 13 twenties

Summary Exercises on Solving Systems of Linear Equations   (page 299)

1. **(a)** Use substitution, since the second equation is solved for y. **(b)** Use elimination, since the coefficients of the y-terms are

opposites. **(c)** Use elimination, since the equations are in standard form with no coefficients of 1 or -1. Solving by substitution

would involve fractions.   **2.** The system on the right is easier to solve by substitution because the second equation is already solved for y.   **3. (a)** $\{(1, 4)\}$ **(b)** $\{(1, 4)\}$ **(c)** Answers will vary.   **4. (a)** $\{(-5, 2)\}$ **(b)** $\{(-5, 2)\}$ **(c)** Answers will vary.

5. $\{(3, 12)\}$ **6.** $\{(-3, 2)\}$ **7.** $\left\{\left(\dfrac{1}{3}, \dfrac{1}{2}\right)\right\}$ **8.** $\emptyset$ **9.** $\{(3, -2)\}$ **10.** $\{(-1, -11)\}$ **11.** $\{(x, y) \mid 2x - 3y = 5\}$

12. $\{(9, 4)\}$ **13.** $\left\{\left(\dfrac{45}{31}, \dfrac{4}{31}\right)\right\}$ **14.** $\{(4, -5)\}$ **15.** $\emptyset$ **16.** $\{(-4, 6)\}$ **17.** $\{(-3, 2)\}$ **18.** $\left\{\left(\dfrac{22}{13}, -\dfrac{23}{13}\right)\right\}$

19. $\{(6, 4)\}$ **20.** $\{(5, 3)\}$ **21.** $\{(2, -3)\}$ **22.** $\{(24, -12)\}$ **23.** $\{(10, -12)\}$ **24.** $\{(3, 2)\}$ **25.** $\{(-4, 2)\}$

Section 4.4   (pages 305–310)

1. D   **3.** B   **5.** D   **7.** C   **9.** the second number; $x - y = 48$; The two numbers are 73 and 25.   **11.** Céline Dion: 154; Prince: 96   **13.** *The Lord of the Rings: The Return of the King*: $361.1 million; *Finding Nemo*: $339.7 million

15. (a) 45 units **(b)** Do not produce; the product will lead to a loss.   **17.** quarters: 24; dimes: 15   **19.** 2 DVDs of *Million Dollar Baby;* 5 Clay Aiken CDs   **21.** $2500 at 4%; $5000 at 5%   **23.** Japan: $17.19; Switzerland: $13.15

25. 40% solution: 80 L; 70% solution: 40 L   **27.** 30 lb at $6 per lb; 60 lb at $3 per lb   **29.** nuts: 40 lb; raisins: 20 lb

31. 60 mph; 50 mph   **33.** 35 mph; 65 mph   **35.** bicycle: 15 mph; car: 55 mph   **37.** boat: 10 mph; current: 2 mph

39. plane: 470 mph; wind: 30 mph   **41.** MacFarlane: 17.5 mph; McGuckian: 12.5 mph

43.   **45.**   **47.**

Section 4.5   (pages 314–315)

1. C   **3.** B   **5.**   **7.**   **9.**   **11.**   **13.**

15.   **17.**   **19.**   **21.** 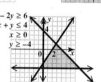   **23.** 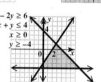   **25.** D

27. A   **29.** 64   **31.** 625   **33.** $\dfrac{8}{27}$ **35.** 128

Chapter 4 Review Exercises   (pages 320–323)

1. yes   **3.** $\{(3, 1)\}$ **5.** $\emptyset$ **7.** No, two lines cannot intersect in exactly three points.   **9.** $\{(3, 5)\}$ **11.** no   **13.** C

15. $\{(7, 1)\}$ **17.** $\{(x, y) \mid 3x - 4y = 9\}$ **19.** $\{(-4, 1)\}$ **21.** $\{(9, 2)\}$ **23.** Answers will vary.   **25.** Subway: 13,247 restaurants; McDonald's: 13,099 restaurants   **27.** U.S. Bank Tower: 73 stories; Aon Center: 62 stories   **29.** length: 27 m; width: 18 m   **31.** 25 lb of $1.30 candy; 75 lb of $0.90 candy   **33.** $7000 at 3%; $11,000 at 4%

35.   **37.**   **39. (a)** years 0 to 6 **(b)** year 6; about $650   **41.** $\{(x, y) \mid x - y = 6\}$

43. $\{(-4, 15)\}$ **45.**   **47.** Statue of Liberty: 3.6 million; National World War II Memorial: 5.4 million

49. small bottles: 102; large bottles: 44

Chapter 4 Test (pages 323–324)

[4.1] **1. (a)** $x = 8$, or 800 parts **(b)** \$3000   **2. (a)** no **(b)** no **(c)** yes   **3.** $\{(4, 1)\}$ [4.2] **4.** $\{(1, -6)\}$
5. $\{(-35, 35)\}$ [4.3] **6.** $\{(5, 6)\}$ **7.** $\{(-1, 3)\}$ **8.** $\{(-1, 3)\}$ **9.** $\emptyset$ **10.** $\{(0, 0)\}$ **11.** $\{(-15, 6)\}$
[4.1–4.3] **12.** $\{(x, y) \mid 3x + 4y = 5\}$ **13.** It has no solution.   [4.4] **14.** Memphis and Atlanta: 394 mi; Minneapolis and

Houston: 1176 mi   **15.** Magic Kingdom: 15.2 million; Disneyland: 13.4 million   **16.** 25% solution: $33\dfrac{1}{3}$ L; 40% solution:

$16\dfrac{2}{3}$ L   **17.** slower car: 45 mph; faster car: 60 mph   [4.5] **18.**   **19.**   **20.** B

Chapters 1–4 Cumulative Review Exercises (pages 325–326)

[1.6] **1.** $-1, 1, -2, 2, -4, 4, -5, 5, -8, 8, -10, 10, -20, 20, -40, 40$ [1.3] **2.** 1   [1.7] **3.** commutative property

4. distributive property   **5.** inverse property   [1.6] **6.** 46   [2.3] **7.** $\left\{-\dfrac{13}{11}\right\}$ **8.** $\left\{\dfrac{9}{11}\right\}$ [2.5] **9.** $T = \dfrac{PV}{k}$

[2.8] **10.** $(-18, \infty)$ **11.** $\left(-\dfrac{11}{2}, \infty\right)$ [2.7] **12.** 2010; 1813; 62.8%; 57.2%   [2.4] **13.** in favor: 78; against: 22

[2.5] **14.** $46°, 46°, 88°$ **15.** width: 8.16 in.; length: 10.74 in.   [3.2] **16.**   **17.**   [3.3] **18.** $-\dfrac{4}{3}$

19. $-\dfrac{1}{4}$ [3.4] **20.** $y = \dfrac{1}{2}x + 3$ **21.** $y = 2x + 1$ **22. (a)** $x = 9$ **(b)** $y = -1$ [4.1–4.3] **23.** $\{(-1, 6)\}$
24. $\{(3, -4)\}$ **25.** $\{(2, -1)\}$ **26.** $\emptyset$ [4.4] **27.** 405 adults and 49 children   **28.** 19 in., 19 in., 15 in.
29. 20% solution: 4 L; 50% solution: 8 L   [4.5] **30.**

CHAPTER 5 Exponents and Polynomials

Section 5.1 (pages 333–336)

1. false; $3^3 = 3 \cdot 3 \cdot 3 = 27$ **3.** false; $(a^2)^3 = a^{2 \cdot 3} = a^6$ **5.** w^6 **7.** $\left(\dfrac{1}{2}\right)^6$ **9.** $(-4)^3$ **11.** $(-7x)^4$ **13.** In $(-3)^4$,
-3 is the base; in -3^4, 3 is the base. $(-3)^4 = 81$; $-3^4 = -81$.   **15.** base: 3; exponent: 5; 243   **17.** base: -3; exponent: 5;
-243 **19.** base: $-6x$; exponent: 4   **21.** base: x; exponent: 4

23. $5^2 + 5^3$ is a sum, not a product; $5^2 + 5^3 = 25 + 125 = 150$. **25.** 5^8 **27.** 4^{12} **29.** $(-7)^9$ **31.** t^{24} **33.** $-56r^7$
35. $42p^{10}$ **37.** $-30x^9$ **39.** The product rule does not apply. **41.** The product rule does not apply. **43.** 4^6 **45.** t^{20}
47. 7^3r^3 **49.** $5^5x^5y^5$ **51.** 5^{12} **53.** -8^{15} **55.** $8q^3r^3$ **57.** $\dfrac{9^8}{5^8}$ **59.** $\dfrac{1}{2^3}$ **61.** $\dfrac{a^3}{b^3}$ **63.** $\dfrac{5^5}{2^5}$ **65.** $\dfrac{9^5}{8^3}$ **67.** $2^{12}x^{12}$
69. -6^5p^5 **71.** $6^5x^{10}y^{15}$ **73.** x^{21} **75.** $4w^4x^{26}y^7$ **77.** $-r^{18}s^{17}$ **79.** $\dfrac{5^3a^6b^{15}}{c^{18}} = \dfrac{125a^6b^{15}}{c^{18}}$ **81.** Using the product rule, it
is simplified as follows: $(10^2)^3 = 10^{2\cdot3} = 10^6 = 1{,}000{,}000$. **83.** $12x^5$ **85.** $6p^7$ **87.** $125x^6$ **89.** $-a^4, -a^3, -(-a)^3,$
$(-a)^4$; One way is to choose a positive number greater than 1 and substitute it for a in each expression. Then arrange the terms
from least to greatest. **91.** \$304.16 **93.** \$1843.88 **95.** $\dfrac{1}{5}$ **97.** -4 **99.** 10 **101.** 3

Section 5.2 (pages 342–344)

1. 1 **3.** 1 **5.** -1 **7.** -1 **9.** 0 **11.** 0 **13.** (a) B (b) C (c) D (d) B (e) E (f) B **15.** 2 **17.** $\dfrac{1}{64}$ **19.** 16
21. $\dfrac{49}{36}$ **23.** $\dfrac{1}{81}$ **25.** $\dfrac{8}{15}$ **27.** $-\dfrac{7}{18}$ **29.** $5^3 = 125$ **31.** $\dfrac{5^3}{3^2} = \dfrac{125}{9}$ **33.** $5^2 = 25$ **35.** x^{15} **37.** $6^3 = 216$ **39.** $2r^4$
41. $\dfrac{5^2}{4^3} = \dfrac{25}{64}$ **43.** $\dfrac{p^5}{q^8}$ **45.** r^9 **47.** $\dfrac{yz^2}{4x^3}$ **49.** $a + b$ **51.** $(x + 2y)^2$ **53.** 1 **54.** $\dfrac{5^2}{5^2}$ **55.** 5^0 **56.** $1 = 5^0$; this
supports the definition of 0 as an exponent. **57.** $7^3 = 343$ **59.** $\dfrac{1}{x^2}$ **61.** $\dfrac{64x}{9}$ **63.** $\dfrac{x^2z^4}{y^2}$ **65.** $6x$ **67.** $\dfrac{1}{m^{10}n^5}$ **69.** $\dfrac{1}{xyz}$
71. x^3y^9 **73.** $\dfrac{a^{11}}{2b^5}$ **75.** $\dfrac{108}{y^5z^3}$ **77.** $\dfrac{9z^2}{400x^3}$ **79.** The student attempted to use the quotient rule with unequal bases. The
correct way to simplify is $\dfrac{16^3}{2^2} = \dfrac{(2^4)^3}{2^2} = \dfrac{2^{12}}{2^2} = 2^{10} = 1024$. **81.** 64,270 **83.** 1230 **85.** 3.4 **87.** 0.237

Summary Exercises on the Rules for Exponents (pages 344–345)

1. $\dfrac{6^{12}x^{24}}{5^{12}}$ **2.** $\dfrac{r^6s^{12}}{729t^6}$ **3.** $10^5x^7y^{14}$ **4.** $-128a^{10}b^{15}c^4$ **5.** $\dfrac{729w^3x^9}{y^{12}}$ **6.** $\dfrac{x^4y^6}{16}$ **7.** c^{22} **8.** $\dfrac{1}{k^4t^{12}}$ **9.** $\dfrac{11}{30}$ **10.** $y^{12}z^3$
11. $\dfrac{x^6}{y^5}$ **12.** 0 **13.** $\dfrac{1}{z^2}$ **14.** $\dfrac{9}{r^2s^2t^{10}}$ **15.** $\dfrac{300x^3}{y^3}$ **16.** $\dfrac{3}{5x^6}$ **17.** x^8 **18.** $\dfrac{y^{11}}{x^{11}}$ **19.** $\dfrac{a^6}{b^4}$ **20.** $6ab$ **21.** $\dfrac{61}{900}$ **22.** 1
23. $\dfrac{343a^6b^9}{8}$ **24.** 1 **25.** -1 **26.** 0 **27.** $\dfrac{27y^{18}}{4x^8}$ **28.** $\dfrac{1}{a^8b^{12}c^{16}}$ **29.** $\dfrac{x^{15}}{216z^9}$ **30.** $\dfrac{q}{8p^6r^3}$ **31.** x^6y^6 **32.** 0 **33.** $\dfrac{343}{x^{15}}$
34. $\dfrac{9}{x^6}$ **35.** $5p^{10}q^9$ **36.** $\dfrac{7}{24}$ **37.** $\dfrac{r^{14}t}{2s^2}$ **38.** 1 **39.** $8p^{10}q$ **40.** $\dfrac{1}{mn^3p^3}$ **41.** -1 **42.** $\dfrac{3}{40}$ **43.** (a) D (b) D (c) E
(d) B (e) J (f) F (g) I (h) B (i) E (j) F

Connections **(page 349)** **1.** The Sumatra earthquake was 10 times as powerful as the Tangshan earthquake. **2.** The
Colombia earthquake had one-hundredth the power of the Alaska earthquake. **3.** The Sumatra earthquake was about 19.95
times as powerful as the California earthquake. **4.** "+3.0" corresponds to a factor of 1000 times stronger; "−1.0" corresponds
to a factor of one-tenth.

Section 5.3 (pages 350–353)

1. (a) C (b) A (c) B (d) D **3.** in scientific notation **5.** not in scientific notation; 5.6×10^6 **7.** not in scientific
notation; 8×10^1 **9.** not in scientific notation; 4×10^{-3} **11.** It is written as the product of a power of 10 and a number

whose absolute value is between 1 and 10 (inclusive of 1). Some examples are 2.3×10^{-4} and 6.02×10^{23}.   **13.** 5.876×10^9

15. 8.235×10^4 **17.** 7×10^{-6} **19.** 2.03×10^{-3} **21.** $750,000$ **23.** $5,677,000,000,000$ **25.** $1,000,000,000,000$

27. 6.21 **29.** 0.00078 **31.** 0.000000005134 **33.** (a) 6×10^{11} (b) $600,000,000,000$ **35.** (a) 1.5×10^7

(b) $15,000,000$ **37.** (a) 6×10^4 (b) $60,000$ **39.** (a) 2.4×10^2 (b) 240 **41.** (a) 6.3×10^{-2} (b) 0.063

43. (a) 3×10^{-4} (b) 0.0003 **45.** (a) 4×10 (b) 40 **47.** (a) 1.3×10^{-5} (b) 0.000013 **49.** (a) 5×10^2 (b) 500

51. $4.7\text{E}^{-}7$ **53.** $2\text{E}7$ **55.** $1\text{E}1$ **57.** 1.04×10^8 **59.** 9.2×10^{-3} **61.** 6×10^9 **63.** 1×10^{10} **65.** $2,000,000,000$

67. 1.341×10^9 **69.** $\$2.524 \times 10^{10}$ **71.** about 2.76×10^{-1}, or 0.276, lb   **73.** 3.59×10^2, or 359, sec   **75.** $\$66.87$

77. $\$30,000$ **79.** 1.5×10^{17} mi   **81.** $\$1,951,700,000,000$ **83.** $\$3554$ **85.** $2x - 36$ **87.** 18 **89.** 63

Connections   (page 357)   The polynomial gives the following: for 3 dog years, 26.4 human years (less than actual); for 5 dog years, 35.7 human years (less than actual, but close); for 7 dog years, 44.4 human years (greater than actual, but close); for 9 dog years, 52.8 human years (greater than actual); for 11 dog years, 60.7 human years (greater than actual); for 15 dog years, 75.2 human years (less than actual).

Section 5.4   (pages 361–364)

1. 7; 5   **3.** 8   **5.** 26   **7.** 0   **9.** 1; 6   **11.** 1; 1   **13.** 2; -19, -1 **15.** 3; 1, 8, 5   **17.** $2m^5$ **19.** $-r^5$

21. It cannot be simplified.   **23.** $-5x^5$ **25.** $5p^9 + 4p^7$ **27.** $-2xy^2$ **29.** already simplified; 4; binomial

31. $11m^4 - 7m^3 - 3m^2$; 4; trinomial   **33.** x^4; 4; monomial   **35.** 7; 0; monomial   **37.** (a) -3 (b) 0   **39.** (a) 14 (b) -19

41. (a) 36 (b) -12 **43.** $5x^2 - 2x$ **45.** $5m^2 + 3m + 2$ **47.** $\dfrac{7}{6}x^2 - \dfrac{2}{15}x + \dfrac{5}{6}$ **49.** $6m^3 + m^2 + 4m - 14$

51. $3y^3 - 11y^2$ **53.** $4x^4 - 4x^2 + 4x$ **55.** $15m^3 - 13m^2 + 8m + 11$ **57.** Answers will vary.   **59.** $5m^2 - 14m + 6$

61. $4x^3 + 2x^2 + 5x$ **63.** $-11y^4 + 8y^2 + y$ **65.** $a^4 - a^2 + 1$ **67.** $5m^2 + 8m - 10$ **69.** $-6x^2 - 12x + 12$

71. -10 **73.** $4b - 5c$ **75.** $6x - xy - 7$ **77.** $-3x^2y - 15xy - 3xy^2$ **79.** $8x^2 + 8x + 6$ **81.** $2x^2 + 8x$

83. (a) $23y + 5t$ (b) $25°, 67°, 88°$ **85.** $-6x^2 + 6x - 7$ **87.** $-7x - 1$

89. $0, -3, -4, -3, 0$ **91.** $7, 1, -1, 1, 7$ **93.** $0, 3, 4, 3, 0$ **95.** $4, 1, 0, 1, 4$ **97.** 9; 63   **98.** 6; $\$27$

99. 2.5; 130   **100.** 60,790   **101.** $-10a^2b$ **103.** $-m^7$ **105.** $5x + 20$ **107.** $8a + 24b$

Section 5.5   (pages 369–372)

1. (a) B (b) D (c) A (d) C   **3.** $15y^{11}$ **5.** $30a^9$ **7.** $15pq^2$ **9.** $-18m^3n^2$ **11.** $6m^2 + 4m$ **13.** $-6p^4 + 12p^3$

15. $-16z^2 - 24z^3 - 24z^4$ **17.** $6y^3 + 4y^4 + 10y^7$ **19.** $28r^5 - 32r^4 + 36r^3$ **21.** $6a^4 - 12a^3b + 15a^2b^2$

23. $21m^5n^2 + 14m^4n^3 - 7m^3n^5$ **25.** $12x^3 + 26x^2 + 10x + 1$ **27.** $81a^3 + 27a^2 + 11a + 2$

29. $20m^4 - m^3 - 8m^2 - 17m - 15$ **31.** $6x^6 - 3x^5 - 4x^4 + 4x^3 - 5x^2 + 8x - 3$ **33.** $5x^4 - 13x^3 + 20x^2 + 7x + 5$

35. $3x^5 + 18x^4 - 2x^3 - 8x^2 + 24x$ **37.** $m^2 + 12m + 35$ **39.** $x^2 - 25$ **41.** $12x^2 + 10x - 12$ **43.** $9x^2 - 12x + 4$

45. $10a^2 + 37a + 7$ **47.** $12 + 8m - 15m^2$ **49.** $20 - 7x - 3x^2$ **51.** $8xy - 4x + 6y - 3$ **53.** $15x^2 + xy - 6y^2$

55. $6y^5 - 21y^4 - 45y^3$ **57.** $-200r^7 + 32r^3$ **59.** (a) $3y^2 + 10y + 7$ (b) $8y + 16$ **61.** $6p^2 - \dfrac{5}{2}pq - \dfrac{25}{12}q^2$

63. $x^2 + 14x + 49$ **65.** $a^2 - 16$ **67.** $4p^2 - 20p + 25$ **69.** $25k^2 + 30kq + 9q^2$ **71.** $m^3 - 15m^2 + 75m - 125$

73. $8a^3 + 12a^2 + 6a + 1$ **75.** $56m^2 - 14m - 21$ **77.** $-9a^3 + 33a^2 + 12a$

79. $81r^4 - 216r^3s + 216r^2s^2 - 96rs^3 + 16s^4$ **81.** $6p^8 + 15p^7 + 12p^6 + 36p^5 + 15p^4$ **83.** $-24x^8 - 28x^7 + 32x^6 + 20x^5$

85. $14x + 49$ **87.** $\pi x^2 - 9$ **89.** $30x + 60$ **90.** $30x + 60 = 600$; 18   **91.** 10 yd by 60 yd   **92.** \$2100   **93.** 140 yd
94. \$1260   **95. (a)** $30kx + 60k$ (dollars) **(b)** $6rx + 32r$ **97.** $9m^2$ **99.** $4r^2$ **101.** $16x^4$

Section 5.6   (pages 375–378)

1. (a) $4x^2$ **(b)** $12x$ **(c)** 9 **(d)** $4x^2 + 12x + 9$ **3.** $m^2 + 4m + 4$ **5.** $r^2 - 6r + 9$ **7.** $x^2 + 4xy + 4y^2$

9. $25p^2 + 20pq + 4q^2$ **11.** $16a^2 + 40ab + 25b^2$ **13.** $49t^2 + 14ts + s^2$ **15.** $36m^2 - \dfrac{48}{5}mn + \dfrac{16}{25}n^2$

17. $9t^3 - 6t^2 + t$ **19.** $-16r^2 + 16r - 4$ **21. (a)** $49x^2$ **(b)** 0 **(c)** $-9y^2$ **(d)** $49x^2 - 9y^2$; because 0 is the identity element
for addition, it is not necessary to write "+ 0."   **23.** three; two   **25.** $k^2 - 25$ **27.** $16 - 9t^2$ **29.** $25x^2 - 4$

31. $36a^2 - p^2$ **33.** $100x^2 - 9y^2$ **35.** $4x^4 - 25$ **37.** $\dfrac{9}{16} - x^2$ **39.** $81y^2 - \dfrac{4}{9}$ **41.** $25q^3 - q$

43. $x^3 + 3x^2 + 3x + 1$ **45.** $t^3 - 9t^2 + 27t - 27$ **47.** $r^3 + 15r^2 + 75r + 125$ **49.** $8a^3 + 12a^2 + 6a + 1$
51. $256x^4 - 256x^3 + 96x^2 - 16x + 1$ **53.** $81r^4 - 216r^3t + 216r^2t^2 - 96rt^3 + 16t^4$ **55.** $(a + b)^2$ **56.** a^2 **57.** $2ab$
58. b^2 **59.** $a^2 + 2ab + b^2$ **60.** They both represent the area of the entire large square.   **61.** 1225

62. $30^2 + 2(30)(5) + 5^2$ **63.** 1225   **64.** They are equal.   **65.** 9999   **67.** 39,999   **69.** $399\dfrac{3}{4}$ **71.** $\dfrac{1}{2}m^2 - 2n^2$

73. $9a^2 - 4$ **75.** $\pi x^2 + 4\pi x + 4\pi$ **77.** $x^3 + 6x^2 + 12x + 8$ **79.** $2p + 1 + \dfrac{4}{p}$ **81.** $\dfrac{m^2}{3} + 3m - 2$

83. $-24k^3 + 36k^2 - 6k$ **85.** $-16k^3 - 10k^2 + 3k + 3$ **87.** $10x^2 - 4x + 11$

Section 5.7   (pages 384–386)

1. $6x^2 + 8$; 2; $3x^2 + 4$ **3.** $3x^2 + 4$; 2 (These may be reversed.); $6x^2 + 8$ **5.** The first is a polynomial divided by a
monomial, covered in Objective 1. This section does not cover dividing a monomial by a polynomial of several terms.

7. $30x^3 - 10x + 5$ **9.** $4m^3 - 2m^2 + 1$ **11.** $4t^4 - 2t^2 + 2t$ **13.** $a^4 - a + \dfrac{2}{a}$ **15.** $4x^3 - 3x^2 + 2x$

17. $-9x^2 + 5x + 1$ **19.** $2x + 8 + \dfrac{12}{x}$ **21.** $\dfrac{4x^2}{3} + x + \dfrac{2}{3x}$ **23.** $9r^3 - 12r^2 + 2r + \dfrac{26}{3} - \dfrac{2}{3r}$ **25.** $-m^2 + 3m - \dfrac{4}{m}$

27. $-3a + 4 + \dfrac{5}{a}$ **29.** $\dfrac{12}{x} - \dfrac{6}{x^2} + \dfrac{14}{x^3} - \dfrac{10}{x^4}$ **31.** $6x^4y^2 - 4xy + 2xy^2 - x^4y$ **33.** $5x^3 + 4x^2 - 3x + 1$

35. 1423   **36.** $(1 \times 10^3) + (4 \times 10^2) + (2 \times 10^1) + (3 \times 10^0)$ **37.** $x^3 + 4x^2 + 2x + 3$ **38.** They are similar in that
the coefficients of powers of 10 are equal to the coefficients of the powers of x. They are different in that one is a constant while
the other is a polynomial. They are equal if $x = 10$ (the base of our decimal system).   **39.** $x + 2$ **41.** $2y - 5$

43. $p - 4 + \dfrac{44}{p + 6}$ **45.** $r - 5$ **47.** $6m - 1$ **49.** $2a - 14 + \dfrac{74}{2a + 3}$ **51.** $4x^2 - 7x + 3$ **53.** $4k^3 - k + 2$

55. $5y^3 + 2y - 3$ **57.** $3k^2 + 2k - 2 + \dfrac{6}{k - 2}$ **59.** $2p^3 - 6p^2 + 7p - 4 + \dfrac{14}{3p + 1}$ **61.** $x^2 + x + 3$

63. $2x^2 - 2x + 3 + \dfrac{-1}{x + 1}$ **65.** $r^2 - 1 + \dfrac{4}{r^2 - 1}$ **67.** $y^2 - y + 1$ **69.** $a^2 + 1$ **71.** $x^2 - 4x + 2 + \dfrac{9x - 4}{x^2 + 3}$

73. $x^3 + 3x^2 - x + 5$ **75.** $\dfrac{3}{2}a - 10 + \dfrac{77}{2a + 6}$ **77.** $(x^2 + x - 3)$ units   **79.** $(5x^2 - 11x + 14)$ hr

81. 1, 2, 3, 6, 9, 18   **83.** 1, 2, 3, 4, 6, 8, 12, 16, 24, 48

Chapter 5 Review Exercises   (pages 391–394)

1. 4^{11} **3.** $-72x^7$ **5.** 19^5x^5 **7.** $5p^4t^4$ **9.** $3^3x^6y^9$ **11.** $6^2x^{16}y^4z^{16}$ **13.** The expression is a *sum* of powers of 7, not a

product.   **15.** -1 **17.** $-\dfrac{1}{49}$ **19.** 5^8 **21.** $\dfrac{3}{4}$ **23.** x^2 **25.** $\dfrac{r^8}{81}$ **27.** $\dfrac{1}{a^3b^5}$ **29.** 4.8×10^7 **31.** 8.24×10^{-8}

33. $78{,}300{,}000$ **35.** 800 **37.** 0.025 **39.** 0.0000000000016 **41.** 9.7×10^4; 5×10^3 **43.** 1×10^3; 2×10^3; 5×10^4;

1×10^5 **45.** $p^3 - p^2 + 4p + 2$; degree 3; none of these   **47.** $-8y^5 - 7y^4 + 9y$; degree 5; trinomial

49. $13x^3y^2 - 5xy^5 + 21x^2$ **51.** $y^2 - 10y + 9$ **53.** $1, 4, 5, 4, 1$ **55.** $a^3 - 2a^2 - 7a + 2$

57. $5p^5 - 2p^4 - 3p^3 + 25p^2 + 15p$ **59.** $6k^2 - 9k - 6$ **61.** $12k^2 - 32kq - 35q^2$ **63.** $a^2 + 8a + 16$ **65.** $36m^2 - 25$
67. $r^3 + 6r^2 + 12r + 8$ **69.** (a) Answers will vary. For example, let $x = 1$ and $y = 2$. $(1 + 2)^2 \neq 1^2 + 2^2$, because $9 \neq 5$.
(b) Answers will vary. For example, let $x = 1$ and $y = 2$. $(1 + 2)^3 \neq 1^3 + 2^3$, because $27 \neq 9$.   **71.** In both cases, $x = 0$ and
$y = 1$ lead to 1 on each side of the inequality. This would not be sufficient to show that, *in general*, the inequality is true. It would
be necessary to choose other values of x and y.   **73.** $\left(\dfrac{4}{3}\pi(x + 1)^3\right)$ in.^{3}, or $\left(\dfrac{4}{3}\pi x^3 + 4\pi x^2 + 4\pi x + \dfrac{4}{3}\pi\right)$ in.^{3}

75. $y^3 - 2y + 3$ **77.** $2mn + 3m^4n^2 - 4n$ **79.** $2r + 7$ **81.** $x^2 + 3x - 4$ **83.** $4x - 5$ **85.** $y^2 + 2y + 4$

87. $2y^2 - 5y + 4 + \dfrac{-5}{3y^2 + 1}$ **89.** 2 **91.** $144a^2 - 1$ **93.** $\dfrac{1}{8^{12}}$ **95.** $\dfrac{2}{3m^3}$ **97.** r^{13} **99.** $-y^2 - 4y + 4$

101. $y^2 + 5y + 1$ **103.** $10p^2 - 3p - 5$ **105.** $49 - 28k + 4k^2$ **107.** (a) $6x - 2$ (b) $2x^2 + x - 6$

Chapter 5 Test   (pages 395–396)

[5.1, 5.2] **1.** $\dfrac{1}{625}$ **2.** 2 **3.** $\dfrac{7}{12}$ **4.** $9x^3y^5$ **5.** 8^5 **6.** x^2y^6 **7.** (a) positive (b) positive (c) negative (d) positive

(e) zero (f) negative   [5.3] **8.** (a) 4.5×10^{10} (b) 0.0000036 (c) 0.00019 **9.** (a) 1×10^3; 5.89×10^{12}
(b) 5.89×10^{15} mi   [5.4] **10.** $-7x^2 + 8x$; 2; binomial   **11.** $4n^4 + 13n^3 - 10n^2$; 4; trinomial
12. $4, -2, -4, -2, 4$ **13.** $-2y^2 - 9y + 17$ **14.** $-21a^3b^2 + 7ab^5 - 5a^2b^2$ **15.** $-12t^2 + 5t + 8$

[5.5] **16.** $-27x^5 + 18x^4 - 6x^3 + 3x^2$ **17.** $t^2 - 5t - 24$ **18.** $8x^2 + 2xy - 3y^2$ [5.6] **19.** $25x^2 - 20xy + 4y^2$

20. $100v^2 - 9w^2$ [5.5] **21.** $2r^3 + r^2 - 16r + 15$ [5.6] **22.** $12x + 36$; $9x^2 + 54x + 81$ [5.7] **23.** $4y^2 - 3y + 2 + \dfrac{5}{y}$

24. $-3xy^2 + 2x^3y^2 + 4y^2$ **25.** $3x^2 + 6x + 11 + \dfrac{26}{x - 2}$

Chapters 1–5 Cumulative Review Exercises   (pages 396–398)

[1.1] **1.** $\dfrac{7}{4}$ **2.** 5 **3.** $\dfrac{19}{24}$ **4.** $-\dfrac{1}{20}$ **5.** $31\dfrac{1}{4}$ yd^3 [1.6] **6.** \$1836   **7.** $1, 3, 5, 9, 15, 45$ **8.** -8 **9.** $\dfrac{1}{2}$

10. -4 [1.7] **11.** associative property   **12.** distributive property   [1.8] **13.** $-10x^2 + 21x - 29$

[2.1–2.3] **14.** $\left\{\dfrac{13}{4}\right\}$ **15.** $\varnothing$ [2.5] **16.** $r = \dfrac{d}{t}$ [2.6] **17.** $\{-5\}$ [2.1–2.3] **18.** $\{-12\}$ **19.** $\{20\}$

20. {all real numbers}   [2.4] **21.** exertion: 9443 calories; regulating body temperature: 1757 calories   **22.** 4

[2.8] **23.** 11 ft and 22 ft   **24.** $[10, \infty)$ **25.** $\left(-\infty, -\dfrac{14}{5}\right)$ **26.** $[-4, 2)$ [3.2] **27.**

[3.3, 3.4] **28.** **(a)** 1 **(b)** $y = x + 6$ [3.5] **29.** no   [3.6] **30.** -1 [4.2] **31.** $\{(-3, -1)\}$ [4.3] **32.** $\{(4, -5)\}$

[5.1, 5.2] **33.** $\dfrac{5}{4}$ **34.** 2   **35.** 1   **36.** $\dfrac{2b}{a^{10}}$ [5.3] **37.** 3.45×10^4 **38.** about 10,800,000 km

[5.4] **39.**   **40.** $11x^3 - 14x^2 - x + 14$ [5.5] **41.** $18x^7 - 54x^6 + 60x^5$ **42.** $63x^2 + 57x + 12$

[5.6] **43.** $25x^2 + 80x + 64$ [5.7] **44.** $2x^2 - 3x + 1$ **45.** $y^2 - 2y + 6$

CHAPTER 6   Factoring and Applications

Section 6.1   (pages 406–408)

1. 4   **3.** 6   **5.** 1   **7.** First, verify that you have factored completely. Then multiply the factors. The product should be the original polynomial.   **9.** 8   **11.** $10x^3$ **13.** $6m^3n^2$ **15.** xy^2 or $-xy^2$ **17.** factored   **19.** not factored   **21.** yes; x^3y^2 **23.** $3m^2$ **25.** $2z^4$ **27.** $2mn^4$ **29.** $y + 2$ **31.** $a - 2$ **33.** $2 + 3xy$ **35.** $9m(3m^2 - 1)$ **37.** $8z^2(2z^2 + 3)$ **39.** $\dfrac{1}{4}d(d - 3)$ **41.** $6x^2(2x + 1)$ **43.** $5y^6(13y^4 + 7)$ **45.** no common factor (except 1)   **47.** $8m^2n^2(n + 3)$ **49.** $13y^2(y^6 + 2y^2 - 3)$ **51.** $9p^3q(4p^3 + 5p^2q^3 + 9q)$ **53.** $a^3(a^2 + 2b^2 - 3a^2b^2 + 4ab^3)$ **55.** $(x + 2)(c - d)$ **57.** $(m + 2n)(m + n)$ **59.** not in factored form; $(7t + 4)(8 + x)$ **61.** in factored form   **63.** not in factored form; $(y + 4)(18x^2 + 7)$ **65.** The quantities in parentheses are not the same, so there is no common factor of the two terms $12k^3(s - 3)$ and $7(s + 3)$.   **67.** $(p + 4)(p + q)$ **69.** $(a - 2)(a + b)$ **71.** $(z + 2)(7z - a)$ **73.** $(3r + 2y)(6r - x)$ **75.** $(a^2 + b^2)(3a + 2b)$ **77.** $(1 - a)(1 - b)$ **79.** $(4m - p^2)(4m^2 - p)$ **81.** $(5 - 2p)(m + 3)$ **83.** $(3r + 2y)(6r - t)$ **85.** $(a^5 - 3)(1 + 2b)$ **87.** commutative property   **88.** $2x(y - 4) - 3(y - 4)$ **89.** No, because it is not a product. It is the difference between $2x(y - 4)$ and $3(y - 4)$.   **90.** $(2x - 3)(y - 4)$; yes   **91.** **(a)** yes **(b)** When either one is multiplied out, the product is $1 - a + ab - b$.   **93.** $x^2 - 3x - 54$ **95.** $x^2 + 9x + 14$ **97.** $2x^4 + 6x^3 + 10x^2$

Section 6.2   (pages 412–414)

1. 1 and 48, -1 and -48, 2 and 24, -2 and -24, 3 and 16, -3 and -16, 4 and 12, -4 and -12, 6 and 8, -6 and -8; the pair with a sum of -19 is -3 and -16.   **3.** 1 and -24, -1 and 24, 2 and -12, -2 and 12, 3 and -8, -3 and 8, 4 and -6, -4 and 6; the pair with a sum of -5 is 3 and -8.   **5.** a and b must have different signs, one positive and one negative.   **7.** A prime polynomial is a polynomial that cannot be factored by using only integers in the factors.   **9.** C   **11.** $p + 6$ **13.** $x + 11$ **15.** $x - 8$ **17.** $y - 5$ **19.** $x + 11$ **21.** $y - 9$ **23.** $(y + 8)(y + 1)$ **25.** $(b + 3)(b + 5)$ **27.** $(m + 5)(m - 4)$ **29.** $(y - 5)(y - 3)$ **31.** prime   **33.** $(z - 7)(z - 8)$ **35.** $(r - 6)(r + 5)$ **37.** $(a + 4)(a - 12)$ **39.** prime   **41.** Factor $8 + 6x + x^2$ directly to get $(2 + x)(4 + x)$. Alternatively, use the commutative property to write the trinomial as $x^2 + 6x + 8$ and factor to get $(x + 2)(x + 4)$, an equivalent answer.   **43.** $(r + 2a)(r + a)$ **45.** $(t + 2z)(t - 3z)$ **47.** $(x + y)(x + 3y)$ **49.** $(v - 5w)(v - 6w)$ **51.** $4(x + 5)(x - 2)$

53. $2t(t + 1)(t + 3)$ **55.** $2x^4(x - 3)(x + 7)$ **57.** $5m^2(m^3 + 5m^2 - 8)$ **59.** $mn(m - 6n)(m - 4n)$

61. $(2x + 4)$ has a common factor of 2 that must be factored out.   **63.** $a^3(a + 4b)(a - b)$ **65.** $yz(y + 3z)(y - 2z)$

67. $z^8(z - 7y)(z + 3y)$ **69.** $(a + b)(x + 4)(x - 3)$ **71.** $(2p + q)(r - 9)(r - 3)$ **73.** $a^2 + 13a + 36$

75. $2y^2 + y - 28$ **77.** $15z^2 - 4z - 4$ **79.** $8p^2 - 10p - 3$

Section 6.3   (pages 419–421)

1. $(2t + 1)(5t + 2)$ **3.** $(3z - 2)(5z - 3)$ **5.** $(2s - t)(4s + 3t)$ **7.** **(a)** 2, 12, 24, 11 **(b)** 3, 8 (Order is irrelevant.)
(c) $3m, 8m$ **(d)** $2m^2 + 3m + 8m + 12$ **(e)** $(2m + 3)(m + 4)$ **(f)** $(2m + 3)(m + 4) = 2m^2 + 11m + 12$ **9.** B   **11.** B
13. A   **15.** $2a + 5b$ **17.** $x^2 + 3x - 4; x + 4, x - 1$ or $x - 1, x + 4$ **19.** The binomial $2x - 6$ cannot be a factor
because it has a common factor of 2, which the polynomial does not have.   **21.** $(3a + 7)(a + 1)$ **23.** $(2y + 3)(y + 2)$
25. $(3m - 1)(5m + 2)$ **27.** $(3s - 1)(4s + 5)$ **29.** $(5m - 4)(2m - 3)$ **31.** $(4w - 1)(2w - 3)$
33. $(4y + 1)(5y - 11)$ **35.** prime   **37.** $2(5x + 3)(2x + 1)$ **39.** $3(4x - 1)(2x - 3)$ **41.** $q(5m + 2)(8m - 3)$
43. $3n^2(5n - 3)(n - 2)$ **45.** $y^2(5x - 4)(3x + 1)$ **47.** $(5a + 3b)(a - 2b)$ **49.** $(4s + 5t)(3s - t)$
51. $m^4n(3m + 2n)(2m + n)$ **53.** $(5 - x)(1 - x)$ **55.** $(4 + 3x)(4 + x)$ **57.** $-5x(2x + 7)(x - 4)$
59. $(12x + 1)(x - 4)$ **61.** $(24y + 7x)(y - 2x)$ **63.** $(18x^2 - 5y)(2x^2 - 3y)$ **65.** $2(24a + b)(a - 2b)$
67. $x^2y^5(10x - 1)(x + 4)$ **69.** $4ab^2(9a + 1)(a - 3)$ **71.** $(12x - 5)(2x - 3)$ **73.** $(8x^2 - 3)(3x^2 + 8)$
75. $(4x + 3y)(6x + 5y)$ **77.** $(xz^2 - 5)(24xz^2 + 7)$ **79.** $-1(x + 7)(x - 3)$ **81.** $-1(3x + 4)(x - 1)$
83. $-1(a + 2b)(2a + b)$ **85.** Yes, $(x + 7)(3 - x)$ is equivalent to $-1(x + 7)(x - 3)$ because
$-1(x - 3) = -x + 3 = 3 - x.$ **87.** $(m + 1)^3(5q - 2)(5q + 1)$ **89.** $(r + 3)^3(5x + 2y)(3x - 8y)$ **91.** $-4, 4$

93. $-11, -7, 7, 11$ **95.** $49p^2 - 9$ **97.** $r^4 - \dfrac{1}{4}$ **99.** $9t^2 + 24t + 16$

Section 6.4   (pages 428–430)

1. 1; 4; 9; 16; 25; 36; 49; 64; 81; 100; 121; 144; 169; 196; 225; 256; 289; 324; 361; 400   **3.** 1; 8; 27; 64; 125; 216; 343; 512;
729; 1000   **5.** **(a)** both of these **(b)** perfect cube **(c)** perfect square **(d)** perfect square   **7.** $(y + 5)(y - 5)$

9. $\left(p + \dfrac{1}{3}\right)\left(p - \dfrac{1}{3}\right)$ **11.** prime   **13.** $(3r + 2)(3r - 2)$ **15.** $\left(6m + \dfrac{4}{5}\right)\left(6m - \dfrac{4}{5}\right)$ **17.** $4(3x + 2)(3x - 2)$

19. $(14p + 15)(14p - 15)$ **21.** $(4r + 5a)(4r - 5a)$ **23.** prime   **25.** $(p^2 + 7)(p^2 - 7)$ **27.** $(x^2 + 1)(x + 1)(x - 1)$
29. $(p^2 + 16)(p + 4)(p - 4)$ **31.** $x^2 - 9$ can be factored as $(x + 3)(x - 3)$. The completely factored form is

$(x^2 + 9)(x + 3)(x - 3).$ **33.** $(w + 1)^2$ **35.** $(x - 4)^2$ **37.** $\left(t + \dfrac{1}{2}\right)^2$ **39.** $(x - 0.5)^2$ **41.** $2(x + 6)^2$

43. $(4x - 5)^2$ **45.** $(7x - 2y)^2$ **47.** $(8x + 3y)^2$ **49.** $2(5h - 2y)^2$ **51.** $k(4k^2 - 4k + 9)$ **53.** $z^2(25z^2 + 5z + 1)$
55. 10   **57.** 9   **59.** $(a - 1)(a^2 + a + 1)$ **61.** $(m + 2)(m^2 - 2m + 4)$ **63.** $(3x - 4)(9x^2 + 12x + 16)$
65. $6(p + 1)(p^2 - p + 1)$ **67.** $5(x + 2)(x^2 - 2x + 4)$ **69.** $2(x - 2y)(x^2 + 2xy + 4y^2)$
71. $(2p + 9q)(4p^2 - 18pq + 81q^2)$ **73.** $(3a + 4b)(9a^2 - 12ab + 16b^2)$ **75.** $(5t + 2s)(25t^2 - 10ts + 4s^2)$
77. $(2x - 5y^2)(4x^2 + 10xy^2 + 25y^4)$ **79.** $(3m^2 + 2n)(9m^4 - 6m^2n + 4n^2)$ **81.** $(x + y)(x^2 - xy + y^2)(x^6 - x^3y^3 + y^6)$

83. $4mn$ **85.** $(m - p + 2)(m + p)$ **87.** $\{4\}$ **89.** $\left\{\dfrac{9}{4}\right\}$ **91.** $\left\{\dfrac{2}{3}\right\}$

Summary Exercises on Factoring   (pages 431–432)

1. $(a - 6)(a + 2)$ **2.** $(a + 8)(a + 9)$ **3.** $6(y - 2)(y + 1)$ **4.** $7y^4(y + 6)(y - 4)$ **5.** $6(a + 2b + 3c)$
6. $(m - 4n)(m + n)$ **7.** $(p - 11)(p - 6)$ **8.** $(z + 7)(z - 6)$ **9.** $(5z - 6)(2z + 1)$ **10.** $2(m - 8)(m + 3)$

11. $17xy(x^2y + 3)$ **12.** $5(3y + 1)$ **13.** $8a^3(a - 3)(a + 2)$ **14.** $(4k + 1)(2k - 3)$ **15.** $(z - 5a)(z + 2a)$

16. $50(z^2 - 2)$ **17.** $(x - 5)(x - 4)$ **18.** $10nr(10nr + 3r^2 - 5n)$ **19.** $(3n - 2)(2n - 5)$ **20.** $(3y - 1)(3y + 5)$

21. $4(4x + 5)$ **22.** $(m + 5)(m - 3)$ **23.** $(3y - 4)(2y + 1)$ **24.** $(m + 9)(m - 9)$ **25.** $(6z + 1)(z + 5)$

26. $(12x - 1)(x + 4)$ **27.** $(2k - 3)^2$ **28.** $(8p - 1)(p + 3)$ **29.** $6(3m + 2z)(3m - 2z)$ **30.** $(4m - 3)(2m + 1)$

31. $(3k - 2)(k + 2)$ **32.** $15a^3b^2(3b^3 - 4a + 5a^3b^2)$ **33.** $7k(2k + 5)(k - 2)$ **34.** $(5 + r)(1 - s)$

35. $(y^2 + 4)(y + 2)(y - 2)$ **36.** $10y^4(2y - 3)$ **37.** $8m(1 - 2m)$ **38.** $(k + 4)(k - 4)$ **39.** $(z - 2)(z^2 + 2z + 4)$

40. $(y - 8)(y + 7)$ **41.** prime **42.** $9p^8(3p + 7)(p - 4)$ **43.** $8m^3(4m^6 + 2m^2 + 3)$

44. $(2m + 5)(4m^2 - 10m + 25)$ **45.** $(4r + 3m)^2$ **46.** $(z - 6)^2$ **47.** $(5h + 7g)(3h - 2g)$ **48.** $5z(z - 7)(z - 2)$

49. $(k - 5)(k - 6)$ **50.** $4(4p - 5m)(4p + 5m)$ **51.** $3k(k - 5)(k + 1)$ **52.** $(y - 6k)(y + 2k)$

53. $(10p + 3)(100p^2 - 30p + 9)$ **54.** $(4r - 7)(16r^2 + 28r + 49)$ **55.** $(2 + m)(3 + p)$ **56.** $(2m - 3n)(m + 5n)$

57. $(4z - 1)^2$ **58.** $5m^2(5m - 3n)(5m - 13n)$ **59.** $3(6m - 1)^2$ **60.** $(10a + 9y)(10a - 9y)$ **61.** prime

62. $(2y + 5)(2y - 5)$ **63.** $8z(4z - 1)(z + 2)$ **64.** $5(2m - 3)(m + 4)$ **65.** $(4 + m)(5 + 3n)$ **66.** $(2 - q)(2 - 3p)$

67. $2(3a - 1)(a + 2)$ **68.** $6y^4(3y + 4)(2y - 5)$ **69.** $(a - b)(a^2 + ab + b^2 + 2)$ **70.** $4(2k - 3)^2$ **71.** $(8m - 5n)^2$

72. $12y^2(6yz^2 + 1 - 2y^2z^2)$ **73.** $(4k - 3h)(2k + h)$ **74.** $(2a + 5)(a - 6)$ **75.** $2(x + 4)(x^2 - 4x + 16)$

76. $(2a - 3)(4a^2 + 6a + 9)$ **77.** $(5y - 6z)(2y + z)$ **78.** $(m - 2)^2$ **79.** $(8a - b)(a + 3b)$

80. $(a^2 + 25)(a + 5)(a - 5)$ **81.** $(x^3 - 1)(x^3 + 1)$ **82.** $(x - 1)(x^2 + x + 1)(x + 1)(x^2 - x + 1)$

83. $(x^2 - 1)(x^4 + x^2 + 1)$ **84.** $(x - 1)(x + 1)(x^4 + x^2 + 1)$ **85.** The result in Exercise 82 is factored completely.

86. Show that $x^4 + x^2 + 1 = (x^2 + x + 1)(x^2 - x + 1)$. **87.** difference of squares

88. $(x - 3)(x^2 + 3x + 9)(x + 3)(x^2 - 3x + 9)$

Section 6.5 (pages 438–441)

1. $ax^2 + bx + c$ **3.** factor **5.** $0; x$ **7.** $\left\{-3, \dfrac{7}{2}, 4\right\}$ **9.** To solve $2x(3x - 4) = 0$, set each *variable* factor equal to 0 to

get $x = 0$ or $3x - 4 = 0$. The *constant* factor 2 does not introduce solutions into the equation. The solution set is $\left\{0, \dfrac{4}{3}\right\}$.

11. $\{-5, 2\}$ **13.** $\left\{3, \dfrac{7}{2}\right\}$ **15.** $\left\{-\dfrac{5}{6}, 0\right\}$ **17.** $\left\{0, \dfrac{4}{3}\right\}$ **19.** $\left\{-\dfrac{1}{2}, \dfrac{1}{6}\right\}$ **21.** $\{-0.8, 2\}$ **23.** $\{9\}$

25. Set each *variable* factor equal to 0 to get $3x = 0$ or $5x - 4 = 0$. The solution set is $\left\{0, \dfrac{4}{5}\right\}$. **27.** $\{-2, -1\}$ **29.** $\{1, 2\}$

31. $\{-8, 3\}$ **33.** $\{-1, 3\}$ **35.** $\{-2, -1\}$ **37.** $\{-4\}$ **39.** $\left\{-2, \dfrac{1}{3}\right\}$ **41.** $\left\{-\dfrac{4}{3}, \dfrac{1}{2}\right\}$ **43.** $\left\{-\dfrac{2}{3}\right\}$ **45.** $\{-3, 3\}$

47. $\left\{-\dfrac{7}{4}, \dfrac{7}{4}\right\}$ **49.** $\{-11, 11\}$ **51.** $\{0, 7\}$ **53.** $\left\{0, \dfrac{1}{2}\right\}$ **55.** $\{2, 5\}$ **57.** $\left\{-4, \dfrac{1}{2}\right\}$ **59.** $\{-17, 4\}$

61. $\left\{-\dfrac{5}{2}, \dfrac{1}{3}, 5\right\}$ **63.** $\left\{-\dfrac{7}{2}, -3, 1\right\}$ **65.** $\left\{-\dfrac{7}{3}, 0, \dfrac{7}{3}\right\}$ **67.** $\{-2, 0, 4\}$ **69.** $\{-5, 0, 4\}$ **71.** $\{-3, 0, 5\}$

73. $\{-1, 3\}$ **75.** $\{-1, 3\}$ **77.** $\{3\}$ **79.** $\left\{-\dfrac{4}{3}, -1, \dfrac{1}{2}\right\}$ **81.** $\left\{-\dfrac{2}{3}, 4\right\}$ **83.** **(a)** 64; 144; 4; 6 **(b)** No time has

elapsed, so the object hasn't fallen (been released) yet. **(c)** Time cannot be negative. **85.** $\{-0.5, 0.1\}$ **87.** $\{-2.5, -1.1\}$

89. Florida: 67 counties; California: 58 counties **91.** 7 m **93.** 9, 10 **95.** 8 in.

Section 6.6   (pages 447–452)

1. Read; variable; equation; Solve; answer; Check, original   **3.** *Step 3:* $45 = (2x + 1)(x + 1)$; *Step 4:* $x = 4$ or $x = -\dfrac{11}{2}$;

Step 5: base: 9 units; height: 5 units; *Step 6:* $9 \cdot 5 = 45$ **5.** *Step 3:* $80 = (x + 8)(x - 8)$; *Step 4:* $x = 12$ or $x = -12$;

Step 5: length: 20 units; width: 4 units; *Step 6:* $20 \cdot 4 = 80$ **7.** length: 14 cm; width: 12 cm   **9.** height: 13 in.; width: 10 in.

11. base: 12 in.; height: 5 in.   **13.** length: 15 in.; width: 12 in.   **15.** mirror: 7 ft; painting: 9 ft   **17.** 20, 21

19. 0, 1, 2 or 7, 8, 9   **21.** 7, 9, 11   **23.** $-2, 0, 2$ or 6, 8, 10   **25.** 12 cm   **27.** 12 mi   **29.** 8 ft   **31. (a)** 1 sec

(b) $\dfrac{1}{2}$ sec and $1\dfrac{1}{2}$ sec **(c)** 3 sec **(d)** The negative solution, -1, does not make sense, since t represents time, which cannot be

negative.   **33. (a)** 45.5 million; the result obtained from the model is a little more than 44 million, the actual number for 1996.

(b) 14 **(c)** 184.3 million; the result is a little more than 182 million. **(d)** 233.9 million   **34.** $107.4 billion; 40%

35. 1997: $148.5 billion; 1999: $230.1 billion; 2000: $270.9 billion   **36.** The answers using the linear equation are not at all

close to the actual data.   **37.** 1997: $111.2 billion; 1999: $266.4 billion; 2000: $399.5 billion   **38.** The answers in

Exercise 37 are fairly close to the actual data. The quadratic equation models the data better.   **39.** $(0, 97.5)$, $(1, 104.3)$,

$(2, 104.7)$, $(3, 164.3)$, $(4, 271.3)$ $(5, 378.7)$ **40.** no   **41.** $776.7 billion

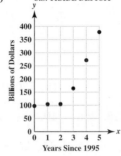

U.S. TRADE DEFICIT

42. (a) The actual deficit is quite a bit less than the prediction. **(b)** No, data for later years might not follow the same pattern.

43. $\dfrac{25}{36}$ **45.** $\dfrac{2}{1}$, or 2   **47.** $\dfrac{16}{-9}$, or $-\dfrac{16}{9}$

Chapter 6 Review Exercises   (pages 457–461)

1. $7(t + 2)$ **3.** $(2y + 3)(x - 4)$ **5.** $(x + 3)(x + 2)$ **7.** $(q + 9)(q - 3)$ **9.** $(r + 8s)(r - 12s)$

11. $8p(p + 2)(p - 5)$ **13.** $p^5(p - 2q)(p + q)$ **15.** $9x^2y(x + 2)(x - 3)$ **17.** r and $6r$, $2r$ and $3r$

19. $(2k - 1)(k - 2)$ **21.** $(3r + 2)(2r - 3)$ **23.** $(v + 3)(8v - 7)$ **25.** $-3(x + 2)(2x - 5)$

27. $4x^2y(3x + y)(4x - y)$ **29.** B   **31.** $(n + 7)(n - 7)$ **33.** $(7y + 5w)(7y - 5w)$ **35.** prime   **37.** $(3t - 7)^2$

39. $(5k + 4x)(25k^2 - 20kx + 16x^2)$ **41.** $(10 - 3x^2)(100 + 30x^2 + 9x^4)$ **43.** $\left\{-\dfrac{3}{4}, 1\right\}$ **45.** $\left\{0, \dfrac{5}{2}\right\}$ **47.** $\{1, 4\}$

49. $\left\{-\dfrac{4}{3}, 5\right\}$ **51.** $\{0, 8\}$ **53.** $\{7\}$ **55.** $\left\{-2, -1, -\dfrac{2}{5}\right\}$ **57.** length: 10 ft; width: 4 ft   **59.** length: 6 m; width: 2 m

61. 6, 7 or $-5, -4$ **63. (a)** $4.51 billion **(b)** Yes, the prediction seems reliable. If eBay revenues in the last half of 2005 are

comparable to those for the first half of the year, annual revenue in 2005 would be about $4.24 billion.   **65.** 192 ft

67. after 8 sec   **69.** The factor $(2x + 8)$ has a common factor of 2. The completely factored form is $2(x + 4)(3x - 4)$.

71. $(3k + 5)(k + 2)$ **73.** $(y^2 + 25)(y + 5)(y - 5)$ **75.** $8abc(3b^2c - 7ac^2 + 9ab)$ **77.** $6xyz(2xz^2 + 2y - 5x^2yz^3)$

79. $(2r + 3q)(6r - 5)$ **81.** $(7t + 4)^2$ **83.** $\{-5, 2\}$ **85. (a)** 603,000 vehicles **(b)** The estimate may be unreliable

because the conditions that prevailed in the years 2001–2004 may have changed, causing either a greater increase or a decrease

in the number of alternative-fueled vehicles.   **87.** length: 6 m; width: 4 m   **89.** 256 ft   **91.** width: 10 m; length: 17 m

Chapter 6 Test   (pages 461–462)

[6.1–6.4] **1.** D   **2.** $6x(2x - 5)$ **3.** $m^2n(2mn + 3m - 5n)$ **4.** $(2x + y)(a - b)$ **5.** $(x + 3)(x - 8)$
6. $(2x + 3)(x - 1)$ **7.** $(5z - 1)(2z - 3)$ **8.** prime   **9.** prime   **10.** $(2 - a)(6 + b)$ **11.** $(3y + 8)(3y - 8)$
12. $(2x - 7y)^2$ **13.** $-2(x + 1)^2$ **14.** $3t^2(2t + 9)(t - 4)$ **15.** $(r - 5)(r^2 + 5r + 25)$ **16.** $8(k + 2)(k^2 - 2k + 4)$
17. $(x^2 + 9)(x + 3)(x - 3)$ **18.** $(3x + 2y)(3x - 2y)(9x^2 + 4y^2)$ **19.** $(3x^3y^2 + 2)^2$ **20.** The product
$(p + 3)(p + 3) = p^2 + 6p + 9$, which does not equal $p^2 + 9$. The binomial $p^2 + 9$ is a prime polynomial.

[6.5] **21.** $\left\{\frac{1}{2}, 6\right\}$ **22.** $\left\{-\frac{2}{5}, \frac{2}{5}\right\}$ **23.** $\{10\}$ **24.** $\{-3, 0, 3\}$ [6.6] **25.** 6 ft by 9 ft   **26.** $-2, -1$ **27.** 17 ft

28. 243

Chapters 1–6 Cumulative Review Exercises   (pages 462–464)

[2.1–2.3] **1.** $\{0\}$ **2.** $\{0.05\}$ **3.** $\{6\}$ [2.5] **4.** $P = \dfrac{A}{1 + rt}$ **5.** $110°$ and $70°$ [2.4] **6.** gold: 11; silver: 12;
bronze: 6   [2.7] **7.** 51.0 million   **8.** 345; 210; 38%; 15%   [3.1] **9.** **(a)** negative, positive **(b)** negative, negative
[3.2] **10.** $\left(-\dfrac{1}{4}, 0\right)$, $(0, 3)$ [3.3] **11.** 12   [3.2] **12.**   [3.3, 3.4] **13.** **(a)** 97; A slope of 97 means that the

number of radio stations increased by about 97 stations per year. **(b)** $(2000, 10{,}375)$ [3.6] **14.** **(a)** 10,569 **(b)** yes

[4.1] **15.** $\{(-1, 2)\}$ **16.** $\emptyset$ [5.1, 5.2] **17.** $\dfrac{16}{9}$ **18.** 256   **19.** $\dfrac{1}{p^2}$ **20.** $\dfrac{1}{m^6}$ [5.4] **21.** $-4k^2 - 4k + 8$
[5.5] **22.** $45x^2 + 3x - 18$ [5.6] **23.** $9p^2 + 12p + 4$ [5.7] **24.** $4x^3 + 6x^2 - 3x + 10$ [5.3] **25.** 5.5×10^4; 2.0×10^6
[6.2, 6.3] **26.** $(2a - 1)(a + 4)$ **27.** $(2m + 3)(5m + 2)$ **28.** $(4t + 3v)(2t + v)$ [6.4] **29.** $(2p - 3)^2$
30. $(5r + 9t)(5r - 9t)$ [6.3] **31.** $2pq(3p + 1)(p + 1)$ [6.5] **32.** $\left\{-\dfrac{2}{3}, \dfrac{1}{2}\right\}$ **33.** $\{0, 8\}$ [6.6] **34.** 5 m, 12 m, 13 m

CHAPTER 7   Rational Expressions and Applications

Connections   (page 472)   **1.** $3x^2 + 11x + 8$ cannot be factored, so this quotient cannot be simplified. By long division, the
quotient is $3x + 5 + \dfrac{-2}{x + 2}$.   **2.** The numerator factors as $(x - 2)(x^2 + 2x + 4)$, so, after simplification, the quotient is $x - 2$.
Long division gives the same quotient.

Section 7.1   (pages 473–475)

1. A rational expression is a quotient of two polynomials, such as $\dfrac{x^2 + 3x - 6}{x + 4}$. One can think of this as an algebraic fraction.

3. **(a)** $\dfrac{7}{10}$ **(b)** $\dfrac{8}{15}$ **5.** **(a)** 0 **(b)** -1 **7.** **(a)** $-\dfrac{64}{15}$ **(b)** undefined   **9.** **(a)** undefined **(b)** $\dfrac{8}{25}$ **11.** Division by 0 is undefined,

so if the denominator of a rational expression equals 0, the expression is undefined.   **13.** $y \neq 0$ **15.** $x \neq 6$ **17.** $x \neq -\dfrac{5}{3}$
19. $m \neq -3, m \neq 2$ **21.** It is never undefined.   **23.** It is never undefined.   **25.** **(a)** numerator: $x^2, 4x$; denominator: $x, 4$

(b) First factor the numerator, getting $x(x + 4)$. Then divide the numerator and denominator by the common factor $x + 4$ to get $\dfrac{x}{1}$, or x.   **27.** $3r^2$ **29.** $\dfrac{2}{5}$ **31.** $\dfrac{x - 1}{x + 1}$ **33.** $\dfrac{7}{5}$ **35.** $\dfrac{6}{7}$ **37.** $m - n$ **39.** $\dfrac{2}{t - 3}$ **41.** $\dfrac{3(2m + 1)}{4}$ **43.** $\dfrac{3m}{5}$

45. $\dfrac{3r - 2s}{3}$ **47.** $k - 3$ **49.** $\dfrac{x - 3}{x + 1}$ **51.** $\dfrac{x + 1}{x - 1}$ **53.** $\dfrac{z - 3}{z + 5}$ **55.** $\dfrac{r + s}{r - s}$ **57.** $\dfrac{m + n}{2}$ **59.** $1 - p + p^2$

61. $-\dfrac{b^2 + ba + a^2}{a + b}$ **63.** $\dfrac{z + 3}{z}$ **65.** B, D   **67.** -1 **69.** $-(m + 1)$ **71.** -1 **73.** It is already in lowest terms.

75. B   *Answers may vary in Exercises 77, 79, and 81.*   **77.** $\dfrac{-(x + 4)}{x - 3}, \dfrac{-x - 4}{x - 3}, \dfrac{x + 4}{-(x - 3)}, \dfrac{x + 4}{-x + 3}$

79. $\dfrac{-(2x - 3)}{x + 3}, \dfrac{-2x + 3}{x + 3}, \dfrac{2x - 3}{-(x + 3)}, \dfrac{2x - 3}{-x - 3}$ **81.** $\dfrac{-(3x - 1)}{5x - 6}, \dfrac{-3x + 1}{5x - 6}, \dfrac{3x - 1}{-(5x - 6)}, \dfrac{3x - 1}{-5x + 6}$ **83.** $x^2 + 3$

85. **(a)** 0 **(b)** 1.6 **(c)** 4.1 **(d)** The waiting time also increases.   **87.** $\dfrac{5}{9}$ **89.** $\dfrac{10}{3}$ **91.** 4

Section 7.2   (pages 480–482)

1. **(a)** B **(b)** D **(c)** C **(d)** A   **3.** $\dfrac{3a}{2}$ **5.** $-\dfrac{4x^4}{3}$ **7.** $\dfrac{2}{c + d}$ **9.** $4(x - y)$ **11.** $\dfrac{t^2}{2}$ **13.** $\dfrac{x + 3}{2x}$ **15.** 5   **17.** $-\dfrac{3}{2t^4}$

19. $\dfrac{1}{4}$ **21.** $-\dfrac{35}{8}$ **23.** $\dfrac{2(x + 2)}{x(x - 1)}$ **25.** $\dfrac{x(x - 3)}{6}$ **27.** Suppose I want to multiply $\dfrac{a^2 - 1}{6} \cdot \dfrac{9}{2a + 2}$. I start by factoring

where possible: $\dfrac{(a + 1)(a - 1)}{6} \cdot \dfrac{9}{2(a + 1)}$. Next, I divide out common factors in the numerator and denominator to get

$\dfrac{a - 1}{2} \cdot \dfrac{3}{2}$. Finally, I multiply numerator by numerator and denominator by denominator to get the final product, $\dfrac{3(a - 1)}{4}$.

29. $\dfrac{10}{9}$ **31.** $-\dfrac{3}{4}$ **33.** $-\dfrac{9}{2}$ **35.** $\dfrac{p + 4}{p + 2}$ **37.** -1 **39.** $\dfrac{(2x - 1)(x + 2)}{x - 1}$ **41.** $\dfrac{(k - 1)^2}{(k + 1)(2k - 1)}$ **43.** $\dfrac{4k - 1}{3k - 2}$

45. $\dfrac{m + 4p}{m + p}$ **47.** $\dfrac{m + 6}{m + 3}$ **49.** $\dfrac{y + 3}{y + 4}$ **51.** $\dfrac{m}{m + 5}$ **53.** $\dfrac{r + 6s}{r + s}$ **55.** $\dfrac{(q - 3)^2(q + 2)^2}{q + 1}$ **57.** $\dfrac{x + 10}{10}$

59. $\dfrac{3 - a - b}{2a - b}$ **61.** $-\dfrac{(x + y)^2(x^2 - xy + y^2)}{3y(y - x)(x - y)}$, or $\dfrac{(x + y)^2(x^2 - xy + y^2)}{3y(x - y)^2}$ **63.** $\dfrac{5xy^2}{4q}$ **65.** $2 \cdot 3^2$ **67.** $2^2 \cdot 3^3$

69. 6   **71.** $6q^3$

Section 7.3   (pages 486–488)

1. C   **3.** C   **5.** 60   **7.** 1800   **9.** x^5 **11.** $30p$ **13.** $180y^4$ **15.** $84r^5$ **17.** $15a^5b^3$ **19.** $12p(p - 2)$

21. $28m^2(3m - 5)$ **23.** $30(b - 2)$ **25.** $18(r - 2)$ **27.** $12p(p + 5)^2$ **29.** $8(y + 2)(y + 1)$ **31.** $c - d$ or $d - c$

33. $m - 3$ or $3 - m$ **35.** $p - q$ or $q - p$ **37.** $k(k + 5)(k - 2)$ **39.** $a(a + 6)(a - 3)$ **41.** $(p + 3)(p + 5)(p - 6)$

43. $(k + 3)(k - 5)(k + 7)(k + 8)$ **45.** 7   **46.** 1   **47.** identity property of multiplication   **48.** 7   **49.** 1

50. identity property of multiplication   **51.** $\dfrac{20}{55}$ **53.** $\dfrac{-45}{9k}$ **55.** $\dfrac{60m^2k^3}{32k^4}$ **57.** $\dfrac{57z}{6z - 18}$ **59.** $\dfrac{-4a}{18a - 36}$

61. $\dfrac{6(k + 1)}{k(k - 4)(k + 1)}$ **63.** $\dfrac{36r(r + 1)}{(r - 3)(r + 2)(r + 1)}$ **65.** $\dfrac{ab(a + 2b)}{2a^3b + a^2b^2 - ab^3}$ **67.** $\dfrac{(t - r)(4r - t)}{t^3 - r^3}$

69. $\dfrac{2y(z - y)(y - z)}{y^4 - z^3y}$, or $\dfrac{-2y(y - z)^2}{y^4 - z^3y}$ **71.** $\dfrac{5}{2}$ **73.** $\dfrac{11}{8}$ **75.** $\dfrac{13}{20}$ **77.** $\dfrac{13}{12}$

Section 7.4 (pages 494–497)

1. E **3.** C **5.** B **7.** G **9.** $\dfrac{11}{m}$ **11.** $\dfrac{4}{y+4}$ **13.** 1 **15.** $\dfrac{m-1}{m+1}$ **17.** b **19.** x **21.** $y-6$ **23.** To add or

subtract rational expressions with the same denominators, combine the numerators and keep the same denominator. For example,

$\dfrac{3x+2}{x-6} + \dfrac{-2x-8}{x-6} = \dfrac{x-6}{x-6}$. Then write the resulting fraction in lowest terms. In this example, the sum simplifies to 1.

25. $\dfrac{3z+5}{15}$ **27.** $\dfrac{10-7r}{14}$ **29.** $\dfrac{-3x-2}{4x}$ **31.** $\dfrac{57}{10x}$ **33.** $\dfrac{x+1}{2}$ **35.** $\dfrac{5x+9}{6x}$ **37.** $\dfrac{7-6p}{3p^2}$ **39.** $\dfrac{-k-8}{k(k+4)}$

41. $\dfrac{x+4}{x+2}$ **43.** $\dfrac{6m^2+23m-2}{(m+2)(m+1)(m+5)}$ **45.** $\dfrac{4y^2-y+5}{(y+1)^2(y-1)}$ **47.** $\dfrac{3}{t}$ **49.** $m-2$ or $2-m$ **51.** $\dfrac{-2}{x-5}$, or $\dfrac{2}{5-x}$

53. -4 **55.** $\dfrac{-5}{x-y^2}$, or $\dfrac{5}{y^2-x}$ **57.** $\dfrac{x+y}{5x-3y}$, or $\dfrac{-x-y}{3y-5x}$ **59.** $\dfrac{-6}{4p-5}$, or $\dfrac{6}{5-4p}$ **61.** $\dfrac{-(m+n)}{2(m-n)}$

63. $\dfrac{-x^2+6x+11}{(x+3)(x-3)(x+1)}$ **65.** $\dfrac{-5q^2-13q+7}{(3q-2)(q+4)(2q-3)}$ **67.** $\dfrac{9r+2}{r(r+2)(r-1)}$ **69.** $\dfrac{2(x^2+3xy+4y^2)}{(x+y)(x+y)(x+3y)}$, or

$\dfrac{2(x^2+3xy+4y^2)}{(x+y)^2(x+3y)}$ **71.** $\dfrac{15r^2+10ry-y^2}{(3r+2y)(6r-y)(6r+y)}$ **73. (a)** $\dfrac{9k^2+6k+26}{5(3k+1)}$ **(b)** $\dfrac{1}{4}$ **75.** $\dfrac{10x}{49(101-x)}$

77. 6 **79.** $\dfrac{3}{25}$

Section 7.5 (pages 502–504)

1. (a) $6; \dfrac{1}{6}$ **(b)** $12; \dfrac{3}{4}$ **(c)** $\dfrac{1}{6} \div \dfrac{3}{4}$ **(d)** $\dfrac{2}{9}$ **3.** Choice D is correct, because every sign has been changed in the fraction. This

means it was multiplied by $\dfrac{-1}{-1} = 1$. **5.** Method 1 indicates that we write the complex fraction as a division problem and then

perform the division. For example, to simplify $\dfrac{\frac{1}{2}}{\frac{2}{3}}$, write $\dfrac{1}{2} \div \dfrac{2}{3}$. Then simplify as $\dfrac{1}{2} \cdot \dfrac{3}{2} = \dfrac{3}{4}$. **7.** -6 **9.** $\dfrac{1}{xy}$

11. $\dfrac{2a^2b}{3}$ **13.** $\dfrac{m(m+2)}{3(m-4)}$ **15.** $\dfrac{2}{x}$ **17.** $\dfrac{8}{x}$ **19.** $\dfrac{a^2-5}{a^2+1}$ **21.** $\dfrac{31}{50}$ **23.** $\dfrac{y^2+x^2}{xy(y-x)}$ **25.** $\dfrac{40-12p}{85p}$ **27.** $\dfrac{5y-2x}{3+4xy}$

29. $\dfrac{a-2}{2a}$ **31.** $\dfrac{z-5}{4}$ **33.** $\dfrac{-m}{m+2}$ **35.** $\dfrac{3m(m-3)}{(m-1)(m-8)}$ **37.** division **39.** $\dfrac{\frac{3}{8}+\frac{5}{6}}{2}$ **40.** $\dfrac{29}{48}$ **41.** $\dfrac{29}{48}$

42. Answers will vary. **43.** $\dfrac{5}{3}$ **45.** $\dfrac{13}{2}$ **47.** $\dfrac{19r}{15}$ **49.** $12x+2$ **51.** $-44p^2+27p$ **53.** $\left\{\dfrac{1}{2}\right\}$ **55.** $\{-5\}$

Section 7.6 (pages 511–514)

1. expression; $\dfrac{43}{40}x$ **3.** equation; $\left\{\dfrac{40}{43}\right\}$ **5.** expression; $-\dfrac{1}{10}x$ **7.** equation; $\{-10\}$ **9.** $x \neq -2, 0$

11. $x \neq -3, 4, -\dfrac{1}{2}$ **13.** $x \neq -9, 1, -2, 2$ **15.** $\dfrac{2}{3x}+\dfrac{1}{5x}$ is an exression, not an equation. Only equaations and inequalities

are "solved." **17.** $\left\{\dfrac{1}{4}\right\}$ **19.** $\left\{-\dfrac{3}{4}\right\}$ **21.** $\{-15\}$ **23.** $\{7\}$ **25.** $\{-15\}$ **27.** $\{-5\}$ **29.** $\{-6\}$ **31.** $\emptyset$

33. $\{5\}$ **35.** $\{4\}$ **37.** $\{1\}$ **39.** $\{4\}$ **41.** $\{5\}$ **43.** $\{-4\}$ **45.** $\{-2, 12\}$ **47.** $\emptyset$ **49.** $\{3\}$ **51.** $\{3\}$ **53.** $\{-3\}$

55. $\left\{-\dfrac{1}{5}, 3\right\}$ **57.** $\left\{-\dfrac{1}{2}, 5\right\}$ **59.** $\{3\}$ **61.** $\left\{-\dfrac{1}{3}, 3\right\}$ **63.** $\{-1\}$ **65.** $\{-6\}$ **67.** $\left\{-6, \dfrac{1}{2}\right\}$ **69.** $\{6\}$

71. Transform the equation so that the terms with k are on one side and the remaining term is on the other.   **73.** $F = \dfrac{ma}{k}$

75. $a = \dfrac{kF}{m}$ **77.** $R = \dfrac{E - Ir}{I}$, or $R = \dfrac{E}{I} - r$ **79.** $A = \dfrac{h(B + b)}{2}$ **81.** $a = \dfrac{2S - ndL}{nd}$, or $a = \dfrac{2S}{nd} - L$ **83.** $y = \dfrac{xz}{x + z}$

85. $t = \dfrac{rs}{rs - 2s - 3r}$, or $t = \dfrac{-rs}{-rs + 2s + 3r}$ **87.** $z = \dfrac{3y}{5 - 9xy}$, or $z = \dfrac{-3y}{9xy - 5}$ **89.** $\dfrac{288}{t}$ mph   **91.** $\dfrac{289}{z}$ hr

Summary Exercises on Rational Expressions and Equations   (page 516)

1. expression; $\dfrac{10}{p}$ **2.** expression; $\dfrac{y^3}{x^3}$ **3.** expression; $\dfrac{1}{2x^2(x + 2)}$ **4.** equation; $\{9\}$ **5.** expression; $\dfrac{y + 2}{y - 1}$

6. expression; $\dfrac{5k + 8}{k(k - 4)(k + 4)}$ **7.** equation; $\{39\}$ **8.** expression; $\dfrac{t - 5}{3(2t + 1)}$ **9.** expression; $\dfrac{13}{3(p + 2)}$

10. equation; $\left\{-1, \dfrac{12}{5}\right\}$ **11.** equation; $\left\{\dfrac{1}{7}, 2\right\}$ **12.** expression; $\dfrac{16}{3y}$ **13.** expression; $\dfrac{7}{12z}$ **14.** equation; $\{13\}$

15. expression; $\dfrac{3m + 5}{(m + 3)(m + 2)(m + 1)}$ **16.** expression; $\dfrac{k + 3}{5(k - 1)}$ **17.** equation; $\emptyset$ **18.** equation; $\emptyset$

19. expression; $\dfrac{t + 2}{2(2t + 1)}$ **20.** equation; $\{-7\}$

Section 7.7   (pages 521–525)

1. (a) the amount **(b)** $5 + x$ **(c)** $\dfrac{5 + x}{6} = \dfrac{13}{3}$ **3.** $\dfrac{12}{18}$ **5.** $\dfrac{12}{3}$ **7.** 12   **9.** $\dfrac{1386}{97}$ **11.** 76.57 sec

13. 338.730 m per min   **15.** 3.699 hr   **17.** $\dfrac{D}{R} = \dfrac{d}{r}$ **19.** $\dfrac{500}{x - 10} = \dfrac{600}{x + 10}$ **21.** 8 mph   **23.** 32 mph

25. 165 mph   **27.** 3 mph   **29.** 18.5 mph   **31.** $\dfrac{1}{10}$ job per hr   **33.** $\dfrac{1}{8}t + \dfrac{1}{6}t = 1$, or $\dfrac{1}{8} + \dfrac{1}{6} = \dfrac{1}{t}$ **35.** $2\dfrac{2}{5}$ hr

37. $5\dfrac{5}{11}$ hr   **39.** 3 hr   **41.** $2\dfrac{7}{10}$ hr   **43.** $9\dfrac{1}{11}$ min   **45.** $\left\{\dfrac{40}{3}\right\}$ **47.** $\{3600\}$ **49.** $k = \dfrac{y}{x}$ **51.** $k = xy$

Section 7.8   (pages 529–532)

1. direct   **3.** direct   **5.** inverse   **7.** inverse   **9.** inverse   **11.** direct   **13.** direct   **15.** inverse   **17. (a)** increases

(b) decreases   **19.** 9   **21.** 250   **23.** 6   **25.** 21   **27.** $\dfrac{16}{5}$ **29.** $\dfrac{4}{9}$ **31.** \$40.32   **33.** $42\dfrac{2}{3}$ in.   **35.** $106\dfrac{2}{3}$ mph

37. $12\dfrac{1}{2}$ amps   **39.** 20 lb   **41.** 52.817 in.^{2} **43.** $14\dfrac{22}{27}$ footcandles   **45.** direct   **47.** inverse   **49.** 64   **51.** -144

53. 169

Chapter 7 Review Exercises   (pages 539–542)

1. (a) $\dfrac{11}{8}$ **(b)** $\dfrac{13}{22}$ **3.** $x \neq 3$ **5.** $k \neq -5, -\dfrac{2}{3}$ **7.** $\dfrac{b}{3a}$ **9.** $\dfrac{-(2x + 3)}{2}$ **11.** (Answers may vary.) $\dfrac{-(4x - 9)}{2x + 3}, \dfrac{-4x + 9}{2x + 3},$

$\dfrac{4x - 9}{-(2x + 3)}, \dfrac{4x - 9}{-2x - 3}$ **13.** $\dfrac{72}{p}$ **15.** $\dfrac{5}{8}$ **17.** $\dfrac{3a - 1}{a + 5}$ **19.** $\dfrac{p + 5}{p + 1}$ **21.** $108y^4$ **23.** $\dfrac{15a}{10a^4}$ **25.** $\dfrac{15y}{50 - 10y}$ **27.** $\dfrac{15}{x}$

29. $\dfrac{4k-45}{k(k-5)}$ **31.** $\dfrac{-2-3m}{6}$ **33.** $\dfrac{7a+6b}{(a-2b)(a+2b)}$ **35.** $\dfrac{5z-16}{z(z+6)(z-2)}$ **37.** (a) $\dfrac{a}{b}$ (b) $\dfrac{a}{b}$ (c) Answers will vary.

39. $\dfrac{4(y-3)}{y+3}$ **41.** $\dfrac{xw+1}{xw-1}$ **43.** It would cause the first and third denominators to equal 0. **45.** $\emptyset$ **47.** $t=\dfrac{Ry}{m}$

49. $m=\dfrac{4+p^2q}{3p^2}$ **51.** $\dfrac{3}{18}$ **53.** $3\dfrac{1}{13}$ hr **55.** inverse **57.** $\dfrac{36}{5}$ **59.** $8p^2$ **61.** 3 **63.** $v=at+w$ **65.** $5\dfrac{1}{11}$ hr

67. 4 **69.** (a) -3 (b) -1 (c) $-3,-1$ **70.** $\dfrac{15}{2x}$ **71.** If $x=0$, the divisor R is equal to 0, and division by 0 is undefined.

72. $(x+3)(x+1)$ **73.** $\dfrac{7}{x+1}$ **74.** $\dfrac{11x+21}{4x}$ **75.** $\emptyset$ **76.** We know that -3 is not allowed, because P and R are

undefined for $x=-3$. **77.** Rate is equal to distance divided by time. Here, distance is 6 mi and time is $x+3$ min, so

rate $=\dfrac{6}{x+3}$, which is the expression for P. **78.** $\dfrac{6}{5},\dfrac{5}{2}$

Chapter 7 Test (pages 542–543)

[7.1] **1.** (a) $\dfrac{11}{6}$ (b) undefined **2.** $x\ne-2,4$ **3.** (Answers may vary.) $\dfrac{-(6x-5)}{2x+3},\dfrac{-6x+5}{2x+3},\dfrac{6x-5}{-(2x+3)},\dfrac{6x-5}{-2x-3}$

4. $-3x^2y^3$ **5.** $\dfrac{3a+2}{a-1}$ [7.2] **6.** $\dfrac{25}{27}$ **7.** $\dfrac{3k-2}{3k+2}$ **8.** $\dfrac{a-1}{a+4}$ [7.3] **9.** $150p^5$ **10.** $(2r+3)(r+2)(r-5)$

11. $\dfrac{240p^2}{64p^3}$ **12.** $\dfrac{21}{42m-84}$ [7.4] **13.** 2 **14.** $\dfrac{-14}{5(y+2)}$ **15.** $\dfrac{-x^2+x+1}{3-x}$, or $\dfrac{x^2-x-1}{x-3}$

16. $\dfrac{-m^2+7m+2}{(2m+1)(m-5)(m-1)}$ [7.5] **17.** $\dfrac{2k}{3p}$ **18.** $\dfrac{-2-x}{4+x}$ [7.6] **19.** $\left\{-\dfrac{1}{2}\right\}$ **20.** $D=\dfrac{dF-k}{F}$, or $D=\dfrac{k-dF}{-F}$

[7.7] **21.** 3 mph **22.** $2\dfrac{2}{9}$ hr [7.8] **23.** 27 **24.** 27 days

Chapters 1–7 Cumulative Review Exercises (pages 544–545)

[1.2, 1.5, 1.6] **1.** 2 [2.3] **2.** $\{17\}$ [2.5] **3.** $b=\dfrac{2A}{h}$ [2.6] **4.** $\left\{-\dfrac{2}{7}\right\}$ [2.8] **5.** $[-8,\infty)$ **6.** $(4,\infty)$

[3.1] **7.** (a) $(-3,0)$ (b) $(0,-4)$ [3.2] **8.** [5.4] **9.** [4.1–4.3] **10.** $\{(-1,3)\}$

11. $\emptyset$ [5.1, 5.2] **12.** $\dfrac{1}{2^4x^7}$ **13.** $\dfrac{1}{m^6}$ **14.** $\dfrac{q}{4p^2}$ [5.4] **15.** k^2+2k+1 [5.1] **16.** $72x^6y^7$ [5.6] **17.** $4a^2-4ab+b^2$

[5.5] **18.** $3y^3+8y^2+12y-5$ [5.7] **19.** $6p^2+7p+1+\dfrac{3}{p-1}$ [5.3] **20.** 1.4×10^5 sec

[6.3] **21.** $(4t+3v)(2t+v)$ **22.** prime [6.4] **23.** $(4x^2+1)(2x+1)(2x-1)$ [6.5] **24.** $\{-3,5\}$ **25.** $\left\{5,-\dfrac{1}{2},\dfrac{2}{3}\right\}$

[6.6] **26.** -2 or -1 **27.** 6 m [7.1] **28.** A **29.** D [7.4] **30.** $\dfrac{4}{q}$ **31.** $\dfrac{3r+28}{7r}$ **32.** $\dfrac{7}{15(q-4)}$

33. $\dfrac{-k - 5}{k(k + 1)(k - 1)}$ [7.2] **34.** $\dfrac{7(2z + 1)}{24}$ [7.5] **35.** $\dfrac{195}{29}$ [7.6] **36.** $x \neq 4, 0$ **37.** $\left\{\dfrac{21}{2}\right\}$ **38.** $\{-2, 1\}$

[7.7] **39.** $1\dfrac{1}{5}$ hr   [7.8] **40.** 32.97 in.

CHAPTER 8   Roots and Radicals

Connections   (page 555)   The area of the large square is $(a + b)^2$, or $a^2 + 2ab + b^2$. The sum of the areas of the smaller square and the four right triangles is $c^2 + 2ab$. Set these equal to each other and subtract $2ab$ from each side to get $a^2 + b^2 = c^2$.

Section 8.1   (pages 555–558)

1. true   **3.** false; zero has only one square root.   **5.** true   **7.** $-3, 3$ **9.** $-8, 8$ **11.** $-12, 12$ **13.** $-\dfrac{5}{14}, \dfrac{5}{14}$

15. $-30, 30$ **17.** 1   **19.** 7   **21.** -11 **23.** $-\dfrac{12}{11}$ **25.** It is not a real number.   **27.** 19   **29.** 19   **31.** $\dfrac{2}{3}$

33. $3x^2 + 4$ **35.** a must be positive.   **37.** a must be negative.   **39.** rational; 5   **41.** irrational; 5.385   **43.** rational; -8
45. irrational; -17.321 **47.** It is not a real number.   **49.** irrational; 34.641   **51.** 9 and 10   **53.** 7 and 8
55. -7 and -6 **57.** 4 and 5   **59.** C   **61.** $c = 17$ **63.** $b \approx 6.708$ **65.** $c \approx 11.705$ **67.** 24 cm   **69.** 80 ft
71. 195 ft   **73.** 11.1 ft   **75.** 9.434   **77.** Answers will vary. For example, if $a = 2$ and $b = 7$, $\sqrt{a^2 + b^2} = \sqrt{2^2 + 7^2} = \sqrt{53}$, while $a + b = 2 + 7 = 9$. Therefore, $\sqrt{a^2 + b^2} \neq a + b$ because $\sqrt{53} \neq 9$.   **79.** 5   **81.** 17   **83.** $\sqrt{13}$ **85.** 5
87. 13   **89.** $\sqrt{13}$ **91.** $\sqrt{2}$ **93.** 1   **95.** 5   **97.** -3 **99.** -2 **101.** 5   **103.** It is not a real number.   **105.** -3
107. -4 **109.** 2.289   **111.** 5.074   **113.** -4.431 **115.** The area of the square on the left is c^2. The small square inside that figure has area $(b - a)^2 = b^2 - 2ba + a^2$. The sum of the areas of the two rectangles in the figure on the right is $2ab$. Since the areas of the two figures are the same, we have $c^2 = 2ab + b^2 - 2ba + a^2$, which simplifies to $c^2 = a^2 + b^2$.
117. $2^3 \cdot 3^2$ **119.** $2^3 \cdot 5$ **121.** 23 is a prime number.

Section 8.2   (pages 564–567)

1. true   **3.** true   **5.** false; $\sqrt{(-6)^2} = \sqrt{36} = 6$, not -6 **7.** $\sqrt{15}$ **9.** $\sqrt{22}$ **11.** $\sqrt{42}$ **13.** $\sqrt{81}$, or 9   **15.** 13
17. $\sqrt{13r}$ **19.** For $x \geq 0$, $\sqrt{x}$ is the nonnegative real number that, when used as a factor twice (i.e., when squared), gives a product of x. Therefore, $\sqrt{x} \cdot \sqrt{x} = x$.   **21.** A   **23.** $3\sqrt{5}$ **25.** $2\sqrt{6}$ **27.** $3\sqrt{10}$ **29.** $5\sqrt{3}$ **31.** $5\sqrt{5}$
33. It cannot be simplified.   **35.** $4\sqrt{10}$ **37.** $-10\sqrt{7}$ **39.** $9\sqrt{3}$ **41.** $25\sqrt{2}$ **43.** $12\sqrt{2}$ **45.** 24   **47.** $6\sqrt{10}$
49. $\sqrt{8} \cdot \sqrt{32} = \sqrt{8 \cdot 32} = \sqrt{256} = 16$. Also, $\sqrt{8} = 2\sqrt{2}$ and $\sqrt{32} = 4\sqrt{2}$, so $\sqrt{8} \cdot \sqrt{32} = 2\sqrt{2} \cdot 4\sqrt{2} = 8 \cdot 2 = 16$. Both methods give the same answer, and the correct answer can always be obtained with either method.   **51.** $\dfrac{4}{15}$ **53.** $\dfrac{\sqrt{7}}{4}$

55. $\dfrac{\sqrt{2}}{5}$ **57.** 5   **59.** $6\sqrt{5}$ **61.** $\dfrac{25}{4}$ **63.** m **65.** y^2 **67.** $6z$ **69.** $20x^3$ **71.** $3x^4\sqrt{2}$ **73.** $3c^7\sqrt{5}$ **75.** $z^2\sqrt{z}$

77. $a^6\sqrt{a}$ **79.** $8x^3\sqrt{x}$ **81.** x^3y^6 **83.** $9m^2n$ **85.** $\dfrac{\sqrt{7}}{x^5}$ **87.** $\dfrac{y^2}{10}$ **89.** $\dfrac{x^2y^3}{13}$ **91.** $2\sqrt[3]{5}$ **93.** $3\sqrt[3]{2}$ **95.** $4\sqrt[3]{2}$

97. $2\sqrt[4]{5}$ **99.** $\dfrac{2}{3}$ **101.** $-\dfrac{6}{5}$ **103.** p **105.** x^3 **107.** $4z^2$ **109.** $7a^3b$ **111.** $2t\sqrt[3]{2t^2}$ **113.** $\dfrac{m^4}{2}$

115. (a) 4.472135955 **(b)** 4.472135955 **(c)** The numerical results are not a proof because both answers are approximations and they might differ if calculated to more decimal places.   **117.** The product rule for radicals requires that both a and b be nonnegative. Otherwise $\sqrt{a}$ and $\sqrt{b}$ would not be real numbers (except when $a = b = 0$).   **119.** 6 cm   **121.** 6 in.

123. D   **125.** $-5x + 19$ **127.** $11x^2y - 7xy$

Section 8.3 (pages 570–571)

1. $7\sqrt{3}$ **3.** $-5\sqrt{7}$ **5.** $2\sqrt{6}$ **7.** $3\sqrt{17}$ **9.** $7\sqrt{3}$ **11.** It cannot be added by the distributive property.   **13.** $3\sqrt{3}$
15. $-20\sqrt{2}$ **17.** $19\sqrt{7}$ **19.** $12\sqrt{6} + 6\sqrt{5}$ **21.** $-2\sqrt{2} - 12\sqrt{3}$ **23.** $4\sqrt{2}$ **25.** $3\sqrt{3} - 2\sqrt{5}$ **27.** $3\sqrt{21}$
29. $5\sqrt{3}$ **31.** $-\sqrt[3]{2}$ **33.** $24\sqrt[3]{3}$ **35.** $10\sqrt[4]{2} + 4\sqrt[4]{8}$ **37.** $2\sqrt{3} + 4\sqrt{3} = (2 + 4)\sqrt{3} = 6\sqrt{3}$ **39.** $\sqrt{2x}$
41. $7\sqrt{3r}$ **43.** $15x\sqrt{3}$ **45.** $2x\sqrt{2}$ **47.** $13p\sqrt{3}$ **49.** $42x\sqrt{5z}$ **51.** $6k^2h\sqrt{6} + 27hk\sqrt{6k}$ **53.** $6\sqrt[3]{p^2}$
55. $21\sqrt[4]{m^3}$ **57.** $-8p\sqrt[4]{p}$ **59.** $-24z\sqrt[3]{4z}$ **61.** 0   **63.** The variables may represent negative numbers, because negative
numbers have cube roots.   **65.** $\sqrt{85}$ **67.** 5   **69.** 7   **71.** 14   **73.** $2\sqrt{2}$ **75.** $-6x^2y$ **76.** $-6(p - 2q)^2(a + b)$
77. $-6a^2\sqrt{xy}$ **78.** The answers are alike because the numerical coefficient of the three answers is the same: -6. Also, the
first variable factor is raised to the second power, and the second variable factor is raised to the first power. The answers are
different because the variables are different: x and y, then $p - 2q$ and $a + b$, and then a and $\sqrt{xy}$.   **79.** 6   **81.** 2
83. $50\sqrt{3}$

Section 8.4 (pages 576–577)

1. $\dfrac{6\sqrt{5}}{5}$ **3.** $\sqrt{5}$ **5.** $\dfrac{2\sqrt{6}}{3}$ **7.** $\dfrac{8\sqrt{15}}{5}$ **9.** $\dfrac{\sqrt{30}}{2}$ **11.** $\dfrac{8\sqrt{3}}{9}$ **13.** $\dfrac{3\sqrt{2}}{10}$ **15.** $\sqrt{2}$ **17.** $\sqrt{2}$ **19.** $\dfrac{2\sqrt{30}}{3}$

21. $\dfrac{\sqrt{2}}{8}$ **23.** $\dfrac{3\sqrt{5}}{5}$ **25.** $\dfrac{-3\sqrt{2}}{10}$ **27.** $\dfrac{21\sqrt{5}}{5}$ **29.** $\dfrac{\sqrt{3}}{3}$ **31.** $-\dfrac{\sqrt{5}}{5}$ **33.** $\dfrac{\sqrt{65}}{5}$ **35.** 1; identity property for

multiplication   **37.** $\dfrac{\sqrt{21}}{3}$ **39.** $\dfrac{3\sqrt{14}}{4}$ **41.** $\dfrac{1}{6}$ **43.** 1   **45.** $\dfrac{\sqrt{15}}{10}$ **47.** $\dfrac{17\sqrt{2}}{6}$ **49.** $\dfrac{\sqrt{3}}{5}$ **51.** $\dfrac{4\sqrt{3}}{27}$ **53.** $\dfrac{\sqrt{6p}}{p}$

55. $\dfrac{\sqrt{3y}}{y}$ **57.** $\dfrac{4\sqrt{m}}{m}$ **59.** $\dfrac{p\sqrt{3q}}{q}$ **61.** $\dfrac{x\sqrt{7xy}}{y}$ **63.** $\dfrac{p\sqrt{2pm}}{m}$ **65.** $\dfrac{x\sqrt{y}}{2y}$ **67.** $\dfrac{3a\sqrt{5r}}{5}$ **69.** B   **71.** $\dfrac{\sqrt[3]{4}}{2}$

73. $\dfrac{\sqrt[3]{2}}{4}$ **75.** $\dfrac{\sqrt[3]{121}}{11}$ **77.** $\dfrac{\sqrt[3]{50}}{5}$ **79.** $\dfrac{\sqrt[3]{196}}{7}$ **81.** $\dfrac{\sqrt[3]{6y}}{2y}$ **83.** $\dfrac{\sqrt[3]{42mn^2}}{6n}$ **85.** $\dfrac{\sqrt[4]{2}}{2}$ **87.** $32x^2 + 44x - 21$
89. $36x^2 - 1$ **91.** $pa - pm + qa - qm$

Section 8.5 (pages 582–585)

1. 13   **3.** 4   **5.** $\sqrt{15} - \sqrt{35}$ **7.** $2\sqrt{10} + 30$ **9.** $4\sqrt{7}$ **11.** $57 + 23\sqrt{6}$ **13.** $81 + 14\sqrt{21}$ **15.** $71 - 16\sqrt{7}$
17. $37 + 12\sqrt{7}$ **19.** $7 + 2\sqrt{6}$ **21.** 23   **23.** 1   **25.** 2   **27.** $2\sqrt{3} - 2 + 3\sqrt{2} - \sqrt{6}$ **29.** $15\sqrt{2} - 15$
31. $\sqrt{30} + \sqrt{15} + 6\sqrt{5} + 3\sqrt{10}$ **33.** $187 - 20\sqrt{21}$ **35.** Because multiplication must be performed before addition, it
is incorrect to add -37 and -2. Since $-2\sqrt{15}$ cannot be simplified, the expression cannot be written in a simpler form, and the
final answer is $-37 - 2\sqrt{15}$.   **37.** $49 + 14\sqrt{x} + x$ **39.** $6t - 3\sqrt{14t} + 2\sqrt{7t} - 7\sqrt{2}$ **41.** $3m - 2n$

43. (a) $\sqrt{5} - \sqrt{3}$ (b) $\sqrt{6} + \sqrt{5}$ **45.** $-2 + \sqrt{5}$ **47.** $-2 - \sqrt{11}$ **49.** $3 - \sqrt{3}$ **51.** $\dfrac{-3 + 5\sqrt{3}}{11}$

53. $\dfrac{\sqrt{6} + \sqrt{2} + 3\sqrt{3} + 3}{2}$ **55.** $\dfrac{-6\sqrt{2} + 12 + \sqrt{10} - 2\sqrt{5}}{2}$ **57.** $\dfrac{-4\sqrt{3} - \sqrt{2} + 10\sqrt{6} + 5}{23}$

59. $\sqrt{21} + \sqrt{14} + \sqrt{6} + 2$ **61.** $-\sqrt{10} + \sqrt{15}$ **63.** $3 - \sqrt{3}$ **65.** $\dfrac{8(4 + \sqrt{x})}{16 - x}$ **67.** $\dfrac{\sqrt{x} - \sqrt{y}}{x - y}$ **69.** $\sqrt{7} - 2$

71. $\dfrac{\sqrt{3} + 5}{4}$ **73.** $\dfrac{6 - \sqrt{10}}{2}$ **75.** $\dfrac{2 + \sqrt{2}}{3}$ **77.** $2 - 3\sqrt[3]{4}$ **79.** $12 + 10\sqrt[4]{8}$ **81.** $-1 + 3\sqrt[3]{2} - \sqrt[3]{4}$ **83.** 1

85. 4 in.   **87.** (a) $\dfrac{9\sqrt{2}}{4}$ sec (b) 3.182 sec   **89.** $30 + 18x$ **90.** They are not like terms.   **91.** $30 + 18\sqrt{5}$

92. They are not like radicals.   **93.** Make the first term $30x$, so that $30x + 18x = 48x$; make the first term $30\sqrt{5}$, so that

$30\sqrt{5} + 18\sqrt{5} = 48\sqrt{5}$.   **94.** When combining like terms, we add (or subtract) the coefficients of the common factors of the terms: $2xy + 5xy = 7xy$. When combining like radicals, we add (or subtract) the coefficients of the common radical terms:

$2\sqrt{ab} + 5\sqrt{ab} = 7\sqrt{ab}$.   **95.** $\left\{\dfrac{1}{2}, \dfrac{3}{4}\right\}$ **97.** $\{-3, -1\}$ **99.** $\{-3, 1\}$

Summary Exercises on Operations with Radicals (pages 585–586)

1. $-3\sqrt{10}$ **2.** $5 - \sqrt{15}$ **3.** $2 - \sqrt{6} + 2\sqrt{3} - 3\sqrt{2}$ **4.** $6\sqrt{2}$ **5.** $73 - 12\sqrt{35}$ **6.** $\dfrac{\sqrt{6}}{2}$ **7.** $-3 - 2\sqrt{2}$

8. $4\sqrt{7} + 4\sqrt{5}$ **9.** -33 **10.** $\dfrac{\sqrt{t} - \sqrt{3}}{t - 3}$ **11.** $2xyz^2\sqrt[3]{y^2}$ **12.** $4\sqrt[3]{3}$ **13.** $\sqrt{6} + 1$ **14.** $\dfrac{\sqrt{6x}}{3x}$ **15.** $\dfrac{3}{5}$

16. $4\sqrt{2}$ **17.** $-2\sqrt[3]{2}$ **18.** $11 - 2\sqrt{30}$ **19.** $3\sqrt{3x}$ **20.** $52 + 30\sqrt{3}$ **21.** $\dfrac{2\sqrt[3]{18}}{9}$ **22.** 1 **23.** $-x^2\sqrt[4]{x}$

24. $3\sqrt[3]{2t^2}$ **25.** 4 **26. (a)** 6 **(b)** $\{-6, 6\}$ **27. (a)** 9 **(b)** $\{-9, 9\}$ **28. (a)** $\{-2, 2\}$ **(b)** -2

29. (a) $\{-3, 3\}$ **(b)** -3 **30. (a)** $\left\{-\dfrac{1}{2}, \dfrac{1}{2}\right\}$ **(b)** $\dfrac{1}{2}$ **31. (a)** $\left\{-\dfrac{1}{7}, \dfrac{1}{7}\right\}$ **(b)** $\dfrac{1}{7}$ **32. (a)** $-\dfrac{4}{5}$ **(b)** $\left\{-\dfrac{4}{5}, \dfrac{4}{5}\right\}$

33. (a) $-\dfrac{7}{10}$ **(b)** $\left\{-\dfrac{7}{10}, \dfrac{7}{10}\right\}$ **34. (a)** $\{-0.2, 0.2\}$ **(b)** 0.2 **35. (a)** $\{-0.4, 0.4\}$ **(b)** 0.4 **36.** $x^2 = 25; x^2 - 25 = 0;$
$(x + 5)(x - 5) = 0; x = -5$ or $x = 5;$ Solution set: $\{-5, 5\}$

Section 8.6 (pages 592–596)

1. $\{49\}$ **3.** $\{7\}$ **5.** $\{85\}$ **7.** $\{-45\}$ **9.** $\left\{-\dfrac{3}{2}\right\}$ **11.** $\emptyset$ **13.** $\{121\}$ **15.** $\{8\}$ **17.** $\{1\}$ **19.** $\{6\}$ **21.** $\emptyset$
23. $\{5\}$ **25.** $\{7\}$ **27.** $\{6\}$ **29.** $\emptyset$ **31.** $x^2 - 14x + 49$ **33.** $\{-2, 1\}$ **35.** $\{12\}$ **37.** $\{-2, -1\}$ **39.** $\{11\}$
41. $\{-1, 3\}$ **43.** $\{9\}$ **45.** $\{8\}$ **47.** $\{9\}$ **49.** $\{2, 11\}$ **51.** Since $\sqrt{x}$ must be greater than or equal to 0 for any

replacement for x, it cannot equal -8, a negative number.   **53.** $\{2\}$ **55.** $\{9\}$ **57.** $\{4, 20\}$ **59.** $\{-5\}$ **61.** $\left\{-\dfrac{2}{3}\right\}$

63. $\{-1, 8\}$ **65.** $\left\{\dfrac{4}{3}, 2\right\}$ **67.** $\{-27, 3\}$ **69.** 21 **71.** 8 **73. (a)** 70.5 mph **(b)** 59.8 mph **(c)** 53.9 mph

75. 158.6 ft   **77.** yes; 26 mi   **79.** 47 mi   **81.** $s = 13$ units   **82.** $6\sqrt{13}$ sq. units   **83.** $h = \sqrt{13}$ units

84. $3\sqrt{13}$ sq. units   **85.** $6\sqrt{13}$ sq. units   **86.** They are both $6\sqrt{13}$.   **87.** 5^6 **89.** $\dfrac{1}{a^3}$ **91.** $\dfrac{3^2}{p^2}$ **93.** c^{13}

Section 8.7 (pages 600–601)

1. A   **3.** C   **5.** 5   **7.** 4   **9.** 2   **11.** 2   **13.** 8   **15.** 9   **17.** 8   **19.** 4   **21.** -4 **23.** -4 **25.** $\dfrac{1}{343}$ **27.** $\dfrac{1}{36}$

29. $-\dfrac{1}{32}$ **31.** $2^{8/3}$ **33.** $\dfrac{1}{6^{1/2}}$ **35.** $\dfrac{1}{15^{1/2}}$ **37.** $11^{1/7}$ **39.** 8^3 **41.** $6^{1/2}$ **43.** $\dfrac{5^3}{2^3}$ **45.** $\dfrac{1}{2^{8/5}}$ **47.** $6^{2/9}$ **49.** $x^{9/5}$

51. $r^{1/9}$ **53.** $m^2n^{1/6}$ **55.** $\dfrac{a^4}{b^{3/2}}$ **57.** $m^{1/6}$ **59.** 2   **61.** 2   **63.** $\sqrt{a}$ **65.** $\sqrt[3]{k^2}$ **67.** 2   **69.** 1.883

71. (a) $d = 1.22x^{1/2}$ **(b)** 211.31 mi   **73.** Because $(7^{1/2})^2 = 7$ and $(\sqrt{7})^2 = 7$, they should be equal, so we define $7^{1/2}$ to be $\sqrt{7}$.

75. $-11, 11$ **77.** $-0.7, 0.7$ **79.** $-\dfrac{1}{2}, \dfrac{1}{2}$ **81.** $-\dfrac{2}{5}, \dfrac{2}{5}$ **83.** $2\sqrt{59}$ **85.** $7\sqrt{3}$

Chapter 8 Review Exercises (pages 605–608)

1. $-7, 7$ **3.** $-14, 14$ **5.** $-15, 15$ **7.** 4 **9.** 10 **11.** It is not a real number. **13.** $\dfrac{7}{6}$ **15.** a must be negative.

17. 48.3 cm **19.** rational; -5 **21.** $5\sqrt{3}$ **23.** $4\sqrt{10}$ **25.** 12 **27.** $16\sqrt{6}$ **29.** $-\dfrac{11}{20}$ **31.** $\dfrac{\sqrt{7}}{13}$ **33.** $\dfrac{2}{15}$ **35.** 8

37. p **39.** r^9 **41.** $a^7 b^{10}\sqrt{ab}$ **43.** Yes, because both approximations are 0.7071067812. **45.** $21\sqrt{3}$ **47.** 0

49. $2\sqrt{3} + 3\sqrt{10}$ **51.** $6\sqrt{30}$ **53.** 0 **55.** $11k^2\sqrt{2n}$ **57.** $\sqrt{5}$ **59.** $\dfrac{\sqrt{30}}{15}$ **61.** $\sqrt{10}$ **63.** $\dfrac{r\sqrt{x}}{4x}$ **65.** $\dfrac{\sqrt[3]{98}}{7}$

67. $-\sqrt{15} - 9$ **69.** $22 - 16\sqrt{3}$ **71.** -2 **73.** $-2 + \sqrt{5}$ **75.** $\dfrac{-2 + 6\sqrt{2}}{17}$ **77.** $\dfrac{-\sqrt{10} + 3\sqrt{5} + \sqrt{2} - 3}{7}$

79. $\dfrac{3 + 2\sqrt{6}}{3}$ **81.** $3 + 4\sqrt{3}$ **83.** $\emptyset$ **85.** $\{1\}$ **87.** $\{6\}$ **89.** $\{-2\}$ **91.** $\{5\}$ **93.** 9 **95.** 7^3, or 343 **97.** $x^{3/4}$

99. 16 **101.** $\dfrac{5 - \sqrt{2}}{23}$ **103.** $5y\sqrt{2}$ **105.** $-\sqrt{10} - 5\sqrt{15}$ **107.** $\dfrac{2 + \sqrt{13}}{2}$ **109.** $7 - 2\sqrt{10}$ **111.** -11

113. $\{7\}$ **115.** $\{8\}$

Chapter 8 Test (pages 608–610)

[8.1] **1.** $-14, 14$ **2.** (a) irrational (b) 11.916 **3.** There is no real number whose square is -5. [8.2] **4.** $-3\sqrt{3}$

5. $\dfrac{8\sqrt{2}}{5}$ **6.** $2\sqrt[3]{4}$ **7.** $4\sqrt{6}$ [8.3] **8.** $9\sqrt{7}$ **9.** $-5\sqrt{3x}$ [8.2] **10.** $2y\sqrt[3]{4x^2}$ [8.5] **11.** 31

12. $6\sqrt{2} + 2 - 3\sqrt{14} - \sqrt{7}$ **13.** $11 + 2\sqrt{30}$ [8.3] **14.** $-6x\sqrt[3]{2x}$ [8.4] **15.** $\dfrac{\sqrt[3]{18}}{3}$ [8.1] **16.** (a) $6\sqrt{2}$ in.

(b) 8.485 in. **17.** 50 ohms [8.4] **18.** $r = \dfrac{\sqrt{\pi S}}{2\pi}$ **19.** $\dfrac{5\sqrt{14}}{7}$ **20.** $\dfrac{\sqrt{6x}}{3x}$ **21.** $-\sqrt[3]{2}$ [8.5] **22.** $\dfrac{-12 - 3\sqrt{3}}{13}$

23. $\dfrac{1 + \sqrt{2}}{2}$ **24.** $\dfrac{\sqrt{3} + 12\sqrt{2}}{3}$ [8.6] **25.** $\emptyset$ [8.6] **26.** $\{3\}$ **27.** $\left\{\dfrac{1}{4}, 1\right\}$ **28.** $\{-4\}$ [8.7] **29.** 16

30. -25 **31.** 5 **32.** $\dfrac{1}{3}$ [8.6] **33.** 12 is not a solution. A check shows that it does not satisfy the original equation. The solution set is $\emptyset$.

Chapters 1–8 Cumulative Review Exercises (pages 610–612)

[1.2] **1.** 54 **2.** 6 [1.4] **3.** 3 [2.3] **4.** $\{3\}$ [2.8] **5.** $[-16, \infty)$ **6.** $(5, \infty)$ [2.4] **7.** 2003: 10.1 billion bushels; 2004: 11.8 billion bushels [3.2] **8.** 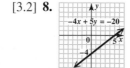 **9.** **10.** [3.5]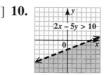

[3.3, 3.4] **11.** (a) 0.46; convention spending increased $\$0.46$ million per year. (b) $y = 0.46x - 906.8$ (c) $\$16.9$ million

[4.1–4.3] **12.** $\{(3, -7)\}$ **13.** $\{(x, y)\mid 2x - y = 6\}$ [4.4] **14.** from Chicago: 61 mph; from Des Moines: 54 mph

[5.1] **15.** $12x^{10}y^2$ [5.2] **16.** $\dfrac{y^{15}}{2^3 \cdot 3^6}$, or $\dfrac{y^{15}}{5832}$ [5.4] **17.** $3x^3 + 11x^2 - 13$ [5.7] **18.** $4t^2 - 8t + 5$

[6.2–6.4] **19.** $(m + 8)(m + 4)$ **20.** $(5t^2 + 6)(5t^2 - 6)$ **21.** $(6a + 5b)(2a - b)$ **22.** $(9z + 4)^2$ [6.5] **23.** $\{3, 4\}$

24. $\{-2, -1\}$ [7.1] **25.** $-7, 2$ [7.2] **26.** $\dfrac{x + 1}{x}$ **27.** $(t + 5)(t + 3)$, or $t^2 + 8t + 15$ [7.4] **28.** $\dfrac{y^2}{(y + 1)(y - 1)}$

29. $\dfrac{-2x - 14}{(x + 3)(x - 1)}$ [7.5] **30.** -21 [7.6] **31.** $\{19\}$ **32.** $B = \dfrac{CD - AD}{AC - 1}$ [7.8] **33.** 405 revolutions per min

[8.3] **34.** $29\sqrt{3}$ [8.5] **35.** $-\sqrt{3} + \sqrt{5}$ [8.2] **36.** $10xy^2\sqrt{2y}$ [8.5] **37.** $21 - 5\sqrt{2}$ [8.6] **38.** $\{16\}$

[8.7] **39.** 32   **40.** $\dfrac{1}{64}$

CHAPTER 9   Quadratic Equations

Section 9.1   (pages 617–619)

1. C   **3.** D   **5.** $\{\pm 9\}$ **7.** $\{\pm\sqrt{14}\}$ **9.** $\{\pm 4\sqrt{3}\}$ **11.** $\emptyset$ **13.** $\left\{\pm\dfrac{5}{2}\right\}$ **15.** $\{\pm 1.5\}$ **17.** $\{\pm\sqrt{3}\}$

19. $\left\{\pm\dfrac{2\sqrt{7}}{7}\right\}$ **21.** $\{\pm 2\sqrt{6}\}$ **23.** $\left\{\pm\dfrac{2\sqrt{5}}{5}\right\}$ **25.** $\{\pm 3\sqrt{3}\}$ **27.** According to the square root property, -9 is

also a solution, so her answer was not completely correct. The solution set is $\{\pm 9\}$.   **29.** $\{-2, 8\}$ **31.** $\emptyset$ **33.** $\{8 \pm 3\sqrt{3}\}$

35. $\left\{-3, \dfrac{5}{3}\right\}$ **37.** $\left\{0, \dfrac{3}{2}\right\}$ **39.** $\left\{\dfrac{5 \pm \sqrt{30}}{2}\right\}$ **41.** $\left\{\dfrac{-1 \pm 3\sqrt{2}}{3}\right\}$ **43.** $\{-10 \pm 4\sqrt{3}\}$ **45.** $\left\{\dfrac{1 \pm 4\sqrt{3}}{4}\right\}$

47. Johnny's first solution, $\dfrac{5 + \sqrt{30}}{2}$, is equivalent to Linda's second solution, $\dfrac{-5 - \sqrt{30}}{-2}$. This can be verified by multiplying

$\dfrac{5 + \sqrt{30}}{2}$ by 1 in the form $\dfrac{-1}{-1}$. Similarly, Johnny's second solution is equivalent to Linda's first one.   **49.** $\{-4.48, 0.20\}$

51. $\{-3.09, -0.15\}$ **53.** about $\dfrac{1}{2}$ sec   **55.** 9 in.   **57.** 5%   **59.** $\dfrac{4 + 4\sqrt{3}}{5}$ **61.** $\dfrac{3 + \sqrt{6}}{4}$ **63.** $(x - 5)^2$

65. $\left(x - \dfrac{7}{2}\right)^2$

Section 9.2   (pages 624–626)

1. $25; (x + 5)^2$ **3.** $100; (z - 10)^2$ **5.** $1; (x + 1)^2$ **7.** $\dfrac{25}{4}; \left(p - \dfrac{5}{2}\right)^2$ **9.** D   **11.** $\{1, 3\}$ **13.** $\{-1 \pm \sqrt{6}\}$

15. $\{4 \pm 2\sqrt{3}\}$ **17.** $\{-3\}$ **19.** $\left\{-\dfrac{3}{2}, \dfrac{1}{2}\right\}$ **21.** $\emptyset$ **23.** $\left\{\dfrac{9 \pm \sqrt{21}}{6}\right\}$ **25.** $\left\{\dfrac{-7 \pm \sqrt{97}}{6}\right\}$ **27.** $\{-4, 2\}$

29. $\{4 \pm \sqrt{3}\}$ **31.** $\{1 \pm \sqrt{6}\}$ **33.** (a) $\left\{\dfrac{3 \pm 2\sqrt{6}}{3}\right\}$ (b) $\{-0.633, 2.633\}$ **35.** (a) $\{-2 \pm \sqrt{3}\}$ (b) $\{-3.732, -0.268\}$

37. 3 sec and 5 sec   **39.** 1 sec and 5 sec   **41.** 75 ft by 100 ft   **43.** 8 mi   **45.** x^2 **46.** $x^2 + 8x$

47. $x^2 + 8x + 16$ **48.** It occurred when we added the 16 squares.   **49.** $\dfrac{4 - 3\sqrt{3}}{3}$ **51.** $\dfrac{2 - \sqrt{5}}{2}$ **53.** $2\sqrt{5}$ **55.** 0

Section 9.3   (pages 631–633)

1. $a = 3, b = 4, c = -8$ **3.** $a = -8, b = -2, c = -3$ **5.** $a = 3, b = -4, c = -2$ **7.** $a = 3, b = 7, c = 0$

9. $a = 1, b = 1, c = -12$ **11.** $a = 9, b = 9, c = -26$ **13.** If a were 0, the equation would be linear, not quadratic.

15. No, because $2a$ should be the denominator for $-b$ as well. The correct formula is $x = \dfrac{-b \pm \sqrt{b^2 - 4ac}}{2a}$.   **17.** $\{2\}$

19. $\{-13, 1\}$ **21.** $\left\{\dfrac{-6 \pm \sqrt{26}}{2}\right\}$ **23.** $\left\{-1, \dfrac{5}{2}\right\}$ **25.** $\{-1, 0\}$ **27.** $\left\{0, \dfrac{12}{7}\right\}$ **29.** $\{\pm 2\sqrt{6}\}$ **31.** $\left\{\pm\dfrac{2}{5}\right\}$

33. $\left\{\dfrac{6 \pm 2\sqrt{6}}{3}\right\}$ **35.** $\emptyset$ **37.** $\emptyset$ **39.** $\left\{\dfrac{-5 \pm \sqrt{61}}{2}\right\}$ **41. (a)** $\left\{\dfrac{-1 \pm \sqrt{11}}{2}\right\}$ **(b)** $\{-2.158, 1.158\}$

43. (a) $\left\{\dfrac{1 \pm \sqrt{5}}{2}\right\}$ **(b)** $\{-0.618, 1.618\}$ **45.** $\left\{-\dfrac{2}{3}, \dfrac{4}{3}\right\}$ **47.** $\left\{\dfrac{-1 \pm \sqrt{73}}{6}\right\}$ **49.** $\emptyset$ **51.** $\{1 \pm \sqrt{2}\}$

53. $\left\{-1, \dfrac{5}{2}\right\}$ **55.** $r = \dfrac{-\pi h \pm \sqrt{\pi^2 h^2 + \pi S}}{\pi}$ **57.** 3.5 ft **59.** $\{16, -8\}$; Only 16 ft is a reasonable answer.

61. (a) 225 **(b)** 169 **(c)** 4 **(d)** 121 **62.** Each is a perfect square. **63. (a)** $(3x - 2)(6x + 1)$ **(b)** $(x + 2)(5x - 3)$
(c) $(8x + 1)(6x + 1)$ **(d)** $(x + 3)(x - 8)$ **64. (a)** 41 **(b)** -39 **(c)** 12 **(d)** 112 **65.** no **66.** If the discriminant is a
perfect square, the trinomial is factorable. **(a)** yes **(b)** yes **(c)** no **67.** $-5 + 8z$ **69.** $-2 + 3k$ **71.** $24 - 2r - 15r^2$

Summary Exercises on Quadratic Equations (pages 633–634)

1. $\{\pm 6\}$ **2.** $\left\{\dfrac{-3 \pm \sqrt{5}}{2}\right\}$ **3.** $\left\{\pm\dfrac{10}{9}\right\}$ **4.** $\left\{\pm\dfrac{7}{9}\right\}$ **5.** $\{1, 3\}$ **6.** $\{-2, -1\}$ **7.** $\{4, 5\}$ **8.** $\left\{\dfrac{-3 \pm \sqrt{17}}{2}\right\}$

9. $\left\{-\dfrac{1}{3}, \dfrac{5}{3}\right\}$ **10.** $\left\{\dfrac{1 \pm \sqrt{10}}{2}\right\}$ **11.** $\{-17, 5\}$ **12.** $\left\{-\dfrac{7}{5}, 1\right\}$ **13.** $\left\{\dfrac{7 \pm 2\sqrt{6}}{3}\right\}$ **14.** $\left\{\dfrac{1 \pm 4\sqrt{2}}{7}\right\}$ **15.** $\emptyset$

16. $\emptyset$ **17.** $\left\{-\dfrac{1}{2}, 2\right\}$ **18.** $\left\{-\dfrac{1}{2}, 1\right\}$ **19.** $\left\{-\dfrac{5}{4}, \dfrac{3}{2}\right\}$ **20.** $\left\{-3, \dfrac{1}{3}\right\}$ **21.** $\{1 \pm \sqrt{2}\}$ **22.** $\left\{\dfrac{-5 \pm \sqrt{13}}{6}\right\}$

23. $\left\{\dfrac{2}{5}, 4\right\}$ **24.** $\{-3 \pm \sqrt{5}\}$ **25.** $\left\{\dfrac{-3 \pm \sqrt{41}}{2}\right\}$ **26.** $\left\{-\dfrac{5}{4}\right\}$ **27.** $\left\{\dfrac{1}{4}, 1\right\}$ **28.** $\left\{\dfrac{1 \pm \sqrt{3}}{2}\right\}$

29. $\left\{\dfrac{-2 \pm \sqrt{11}}{3}\right\}$ **30.** $\left\{\dfrac{-5 \pm \sqrt{41}}{8}\right\}$ **31.** $\left\{\dfrac{-7 \pm \sqrt{5}}{4}\right\}$ **32.** $\left\{-\dfrac{8}{3}, -\dfrac{6}{5}\right\}$ **33.** $\left\{\dfrac{8 \pm 8\sqrt{2}}{3}\right\}$

34. $\left\{\dfrac{-5 \pm \sqrt{5}}{2}\right\}$ **35.** $\emptyset$ **36.** $\emptyset$ **37.** $\left\{-\dfrac{2}{3}, 2\right\}$ **38.** $\left\{-\dfrac{1}{4}, \dfrac{2}{3}\right\}$ **39.** $\left\{-4, \dfrac{3}{5}\right\}$ **40.** $\{-3, 5\}$

41. $\left\{-\dfrac{2}{3}, \dfrac{2}{5}\right\}$ **42.** $\{-4, 6\}$ **43.** There are no real solutions.

Section 9.4 (pages 639–640)

1. $3i$ **3.** $2i\sqrt{5}$ **5.** $3i\sqrt{2}$ **7.** $5i\sqrt{5}$ **9.** $5 + 3i$ **11.** $6 - 9i$ **13.** $6 - 7i$ **15.** $14 + 5i$ **17.** $7 - 22i$ **19.** 45

21. $3 - i$ **23.** $2 - 6i$ **25.** $-\dfrac{3}{25} + \dfrac{4}{25}i$ **27.** $\{-1 \pm 2i\}$ **29.** $\{3 \pm i\sqrt{5}\}$ **31.** $\left\{-\dfrac{2}{3} \pm i\sqrt{2}\right\}$ **33.** $\{1 \pm i\}$

35. $\left\{-\dfrac{3}{4} \pm \dfrac{\sqrt{31}}{4}i\right\}$ **37.** $\left\{\dfrac{3}{2} \pm \dfrac{\sqrt{7}}{2}i\right\}$ **39.** $\left\{\dfrac{1}{5} \pm \dfrac{\sqrt{14}}{5}i\right\}$ **41.** $\left\{-\dfrac{1}{2} \pm \dfrac{\sqrt{13}}{2}i\right\}$ **43.** $\left\{\dfrac{1}{2} \pm \dfrac{\sqrt{11}}{2}i\right\}$

45. If the discriminant $b^2 - 4ac$ is negative, the equation will have solutions that are not real. **47.** true **49.** false; for
example, $3 + 2i$ is a complex number but it is not real. **51.** **53.** **55.** 16

Section 9.5   (pages 646–649)

1. The vertex of a parabola is the lowest or highest point on the graph.   **3.** $(0, -6)$ **5.** $(-3, 0)$

7. $(-1, 2)$ **9.** $(4, 0)$ **11.** $(3, 4)$ **13.** one real solution; $\{2\}$

15. two real solutions; $\{\pm 2\}$ **17.** no real solutions; $\emptyset$ **19.** If $a > 0$, it opens upward, and if $a < 0$, it opens downward.

21. $\{-2, 3\}$ **23.** $\{-1, 1.5\}$ **25.** $(-\infty, \infty); [0, \infty)$ **27.** $(-\infty, \infty); (-\infty, 4]$ **29.** $(-\infty, \infty); [1, \infty)$ **31.** 3   **33.** 21

35. $y = \dfrac{11}{5625} x^2$ **37.** 40 and 40   **39.** In each case, there is a vertical "stretch" of the parabola. It becomes narrower as the

coefficient gets larger.   **40.** In each case, there is a vertical "shrink" of the parabola. It becomes wider as the coefficient gets

smaller.   **41.** The graph of Y_2 is obtained by reflecting the graph of Y_1 across the x-axis.   **42.** When the coefficient of x^2 is

negative, the parabola opens downward.   **43.** By adding a positive constant k, the graph is shifted k units upward. By

subtracting a positive constant k, the graph is shifted k units downward.   **44.** Adding a positive constant k before squaring

moves the graph k units to the left. Subtracting a positive constant k before squaring moves the graph k units to the right.

Chapter 9 Review Exercises   (pages 655–656)

1. $\{\pm 12\}$ **3.** $\{\pm 8\sqrt{2}\}$ **5.** $\{3 \pm \sqrt{10}\}$ **7.** $\emptyset$ **9.** $\{-5, -1\}$ **11.** $\{-1 \pm \sqrt{6}\}$ **13.** $\left\{-\dfrac{2}{5}, 1\right\}$ **15.** 2.5 sec

17. $\left(\dfrac{3}{2}\right)^2$, or $\dfrac{9}{4}$ **19.** $\{1 \pm \sqrt{5}\}$ **21.** $\left\{\dfrac{2 \pm \sqrt{10}}{2}\right\}$ **23.** $\left\{\dfrac{-3 \pm \sqrt{41}}{2}\right\}$ **25.** The term $-b$ should be above the

fraction bar.   **27.** $-6 - 5i$ **29.** 13   **31.** $\dfrac{28}{13} - \dfrac{3}{13} i$ **33.** No, the product $(a + bi)(a - bi) = a^2 + b^2$ will always be the

sum of the squares of two real numbers, which is a real number.   **35.** $\left\{\dfrac{2}{3} \pm \dfrac{2\sqrt{2}}{3} i\right\}$ **37.** $\left\{-\dfrac{3}{2} \pm \dfrac{\sqrt{23}}{2} i\right\}$

39. $\left\{-\dfrac{1}{9} \pm \dfrac{2\sqrt{2}}{9} i\right\}$ **41.** vertex: $(0, 5)$ **43.** vertex: $(1, 0)$

45. vertex: $(-2, -2)$ **47.** one; $\{2\}$ **49.** Exercise 46, domain: $(-\infty, \infty)$; range: $[-2, \infty)$; Exercise 47, domain: $(-\infty, \infty)$; range: $(-\infty, 0]$; Exercise 48, domain: $(-\infty, \infty)$; range: $[1, \infty)$

51. $\left\{-\dfrac{11}{2}, 5\right\}$ **53.** $\left\{\dfrac{-1 \pm \sqrt{21}}{2}\right\}$ **55.** $\left\{\dfrac{-5 \pm \sqrt{17}}{2}\right\}$ **57.** $\emptyset$ **59.** $\left\{-\dfrac{5}{3}\right\}$ **61.** $\{-2 \pm \sqrt{5}\}$ **63.** 400 or 800

Chapter 9 Test (pages 657–658)

[9.1] **1.** $\left\{\pm\sqrt{39}\right\}$ **2.** $\{-11, 5\}$ **3.** $\left\{\dfrac{-3 \pm 2\sqrt{6}}{4}\right\}$ [9.2] **4.** $\left\{2 \pm \sqrt{10}\right\}$ **5.** $\left\{\dfrac{-6 \pm \sqrt{42}}{2}\right\}$

[9.3] **6.** $\left\{0, -\dfrac{2}{5}\right\}$ **7.** $\left\{-3, \dfrac{1}{2}\right\}$ **8.** $\left\{\dfrac{3 \pm \sqrt{3}}{3}\right\}$ [9.4] **9.** $\left\{-1 \pm \dfrac{\sqrt{7}}{2}i\right\}$ [9.3] **10.** $\left\{\dfrac{5 \pm \sqrt{13}}{6}\right\}$

[9.1–9.3] **11.** $\left\{1 \pm \sqrt{2}\right\}$ **12.** $\left\{\dfrac{-1 \pm 3\sqrt{2}}{2}\right\}$ **13.** $\left\{\dfrac{11 \pm \sqrt{89}}{4}\right\}$ **14.** $\{5\}$ **15.** 2 sec **16.** 12, 16, 20

[9.4] **17.** $-5 + 5i$ **18.** $-17 - 4i$ **19.** 73 **20.** $2 - i$ [9.5] **21.** vertex: $(3, 0)$

22. vertex: $(-1, -3)$ **23.** vertex: $(-3, -2)$ **24.** (a) two (b) $\left\{-3 \pm \sqrt{2}\right\}$

(c) $-3 - \sqrt{2} \approx -4.414$ and $-3 + \sqrt{2} \approx -1.586$ **25.** 200 and 200

Chapters 1–9 Cumulative Review Exercises (pages 658–660)

[1.2] **1.** 15 [1.5] **2.** -2 [1.6] **3.** 5 [1.8] **4.** $-r + 7$ **5.** $-2k$ **6.** $19m - 17$ [2.1–2.3] **7.** $\{18\}$ **8.** $\{5\}$

9. $\left\{\dfrac{8}{3}\right\}$ **10.** $\{2\}$ [2.5] **11.** $100°, 80°$ **12.** width: 50 ft; length: 94 ft **13.** $L = \dfrac{P - 2W}{2}$, or $L = \dfrac{P}{2} - W$

[2.8] **14.** $(-2, \infty)$  **15.** $(-\infty, 4]$ [3.2] **16.** **17.**

[3.5] **18.** [3.3] **19.** $-\dfrac{1}{3}$ [3.4] **20.** $2x - y = -3$ [4.1–4.3] **21.** $\{(-3, 2)\}$ **22.** $\emptyset$

[4.4] **23.** AT&T: $12.99; GE: $29.99 [4.5] **24.** 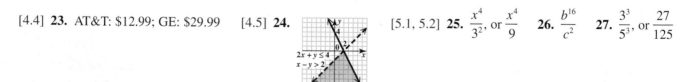 [5.1, 5.2] **25.** $\dfrac{x^4}{3^2}$, or $\dfrac{x^4}{9}$ **26.** $\dfrac{b^{16}}{c^2}$ **27.** $\dfrac{3^3}{5^3}$, or $\dfrac{27}{125}$

[5.4] **28.** $8x^5 - 17x^4 - x^2$ [5.5] **29.** $2x^4 + x^3 - 19x^2 + 2x + 20$ [5.7] **30.** $3x^2 - 2x + 1$ [5.3] **31.** (a) 6.35×10^9

(b) 0.00023 [6.1] **32.** $16x^2(x - 3y)$ [6.3] **33.** $(2a + 1)(a - 3)$ [6.4] **34.** $(4x^2 + 1)(2x + 1)(2x - 1)$

35. $(5m - 2)^2$ [6.5] **36.** $\{-9, 6\}$ [6.6] **37.** 50 m [7.2] **38.** $\dfrac{4}{5}$ [7.4] **39.** $\dfrac{-k - 1}{k(k - 1)}$ **40.** $\dfrac{5a + 2}{(a - 2)^2(a + 2)}$

[7.5] **41.** $\dfrac{b + a}{b - a}$ [7.6] **42.** $\left\{-\dfrac{15}{7}, 2\right\}$ [7.7] **43.** 3 mph [8.1] **44.** 10 [8.4] **45.** $\dfrac{6\sqrt{30}}{5}$ **46.** $\dfrac{\sqrt[3]{28}}{4}$

[8.3] **47.** $4\sqrt{5}$ **48.** $-ab\sqrt[3]{2b}$ [8.6] **49.** $\{7\}$ [8.7] **50.** (a) 4 (b) -2 [9.1] **51.** $\left\{\dfrac{-2 \pm 2\sqrt{3}}{3}\right\}$

[9.2] **52.** $\left\{-1 \pm \sqrt{6}\right\}$ [9.3] **53.** $\left\{\dfrac{2 \pm \sqrt{10}}{2}\right\}$ [9.2, 9.3] **54.** $\emptyset$ **55.** 3, 4, 5   [9.4] **56.** **(a)** $8i$ **(b)** $5 + 2i$

57. $\left\{-\dfrac{1}{2} \pm \dfrac{\sqrt{17}}{2}i\right\}$ [9.5] **58.** vertex: $(-1, 2)$; domain: $(-\infty, \infty)$; range: $(-\infty, 2]$

APPENDIXES

Appendix B (pages 669–670)

1. 117.385   **3.** 13.21   **5.** 150.49   **7.** 96.101   **9.** 4.849   **11.** 166.32   **13.** 164.19   **15.** 1.344   **17.** 4.14
19. 2.23   **21.** 4800   **23.** 0.53   **25.** 1.29   **27.** 0.96   **29.** 0.009   **31.** 80%   **33.** 0.7%   **35.** 67%   **37.** 12.5%
39. 109.2   **41.** 238.92   **43.** 5%   **45.** 110%   **47.** 25%   **49.** 148.44   **51.** 7839.26   **53.** 7.39%   **55.** $2760
57. 805 mi   **59.** 63 mi   **61.** $2400

Appendix C (pages 674–675)

1. $\{1, 2, 3, 4, 5, 6, 7\}$ **3.** $\{$winter, spring, summer, fall$\}$ **5.** $\emptyset$ **7.** $\{L\}$ **9.** $\{2, 4, 6, 8, 10, \dots\}$ **11.** The sets in
Exercises 9 and 10 are infinite sets.   **13.** true   **15.** false   **17.** true   **19.** true   **21.** true   **23.** true   **25.** true
27. false   **29.** true   **31.** true   **33.** false   **35.** true   **37.** true   **39.** false   **41.** false   **43.** true   **45.** $\{g, h\}$
47. $\{b, c, d, e, g, h\}$ **49.** $\{a, c, e\} = B$ **51.** $\{d\} = D$ **53.** $\{a\}$ **55.** $\{a, c, d, e\}$ **57.** $\{a, c, e, f\}$ **59.** $\emptyset$
61. B and D; C and D

Appendix D (pages 681–682)

1. 7   **3.** 69.8 (rounded)   **5.** $39,622   **7.** $58.24   **9.** $35,500   **11.** 6.1   **13.** 17.2   **15.** 2.60   **17.** **(a)** 2.80
(b) 2.93 (rounded) **(c)** 3.13 (rounded)   **19.** 15   **21.** 516   **23.** 48   **25.** 8   **27.** 68 and 74; bimodal   **29.** no mode

Appendix E (page 686)

1. C   **3.** **(a)** true **(b)** true **(c)** false **(d)** true   **5.** $(-3, 3)$ **7.** $(-\infty, -6] \cup [7, \infty)$

9. $(-\infty, -3) \cup (-2, \infty)$ **11.** $[-1, 5]$ **13.** $\left(-1, \dfrac{2}{5}\right)$

15. $\left(-\dfrac{1}{2}, \dfrac{4}{3}\right)$ **17.** $(1, 6)$ **19.** $\left(-\infty, -\dfrac{1}{2}\right) \cup \left(\dfrac{1}{3}, \infty\right)$

21. $\left(-\dfrac{2}{3}, -\dfrac{1}{4}\right)$ **23.** $(-\infty, -2) \cup (2, \infty)$

25. $(-\infty, -4) \cup (4, \infty)$ **27.** $\left[-2, \dfrac{1}{3}\right] \cup [4, \infty)$

29. $(-\infty, -1) \cup (2, 4)$

Glossary

A

absolute value The absolute value of a number is the distance between 0 and the number on a number line. (Section 1.4)

addition property of equality The addition property of equality states that the same number can be added to (or subtracted from) both sides of an equation to obtain an equivalent equation. (Section 2.1)

addition property of inequality The addition property of inequality states that the same number can be added to (or subtracted from) both sides of an inequality without changing the solution set. (Section 2.8)

additive inverse (opposite) Two numbers that are the same distance from, but on opposite sides, of 0 on a number line are called additive inverses. (Section 1.4)

algebraic expression An algebraic expression is a sequence of numbers, variables, operation symbols, and/or grouping symbols (such as parentheses) formed according to the rules of algebra. (Section 1.3)

area Area is a measure of the surface covered by a two-dimensional (flat) figure. (Section 2.5)

associative property of addition The associative property of addition states that the way in which numbers being added are grouped does not change the sum. (Section 1.7)

associative property of multiplication The associative property of multiplication states that the way in which numbers being multiplied are grouped does not change the product. (Section 1.7)

axis (axis of symmetry) The axis of a parabola is the vertical or horizontal line (depending on the orientation of the graph) through the vertex of the parabola. (Sections 5.4, 9.5)

B

bar graph A bar graph is a series of bars (or simulations of bars) arranged either vertically or horizontally to show comparisons of data. (Section 3.1)

base The base is the number that is a repeated factor when written with an exponent. (Sections 1.2, 5.1)

bimodal A group of numbers with two modes is bimodal. (Appendix D)

binomial A binomial is a polynomial with exactly two terms. (Section 5.4)

boundary line In the graph of a linear inequality, the boundary line separates the region that satisfies the inequality from the region that does not satisfy the inequality. (Section 3.5)

C

circle graph (pie chart) A circle graph (or pie chart) is a circle divided into sectors, or wedges, whose sizes show the relative magnitudes of the categories of data being represented. (Section 1.1)

coefficient (numerical coefficient) A coefficient is the numerical factor of a term. (Sections 1.8, 5.4)

combining like terms Combining like terms is a method of adding or subtracting like terms by using the properties of real numbers. (Section 1.8)

common factor An integer that is a factor of two or more integers is called a common factor of those integers. (Section 6.1)

commutative property of addition The commutative property of addition states that the order of numbers in an addition problem can be changed without changing the sum. (Section 1.7)

commutative property of multiplication The commutative property of multiplication states that the product in a multiplication problem remains the same regardless of the order of the factors. (Section 1.7)

complement of a set The set of elements in the universal set that are not in a set A is the complement of A, written A'. (Appendix C)

complementary angles (complements) Complementary angles are angles whose measures have a sum of 90°. (Section 2.4)

completing the square The process of adding to a binomial the number that makes it a perfect square trinomial is called completing the square. (Section 9.2)

complex fraction A complex fraction is an expression with one or more fractions in the numerator, denominator, or both. (Section 7.5)

complex number A complex number is any number that can be written in the form $a + bi$, where a and b are real numbers. (Section 9.4)

components In an ordered pair (x, y), x and y are called the components of the ordered pair. (Section 3.6)

composite number A composite number has at least one factor other than itself and 1. (Section 1.1)

conditional equation A conditional equation is true for some replacements of the variable and false for others. (Section 2.3)

conjugate The conjugate of $a + b$ is $a - b$. (Section 8.5)

conjugate of a complex number The conjugate of a complex number $a + bi$ is $a - bi$. (Section 9.4)

consecutive integers Two integers that differ by 1 are called consecutive integers. (Section 2.4)

consistent system A system of equations with a solution is called a consistent system. (Section 4.1)

constant of variation In the equation $y = kx$, the number k is called the constant of variation. (Section 7.8)

contradiction A contradiction is an equation that is never true. It has no solutions. (Section 2.3)

coordinate on a number line Each number on a number line is called the coordinate of the point that it labels. (Section 1.4)

coordinates of a point The numbers in an ordered pair are called the coordinates of the corresponding point. (Section 3.1)

cross products The cross products in the proportion $\frac{a}{b} = \frac{c}{d}$ are ad and bc. (Section 2.6)

cube root A number b is a cube root of a if $b^3 = a$. (Section 8.1)

D

decimal A decimal is a number written with a decimal point. (Appendix B)

degree A degree is a basic unit of measure for angles in which one degree (1°) is $\frac{1}{360}$ of a complete revolution. (Section 2.4)

degree of a polynomial The degree of a polynomial is the greatest degree of any of the terms in the polynomial. (Section 5.4)

degree of a term The degree of a term is the sum of the exponents on the variables in the term. (Section 5.4)

denominator The number below the fraction bar in a fraction is called the denominator. It shows the number of equal parts in a whole. (Section 1.1)

dependent equations Equations of a system that have the same graph (because they are different forms of the same equation) are called dependent equations. (Section 4.1)

descending powers A polynomial in one variable is written in descending powers of the variable if the exponents on the terms of the polynomial decrease from left to right. (Section 5.4)

difference The answer to a subtraction problem is called the difference. (Section 1.1)

difference of cubes The difference of cubes, $x^3 - y^3$, can be factored as $x^3 - y^3 = (x - y)(x^2 + xy + y^2)$. (Section 6.4)

difference of squares The difference of squares, $x^2 - y^2$, can be factored as the product of the sum and difference of two terms, or $x^2 - y^2 = (x + y)(x - y)$. (Section 6.4)

direct variation y varies directly as x if there exists a real number (constant) k such that $y = kx$. (Section 7.8)

discriminant The discriminant is the quantity $b^2 - 4ac$ under the radical in the quadratic formula. (Section 9.3)

disjoint sets Sets that have no elements in common are disjoint sets. (Appendix C)

distributive property For any real numbers a, b, and c, the distributive property states that $a(b + c) = ab + ac$ and $(b + c)a = ba + ca$. (Section 1.7)

domain The set of all first components (x-values) in the ordered pairs of a relation is the domain. (Section 3.6)

E

elements (members) Elements are the objects that belong to a set. (Section 1.3, Appendix C)

elimination method The elimination method is an algebraic method used to solve a system of equations in which the equations of the system are combined so that one or more variables is eliminated. (Section 4.3)

empty set (null set) The empty set, denoted by { } or ∅, is the set containing no elements. (Section 2.3, Appendix C)

equation An equation is a statement that two algebraic expressions are equal. (Section 1.3)

equivalent equations Equivalent equations are equations that have the same solution set. (Section 2.1)

exponent (power) An exponent is a number that indicates how many times a factor is repeated. (Sections 1.2, 5.1)

exponential expression A number or letter (variable) written with an exponent is an exponential expression. (Sections 1.2, 5.1)

extraneous solution A solution to a new equation that does not satisfy the original equation is called an extraneous solution. (Section 8.6)

extremes of a proportion In the proportion $\frac{a}{b} = \frac{c}{d}$, the a- and d-terms are called the extremes. (Section 2.6)

F

factor A factor of a given number is any number that divides evenly (without remainder) into the given number. (Sections 1.1, 6.1)

factored A number is factored by writing it as the product of two or more numbers. (Section 1.1)

factored form An expression is in factored form when it is written as a product. (Section 6.1)

factoring Writing a polynomial as the product of two or more simpler polynomials is called factoring. (Section 6.1)

factor by grouping Factoring by grouping is a method for grouping the terms of a polynomial in such a way that the polynomial can be factored even though its greatest common factor is 1. (Section 6.1)

factor out the greatest common factor Factoring out the greatest common factor is the process of using the distributive property to write a polynomial as a product of the greatest common factor and a simpler polynomial. (Section 6.1)

FOIL FOIL is a method for multiplying two binomials $(A + B)(C + D)$. Multiply **F**irst terms AC, **O**uter terms AD, **I**nner terms BC, and **L**ast terms BD. Then combine like terms. (Section 5.5)

formula A formula is an equation in which letters are used to describe a relationship. (Section 2.5)

fourth root A number b is a fourth root of a if $b^4 = a$. (Section 8.1)

function A function is a set of ordered pairs (a relation) in which each value of the first component x corresponds to exactly one value of the second component y. (Section 3.6)

function notation Function notation $f(x)$ represents the value of the function at x—that is, the y-value which corresponds to x. (Section 3.6)

G

graph of a number The point on a number line that corresponds to a number is its graph. (Section 1.4)

graph of an equation The graph of an equation is the set of all points that correspond to all of the ordered pairs that satisfy the equation. (Section 3.2)

graphing method The graphing method for solving a system of equations requires graphing all equations of the system on the same axes and locating the ordered pair(s) at their intersection. (Section 4.1)

greatest common factor (GCF) The greatest common factor of a list of integers is the largest common factor of those integers. The greatest common factor of a polynomial is the largest term that is a factor of all terms in the polynomial. (Sections 1.1, 6.1)

grouping symbols Grouping symbols are parentheses ( ), brackets [ ], or fraction bars. (Section 1.2)

H

hypotenuse   The hypotenuse is the longest side in a right triangle. It is the side opposite the right angle. (Section 6.6)

I

identity   An identity is an equation that is true for all replacements of the variable. It has an infinite number of solutions. (Section 2.3)

identity element for addition   Since adding 0 to a number does not change the number, 0 is called the identity element for addition. (Section 1.7)

identity element for multiplication   Since multiplying a number by 1 does not change the number, 1 is called the identity element for multiplication. (Section 1.7)

identity property   The identity properties state that the sum of 0 and any number equals the number, and the product of 1 and any number equals the number. (Section 1.7)

imaginary part   The imaginary part of the complex number $a + bi$ is b. (Section 9.4)

inconsistent system   An inconsistent system of equations is a system with no solution. (Section 4.1)

independent equations   Equations of a system that have different graphs are called independent equations. (Section 4.1)

index (order)   In a radical of the form $\sqrt[n]{a}$, n is called the index or order. (Section 8.1)

inequality   An inequality is a statement that two expressions are not equal. (Section 1.2)

inner product   When using the FOIL method to multiply two binomials $(A + B)(C + D)$, the inner product is BC. (Section 5.5)

integers   The set of integers is $\{\dots, -3, -2, -1, 0, 1, 2, 3, \dots\}$. (Section 1.4)

intersection   The intersection of two sets A and B, written $A \cap B$, is the set of elements that belong to both A and B. (Appendix C)

interval   An interval is a portion of a number line. (Section 2.8)

interval notation   Interval notation is a simplified notation that uses parentheses ( ) and/or brackets [ ] to describe an interval on a number line. (Section 2.8)

inverse property   The inverse properties state that a number added to its opposite is 0 and a number multiplied by its reciprocal is 1. (Section 1.7)

inverse variation y varies inversely as x if there exists a real number (constant) k such that $y = \frac{k}{x}$. (Section 7.8)

irrational number   An irrational number cannot be written as the quotient of two integers but can be represented by a point on the number line. (Section 1.4)

L

least common denominator (LCD)   Given several denominators, the smallest expression that is divisible by all the denominators is called the least common denominator. (Sections 1.1, 7.3)

legs of a right triangle   The two shorter sides of a right triangle are called the legs. (Section 6.6)

like radicals   Like radicals are terms that have multiples of the same root of the same number or expression. (Section 8.3)

like terms   Terms with exactly the same variables raised to exactly the same powers are called like terms. (Sections 1.8, 5.4)

line graph   A line graph is a series of line segments that connect points representing data. (Section 3.1)

linear equation in one variable   A linear equation in one variable can be written in the form $Ax + B = C$, where A, B, and C are real numbers, with $A \neq 0$. (Section 2.1)

linear equation in two variables   A linear equation in two variables is an equation that can be written in the form $Ax + By = C$, where A, B, and C are real numbers and A and B are not both 0. (Section 3.1)

linear inequality in one variable   A linear inequality in one variable can be written in the form $Ax + B < C$, $Ax + B \leq C$, $Ax + B > C$, or $Ax + B \geq C$, where A, B, and C are real numbers, with $A \neq 0$. (Section 2.8)

linear inequality in two variables   A linear inequality in two variables can be written in the form $Ax + By < C$ or $Ax + By > C$ (or with $\leq$ or $\geq$), where A, B, and C are real numbers and A and B are not both 0. (Section 3.5)

line of symmetry   The axis of a parabola is a line of symmetry for the graph. It is a line that can be drawn through the graph in such a way that the part of the graph on one side of the line is an exact reflection of the part on the opposite side. (Sections 5.4, 9.5)

lowest terms   A fraction is in lowest terms when there are no common factors in the numerator and denominator (except 1). (Sections 1.1, 7.1)

M

mean   The mean (average) of a group of numbers is the sum of the numbers, divided by the number of numbers. (Appendix D)

means of a proportion   In the proportion $\frac{a}{b} = \frac{c}{d}$, the b- and c-terms are called the means. (Section 2.6)

median   The median divides a group of numbers, listed in order from least to greatest, in half; that is, half the numbers lie above the median and half lie below the median. (Appendix D)

mixed number   A mixed number includes a whole number and a fraction written together and is understood to be the sum of the whole number and the fraction. (Section 1.1)

mode   The mode is the number (or numbers) that occurs most often in a group of numbers. (Appendix D)

monomial   A monomial is a polynomial with only one term. (Section 5.4)

multiplication property of equality   The multiplication property of equality states that the same nonzero number can be multiplied by (or divided into) both sides of an equation to obtain an equivalent equation. (Section 2.2)

multiplication property of inequality   The multiplication property of inequality states that both sides of an inequality may be multiplied (or divided) by a positive number without changing the direction of the inequality symbol. Multiplying (or dividing) by a negative number reverses the inequality symbol. (Section 2.8)

multiplicative inverse (reciprocal)   The multiplicative inverse of a nonzero real number a is $\frac{1}{a}$. (Section 1.6)

N

natural numbers   The set of natural numbers consists of the numbers used for counting: $\{1, 2, 3, 4, \dots\}$. (Sections 1.1, 1.4)

negative number A negative number is located to the left of 0 on a number line. (Section 1.4)

number line A number line is a line with a scale that is used to show how numbers relate to each other. (Section 1.4)

numerator The number above the fraction bar in a fraction is called the numerator. It shows how many of the equivalent parts are being considered. (Section 1.1)

numerical coefficient The numerical factor in a term is its numerical coefficient. (Sections 1.8, 5.4)

O

ordered pair An ordered pair is a pair of numbers written within parentheses in which the order of the numbers is important. (Section 3.1)

origin The point at which the x-axis and y-axis of a rectangular coordinate system intersect is called the origin. (Section 3.1)

outer product When using the FOIL method to multiply two binomials $(A + B)(C + D)$, the outer product is AD. (Section 5.5)

P

parabola The graph of a second-degree (quadratic) equation in two variables is called a parabola. (Sections 5.4, 9.5)

parallel lines Parallel lines are two lines in the same plane that never intersect. (Section 3.3)

percent Percent, written with the sign %, means per one hundred. (Appendix B)

percentage A percentage is a part of a whole. (Appendix B)

perfect cube A perfect cube is a number with a rational cube root. (Section 8.1)

perfect square A perfect square is a number with a rational square root. (Section 8.1)

perfect square trinomial A perfect square trinomial is a trinomial that can be factored as the square of a binomial. (Section 6.4)

perimeter The perimeter of a two-dimensional figure is a measure of the distance around the outside edges of the figure—that is, the sum of the lengths of its sides. (Section 2.5)

perpendicular lines Perpendicular lines are two lines that intersect to form a right (90°) angle. (Section 3.3)

plot To plot an ordered pair is to locate it on a rectangular coordinate system. (Section 3.1)

point–slope form A linear equation is written in point–slope form if it is in the form $y - y_1 = m(x - x_1)$, where m is the slope and (x_1, y_1) is a point on the line. (Section 3.4)

polynomial A polynomial is a term or a finite sum of terms in which all coefficients are real, all variables have whole number exponents, and no variables appear in denominators. (Section 5.4)

polynomial in x A polynomial whose only variable is x is called a polynomial in x. (Section 5.4)

positive number A positive number is located to the right of 0 on a number line. (Section 1.4)

prime factors The prime factors of a number are those factors which are prime numbers. (Section 1.1)

prime number A natural number (except 1) is prime if it has only 1 and itself as factors. (Section 1.1)

prime polynomial A prime polynomial is a polynomial that cannot be factored into factors having only integer coefficients. (Section 6.2)

principal root (principal nth root) For even indexes, the symbols $\sqrt{}$, $\sqrt[4]{}$, $\sqrt[6]{}$, . . . , $\sqrt[n]{}$, are used for nonnegative roots, which are called principal roots. (Section 8.1)

product The answer to a multiplication problem is called the product. (Section 1.1)

product of the sum and difference of two terms The product of the sum and difference of two terms is the difference of the squares of the terms, or $(x + y)(x - y) = x^2 - y^2$. (Section 5.6)

proportion A proportion is a statement that two ratios are equal. (Section 2.6)

pure imaginary number If $a = 0$ and $b \neq 0$ in the complex number $a + bi$, the complex number is said to be a pure imaginary number. (Section 9.4)

Pythagorean formula The Pythagorean formula states that the square of the length of the hypotenuse of a right triangle equals the sum of the squares of the lengths of the two legs. (Section 6.6)

Q

quadrant A quadrant is one of the four regions in the plane determined by a rectangular coordinate system. (Section 3.1)

quadratic equation A quadratic equation is an equation that can be written in the form $ax^2 + bx + c = 0$, where a, b, and c are real numbers, with $a \neq 0$. (Sections 6.5, 9.1)

quadratic formula The quadratic formula is a general formula used to solve any quadratic equation. (Section 9.3)

quadratic function A function defined by an equation of the form $f(x) = ax^2 + bx + c$, for real numbers a, b, and c, with $a \neq 0$, is a quadratic function. (Section 9.5)

quadratic inequality A quadratic inequality can be written in the form $ax^2 + bx + c < 0$ or $ax^2 + bx + c > 0$ (or with $\leq$ or $\geq$), where a, b, and c are real numbers, with $a \neq 0$. (Appendix E)

quotient The answer to a division problem is called the quotient. (Section 1.1)

R

radical An expression consisting of a radical sign with a radicand underneath is called a radical. (Section 8.1)

radical equation A radical equation is an equation with a variable in the radicand. (Section 8.6)

radical expression A radical expression is an algebraic expression that contains radicals. (Section 8.1)

radical sign The symbol $\sqrt{}$ is called a radical sign. (Section 8.1)

radicand The number or expression under a radical sign is called the radicand. (Section 8.1)

range The set of all second components (y-values) in the ordered pairs of a relation is the range. (Section 3.6)

ratio A ratio is a comparison of two quantities with the same units. (Section 2.6)

rational expression The quotient of two polynomials with denominator not 0 is called a rational expression. (Section 7.1)

rationalizing the denominator The process of removing radicals from a denominator so that the denominator contains only rational numbers is called rationalizing the denominator. (Section 8.4)

rational numbers Rational numbers can be written as the quotient of two integers, with denominator not 0. (Section 1.4)

real numbers Real numbers include all numbers that can be represented by points on the number line—that is, all rational and irrational numbers. (Section 1.4)

real part The real part of a complex number $a + bi$ is a. (Section 9.4)

reciprocal Pairs of numbers whose product is 1 are called reciprocals of each other. (Section 1.1)

rectangular (Cartesian) coordinate system The x-axis and y-axis placed at a right angle at their zero points form a rectangular coordinate system, also called the Cartesian coordinate system. (Section 3.1)

relation A relation is a set of ordered pairs. (Section 3.6)

right angle A right angle measures $90°$. (Section 2.4)

rise Rise is the vertical change between two points on a line—that is, the change in y-values. (Section 3.3)

run Run is the horizontal change between two points on a line—that is, the change in x-values. (Section 3.3)

S

scatter diagram A scatter diagram is a graph of ordered pairs of data. (Section 3.1)

scientific notation A number is written in scientific notation when it is expressed in the form $a \times 10^n$, where $1 \le |a| < 10$ and n is an integer. (Section 5.3)

set A set is a collection of objects. (Section 1.3, Appendix C)

set-builder notation Set-builder notation is used to describe a set of numbers without actually having to list all of the elements. (Section 1.4)

signed numbers Signed numbers are numbers that can be written with a positive or negative sign. (Section 1.4)

simplified radical A simplified radical meets three conditions:

1. The radicand has no factor (except 1) that is a perfect square (if the radical is a square root), a perfect cube (if the radical is a cube root), and so on.

2. The radicand has no fractions.

3. No denominator contains a radical.

(Section 8.4)

slope The ratio of the change in y to the change in x along a line is called the slope of the line. (Section 3.3)

slope–intercept form A linear equation is written in slope–intercept form if it is in the form $y = mx + b$, where m is the slope and $(0, b)$ is the y-intercept. (Section 3.4)

solution of an equation A solution of an equation is any replacement for the variable that makes the equation true. (Section 1.3)

solution of a system A solution of a system of two equations is an ordered pair (x, y) that makes both equations true at the same time. (Section 4.1)

solution set The solution set is the set of all solutions of a particular equation. (Section 2.1)

solution set of a linear system The solution set of a linear system of equations includes all ordered pairs that satisfy all the equations of the system at the same time. (Section 4.1)

solution set of a system of linear inequalities The solution set of a system of linear inequalities includes all ordered pairs that make all inequalities of the system true at the same time. (Section 4.5)

square of a binomial The square of a binomial is the sum of the square of the first term, twice the product of the two terms, and the square of the last term—that is, $(x + y)^2 = x^2 + 2xy + y^2$ or $(x - y)^2 = x^2 - 2xy + y^2$. (Section 5.6)

square root The opposite of squaring a number is called taking its square root; that is, a number a is a square root of k if $a^2 = k$. (Section 8.1)

square root property The square root property states that if $x^2 = k$, with $k > 0$, then $x = \sqrt{k}$ or $x = -\sqrt{k}$. (Section 9.1)

squaring property If each side of a given equation is squared, then all solutions of the given equation are *among* the solutions of the squared equation. (Section 8.6)

standard form of a complex number The standard form of a complex number is $a + bi$. (Section 9.4)

standard form of a linear equation A linear equation in two variables written in the form $Ax + By = C$, with A and B not both 0, is in standard form. (Section 3.4)

standard form of a quadratic equation A quadratic equation written in the form $ax^2 + bx + c = 0$, where $a \ne 0$, is in standard form. (Section 6.5)

straight angle A straight angle measures $180°$. (Section 2.4)

subscript notation Subscript notation is a way of indicating nonspecific values, such as x_1 and x_2. (Section 3.3)

subset If all elements of set A are in set B, then A is a subset of B, written $A \subseteq B$. (Appendix C)

substitution method The substitution method is an algebraic method for solving a system of equations in which one equation is solved for one of the variables and the result is substituted into the other equation. (Section 4.2)

sum The answer to an addition problem is called the sum. (Section 1.1)

sum of cubes The sum of cubes, $x^3 + y^3$, can be factored as $x^3 + y^3 = (x + y) \cdot (x^2 - xy + y^2)$. (Section 6.4)

supplementary angles (supplements) Supplementary angles are angles whose measures have a sum of $180°$. (Section 2.4)

system of linear equations (linear system) A system of linear equations consists of two or more linear equations to be solved at the same time. (Section 4.1)

system of linear inequalities A system of linear inequalities consists of two or more linear inequalities to be solved at the same time. (Section 4.5)

T

table of values A table of values is an organized way of displaying ordered pairs. (Section 3.1)

term A term is a number, a variable, or the product or quotient of a number and one or more variables raised to powers. (Section 1.8)

terms of a proportion The terms of the proportion $\frac{a}{b} = \frac{c}{d}$ are a, b, c, and d. (Section 2.6)

three-part inequality An inequality that says that one number is between two other numbers is called a three-part inequality. (Section 2.8)

trinomial A trinomial is a polynomial with exactly three terms. (Section 5.4)

U

union The union of two sets A and B, written $A \cup B$, is the set that includes all elements of A together with all elements of B. (Appendix C)

universal set The set that includes all elements under consideration is the universal set, written U. (Appendix C)

unlike terms Unlike terms are terms that do not have the same variable or terms with the same variables but whose variables are not raised to the same powers. (Section 1.8)

V

variable A variable is a symbol, usually a letter, used to represent an unknown number. (Section 1.3)

vary directly (is proportional to) y varies directly as x if there exists a nonzero number (constant) k such that $y = kx$. (Section 7.8)

vary inversely y varies inversely as x if there exists a real number (constant) k such that $y = \frac{k}{x}$. (Section 7.8)

Venn diagram A Venn diagram represents the relationships between sets. (Appendix C)

vertex The point on a parabola that has the smallest y-value (if the parabola opens upward) or the largest y-value (if the parabola opens downward) is called the vertex of the parabola. (Sections 5.4, 9.5)

vertical angles When two intersecting lines are drawn, the angles that lie opposite each other have the same measure and are called vertical angles. (Section 2.5)

vertical line test The vertical line test states that any vertical line drawn through the graph of a function must intersect the graph in at most one point. (Section 3.6)

volume The volume of a three-dimensional figure is a measure of the space occupied by the figure. (Section 2.5)

W

weighted mean A weighted mean is a mean in which each number is weighted by multiplying it by the number of times it occurs. (Appendix D)

whole numbers The set of whole numbers is $\{0, 1, 2, 3, 4, \ldots\}$. (Sections 1.1, 1.4)

X

x-axis The horizontal number line in a rectangular coordinate system is called the x-axis. (Section 3.1)

x-intercept A point where a graph intersects the x-axis is called an x-intercept. (Section 3.2)

Y

y-axis The vertical number line in a rectangular coordinate system is called the y-axis. (Section 3.1)

y-intercept A point where a graph intersects the y-axis is called a y-intercept. (Section 3.2)

Z

zero-factor property The zero-factor property states that if two numbers have a product of 0, then at least one of the numbers is 0. (Section 6.5)

Index

Formulas

Figure	Formulas	Illustration
Square	Perimeter: $P = 4s$ Area: $A = s^2$	
Rectangle	Perimeter: $P = 2L + 2W$ Area: $A = LW$	
Triangle	Perimeter: $P = a + b + c$ Area: $A = \frac{1}{2}bh$	
Parallelogram	Perimeter: $P = 2a + 2b$ Area: $A = bh$	
Trapezoid	Perimeter: $P = a + b + c + B$ Area: $A = \frac{1}{2}h(b + B)$	
Circle	Diameter: $d = 2r$ Circumference: $C = 2\pi r$ $C = \pi d$ Area: $A = \pi r^2$	